An Introduction to Dynamical Systems and Chaos

G.C. Layek

An Introduction
to Dynamical Systems
and Chaos

 Springer

G.C. Layek
Department of Mathematics
The University of Burdwan
Burdwan, West Bengal
India

ISBN 978-81-322-3794-5   ISBN 978-81-322-2556-0   (eBook)
DOI 10.1007/978-81-322-2556-0

Springer New Delhi Heidelberg New York Dordrecht London
© Springer India 2015
Softcover re-print of the Hardcover 1st edition 2015

Printed on acid-free paper

Springer (India) Pvt. Ltd. is part of Springer Science+Business Media (www.springer.com)

Dedicated to my father
Late Bijoychand Layek
for his great interest in my education.

Preface

This book is the outcome of my teaching and research on dynamical systems, chaos, fractals, and fluid dynamics for the past two decades in the Department of Mathematics, University of Burdwan, India. There are a number of excellent books on dynamical systems that cover different aspects and approaches to nonlinear dynamical systems and chaos theory. However, there lies a gap among mathematical theories, intuitions, and perceptions of nonlinear science and chaos. There is a need for amalgamation among theories, intuitions, and perceptions of the subject and it is also necessary for systematic, sequential, and logical developments in the subject, which helps students at the undergraduate and postgraduate levels. Teachers and researchers in this discipline will be benefitted from this book. Readers are presumed to have a basic knowledge in linear algebra, mathematical analysis, topology, and differential equations.

Over the past few decades an unprecedented interest and progress in nonlinear systems, chaos theory, and fractals have been noticed, which are reflected in the undergraduate and postgraduate curriculum of science and engineering. The essence of writing this book is to provide the basic ideas and the recent developments in the field of nonlinear dynamical systems and chaos theory; their mathematical theories and physical examples. Nonlinearity is a driving mechanism in most physical and biological phenomena. Scientists are trying to understand the inherent laws underlying these phenomena over the centuries through mathematical modeling. We know nonlinear equations are harder to solve analytically, except for a few special equations. The superposition principle does not hold for nonlinear equations. Scientists are now convinced about the power of geometric and qualitative approaches in analyzing the dynamics of a system that governs nonlinearly. Using these techniques, some nonlinear intractable problems had been analyzed from an analytical point of view and the results were found to be quite interesting. Solutions of nonlinear system may have extremely complicated geometric structure. Historically, these types of solutions were known to both Henri Poincaré (1854–1912), father of nonlinear dynamics, and George David Brikhoff (1884–1944) in the late nineteenth and early twentieth centuries. In the year 1963,

Edward Lorenz published a paper entitled "Deterministic Nonperiodic Flow" that described numerical results obtained by integrating third-order nonlinear ordinary differential equations, which was nothing but a simplified version of convection rolls in atmosphere. This work was most influential and the study of chaotic systems began. Throughout the book, emphasis has been given to understanding the subject mathematically and then explaining the dynamics of systems physically. Some mathematical theorems are given so that the reader can follow the logical steps easily and, also, for further developments in the subject. In this book, continuous and discrete time systems are presented separately, which will help beginners. Discrete-time systems and chaotic maps are given more emphasis. Conjugacy/semi-conjugacy relations among maps and their properties are also described. Mathematical theories for chaos are needed for proper understanding of chaotic motion. The concept and theories are lucidly explained with many worked-out examples, including exercises.

Bankura, India   G.C. Layek
October 2015

Acknowledgments

I would like to acknowledge the academic help and cooperation from my two Ph.D. students, Naresh Chandra Pati and Sunita, during the preparation of this book. I must thank both of them. As mentioned earlier, this book is an amalgamation of materials from different books at the advanced level. I appreciate all authors for their contributions to nonlinear dynamics and chaos theory. I would like to give special thanks to my colleagues Prof. Mantu Saha and Dr. Swati Mukhophadhay for their encouragement, help, and appreciation. It is indeed a great pleasure for me to mention Prof. Prashanta Chatterjee, Dr. Anjan Kumar Bhunia of Visva Bharati University, Dr. Tanmoy Banerjee, Department of Physics, Prof. Goutam Mitra, MBA Department and Dr. Bibhus Bagchi, Department of Sociology, Burdwan University, Prof. H.P. Mazumdar, Indian Statistical Institute, Kolkata, Dr. Rajib Basu, Nistarini College, Purulia, Dr. Subhasis Banerjee, Bankura Christian College, who helped me in various ways while preparing this book. I would like to express my sincere thanks to Mr. Shamim Ahmad, Editor, Mathematics & Statistics series, Springer, and his staff. I must appreciate their patience and cooperation in all respects during the preparation of the final version of the manuscripts. I thank my wife Atasi, my children Soumita and Soumyadeep, and express my gratitude to my mother, sisters, my father-in law, mother-in law, and other family members and relatives for their support and encouragement.

Contents

About the Book

The materials of the book have been assembled from different articles and books published over the past 50 years along with my thinking and research experience. The book contains 13 chapters covering all aspects of nonlinear dynamical systems at the basic and advanced levels. The book is self-contained as the initial three chapters cover mainly ordinary differential equations and the concept of flows. The first chapter contains an introduction followed by a brief history of nonlinear science and discussions of one-dimensional continuous systems. Flows and their mathematical basis, qualitative approach, analysis of one-dimensional flows with examples, some important definitions, and conservative-dissipative systems are discussed in this chapter. Chapter 2 presents the solution technique of homogeneous linear systems using eigenvalue–eigenvector method and the fundamental matrix method. Discussions and theories on linear systems are presented. The solutions of a linear system form a vector space. The solution technique for higher dimensional systems and properties of exponential matrices are given in detail. The solution technique for nonhomogeneous linear equations using fundamental matrix is also given in this chapter. Flows in $\mathbb{R}^2$ that is, phase plane analysis, the equilibrium points and their stability characters, linearization of nonlinear systems, and its limitations are subject matters in Chap. 3. Mathematical pendulum problems and linear oscillators are also discussed in this chapter. Chapter 4 gives the theory of stability of linear and nonlinear systems. It also contains the notion of hyperbolicity, stable and unstable subspaces, Hartman–Grobman theorem, stable manifold theorem, and their applications. The most important contribution to the history of nonlinear dynamical systems is the theory of nonlinear oscillations. The problem of constructing mathematical tools for the study of nonlinear oscillations was first formulated by Mandelstham around 1920, in connection with the study of dynamical systems in radio-engineering. In 1927, Andronov, the most famous student of Mandelstham presented his thesis on a topic "Poincare's limit cycles and the theory of oscillations." Subsequently, van der Pol and Liénard made significant contributions with practical applications of nonlinear oscillations. Chapter 5 deals with linear and nonlinear oscillations with some important theorems and physical applications. Bifurcation is the study of possible changes in the structure of the orbit of a dynamical system depending on the changing values of the parameters. Chapter 6 presents the bifurcations in one-dimensional and two-dimensional systems. Lorenz system and its properties, for example in fluid system, are also given in this chapter. Hamiltonian systems are elegant and beautiful concepts in classical mechanics. Chapter 7 discusses the basics of Lagranergian and Hamiltonian systems and their derivations. Hamiltonian flows, their properties, and a number of worked-out examples are presented in this chapter. Symmetry is an inherent character in many physical phenomena. Symmetry analysis is one of the important discoveries of the ninetieth century. This is based on a continuous group of transformations discovered by the great Norwegian mathematician Sophus Lie (1842–1899). Symmetry groups or simply symmetries are invariant transformations that do not change the structural form of the equation under consideration. Knowledge of symmetries of a system definitely enhances our understanding of complex physical phenomena and their inherent laws. It has been presumed that students must be familiar with symmetry analysis of simple nonlinear systems for understanding natural phenomena in-depth. With this motivation we introduce the Lie symmetry under continuous group of transformations, invariance principle, and systematic calculation of symmetries for ordinary and partial differential equations in Chap. 8. Maps and their compositions have a vast dynamics with immense applications. Chapter 9 discusses maps, their iterates, fixed points and their stabilities, periodic cycles, and some important theorems. In Chap. 10 some important maps, namely tent map, logistic map, shift map, Hénon map, etc., are discussed elaborately. Chapter 11 deals with conjugacy/semi-conjugacy relations among maps, their properties, and proofs of some important theorems. In the twenty-first century, chaos and its mathematical foundation are crucially important. The chaos theory is an emergent area in twenty-first century science. The chaotic motion was first formulated by the French mathematician Henri Poincare in his paper on the stability of the solar system. What kinds of systems exhibit chaotic motion? Is there any universal quantifying feature of chaos? Chapter 12 contains a brief history of chaos and its mathematical theory. Emphasis has been given to establish mathematical theories on chaotic systems, quantifying chaos and universality. Routes of chaos, chaotic maps, Sharkovskii ordering, and theory are discussed in this chapter. The term 'fractal' was coined by Benoit Mandelbrot. It appeared as mathematical curiosities at the end of the twentieth century and its connection with chaotic orbit. Fractals are complex geometric shapes with fine structure at arbitrarily small scales. The self-similarity property is evidenced in most fractal objects. The dimension of a fractal object is not an integer. Chaotic orbit may be represented by fractals. Chapter 13 is devoted to the study of fractals, their self-similarities, scaling, and dimensions of fractal objects with many worked-out examples.

Chapter 1
Continuous Dynamical Systems

Dynamics is a time-evolutionary process. It may be deterministic or stochastic. Long-term predictions of some systems often become impossible. Even their trajectories cannot be represented by usual geometry. In many natural and social phenomena there is unpredictability. Unpredictability is an intrinsic property which is present in the phenomenon itself. It has great impact on human civilization as well as scientific thoughts. There are numerous questions in human mind e.g., how can a deterministic trajectory be unpredictable? What are the causes in formation of symmetric crystals and snowflakes in Nature? How can one find chaotic trajectories? Can a deterministic trajectory be random? How can one define and explain turbulence in fluid motion? Is there any local symmetry in chaos? How can one relate chaotic dynamics with fractal object? For answering these questions we have no way but to study nonlinear dynamics.

Dynamical systems are generally described by differential or difference equations. Studies of differential equations in mathematics were devoted mainly of finding analytical solutions of equations for more than two centuries. But the dynamical behaviors of a system may not always be determined by analytical or closed-form solutions. Moreover, analytical solution of nonlinear equations is difficult to obtain except in a few special cases. The subject dynamical systems had evolved at the end of nineteenth century and made significant contributions to understanding some nonlinear phenomena. The dynamics of a system may be expressed either as a continuous-time or as a discrete-time-evolutionary process. The simplest mathematical models of continuous systems are those consisting of first-order differential equations. In first-order autonomous system (explicit in time), the dynamics is a very restrictive class of system since its motion is in the real line. In simple nonautonomous cases, on the other hand, the dynamics is very rich.

Nonlinear science and its dynamics have been a matter of great importance in the field of natural and social sciences. Examples include physical science (e.g., earth's atmosphere, laser, electronic circuit, superconductivity, fluid turbulence, etc.), chemistry (Belousov–Zhabotinsky reaction, Brusselator model, etc.), biology (neural and cardiac systems, biochemical processes), ecology and social sciences (spreading of fading, spreading diseases, price fluctuations of markets and stock markets, etc.), to mention a few. Nonlinear systems are harder (if not sometime impossible) to solve than linear systems, because the latter follow the superposition

© Springer India 2015
G.C. Layek, *An Introduction to Dynamical Systems and Chaos*,
DOI 10.1007/978-81-322-2556-0_1

principle and can be divided into parts. Each part can be solved individually and adding them all provides the final result. However, solutions of linear systems are helpful for the analysis of nonlinear systems.

In this chapter we discuss some important definitions, concept of flows, their properties, examples, and analysis of one-dimensional flows for an easy way to understand the nonlinear dynamical systems.

1.1 Dynamics: A Brief History

The explicit time behaviors of a system and its dependency on initial conditions of solutions began after the 1880s. It is well known that analytical or closed-form solutions of nonlinear equations cannot be obtained except for very few special forms. Moreover, the solution behaviors at different initial conditions or their asymptotic characters are sometimes cumbersome to determine from closed-form solutions. In this situation scientists felt the necessity for developing a method that determines the qualitative features of a system rather than the quantitative analysis. The French mathematician *Henri Poincaré* (1854–1912) pioneered the qualitative approach, a combination of analysis and geometry which was proved to be a powerful approach for analyzing behaviors of a system and brought *Poincaré* recognition as the "father of nonlinear dynamics." The time-evolutionary process governed either by linear or nonlinear equations gives the dynamical system. Dynamics and its representations are inextricably tied with mathematics. The subject initiated informally from the different views of mathematicians and physicists. Studies began in the mid-1600s when *Newton* (1643–1727) invented calculus, differential equations, the laws of motion, and universal gravitation. With the help of Newton's discoveries the laws of planetary motions, already postulated by *Jonaesh Kepler*, a German astrologist (1609, 1619) were established mathematically and the study of dynamical systems commenced. In the qualitative approach, the local and asymptotic behaviors of an equation could be explained. Unfortunately, the qualitative study was restricted to mathematicians only. However, the power and necessity of the qualitative approach for analyzing the dynamical evolution of a system were subsequently enriched by *A.M. Lyapunov* (1857–1918), *G.D. Birkhoff* (1908–1944) and a group of mathematicians from the Russian schools, viz. *A.A. Andronov, V.I. Arnold,* and *co-workers* (1937, 1966, 1971, 1973).

In fact, Poincaré studied continuous systems in connection with an international competition held in honor of the 60th birthday of King Oscar II of Sweden and Norway. Of the four questions announced in the competition, he opted for the stability of the solar system. He won the prize. But the published memoir differed significantly from the original due to an error. In the study of dynamics he found it convenient to replace a continuous flow of time with a discrete analog. In celestial mechanics, Newton solved two-body problems: the motion of the Earth around the Sun. This is the famous inverse-square law: F(gravitational force) $\propto$

(distance between two bodies)$^{-2}$. Many great mathematicians and physicists tried to extend Newton's analytical method to the three-body problem (Sun, Earth, and Moon), but three or more than three-body problems were found to be remarkably difficult. At this juncture the situation seemed completely hopeless. This means that instead of asking about the exact positions of the planets always, one may ask "Is the solar system stable forever?" Answering this question Poincaré devised a new way of analysis which emphasized the qualitative approach. This eventually gave birth to the subject of '*Dynamical Systems.*' The Russian Schools, viz. Nonlinear Mechanics and the Gorki (*Andronov* or *Mandelstham Andronov*) contributed immensely to the mathematical theories for dynamical systems. In the dynamic evolution stability of a system is an important property. The Russian academician *A.M. Lyapunov* made a significant contribution to the stability/instability of a system. The mathematical definition of stability, construction of Lyapunov function, and Lyapunov theorem are extensively used for analyzing the stability of a particular class of systems. Also, Lyapunov exponent, assuming the exponential growth/decay with time of nearby orbits are applied for quantifying in chaotic motions.

One of the most remarkable breakthroughs in the early nineteenth century was the discovery of solitary waves in shallow water. Solitary waves are disturbances occurring on the surface of a fluid. They are dispersive in nature and form a single hump above the surface by displacing an equal amount of fluid, creating a bore at the place. Furthermore, these waves spread while propagating without changing their shape and velocity. The speed of these waves is proportional to the fluid depth, which causes large amplitude of the wave. Consequently, the speed of the wave increases with increase in the height of the wave. When a high amplitude solitary wave is formed behind a low amplitude wave, the former overtakes the latter keeping its shape unchanged with only a shift in position. This preservation of shape and velocity after collision suggests a particle like character of these waves and therefore called as solitary wave or solition, coined by Zabusky and Kruskal relevant with photon, proton, etc. John Scott Russel, the Scottish naval engineer first observed solitary wave on the Edinburgh-Glasgow canal in 1834 and he called it the '*great wave of translation.*' Russel reported his observations to the British Association in 1844 as 'Report on waves.' The mathematical form of these waves was given by Boussinesque in 1871 and subsequently by Lord Rayleigh in 1876. The equation for solitary wave was later derived by Korteweg and de Vries in 1895 and was popularly known as the KdV equation. This is a nonlinear equation with a balance between the nonlinear advection term and dispersion resulting in the propagation of solitary waves in an inviscid fluid.

In the first half of the twentieth century nonlinear dynamics was mainly concerned with nonlinear oscillations and their applications in physics, electrical circuits, mechanical engineering, and biological science. Oscillations occur widely in nature and are exploited in many manmade devices. Many great scientists, viz. *van der Pol* (1889–1959), *Alfred-Marie Liénard* (1869–1958), *Georg Duffing* (1861–1944), *John Edensor Littlewood* (1885–1977), *A.A. Andronov* (1901-1952), *M.L. Cartwright* (1900–1998), *N. Levinson* (1912–1975), and others made mathematical formulations and analyzed different aspects of nonlinear oscillations. Balthasar van der Pol had made significant contributions to areas such as limit cycles (isolated closed trajectory but neighboring trajectories are not closed either as they spiral toward the closed trajectory or away from it), relaxation oscillations (limiting cycles exhibit an extremely slow buildup followed by a sudden discharge, and then followed by another slow buildup and sudden discharge, and so on) of nonlinear electrical circuits, forced oscillators hysteresis and bifurcation phenomena. The well-known van der Pol equation first appeared in his paper entitled "On relaxation oscillations" published in the Philosophical Magazine in the year 1926. The van der Pol oscillator in a triode circuit is a simple example of a system with a limit cycle. He and var der Mark used van der Pol nonlinear equation to describe the heartbeat and an electrical model of the heart. Limit cycles were found later in mechanical and biological systems. The existence of limit cycle of a system is important scientifically and stable limit cycle exhibits self-sustained oscillations.

Species live in harmony in Nature. The existence of one species depends on the other, otherwise, one of the species would become extinct. Coexistence and sometimes mutual exclusion occur in reality in which one of the species becomes extinct. *Alfred James Lotka* (1880–1949), *Vito Volterra* (1860–1940), *Ronald Fisher* (1890–1962) and *Nicols Rashevsky* (1899–1972), and many others had explored the area of mathematical biology. The interaction dynamics of species, its mathematical model, and their asymptotic behaviors are useful tools in population dynamics of interacting species. Interaction dynamics among species have a great impact on the ecology and environment. The two-species predator–prey model in which one species preys on another was formulated by Lotka in 1910 and later by Volterra in 1926. This is known as the Lotka–Volterra model. In reality, the predator–prey populations rise and fall periodically and the maximum and minimum values (amplitudes) are relatively constant. However, this is not true for the Lotka–Volterra model. Different initial conditions can have solutions with different amplitudes. Holling and Tanner (1975) constructed a mathematical model for predator–prey populations whose solutions have the same amplitudes in the long time irrespective of the initial populations. The mathematical ecologist *Robert May* (1972) and many other scientists formulated several realistic population models that are useful in analyzing the population dynamics.

The perception of unpredictability in natural and social phenomena has a great impact on human thoughts and also in scientific evolutions. The conflict between determinism and free-will has been a long-standing continuing debate in philosophy. Nature is our great teacher. In the nineteenth century, the French engineer *Joseph Fourier* (1770–1830) wrote *"The study of Nature is the most productive source of mathematical discoveries. By offering a specific objective, it provides the advantage of excluding vague problems and unwieldy calculations. It is also a means to formulate mathematical analysis, and to isolate the most important aspects to know and to conserve. These fundamental elements are those which appear in all natural effects."*

Newtonian mechanics gives us a deterministic view of an object in which the future is determined from the laws of force and the initial conditions. There is no question of unpredictability or free-will in the Newtonian setup. In the beginning of the twentieth century experimental evidence, logical description, and also philosophical perception of physical phenomena, both in microscopic and macroscopic, made a breakthrough in science as a whole. The perception of infinity, how we approach the stage of infinitum, was a matter of great concern in the scientific community of the twentieth century. In the macroscopic world, studies particularly in oscillations in electrical, mechanical, and biological systems and the emergence of statistical mechanics either in fluid system or material body established the role and consequence of nonlinearity on their dynamics.

The existence of a chaotic orbit for a forced van der Pol equation (nonlinear equation) was proved mathematically by *M.L. Cartwright*, *J.E. Littlewood* about the 1950s. During this period mathematician *N. Levinson* showed that a physical model had a family of solutions that is unpredictable in nature. On the other hand, the turbulence in fluid flows is an unsolved and challenging problem in classical mechanics. The Soviet academician *A.N. Kolmogorov* (1903–1987), the greatest probabilist of the twentieth century and his co-workers made significant contributions to isotropic turbulence in fluids, the famous Kolmogorov-5/3 law (K41 theory) in the statistical equilibrium range. Kolmogorov's idea was based on the assumption of statistical equilibrium in an isotropic fluid turbulence. In turbulent motion large unstable eddies form and decay spontaneously into smaller unstable eddies, so that the energy-eddy cascade continues until the eddies reach a size so small that the cascade is damped effectively by fluid viscosity. *Geoffrey Ingram Taylor* (1886–1975), *von Karman* (1881–1963), and co-workers made significant contributions to the statistical description of turbulent motion. Yet, till today the nature of turbulent flow and universal law remain elusive. In nonlinear dynamics the well-known Kolmogorov–Arnold–Moser (KAM) theorem proves the existence of a positive measure set of quasi-periodic motions lying on invariant tori for Hamiltonian flows that are sufficiently close to completely integrable systems. This is the condition of weak chaotic motion in conservative systems. In chemistry, oscillation in chemical reaction such as the Belousov–Zhabotinsky reaction provided a wonderful example of relaxation oscillation. The experiment was conducted by the Russian biochemist Boris Belousov around the 1950s. However, he could not publish his discovery as in those days it was believed that chemical reagents must go monotonically to equilibrium solution, no oscillatory motion. Later, Zhabotinsky confirmed Belousov's results and brought this discovery to the notice of the scientific community at an international conference in Prague in the year 1968. For the progress of nonlinear science in the twentieth century both in theory and experiments such as hydrodynamic (water, helium, liquid mercury), electronic, biological (heart muscles), chemical, etc., scientists believed that simple looking systems can display highly complex seemingly random behavior. It was Henri Poincaré who first reported the notion of sensitivity to initial conditions in his work. The quotation from his essay on Science and Method is relevant here: *"It may happen that small differences in the initial produce very great ones in the final phenomena. A small error in the former will produce an enormous error in the later prediction becomes impossible."* Perhaps the most intriguing characteristic of a chaotic system is the extreme sensitivity to initial conditions. Naturally, there is a need to develop the science of the unpredictable. The real breakthrough came from the computational result of a simple nonlinear system. In the year 1963, *Edward Lorenz* (1917–2008) published a paper entitled *"Deterministic Non-periodic Flow."* In this paper he derived equations for thermal convection in a simplified model of atmospheric flow and noticed a very strange thing that the solutions of the equations could be unpredictable and irregular despite being deterministic. The sensitive dependence of the evolution of a system for an infinitesimal change of initial conditions is called the butterfly effect. Deterministic systems may exhibit a regular behavior for some values of their control parameters and irregular behavior for other values. Deterministic systems can give rise to motions that are essentially random and the long-term prediction is impossible. Another paper from the discrete system *"Differential Dynamical Systems"* published by *Stephen Smale* proved mathematically the existence of chaotic solutions and gave a geometric description of the chaotic set, the Smale horseshoe map. Mathematicians/physicists such as *Lev. D. Laudau* (fluid dynamics and stability), *James Yorke* ("Period three implies chaos"), *Robert May* (mathematical biology), *Enrico Fermi* (ergodicity), *Stanislaw Ulam* (the growth of patterns in cellular automata, lattice dynamics), *J.G. Senai* (ergodic theory), *Sarkovskii* (ordering of infinitely many periodic points of a map), *Ruelle* and *Takens* (fluid turbulence and 'strange attractor'), *A. Libchaber,* and *J. Maurer* (intermittency as a route to fluid turbulence) and many others are the great contributors to the development of nonlinear science and chaos theory. In the mid-1970s a remarkable discovery was made by *Mitchell Feigenbaum*: the universality in chaotic regime for unimodal maps undergoing period doubling bifurcation.

The concept of fractal geometry or fractal objects is about 50 years old and was first introduced by the Polish–French–American mathematician *Benoit Mandelbrot* (1924–2010) in 1975. Fractals are structures that are irregular, erratic and self-similarity is intrinsic in most of these objects. Fractal objects consist of self-similarity between scales, that is, the patterns observed in larger scales repeat in ever decreasing smaller and smaller scales. In short, a fractal object is made up of parts similar to the whole in some way but lacks a characteristic smallest scale to measure. Fractal geometry is different from Euclidean geometry, and finds order in chaotic shapes and processes. Chaotic orbit can be expressed in terms of fractal object. Scaling and self-similarity are important features in most natural and manmade fractal objects. There exist numerous examples of fractals in natural and physical sciences. One can also find a number of examples of fractals in the human anatomy. For instance, lungs, heart, and many other anatomical structures are either fractal or fractal-like. Moreover, in recent years the idea of fractals is being exploited to find applications in medical science to curb fatal diseases. Mandelbrot and other researchers have shown how fractals could be explored in different areas and chaos in particular. The phenomenon of chaos is a realistic phenomenon and therefore one has to understand and realize chaos in usual incidents happening in our everyday life. The study of chaotic phenomena has begun in full length nowadays and is widely applied in different areas. The theory of chaos is now applied in computer security, digital watermarking, secure data aggregation and video surveillance successfully. Thus, chaotic phenomena are not only destructive as in tsunami, tornado, etc., but can also be effectively utilized for the welfare of human beings. Chaos has been considered as the third greatest discovery, after relativity and quantum mechanics in the twentieth century science and philosophy. In the past 20 years scientists and technologists have been realizing the potential use of chaos in natural and technological sciences.

1.2 Dynamical Systems

Dynamics is primarily the study of the time-evolutionary process and the corresponding system of equations is known as dynamical system. Generally, a system of n first-order differential equations in the space $\mathbb{R}^n$ is called a dynamical system of dimension n which determines the time behavior of evolutionary process. Evolutionary processes may possess the properties of determinacy/non-determinacy, finite/infinite dimensionality, and differentiability. A process is called **deterministic** if its entire future course and its entire past are *uniquely* determined by its state at the present time. Otherwise, the process is called **nondeterministic**. However, the process may be **semi-deterministic** (determined, but not uniquely). In classical mechanics the motion of a system whose future and past are uniquely determined by the initial positions and the initial velocities is an example of a deterministic dynamical system. The evolutionary process may describe, viz. (i) a continuous-time process and (ii) a discrete-time process. The continuous-time process is represented by differential equations, whereas the discrete-time process is by difference equations (or maps). The continuous-time dynamical systems may be described mathematically as follows:

Let $\underset{\sim}{x} = \underset{\sim}{x}(t) \in \mathbb{R}^n$, $t \in I \subseteq \mathbb{R}$ be the vector representing the dynamics of a continuous system (continuous-time system). The mathematical representation of the system may be written as

$$\frac{d\underset{\sim}{x}}{dt} = \dot{\underset{\sim}{x}} = \underset{\sim}{f}(\underset{\sim}{x}, t) \tag{1.1}$$

where $\underset{\sim}{f}(\underset{\sim}{x}, t)$ is a sufficiently smooth function defined on some subset $U \subset \mathbb{R}^n \times \mathbb{R}$. Schematically, this can be shown as

$$\underset{\text{(state space)}}{\mathbb{R}^n} \times \underset{\text{(time)}}{\mathbb{R}} = \underset{\text{(space of motions)}}{\mathbb{R}^{n+1}}$$

The variable t is usually interpreted as time and the function $\underset{\sim}{f}(\underset{\sim}{x}, t)$ is generally nonlinear. The time interval may be finite, semi-finite or infinite. On the other hand, the discrete system is related to a discrete map (given only at equally spaced points of time) such that from a point x_0, one can obtain a point x_1 which in turn maps into

x_2, and so on. In other words, $x_{n+1} = g(x_n) = g(g(x_{n-1}))$, etc. This is also written in the form $x_{n+1} = g(x_n) = g^2(x_{n-1}) = \cdots$. The discrete system will be discussed in the later chapters.

If the right-hand side of Eq. (1.1) is explicitly time independent then the system is called **autonomous**. The trajectories of such a system do not change in time. On the other hand, if the right-hand side of Eq. (1.1) has explicit dependence on time then the system is called **nonautonomous**. An n-dimensional nonautonomous system can be converted into autonomous form by introducing a new dependent variable x_{n+1} such that $x_{n+1} = t$. In general, the solution of Eq. (1.1) is difficult or sometimes impossible to obtain when the function $f(\underset{\sim}{x}, t)$ is nonlinear, except in some special cases. Examples of autonomous and nonautonomous systems are given below.

(i) **Autonomous systems**

    (a) $\ddot{x} + \alpha\dot{x} + \beta x = 0, \alpha, \beta > 0$. This is a damped linear harmonic oscillator. The parameters α and β are, respectively, the strength of damping and the strength of linear restoring force.

    (b) $\ddot{x} + \omega^2 \sin x = 0, \ \omega = \sqrt{g/L}$. g is the gravitational acceleration, L the string length. This is a simple undamped nonlinear oscillator (pendulum).

    (c) $\left.\begin{array}{l} \dot{x} = \alpha x - \beta xy \\ \dot{y} = -\gamma y + \delta xy \end{array}\right\}$. This is the well-known Lotka–Volterra predator–prey

        model, where $\alpha, \beta, \gamma, \delta$ are all positive constants.

    (d) $\ddot{x} - \mu(1 - x^2)\dot{x} + \beta x = 0, \mu > 0$. This is the well-known van der Pol oscillator.

(ii) **Nonautonomous systems**

    (a) $\ddot{x} + \alpha\dot{x} + \beta x = f \cos \omega t, \alpha, \beta > 0$. This is an example of linear oscillator with external time-dependent force. f and ω are the amplitude and frequency of driving force, respectively.

    (b) $\ddot{x} + \alpha\dot{x} + \omega_0^2 x + \beta x^3 = f \sin \omega t$. This is a Duffing nonlinear oscillator with cubic restoring force. α is the strength of damping, ω_0 is the natural frequency and β is the strength of the nonlinear restoring force.

    (c) $\ddot{x} - \mu(1 - x^2)\dot{x} + \beta x = f \cos \omega t, \mu > 0$. This is a van der Pol nonlinear forced oscillator.

    (d) $\ddot{x} - \mu(1 - x^2)\dot{x} + \omega_0^2 x + \beta x^3 = f \cos \omega t$. This is a Duffing-van der Pol nonlinear forced oscillator.

Some examples of dynamical systems

(a) The most common example of a dynamical system is Newtonian systems governed by Newton's law of motion. This law states that the acceleration of a particle is determined by the force per unit mass. The force can be a function of the velocity $(\dot{x})$ and the position (x) and so the Newtonian systems take the form

$$m\ddot{x} = F(x, \dot{x}), \quad (m = \text{mass}, F = \text{force}). \tag{1.2}$$

Equation (1.2) may be written as a system of two first-order differential equations as

$$\dot{x} = y, \text{ and } \dot{y} = F(x, y) \tag{1.3}$$

System (1.3) may be viewed as a dynamical system of dimension two in the xy-plane and the dynamics is a set of trajectories giving time evolution of motion.

(b) The simple exponential growth model for a single population is expressed mathematically as

$$\frac{dx}{dt} = rx \text{ with } x = x_0 \text{ at } t = 0, \tag{1.4}$$

where $r > 0$ is the population growth parameter. The solution of (1.4) is $x(t) = x_0 e^{rt}$. This solution expresses the simplest model for population growth with time in unrestricted resources and the population $x(t) \to \infty$ as $t \to \infty$. Obviously, this model does not obey realistic population growth of any species.

The simple population growth model, considering effects like intraspecies competitions, depletion of resources with population growth is given as

$$\frac{dx}{dt} = (r - bx)x \tag{1.5}$$

with the condition $x = x_0$ at $t = 0$. The solution of (1.5) is given as

$$x(t) = \frac{\left(\frac{r}{b}\right)x_0}{x_0 + \left(\frac{r}{b} - x_0\right)e^{-rt}} \tag{1.6}$$

Clearly $x \to \frac{r}{b}$ as $t \to \infty$ for both the cases $x_0 > \frac{r}{b}$ and $x_0 < \frac{r}{b}$.

This growth model is known as the **Logistic growth model** of population. The graphical representation of the above solution is shown in Fig. 1.1.

This simple model shows that population $x(t)$ is of constant growth rate after some time t.

(c) Populations of two competing species (predator and prey populations) could be modeled mathematically. The predator–prey population model was first formulated by *Alfred J. Lotka* (1880–1949) in the year 1910 and later by *Vito Volterra* (1860–1940) in the year 1926. This is known as Lotka–Volterra predator–prey model. In this model the fox population preys on the rabbit population. The population density of rabbit affects the population density of fox, since the latter relies on the former for food. If the density of rabbit is high, the fox population decreases, while when the fox population increases,

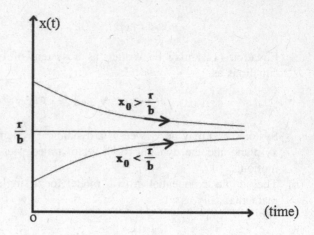

Fig. 1.1 Graphical representation of population growth model

the rabbit population decreases. When the rabbit population falls, the fox population also falls. When the fox population drops, the rabbits can multiply again and so on. The growth or decrease of two populations could be analyzed using dynamical system principles. The dynamical equations for predator–prey model are given as

$$\left. \begin{array}{l} \dot{x} = \alpha x - \beta xy \\ \dot{y} = -\gamma y + \delta xy \end{array} \right\} \tag{1.7}$$

where x denotes the population density of the prey and y, the population density of the predator. The parameter α represents the growth rate of the prey in the absence of intersection with the predators whereas the parameter γ represents the death rate of the predators in the absence of interaction with the prey and β, δ are the interaction parameters and are all constants (for simple model). Using the dynamical principle one can obtain a necessary condition for coexistence of the two species. In this model the survival of the predators depends entirely on the population of the prey. If initially $x = 0$, then $\dot{y} = -\gamma y$, that is, $y(t) = y(0)e^{-\gamma t}$ and $y(t) \to 0$ as $t \to \infty$ (see the book Arrowsmith and Place [1]).

(d) Suppose we have an LCR circuit consisting of a resistor of resistance R, a capacitor of capacitance C, and an inductor of inductance L. In a simple electrical circuit the values of R, C, and L are always nonnegative and are independent of time t. Kirchhoff's current law (the sum of the currents flowing into a node is equal to the sum of the currents flowing out of it) is satisfied if we pass a current I to the closed loop as shown in Fig. 1.2.

According to Kirchhoff's voltage law of the circuit (the sum of the potential differences around any closed loop in a circuit is zero), we have the equation

Fig. 1.2 Schematic of an
LCR circuit

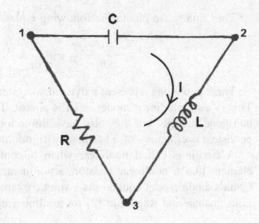

$$V_{12} + V_{23} + V_{31} = 0 \qquad (1.8)$$

Here V_{ij} denotes the voltage difference between node i and node j.
From Ohm's law, we get the relation

$$V_{31} = IR \qquad (1.9)$$

Also, from the definition of capacitance C, we have

$$C \frac{dV_{12}}{dt} = I \qquad (1.10)$$

Again, Faraday's law of inductance gives

$$L \frac{dI}{dt} = V_{23} \qquad (1.11)$$

Substituting (1.8) and (1.10) into (1.11) and writing $V_{12} = V$, we get

$$V + L \frac{dI}{dt} + IR = 0$$

or,

$$\frac{dI}{dt} = -\frac{R}{L} I - \frac{V}{L} \qquad (1.12)$$

Again, from (1.10)

$$\frac{dV}{dt} = \frac{I}{C}. \qquad (1.13)$$

Thus finally we obtain the following equations

$$\frac{dV}{dt} = \frac{I}{C} \text{ and } \frac{dI}{dt} = -\frac{R}{L}I - \frac{V}{L}. \tag{1.14}$$

These equations represent a dynamical system of dimension two in the VI plane. This is a simple linear model in LCR circuit. The linear models have undoubtedly had good success, but they also have limitations. Linear models can only produce persistent oscillations of a harmonic (trigonometric) type.

A circuit is called nonlinear when it contains at least one nonlinear circuit element like a nonlinear resistor, a nonlinear capacitor or a nonlinear inductor. Chua's diode model equation is a simple example of nonlinear electric circuit (see Lakshmanan and Rajasekar [2] for nonlinear electrical circuit).

1.3 Flows

The time-evolutionary process may be described as a flow of a vector field. Generally, flow is frequently used for discussing the dynamics as a whole rather than the evolution of a system at a particular point. The solution $x\,(t)$ of a system $\dot{x} = f(x)$ which satisfies $x\,(t_0) = x_0$ gives the past $(t < t_0)$ and future $(t > t_0)$ evolutions of the system. Mathematically, the flow is defined by $\phi_t(x) : U \to \mathbb{R}^n$ where $\phi_t(x) = \phi(t, x)$ is a smooth vector function of $x \in U \subseteq \mathbb{R}^n$ and $t \in I \subseteq \mathbb{R}$ satisfying the equation

$$\frac{d}{dt}\phi_t(x) = f(\phi_t(x))$$

for all t such that the solution through x exists and $\phi(0, x) = x$. The flow $\phi_t(x)$ satisfies the following properties:

(a) $\phi_o = I_d$, (b) $\phi_{t+s} = \phi_t \circ \phi_s$.

Some flows may also satisfy the property

$$\phi(t + s, \; x) = \phi(t, \phi(s, \; x)) = \phi(s, \phi(t, \; x)) = \phi(s + t, x).$$

Flows in $\mathbb{R}$: Consider a one-dimensional autonomous system represented by $\dot{x} = f(x), x \in \mathbb{R}$. We can imagine that a fluid is flowing along the real line with local velocity $f(x)$. This imaginary fluid is called the **phase fluid** and the real line is called the **phase line**. For solution of the system $\dot{x} = f(x)$ starting from an arbitrary initial position x_0, we place an imaginary particle, called a **phase point**, at x_0 and watch how it moves along with the flow in phase line in varying time t. As time goes on, the phase point (x, t) in the one-dimensional system $\dot{x} = f(x)$ with $x(0) = x_0$ moves along the x-axis according to some function $\phi(t, x_0)$. The function $\phi(t, x_0)$ is called the **trajectory** for a given initial state x_0, and the set $\{\phi(t, x_0) | t \in I \subseteq \mathbb{R}\}$

is the orbit of $x_0 \in \mathbb{R}$. The set of all qualitative trajectories of the system is called **phase portrait**.

Flows in $\mathbb{R}^2$: Consider a two-dimensional system represented by the following equations $\dot{x} = f(x, y)$, $\dot{y} = g(x, y)$, $(x, y) \in \mathbb{R}^2$. An imaginary fluid particle flows in the plane $\mathbb{R}^2$, known as phase plane of the system. The succession of states given parametrically by $x = x(t)$, $y = y(t)$ trace out a curve through some initial point $P(x(t_0), y(t_0))$ is called a **phase path**. The set $\{\phi(t, \underset{\sim}{x_0}) | t \in I \subseteq \mathbb{R}\}$ is the orbit of $\underset{\sim}{x_0}$ in $\mathbb{R}^2$. There are an infinite number of trajectories that would fill the phase plane when they are plotted. But the qualitative behavior can be determined by plotting a few trajectories with different initial conditions. The phase portrait displays how the qualitative behavior of a system is changing as x and y varies with time t. An orbit is called periodic if $x(t + p) = x(t)$ for some $p > 0$, for all t. The smallest integer p for which the relation is satisfied is called the prime period of the orbit. Flows in $\mathbb{R}$ cannot have oscillatory or closed path.

Flows in $\mathbb{R}^n$: Let us now define an autonomous system representing n ordinary differential equations as

$$\left. \begin{array}{l} \dot{x}_1 = f_1(x_1, x_2, \ldots, x_n) \\ \dot{x}_2 = f_2(x_1, x_2, \ldots, x_n) \\ \vdots \\ \dot{x}_n = f_n(x_1, x_2, \ldots, x_n) \end{array} \right\}$$

which can also be written in symbolic notation as $\dot{\underset{\sim}{x}} = f(\underset{\sim}{x})$, where $\underset{\sim}{x} = (x_1, x_2, \ldots, x_n)$ and $f = (f_1, f_2, \ldots, f_n)$. The solution of this system with the initial condition $\underset{\sim}{x}(0) = \underset{\sim}{x_0}$ can be thought as a continuous curve in the phase space $\mathbb{R}^n$ parameterized by time $t \in I \subseteq \mathbb{R}$. So the set of all states of the evolutionary process is represented by an n-valued vector field in $\mathbb{R}^n$. The solutions of the system with different initial conditions describe a family of phase curves in the phase space, called the phase portrait of the system. The vector field $f(\underset{\sim}{x})$ is everywhere tangent to these curves and their orientation is directed by the direction of the tangent vector of $f(\underset{\sim}{x})$.

1.4 Evolution

Consider a system $\dot{\underset{\sim}{x}} = f(\underset{\sim}{x})$, $\underset{\sim}{x} \in \mathbb{R}^n$ with initial conditions $\underset{\sim}{x}(t_0) = \underset{\sim}{x_0}$. Let $E \subset \mathbb{R}^n$ be an open set and $f \in C^1(E)$. For $\underset{\sim}{x_0} \in E$, let $\phi(t, \underset{\sim}{x_0})$ be a solution of the above system on the maximum interval of existence $I(\underset{\sim}{x_0}) \subset \mathbb{R}$. The mapping $\phi_t : \mathbb{R}^n \to \mathbb{R}^n$ defined by $\phi_t(\underset{\sim}{x_0}) = \phi(t, \underset{\sim}{x_0})$ is known as **evolution operator** of the system. The linear flow for the system $\dot{\underset{\sim}{x}} = A\underset{\sim}{x}$ with $\underset{\sim}{x}(t_0) = \underset{\sim}{x_0}$, is defined by $\phi_t : \mathbb{R}^n \to \mathbb{R}^n$ and $\phi_t = e^{At}$, the exponential matrix. The mappings ϕ_t for both linear and nonlinear systems satisfy the following properties:

(i) $\phi_0(\underset{\sim}{x}) = \underset{\sim}{x}$

(ii) $\phi_s(\phi_t(\underset{\sim}{x})) = \phi_{s+t}(\underset{\sim}{x}), \forall s, t \in \mathbb{R}$

(iii) $\phi_t(\phi_{-t}(\underset{\sim}{x})) = \phi_{-t}(\phi_t(\underset{\sim}{x})) = \underset{\sim}{x}, \forall t \in \mathbb{R}$

In general a dynamical system may be viewed as group of nonlinear / linear operators evolving as $\{\phi_t(\underset{\sim}{x}), t \in \mathbb{R}, \underset{\sim}{x} \in \mathbb{R}^n\}$. The following dynamical group properties hold good:

(i) $\phi_t \phi_s \in \{\phi_t(\underset{\sim}{x}), t \in \mathbb{R}, \underset{\sim}{x} \in \mathbb{R}^n\}$ (Closure property)

(ii) $\phi_t(\phi_s \phi_r) = (\phi_t \phi_s)\phi_r$ (Associative property)

(iii) $\phi_0(\underset{\sim}{x}) = \underset{\sim}{x}$, ϕ_0 being the Identity operator.

(iv) $\phi_t \phi_{-t} = \phi_{-t} \phi_t = \phi_0$, where ϕ_{-t} is the Inverse of ϕ_t.

For some cases the flow satisfies the commutative property $\phi_t \phi_s = \phi_s \phi_t$.

1.5   Fixed Points of a System

The notion of fixed point is important in analyzing the local behavior of a system. The fixed point is nothing but a constant or equilibrium or invariant solution of a system. A point is a fixed point of the flow generated by an autonomous system $\dot{x} = f(\underset{\sim}{x})$, $\underset{\sim}{x} \in \mathbb{R}^n$ if and only if $\phi(t, \underset{\sim}{x}) = \underset{\sim}{x}$ for all $t \in \mathbb{R}$. Consequently in continuous system, this gives $\dot{x} = 0 \Rightarrow f(\underset{\sim}{x}) = 0$. For nonautonomous systems fixed point can be defined for a fixed time interval. A fixed point is also known as a **critical point** or an **equilibrium point** or a **stationary point**. This point is also called **stagnation point** with respect to the flow ϕ_t in $\mathbb{R}^n$. Flows on line may have no fixed points, only one fixed point, finite number of fixed points, and infinite number of fixed points. For example, the flow $\dot{x} = 5$ (no fixed points), $\dot{x} = x$ (only one fixed point), $\dot{x} = x^2 - 1$ (two fixed points), and $\dot{x} = \sin x$ (infinite number of fixed points).

1.6   Linear Stability Analysis

A fixed point, say $\underset{\sim}{x}_0$ is said to be stable if for a given $\varepsilon > 0$, there exists a $\delta > 0$ depending upon ε such that for all $t \geq t_0$, $\| \underset{\sim}{x}(t) - \underset{\sim}{x}_0(t) \| < \varepsilon$, whenever $\| \underset{\sim}{x}(t_0) - \underset{\sim}{x}_0(t_0) \| < \delta$, where $\|.\| : \mathbb{R}^n \to \mathbb{R}$ denotes the norm of a vector in $\mathbb{R}^n$. Otherwise, the fixed point is called unstable. In linear stability analysis the quadratic and higher order terms in the Taylor series expansion about a fixed point x^* of a system $\dot{x} = f(x)$, $x \in \mathbb{R}$ are neglected due to the smallness of the terms. Consider a small perturbation quantity $\xi(t)$, away from the fixed point x^*, such that $x(t) = x^* + \xi(t)$. We see whether the perturbation grows or decays as time goes on. So we get the perturbation equation as

$$\dot{\xi} = \dot{x} = f(x) = f(x^* + \xi).$$

Taylor series expansion of $f(x^* + \xi)$ gives

$$\dot{\xi} = f(x^*) + \xi f'(x^*) + \frac{\xi^2}{2} f''(x^*) + \cdots$$

According to linear stability analysis, we get

$$\dot{\xi} = \xi f'(x^*) [\because f(x^*) = 0]$$

Assuming $f'(x^*) \neq 0$, the perturbation $\xi(t)$ grows exponentially if $f'(x^*) > 0$ and decays exponentially if $f'(x^*) < 0$. Linear theory fails if $f'(x^*) = 0$ and then higher order derivatives must be considered in the neighborhood of fixed point for stability analysis of the system.

Example 1.1 Find the evolution operator ϕ_t for the one-dimensional flow $\dot{x} = -x^2$. Show that ϕ_t forms a dynamical group. Is it a commutative group?

Solution The solutions of the given system are obtained as below:

$$\dot{x} = \frac{dx}{dt} = -x^2 \Rightarrow \frac{1}{x} = t + A \Rightarrow x(t) = \frac{1}{t+A}$$

in any interval of $\mathbb{R}$ that does not contain the point $x = 0$, where A is a constant.
If we take starting point $x(0) = x_0$, then $A = 1/x_0$ and so we get

$$x(t) = \frac{x_0}{1 + x_0 t}, \quad t \neq -1/x_0.$$

The point $x = 0$ is not included in this solution. But it is the fixed point of the given system, because $\dot{x} = 0 \Leftrightarrow x = 0$. Therefore, $\phi_t(0) = 0$ for all $t \in \mathbb{R}$. So the evolution operator of the system is given as $\phi_t(x) = \frac{x}{1+xt}$ for all $x \in \mathbb{R}$.

The evolution operator ϕ_t is not defined for all $t \in \mathbb{R}$. For example, if $t = -1/x$, $x \neq 0$, then ϕ_t is undefined. Thus we see that the interval in which ϕ_t is defined is completely dependent on x.

We shall now examine the group properties of the evolution operator ϕ_t below:

(i) $\phi_r \phi_s \in \{\phi_t(x), t \in \mathbb{R}, x \in \mathbb{R}\} \ \forall r, s \in \mathbb{R}$ (Closure property)
Now,

$$\phi_r(y) = \frac{y}{1+yr}. \text{ Take } y = \frac{x}{1+xs}$$

$$= \frac{x/1+sx}{1 + \frac{xr}{1+xs}} = \frac{x}{1+xs+xr} = \frac{x}{1+x(s+r)}$$

$$= \phi_{s+r} \in \{\phi_t(x), t \in \mathbb{R}, x \in \mathbb{R}\}$$

(ii) $\phi_t(\phi_s\phi_r) = (\phi_t\phi_s)\phi_r$ (Associative property)

$$\text{L.H.S.} = \phi_t((\phi_s\phi_r)(x)) = \phi_t(y) = \frac{y}{1+yt} = \frac{z}{1+zs} = \frac{x}{1+x(r+s)}, y = \phi_s(\phi_r(x))$$

$$\left(\text{where } y = \phi_s(z), z = \phi_r(x) = \frac{x}{1+rx}\right)$$

$$\therefore \text{L.H.S} = \frac{x}{1+x(t+r+s)} = \phi_{t+r+s}(x)$$

$$\text{R.H.S.} = ((\phi_t\phi_s)\phi_r(x))$$

Now,

$$\phi_t(y) = \frac{y}{1+yt}, y = \phi_s(x) = \frac{x}{1+sx}$$

$$= \frac{x}{1+x(t+s)} = \phi_{t+s}(x)$$

$$\phi_{t+s}(\phi_r)(x) = \phi_{t+s}(z) = \frac{z}{1+z(t+s)}, z = \phi_r(x) = \frac{x}{1+rx}$$

$$\phi_{t+s}(\phi_r)(x) = \frac{x}{1+x(t+s+r)} = \phi_{t+s+r}(x)$$

Hence, $\phi_t(\phi_s\phi_r)(x) = (\phi_s\phi_r)\phi_t(x), \forall x \in \mathbb{R}$.

(iii) $\phi_0(x) = \frac{x}{1+x\cdot0} = x$, ϕ_0 is the identity operator.

$$\phi_t\phi_{-t}(x) = \phi_t(y) = \frac{y}{1+ty}, y = \phi_{-t}(x) = \frac{x}{1-tx}$$

(iv)

$$= \frac{x}{1-tx+tx} = x = \phi_0(x) \quad (\phi_{-t} \text{ is the inverse of } \phi_t)$$

Hence the flow evolution operator forms a dynamical group.

(v) $\phi_t\phi_s = \phi_s\phi_t$

Now,

$$(\phi_t\phi_s)(x) = \phi_t(y) = \frac{y}{1+ty}, y = \phi_s(x) = \frac{x}{1+xs}$$

$$= \frac{x}{1+x(t+s)} = \phi_{t+s}(x)$$

$$\phi_s\phi_t(x) = \phi_s(z) = \frac{z}{1+sz}, z = \phi_t(x) = \frac{x}{1+tx}$$

$$= \frac{x}{1+tx+sx} = \frac{x}{1+(s+t)x} = \phi_{s+t}(x)$$

So, $\phi_t\phi_s = \phi_s\phi_t$ (Commutative property).

Thus, the evolution operator ϕ_t forms a commutative group.

Example 1.2 Find the evolution operator ϕ_t for the system $\dot{x} = x^2 - 1$ and also verify that $\phi_t(\phi_s(x)) = \phi_{t+s}(x)$ for all $s, t \in \mathbb{R}$. Show that the evolution operator forms a dynamical group. Examine whether it is commutative dynamical group.

Solution The solutions of the system satisfy the equation

$$\frac{dx}{dt} = x^2 - 1 \Rightarrow \frac{1}{2} \log \left| \frac{x-1}{x+1} \right| = t + A, \quad x \neq 1, -1$$

$$\Rightarrow \frac{x-1}{x+1} = Be^{2t}, \quad B = \pm e^{2A}$$

$$\Rightarrow x(t) = \frac{Be^{2t} + 1}{1 - Be^{2t}}$$

If we take $x(0) = x_0$, then from the above relation we get $B = (x_0 - 1)/(x_0 + 1)$ and so the solution can be written as

$$x(t) = \frac{(x_0 - 1)e^{2t} + x_0 + 1}{x_0 + 1 - (x_0 - 1)e^{2t}}.$$

This solution is defined for all $t \in \mathbb{R}$ and for $x_0 \neq -1, 1$. But the points $x = -1, 1$ are the fixed points of the system. Therefore, the evolution operator of the given one-dimensional system is

$$\phi_t(x) = \frac{(x-1)e^{2t} + x + 1}{x + 1 - (x-1)e^{2t}} \text{ for all } x, t \in \mathbb{R}.$$

Now for all $x, t, s \in \mathbb{R}$, we have

$$\phi_t(\phi_s(x)) = \phi_t(y) = \frac{(y-1)e^{2t} + y + 1}{y + 1 - (y-1)e^{2t}}$$

where $y = \phi_s(x) = \frac{(x-1)e^{2s} + x + 1}{x + 1 - (x-1)e^{2s}}$

Substituting this value of y into the above expression we get

$$\phi_t(\phi_s(x)) = \frac{(x-1)e^{2(t+s)} + x + 1}{x + 1 - (x-1)e^{2(t+s)}} = \phi_{t+s}(x).$$

(ii) $\phi_0(x) = \frac{x - 1 + x + 1}{x + 1 - x + 1} = x$, the identity operator

(iii) $\phi_t\phi_{-t}(x) = \phi_t(y) = \dfrac{(y-1)e^{2t}+y+1}{y+1-(y-1)e^{2t}}, y = \phi_{-t}(x) = \dfrac{(x-1)e^{-2t}+x+1}{x+1-(x-1)e^{-2t}}$

$$= \dfrac{\left(\frac{(x-1)e^{-2t}+x+1}{(x+1)-(x-1)e^{-2t}}-1\right)e^{2t}+\left(\frac{(x-1)e^{-2t}+x+1}{(x+1)-(x-1)e^{-2t}}+1\right)}{\left(\frac{(x-1)e^{-2t}+x+1}{(x+1)-(x-1)e^{-2t}}+1\right)-\left(\frac{(x-1)e^{-2t}+x+1}{(x+1)-(x-1)e^{-2t}}-1\right)e^{2t}}$$

$$= \dfrac{\left(\frac{(x-1)e^{-2t}+x+1-(x+1)+(x-1)e^{-2t}}{(x+1)-(x-1)e^{-2t}}\right)e^{2t}+\left(\frac{(x-1)e^{-2t}+x+1+(x+1)-(x-1)e^{-2t}}{(x+1)-(x-1)e^{-2t}}\right)}{\left(\frac{(x-1)e^{-2t}+x+1+(x+1)-(x-1)e^{-2t}}{(x+1)-(x-1)e^{-2t}}\right)-\left(\frac{(x-1)e^{-2t}+x+1-(x+1)+(x-1)e^{-2t}}{(x+1)-(x-1)e^{-2t}}\right)e^{2t}}$$

$$= \dfrac{2(x-1)+2(x+1)}{2(x+1)-2(x-1)} = x$$

Again,

$$\phi_{-t}\phi_t(x) = \phi_{-t}(y) = \dfrac{(y-1)e^{-2t}+y+1}{y+1-(y-1)e^{-2t}},$$

$$y = \phi_t(x) = \dfrac{(x-1)e^{2t}+x+1}{x+1-(x-1)e^{2t}}$$

$$= \dfrac{\left(\frac{(x-1)e^{2t}+(x+1)}{(x+1)-(x-1)e^{2t}}-1\right)e^{-2t}+\left(\frac{(x-1)e^{2t}+(x+1)}{(x-1)-(x-1)e^{2t}}+1\right)}{\left(\frac{(x-1)e^{2t}+(x+1)}{(x+1)e^{2t}-(x-1)e^{2t}}+1\right)-\left(\frac{(x-1)e^{2t}+(x+1)}{(x+1)-(x-1)e^{2t}}-1\right)e^{-2t}}$$

$$= \dfrac{\left(\frac{(x-1)e^{2t}+(x+1)-(x+1)-(x-1)e^{2t}}{(x+1)-(x-1)e^{2t}}\right)e^{-2t}+\left(\frac{(x-1)e^{2t}+(x+1)+(x+1)-(x-1)e^{2t}}{(x-1)-(x-1)e^{2t}}\right)}{\left(\frac{(x-1)e^{2t}+(x+1)+(x+1)-(x-1)e^{2t}}{(x+1)e^{2t}-(x-1)e^{2t}}\right)-\left(\frac{(x-1)e^{2t}+(x+1)-(x+1)-(x-1)e^{2t}}{(x+1)-(x-1)e^{2t}}\right)e^{-2t}}$$

$$= \dfrac{2(x-1)+2(x+1)}{2(x+1)-2(x-1)} = x$$

Hence, the evolution operator forms a dynamical group.

(iv) $\phi_t(\phi_s\phi_r) = (\phi_t\phi_s)\phi_r$ (Associative property)

L.H.S. $\phi_t(y) = \frac{(y-1)e^{2t}+y+1}{y+1-(y-1)e^{2t}}$
where,

$$y = \phi_s(z), z = \phi_r(x) = \dfrac{(x-1)e^{2r}+x+1}{x+1-(x-1)e^{2r}}$$

$$= \dfrac{(z-1)e^{2s}+z+1}{z+1-(z-1)e^{2s}} = \dfrac{(x-1)e^{2(r+s)}+x+1}{x+1-(x-1)e^{2(r+s)}}$$

L.H.S. $= \dfrac{(x-1)e^{2(r+s+t)}+x+1}{x+1-(x-1)e^{2(r+s+t)}} = \phi_{t+r+s}(x)$

R.H.S. $= (\phi_t\phi_s)\phi_r$

Now,

$$\phi_t(y) = \frac{(y-1)e^{2t}+y+1}{y+1-(y-1)e^{2t}}, y = \phi_s(x) = \frac{(x-1)e^{2s}+x+1}{x+1-(x-1)e^{2s}}$$

$$= \frac{(x-1)e^{2(t+s)}+x+1}{x+1-(x-1)e^{2(t+s)}} = \phi_{t+s}(x)$$

$$\phi_{t+s}(\phi_r)(x) = \phi_{t+s}(z) = \frac{(z-1)e^{2t}+z+1}{z+1-(z-1)e^{2t}}, z = \phi_r(x) = \frac{(x-1)e^{2r}+x+1}{x+1-(x-1)e^{2r}}$$

$$\phi_{t+s}(\phi_r)(x) = \frac{(x-1)e^{2(t+s+r)}+x+1}{x+1-(x-1)e^{2(t+s+r)}} = \phi_{t+s+r}(x)$$

$$\Rightarrow \phi_t(\phi_s\phi_r) = (\phi_s\phi_r)\phi_t$$

(v) $\phi_t\phi_s = \phi_s\phi_t$
   Now,

$$\phi_t(y) = \frac{(y-1)e^{2t}+y+1}{y+1-(y-1)e^{2t}}, y = \phi_s(x) = \frac{(x-1)e^{2s}+x+1}{x+1-(x-1)e^{2s}}$$

$$= \frac{\left(\frac{(x-1)e^{2s}+x+1}{x+1-(x-1)e^{2s}}-1\right)e^{2t}+\left(\frac{(x-1)e^{2s}+x+1}{x+1-(x-1)e^{2s}}+1\right)}{\left(\frac{(x-1)e^{2s}+x+1}{x+1-(x-1)e^{2s}}+1\right)-\left(\frac{(x-1)e^{2s}+x+1}{x+1-(x-1)e^{2s}}-1\right)e^{2t}}$$

$$= \frac{2(x-1)e^{2(s+t)}+2(x+1)}{2(x+1)-2(x-1)e^{2(s+t)}}$$

$$= \frac{(x-1)e^{2(s+t)}+(x+1)}{(x+1)-(x-1)e^{2(s+t)}} = \phi_{s+t}(x)$$

$$\phi_s\phi_t(x) = \phi_s(z) = \frac{(z-1)e^{2s}+(z+1)}{(z+1)-(z-1)e^{2s}}, z = \phi_t(x) = \frac{(x-1)e^{2t}+(x+1)}{(x+1)-(x-1)e^{2t}}$$

$$= \frac{\left(\frac{(x-1)e^{2t}+x+1}{x+1-(x-1)e^{2t}}-1\right)e^{2s}+\left(\frac{(x-1)e^{2t}+x+1}{x+1-(x-1)e^{2t}}+1\right)}{\left(\frac{(x-1)e^{2t}+x+1}{x+1-(x-1)e^{2t}}+1\right)-\left(\frac{(x-1)e^{2t}+x+1}{x+1-(x-1)e^{2t}}-1\right)e^{2s}}$$

$$= \frac{2(x-1)e^{2(t+s)}+2(x+1)}{2(x+1)-2(x-1)e^{2(t+s)}} = \frac{(x+1)+(x-1)e^{2(t+s)}}{(x+1)-(x-1)e^{2(t+s)}} = \phi_{t+s}(x)$$

So, $\phi_t\phi_s = \phi_s\phi_t$ (commutative property).
Hence the flow evolution operator ϕ_t forms a commutative dynamical group.

Example 1.3 Find the maximal interval of existence for unique solution of the following systems

$$\text{(i) } \dot{x}(t) = x^2 + \cos^2 t, t > 0, x(0) = 0$$
$$\text{(ii) } \dot{x} = x^2, x(0) = 1$$

Solution (i) By maximal interval of existence of solution we mean the largest interval for which the solution of the equation exists. The given system is nonautonomous and $f(t,x) = x^2 + \cos^2 t$. Consider the rectangle $R = \{(t,x) : 0 \le t \le a, |x| \le b, a > 0, b > 0\}$ containing the point $(0,0)$. Clearly, $f(t,x)$ is continuous and $\frac{\partial f}{\partial x} = 2x$ is bounded on R. The Lipschitz condition $|f(t,x_1) - f(t,x_2)| \le K|x_1 - x_2|, \forall (t,x_1), (t,x_2) \in R$, K being the Lipschitz constant, is satisfied on R. Since $|f(t,x)| = |x^2 + \cos^2 t| \le |x^2| + |\cos^2 t| \le |x^2| + 1$, and $M = \max|f(t,x)| = 1 + b^2$ in R. Therefore, from Picard's theorem (if $f(t,x)$ is a continuous function in a rectangle $R = \{(t,x) : |t - t_0| \le a, |x - x_0| \le b, a > 0, b > 0\}$ and satisfies Lipschitz condition therein, then the initial value problem $\dot{x} = f(t,x), x(t_0) = x_0$ has a unique solution in the rectangle $R' = \{(t,x) : |t - t_0| \le h, |x - x_0| \le b\}$, where $h = \min\{a, b/M\}$, $M = \max|f(t,x)|$ for all $(t,x) \in R$, see the books Coddington and Levinson [3], Arnold [4]). Now $h = \min\{a, \frac{b}{M}\} = \min\{a, \frac{b}{1+b^2}\}$. We now determine the maximum/minimum value(s) of $b/(1+b^2)$. Let $g(b) = \frac{b}{1+b^2}$. Then $g'(b) = \frac{1-b^2}{(1+b^2)^2}$ and $g''(b) = \frac{2b(b^2-3)}{(1+b^2)^3}$. For max or min value(s) of $g(b)$, $g'(b) = 0$. This gives $b = 1$. Since $g''(1) = -1/2 < 0$, $g(b)$ is maximum at $b = 1$ and the maximum value is given by $g(1) = \frac{1}{2}$. Now, if $a \ge 1/2$, then $h = \frac{b}{1+b^2} \le 1/2$ and if $a < 1/2$, then $h < 1/2$. Thus we must have $h \le 1/2$. Hence the maximum interval of existence of the solution of the given system is $0 \le t \le 1/2$.

(ii) Here $f(t,x) = x^2$. Consider the rectangle $R = \{(t,x) : |t| \le a, |x - 1| \le b, a > 0, b > 0\}$ containing the point $(0,1)$. Clearly, $f(t,x)$ is continuous and $\frac{\partial f}{\partial x} = 2x$ is bounded on R. Hence the Lipschitz condition is satisfied on R. Also in R, $M = \max|f(t,x)| = (1+b)^2$. Therefore, $h = \min\{a, \frac{b}{M}\} = \min\{a, \frac{b}{(1+b)^2}\}$. It can be shown, as earlier, that $g(b) = \frac{b}{(1+b)^2}$ is maximum at $b = 1$ and the maximum value is $g(1) = 1/4$. Now if $a \ge 1/4$, then $h = \frac{b}{(1+b)^2} \le 1/4$ and if $a < 1/4$, then $h < 1/4$. Thus we must have $h \le 1/4$. Hence the maximum interval of existence of solution of the given system is $|t| \le 1/4$, that is, $-1/4 \le t \le 1/4$. Note that the Picard's theorem gives the local region of existence of unique solution for a system.

Example 1.4 Using linear stability analysis determine the stability of the critical points for the following systems:

$$\text{(i)}\,\dot{x} = \sin x, \quad \text{(ii)}\,\dot{x} = x^2.$$

Solution (i) The given system has infinite numbers of critical points. The critical points are $x_n^* = n\pi, n = 0, \pm 1, \pm 2, \ldots$. When n is even, $f'(x_n^*) = \cos(x_n^*) = \cos(n\pi) = (-1)^n = 1 > 0$. So, these critical points are unstable. When n is odd, $f'(x_n^*) = -1 < 0$, and so these critical points are stable.

(ii) The critical point of the system is at $x^* = 0$. Now, $f'(x^*) = 0$ and $f''(x^*) = 2 > 0$. Hence, x^* is attracting when $x < 0$ and repelling when $x > 0$. Actually, the critical point is semi-stable in nature.

1.7 Analysis of One-Dimensional Flows

As we know qualitative approach is the combination of analysis and geometry and is a powerful tool for analyzing solution behaviors of a system qualitatively. By drawing trajectories in phase line/plane/space, the behaviors of phase points may be found easily. In qualitative analysis we mainly look for the following solution behaviors:

(i)   Local stabilities of fixed points for a system;
(ii) Analyzing the existence of periodic/quasi-periodic solutions, limit cycle, relaxation oscillation, hysteresis, etc.;
(iii) Local and asymptotic solution behaviors of a system;
(iv) Topological features of flows such as bifurcations, catastrophe, topological equivalence, transitiveness, etc.

We shall now analyze a simple one-dimensional system as follows.

Consider a one-dimensional system represented as $\dot{x}(t) = \sin x$ with the initial condition $x(t = 0) = x(0) = x_0$. The characteristic features of the system are (i) it is a one-dimensional system, (ii) nonlinear system (iii) autonomous system, and (iv) its closed-form solution (analytical solution) exists. This is a one-dimensional flow and we analyze the system on the basis of flow. The analytical solution of the system is obtained easily

$$\frac{dx}{dt} = \sin x \Rightarrow dt = \text{cosec} \, (x) dx$$

Integrating, we get

$$t = \int \text{cosec}(x) \, dx$$
$$= -\log|\text{cosec}(x) + \cot(x)| + c,$$

where c is an integrating constant. Using the initial condition $x(0) = x_0$, we get the integrating constant c as

$$c = \log|\text{cosec}(x_0) + \cot(x_0)|.$$

Thus the solution of the system is given as

$$t = \log \left| \frac{\text{cosec} \, (x_0) + \cot(x_0)}{\text{cosec} \, (x) + \cot(x)} \right|.$$

From this closed-form solution, the behaviors of solutions for any initial conditions are difficult to analyse. Moreover, the asymptotic values of the system are also difficult to obtain. The qualitative approach can give better dynamical behavior about this simple system.

We consider t as time, x as the position of an imaginary particle moving along the flow in real line and $\dot{x}$ as the velocity of that particle. The differential equation $\dot{x} = \sin x$ represents a vector field on the line. It gives the velocity vector $\dot{x}$ at each position x. The arrows point to the right when $\dot{x} > 0$ and to the left when $\dot{x} < 0$. We shall draw the graph of $\sin x$ versus x in $x\dot{x}$- plane which gives the flow in the x-axis (see Fig. 1.3).

We may imagine that fluid is flowing steadily along the x-axis with a velocity $\dot{x}$ which varies from place to place, according to equation $\dot{x} = \sin x$. At points $\dot{x} = 0$, there is no flow and such points are called equilibrium points (fixed points). According to the definition of fixed point, the equilibrium points of this system are obtained as $\sin x = 0 \Rightarrow x = n\pi (n = 0, \pm1, \pm2, \ldots)$. This simple looking autonomous system has infinite numbers of equilibrium points in $\mathbb{R}$. We can see that there are two kinds of equilibrium points. The equilibrium point where the flow is toward the point is called **sink** or **attractor** (neighboring trajectories approach asymptotically to the point as $t \to \infty$). On the other hand, when the flow is away from the point, the point is called **source** or **repellor** (neighboring trajectories move away from the point as $t \to \infty$). From the above figure the solid circles represent the sinks that are stable equilibrium points and the open circles are the sources, which are unstable equilibrium points. The names are given because the sinks and sources are common in fluid flow problems. From the geometric approach one can get local stability behavior of the equilibrium points of the system easily and is valid for all time. We shall now re-look the analytical solution of the system. The analytical solution can be expressed as

$$t = \log|\tan(x/2)| + c \Rightarrow x(t) = 2\tan^{-1}(Ae^t)$$

where A is an integrating constant.

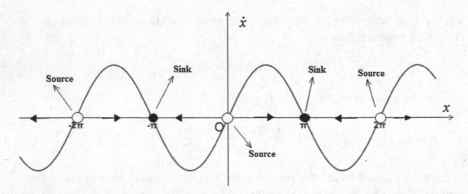

Fig. 1.3 Graphical representation of flow generated by $\sin(x)$

Let the initial condition be $x_0 = x(0) = \pi/4$. Then from the above solution we obtain

$$A = \tan(\pi/8) = -1 + \sqrt{2} = 1/\left(1 + \sqrt{2}\right).$$

So the solution is expressed as

$$x(t) = 2\tan^{-1}\left(\frac{e^t}{1 + \sqrt{2}}\right).$$

We see that the solution $x(t) \to \pi$ and $t \to \infty$.

Without using analytical solution for this particular initial condition the same result can be found by drawing the graph of x versus t. So the solution's behavior at any initial condition can be obtained easily by geometric approach. This simple one-dimensional system also has an interesting application. For a slow motion of a spring immersed in a highly viscous fluid such as grease or viscoelastic fluid (the combined effects of fluid viscosity and elasticity for example, synovial fluid in the joints of human bones), the viscous damping force is very strong compared to the inertia of motion. So one can neglect acceleration term (that is, inertia) and the spring-mass system may be governed by the equation $\alpha\dot{x} = \sin x$, where $\alpha > 0$ (string constant) is a real number and the dynamics can be obtained using this approach for different values of α (see the book Strogatz [5] for more physical examples and explanations).

We shall discuss a few worked out examples presented below.

Example 1.5 With the help of flow concept discuss the local stability of the fixed points of $\dot{x} = f(x) = (x^2 - 1)$.

Solution The fixed points of the given autonomous system are given by setting $f(x) = 0$. This gives $x = \pm 1$. So the fixed points of the system are 1 and -1. For the local stability of the system about these fixed points we plot the graph of the function $f(x)$ and then sketch the vector field. The flow is to the right direction, indicated by the symbol '$\rightarrow$', where the velocity $\dot{x} > 0$, that is, where $(x^2 - 1) > 0$ and to the left direction, indicated by the symbol '$\leftarrow$', where $\dot{x} < 0$, that is, $(x^2 - 1) < 0$. We also use solid circles to represent stable fixed points and open circles for unstable fixed points.

In Fig. 1.4 the arrows indicate the flow of the system. From the figure, we see that the fixed point $x = 1$ is unstable, since it acts as a source point and the fixed point $x = -1$ is stable, since it acts as a sink point.

Example 1.6 Discuss the stability character of the fixed points for the system $\dot{x} = x(1 - x)$ using the concept of flow.

Fig. 1.4 Graphical
representation of
$f(x) = (x^2 - 1)$

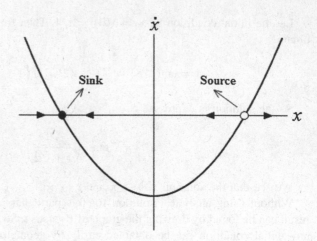

Solution Here $f(x) = x(1 - x)$. Then for the fixed points, we have

$$f(x) = 0 \Rightarrow x(1 - x) = 0 \Rightarrow x = 0, 1.$$

Thus the fixed points are 0-and 1. To discuss the stability of these fixed points we plot the system (x versus $\dot{x}$) and then sketch the vector field. The flow is to the right direction, indicated by the symbol '$\rightarrow$', when the velocity $\dot{x} > 0$, and to the left direction, indicated by the symbol '$\leftarrow$', when $\dot{x} < 0$. We also use solid circle to represent stable fixed point and open circle to represent unstable fixed point.

From Fig. 1.5 we see that the fixed point $x = 1$ is stable whereas the fixed point $x = 0$ is unstable.

Example 1.7 Find the fixed points and analyze the local stability of the following systems (i) $\dot{x} = x + x^3$ (ii) $\dot{x} = x - x^3$ (iii) $\dot{x} = -x - x^3$

Solution (i) Here $f(x) = x + x^3$. Then for fixed points $f(x) = 0 \Rightarrow x + x^3 = 0 \Rightarrow x = 0$, as $x \in \mathbb{R}$. So, 0 is the only fixed point of the system. We now see that when $x > 0$, $\dot{x} > 0$ and when $x < 0$, $\dot{x} < 0$. Hence the fixed point $x = 0$ is unstable. The graphical representation of the flow generated by the system is displayed in Fig. 1.6.

(ii) Here $f(x) = x - x^3$. Then $f(x) = 0 \Rightarrow x - x^3 = 0 \Rightarrow x = 0, 1, -1$. Therefore, the fixed points of the system are $0, 1, -1$. We now see that

(a)   when $x < -1$, then $\dot{x} > 0$
(b)   when $-1 < x < 0$, $\dot{x} < 0$
(c)   when $0 < x < 1$, $\dot{x} > 0$
(d)   when $x > 1$, then $\dot{x} < 0$.

This shows that the fixed points 1 and -1 are stable whereas the fixed point 0 is unstable (Fig. 1.7).

Fig. 1.5 Pictorial representation of $f(x) = x(1-x)$

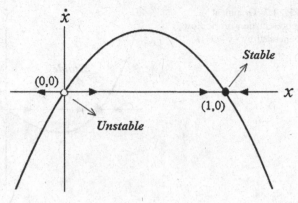

Fig. 1.6 Graphical representation of $f(x) = (x + x^3)$

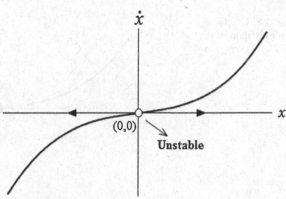

(iii) Here $f(x) = -x - x^3$. Then $f(x) = 0 \Rightarrow -x - x^3 = 0 \Rightarrow x = 0$, as $x \in \mathbb{R}$. So $x = 0$ is the only fixed point of the system. We now see that $\dot{x} > 0$ when $x < 0$ and $\dot{x} < 0$ when $x > 0$. This shows that the fixed point $x = 0$ is stable. The graphical representation of the flow generated by the system is displayed in Fig. 1.8.

Example 1.8 Determine the equilibrium points and sketch the phase diagram in the neighborhood of the equilibrium points for the system represented as $\dot{x} + x\,\mathrm{sgn}(x) = 0$.

Solution Given system is $\dot{x} + x\,\mathrm{sgn}(x) = 0$, that is, $\dot{x} = -x\,\mathrm{sgn}(x)$, where the function $\mathrm{sgn}(x)$ is defined as

$$\mathrm{sgn}(x) = \begin{cases} 1, & x > 0 \\ 0, & x = 0 \\ -1, & x < 0 \end{cases}$$

For equilibrium points, we have

$$\dot{x} = 0 \Rightarrow x\,\mathrm{sgn}\,x = 0 \Rightarrow x = 0.$$

Fig. 1.7 Graphical representation of the flow generated by $(x - x^3)$

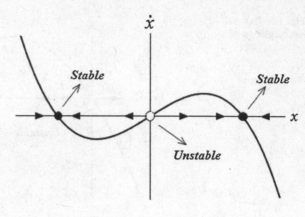

Fig. 1.8 Graphical representation of $f(x) = (-x - x^3)$ versus x

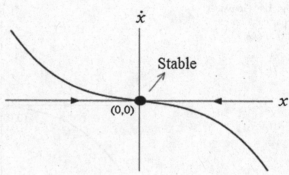

This shows that the system has only one equilibrium point at $x = 0$. In flow analysis we see that the velocity $\dot{x} < 0$ for all $x \neq 0$. The flow is to the right direction, when $\dot{x} > 0$, in the negative x-axis and to the left direction, when $\dot{x} < 0$, in the positive x-axis. This is shown in the phase diagram depicted in Fig. 1.9, which shows that the fixed point origin is semi-stable.

1.8 Conservative and Dissipative Dynamical Systems

The dichotomy of dynamical systems in conservative versus dissipative is very important. They have some fundamental properties. Particularly, conservative systems are the integral part of Hamiltonian mechanics. We give here only the formal definitions of conservative and dissipative systems. Consider an autonomous system represented as

$$\dot{x} = f(x), x \in \mathbb{R}^n. \tag{1.15}$$

Fig. 1.9 Graphical representation of the flow $\dot{x} = -x\,\mathrm{sgn}\,x$

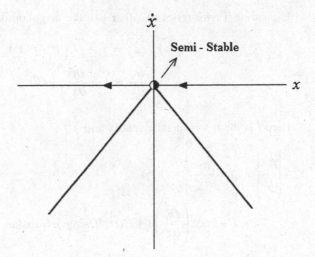

The conservative and dissipative systems are defined with respect to the divergence of the corresponding vector field, which in turn refers to the conservation of volume or area in their state space or phase plane, respectively as follows:

A system is said to be **conservative** if the divergence of its vector field is zero. On the other hand, it is said to be **dissipative** if its vector field has negative divergence. The phase volume in a conservative system is constant under the flow while for a dissipative system the phase volume occupied by the system is gradually decreased as the time t increases and shrinks to zero as $t \to \infty$. When divergence of vector field is positive, the phase volume is gradually expanding. We shall discuss it in a later chapter. We state a lemma below which gives the change of volume in a phase space for an autonomous system.

Sometimes, it is useful to find the evolution of volume in the phase space of a system $\dot{x} = f(x), x \in \mathbb{R}^n$. The system generates a flow $\phi(t, x)$. We give Liouville's theorem which describes the time evolution of volume under the flow $\phi(t, x)$. Before this we now give the following lemma.

Lemma 1.1 *Consider an autonomous vector field $\dot{x} = f(x), x \in \mathbb{R}^n$ and generates a flow $\phi_t(x)$. Let D_0 be a domain in $\mathbb{R}^n$ and $\phi_t(D_0)$ be its evolution under the flow. If $V(t)$ is the volume of D_t, then the time rate of change of volume is given as*

$$\left.\frac{dV}{dt}\right|_{t=0} = \int_{D_0} \nabla \cdot f\, dx.$$

Proof The volume $V(t)$ can be expressed in the following form using the definition of the Jacobian of a transformation as

$$V(t) = \int_{D_0} \left| \frac{\partial \phi(t, x)}{\partial x} \right| dx$$

Expanding Taylor series of $\phi(t, \underset{\sim}{x})$ in the neighborhood of $t = 0$, we get

$$\phi(t, \underset{\sim}{x}) = \underset{\sim}{x} + f(\underset{\sim}{x})t + O(t^2)$$

$$\Rightarrow \frac{\partial \phi}{\partial \underset{\sim}{x}} = I + \frac{\partial f}{\partial \underset{\sim}{x}}t + O(t^2)$$

Here I is the $n \times n$ identity matrix and

$$\left| \frac{\partial \phi}{\partial \underset{\sim}{x}} \right| = \left| I + \frac{\partial f}{\partial \underset{\sim}{x}}t \right| + O(t^2)$$

$$= 1 + \text{trace} \left(\frac{\partial f}{\partial \underset{\sim}{x}} \right) t + O(t^2) \text{ [Using expansion of the determinant]}$$

Now, $\text{trace} \left(\frac{\partial f}{\partial \underset{\sim}{x}} \right) = \nabla \cdot f$, so we have

$$V(t) = V(0) + \int_{D_0} t \nabla \cdot f \, d\underset{\sim}{x} + O(t^2).$$

This gives $\frac{\mathrm{d}V}{\mathrm{d}t} \Big|_{t=0} = \int_{D_0} \nabla \cdot f \, d\underset{\sim}{x}$.

Theorem 1.1 (Liouville's Theorem) *Suppose $\nabla \cdot f = 0$ for a vector field f. Then for any region $D_0 \subseteq \mathbb{R}^n$, the volume $V(t)$ generated by the flow $\phi(t, \underset{\sim}{x})$ is $V(t) = V(0)$, $V(0)$ being the volume of D_0.*

Proof Suppose the divergence of the vector field f is everywhere constant, that is, $\nabla \cdot f = c$. For arbitrary time t_0 the evolution equation for the volume is given as $\dot{V} = cV$. This gives $V(t) = V(0)e^{ct}$. When the vector field is divergence free, that is, $c = 0$, we get the result $\dot{V} = 0 \Rightarrow V(t) = V(0) = $ constant. Thus we can say that the flow generated by a time independent system is volume preserving.

Examples of conservative and dissipative systems are presented below.

(a) Consider a linear and undamped pendulum represented as $\ddot{x} + x = 0$. This is an example of a conservative system. Setting $\dot{x} = y$, we can write it as a system of equations

$$\left. \begin{array}{l} \dot{x} = y \\ \dot{y} = -x \end{array} \right\}$$

The system may also be written in the compact form $\dot{\underset{\sim}{x}} = f(\underset{\sim}{x})$, where $f(\underset{\sim}{x}) = \begin{pmatrix} y \\ -x \end{pmatrix}$. The divergence of the vector field $\underset{\sim}{f}$ is given by $\vec{\nabla} \cdot \underset{\sim}{f} = \frac{\partial}{\partial x}(y) + \frac{\partial}{\partial y}(-x) = 0$. According to the definition, the system is conservative and the area occupied in the xy-phase plane is constant.

(b)   The damped pendulum governed by $\ddot{x} + \alpha\dot{x} + \beta x = 0, \alpha, \beta > 0$ is an example of a dissipative system. Setting $\dot{x} = y$, we can write the system as

$$\left. \begin{array}{l} \dot{x} = y \\ \dot{y} = -\alpha y - \beta x \end{array} \right\}$$

The vector field is then expressed as $f(\underset{\sim}{x}) = \begin{pmatrix} y \\ -\alpha y - \beta x \end{pmatrix}$.

Now, $\vec{\nabla} \cdot \underset{\sim}{f} = \frac{\partial}{\partial x}(y) + \frac{\partial}{\partial y}(-\alpha y - \beta x) = -\alpha < 0$, since $\alpha > 0$.

This shows that the divergence of the vector field is negative. So the system is dissipative in nature and the area in the phase plane is decreasing as time goes on. This is the simplest linear oscillator with linear damping. It describes a spring-mass system with a damper in parallel. The spring force is proportional to the extension x of the spring and the damping or frictional force is proportional to the velocity $\dot{x}$. The two constants α and β are related to the stiffness of the spring and the degrees of friction in the damper, respectively. According to the above lemma, the change in phase area is given by

$$A(t) = cA(0)e^{-\alpha t}, \alpha > 0 \text{ as } t \to \infty, c \text{ being a constant.}$$

Example 1.9 Find the phase volume element for the systems (i) $\dot{x} = -x$, (ii) $\dot{x} = ax - bxy, \dot{y} = bxy - cy$ where $x, y \geq 0$ and a, b, c are positive constants.

Solution (i) The flow of the system $\dot{x} = -x$ is attracted toward the point $x = 0$. The time rate of change of volume element $V(t)$ under the flow is given as

$$\left. \frac{dV}{dt} \right|_{t=0} = -\int_{D(0)} dx = -V(0)$$

or, $V(t) = V(0)e^{-t} \to 0$ as $t \to \infty$.

Hence the phase volume element $V(t)$ shrinks exponentially.

   (ii) The given system is a Lotka-Volterra predator-prey population model. The rate of change in phase area $A(t)$ is given as

$$\frac{dV}{dt} = - \int \vec{\nabla} \cdot \underset{\sim}{f} \, dxdy$$

$$= - \int (a - c - by + bx)dxdy$$

This shows that a phase area periodically shrinks and expands.

1.9 Some Definitions

In this section we give some important preliminary definitions relating to flow of a system. The definitions given here are elaborately discussed in the later chapters for higher dimensional systems.

Invariant set A set $D \subset \mathbb{R}^n$ is said to be an invariant set under the flow ϕ_t if for any point $p \in D$, $\phi_t(p) \in D$ for all $t \in \mathbb{R}$. The set D is said to be positively invariant if $\phi_t(p) \in D$ for $t \geq 0$. Trajectories starting in an invariant set remain in the set for all times. An interval is called trapping if it is mapped into itself and is said to be invariant if it is mapped exactly onto itself. Moreover, if a bounded interval is trapping, then all of its trajectories are trapped inside and must converge to a closed, invariant, and bounded limit set. Basically these limit sets are the attractors of a system. So the periodic orbits are examples of invariant sets. We now define two limiting topological concepts which are relevant to the orbits of dynamical systems.

Limit points (ω- and α-limit points)

The asymptotic behavior of a trajectory may be related with limit points/sets or cycles and are termed as ω- and α-limit points/sets or cycles. We now give the definitions.

A point $p \in \mathbb{R}^n$ is called an ω-(resp. a α-) limit point if there exists a sequence $\{t_i\}$ with $t_i \to \infty$ (resp. $t_i \to -\infty$) such that $\phi(t_i, x) \to p$ as $i \to \infty$. The ω-limit set(cycle) is denoted by $\Lambda(\underset{\sim}{x})$ and is defined as

$$\Lambda(\underset{\sim}{x}) = \left\{ \underset{\sim}{x} \in \mathbb{R}^n | \exists \{t_i\} \text{ with } t_i \to \infty \text{ and } \phi(t_i, \underset{\sim}{x}) \to p \text{ as } i \to \infty \right\}.$$

Similarly, the α-limit set (cycle), $\mu(\underset{\sim}{x})$, is defined as

$$\mu(\underset{\sim}{x}) = \left\{ \underset{\sim}{x} \in \mathbb{R}^n | \exists \{t_i\} \text{ with } t_i \to -\infty \text{ and} \phi(t_i, \underset{\sim}{x}) \to p \text{ as } i \to \infty \right\}.$$

For example, consider a flow $\phi(t, x)$ on $\mathbb{R}^2$ generated by the system $\dot{r} = cr(1 - r)$, $\dot{\theta} = 1$, c being a positive constant. For $x \neq 0$, let p be any point of the closed orbit C and take $\{t_i\}_{i=1}^{\infty}$ to be the sequence of $t > 0$. The trajectory through x crosses the radial line through p. So, $t_i \to \infty$ as $i \to \infty$ and

$\phi(t_i, x) \to p$ as $i \to \infty$. If $\underset{\sim}{x}$ lies in the closed orbit C, then $\phi(t_i, x) = p$ for each i. Hence every point of C is a ω-limit point of $\underset{\sim}{x}$ and so $\Lambda(x) = C$ for every $x \neq 0$. When $|x| \leq 1$, the sequence $\{t_i\}_{i=1}^{\infty}$ with $t < 0$ gives the α-limit set

$$\mu(x) = \begin{cases} \{0\} & \text{for } |x| < 1 \\ \text{closed orbit} & \text{for } |x| = 1 \end{cases}.$$

When $|x| > 1$, there is no sequence $\{t_i\}_{i=1}^{\infty}$, with $t_i \to \infty$ as $i \to \infty$, such that $\phi(t_i, x)$ exists as $i \to \infty$. So, $\mu(x)$ is empty when $|x| > 1$. The closed orbit C is called a limit cycle of the system.

The trajectory of a system through a point $\underset{\sim}{x}$ is the set $\gamma(\underset{\sim}{x}) = \bigcup_{t \in \mathbb{R}} \phi(t, \underset{\sim}{x})$ and the corresponding positive semi-trajectory $\gamma^+(\underset{\sim}{x})$ and negative semi-trajectory $\gamma^-(\underset{\sim}{x})$ are defined as follows:

$$\gamma^+(\underset{\sim}{x}) = \bigcup_{t \geq 0} \phi(t, \underset{\sim}{x}) \text{ and } \gamma^-(\underset{\sim}{x}) = \bigcup_{t \leq 0} \phi(t, \underset{\sim}{x}).$$

We now state two lemmas below. Interested readers can try for proofs (see the book Glendinning [6]).

Lemma 1.2

(a) *The set D is invariant if and only if $\gamma(\underset{\sim}{x}) \subset D$ for all $\underset{\sim}{x} \in D$.*
(b) *D is invariant if and only if $\mathbb{R}^n \backslash D$ is invariant.*
(c) *Let (D_i) be a countable collection of invariant subsets of $\mathbb{R}^n$. Then $\bigcup D_i$ and $\bigcap_i D_i$ are also invariant subsets of $\mathbb{R}^n$.*

Lemma 1.3 *The set $\Lambda(\underset{\sim}{x}) = \bigcap_{x \in \gamma(\underset{\sim}{x})} \text{cl}\left(\gamma^+(\underset{\sim}{x})\right)$, where cl denotes the closure of the set, is the ω-limit set.*

Non-wandering point A point p is called a non-wandering point if for any neighborhood U of p and for any $T > 0$, there exists some $|t| > T$ such that $\phi(t, U) \cap U \neq \varphi$. The nonwandering set, denoted by Ω, contains all such points $p \in U$ and it is closed. Non-wandring points give asymptotic behavior of the orbit. In the above definition, if $\phi(t, U) \cap U = \varphi$, then the point p is called a wandering point.

The examples of non-wandering points are fixed points and periodic orbits of a system. For the undamped oscillator $(\ddot{x} + x = 0)$, all points are non-wandering in $x\dot{x}$ phase plane while for the damped oscillator $(\ddot{x} + \alpha\dot{x} + x = 0)$, origin is the only non-wandering point.

Attracting set A closed invariant set $D \subset \mathbb{R}^n$ for a flow ϕ_t is said to be an attracting set if there exists some neighborhood U in D such that $\forall t \geq 0$, $\phi(t, U) \subset U$ and $\bigcap_{t > 0} \phi(t, U) = D$.

Absorbing set A positive invariant compact subset $B \subseteq \mathbb{R}^n$ is said to be an absorbing set if there exists a bounded subset C of $\mathbb{R}^n$ with $C \supset B$ such that $t_C > 0 \Rightarrow \phi(t, C) \subset B \forall t \geq t_C$ (see the book by Wiggins [7] for details).

Trapping zone An open set U in an invariant set $D \subset \mathbb{R}^n$ in an attracting set for a flow generated by a system is called a trapping zone. Let a set A be closed and invariant. The set A is said to be stable if and only if every neighborhood of A contains a neighborhood U of A which is trapping.

Basin of attraction The domain (called as basin of attraction) of an attracting set D is defined as $\bigcup_{t \leq 0} \phi(t, U)$ where U is any open set in $D \subset \mathbb{R}^n$.

We now give an example from Ruelle [8]. This example is also discussed in the book by Guckenheimer and Holmes [9]. Consider the one-dimensional system $\dot{x} = -x^4 \sin(\pi/x)$. It has countably infinite set of fixed points at $x^* = 0, \pm\frac{1}{n}, n = 1, 2, 3, \ldots$. Now,

$$f(x) = -x^4 \sin(\pi/x) \Rightarrow f'(x) = -4x^3 \sin(\pi/x) + \pi x^2 \cos(\pi/x)$$

$$\Rightarrow f'(x^*)|_{x^* = \pm\frac{1}{n}} = \frac{\pi}{n^2} \cos(n\pi) = \frac{\pi}{n^2}(-1)^n.$$

The fixed point $x^* = 0$ is neither attracting nor repelling. The interval $[-1, 1]$ is an attracting set of the given system. The fixed points $x^* = \pm\frac{1}{2n}$, $n = 1, 2, \ldots$ are repelling while the fixed points $x^* = \pm\frac{1}{(2n-1)}$, $n = 1, 2, \ldots$ are attracting.

1.10 Exercises

1.(a) What is a dynamical system? Write its importance.
  (b) Discuss continuous and discrete dynamical systems with examples.
  (c) Explain deterministic, semi-deterministic and nondeterministic dynamical processes with examples.
  (d) What do you mean by 'qualitative study' of a nonlinear system? Write it importance in nonlinear dynamics.

2.(a) Give the mathematical definition of flow. Discuss the concept related to 'a flow and its orbit'. Also indicate its implication on uniqueness theorem of differential equation.
  (b) Show that the initial value problem $\dot{x} = x^{\frac{1}{3}}, x(0) = 0$ has an infinite number of solutions. How would you explain it in the context of flow?
  (c) Consider a system $\dot{x} = |x|^{p/q}$ with $x(0) = 0$ where p and q are prime to each other. Show that the system has an infinite number of solutions if $p < q$, and it has a unique solution if $p > q$.

3. Find the maximum interval of existence for the solutions of the following equations
  (a) $\dot{x} = x^2, x(0) = x_0$, (b) $\dot{x} = tx^2, x(0) = x_0$, (c) $\dot{x} = x^2 + t^2, x(0) = x_0$,
  (d) $\dot{x} = x(x - 2), x(0) = 3$

4. For what values of t_0 and x_0 does the initial value problem $\dot{x} = \sqrt{x}, x(t_0) = x_0$ have a unique solution?

5. Show that the initial value problem $\dot{x} = 3x^{2/3}$ with $x(0) = 0$ has two solutions passing through the point $(0, 0)$. How do you explain the context of flow in this?

6. Show that the initial value problem $\dot{x} = |x|^{1/2}, x(0) = 0$ has four different solutions through the point $(0,0)$. Sketch these solutions in the $t-x$ plane. Explain this from Picard's theorem.

7. Prove that the system $\dot{x} = x^3$ with the initial condition $x(0) = 2$ has a solution on an interval $(-\infty, c), c \in \mathbb{R}$. Sketch the solution $x(t)$ in the $t - x$ plane and find the limiting behavior of solution $x(t)$ as $t \to c-$.

8. Prove that the solutions of the initial value problem $\dot{x} = \begin{cases} 0 & \text{when } x \leq 0 \\ x^{1/n} & \text{when } x > 0 \end{cases}$ with $x(0) = 0$ are not unique for $n = 2, 3, 4, \ldots$.

9. What do you mean by fixed point of a system? Determine the fixed points of the system $\dot{x} = x^2 - x, x \in \mathbb{R}$. Show that solutions exist for all time and become unbounded in finite time.

10. Give mathematical definitions of 'flow evolution operator' of a system. Write the basic properties of an evolution operator of a flow.

11. Show that the dynamical system (or evolution) forms a dynamical group. What can you say about commutative/non-commutative group of a system? Give reasons in support of your answer.

12. Find the evolution operators for the following systems: (i) $\dot{x} = x - x^2$, (ii) $\dot{x} = x^2$, (iii) $\dot{x} = x \ln x, x > 0$, (iv) $\dot{x} = \tanh(x)$, (v) $\dot{x} = x - x^3$, (vi) $\dot{x} = f \cos \omega t$, (vii) $\dot{x} = f \sin \omega t$.
Verify that $\phi_t(\phi_s(x)) = \phi_{t+s}(x) \ \forall x, t, s \in \mathbb{R}$ for all cases. Also, show that the evolution operator ϕ_t for the system $\dot{x} = x^2$ forms a dynamical group.

13. Define fixed point of a system in the context of flow. Give its geometrical interpretation. How do you relate this concept with the usual notion of fixed point in a continuous dynamical system?

14.(a) Define source and sink for a one-dimensional flow. Illustrate them with examples.

  (b) Locate the source and sink for the system $\dot{x} = (x^2 - 1), x \in \mathbb{R}$.

15. Consider the one-dimensional system represented as $\dot{x} = ax + b$, where b is a constant and a is a nonzero parameter. Find all fixed points of the system and discuss their stability for different values of a.

16.(a) Sketch the region of the flow generated by the one-dimensional system $\dot{x} = 1/x, x > 0$ with the starting condition $x(0) = x_0$.

  (b) What do you mean by oscillating solution of a system? Explain with examples. Show that one-dimensional flow cannot oscillate.

17.(a) Find the critical points of the following systems: (i) $\dot{x} = e^x - 1$, (ii) $\dot{x} = x^2 - x - 1$, (iii) $\dot{x} = \sin(\pi x)$, (iv) $\dot{x} = \cos x$, (v) $\dot{x} = \sinh x$, (vi) $\dot{x} = rx - x^2$ for $r < 0$, $r = 0$, $r > 0$, (vii) $\dot{x} = x - \ln(1 + x) + r$, $r > 0$, (viii) $\dot{x} = x - x^3/6$.

(b) Using linear stability analysis determine the stabilities/instabilities of the following systems about their critical points:
   (i) $\dot{x} = x(x-1)(x-2)$,   (ii) $\dot{x} = (x-2)(x-3)$,   (iii) $\dot{x} = \log x$,
   (iv) $\dot{x} = \cos x$, (v) $\dot{x} = \tan x$, (vi) $\dot{x} = 2 + \sin x$, (vii) $\dot{x} = x - x^3/6$, (viii) $\dot{x} = 1 - x^2/2 + x^4/24$.

18. Sketch the family of solutions of the differential equation $\dot{x} = ax - bx^2$, $x > 0$ and $a, b > 0$. How does the velocity vector $\dot{x}$ behave when $(a/b) < x < \infty$?

19. Find the critical points and analyze the local stability about the critical points of each of the following systems: (i)$\dot{x} = x^4 - x^3 - 2x^2$, (ii) $\dot{x} = \sinh(x^2)$ (iii) $\dot{x} = \cos x - 1$, (iv) $\dot{x} = (x - a)^2$, (v) $\dot{x} = (x+1)(x+2)$, (vi) $\dot{x} = \tan x$, (vii) $\dot{x} = \log x$,(viii) $\dot{x} = e^x - x - 1$.

20. Classify all possible flows in $\mathbb{R}$ of the system $\dot{x} = a_0 + a_1 x + a_2 x^2 + x^3$, where $a_0, a_1, a_2 \in \mathbb{R}$.

21. Consider the one-dimensional system $\dot{x} = \mu x + x^3$, $\mu \geq 0$. Using geometric approach find the solution behavior for any initial condition $x_0 (\neq 0)$.

22. When is a flow called conservative? Give an example of conservative flow.

23. Prove that the phase volume of a conservative system is constant. Is the converse true? Give reasons in support of your answer.

24. What can you say about time rate of change of phase volume element in a dissipative dynamical system? Explain it geometrically. Give an example of a dissipative system.

25. Prove that the α- and ω-limit sets of a flow $\phi_t(x)$ are contained in the non-wandering set of the flow $\phi_t(x)$.

26. Define absorbing set of a flow. Write down the relation between trapping zones and absorbing sets. Prove that for an absorbing set A, $\bigcap_{t \geq 0} \phi(t, A)$ forms an attracting set.

27. Give the definition of invariant set of a flow. Write its importance in dynamical evolution of a system. Prove that the ω-limit set, $\Lambda(\underset{\sim}{x})$, is invariant and it is nonempty and compact if the positive orbit $\gamma^+(\underset{\sim}{x})$ of $\underset{\sim}{x}$ is bounded.

28. If two orbits $\gamma(x)$ and $\gamma(y)$ of autonomous systems satisfy $\gamma(x) \cap \gamma(y) \neq \varphi$, prove that both the orbits are coinciding.

References

1. Arrowsmith, D.K., Place, L.M.: Dynamical Systems: Differential equations, Maps, and Chaotic Behaviour. Chapman and Hall/CRC (1992)
2. Lakshmanan, M., Rajasekar, S.: Nonlinear Dynamics: Integrability, Chaos and Patterns. Springer, 2003.
3. Coddington, E.A., Levinson, N.: Theory of Ordinary Differential Equations. McGraw Hill, New York (1955)
4. Arnold, V.I.: Ordinary Differential Equations. MIT Press, Cambridge, MA (1973)
5. Strogatz, S.H.: Nonlinear Dynamics and Chaos with application to physics, biology, chemistry and engineering. Perseus books, L.L.C, Massachusetts (1994)

6. Glendinning, P.: Stability, Instability and Chaos: An Introduction to the Theory of Nonlinear Differential Equations. Cambridge University Press. 1994
7. Wiggins, S.: Introduction to Applied Nonlinear Dynamical Systems and Chaos, 2nd edn. Springer (2003)
8. Ruelle, D.: Elements of Differentiable dynamics and Bifurcation Theory. Academic Press, New York (1989)
9. Gluckenheimer, J., Holmes, P.: Nonlinear Oscillations, Dynamical Systems, and Bifurcations of Vector Fields. Springer (1983)

Chapter 2
Linear Systems

This chapter deals with linear systems of ordinary differential equations (ODEs), both homogeneous and nonhomogeneous equations. Linear systems are extremely useful for analyzing nonlinear systems. The main emphasis is given for finding solutions of linear systems with constant coefficients so that the solution methods could be extended to higher dimensional systems easily. The well-known methods such as eigenvalue–eigenvector method and the fundamental matrix method have been described in detail. The properties of fundamental matrix, the fundamental theorem, and important properties of exponential matrix function are given in this chapter. It is important to note that the set of all solutions of a linear system forms a vector space. The eigenvectors constitute the solution space of the linear system. The general solution procedure for linear systems using fundamental matrix, the concept of generalized eigenvector, solutions of multiple eigenvalues, both real and complex, are discussed.

2.1 Linear Systems

Consider a linear system of ordinary differential equations as follows:

$$
\left.
\begin{aligned}
\frac{dx_1}{dt} &= \dot{x}_1 = a_{11}x_1 + a_{12}x_2 + \cdots + a_{1n}x_n + b_1 \\
\frac{dx_2}{dt} &= \dot{x}_2 = a_{21}x_1 + a_{22}x_2 + \cdots + a_{2n}x_n + b_2 \\
&\vdots \\
\frac{dx_n}{dt} &= \dot{x}_n = a_{n1}x_1 + a_{n2}x_2 + \cdots + a_{nn}x_n + b_n
\end{aligned}
\right\}
\tag{2.1}
$$

where $a_{ij}, b_j (i,j = 1, 2, \ldots, n)$ are all given constants. The system (2.1) can be written in matrix notation as

© Springer India 2015
G.C. Layek, *An Introduction to Dynamical Systems and Chaos*,
DOI 10.1007/978-81-322-2556-0_2

$$\dot{\underset{\sim}{x}} = A\underset{\sim}{x} + \underset{\sim}{b} \tag{2.2}$$

where $\underset{\sim}{x}(t) = (x_1(t), x_2(t), \ldots, x_n(t))^t$, $\underset{\sim}{b} = (b_1, b_2, \ldots, b_n)^t$ are the column vectors and $A = [a_{ij}]_{n \times n}$ is the square matrix of order n, known as the coefficient matrix of the system. The system (2.2) is said to be homogeneous if $\underset{\sim}{b} = \underset{\sim}{0}$, that is, if all b_i's are identically zero. On the other hand, if $\underset{\sim}{b} \neq \underset{\sim}{0}$, that is, if at least one b_i is nonzero, then the system is called nonhomogeneous. We consider first linear homogeneous system as

$$\dot{\underset{\sim}{x}} = A\underset{\sim}{x} \tag{2.3}$$

A differentiable function $\underset{\sim}{x}(t)$ is said to be a solution of (2.3) if it satisfies the equation $\dot{\underset{\sim}{x}} = A\underset{\sim}{x}$. Let $\underset{\sim}{x}_1(t)$ and $\underset{\sim}{x}_2(t)$ be two solutions of (2.3). Then any linear combination $\underset{\sim}{x}(t) = c_1 \underset{\sim}{x}_1(t) + c_2 \underset{\sim}{x}_2(t)$ of $\underset{\sim}{x}_1(t)$ and $\underset{\sim}{x}_2(t)$ is also a solution of (2.3). This can be shown very easily as below.

$$\dot{\underset{\sim}{x}} = c_1 \dot{\underset{\sim}{x}}_1 + c_2 \dot{\underset{\sim}{x}}_2$$

and so

$$A\underset{\sim}{x} = A(c_1 \underset{\sim}{x}_1 + c_2 \underset{\sim}{x}_2) = c_1 A\underset{\sim}{x}_1 + c_2 A\underset{\sim}{x}_2 = c_1 \dot{\underset{\sim}{x}}_1 + c_2 \dot{\underset{\sim}{x}}_2 = \dot{\underset{\sim}{x}}.$$

The solution $\underset{\sim}{x} = c_1 \underset{\sim}{x}_1 + c_2 \underset{\sim}{x}_2$ is known as **general solution** of the system (2.3). Thus the general solution of a system is the linear combination of the set of all solutions of that system (superposition principle). Since the system is linear, we may consider a nontrivial solution of (2.3) as

$$\underset{\sim}{x}(t) = \underset{\sim}{\alpha} e^{\lambda t} \tag{2.4}$$

where $\underset{\sim}{\alpha}$ is a column vector with components $\underset{\sim}{\alpha} = (\alpha_1, \alpha_2, \ldots, \alpha_n)^t$ and λ is a number. Substituting (2.4) into (2.3) we obtain

$$\lambda \underset{\sim}{\alpha} e^{\lambda t} = A\underset{\sim}{\alpha} e^{\lambda t}$$

$$\text{or, } (A - \lambda I)\underset{\sim}{\alpha} = \underset{\sim}{0} \tag{2.5}$$

where I is the identity matrix of order n. Equation (2.5) gives a nontrivial solution if and only if

$$\det(A - \lambda I) = 0 \tag{2.6}$$

On expansion, Eq. (2.6) gives a polynomial equation of degree n in λ, known as the *characteristic equation* of matrix A. The roots of the characteristic equation (2.6) are called the **characteristic roots** or **eigenvalues** or **latent roots** of A. The vector $\underset{\sim}{\alpha}$, which is a nontrivial solution of (2.5), is known as an **eigenvector** of A corresponding to the eigenvalue λ. If $\underset{\sim}{\alpha}$ is an eigenvector of a matrix A corresponding to an eigenvalue λ, then $\underset{\sim}{x}(t) = e^{\lambda t}\underset{\sim}{\alpha}$ is a solution of the system $\dot{\underset{\sim}{x}} = A\underset{\sim}{x}$. The set of linearly independent eigenvectors constitutes a solution space

of the linear homogeneous ordinary differential equations which is a vector space. All properties of vector space hold good for the solution space. We now discuss the general solution of a linear system below.

2.2 Eigenvalue–Eigenvector Method

As we know, the solution of a linear system constitutes a linear space and the solution is formed by the eigenvectors of the matrix. There may have four possibilities according to the eigenvalues and corresponding eigenvectors of matrix A. We proceed now case-wise as follows.

Case I: Eigenvalues of A are real and distinct

If the coefficient matrix A has real distinct eigenvalues, then it has linearly independent (L.I.) eigenvectors. Let $\underset{\sim}{\alpha}_1, \underset{\sim}{\alpha}_2, \ldots, \underset{\sim}{\alpha}_n$ be the eigenvectors corresponding to the eigenvalues $\lambda_1, \lambda_2, \ldots \lambda_n$ of matrix A. Then each $\underset{\sim}{x}_j(t) = \underset{\sim}{\alpha}_j e^{\lambda_j t}, j = 1, 2, \ldots, n$ is a solution of $\dot{\underset{\sim}{x}} = A\underset{\sim}{x}$. The general solution is a linear combination of the solutions $\underset{\sim}{x}_j(t)$ and is given by

$$\underset{\sim}{x}(t) = \sum_{j=1}^{n} c_j \underset{\sim}{x}_j(t)$$

where $c_1, c_2, \ldots, c_n$ are arbitrary constants. In $\mathbb{R}^2$, the solution can be written as

$$\underset{\sim}{x}(t) = \sum_{j=1}^{2} c_j \underset{\sim}{\alpha}_j e^{\lambda_j t} = c_1 \underset{\sim}{\alpha}_1 e^{\lambda_1 t} + c_2 \underset{\sim}{\alpha}_2 e^{\lambda_2 t}.$$

Case II: Eigenvalues of A are real but repeated

In this case matrix A may have either n linearly independent eigenvectors or only one or many ($<n$) linearly independent eigenvectors corresponding to the repeated

eigenvalues. The generalized eigenvectors have been used for linearly independent eigenvectors. We discuss this case in the following two sub-cases.

Sub-case 1: Matrix A has linearly independent eigenvectors

Let $\underset{\sim}{\alpha}_1, \underset{\sim}{\alpha}_2, \ldots, \underset{\sim}{\alpha}_n$ be n linearly independent eigenvectors corresponding to the repeated real eigenvalue λ of matrix A. In this case the general solution of the linear system is given by

$$\underset{\sim}{x}(t) = \sum_{i=1}^{n} c_i \underset{\sim}{\alpha}_i e^{\lambda t}.$$

Sub-case 2. Matrix A has only one or many ($<n$) linearly independent eigenvectors

First, we give the definition of generalized eigenvector of A. Let λ be an eigenvalue of the $n \times n$ matrix A of multiplicity $m \leq n$. Then for $k = 1, 2, \ldots, m$, any nonzero solution of the equation $(A - \lambda I)^k \underset{\sim}{v} = \underset{\sim}{0}$ is called a **generalized eigenvector** of A. For simplicity consider a two dimensional system. Let the eigenvalues be repeated but only one eigenvector, say $\underset{\sim}{\alpha}_1$ be linearly independent. Let $\underset{\sim}{\alpha}_2$ be a generalized eigenvector of the 2×2 matrix A. Then $\underset{\sim}{\alpha}_2$ can be obtained from the relation $(A - \lambda I)\underset{\sim}{\alpha}_2 = \underset{\sim}{\alpha}_1 \Rightarrow A\underset{\sim}{\alpha}_2 = \lambda \underset{\sim}{\alpha}_2 + \underset{\sim}{\alpha}_1$. So the general solution of the system is given by

$$\underset{\sim}{x}(t) = c_1 \underset{\sim}{\alpha}_1 e^{\lambda t} + c_2 (t \underset{\sim}{\alpha}_1 e^{\lambda t} + \underset{\sim}{\alpha}_2 e^{\lambda t}).$$

Similarly, for an $n \times n$ matrix A, the general solution may be written as $\underset{\sim}{x}(t) = \sum_{i=1}^{n} c_i \underset{\sim}{x}_i(t)$, where

$$\underset{\sim}{x}_1(t) = \underset{\sim}{\alpha}_1 e^{\lambda t},$$

$$\underset{\sim}{x}_2(t) = t \underset{\sim}{\alpha}_1 e^{\lambda t} + \underset{\sim}{\alpha}_2 e^{\lambda t},$$

$$\underset{\sim}{x}_3(t) = \frac{t^2}{2!} \underset{\sim}{\alpha}_1 e^{\lambda t} + t \underset{\sim}{\alpha}_2 e^{\lambda t} + \underset{\sim}{\alpha}_3 e^{\lambda t},$$

$$\vdots$$

$$\underset{\sim}{x}_n(t) = \frac{t^{n-1}}{(n-1)!} \underset{\sim}{\alpha}_1 e^{\lambda t} + \cdots + \frac{t^2}{2!} \underset{\sim}{\alpha}_{n-2} e^{\lambda t} + t \underset{\sim}{\alpha}_{n-1} e^{\lambda t} + \underset{\sim}{\alpha}_n e^{\lambda t}.$$

Case III: Matrix A has non-repeated complex eigenvalues

Suppose the real $n \times n$ matrix A has m-pairs of complex eigenvalues $a_j \pm ib_j, j = 1, 2, \ldots, m$. Let $\underset{\sim}{\alpha}_j \pm i\underset{\sim}{\beta}_j, j = 1, 2, \ldots, m$ denote the corresponding eigenvectors. Then the solution of the system $\underset{\sim}{\dot{x}}(t) = A\underset{\sim}{x}(t)$ for these complex eigenvalues is given by

$$\underset{\sim}{x}(t) = \sum_{j=1}^{m} c_j \underset{\sim}{u}_j + d_j \underset{\sim}{v}_j$$

where $\underset{\sim}{u}_j = \exp(a_j t)\{\underset{\sim}{\alpha}_j \cos(b_j t) - \underset{\sim}{\beta}_j \sin(b_j t)\}$, $\underset{\sim}{v}_j = \exp(a_j t)\{\underset{\sim}{\alpha}_j \sin(b_j t) + \underset{\sim}{\beta}_j$
$\cos(b_j t)\}$ and $c_j, d_j (j = 1, 2, \ldots, m)$ are arbitrary constants. We discuss each of the
above cases through specific examples below.

Example 2.1 Find the general solution of the following linear homogeneous system
using eigenvalue-eigenvector method:

$$\dot{x} = 5x + 4y$$
$$\dot{y} = x + 2y.$$

Solution In matrix notation, the system can be written as $\dot{\underset{\sim}{x}} = A\underset{\sim}{x}$, where $\underset{\sim}{x} = \begin{pmatrix} x \\ y \end{pmatrix}$ and $A = \begin{pmatrix} 5 & 4 \\ 1 & 2 \end{pmatrix}$. The eigenvalues of A satisfy the equation

$$\det(A - \lambda I) = 0$$
$$\Rightarrow \begin{vmatrix} 5 - \lambda & 4 \\ 1 & 2 - \lambda \end{vmatrix} = 0$$
$$\Rightarrow (5 - \lambda)(2 - \lambda) - 4 = 0$$
$$\Rightarrow \lambda^2 - 7\lambda + 6 = 0.$$

The roots of the characteristic equation $\lambda^2 - 7\lambda + 6 = 0$ are $\lambda = 1, 6$. So the
eigenvalues of A are real and distinct. We shall now find the eigenvectors corresponding to these eigenvalues.

Let $\underset{\sim}{e} = \begin{pmatrix} e_1 \\ e_2 \end{pmatrix}$ be the eigenvector corresponding to the eigenvalue $\lambda_1 = 1$.
Then

$$(A - I)\underset{\sim}{e} = \underset{\sim}{0}$$
$$\Rightarrow \begin{pmatrix} 5 - 1 & 4 \\ 1 & 2 - 1 \end{pmatrix} \begin{pmatrix} e_1 \\ e_2 \end{pmatrix} = \begin{pmatrix} 0 \\ 0 \end{pmatrix}$$
$$\Rightarrow \begin{pmatrix} 4e_1 + 4e_2 \\ e_1 + e_2 \end{pmatrix} = \begin{pmatrix} 0 \\ 0 \end{pmatrix}$$
$$\Rightarrow 4e_1 + 4e_2 = 0, \ e_1 + e_2 = 0.$$

We can choose $e_1 = 1$, $e_2 = -1$. So, the eigenvector corresponding to the eigenvalue $\lambda_1 = 1$ is $\underset{\sim}{e} = \begin{pmatrix} 1 \\ -1 \end{pmatrix}$.

Again, let $\underset{\sim}{e}' = \begin{pmatrix} e_1' \\ e_2' \end{pmatrix}$ be the eigenvector corresponding to the eigenvalue $\lambda_2 = 6$. Then

$$(A - 6I)\underset{\sim}{e}' = \underset{\sim}{0}$$

$$\Rightarrow \quad \begin{pmatrix} 5-6 & 4 \\ 1 & 2-6 \end{pmatrix} \begin{pmatrix} e_1' \\ e_2' \end{pmatrix} = \begin{pmatrix} 0 \\ 0 \end{pmatrix}$$

$$\Rightarrow \quad \begin{pmatrix} -e_1' + 4e_2' \\ e_1' - 4e_2' \end{pmatrix} = \begin{pmatrix} 0 \\ 0 \end{pmatrix}$$

$$\Rightarrow \quad -e_1' + 4e_2' = 0 \, e_1' - 4e_2' = 0.$$

We can choose $e_1' = 4, e_2' = 1$. So, the eigenvector corresponding to theeigenvalue $\lambda_2 = 6$ is $\underset{\sim}{e}' = \begin{pmatrix} 4 \\ 1 \end{pmatrix}$. The eigenvectors $\underset{\sim}{e}$, $\underset{\sim}{e}'$ are linearly independent. Hence the general solution of the system is given as

$$\underset{\sim}{x}(t) = c_1 \underset{\sim}{e} \, e^t + c_2 \underset{\sim}{e}' e^{6t} = c_1 \begin{pmatrix} 1 \\ -1 \end{pmatrix} e^t + c_2 \begin{pmatrix} 4 \\ 1 \end{pmatrix} e^{6t}$$

or, $\left. \begin{array}{l} x(t) = c_1 e^t + 4c_2 e^{6t} \\ y(t) = -c_1 e^t + c_2 e^{6t} \end{array} \right\}$, where c_1, c_2 are arbitrary constants.

Example 2.2 Find the general solution of the linear system

$$\frac{d}{dt} \begin{pmatrix} x \\ y \end{pmatrix} = \begin{pmatrix} 3 & 0 \\ 0 & 3 \end{pmatrix} \begin{pmatrix} x \\ y \end{pmatrix}$$

Solution The characteristic equation of matrix A is

$$\det(A - \lambda I) = 0$$

or, $\begin{vmatrix} 3 - \lambda & 0 \\ 0 & 3 - \lambda \end{vmatrix} = 0$

or, $(3 - \lambda)^2 = 0$

or, $\lambda = 3, 3.$

So, the eigenvalues of A are 3, 3, which are real and repeated. Clearly, $\underset{\sim}{e}_1 = \begin{pmatrix} 1 \\ 0 \end{pmatrix}$ and $\underset{\sim}{e}_2 = \begin{pmatrix} 0 \\ 1 \end{pmatrix}$ are two linearly independent eigenvectors corresponding to the repeated eigenvalue $\lambda = 3$. Thus, the general solution of the system is

$$\underset{\sim}{x}(t) = c_1 \underset{\sim}{e}_1 e^{\lambda t} + c_2 \underset{\sim}{e}_2 e^{\lambda t}$$

$$\Rightarrow \quad \begin{pmatrix} x(t) \\ y(t) \end{pmatrix} = c_1 \begin{pmatrix} 1 \\ 0 \end{pmatrix} e^{3t} + c_2 \begin{pmatrix} 0 \\ 1 \end{pmatrix} e^{3t} = \begin{pmatrix} c_1 e^{3t} \\ c_2 e^{3t} \end{pmatrix}.$$

$\Rightarrow \quad x(t) = c_1 e^{3t}, \ y(t) = c_2 e^{3t},$ where c_1, c_2 are arbitrary constants.

Example 2.3 Find the general solution of the system

$$\dot{x} = 3x - 4y$$
$$\dot{y} = x - y$$

using eigenvalue-eigenvector method.

Solution The characteristic equation of matrix A is

$$\det(A - \lambda I) = 0$$

$$\Rightarrow \quad \begin{vmatrix} 3 - \lambda & -4 \\ 1 & -1 - \lambda \end{vmatrix} = 0$$

$$\Rightarrow \quad \lambda^2 - 2\lambda + 1 = 0$$

$$\Rightarrow \quad \lambda = 1, 1$$

So matrix A has repeated real eigenvalues $\lambda = 1, 1$.

Let $\underset{\sim}{e} = \begin{pmatrix} e_1 \\ e_2 \end{pmatrix}$ be the eigenvector corresponding to the eigenvalue $\lambda = 1$. Then

$$(A - I)\underset{\sim}{e} = \underset{\sim}{0}$$

$$\Rightarrow \quad \begin{pmatrix} 3 - 1 & -4 \\ 1 & -1 - 1 \end{pmatrix} \begin{pmatrix} e_1 \\ e_2 \end{pmatrix} = \begin{pmatrix} 0 \\ 0 \end{pmatrix}$$

$$\Rightarrow \quad \begin{pmatrix} 2e_1 - 4e_2 \\ e_1 - 2e_2 \end{pmatrix} = \begin{pmatrix} 0 \\ 0 \end{pmatrix}$$

$$\Rightarrow \quad 2e_1 - 4e_2 = 0, e_1 - 2e_2 = 0$$

We can choose $e_1 = 2, e_2 = 1$. Therefore, $\underset{\sim}{e} = \begin{pmatrix} 2 \\ 1 \end{pmatrix}$.

Let $\underset{\sim}{g} = \begin{pmatrix} g_1 \\ g_2 \end{pmatrix}$ be the generalized eigenvector corresponding to the eigenvalue $\lambda = 1$. Then

$$(A - I)\underset{\sim}{g} = \underset{\sim}{e}$$

$$\Rightarrow \begin{pmatrix} 3 - 1 & -4 \\ 1 & -1 - 1 \end{pmatrix} \begin{pmatrix} g_1 \\ g_2 \end{pmatrix} = \begin{pmatrix} 2 \\ 1 \end{pmatrix}$$

$$\Rightarrow \begin{pmatrix} 2g_1 - 4g_2 \\ g_1 - 2g_2 \end{pmatrix} = \begin{pmatrix} 2 \\ 1 \end{pmatrix}$$

$$\Rightarrow \quad 2g_1 - 4g_2 = 2, g_1 - 2g_2 = 1$$

We can choose $g_2 = 1, g_1 = 3$. Therefore $\underset{\sim}{g} = \begin{pmatrix} 3 \\ 1 \end{pmatrix}$.

Therefore the general solution of the system is

$$\underset{\sim}{x}(t) = c_1 \underset{\sim}{e} \, e^t + c_2 \left(\underset{\sim}{e} \, t e^t + \underset{\sim}{g} \, e^t \right)$$

or, $\begin{pmatrix} x(t) \\ y(t) \end{pmatrix} = c_1 \begin{pmatrix} 2 \\ 1 \end{pmatrix} e^t + c_2 \begin{pmatrix} 2 \\ 1 \end{pmatrix} t e^t + c_2 \begin{pmatrix} 3 \\ 1 \end{pmatrix} e^t$

or, $\left. \begin{array}{l} x(t) = \{2c_1 + (2t+3)c_2\}e^t \\ y(t) = \{c_1 + (t+1)c_2\}e^t \end{array} \right\}$, where c_1 and c_2 are arbitrary constants.

Example 2.4 Find the general solution of the linear system

$$\dot{x} = 10x - y$$
$$\dot{y} = 25x + 2y$$

Solution Given system can be written as

$$\underset{\sim}{\dot{x}} = A\underset{\sim}{x}, \text{ where } A = \begin{pmatrix} 10 & -1 \\ 25 & 2 \end{pmatrix} \text{ and } \underset{\sim}{x} = \begin{pmatrix} x \\ y \end{pmatrix}.$$

The characteristic equation of matrix A is

$$\det(A - \lambda I) = 0$$

$$\Rightarrow \quad \begin{vmatrix} 10 - \lambda & -1 \\ 25 & 2 - \lambda \end{vmatrix} = 0$$

$$\Rightarrow \quad \lambda^2 - 12\lambda + 45 = 0$$

$$\Rightarrow \quad \lambda = 6 \pm 3i.$$

Therefore, matrix A has a pair of complex conjugate eigenvalues $6 \pm 3i$.

Let $\underset{\sim}{e} = \begin{pmatrix} e_1 \\ e_2 \end{pmatrix}$ be the eigenvector corresponding to the eigenvalue $\lambda = 6 + 3i$. Then

$$(A - (6+3i)I)\underset{\sim}{e} = \underset{\sim}{0}$$

$$\Rightarrow \begin{pmatrix} 10-6-3i & -1 \\ 25 & 2-6-3i \end{pmatrix} \begin{pmatrix} e_1 \\ e_2 \end{pmatrix} = \begin{pmatrix} 0 \\ 0 \end{pmatrix}$$

$$\Rightarrow \begin{pmatrix} (4-3i)e_1 - e_2 \\ 25e_1 - (4+3i)e_2 \end{pmatrix} = \begin{pmatrix} 0 \\ 0 \end{pmatrix}$$

$$\Rightarrow \quad (4-3i)e_1 - e_2 = 0, 25e_1 - (4+3i)e_2 = 0.$$

A nontrivial solution of this system is

$$e_1 = 1, \ e_2 = 4 - 3i.$$

Therefore $\underset{\sim}{e} = \begin{pmatrix} 1 \\ 4-3i \end{pmatrix} = \begin{pmatrix} 1 \\ 4 \end{pmatrix} + i \begin{pmatrix} 0 \\ -3 \end{pmatrix} = \underset{\sim}{\alpha}_1 + i\underset{\sim}{\alpha}_2$, where $\underset{\sim}{\alpha}_1 = \begin{pmatrix} 1 \\ 4 \end{pmatrix}$

and $\underset{\sim}{\alpha}_2 = \begin{pmatrix} 0 \\ -3 \end{pmatrix}$.

Similarly, the eigenvector corresponding to the eigenvalue $\lambda = 6 - 3i$ is
$\underset{\sim}{e}' = \begin{pmatrix} 1 \\ 4+3i \end{pmatrix} = \underset{\sim}{\alpha}_1 - i\underset{\sim}{\alpha}_2$. Therefore,

$$\underset{\sim}{u}_1 = e^{at}\left(\underset{\sim}{\alpha}_1 \cos bt - \underset{\sim}{\alpha}_2 \sin bt \right) = e^{6t}\left\{ \begin{pmatrix} 1 \\ 4 \end{pmatrix} \cos 3t - \begin{pmatrix} 0 \\ -3 \end{pmatrix} \sin 3t \right\}$$

and

$$\underset{\sim}{v}_1 = e^{at}\left(\underset{\sim}{\alpha}_1 \sin bt + \underset{\sim}{\alpha}_2 \cos bt \right) = e^{6t}\left\{ \begin{pmatrix} 1 \\ 4 \end{pmatrix} \sin 3t + \begin{pmatrix} 0 \\ -3 \end{pmatrix} \cos 3t \right\}.$$

Therefore, the general solution is

$$\underset{\sim}{x}(t) = c_1\underset{\sim}{u}_1 + d_1\underset{\sim}{v}_1$$

$$= e^{6t}\left\{ \begin{pmatrix} 1 \\ 4 \end{pmatrix} c_1 \cos 3t - \begin{pmatrix} 0 \\ -3 \end{pmatrix} c_1 \sin 3t \right\}$$

$$+ e^{6t}\left\{ \begin{pmatrix} 1 \\ 4 \end{pmatrix} d_1 \sin 3t + \begin{pmatrix} 0 \\ -3 \end{pmatrix} d_1 \cos 3t \right\}$$

$$= e^{6t}\begin{pmatrix} c_1 \cos 3t + d_1 \sin 3t \\ (4c_1 - 3d_1) \cos 3t + (3c_1 + 4d_1) \sin 3t \end{pmatrix}$$

$$\Rightarrow \quad x(t) = e^{6t}(c_1 \cos 3t + d_1 \sin 3t),$$

$$y(t) = e^{6t}[(4c_1 - 3d_1) \cos 3t + (3c_1 + 4d_1) \sin 3t]$$

where c_1 and d_1 are arbitrary constants.

Example 2.5 Find the solution of the system

$$\dot{x} = x - 5y, \dot{y} = x - 3y$$

satisfying the initial condition $x(0) = 1$, $y(0) = 1$. Describe the behavior of the solution as $t \to \infty$.

Solution The characteristic equation of matrix A is

$$\det(A - \lambda I) = 0$$

$$\Rightarrow \quad \begin{vmatrix} 1 - \lambda & -5 \\ 1 & -3 - \lambda \end{vmatrix} = 0$$

$$\Rightarrow \quad \lambda^2 + 2\lambda + 2 = 0$$

$$\Rightarrow \quad \lambda = -1 \pm i.$$

So, matrix A has a pair of complex conjugate eigenvalues $(-1 \pm i)$.

Let $\underset{\sim}{e} = \begin{pmatrix} e_1 \\ e_2 \end{pmatrix}$ be the eigenvector corresponding to the eigenvalue $\lambda = -1 + i$. Then

$$(A - (-1 + i)I)\underset{\sim}{e} = \underset{\sim}{0}$$

$$\Rightarrow \quad \begin{pmatrix} 2 - i & -5 \\ 1 & -2 - i \end{pmatrix} \begin{pmatrix} e_1 \\ e_2 \end{pmatrix} = \begin{pmatrix} 0 \\ 0 \end{pmatrix}$$

$$\Rightarrow \quad \begin{pmatrix} (2 - i)e_1 - 5e_2 \\ e_1 - (2 + i)e_2 \end{pmatrix} = \begin{pmatrix} 0 \\ 0 \end{pmatrix}$$

$$\Rightarrow \quad (2 - i)e_1 - 5e_2 = 0, e_1 - (2 + i)e_2 = 0.$$

A nontrivial solution of this system is

$$e_1 = 2 + i, e_2 = 1.$$

Therefore $\underset{\sim}{e} = \begin{pmatrix} 2 + i \\ 1 \end{pmatrix} = \begin{pmatrix} 2 \\ 1 \end{pmatrix} + i \begin{pmatrix} 1 \\ 0 \end{pmatrix} = \underset{\sim}{\alpha}_1 + i\underset{\sim}{\alpha}_2$, where $\underset{\sim}{\alpha}_1 = \begin{pmatrix} 2 \\ 1 \end{pmatrix}$ and $\underset{\sim}{\alpha}_2 = \begin{pmatrix} 1 \\ 0 \end{pmatrix}$.

Similarly, the eigenvector corresponding to the eigenvalue $\lambda = -1 - i$ is $\underset{\sim}{e}' = \begin{pmatrix} 2 - i \\ 1 \end{pmatrix} = \underset{\sim}{\alpha}_1 - i\underset{\sim}{\alpha}_2$.

$$\therefore \underset{\sim}{u}_1 = e^{at}\left(\underset{\sim}{\alpha}_1 \cos bt - \underset{\sim}{\alpha}_2 \sin bt\right) = e^{-t}\left\{\binom{2}{1}\cos t - \binom{1}{0}\sin t\right\} \text{ and}$$

$$\underset{\sim}{v}_1 = e^{at}\left(\underset{\sim}{\alpha}_1 \sin bt + \underset{\sim}{\alpha}_2 \cos bt\right) = e^{-t}\left\{\binom{2}{1}\sin t + \binom{1}{0}\cos t\right\}$$

Therefore, the solution of the system is

$$\underset{\sim}{x}(t) = x(0)\underset{\sim}{u}_1 + y(0)\underset{\sim}{v}_1$$

$$= e^{-t}\left\{\binom{2}{1}\cos t - \binom{1}{0}\sin t\right\} + e^{-t}\left\{\binom{2}{1}\sin t + \binom{1}{0}\cos t\right\}$$

$$= e^{-t}\left\{\binom{2}{1}(\cos t + \sin t) + \binom{1}{0}(\cos t - \sin t)\right\}.$$

When $t \to \infty$, $e^{-t} \to 0$. So, in this case $\underset{\sim}{x}(t) \to \underset{\sim}{0}$, that is, the solution of the system is stable in the usual sense.

Example 2.6 Find the solution of the system $\underset{\sim}{\dot{x}} = A\underset{\sim}{x}$, where

$$A = \begin{pmatrix} -1 & 2 & 3 \\ 0 & -2 & 1 \\ 0 & 3 & 0 \end{pmatrix}.$$

Solution The characteristic equation of A is

$$\det(A - \lambda I) = 0$$

$$\Rightarrow \begin{vmatrix} -1-\lambda & 2 & 3 \\ 0 & -2-\lambda & 1 \\ 0 & 3 & -\lambda \end{vmatrix} = 0$$

$$\Rightarrow (\lambda+1)(\lambda-1)(\lambda+3) = 0$$

$$\Rightarrow \lambda = -1, 1, -3$$

Therefore the eigenvalues of matrix A are $\lambda = -1, 1, -3$.

We shall now find the eigenvector corresponding to each of the eigenvalues.

Let $\underset{\sim}{e} = \begin{pmatrix} e_1 \\ e_2 \\ e_3 \end{pmatrix}$ be the eigenvector corresponding to the eigenvalue $\lambda = -1$.

Then

$$(A+1)\underset{\sim}{e} = \underset{\sim}{0}$$

$$\Rightarrow \begin{pmatrix} -1+1 & 2 & 3 \\ 0 & -2+1 & 1 \\ 0 & 3 & 1 \end{pmatrix} \begin{pmatrix} e_1 \\ e_2 \\ e_3 \end{pmatrix} = \begin{pmatrix} 0 \\ 0 \\ 0 \end{pmatrix}$$

$$\Rightarrow \quad 2e_2 + 3e_3 = 0, \quad -e_2 + e_3 = 0, \quad 3e_2 + e_3 = 0$$

$$\Rightarrow e_2 = e_3 = 0 \text{ and } e_1 \text{ is } \text{ arbitrary.}$$

We choose $e_1 = 1$. Therefore, the eigenvector corresponding to the eigenvalue $\lambda = -1$ is $\underset{\sim}{e} = \begin{pmatrix} 1 \\ 0 \\ 0 \end{pmatrix}$. Similarly, the eigenvectors corresponding to $\lambda = 1$ and $\lambda = -3$

are, respectively, $\underset{\sim}{g} = \begin{pmatrix} 11/2 \\ 1 \\ 3 \end{pmatrix}$ and $\underset{\sim}{\alpha} = \begin{pmatrix} 1/2 \\ 1 \\ -1 \end{pmatrix}$. Therefore the general solution

is

$$\underset{\sim}{x}(t) = c_1 \underset{\sim}{e} \, e^{-t} + c_2 \underset{\sim}{g} \, e^t + c_3 \underset{\sim}{\alpha} \, e^{-3t}$$

$$= c_1 \begin{pmatrix} 1 \\ 0 \\ 0 \end{pmatrix} e^{-t} + c_2 \begin{pmatrix} 11/2 \\ 1 \\ 3 \end{pmatrix} e^t + c_3 \begin{pmatrix} 1/2 \\ 1 \\ -1 \end{pmatrix} e^{-3t}$$

where c_1, c_2 and c_3 are arbitrary constants.

Example 2.7 Solve the system $\underset{\sim}{\dot{x}} = A\underset{\sim}{x}$, where

$$A = \begin{pmatrix} 1 & -3 & 3 \\ 3 & -5 & 3 \\ 6 & -6 & 4 \end{pmatrix}.$$

Solution The characteristic equation of matrix A is

$$\det(A - \lambda I) = 0$$

$$\Rightarrow \begin{vmatrix} 1-\lambda & -3 & 3 \\ 3 & -5-\lambda & 3 \\ 6 & -6 & 4-\lambda \end{vmatrix} = 0$$

$$\Rightarrow \quad \lambda = 4, -2, -2.$$

So (-2) is a repeated eigenvalue of A. The eigenvector for the eigenvalue $\lambda_1 = 4$ is given as $\begin{pmatrix} 1 \\ 1 \\ 2 \end{pmatrix}$. The eigenvector corresponding to the repeated eigenvalue $\lambda_2 = \lambda_3 = -2$ is $(e_1 \quad e_2 \quad e_3)^T$ such that

$$\begin{pmatrix} 3 & -3 & 3 \\ 3 & -3 & 3 \\ 6 & -6 & 6 \end{pmatrix} \begin{pmatrix} e_1 \\ e_2 \\ e_3 \end{pmatrix} = \begin{pmatrix} 0 \\ 0 \\ 0 \end{pmatrix}$$

which is equivalent to

$$3e_1 - 3e_2 + 3e_3 = 0, \quad 3e_1 - 3e_2 + 3e_3 = 0, \quad 6e_1 - 6e_2 + 6e_3 = 0,$$

that is, $e_1 - e_2 + e_3 = 0$.

We can choose $e_1 = 1$, $e_2 = 1$ and $e_3 = 0$, and so we can take one eigenvector as $\begin{pmatrix} 1 \\ 1 \\ 0 \end{pmatrix}$. Again, we can choose $e_1 = 0$, $e_2 = 1$ and $e_3 = 1$. Then we obtain another eigenvector $\begin{pmatrix} 0 \\ 1 \\ 1 \end{pmatrix}$. Clearly, these two eigenvectors are linearly independent. Thus, we have two linearly independent eigenvectors corresponding to the repeated eigenvalue -2. Hence, the general solution of the system is given by

$$\underset{\sim}{x}(t) = c_1 \begin{pmatrix} 1 \\ 1 \\ 2 \end{pmatrix} e^{4t} + c_2 \begin{pmatrix} 1 \\ 1 \\ 0 \end{pmatrix} e^{-2t} + c_3 \begin{pmatrix} 0 \\ 1 \\ 1 \end{pmatrix} e^{-2t}$$

where c_1, c_2 and c_3 are arbitrary constants.

Example 2.8 Solve the system $\underset{\sim}{\dot{x}} = A\underset{\sim}{x}$ where

$$A = \begin{bmatrix} -1 & -1 & 0 & 0 \\ 1 & -1 & 0 & 0 \\ 0 & 0 & 0 & -2 \\ 0 & 0 & 1 & 2 \end{bmatrix}$$

Solution Here matrix A has two pair of complex conjugate eigenvalues $\lambda_1 = -1 \pm i$ and $\lambda_2 = 1 \pm i$. The corresponding pair of eigenvectors is

$$\underset{\sim}{w}_1 = \underset{\sim}{\alpha}_1 \pm i\underset{\sim}{\beta}_1 = \begin{pmatrix} \pm i \\ 1 \\ 0 \\ 0 \end{pmatrix} = \begin{pmatrix} 0 \\ 1 \\ 0 \\ 0 \end{pmatrix} \pm i \begin{pmatrix} 1 \\ 0 \\ 0 \\ 0 \end{pmatrix} \quad \text{and}$$

$$\underset{\sim}{w}_2 = \underset{\sim}{\alpha}_2 \pm i\underset{\sim}{\beta}_2 = \begin{pmatrix} 0 \\ 0 \\ -1 \pm i \\ 1 \end{pmatrix} = \begin{pmatrix} 0 \\ 0 \\ -1 \\ 1 \end{pmatrix} \pm i \begin{pmatrix} 0 \\ 0 \\ 1 \\ 0 \end{pmatrix}$$

Therefore, the general solution of the system is expressed as

$$\underset{\sim}{x}(t) = \sum_{j=1}^{2} c_j \underset{\sim}{u}_j + d_j \underset{\sim}{v}_j$$

$$= c_1 e^{-t} \left\{ \begin{pmatrix} 0 \\ 1 \\ 0 \\ 0 \end{pmatrix} \cos t - \begin{pmatrix} 1 \\ 0 \\ 0 \\ 0 \end{pmatrix} \sin t \right\} + c_2 e^{t} \left\{ \begin{pmatrix} 0 \\ 0 \\ -1 \\ 1 \end{pmatrix} \cos t - \begin{pmatrix} 0 \\ 0 \\ 1 \\ 0 \end{pmatrix} \sin t \right\}$$

$$+ d_1 e^{-t} \left\{ \begin{pmatrix} 0 \\ 1 \\ 0 \\ 0 \end{pmatrix} \sin t + \begin{pmatrix} 1 \\ 0 \\ 0 \\ 0 \end{pmatrix} \cos t \right\}$$

$$+ d_2 e^{t} \left\{ \begin{pmatrix} 0 \\ 0 \\ -1 \\ 1 \end{pmatrix} \sin t + \begin{pmatrix} 0 \\ 0 \\ 1 \\ 0 \end{pmatrix} \cos t \right\}$$

$$= \begin{pmatrix} e^{-t}(d_1 \cos t - c_1 \sin t) \\ e^{-t}(c_1 \cos t + d_1 \sin t) \\ e^{t}\{(d_2 - c_2)\cos t - (d_2 + c_2)\sin t\} \\ e^{t}(c_2 \cos t + d_2 \sin t) \end{pmatrix}$$

where $c_j, d_j (j = 1, 2)$ are arbitrary constants.

2.3   Fundamental Matrix

A set $\{x_1(t), x_2(t), \ldots, x_n(t)\}$ of solutions of a linear homogeneous system $\dot{x} = Ax$ is said to be a **fundamental set** of solutions of that system if it satisfies the following two conditions:

(i)   The set $\{x_1(t), x_2(t), \ldots, x_n(t)\}$ is linearly independent, that is, for $c_1, c_2, \ldots, c_n \in \mathbb{R}$, $c_1 x_1 + c_2 x_2 + \cdots + c_n x_n = 0 \Rightarrow c_1 = c_2 = \cdots = c_n = 0$.

(ii) For any solution $x(t)$ of the system $\dot{x} = Ax$, there exist $c_1, c_2, \ldots, c_n \in \mathbb{R}$ such that $x(t) = c_1 x_1(t) + c_2 x_2(t) + \cdots + c_n x_n(t), \forall t \in \mathbb{R}$.

The solution, expressed as a linear combination of a fundamental set of solutions of a system, is called a **general solution** of the system.

Let $\{x_1(t), x_2(t), \ldots, x_n(t)\}$ be a fundamental set of solutions of the system $\dot{x} = Ax$ for $t \in I = [a, b]$; $a, b \in \mathbb{R}$. Then the matrix

$$\Phi(t) = \left(x_1(t), x_2(t), \ldots, x_n(t) \right)$$

is called a **fundamental matrix** of the system $\dot{x} = Ax$, $x \in \mathbb{R}^n$. Since the set $\{x_1(t), x_2(t), \ldots, x_n(t)\}$ is linearly independent, the fundamental matrix $\Phi(t)$ is nonsingular. Now the general solution of the system is

$$x(t) = c_1 x_1(t) + c_2 x_2(t) + \cdots + c_n x_n(t)$$

$$= \left(x_1(t), x_2(t), \ldots, x_n(t) \right) \begin{pmatrix} c_1 \\ c_2 \\ \vdots \\ c_n \end{pmatrix}$$

$$= \Phi(t) c$$

where $c = (c_1, c_2, \ldots c_n)^t$ is a constant column vector. If the initial condition is $x(0) = x_0$, then

$$\Phi(0) c = x_0$$

$$\Rightarrow c = \Phi^{-1}(0) x_0 [\text{Since } \Phi(t) \text{ is nonsingular for all } t].$$

Thus the solution of the initial value problem $\dot{x} = A\underset{\sim}{x}$ with the initial conditions $\underset{\sim}{x}(0) = \underset{\sim}{x}_0$ can be expressed in terms of the fundamental matrix $\Phi(t)$ as

$$\underset{\sim}{x}(t) = \Phi(t)\Phi^{-1}(0)\underset{\sim}{x}_0 \qquad (2.7)$$

Note that two different homogeneous systems cannot have the same fundamental matrix. Again, if $\Phi(t)$ is a fundamental matrix of $\dot{x} = A\underset{\sim}{x}$, then for any constant C, $C\Phi(t)$ is also a fundamental matrix of the system.

Example 2.9 Find the fundamental matrix of the system $\dot{x} = A\underset{\sim}{x}$, where $A = \begin{pmatrix} 1 & -2 \\ -3 & 2 \end{pmatrix}$. Hence find its solution.

Solution The characteristic equation of matrix A is

$$|A - \lambda I| = 0$$
$$\Rightarrow \quad \begin{vmatrix} 1-\lambda & -2 \\ -3 & 2-\lambda \end{vmatrix} = 0$$
$$\Rightarrow \quad (1-\lambda)(2-\lambda) - 6 = 0$$
$$\Rightarrow \quad \lambda^2 - 3\lambda - 4 = 0$$
$$\Rightarrow \quad \lambda = -1, 4.$$

So, the eigenvalues of matrix A are $-1, 4$, which are real and distinct.

Let $\underset{\sim}{e} = \begin{pmatrix} e_1 \\ e_2 \end{pmatrix}$ be the eigenvector corresponding to the eigenvalue $\lambda_1 = -1$. Then

$$(A+I)\underset{\sim}{e} = \underset{\sim}{0}$$
$$\Rightarrow \quad \begin{pmatrix} 1+1 & -2 \\ -3 & 2+1 \end{pmatrix}\begin{pmatrix} e_1 \\ e_2 \end{pmatrix} = \begin{pmatrix} 0 \\ 0 \end{pmatrix}$$
$$\Rightarrow \quad 2e_1 - 2e_2 = 0, -3e_1 + 3e_2 = 0.$$

A nontrivial solution of this system is $e_1 = 1, e_2 = 1$.

$$\therefore \underset{\sim}{e} = \begin{pmatrix} 1 \\ 1 \end{pmatrix}.$$

Again, let $\underset{\sim}{g} = \begin{pmatrix} g_1 \\ g_2 \end{pmatrix}$ be the eigenvector corresponding to the eigenvalue $\lambda_2 = 4$. Then

$$(A - 4I)g = 0$$

$$\Rightarrow \begin{pmatrix} 1-4 & -2 \\ -3 & 2-4 \end{pmatrix} \begin{pmatrix} g_1 \\ g_2 \end{pmatrix} = \begin{pmatrix} 0 \\ 0 \end{pmatrix}$$

$$\Rightarrow \quad 3g_1 + 2g_2 = 0$$

Choose $g_1 = 2, g_2 = -3$. Therefore, $g = \begin{pmatrix} 2 \\ -3 \end{pmatrix}$.

Therefore the eigenvectors corresponding to the eigenvalues $\lambda = -1, 4$ are respectively $\begin{pmatrix} 1 \\ 1 \end{pmatrix}$ and $\begin{pmatrix} 2 \\ -3 \end{pmatrix}$, which are linearly independent. So two fundamental solutions of the system are

$$x_1(t) = \begin{pmatrix} 1 \\ 1 \end{pmatrix} e^{-t}, x_2(t) = \begin{pmatrix} 2 \\ -3 \end{pmatrix} e^{4t}$$

and a fundamental matrix of the system is

$$\Phi(t) = \begin{pmatrix} x_1(t) & x_2(t) \end{pmatrix} = \begin{pmatrix} e^{-t} & 2e^{4t} \\ e^{-t} & -3e^{4t} \end{pmatrix}.$$

Now $\Phi(0) = \begin{pmatrix} 1 & 2 \\ 1 & -3 \end{pmatrix}$ and so $\Phi^{-1}(0) = \frac{1}{5} \begin{pmatrix} 3 & 2 \\ 1 & -1 \end{pmatrix}$.

Therefore the general solution of the system is given by

$$x(t) = \Phi(t)\Phi^{-1}(0) x_0 = \frac{1}{5} \begin{pmatrix} e^{-t} & 2e^{4t} \\ e^{-t} & -3e^{4t} \end{pmatrix} \begin{pmatrix} 3 & 2 \\ 1 & -1 \end{pmatrix} x_0$$

$$= \frac{1}{5} \begin{pmatrix} 3e^{-t} + 2e^{4t} & 2e^{-t} - 2e^{4t} \\ 3e^{-t} - 3e^{4t} & 2e^{-t} + 3e^{4t} \end{pmatrix} x_0.$$

2.3.1 General Solution of Linear Systems

Consider a simple linear equation

$$\dot{x} = ax \tag{2.8}$$

with initial condition $x(0) = x_0$, where a and x_0 are certain constants. The solution of this initial value problem (IVP) is given as $x(t) = x_0 e^{at}$. Then we may expect that the solution of the initial value problem for $n \times n$ system

$$\dot{\underset{\sim}{x}} = A\underset{\sim}{x} \text{ with } \underset{\sim}{x}(0) = \underset{\sim}{x}_0 \qquad (2.9)$$

can be expressed in term of exponential matrix function as

$$\underset{\sim}{x}(t) = e^{At}\underset{\sim}{x}_0 \qquad (2.10)$$

where A is an $n \times n$ matrix. Comparing (2.10) with the solution obtained by the fundamental matrix, we have the relation

$$e^{At} = \Phi(t)\Phi^{-1}(0) \qquad (2.11)$$

Thus we see that if $\Phi(t)$ is a fundamental matrix of the system $\dot{\underset{\sim}{x}} = A\underset{\sim}{x}$, then $\Phi(0)$ is invertible and $e^{At} = \Phi(t)\Phi^{-1}(0)$. Note that if $\Phi(0) = I$, then $\Phi^{-1}(0) = I$ and so, $e^{At} = \Phi(t)I = \Phi(t)$.

Example 2.10 Does $\Phi(t) = \begin{pmatrix} 2e^t & -e^{-3t} \\ -4e^t & 2e^{-3t} \end{pmatrix}$ a fundamental matrix for a system $\dot{\underset{\sim}{x}} = A\underset{\sim}{x}$?

Solution We know that if $\Phi(t)$ is a fundamental matrix, then $\Phi(0)$ is invertible. Here $\Phi(t) = \begin{pmatrix} 2e^t & -e^{-3t} \\ -4e^t & 2e^{-3t} \end{pmatrix}$. So, $\Phi(0) = \begin{pmatrix} 2 & -1 \\ -4 & 2 \end{pmatrix}$.

Since $\det(\Phi(0)) = 4 - 4 = 0$, $\Phi(0)$ is not invertible and hence the given matrix is not a fundamental matrix for the system $\dot{\underset{\sim}{x}} = A\underset{\sim}{x}$.

Example 2.11 Find e^{At} for the system $\dot{\underset{\sim}{x}} = A\underset{\sim}{x}$, where $A = \begin{pmatrix} 1 & 1 \\ 4 & 1 \end{pmatrix}$.

Solution The characteristic equation of A is

$$|A - \lambda I| = 0$$

$$\Rightarrow \quad \begin{vmatrix} 1 - \lambda & 1 \\ 4 & 1 - \lambda \end{vmatrix} = 0$$

$$\Rightarrow \quad (\lambda - 1)^2 - 4 = 0$$

$$\Rightarrow \quad \lambda = 3, -1.$$

So, the eigenvalue of A are $\lambda = 3, -1$. The eigenvector corresponding to the eigenvalues $\lambda = 3, -1$ are, respectively, $\begin{pmatrix} 1 \\ 2 \end{pmatrix}$ and $\begin{pmatrix} 1 \\ -2 \end{pmatrix}$, which are linearly independent. So, two fundamental solutions of the system are $\underset{\sim}{x}_1(t) = \begin{pmatrix} 1 \\ 2 \end{pmatrix} e^{3t}, \underset{\sim}{x}_2(t) = \begin{pmatrix} 1 \\ -2 \end{pmatrix} e^{-t}$. Therefore a fundamental matrix of the system is

$$\Phi(t) = \begin{pmatrix} \underset{\sim}{x}_1(t) & \underset{\sim}{x}_2(t) \end{pmatrix} = \begin{pmatrix} e^{3t} & e^{-t} \\ 2e^{3t} & -2e^{-t} \end{pmatrix}.$$

Now, $\Phi(0) = \begin{pmatrix} 1 & 1 \\ 2 & -2 \end{pmatrix}$ and $\Phi^{-1}(0) = -\frac{1}{4}\begin{pmatrix} -2 & -1 \\ -2 & 1 \end{pmatrix} = \begin{pmatrix} \frac{1}{2} & \frac{1}{4} \\ \frac{1}{2} & -\frac{1}{4} \end{pmatrix}.$
Therefore,

$$e^{At} = \Phi(t)\Phi^{-1}(0)$$

$$= \begin{pmatrix} e^{3t} & e^{-t} \\ 2e^{3t} & -2e^{-t} \end{pmatrix}\begin{pmatrix} \frac{1}{2} & \frac{1}{4} \\ \frac{1}{2} & -\frac{1}{4} \end{pmatrix} = \begin{pmatrix} \frac{1}{2}(e^{3t}+e^{-t}) & \frac{1}{4}(e^{3t}-e^{-t}) \\ (e^{3t}-e^{-t}) & \frac{1}{2}(e^{3t}+e^{-t}) \end{pmatrix}.$$

2.3.2 Fundamental Matrix Method

The fundamental matrix can be used to obtain the general solution of a linear system. The fundamental theorem gives the existence and uniqueness of solution of a linear system $\dot{\underset{\sim}{x}} = A\underset{\sim}{x}$, $\underset{\sim}{x} \in \mathbb{R}^n$ subject to the initial conditions $\underset{\sim}{x}_0 \in \mathbb{R}^n$. We now present the fundamental theorem.

Theorem 2.1 (Fundamental theorem) *Let A be an $n \times n$ matrix. Then for given any initial condition $\underset{\sim}{x}_0 \in \mathbb{R}^n$, the initial value problem $\dot{\underset{\sim}{x}} = A\underset{\sim}{x}$ with $\underset{\sim}{x}(0) = \underset{\sim}{x}_0$ has the unique solution $\underset{\sim}{x}(t) = e^{At}\underset{\sim}{x}_0$.*

Proof The initial value problem is

$$\dot{\underset{\sim}{x}} = A\underset{\sim}{x}, \quad \underset{\sim}{x}(0) = \underset{\sim}{x}_0 \tag{2.12}$$

We have

$$e^{At} = I + At + \frac{A^2 t^2}{2!} + \frac{A^3 t^3}{3!} + \cdots \tag{2.13}$$

Differentiating (2.13) w.r.to t,

$$\frac{d}{dt}(e^{At}) = \frac{d}{dt}\left(I + At + \frac{A^2 t^2}{2!} + \frac{A^3 t^3}{3!} + \cdots\right)$$

$$= \frac{d}{dt}(I) + \frac{d}{dt}(At) + \frac{d}{dt}\left(\frac{A^2 t^2}{2!}\right) + \frac{d}{dt}\left(\frac{A^3 t^3}{3!}\right) + \cdots$$

The term by term differentiation is valid because the series of e^{At} is convergent for all t under the operator.

$$\text{or, } \frac{d}{dt}\left(e^{At}\right) = \varphi + A + A^2 t + \frac{A^3 t^2}{2!} + \frac{A^4 t^3}{3!} + \cdots$$

$$= A\left(I + At + \frac{A^2 t^2}{2!} + \frac{A^3 t^3}{3!} + \cdots\right)$$

$$= Ae^{At}.$$

Therefore,

$$\frac{d}{dt}\left(e^{At}\right) = Ae^{At} \tag{2.14}$$

This shows that the matrix $\underset{\sim}{x} = e^{At}$ is a solution of the matrix differential equation $\underset{\sim}{\dot{x}} = A\underset{\sim}{x}$. The matrix e^{At} is known as the fundamental matrix of the system (2.12). Now using (2.14)

$$\frac{d}{dt}\left(e^{At}\underset{\sim}{x}_0\right) = \frac{d}{dt}\left(e^{At}\right)\underset{\sim}{x}_0 = Ae^{At}\underset{\sim}{x}_0$$

$$\Rightarrow \quad \underset{\sim}{\dot{x}} = \frac{d}{dt}(\underset{\sim}{x}) = A\underset{\sim}{x},$$

where $\underset{\sim}{x} = e^{At}\underset{\sim}{x}_0$.

Also, $\underset{\sim}{x}(0) = \left[e^{At}\underset{\sim}{x}_0\right]_{t=0} = [e^{At}]_{t=0}\underset{\sim}{x}_0 = I\underset{\sim}{x}_0 = \underset{\sim}{x}_0$. Thus $\underset{\sim}{x}(t) = e^{At}\underset{\sim}{x}_0$ is a solution of (2.12). We prove the uniqueness of solution as follows. Let $\underset{\sim}{x}(t)$ be a solution of (2.12) and $\underset{\sim}{y}(t) = e^{-At}\underset{\sim}{x}(t)$ be its another solution. Then

$$\underset{\sim}{\dot{y}}(t) = -Ae^{-At}\underset{\sim}{x}(t) + e^{-At}\underset{\sim}{\dot{x}}(t)$$

$$= -Ae^{-At}\underset{\sim}{x}(t) + Ae^{-At}\underset{\sim}{x}(t) = 0.$$

This implies $\underset{\sim}{y}(t)$ is constant. At $t = 0$, for $t \in \mathbb{R}$, it shows that $\underset{\sim}{y}(t) = \underset{\sim}{x}_0$. Therefore any solution of the IVP (2.12) is given as $\underset{\sim}{x}(t) = e^{At}\underset{\sim}{y}(t) = e^{At}\underset{\sim}{x}_0$. This completes the proof.

2.3.3   Matrix Exponential Function

From the fundamental theorem, the general solution of a linear system can be obtained using the exponential matrix function. The exponential matrix function has some interesting properties in which the general solution can be obtained easily. For an $n \times n$ matrix A, the **matrix exponential function** e^A of A is defined as

$$e^A = \sum_{n=0}^{\infty} \frac{A^n}{n!} = I + A + \frac{A^2}{2!} + \cdots \tag{2.15}$$

Note that the infinite series (2.15) converges for all $n \times n$ matrix A. If $A = [a]$, a 1×1 matrix, then $e^A = [e^a]$ (see the book by L. Perko [1]). We now discuss some of the important properties of matrix exponential function e^A.

Property 1 *If $A = \varphi$, the null matrix, then $e^{At} = I$.*

Proof By definition

$$e^{At} = I + At + \frac{A^2 t^2}{2!} + \frac{A^3 t^3}{3!} + \cdots$$

$$= I + \varphi t + \frac{\varphi^2 t^2}{2!} + \frac{\varphi^3 t^3}{3!} + \cdots$$

$$= I.$$

So, $e^{At} = I$ for $A = \varphi$.

Property 2 *Let $A = I$, the identity matrix. Then*

$$e^{At} = \begin{bmatrix} e^t & 0 \\ 0 & e^t \end{bmatrix} = Ie^t.$$

Proof We know that $e^{At} = I + At + \frac{A^2 t^2}{2!} + \frac{A^3 t^3}{3!} + \cdots$. Therefore

$$e^{It} = I + It + \frac{I^2 t^2}{2!} + \frac{I^3 t^3}{3!} + \cdots$$

$$= I + It + \frac{It^2}{2!} + \frac{It^3}{3!} + \cdots$$

$$= I\left(1 + t + \frac{t^2}{2!} + \frac{t^3}{3!} + \cdots\right)$$

$$= Ie^t = e^t \begin{bmatrix} 1 & 0 \\ 0 & 1 \end{bmatrix} = \begin{bmatrix} e^t & 0 \\ 0 & e^t \end{bmatrix}.$$

Note If $A = \alpha I$, α being a scalar, then

$$e^{At} = e^{\alpha It} = Ie^{\alpha t} = \begin{bmatrix} e^{\alpha t} & 0 \\ 0 & e^{\alpha t} \end{bmatrix}.$$

Property 3 *Suppose* $D = \begin{bmatrix} \lambda_1 & 0 \\ 0 & \lambda_2 \end{bmatrix}$, *a diagonal matrix. Then*

$$e^{Dt} = \begin{bmatrix} e^{\lambda_1 t} & 0 \\ 0 & e^{\lambda_2 t} \end{bmatrix}$$

Proof By definition

$$
\begin{aligned}
e^{Dt} &= I + Dt + \frac{D^2 t^2}{2!} + \frac{D^3 t^3}{3!} + \cdots \\
&= \begin{bmatrix} 1 & 0 \\ 0 & 1 \end{bmatrix} + \begin{bmatrix} \lambda_1 & 0 \\ 0 & \lambda_2 \end{bmatrix} t + \begin{bmatrix} \lambda_1 & 0 \\ 0 & \lambda_2 \end{bmatrix}^2 \frac{t^2}{2!} + \cdots \\
&= \begin{bmatrix} 1 & 0 \\ 0 & 1 \end{bmatrix} + \begin{bmatrix} \lambda_1 & 0 \\ 0 & \lambda_2 \end{bmatrix} t + \begin{bmatrix} \lambda_1^2 & 0 \\ 0 & \lambda_2^2 \end{bmatrix} \frac{t^2}{2!} + \cdots \\
&= \begin{bmatrix} 1 + \lambda_1 t + \frac{\lambda_1^2 t^2}{2!} + \cdots & 0 \\ 0 & 1 + \lambda_2 t + \frac{\lambda_2^2 t^2}{2!} + \cdots \end{bmatrix} \\
&= \begin{bmatrix} e^{\lambda_1 t} & 0 \\ 0 & e^{\lambda_2 t} \end{bmatrix}.
\end{aligned}
$$

Property 4 *Let* $P^{-1}AP = D$, *D being a diagonal matrix. Then*

$$e^{At} = Pe^{Dt}P^{-1} = P \begin{bmatrix} e^{\lambda_1 t} & 0 \\ 0 & e^{\lambda_2 t} \end{bmatrix} P^{-1}, \text{ where } D = \begin{bmatrix} \lambda_1 & 0 \\ 0 & \lambda_2 \end{bmatrix}.$$

Proof We have

$$
\begin{aligned}
e^{At} &= \lim_{n \to \infty} \sum_{k=0}^{n} \frac{A^k t^k}{k!} \\
&= \lim_{n \to \infty} \sum_{k=0}^{n} \frac{(PDP^{-1})^k t^k}{k!} \left[\because D = P^{-1}AP, \text{ so } A = PDP^{-1} \right] \\
&= \lim_{n \to \infty} \sum_{k=0}^{n} \frac{(PD^k P^{-1}) t^k}{k!} \begin{bmatrix} (PDP^{-1})^k = (PDP^{-1})(PDP^{-1}) \cdots (PDP^{-1}) \\ = PD(P^{-1}P)D(P^{-1}P) \cdots (P^{-1}P)DP^{-1} \\ = PD^k P^{-1} \end{bmatrix} \\
&= P \left(\lim_{n \to \infty} \sum_{k=0}^{n} \frac{D^k t^k}{k!} \right) P^{-1} \\
&= Pe^{Dt} P^{-1} \\
&= P \begin{bmatrix} e^{\lambda_1 t} & 0 \\ 0 & e^{\lambda_2 t} \end{bmatrix} P^{-1}
\end{aligned}
$$

Property 5 *Let N be a nilpotent matrix of order k. Then e^{Nt} is a series containing finite terms only.*

Proof A matrix N is said to be a nilpotent matrix of order or index k if k is the least positive integer such that $N^k = \varphi$ but $N^{k-1} \neq \varphi$, φ being the null matrix.

Since N is a nilpotent matrix of order k, $N^{k-1} \neq \varphi$ but $N^k = \varphi$.

Therefore

$$e^{Nt} = I + Nt + \frac{N^2 t^2}{2!} + \frac{N^3 t^3}{3!} + \cdots + \frac{N^{k-1} t^{k-1}}{(k-1)!} + \frac{N^k t^k}{k!} + \cdots$$

$$= I + Nt + \frac{N^2 t^2}{2!} + \frac{N^3 t^3}{3!} + \cdots + \frac{N^{k-1} t^{k-1}}{(k-1)!}$$

which is a series of finite terms only.

Property 6 *If $A = \begin{bmatrix} a & -b \\ b & a \end{bmatrix}$, then $e^{At} = e^{alt}[I \cos(bt) + J \sin(bt)]$, where $I = \begin{bmatrix} 1 & 0 \\ 0 & 1 \end{bmatrix}$ and $J = \begin{bmatrix} 0 & -1 \\ 1 & 0 \end{bmatrix}$.*

Proof We have

$$A = \begin{bmatrix} a & -b \\ b & a \end{bmatrix} = a \begin{bmatrix} 1 & 0 \\ 0 & 1 \end{bmatrix} + b \begin{bmatrix} 0 & -1 \\ 1 & 0 \end{bmatrix} = aI + bJ, \text{ where } I = \begin{bmatrix} 1 & 0 \\ 0 & 1 \end{bmatrix},$$

$$J = \begin{bmatrix} 0 & -1 \\ 1 & 0 \end{bmatrix}.$$

Therefore

$$e^{At} = e^{aIt + bJt}$$

$$= e^{aIt} \cdot e^{bJt} = e^{aIt} \left[I + bJt + \frac{(bJt)^2}{2!} + \frac{(bJt)^3}{3!} + \cdots \right]$$

$$= e^{aIt} \left[I \left(1 - \frac{b^2 t^2}{2!} + \frac{b^4 t^4}{4!} + \cdots \right) + J \left(bt - \frac{b^3 t^3}{3!} + \cdots \right) \right]$$

$$= e^{aIt}[I \cos(bt) + J \sin(bt)] \quad \begin{bmatrix} \because J^2 = \begin{bmatrix} 0 & -1 \\ 1 & 0 \end{bmatrix} \begin{bmatrix} 0 & -1 \\ 1 & 0 \end{bmatrix} \\ = \begin{bmatrix} -1 & 0 \\ 0 & -1 \end{bmatrix} = -I \\ J^3 = J^2 J = (-I)J = -J \\ J^4 = J^3 J = (-J)J = -J^2 = I \\ \text{etc.} \ldots \end{bmatrix}$$

Property 7 $e^{A+B} = e^A e^B$, *provided* $AB = BA$.

Proof Suppose $AB = BA$. Then by Binomial theorem,

$$(A+B)^n = \sum_{k=0}^{n} \frac{n!}{(n-k)!k!} A^{n-k} B^k = n! \sum_{j+k=n} \frac{A^j B^k}{j!k!}.$$

Therefore

$$e^{A+B} = \sum_{n=0}^{\infty} \frac{(A+B)^n}{n!} = \sum_{n=0}^{\infty} \sum_{j+k=n} \frac{A^j B^k}{j!k!}$$

$$= \sum_{j=0}^{\infty} \frac{A^j}{j!} \sum_{k=0}^{\infty} \frac{B^k}{k!}$$

$$= e^A e^B.$$

It is true that $e^{A+B} = e^A e^B$ if $AB = BA$. But in general $e^{A+B} \neq e^A e^B$.

Property 8 *For any* $n \times n$ *matrix* A, $\frac{d}{dt}(e^{At}) = Ae^{At}$.

Proof By definition

$$e^{At} = I + At + \frac{A^2 t^2}{2!} + \frac{A^3 t^3}{3!} + \cdots$$

$$\therefore \frac{d}{dt}(e^{At}) = \frac{d}{dt}\left(I + At + \frac{A^2 t^2}{2!} + \frac{A^3 t^3}{3!} + \cdots\right)$$

$$= \frac{d}{dt}(I) + \frac{d}{dt}(At) + \frac{d}{dt}\left(\frac{A^2 t^2}{2!}\right) + \frac{d}{dt}\left(\frac{A^3 t^3}{3!}\right) + \cdots$$

The term by term differentiation is valid because the series of e^{At} is convergent for all t under the operator.

$$\text{or, } \frac{d}{dt}(e^{At}) = \varphi + A + A^2 t + \frac{A^3 t^2}{2!} + \frac{A^4 t^3}{3!} + \cdots$$

$$= A\left(I + At + \frac{A^2 t^2}{2!} + \frac{A^3 t^3}{3!} + \cdots\right)$$

$$= Ae^{At}.$$

Therefore, $\frac{d}{dt}(e^{At}) = Ae^{At}$.

We now establish the important result below.

Result Multiplying both sides of $\dfrac{d}{dt}\left(e^{At}\right) = Ae^{At}$ by $\Phi(0)$ in right, we have

$$\frac{d}{dt}\left(e^{At}\right)\Phi(0) = Ae^{At}\Phi(0)$$

$$\Rightarrow \frac{d}{dt}\left(e^{At}\Phi(0)\right) = Ae^{At}\Phi(0)$$

$$\Rightarrow \frac{d}{dt}\left(\Phi(t)\Phi^{-1}(0)\Phi(0)\right) = A\Phi(t)\Phi^{-1}(0)\Phi(0) \text{ [since } e^{At} = \Phi(t)\Phi^{-1}(0)]$$

$$\Rightarrow \frac{d}{dt}\left(\Phi(t)\right) = \dot{\Phi}(t) = A\Phi(t).$$

This shows that the fundamental matrix $\Phi(t)$ must satisfy the system $\dot{\underset{\sim}{x}} = A\underset{\sim}{x}$. This is true for all t. So, it is true for $t = 0$. Putting $t = 0$ in $\dot{\Phi}(t) = A\Phi(t)$, we get

$$\dot{\Phi}(0) = A\Phi(0) \Rightarrow A = \dot{\Phi}(0)\Phi^{-1}(0).$$

This gives that the coefficient matrix A can be expressed in terms of the fundamental matrix $\Phi(t)$.

Example 2.12 Does $\Phi(t) = \begin{pmatrix} e^t & e^{-2t} \\ 2e^t & 3e^{-2t} \end{pmatrix}$ a fundamental matrix for the system $\dot{\underset{\sim}{x}} = A\underset{\sim}{x}$? If so, then find the matrix A.

Solution We know that if $\Phi(t)$ is a fundamental matrix, then $\Phi(0)$ is invertible. Here $\Phi(t) = \begin{pmatrix} e^t & e^{-2t} \\ 2e^t & 3e^{-2t} \end{pmatrix}$. So, $\Phi(0) = \begin{pmatrix} 1 & 1 \\ 2 & 3 \end{pmatrix}$.

Since $\det(\Phi(0)) = 3 - 2 = 1 \neq 0$, $\Phi(0)$ is invertible. Hence the given matrix is a fundamental matrix for the system $\dot{\underset{\sim}{x}} = A\underset{\sim}{x}$. We shall now find the coefficient matrix A.

We have $\Phi(0) = \begin{pmatrix} 1 & 1 \\ 2 & 3 \end{pmatrix}$. So $\Phi^{-1}(0) = \begin{pmatrix} 3 & -1 \\ -2 & 1 \end{pmatrix}$.

Also $\dot{\Phi}(t) = \begin{pmatrix} e^t & -2e^{-2t} \\ 2e^t & -6e^{-2t} \end{pmatrix}$, and $\dot{\Phi}(0) = \begin{pmatrix} 1 & -2 \\ 2 & -6 \end{pmatrix}$.

Therefore the matrix A is

$$A = \dot{\Phi}(0)\Phi^{-1}(0) = \begin{pmatrix} 1 & -2 \\ 2 & -6 \end{pmatrix}\begin{pmatrix} 3 & -1 \\ -2 & 1 \end{pmatrix} = \begin{pmatrix} 7 & -3 \\ 18 & -8 \end{pmatrix}.$$

Example 2.13 Find e^{At} for the matrix $A = \begin{pmatrix} 3 & 1 \\ 1 & 3 \end{pmatrix}$. Hence find the solution of the system $\dot{\underset{\sim}{x}} = A\underset{\sim}{x}$.

Solution We see that the eigenvectors corresponding to the eigenvalues $\lambda = 2, 4$ of A are respectively $\underset{\sim}{e} = \begin{pmatrix} 1 \\ -1 \end{pmatrix}$ and $\underset{\sim}{g} = \begin{pmatrix} 1 \\ 1 \end{pmatrix}$, which are linearly independent.

Therefore, two fundamental solutions of the system are $\underset{\sim}{x}_1(t) = \begin{pmatrix} 1 \\ -1 \end{pmatrix} e^{2t}$ and $\underset{\sim}{x}_2(t) = \begin{pmatrix} 1 \\ 1 \end{pmatrix} e^{4t}$. So a fundamental matrix of the system is

$$\Phi(t) = \begin{pmatrix} \underset{\sim}{x}_1(t) & \underset{\sim}{x}_2(t) \end{pmatrix} = \begin{pmatrix} e^{2t} & e^{4t} \\ -e^{2t} & e^{4t} \end{pmatrix}.$$

We find $\Phi(0) = \begin{pmatrix} 1 & 1 \\ -1 & 1 \end{pmatrix}$ and $\Phi^{-1}(0) = \frac{1}{2}\begin{pmatrix} 1 & -1 \\ 1 & 1 \end{pmatrix}$. Therefore

$$e^{At} = \Phi(t)\Phi^{-1}(0) = \frac{1}{2}\begin{pmatrix} e^{2t} & e^{4t} \\ -e^{2t} & e^{4t} \end{pmatrix}\begin{pmatrix} 1 & -1 \\ 1 & 1 \end{pmatrix} = \frac{1}{2}\begin{pmatrix} e^{2t}+e^{4t} & e^{4t}-e^{2t} \\ e^{4t}-e^{2t} & e^{2t}+e^{4t} \end{pmatrix}.$$

By fundamental theorem, the solution of the system $\underset{\sim}{\dot{x}} = A\underset{\sim}{x}$ is

$$\underset{\sim}{x}(t) = e^{At}\underset{\sim}{x}_0 = \frac{1}{2}\begin{pmatrix} e^{2t}+e^{4t} & e^{4t}-e^{2t} \\ e^{4t}-e^{2t} & e^{2t}+e^{4t} \end{pmatrix}\begin{pmatrix} c_1 \\ c_2 \end{pmatrix}$$

where $\underset{\sim}{x}_0 = \begin{pmatrix} c_1 \\ c_2 \end{pmatrix}$ is an arbitrary constant column vector.

2.4   Solution Procedure of Linear Systems

The general solution of a linear homogeneous system can be easily deduced from the fundamental theorem. According to this theorem the solution of $\underset{\sim}{\dot{x}} = A\underset{\sim}{x}$ with $\underset{\sim}{x}(0) = \underset{\sim}{x}_0$ is given as $\underset{\sim}{x}(t) = e^{At}\underset{\sim}{x}_0$ and this solution is unique.

For a simple change of coordinates $\underset{\sim}{x} = P\underset{\sim}{y}$ where P is an invertible matrix, the equation $\underset{\sim}{\dot{x}} = A\underset{\sim}{x}$ is transformed as

$$\underset{\sim}{\dot{x}} = A\underset{\sim}{x}$$

$$\Rightarrow P\underset{\sim}{\dot{y}} = AP\underset{\sim}{y}$$

$$\Rightarrow \underset{\sim}{\dot{y}} = P^{-1}AP\underset{\sim}{y}$$

$$\Rightarrow \underset{\sim}{\dot{y}} = C\underset{\sim}{y}, \text{ where } C = P^{-1}AP.$$

The initial conditions $\underset{\sim}{x}(0) = \underset{\sim}{x}_0$ become $\underset{\sim}{y}(0) = P^{-1}\underset{\sim}{x}(0) = P^{-1}\underset{\sim}{x}_0 = \underset{\sim}{y}_0$.
So, the new system is $\dot{\underset{\sim}{y}} = C\underset{\sim}{y}$ with $\underset{\sim}{y}(0) = \underset{\sim}{y}_0$, where $C = P^{-1}AP$.

It has the solution

$$\underset{\sim}{y}(t) = e^{Ct}\underset{\sim}{y}_0.$$

Hence the solution of the original system is

$$\underset{\sim}{x}(t) = P\underset{\sim}{y}(t) = Pe^{Ct}\underset{\sim}{y}_0 = Pe^{Ct}P^{-1}\underset{\sim}{x}_0.$$

We see that $e^{At} = Pe^{Ct}P^{-1}$. The matrix P is chosen in such a way that matrix C takes a simple form. We now discuss three cases.

(i) **Matrix A has distinct real eigenvalues**

Let $P = \left(\underset{\sim}{\alpha}_1, \underset{\sim}{\alpha}_2, \ldots, \underset{\sim}{\alpha}_n\right)$ so that, P^{-1} exists. The matrix C is obtained as $C = P^{-1}AP$ which is a diagonal matrix. Hence the exponential function of C becomes

$$e^{Ct} = \text{diag}(e^{\lambda_1 t}, e^{\lambda_2 t}, \ldots, e^{\lambda_n t}).$$

Therefore we can write the solution of $\dot{\underset{\sim}{x}} = A\underset{\sim}{x}$ with $\underset{\sim}{x}(0) = \underset{\sim}{x}_0$ as $\underset{\sim}{x}(t) = e^{At}\underset{\sim}{x}_0 = Pe^{Ct}P^{-1}\underset{\sim}{x}_0$. So

$$\underset{\sim}{x}(t) = P\text{diag}(e^{\lambda_1 t}, e^{\lambda_2 t}, \ldots, e^{\lambda_n t})P^{-1}\underset{\sim}{x}_0$$

where $\underset{\sim}{x}_0 = (c_1, c_2, \ldots, c_n)^t$ is an arbitrary constant.

(ii) **Matrix A has real repeated eigenvalues**

In this case the following theorems are relevant (proofs are available in the book Hirsch and Smale [2]) for finding general solution of a linear system when matrix A has repeated eigenvalues.

Theorem 2.2 *Let the $n \times n$ matrix A have real eigenvalues $\lambda_1, \lambda_2, \ldots, \lambda_n$ repeated according to their multiplicity. Then there exists a basis of generalized eigenvectors $\{\underset{\sim}{\alpha}_1, \underset{\sim}{\alpha}_2, \ldots, \underset{\sim}{\alpha}_n\}$ such that the matrix $P = (\underset{\sim}{\alpha}_1, \underset{\sim}{\alpha}_2, \ldots, \underset{\sim}{\alpha}_n)$ is invertible and $A = S + N$, where $P^{-1}SP = \text{diag}(\lambda_1, \lambda_1, \ldots, \lambda_n)$ and $N(=A - S)$ is nilpotent of order $k \leq n$, and S and N commute.*

Using the theorem the linear system subject to the initial conditions $\underset{\sim}{x}(0) = \underset{\sim}{x}_0$ has the solution

$$\underset{\sim}{x}(t) = P\mathrm{diag}(e^{\lambda_j t})P^{-1}\left[I+Nt+\cdots+\frac{N^{k-1}t^{k-1}}{(k-1)!}\right]\underset{\sim}{x}_0.$$

(iii) Matrix A has complex eigenvalues

Theorem 2.3 *Let A be a $2n \times 2n$ matrix with complex eigenvalues $a_j \pm ib_j, j = 1, 2,$ ..., n. Then there exists generalized complex eigenvectors $(\underset{\sim}{\alpha}_j \pm i\underset{\sim}{\beta}_j), j = 1, 2\ldots, n$ such that the matrix $P = (\underset{\sim}{\beta}_1, \underset{\sim}{\alpha}_1, \underset{\sim}{\beta}_2, \underset{\sim}{\alpha}_2, \ldots, \underset{\sim}{\beta}_n, \underset{\sim}{\alpha}_n)$ is invertible and $A = S + N$, where $P^{-1}SP = \mathrm{diag}\begin{bmatrix} a_j & -b_j \\ b_j & a_j \end{bmatrix}$, and $N(=A - S)$ is a nilpotent matrix of order $k \le 2n$, and S and N commute.*

Using the theorem the linear system of equations subject to the initial conditions $\underset{\sim}{x}(0) = \underset{\sim}{x}_0$ has the solution

$$\underset{\sim}{x}(t) = P\mathrm{diag}(e^{a_j t})\begin{bmatrix} \cos(b_j t) & -\sin(b_j t) \\ \sin(b_j t) & \cos(b_j t) \end{bmatrix}P^{-1}\left[I+Nt+\cdots+\frac{N^k t^k}{k!}\right]\underset{\sim}{x}_0.$$

For a 2×2 matrix A with complex eigenvalues $(\alpha \pm i\beta)$ the solution is given by

$$\underset{\sim}{x}(t) = Pe^{\alpha t}\begin{pmatrix} \cos\beta t & -\sin\beta t \\ \sin\beta t & \cos\beta t \end{pmatrix}P^{-1}\underset{\sim}{x}_0.$$

Example 2.14 Solve the initial value problem

$$\dot{x} = x+y, \dot{y} = 4x - 2y$$

with initial condition $\underset{\sim}{x}(0) = \begin{pmatrix} 2 \\ -3 \end{pmatrix}.$

Solution The characteristic equation of matrix A is

$$|A - \lambda I| = 0$$
$$\Rightarrow \begin{vmatrix} 1-\lambda & 1 \\ 4 & -2-\lambda \end{vmatrix} = 0$$
$$\Rightarrow (\lambda - 1)(\lambda + 2) - 4 = 0$$
$$\Rightarrow \lambda^2 + \lambda - 6 = 0$$
$$\Rightarrow \lambda = 2, -3$$

So the eigenvalues of matrix A are 2, −3, which are real and distinct.

Let $\underset{\sim}{e} = \begin{pmatrix} e_1 \\ e_2 \end{pmatrix}$ be the eigenvector corresponding to the eigenvalue $\lambda_1 = 2$. Then

$$(A - 2I)\underset{\sim}{e} = 0$$

$$\Rightarrow \begin{pmatrix} 1-2 & 1 \\ 4 & -2-2 \end{pmatrix} \begin{pmatrix} e_1 \\ e_2 \end{pmatrix} = \begin{pmatrix} 0 \\ 0 \end{pmatrix}$$

$$\Rightarrow \quad -e_1 + e_2 = 0, 4e_1 - 4e_2 = 0$$

A nontrivial solution of this system is $e_1 = 1, e_2 = 1$.

$$\therefore \underset{\sim}{e} = \begin{pmatrix} 1 \\ 1 \end{pmatrix}.$$

Again let $\underset{\sim}{g} = \begin{pmatrix} g_1 \\ g_2 \end{pmatrix}$ be the eigenvector corresponding to the eigenvalue $\lambda_2 = -3$. Then

$$(A + 3I)\underset{\sim}{g} = 0$$

$$\Rightarrow \begin{pmatrix} 1+3 & 1 \\ 4 & -2+3 \end{pmatrix} \begin{pmatrix} g_1 \\ g_2 \end{pmatrix} = \begin{pmatrix} 0 \\ 0 \end{pmatrix}$$

$$\Rightarrow \quad 4g_1 + g_2 = 0, 4g_1 + g_2 = 0$$

A nontrivial solution of this system is $g_1 = 1, g_2 = -4$.

$$\therefore \underset{\sim}{g} = \begin{pmatrix} 1 \\ -4 \end{pmatrix}.$$

Let $P = \begin{pmatrix} \underset{\sim}{e} , \underset{\sim}{g} \end{pmatrix} = \begin{pmatrix} 1 & 1 \\ 1 & -4 \end{pmatrix}$. Then $P^{-1} = -\frac{1}{5} \begin{pmatrix} -4 & -1 \\ -1 & 1 \end{pmatrix} = \frac{1}{5} \begin{pmatrix} 4 & 1 \\ 1 & -1 \end{pmatrix}$

$$\therefore C = P^{-1}AP = \frac{1}{5} \begin{pmatrix} 4 & 1 \\ 1 & -1 \end{pmatrix} \begin{pmatrix} 1 & 1 \\ 4 & -2 \end{pmatrix} \begin{pmatrix} 1 & 1 \\ 1 & -4 \end{pmatrix}$$

$$= \frac{1}{5} \begin{pmatrix} 8 & 2 \\ -3 & 3 \end{pmatrix} \begin{pmatrix} 1 & 1 \\ 1 & -4 \end{pmatrix} = \frac{1}{5} \begin{pmatrix} 10 & 0 \\ 0 & -15 \end{pmatrix} = \begin{pmatrix} 2 & 0 \\ 0 & -3 \end{pmatrix}$$

$$\therefore e^{Ct} = \begin{pmatrix} e^{2t} & 0 \\ 0 & e^{-3t} \end{pmatrix}$$

Therefore by the fundamental theorem, the solution of the system is

$$x(t) = e^{At}x_0 = Pe^{Ct}P^{-1}x_0$$

$$= \begin{pmatrix} 1 & 1 \\ 1 & -4 \end{pmatrix} \begin{pmatrix} e^{2t} & 0 \\ 0 & e^{-3t} \end{pmatrix} \frac{1}{5} \begin{pmatrix} 4 & 1 \\ 1 & -1 \end{pmatrix} x_0$$

$$= \frac{1}{5} \begin{pmatrix} 4e^{2t} + e^{-3t} & e^{2t} - e^{-3t} \\ 4e^{2t} - 4e^{-3t} & e^{2t} + 4e^{-3t} \end{pmatrix} x_0$$

$$\Rightarrow \begin{pmatrix} x(t) \\ y(t) \end{pmatrix} = \begin{pmatrix} \frac{4}{5}e^{2t} + \frac{1}{5}e^{-3t} & \frac{1}{5}e^{2t} - \frac{1}{5}e^{-3t} \\ \frac{4}{5}e^{2t} - \frac{4}{5}e^{-3t} & \frac{1}{5}e^{2t} + \frac{4}{5}e^{-3t} \end{pmatrix} \begin{pmatrix} 2 \\ -3 \end{pmatrix} = \begin{pmatrix} e^{2t} + e^{-3t} \\ e^{2t} - 4e^{-3t} \end{pmatrix}$$

$$\Rightarrow x(t) = e^{2t} + e^{-3t}, y(t) = e^{2t} - 4e^{-3t}.$$

Example 2.15 Solve the system

$$\dot{x}_1 = -x_1 - 3x_2, \dot{x}_2 = 2x_2.$$

Also sketch the phase portrait.

Solution The characteristic equation of matrix A is

$$|A - \lambda I| = 0$$

$$\Rightarrow \begin{vmatrix} -1 - \lambda & -3 \\ 0 & 2 - \lambda \end{vmatrix} = 0$$

$$\Rightarrow (\lambda - +)(\lambda - 2) = 0$$

$$\Rightarrow \lambda = -1, 2$$

The eigenvalues of matrix A are $-1, 2$, which are real and distinct.

Let $e = \begin{pmatrix} e_1 \\ e_2 \end{pmatrix}$ be the eigenvector corresponding to the eigenvalue $\lambda_1 = -1$.
Then

$$(A + I)e = 0$$

$$\Rightarrow \begin{pmatrix} -1 + 1 & -3 \\ 0 & 2 + 1 \end{pmatrix} \begin{pmatrix} e_1 \\ e_2 \end{pmatrix} = \begin{pmatrix} 0 \\ 0 \end{pmatrix}$$

$$\Rightarrow -3e_2 = 0, 3e_2 = 0$$

$$\Rightarrow e_2 = 0 \text{ and } e_1 \text{ is arbitrary.}$$

Choose $e_1 = 1$ so that $e = \begin{pmatrix} 1 \\ 0 \end{pmatrix}$.

Again, let $g = \begin{pmatrix} g_1 \\ g_2 \end{pmatrix}$ be the eigenvector corresponding to the eigenvalue $\lambda_2 = 2$. Then

$$(A - 2I)g = 0$$

$$\Rightarrow \quad \begin{pmatrix} -1-2 & -3 \\ 0 & 2-2 \end{pmatrix} \begin{pmatrix} g_1 \\ g_2 \end{pmatrix} = \begin{pmatrix} 0 \\ 0 \end{pmatrix}$$

$$\Rightarrow \quad g_1 + g_2 = 0$$

Choose $g_1 = 1, g_2 = -1$. Then $g = \begin{pmatrix} 1 \\ -1 \end{pmatrix}$.

Let $P = \left(e, g \right) = \begin{pmatrix} 1 & 1 \\ 0 & -1 \end{pmatrix}$. Then $P^{-1} = \begin{pmatrix} 1 & 1 \\ 0 & -1 \end{pmatrix}$

Therefore

$$C = P^{-1}AP = \begin{pmatrix} 1 & 1 \\ 0 & -1 \end{pmatrix} \begin{pmatrix} -1 & -3 \\ 0 & 2 \end{pmatrix} \begin{pmatrix} 1 & 1 \\ 0 & -1 \end{pmatrix}$$

$$= \begin{pmatrix} -1 & -1 \\ 0 & -2 \end{pmatrix} \begin{pmatrix} 1 & 1 \\ 0 & -1 \end{pmatrix} = \begin{pmatrix} -1 & 0 \\ 0 & 2 \end{pmatrix}$$

and so $e^{Ct} = \begin{pmatrix} e^{-t} & 0 \\ 0 & e^{2t} \end{pmatrix}$.

Therefore by fundamental theorem, the solution of the system is

$$x(t) = e^{At}x_0 = Pe^{Ct}P^{-1}x_0$$

$$= \begin{pmatrix} 1 & 1 \\ 0 & -1 \end{pmatrix} \begin{pmatrix} e^{-t} & 0 \\ 0 & e^{2t} \end{pmatrix} \begin{pmatrix} 1 & 1 \\ 0 & -1 \end{pmatrix} x_0$$

$$= \begin{pmatrix} e^{-t} & e^{-t} - e^{2t} \\ 0 & e^{2t} \end{pmatrix} \begin{pmatrix} c_1 \\ c_2 \end{pmatrix}$$

$$\Rightarrow \quad \begin{pmatrix} x_1(t) \\ x_2(t) \end{pmatrix} = \begin{pmatrix} e^{-t} & e^{-t} - e^{2t} \\ 0 & e^{2t} \end{pmatrix} \begin{pmatrix} c_1 \\ c_2 \end{pmatrix} = \begin{pmatrix} (c_1 + c_2)e^{-t} - c_2e^{2t} \\ c_2e^{2t} \end{pmatrix}$$

$$\Rightarrow \quad x_1(t) = c_1e^{-t} + c_2(e^{-t} - e^{2t}), x_2(t) = c_2e^{2t}$$

where c_1, c_2 are arbitrary constants. The phase diagram is presented in Fig. 2.1.

Example 2.16 Solve the following system using the fundamental theorem.

$$\dot{x} = 5x + 4y$$
$$\dot{y} = -x + y$$

Fig. 2.1 A typical phase portrait of the system

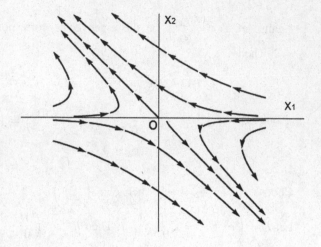

Solution The characteristic equation of matrix A is

$$|A - \lambda I| = 0$$

$$\Rightarrow \quad \begin{vmatrix} 5 - \lambda & 4 \\ -1 & 1 - \lambda \end{vmatrix} = 0$$

$$\Rightarrow \quad (\lambda - 1)(\lambda - 5) + 4 = 0$$

$$\Rightarrow \quad \lambda^2 - 6\lambda + 9 = 0$$

$$\Rightarrow \quad \lambda = 3, 3.$$

This shows that matrix A has an eigenvalue $\lambda = 3$ of multiplicity 2. Then $S = \begin{bmatrix} 3 & 0 \\ 0 & 3 \end{bmatrix}$ and $N = A - S = \begin{bmatrix} 2 & 4 \\ -1 & -2 \end{bmatrix}$. Clearly, matrix N is a nilpotent matrix of order 2. So, the general solution of the system is given by

$$\underset{\sim}{x}(t) = e^{At}\underset{\sim}{x}_0 = e^{(S+N)t}\underset{\sim}{x}_0 = e^{St}e^{Nt}\underset{\sim}{x}_0 = \begin{bmatrix} e^{3t} & 0 \\ 0 & e^{3t} \end{bmatrix}[I + Nt]\underset{\sim}{x}_0$$

$$= \begin{bmatrix} e^{3t} & 0 \\ 0 & e^{3t} \end{bmatrix}\begin{bmatrix} 1 + 2t & 4t \\ -t & 1 - 2t \end{bmatrix}\underset{\sim}{x}_0.$$

Example 2.17 Find the general solution of the system of linear equations

$$\dot{x} = 4x - 2y$$
$$\dot{y} = 5x + 2y$$

Solution The characteristic equation of matrix A is

$$|A - \lambda I| = 0$$

$$\Rightarrow \begin{vmatrix} 4-\lambda & -2 \\ 5 & 2-\lambda \end{vmatrix} = 0$$

$$\Rightarrow (\lambda - 4)(\lambda - 2) + 10 = 0$$

$$\Rightarrow \lambda^2 - 6\lambda + 18 = 0$$

$$\Rightarrow \lambda = \frac{6 \pm \sqrt{36 - 72}}{2} = 3 \pm 3i.$$

So matrix A has a pair of complex conjugate eigenvalues $3 \pm 3i$

Let $\underset{\sim}{e} = \begin{pmatrix} e_1 \\ e_2 \end{pmatrix}$ be the eigenvector corresponding to the eigenvalue $\lambda_1 = 3 + 3i$.
Then

$$(A - (3 + 3i)I)\underset{\sim}{e} = \underset{\sim}{0}$$

$$\Rightarrow \begin{pmatrix} 4 - (3+3i) & -2 \\ 5 & 2 - (3+3i) \end{pmatrix} \begin{pmatrix} e_1 \\ e_2 \end{pmatrix} = \begin{pmatrix} 0 \\ 0 \end{pmatrix}$$

$$\Rightarrow \begin{pmatrix} 1 - 3i & -2 \\ 5 & -1 - 3i \end{pmatrix} \begin{pmatrix} e_1 \\ e_2 \end{pmatrix} = \begin{pmatrix} 0 \\ 0 \end{pmatrix}$$

$$\Rightarrow (1 - 3i)e_1 - 2e_2 = 0,\ 5e_1 + (1 + 3i)e_2 = 0$$

A nontrivial solution of this system is $e_1 = 2$, $e_2 = 1 - 3i$.

$$\therefore \underset{\sim}{e} = \begin{pmatrix} 2 \\ 1 - 3i \end{pmatrix}.$$

Similarly, the eigenvector corresponding to the eigenvalue $\lambda_2 = 3 - 3i$ is
$\underset{\sim}{g} = \begin{pmatrix} 2 \\ 1 + 3i \end{pmatrix}$.

Let $P = \begin{pmatrix} 0 & 2 \\ -3 & 1 \end{pmatrix}$. Then $P^{-1} = \frac{1}{6} \begin{pmatrix} 1 & -2 \\ 3 & 0 \end{pmatrix}$.

Let $C = P^{-1}AP$. Then $C = P^{-1}AP = \frac{1}{6} \begin{pmatrix} 1 & -2 \\ 3 & 0 \end{pmatrix} \begin{pmatrix} 4 & -2 \\ 5 & 2 \end{pmatrix} \begin{pmatrix} 0 & 2 \\ -3 & 1 \end{pmatrix} = \begin{pmatrix} 3 & -3 \\ 3 & 3 \end{pmatrix}$.

So,

$$e^{Ct} = e^{3t} \begin{pmatrix} \cos 3t & -\sin 3t \\ \sin 3t & \cos 3t \end{pmatrix}.$$

Therefore, the solution of the system is

$$\underset{\sim}{x}(t) = e^{At}\underset{\sim}{x_0} = Pe^{Ct}P^{-1}\underset{\sim}{x_0}$$

$$= \frac{1}{6}e^{3t}\begin{pmatrix} 0 & 2 \\ -3 & 1 \end{pmatrix}\begin{pmatrix} \cos 3t & -\sin 3t \\ \sin 3t & \cos 3t \end{pmatrix}\begin{pmatrix} 1 & -2 \\ 3 & 0 \end{pmatrix}\underset{\sim}{x}_0.$$

Example 2.18 Solve the initial value problem $\dot{\underset{\sim}{x}} = A\underset{\sim}{x}$, with $\underset{\sim}{x}(0) = \begin{pmatrix} 1 \\ 0 \end{pmatrix}$, where

$A = \begin{pmatrix} -2 & -1 \\ 1 & -2 \end{pmatrix}, \underset{\sim}{x} = \begin{pmatrix} x \\ y \end{pmatrix}$. Also sketch the solution curve in the phase plane $\mathbb{R}^2$.

Solution The characteristic equation of matrix A is

$$|A - \lambda I| = 0$$

$$\Rightarrow \quad \begin{vmatrix} -2-\lambda & -1 \\ 1 & -2-\lambda \end{vmatrix} = 0$$

$$\Rightarrow \quad (\lambda+2)^2 + 1 = 0$$

$$\Rightarrow \quad \lambda^2 + 4\lambda + 5 = 0$$

$$\Rightarrow \quad \lambda = \frac{-4 \pm \sqrt{16-20}}{2} = -2 \pm i.$$

So matrix A has a pair of complex conjugate eigenvalues $-2 \pm i$

Let $\underset{\sim}{e} = \begin{pmatrix} e_1 \\ e_2 \end{pmatrix}$ be the eigenvector corresponding to the eigenvalue $\lambda_1 = -2+i$. Then

$$(A - (-2+i)I)\underset{\sim}{e} = 0$$

$$\Rightarrow \quad \begin{pmatrix} -2-(-2+i) & -1 \\ 1 & -2-(-2+i) \end{pmatrix}\begin{pmatrix} e_1 \\ e_2 \end{pmatrix} = \begin{pmatrix} 0 \\ 0 \end{pmatrix}$$

$$\Rightarrow \quad \begin{pmatrix} -i & -1 \\ 1 & -i \end{pmatrix}\begin{pmatrix} e_1 \\ e_2 \end{pmatrix} = \begin{pmatrix} 0 \\ 0 \end{pmatrix}$$

$$\Rightarrow \quad -ie_1 - e_2 = 0, e_1 - ie_2 = 0$$

A nontrivial solution of this system is $e_1 = 1, e_2 = -i$.

$\therefore \underset{\sim}{e} = \begin{pmatrix} 1 \\ -i \end{pmatrix}$. Similarly, the eigenvector corresponding to the eigenvalue $\lambda_2 = -2 - i$ is $\underset{\sim}{g} = \begin{pmatrix} 1 \\ i \end{pmatrix}$. Let $P = \begin{pmatrix} 0 & 1 \\ -1 & 0 \end{pmatrix}$. Then $P^{-1} = \begin{pmatrix} 0 & -1 \\ 1 & 0 \end{pmatrix}$ and

$$C = P^{-1}AP = \begin{pmatrix} 0 & -1 \\ 1 & 0 \end{pmatrix}\begin{pmatrix} -2 & -1 \\ 1 & -2 \end{pmatrix}\begin{pmatrix} 0 & 1 \\ -1 & 0 \end{pmatrix} = \begin{pmatrix} -2 & -1 \\ 1 & -2 \end{pmatrix}.$$

So,

$$e^{Ct} = e^{-2t}\begin{pmatrix} \cos t & -\sin t \\ \sin t & \cos t \end{pmatrix}.$$

Hence the solution of the system is

$$\underset{\sim}{x}(t) = e^{At}\underset{\sim}{x}_0 = Pe^{Ct}P^{-1}\underset{\sim}{x}_0$$

$$= e^{-2t}\begin{pmatrix} 0 & 1 \\ -1 & 0 \end{pmatrix}\begin{pmatrix} \cos t & -\sin t \\ \sin t & \cos t \end{pmatrix}\begin{pmatrix} 0 & -1 \\ 1 & 0 \end{pmatrix}\underset{\sim}{x}_0.$$

$$= e^{-2t}\begin{pmatrix} \sin t & \cos t \\ -\cos t & \sin t \end{pmatrix}\begin{pmatrix} 0 & -1 \\ 1 & 0 \end{pmatrix}\underset{\sim}{x}_0$$

$$= e^{-2t}\begin{pmatrix} \cos t & -\sin t \\ \sin t & \cos t \end{pmatrix}\begin{pmatrix} 1 \\ 0 \end{pmatrix}$$

$$= e^{-2t}\begin{pmatrix} \cos t \\ \sin t \end{pmatrix}$$

$$\therefore x(t) = e^{-2t}\cos t, y(t) = e^{-2t}\sin t.$$

Phase Portrait The phase portrait of the solution curve is shown in Fig. 2.2.

Example 2.19 Solve the system $\dot{\underset{\sim}{x}} = A\underset{\sim}{x}$ with $\underset{\sim}{x}(0) = \underset{\sim}{x}_0$, where

$$A = \begin{pmatrix} 2 & 1 & 3 & -1 \\ 0 & 2 & 2 & -1 \\ 0 & 0 & 2 & -5 \\ 0 & 0 & 0 & 2 \end{pmatrix}.$$

Fig. 2.2 Phase portrait of the solution curve

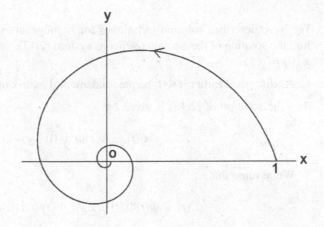

Solution Clearly, matrix A has the eigenvalue $\lambda = 2$ with multiplicity 4. Therefore,

$$S = \begin{pmatrix} 2 & 0 & 0 & 0 \\ 0 & 2 & 0 & 0 \\ 0 & 0 & 2 & 0 \\ 0 & 0 & 0 & 2 \end{pmatrix} \text{ and } N = A - S = \begin{pmatrix} 0 & 1 & 3 & -1 \\ 0 & 0 & 2 & -1 \\ 0 & 0 & 0 & -5 \\ 0 & 0 & 0 & 0 \end{pmatrix}.$$

It is easy to check that the matrix N is nilpotent of order 4. Therefore, the solution of the system is

$$\underset{\sim}{x}(t) = e^{St}\left(I + Nt + \frac{N^2 t^2}{2!} + \frac{N^3 t^3}{3!}\right)\underset{\sim}{x}_0.$$

2.5   Nonhomogeneous Linear Systems

The most general form of a nonhomogeneous linear system is given as

$$\underset{\sim}{\dot{x}}(t) = A(t)\underset{\sim}{x}(t) + \underset{\sim}{b}(t) \tag{2.16}$$

where $A(t)$ is an $n \times n$ matrix, usually depends on time and $\underset{\sim}{b}(t)$ is a time dependent column vector. Here we consider matrix $A(t)$ to be time independent, that is, $A(t) \equiv A$. Then (2.16) becomes

$$\underset{\sim}{\dot{x}}(t) = A\underset{\sim}{x}(t) + \underset{\sim}{b}(t) \tag{2.17}$$

The corresponding homogeneous system is given as

$$\underset{\sim}{\dot{x}}(t) = A\underset{\sim}{x}(t) \tag{2.18}$$

We have described solution techniques for homogeneous system (2.18). We now find the solution of the nonhomogeneous system (2.17), subject to initial conditions $\underset{\sim}{x}(0) = \underset{\sim}{x}_0$.

As discussed earlier if $\Phi(t)$ be the fundamental matrix of (2.18) with $\underset{\sim}{x}(0) = \underset{\sim}{x}_0$, then the solution of (2.18) is given by

$$\underset{\sim}{x}(t) = \Phi(t)\Phi^{-1}(0)\underset{\sim}{x}_0$$

We assume that

$$\underset{\sim}{x}(t) = \Phi(t)\Phi^{-1}(0)\underset{\sim}{x}_0 + \Phi(t)\Phi^{-1}(0)\underset{\sim}{u}(t) \tag{2.19}$$

be the solution of the nonhomogeneous linear system (2.17). Then the initial conditions are obtained as $\underset{\sim}{u}(0) = 0$. Differentiating (2.19) with respect to t, we get

$$\dot{\underset{\sim}{x}}(t) = \dot{\Phi}(t)\Phi^{-1}(0)\underset{\sim}{x}_0 + \dot{\Phi}(t)\Phi^{-1}(0)\underset{\sim}{u}(t) + \Phi(t)\Phi^{-1}(0)\dot{\underset{\sim}{u}}(t) \qquad (2.20)$$

Substituting (2.20) and (2.19) into (2.17),

$$\dot{\Phi}(t)\Phi^{-1}(0)\underset{\sim}{x}_0 + \dot{\Phi}(t)\Phi^{-1}(0)\underset{\sim}{u}(t) + \Phi(t)\Phi^{-1}(0)\dot{\underset{\sim}{u}}(t)$$
$$= A\Phi(t)\Phi^{-1}(0)\underset{\sim}{x}_0 + A\Phi(t)\Phi^{-1}(0)\underset{\sim}{u}(t) + \underset{\sim}{b}(t) \qquad (2.21)$$

Since $\Phi(t)$ is a fundamental matrix solution of (2.18),

$$\dot{\Phi}(t) = A\Phi(t).$$

Using this in (2.21), we get

$$\Phi(t)\Phi^{-1}(0)\dot{\underset{\sim}{u}}(t) = \underset{\sim}{b}(t)$$
$$\Rightarrow \dot{\underset{\sim}{u}}(t) = \Phi(0)\dot{\Phi}^{-1}(t)\underset{\sim}{b}(t)$$

Integrating w.r.to t and using $\underset{\sim}{u}(0) = 0$, we get

$$\underset{\sim}{u}(t) = \int\limits_0^t \Phi(0)\Phi^{-1}(t)\underset{\sim}{b}(t)\mathrm{d}t.$$

Hence the general solution of the nonhomogeneous system (2.17) subject to $\underset{\sim}{x}(0) = \underset{\sim}{x}_0$ is given by

$$\underset{\sim}{x}(t) = \Phi(t)\Phi^{-1}(0)\underset{\sim}{x}_0 + \Phi(t)\int\limits_0^t \Phi^{-1}(\alpha)\underset{\sim}{b}(\alpha)d\alpha \qquad (2.22)$$

Example 2.20 Find the solution of the nonhomogeneous system $\dot{x} = x + y + t$, $\dot{y} = -y + 1$ with the initial conditions $x(0) = 1$, $y(0) = 0$.

Solution In matrix notation, the system takes the form $\dot{\underset{\sim}{x}}(t) = A\underset{\sim}{x}(t) + \underset{\sim}{b}(t)$, where $A = \begin{pmatrix} 1 & 1 \\ 0 & -1 \end{pmatrix}$ and $\underset{\sim}{b}(t) = \begin{pmatrix} t \\ 1 \end{pmatrix}$.

The initial conditions become $x(0) = x_0$, where $x_0 = \begin{pmatrix} 1 \\ 0 \end{pmatrix}$. Matrix A has

eigenvalues $\lambda_1 = 1$, $\lambda_2 = -1$ with corresponding eigenvectors $\begin{pmatrix} 1 \\ 0 \end{pmatrix}$ and $\begin{pmatrix} 1 \\ -2 \end{pmatrix}$.

Therefore

$$\Phi(t) = \begin{pmatrix} e^t & e^{-t} \\ 0 & -2e^{-t} \end{pmatrix}.$$

This gives

$$\Phi^{-1}(t) = \frac{1}{2}\begin{pmatrix} 2e^{-t} & e^{-t} \\ 0 & -e^t \end{pmatrix}, \ \Phi(0) = \begin{pmatrix} 1 & 1 \\ 0 & -2 \end{pmatrix} \text{and } \Phi^{-1}(0) = \frac{1}{2}\begin{pmatrix} 2 & 1 \\ 0 & -1 \end{pmatrix}.$$

Therefore the required solution is

$$x(t) = \Phi(t)\Phi^{-1}(0)x_0 + \Phi(t)\int_0^t \Phi^{-1}(\alpha)b(\alpha)d\alpha$$

$$= \frac{1}{2}\Phi(t)\left\{\begin{pmatrix} 2 & 1 \\ 0 & -1 \end{pmatrix}\begin{pmatrix} 1 \\ 0 \end{pmatrix} + \int_0^t \begin{pmatrix} 2e^{-\alpha} & e^{-\alpha} \\ 0 & -e^{\alpha} \end{pmatrix}\begin{pmatrix} \alpha \\ 1 \end{pmatrix}d\alpha\right\}$$

$$= \frac{1}{2}\Phi(t)\left\{\begin{pmatrix} 2 \\ 0 \end{pmatrix} + \begin{pmatrix} 3 - (2t+3)e^{-t} \\ 1 - e^t \end{pmatrix}\right\}$$

$$= \frac{1}{2}\begin{pmatrix} e^t & e^{-t} \\ 0 & -2e^{-t} \end{pmatrix}\begin{pmatrix} 5 - (2t+3)e^{-t} \\ 1 - e^t \end{pmatrix} = \frac{1}{2}\begin{pmatrix} 5e^t - 2t - 4 + e^{-t} \\ 2 - 2e^{-2t} \end{pmatrix}.$$

Example 2.21 Prove that the flow evolution operator $\phi_t(x) = e^{At}x$ satisfies the following properties:

(i) $\phi_0(x) = x$,

(ii) $\phi_{-t} \circ \phi_t(x) = x$,

(iii) $\phi_t \circ \phi_s(x) = \phi_{t+s}(x)$

for all s, $t \in \mathbb{R}$ and $x \in \mathbb{R}^n$. Is $\phi_t \circ \phi_s = \phi_s \circ \phi_t$?

Solution We have

(i) $\phi_0(x) = e^{A\cdot 0}x = x$.

(ii) $\phi_{-t} \circ \phi_t(x) = \phi_{-t}(y) = e^{-At}y = e^{-At}e^{At}x = x$, where $y = e^{At}x$.

(iii) $\phi_t \circ \phi_s(x) = \phi_t(y) = e^{At}y = e^{At}e^{As}x = e^{A(t+s)}x = \phi_{t+s}(x)$.

Now,

$$\phi_t \circ \phi_s(\underset{\sim}{x}) = \phi_t(\underset{\sim}{y}) = e^{At}\underset{\sim}{y} = e^{At}e^{As}\underset{\sim}{x} = e^{As}e^{At}\underset{\sim}{x} = \phi_s(\underset{\sim}{z}) = \phi_s \circ \phi_t(\underset{\sim}{x})$$

for all $\underset{\sim}{x} \in \mathbb{R}^n$, where $\underset{\sim}{z} = e^{As}\underset{\sim}{x}$.

Hence $\phi_t \circ \phi_s = \phi_s \circ \phi_t$. This indicates that the given flow evolution operator is commutative.

2.6 Exercises

1. Prove that for a square matrix A of order n, the set of solutions of the linear homogeneous system $\dot{\underset{\sim}{x}} = A\underset{\sim}{x}$ in $\mathbb{R}^n$ forms an n-dimensional vector space.

2. Find the eigenvalues and the corresponding eigenvectors of the following matrices:

(i) $\begin{pmatrix} \dfrac{1}{2} & \dfrac{1}{2} \\[2mm] \dfrac{1}{2} & \dfrac{1}{2} \end{pmatrix}$ (ii) $\begin{pmatrix} 1 & 2 \\ -1 & 2 \end{pmatrix}$ (iii) $\begin{pmatrix} 2 & 7 \\ 5 & -10 \end{pmatrix}$ (iv) $\begin{pmatrix} \alpha & \beta \\ 0 & \gamma \end{pmatrix}$ (v) $\begin{bmatrix} 1 & 3 \\ \sqrt{2} & 3\sqrt{2} \end{bmatrix}$

(vi) $\begin{pmatrix} 1 & -2 & 5 \\ 0 & 6 & -1 \\ 3 & -2 & 1 \end{pmatrix}$

3. (a) Consider the matrix $A = \begin{pmatrix} p & 0 \\ 1 & 1 \end{pmatrix}$. Find the value(s) of p for which the matrix A has repeated eigenvalues.

   (b) Find the 2×2 matrix A whose eigenvalues are 1, 4 and the corresponding eigenvectors are $\begin{pmatrix} 1 \\ -1 \end{pmatrix}$ and $\begin{pmatrix} 2 \\ 1 \end{pmatrix}$.

   (c) Find all 2×2 matrices A whose eigenvalues are 0 and 1.

4. Consider the linear homogeneous system

   $$\dot{x} = -4x + y, \dot{y} = -2x - y.$$

   (a) Write the system as $\dot{\underset{\sim}{x}} = A\underset{\sim}{x}$.

   (b) Show that the characteristic polynomial is $\lambda^2 + 5\lambda + 6$

   (c) Find the eigenvalues and the corresponding eigenvectors of the matrix A.

   (d) Find the general solution of the system.

   (e) Solve the system subject to the initial condition $\underset{\sim}{x}(0) = \begin{pmatrix} 1 \\ 2 \end{pmatrix}$.

5.   Find the general solution to each of the following system of homogeneous linear equations:

(i) $\dot{x} = x + 3y, \dot{y} = x - y$

(ii) $\begin{pmatrix} \dot{x} \\ \dot{y} \end{pmatrix} = \begin{pmatrix} 1 & i \\ -i & 1 \end{pmatrix} \begin{pmatrix} x \\ y \end{pmatrix}, i = \sqrt{-1}$

(iii) $\dot{\underline{x}} = A\underline{x}$ where $A = \begin{bmatrix} -4 & 2 \\ -3 & 1 \end{bmatrix}$

(iv) $\dot{\underline{x}}(t) = A\underline{x}(t)$ where $A = \begin{pmatrix} 5 & 4 \\ -1 & 0 \end{pmatrix}$

(v) $\dot{\underline{x}} = A\underline{x}$, where $A = \begin{pmatrix} 3 & 1 \\ -2 & 1 \end{pmatrix}$

(vi) $\dot{x} = -5x, \dot{y} = -5y$

(vii) $\dot{\underline{x}} = \begin{pmatrix} a & b \\ c & a \end{pmatrix} \underline{x}$, where $bc > 0$.

(viii) $\dfrac{d}{dt} \begin{pmatrix} x(t) \\ y(t) \end{pmatrix} = \begin{pmatrix} \lambda & 1 \\ 0 & \lambda \end{pmatrix} \begin{pmatrix} x(t) \\ y(t) \end{pmatrix}$

(ix) $\dot{\underline{x}} = A\underline{x}$ where $A = \begin{pmatrix} 1 & 1 & 1 \\ 0 & 2 & 2 \\ 0 & 0 & 3 \end{pmatrix}$

(x) $\dfrac{d}{dt} \begin{pmatrix} x(t) \\ y(t) \\ z(t) \end{pmatrix} = \begin{pmatrix} 1 & 2 & -1 \\ 0 & 1 & 1 \\ 0 & -1 & 1 \end{pmatrix} \begin{pmatrix} x(t) \\ y(t) \\ z(t) \end{pmatrix}$

(xi) $\dot{x} = y, \dot{y} = z, \dot{z} = x + y - z$

(xii) $\dot{x} = x + 2y - z, \dot{y} = y + z, \dot{z} = -y + z$

(xiii) $\dot{x} = x, \dot{y} = 2y - 3z, \dot{z} = x + 3y + 2z$

(xiv) $\begin{pmatrix} \dot{x}(t) \\ \dot{y}(t) \\ \dot{z}(t) \end{pmatrix} = \begin{pmatrix} 0 & 1 & 1 \\ 1 & 0 & 1 \\ 1 & 1 & 0 \end{pmatrix} \begin{pmatrix} x(t) \\ y(t) \\ z(t) \end{pmatrix}$

(xv) $\dot{\underline{x}} = A\underline{x}$, where $A = \begin{pmatrix} 1 & -1 & 0 & 0 \\ 1 & 1 & 0 & 0 \\ 0 & 0 & 3 & -2 \\ 0 & 0 & 1 & 1 \end{pmatrix}$

6. Solve the following initial value problems:

    (i) $\dot{x} = 9x + 5y, \dot{y} = -6x - 2y \, ; x(0) = 1, y(0) = 0$.

    (ii) $\dot{\underset{\sim}{x}} = \begin{pmatrix} 3 & -1 \\ 1 & 5 \end{pmatrix} \underset{\sim}{x}, \; \underset{\sim}{x}(0) = \begin{pmatrix} 1 \\ 2 \end{pmatrix}$.

    (iii) $\dot{\underset{\sim}{x}} = \begin{pmatrix} 1 & 0 \\ 0 & 1 \end{pmatrix} \underset{\sim}{x}, \; \underset{\sim}{x}(0) = \begin{pmatrix} 1 \\ -1 \end{pmatrix}$.

    (iv) $\dot{\underset{\sim}{x}} = A\underset{\sim}{x}, \underset{\sim}{x}(0) = \begin{pmatrix} 1 \\ -2 \end{pmatrix}$, where $A = \begin{pmatrix} -3 & 2 \\ -1 & -1 \end{pmatrix}$.

    (v) $\begin{pmatrix} \dot{x}(t) \\ \dot{y}(t) \end{pmatrix} = \begin{pmatrix} -3 & -1 \\ 2 & -1 \end{pmatrix} \begin{pmatrix} x(t) \\ y(t) \end{pmatrix}, \; \underset{\sim}{x}(\pi/2) = \begin{pmatrix} 0 \\ 1 \end{pmatrix}$

    (vi) $\dot{\underset{\sim}{x}}(t) = \begin{pmatrix} 1 & 2 & -1 \\ 1 & 0 & 1 \\ 4 & -4 & 5 \end{pmatrix} \underset{\sim}{x}(t), \; \underset{\sim}{x}(0) = \begin{pmatrix} -1 \\ 0 \\ 0 \end{pmatrix}$

7. Find the solution of the IVP $\dot{\underset{\sim}{x}} = A\underset{\sim}{x}$ subject to the initial condition $\underset{\sim}{x}(0) = \begin{pmatrix} 2 \\ 4 \end{pmatrix}$, where

    $A = \begin{pmatrix} 3 & 9 \\ -1 & -3 \end{pmatrix}$ and $\underset{\sim}{x}(t) = \begin{pmatrix} x(t) \\ y(t) \end{pmatrix}$. Also draw the diagram for the solution set.

8. Convert the second order differential equation $\ddot{x} + a\dot{x} + bx = 0$ to a system of two first order differential equations. Find all values of a and b for which the system has real, distinct eigenvalues. Also find the general solution of the system. Find the solution of the system that satisfies the initial condition $\begin{pmatrix} 0 \\ 1 \end{pmatrix}$. Draw the diagram for the solution set.

9. Find the general solution of the system $\begin{pmatrix} \dot{x}(t) \\ \dot{y}(t) \end{pmatrix} = \begin{pmatrix} a & b \\ c & d \end{pmatrix} \begin{pmatrix} x(t) \\ y(t) \end{pmatrix}$, where $a + d \neq 0$ and

    $ad - bc = 0$. Also sketch the diagram.

10. Consider the system $\dot{\underset{\sim}{x}} = \begin{pmatrix} 0 & 1 \\ -k & -b \end{pmatrix} \underset{\sim}{x}$, where $b \geq 0, k > 0$.

    a) For what values of k and b does the system has
        (i) Complex conjugate eigenvalues?
        (ii) Repeated real eigenvalues?
        (iii) Real and distinct eigenvalues?
    b) Find the general solution of the system in each case.

11. Solve the following second order differential equation after reducing them into a system of two first order differential equations:

(i) $\ddot{x} + x = 0$ with $x(0) = 1, \dot{x}(0) = 0$

(ii) $\ddot{x} + 3\dot{x} + 5x = 0$ with $x(0) = 1, \dot{x}(0) = -1$.

12. Find the general solution of the system below and determine the possible values of α, β so that the initial value problem has a solution that tends to zero as $t \to \infty$

$$\dot{x} = \begin{pmatrix} 5 & -1 \\ 7 & 3 \end{pmatrix} x, \ x(0) = \begin{pmatrix} \alpha \\ \beta \end{pmatrix}.$$

13. (a) What do you mean by a fundamental matrix of a homogeneous system of linear equations?

(b) Show that two different homogeneous systems cannot have the same fundamental matrix.

(c) Let $\Phi(t)$ be a fundamental matrix of the system $\dot{x} = Ax$. Show that for any non-zero constant k, $k\Phi(t)$ is also a fundamental matrix of the system.

14. Find the fundamental matrix of the following systems and hence find the solution of each system:

(i) $\dot{x} = \begin{pmatrix} 1 & -2 \\ -3 & 2 \end{pmatrix} x$

(ii) $\begin{pmatrix} \dot{x} \\ \dot{y} \end{pmatrix} = \begin{pmatrix} 3 & 1 \\ 0 & 2 \end{pmatrix} \begin{pmatrix} x \\ y \end{pmatrix}$

(iii) $\dot{x} = \begin{pmatrix} 1 & 1 \\ 4 & 1 \end{pmatrix} x$

(iv) $\dfrac{d}{dt} \begin{pmatrix} x \\ y \end{pmatrix} = \begin{pmatrix} 9 & 5 \\ -6 & -2 \end{pmatrix} \begin{pmatrix} x \\ y \end{pmatrix}$

(v) $\dot{x} = Ax$, where $A = \begin{pmatrix} 3 & -1 \\ 1 & 5 \end{pmatrix}$

(vi) $\dot{x} = x + y, \dot{y} = -5x - 3y$

(vii) $\dot{x} = \begin{pmatrix} 5 & 4 \\ -1 & 0 \end{pmatrix} x$

15. Find the fundamental matrix of the system

$$\dot{x} = \begin{pmatrix} 2 & -1 \\ -4 & 2 \end{pmatrix} x$$

and use it to find the solution of the system satisfying the initial condition $x(0) = \begin{pmatrix} 1 \\ -3 \end{pmatrix}$.

16. Find a fundamental matrix of the system

$$\dot{x} = 2x - y, \dot{y} = 3x - 2y.$$

Also, find the fundamental matrix $\Phi(t)$ satisfying $\Phi(0) = I$. Find the solution of the system satisfying the initial condition $x(0) = -1, y(0) = 1$.

17. Does $\Phi(t) = \begin{pmatrix} 2e^{3t} & 2e^{-2t} \\ 3e^{3t} & 5e^{-2t} \end{pmatrix}$ a fundamental matrix of the system $\dot{x} = A\underset{\sim}{x}$? If yes, then find the coefficient matrix A.

18. Does $\Phi(t) = \begin{pmatrix} 2e^{4t} & -2 \\ -e^{4t} & 1 \end{pmatrix}$ a fundamental solution of a system $\dot{x} = A\underset{\sim}{x}$?

19. Find e^{At} and then solve the linear system $\dot{x} = A\underset{\sim}{x}$ for

(i) $A = \begin{pmatrix} 9 & -5 \\ 4 & 5 \end{pmatrix}$ (ii) $A = \begin{pmatrix} 1 & 3 \\ 3 & 1 \end{pmatrix}$ (iii) $A = \begin{pmatrix} -4 & 12 \\ -3 & 8 \end{pmatrix}$ (iv) $A = \begin{pmatrix} 1 & 1 \\ 0 & 2 \end{pmatrix}$

20. Compute the exponentials of the following matrices:

(i) $\begin{pmatrix} 0 & 1 \\ 0 & 0 \end{pmatrix}$ (ii) $\begin{pmatrix} a & b \\ 0 & a \end{pmatrix}$, $a, b \in \mathbb{R}$ (iii) $\begin{pmatrix} a & 0 \\ 0 & b \end{pmatrix}$, $a, b \in \mathbb{R}$ (iv) $\begin{pmatrix} 2 & 0 \\ 3 & 2 \end{pmatrix}$

(v) $\begin{pmatrix} 5 & -6 \\ 3 & -4 \end{pmatrix}$

21. If $A = \begin{pmatrix} a & -b \\ b & a \end{pmatrix}$ then prove that $e^A = e^a \begin{pmatrix} \cos b & -\sin b \\ \sin b & \cos b \end{pmatrix}$

22. If $AB = BA$, then show that

(i) $e^A e^B = e^B e^A$ (ii) $Ae^B = Be^A$ (iii) $e^{(A+B)t} = e^{At} e^{Bt}$.

23. If $\underset{\sim}{\alpha}$ is an eigenvector of the matrix A corresponding to the eigenvalue λ, then show that $\underset{\sim}{\alpha}$ is also an eigenvector of the matrix e^A corresponding to the eigenvalue e^{λ}.

24. Consider the matrix $A = \begin{pmatrix} 1 & 1 \\ 0 & 2 \end{pmatrix}$.

(i) Compute e^A directly from the expression.

(ii) Compute e^A by diagonalizing the matrix A.

25. Find the solution of the following systems using fundamental theorem:

(i) $\dot{x} = \begin{pmatrix} 0 & 1 \\ 1 & 0 \end{pmatrix} x$

(ii) $\dot{x} = Ax$, where $A = \begin{pmatrix} 5 & -3 \\ 3 & -1 \end{pmatrix}$

(iii) $\dot{x} = \begin{pmatrix} 2 & 1 \\ -2 & 0 \end{pmatrix} x$

(iv) $\dot{x} = Ax$, where $A = \begin{bmatrix} 2 & 0 & 0 \\ 0 & 0 & 2 \\ 0 & 2 & 0 \end{bmatrix}$

(v) $\dot{x} = Ax$, where $A = \begin{bmatrix} 2 & 0 & 0 \\ 0 & 0 & -2 \\ 0 & 2 & 0 \end{bmatrix}$

(vi) $\dot{x} = Ax$, where $A = \begin{bmatrix} 0 & -2 & -1 & -1 \\ 1 & 2 & 1 & 1 \\ 0 & 1 & 1 & 0 \\ 0 & 0 & 0 & 1 \end{bmatrix}$

(vii) $\dot{x} = Ax$, where $A = \begin{bmatrix} 0 & -2 & -1 & -1 \\ 1 & -2 & 1 & 1 \\ 0 & 1 & 1 & 0 \\ 0 & 0 & 0 & 1 \end{bmatrix}$

26. Solve the following system and sketch its phase portrait

$$\frac{d}{dt}\begin{pmatrix} x(t) \\ y(t) \end{pmatrix} = \begin{pmatrix} -1 & -1 \\ 1 & -1 \end{pmatrix}\begin{pmatrix} x(t) \\ y(t) \end{pmatrix}.$$

27. Solve the initial value problem $\dot{x} = \begin{pmatrix} -2 & -1 \\ 1 & 2 \end{pmatrix} x,\ x(0) = \begin{pmatrix} 1 \\ 0 \end{pmatrix}$ and sketch the solution curve in the phase plane $\mathbb{R}^2$.

28. Find the solution of the problem
$$\ddot{x} + \alpha \dot{x} + \beta x = f(t), \quad x(0) = 1, \dot{x}(0) = 0$$
where $\alpha, \beta > 0$ are constants and $f(t)$ is a function of t.

29. Find the solution curve of the system $\dot{x} = x + y + 1, \dot{y} = x + y$ subject to the initial condition $x(0) = a,\ y(0) = b$, where a, b are some constants.

30. Consider the non-homogeneous linear system
$$\dot{x}(t) = Ax(t) + b(t)$$
Now apply the co-ordinate transformation $x = Py$, $x, y \in \mathbb{R}^n$, where P is a $n \times n$ non-singular matrix. Find the transformed system. Hence show that every non-homogeneous system in $\mathbb{R}^2$ can be transformed into a non-homogeneous system with a Jordan matrix.

31. Does the translation property always hold for non-autonomous system of equations? Justify your answer.

32. Show that if the coefficient matrix A of a non-homogeneous system $\dot{x}(t) = Ax(t) + b(t)$ in $\mathbb{R}^2$ has two real distinct eigenvalues, then the system can be decomposed.

33. Show that $x(t) = x_0 e^t$ is a trajectory passing through the point x_0 of a linear vector field $\dot{x} = Ax$ where A is a constant matrix.

34. Show that $x(t + \tau) = x_0 e^{A(t+\tau)}, t, \tau \in \mathbb{R}$ is also a solution of $\dot{x} = Ax$ subject to the initial condition $x(0) = x_0$. Does it violate the uniqueness of solution? Justify.

35. Define a flow in $\mathbb{R}^2$. Write the properties of flow $\phi(t, x)$. Show that $\phi(t, x) = e^{At} x$ satisfies all properties of flow.

36. Find the flow evolution operator $\phi(t, x)$ for the following systems:

    (i) $\dot{x} = -x, \dot{y} = -2y$,

    (ii) $\dot{x} = xy, \dot{y} = y^2$,

    (iii) $\dot{r} = r(1-r), \dot{\theta} = 1$,

    (iv) $\ddot{x} + \dot{x} + x = 0$,

    (v) $\dot{x} = y, \dot{y} = (x - y)$.

References

1. Perko, L.: Differential Equations and Dynamical Systems, 3rd edn. Springer, New York (2001)
2. Hirsch, M.W., Smale, S.: Differential Equations, Dynamical Systems and Linear Algebra. Academic Press, London (1974)
3. Robinson, R.C.: An Introduction to Dynamical Systems: Continuous and Discrete. American Mathematical Society (2012)
4. Jordan, D.W., Smith, P.: Non-linear Ordinary Differential Equations. Oxford University Press, Oxford (2007)
5. Friedberg, S.H., Insel, A.J., Spence, L.E.: Linear Algebra. Prentice Hall (2003)
6. Hoffman, K., Kunze, R.: Linear Algebra. Prentice Hall (1971)

Chapter 3
Phase Plane Analysis

We have so far discussed one-dimensional systems and solution methods for linear systems. As we know analytical solutions of nonlinear systems are very difficult to obtain except for some special nonlinear equations. The essence of this chapter is to give on finding the local solution behaviors of nonlinear systems, known as local analysis. We shall emphasize on qualitative properties of linear and nonlinear systems rather than quantitative analysis or closed-form solution of a system. The qualitative analysis for two-dimensional system is known as phase plane analysis. Two-dimensional systems have a vast and important dynamics with enormous applications and we study now to explore some of them in this chapter.

3.1 Plane Autonomous Systems

We have discussed autonomous and nonautonomous systems in Chap. 1. Consider an autonomous system of two first-order differential equations in the xy-plane represented by

$$\left.\begin{array}{l} \dot{x} = P(x, y) \\ \dot{y} = Q(x, y) \end{array}\right\} \tag{3.1}$$

where $P(x, y)$ and $Q(x, y)$ are continuous and have continuous first-order partial derivatives throughout the xy-plane. The solutions $(x(t), y(t))$ of (3.1) may be represented on the xy-plane which is called the phase plane of the system. As time t increases, $(x(t), y(t))$ traces out a directed curve on the phase plane. This directed curve is known as the phase path or phase trajectory or simply the path of the system. The phase portrait is the set of all qualitatively different trajectories of the system drawn in the phase plane.

A point $\underset{\sim}{x}^* = (x^*, y^*)$ in a two-dimensional flow $\phi(t, \underset{\sim}{x})$ is said to be a fixed point, also known as critical point or equilibrium point or stationary point of the

© Springer India 2015
G.C. Layek, *An Introduction to Dynamical Systems and Chaos*,
DOI 10.1007/978-81-322-2556-0_3

system (3.1) if and only if $\phi(t, \underset{\sim}{x}^*) = \underset{\sim}{x}^*$. This gives $\dot{x} = 0, \dot{y} = 0 \Rightarrow P(x^*, y^*) = 0$ and $Q(x^*, y^*) = 0$ hold simultaneously. This implies that the fixed points of a system are the constant or equilibrium solution of a system.

3.2   Phase Plane Analysis

Consider the plane autonomous system represented by (3.1). Let (x_0, y_0) be a fixed point of (3.1). Expanding $P(x, y)$ and $Q(x, y)$ in Taylor series in the neighborhood of the point (x_0, y_0), we have

$$P(x, y) = P(x_0, y_0) + x\left(\frac{\partial P}{\partial x}\right)_0 + y\left(\frac{\partial P}{\partial y}\right)_0 + \frac{1}{2!}\left(x\frac{\partial}{\partial x} + y\frac{\partial}{\partial y}\right)_0 P + \cdots$$

$$Q(x, y) = Q(x_0, y_0) + x\left(\frac{\partial Q}{\partial x}\right)_0 + y\left(\frac{\partial Q}{\partial y}\right)_0 + \frac{1}{2!}\left(x\frac{\partial}{\partial x} + y\frac{\partial}{\partial y}\right)_0 Q + \cdots$$

Since (x_0, y_0) is a fixed point of (3.1), $P(x_0, y_0) = 0$ and $Q(x_0, y_0) = 0$. The lower suffixes indicate that the quantities are evaluated at the point (x_0, y_0). Using these and neglecting square and higher order terms in the above expansions, we have

$$\left.\begin{array}{l} P(x, y) = ax + by \\ Q(x, y) = cx + dy \end{array}\right\} \tag{3.2}$$

where each of the quantities a, b, c, and d is evaluated at (x_0, y_0) as given below:

$$a = \left(\frac{\partial P}{\partial x}\right)_0, b = \left(\frac{\partial P}{\partial y}\right)_0, c = \left(\frac{\partial Q}{\partial x}\right)_0 \text{ and } d = \left(\frac{\partial Q}{\partial y}\right)_0 \tag{3.3}$$

Finally, we get a linear system corresponding to (3.1) as

$$\left.\begin{array}{l} \dot{x} = ax + by \\ \dot{y} = cx + dy \end{array}\right\} \tag{3.4}$$

which can also be written in matrix notation as $\underset{\sim}{\dot{x}} = A\underset{\sim}{x}$, where $A = \begin{pmatrix} a & b \\ c & d \end{pmatrix}$ and $\underset{\sim}{x} = \begin{pmatrix} x \\ y \end{pmatrix}$. We assume that $\det(A) = (ad - bc) \neq 0$. Then the solution of (3.4) can be obtained by finding the eigenvalues and the corresponding eigenvectors of the coefficient matrix A. By choosing a matrix P such that $|P| = \det(P) \neq 0$, we apply the similarity transformation

$$\underset{\sim}{x} = P\underset{\sim}{y} \tag{3.5}$$

Under this transformation the system $\dot{\underset{\sim}{x}} = A\underset{\sim}{x}$ reduces to

$$P\dot{\underset{\sim}{y}} = AP\underset{\sim}{y}$$

$$\text{or,} \quad \dot{\underset{\sim}{y}} = P^{-1}AP\underset{\sim}{y} = C\underset{\sim}{y} \tag{3.6}$$

where $C = P^{-1}AP$ is a real **Jordan canonical form** of A (see the books Arrowsmith and Place [1] and Perko [2]). The system (3.6) is known as the canonical form of system (3.4). We also see that

$$|C - \lambda I| = |P^{-1}AP - P^{-1}\lambda IP|$$
$$= |P^{-1}(A - \lambda I)P| = |A - \lambda I|$$
$$= \begin{vmatrix} a - \lambda & b \\ c & d - \lambda \end{vmatrix}$$
$$= \lambda^2 - \tau\lambda + \Delta$$

where $\tau = (a + d) = \text{tr}(A) = \text{tr}(C)$ and $\Delta = (ad - bc) = \det(A) = \det(C)$.

Thus the characteristic equations of matrices A and C are same. So instead of taking matrix A of the original system, we take matrix C for analyzing of the system $\dot{\underset{\sim}{x}} = A\underset{\sim}{x}$, Note that similarity transformations do not change the qualitativeare properties of the systems. However, the orientations of the solution trajectories not preserved. It is known from algebraic theory that matrix P can be chosen in such a way that matrix C takes one of the several canonical forms which, in general are simpler than A. The particular form depends on the nature of the eigenvalues of A. We now give a theorem concerning the possible canonical forms of matrix A.

Theorem 3.1 *Let A be a real 2×2 matrix. Then there exists a real, non-singular 2×2 matrix P such that $C = P^{-1}AP$ is one of the following forms*:

(a) $\begin{bmatrix} \lambda_1 & 0 \\ 0 & \lambda_2 \end{bmatrix}$ (b) $\begin{bmatrix} \lambda & 0 \\ 0 & \lambda \end{bmatrix}$ (c) $\begin{bmatrix} \lambda & 1 \\ 0 & \lambda \end{bmatrix}$, (d) $\begin{bmatrix} \alpha & \beta \\ -\beta & \alpha \end{bmatrix}$ *or* $\begin{bmatrix} \alpha & -\beta \\ \beta & \alpha \end{bmatrix}$ *where* $\lambda_1, \lambda_2, \lambda, \alpha, \beta$ *are all real numbers.*

Proof Eigenvalues of matrix A are the roots of the equation

$$\lambda^2 - \text{tr}(A)\lambda + \det(A) = 0 \tag{3.7}$$

where $\text{tr}(A)$ and $\det(A)$ are, respectively, the trace and determinant of A. Solving the Eq. (3.7), we get the eigenvalues of A as

$$\lambda_{1,2} = \frac{\operatorname{tr}(A) \pm \sqrt{D}}{2} \tag{3.8}$$

where

$$D = (\operatorname{tr}(A))^2 - 4\det(A) \tag{3.9}$$

The nature of the eigenvalues depends on the sign of D. If $D > 0$, the eigenvalues are real and distinct. If $D = 0$, they are equal and if $D < 0$, then the eigenvalues are complex.

Case I: $D > 0$

In this case matrix A has two real, distinct eigenvalues, say λ_1, λ_2. Let $\underset{\sim}{x}_1, \underset{\sim}{x}_2$ be the corresponding eigenvectors. Then from linear algebra, $\underset{\sim}{x}_1, \underset{\sim}{x}_2$ are linearly independent and

$$A\underset{\sim}{x}_1 = \lambda_1 \underset{\sim}{x}_1, \quad A\underset{\sim}{x}_2 = \lambda_2 \underset{\sim}{x}_2 \tag{3.10}$$

We assume that $\lambda_1 > \lambda_2$. Take $P = \begin{pmatrix} \underset{\sim}{x}_1 & \underset{\sim}{x}_2 \end{pmatrix}$, a real 2×2 matrix with the eigenvectors $\underset{\sim}{x}_1$ and $\underset{\sim}{x}_2$ as columns. Since the eigenvectors are linearly independent, matrix P is non-singular and hence invertible. We see that

$$AP = A\begin{pmatrix} \underset{\sim}{x}_1 & \underset{\sim}{x}_2 \end{pmatrix} = \begin{pmatrix} A\underset{\sim}{x}_1 & A\underset{\sim}{x}_2 \end{pmatrix} = \begin{pmatrix} \lambda_1\underset{\sim}{x}_1 & \lambda_2\underset{\sim}{x}_2 \end{pmatrix}$$

$$= \begin{pmatrix} \underset{\sim}{x}_1 & \underset{\sim}{x}_2 \end{pmatrix}\begin{pmatrix} \lambda_1 & 0 \\ 0 & \lambda_2 \end{pmatrix}$$

$$= PC$$

$$\Rightarrow \quad P^{-1}AP = C = \begin{pmatrix} \lambda_1 & 0 \\ 0 & \lambda_2 \end{pmatrix}.$$

Case II: $D = 0$

In this case matrix A has two real repeated eigenvalues given by

$$\lambda_1 = \lambda_2 = \frac{1}{2}\operatorname{tr}(A) = \lambda(\text{say}).$$

We now have the following two possibilities.

(i) Matrix A has two linearly independent eigenvectors, say $\underset{\sim}{x}_1, \underset{\sim}{x}_2$ corresponding to the repeated eigenvalue λ. This is possible only when

$$A = \begin{pmatrix} \lambda & 0 \\ 0 & \lambda \end{pmatrix} = \lambda I. \tag{3.11}$$

We also have $A\underset{\sim}{x}_1 = \lambda \underset{\sim}{x}_1$, and $A\underset{\sim}{x}_2 = \lambda \underset{\sim}{x}_2$.

Let $P = \begin{pmatrix} \underset{\sim}{x}_1 & \underset{\sim}{x}_2 \end{pmatrix}$. Then P is non-singular. Therefore,

$$AP = A\begin{pmatrix} \underset{\sim}{x}_1 & \underset{\sim}{x}_2 \end{pmatrix} = \begin{pmatrix} A\underset{\sim}{x}_1 & A\underset{\sim}{x}_2 \end{pmatrix} = \begin{pmatrix} \lambda \underset{\sim}{x}_1 & \lambda \underset{\sim}{x}_2 \end{pmatrix}$$

$$= \begin{pmatrix} \underset{\sim}{x}_1 & \underset{\sim}{x}_2 \end{pmatrix}\begin{pmatrix} \lambda & 0 \\ 0 & \lambda \end{pmatrix}$$

$$= PA$$

$$\Rightarrow \quad P^{-1}AP = A = C = \begin{pmatrix} \lambda & 0 \\ 0 & \lambda \end{pmatrix}$$

(ii) The matrix A has only one linearly independent eigenvector, say $\underset{\sim}{x}$ corresponding to the repeated eigenvalue λ. Then $A\underset{\sim}{x} = \lambda\underset{\sim}{x}$. We now choose a generalized eigenvector $\underset{\sim}{y}$ such that the matrix $P = \begin{pmatrix} \underset{\sim}{x} & \underset{\sim}{y} \end{pmatrix}$ is non-singular. Then

$$AP = A\begin{pmatrix} \underset{\sim}{x} & \underset{\sim}{y} \end{pmatrix} = \begin{pmatrix} A\underset{\sim}{x} & A\underset{\sim}{y} \end{pmatrix}$$

$$= \begin{pmatrix} \lambda P\underset{\sim}{e}_1 & PP^{-1}A\underset{\sim}{y} \end{pmatrix}$$

$$= P\begin{pmatrix} \lambda\underset{\sim}{e}_1 & P^{-1}A\underset{\sim}{y} \end{pmatrix}$$

$$\Rightarrow \quad P^{-1}AP = \begin{pmatrix} \lambda\underset{\sim}{e}_1 & P^{-1}A\underset{\sim}{y} \end{pmatrix}$$

where $\underset{\sim}{e}_1$ is the first column of the identity matrix I_2.

Now the matrices A and $P^{-1}AP$ have the same eigenvalues, and so

$$P^{-1}AP = \begin{pmatrix} \lambda & k \\ 0 & \lambda \end{pmatrix}$$

for some nonzero real k. If we consider the modified matrix

$$Q = P\begin{pmatrix} 1 & 0 \\ 0 & k^{-1} \end{pmatrix},$$

then using some elementary properties of matrix multiplication and inverse, we obtain

$$Q^{-1}AQ = \begin{pmatrix} \lambda & 1 \\ 0 & \lambda \end{pmatrix}.$$

This gives the result (c).

Case III: $D < 0$

In this case matrix A has a pair of complex eigenvalues $(\alpha \pm i\beta)$, where $\alpha = \text{tr}(A)/2$ and $\beta = \sqrt{(-D)}/2$. Let $\lambda = \alpha + i\beta$. Then the other eigenvalue is $\bar{\lambda} = \alpha - i\beta$, where '$-$' denotes the complex conjugate. Let $\underset{\sim}{x} = \underset{\sim}{u} + i\underset{\sim}{v}$ be the eigenvector corresponding to $\lambda = \alpha + i\beta$. Then the eigenvector corresponding to $\bar{\lambda}$ is $\bar{\underset{\sim}{x}} = \underset{\sim}{u} - i\underset{\sim}{v}$. Take $P = \begin{pmatrix} \underset{\sim}{u} & \underset{\sim}{v} \end{pmatrix}$. Then P is non-singular. Now,

$$\begin{aligned} AP = A\begin{pmatrix} \underset{\sim}{u} & \underset{\sim}{v} \end{pmatrix} &= \begin{pmatrix} A\underset{\sim}{u} & A\underset{\sim}{v} \end{pmatrix} \\ &= \begin{pmatrix} \alpha\underset{\sim}{u} - \beta\underset{\sim}{v} & \beta\underset{\sim}{u} + \alpha\underset{\sim}{v} \end{pmatrix} [\because A\underset{\sim}{x} = \lambda\underset{\sim}{x}] \\ &= \begin{pmatrix} \underset{\sim}{u} & \underset{\sim}{v} \end{pmatrix}\begin{pmatrix} \alpha & \beta \\ -\beta & \alpha \end{pmatrix} \\ &= PC \\ \Rightarrow \quad P^{-1}AP = C &= \begin{pmatrix} \alpha & \beta \\ -\beta & \alpha \end{pmatrix}. \end{aligned}$$

If we take $P(\underset{\sim}{v} \ \underset{\sim}{u})$, then it is easy to prove that

$$P^{-1}AP = \begin{pmatrix} \alpha & -\beta \\ \beta & \alpha \end{pmatrix}.$$

(See the books Arrowsmith and Place [1], and Perko[2])
This completes the proof.

We discuss four cases in the phase plane case-wise.

Case I: If matrix A has real, distinct eigenvalues λ_1, λ_2, then $C = \begin{pmatrix} \lambda_1 & 0 \\ 0 & \lambda_2 \end{pmatrix}$.

Case II: (a) If $A = \begin{pmatrix} \lambda & 0 \\ 0 & \lambda \end{pmatrix}$, where λ is real, then $C = \begin{pmatrix} \lambda & 0 \\ 0 & \lambda \end{pmatrix}$.

(b) If A has real repeated eigenvalues λ and $A \neq \lambda I$, then $C = \begin{pmatrix} \lambda & 1 \\ 0 & \lambda \end{pmatrix}$.

Case III: If A has complex eigenvalues $\lambda_1 = \alpha + i\beta$ and $\lambda_2 = \alpha - i\beta$ with $\alpha, \beta \in \mathbb{R}$, then either $C = \begin{pmatrix} \alpha & \beta \\ -\beta & \alpha \end{pmatrix}$ or $C = \begin{pmatrix} \alpha & -\beta \\ \beta & \alpha \end{pmatrix}$ depending upon the choice of the similarity matrix P.

Case IV: One or both eigenvalues of A are zero.

This is the degenerate case.

Case I: The eigenvalues of A are real, distinct, and nonzero

The critical point of the system is $(0, 0)$. Since λ_1, λ_2 are nonzero real, distinct eigenvalues of A, the canonical form of A is

$$C = \begin{pmatrix} \lambda_1 & 0 \\ 0 & \lambda_2 \end{pmatrix}.$$

Therefore the canonical form of system (3.4) is

$$\dot{\underset{\sim}{x}} = C\underset{\sim}{x}$$

or, $\begin{pmatrix} \dot{x} \\ \dot{y} \end{pmatrix} = \begin{pmatrix} \lambda_1 & 0 \\ 0 & \lambda_2 \end{pmatrix} \begin{pmatrix} x \\ y \end{pmatrix}$

or, $\left.\begin{aligned} \dot{x} &= \lambda_1 x \\ \dot{y} &= \lambda_2 y \end{aligned}\right\}$

Now, $\dfrac{dy}{dx} = \dfrac{\dot{y}}{\dot{x}} = \dfrac{\lambda_2 y}{\lambda_1 x} = \left(\dfrac{\lambda_2}{\lambda_1}\right)\left(\dfrac{y}{x}\right)$

which we can integrate immediately to obtain the trajectories as

$$\log(y) = \left(\frac{\lambda_2}{\lambda_1}\right)\log(x) + \log(B)$$

or, $y = Bx^{(\lambda_2/\lambda_1)}$

where B is a positive constant.

There may arise three cases, viz. (i) the eigenvalues λ_1, λ_2 are of same sign, say positive, the critical point origin is called node; (ii) both λ_1, λ_2 are negative in sign, the origin is called node but it is stable; and (iii) when λ_1, λ_2 are of opposite signs, the origin is called saddle but it is unstable or semi-stable in nature. The saddle has one incoming trajectory toward the critical point for negative eigenvalue and one outgoing trajectory away from the critical point for positive eigenvalue. The integral curves for the first two cases resemble a family of parabolas having the tangent at the origin. A sketch of the integral curves or phase diagrams for $\lambda_1 > \lambda_2 > 0$ and $-\lambda_1 > -\lambda_2 > 0$ are, respectively, given in Fig. 3.1a, b.

The directions of arrows are easy to obtain as both $|x|$ and $|y|$ increase with time (since $x(t) = ke^{\lambda_1 t}$, k being a constant). This is called an unstable node (Fig. 3.1a). If

(a)

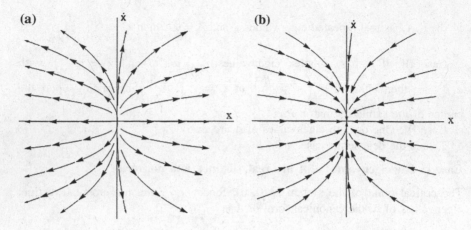

(b)

Fig. 3.1 Phase diagrams for nonzero real, distinct eigenvalues of same sign, **a** unstable node, **b** stable node

Fig. 3.2 Phase diagram for nonzero real, distinct eigenvalues of opposite signs

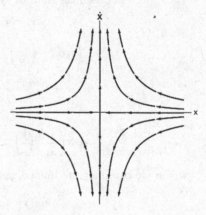

$-\lambda_1 > -\lambda_2 > 0$, we can describe the flow in exactly the same way by reversing the arrow directions to give a stable node (Fig. 3.1b). The stable node is an asymptotically fixed point, that is, the trajectories approach to the fixed point as $t \rightarrow \infty$.

When λ_1, λ_2 are of opposite signs, the integral curves resemble a family of hyperbolas with the axes as asymptotes and the equilibrium point origin is called saddle. It is shown in Fig. 3.2. This type of equilibrium point is also called hyperbolic type.

Case II: (a) The matrix A has real repeated eigenvalues

In this case, the matrix A has either two linearly independent eigenvectors or only one linearly independent eigenvector corresponding to the repeated real eigenvalue λ. We now discuss two sub-cases below.

Suppose that matrix A has two linearly independent eigenvectors corresponding to the repeated real eigenvalue λ.

The canonical form of (3.4) is then given as

$$\dot{\underset{\sim}{x}} = C \underset{\sim}{x}$$

$$\text{or,} \quad \begin{pmatrix} \dot{x} \\ \dot{y} \end{pmatrix} = \begin{pmatrix} \lambda & 0 \\ 0 & \lambda \end{pmatrix} \begin{pmatrix} x \\ y \end{pmatrix}$$

$$\text{or,} \quad \dot{x} = \lambda x, \dot{y} = \lambda y, \text{ where } \lambda \neq 0.$$

So, $\frac{dy}{dx} = \frac{\dot{y}}{\dot{x}} = \frac{\lambda y}{\lambda x} = \frac{y}{x}$. This gives the integral curves as $y = kx$, k being an arbitrary constant. The phase diagrams in xy-plane for $\lambda < 0$ and $\lambda > 0$ are presented in Fig. 3.3.

The flow is toward the origin when $\lambda < 0$. The origin is then called a star node or a proper node and it is asymptotically stable (see Fig. 3.3a). It is an attracting stable fixed point and is also called a sink in which flow is coming toward the equilibrium point as $t \to \infty$. When $\lambda > 0$, the phase diagram is exactly the same except that the flow is away from the origin and it is an unstable star node (Fig. 3.3b). In such a case the equilibrium point origin is known as a source in the context of flow.

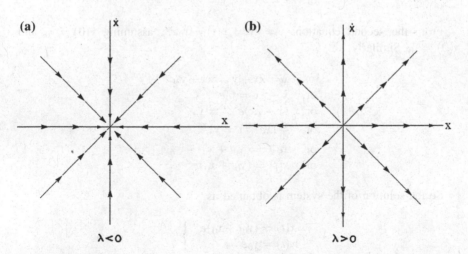

Fig. 3.3 Phase diagrams for repeated real eigenvalues

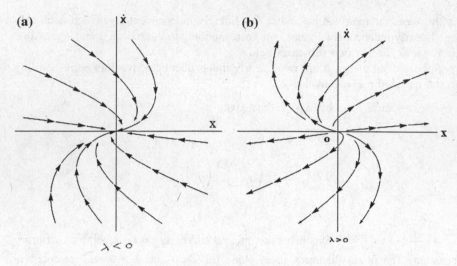

Fig. 3.4 **a** Phase portrait for $\lambda < 0$. **b** Phase portrait for $\lambda > 0$

(b) The matrix A has only one linearly independent eigenvector corresponding to the repeated real eigenvalue λ. The canonical form of (3.4) is then expressed as

$$\dot{\underset{\sim}{x}} = C \underset{\sim}{x}$$

or, $\quad \begin{pmatrix} \dot{x} \\ \dot{y} \end{pmatrix} = \begin{pmatrix} \lambda & 1 \\ 0 & \lambda \end{pmatrix} \begin{pmatrix} x \\ y \end{pmatrix}$

or, $\quad \dot{x} = \lambda x + y, \dot{y} = \lambda y$

From the second equation, $\dot{y} = \lambda y \Rightarrow y(t) = y_0 e^{\lambda t}$, assuming $y(0) = y_0$ and $x(0) = x_0$. Similarly,

$$\dot{x} = \lambda x + y = \lambda x + y_0 e^{\lambda t}$$

or, $\quad \dot{x} - \lambda x = y_0 e^{\lambda t}$

or, $\quad (\dot{x} - \lambda x) e^{-\lambda t} = y_0$

or, $\quad \frac{\mathrm{d}}{\mathrm{d}t}\left(x e^{-\lambda t} \right) = y_0$

or, $\quad x e^{-\lambda t} = y_0 t + x_0$

or, $\quad x(t) = (y_0 t + x_0) e^{\lambda t}$

So the solution of the system is obtained as

$$\left. \begin{array}{l} x(t) = (y_0 t + x_0) e^{\lambda t} \\ y(t) = y_0 e^{\lambda t} \end{array} \right\}$$

When $\lambda < 0$, $x(t) \to 0, y(t) \to 0$ as $t \to \infty$. Also the second equation shows that $y(t)$ has the same sign as that of y_0 for all t. We also see that

$$\frac{y}{x} = \frac{y_0}{(y_0 t + x_0)} = \frac{1}{\left(t + \frac{x_0}{y_0}\right)} \to 0 \text{ as } t \to \infty$$

while remaining positive sign. The phase portrait for $\lambda < 0$ is presented in Fig. 3.4a.

In this case the origin is said to be an imperfect node and it is stable. A **separatrix** is a line or a curve in the orbit that divides the phase plane into distinct types of qualitative behaviors of trajectories. The equilibrium point is called sink and only one separatrix straight line which is the x-axis, the eigen axis of the system. When $\lambda > 0$, $x(t) \to 0$, $y(t) \to 0$ as $t \to -\infty$ and $y/x \to 0$ as $t \to -\infty$ while remaining negative sign. In this case the equilibrium point is called an imperfect node but it is unstable. The phase diagram is shown in Fig. 3.4b.

Case III: The matrix A has complex eigenvalues

Suppose that matrix A has complex eigenvalues, say $(\alpha \pm i\beta)$, $\alpha, \beta \in \mathbb{R}$. Here the canonical form of A is expressed either

$$C = \begin{pmatrix} \alpha & \beta \\ -\beta & \alpha \end{pmatrix} \text{ or } C = \begin{pmatrix} \alpha & -\beta \\ \beta & \alpha \end{pmatrix}$$

depending upon the choice of the similarity matrix P. Therefore the canonical form of (3.4) is expressed as

$$\dot{\underset{\sim}{x}} = C\underset{\sim}{x}$$

$$\text{either} \quad \left. \begin{array}{l} \dot{x} = \alpha x + \beta y \\ \dot{y} = -\beta x + \alpha y \end{array} \right\} \text{ or } \left. \begin{array}{l} \dot{x} = \alpha x - \beta y \\ \dot{y} = \beta x + \alpha y \end{array} \right\}$$

We now transform the system into polar coordinates (r, θ) using the relations $x = r\cos\theta$, $y = r\sin\theta$, where $r^2 = x^2 + y^2$ and $\tan\theta = y/x$.

Differentiating $r^2 = x^2 + y^2$ with respect to t, we have

$$r\dot{r} = x\dot{x} + y\dot{y}$$
$$\text{or,} \quad r\dot{r} = x(\alpha x \pm \beta y) + y(\mp \beta x + \alpha y) = \alpha(x^2 + y^2) = \alpha r^2$$
$$\text{or,} \quad \dot{r} = \alpha r$$

Similarly, differentiating $\tan\theta = y/x$ with respect to t, we get

$$\sec^2(\theta)\dot\theta = \frac{x\dot y - y\dot x}{x^2}$$
$$\Rightarrow \quad (1+\tan^2\theta)\dot\theta = \frac{x\dot y - y\dot x}{x^2}$$
$$\Rightarrow \quad \left(1+\frac{y^2}{x^2}\right)\dot\theta = \frac{x\dot y - y\dot x}{x^2}$$
$$\Rightarrow \quad (x^2+y^2)\dot\theta = x\dot y - y\dot x$$
$$\Rightarrow \quad r^2\dot\theta = x(\mp\beta x + \alpha y) - y(\alpha x \pm \beta y)$$
$$\Rightarrow \quad r^2\dot\theta = \mp\beta(x^2+y^2) = \mp\beta r^2$$
$$\Rightarrow \quad \dot\theta = \mp\beta.$$

Thus the system in polar coordinate becomes

$$\left.\begin{array}{c} \dot r = \alpha r \\ \dot\theta = \mp\beta \end{array}\right\} \tag{3.12}$$

Solutions of (3.12) are given by

$$r(t) = r(0)e^{\alpha t} \tag{3.13}$$

$$\theta(t) = \mp\beta t + \theta(0) \tag{3.14}$$

We first consider the case where $\theta(t) = -\beta t + \theta(0)$. In this case, when $\beta < 0$, θ increases linearly, that is, it corresponds to anti-clockwise rotation. When $\beta > 0$, θ decreases linearly and it corresponds to clockwise rotation. For the other case, $\theta(t) = \beta t + \theta(0)$, $\beta < 0$ gives clockwise motion and $\beta > 0$ gives anti-clockwise motion. Remember that similarity transformation does not preserve orientation of trajectory. The two canonical forms will retain their qualitative properties but the orientation will be opposite in directions. We display the phase portraits for first case only.

When $\alpha < 0$, from (3.13) $r(t) \to 0$ as $t \to \infty$. The equilibrium point origin is then called a stable focus (or a stable spiral). The trajectories wind spirally around the equilibrium point a number of times before reaching asymptotically at origin.

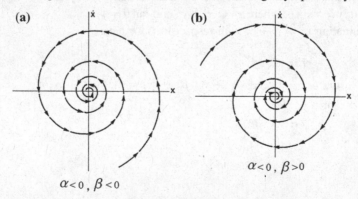

Fig. 3.5 Phase portraits for complex eigenvalues with negative real part

The stable focuses are basically asymptotically stable equilibrium points. Figure 3.5 displays the phase diagrams for two cases.

On the other hand, when $\alpha > 0$, $r(t) \to 0$ as $t \to -\infty$ and the flow becomes away from the origin and the equilibrium point $(0, 0)$ is known as source. This is basically a repelling equilibrium point and hence it is an unstable focus. It is shown graphically in Fig. 3.6.

When $\alpha = 0$, this case is very interesting and we have $r = r(0) =$ constant. The trajectories never reach the equilibrium point as $t \to \infty$. The equilibrium point $(0, 0)$ is called a center or sometimes called an **elliptic equilibrium point** and it is neutrally stable. This gives a closed path of the system and the system exhibits periodic solution. On the other hand, periodic solution represents the closed path. The phase path is shown in Fig. 3.7. The direction of motion of the closed path is determined by the sign of β.

Finally we say that when $\alpha \neq 0$, the phase diagrams are spirals and the spiral toward the equilibrium point origin if $\alpha < 0$ and the equilibrium point is called stable

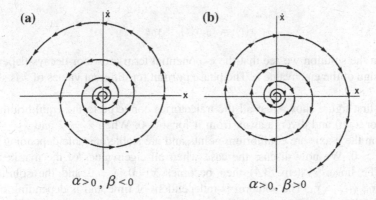

Fig. 3.6 Phase portraits for complex eigenvalues with positive real part

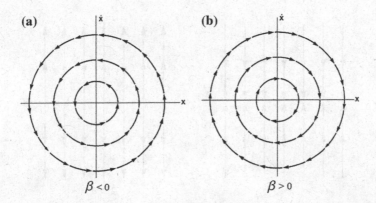

Fig. 3.7 Phase portraits for purely imaginary eigenvalues

focus. And it is asymptotically stable. If $\alpha > 0$, the spiral is winding away from the origin and the equilibrium point is called unstable focus. When $\alpha = 0$, the equilibrium point is called center, representing the closed path which is neutrally stable but not asymptotically.

Case IV: One or both eigenvalues are zero

We now discuss the degenerate cases where either $\lambda_1 = 0$ or $\lambda_2 = 0$ or both $\lambda_1 = \lambda_2 = 0$. This case arises when $|A| = 0$, that is, at least one of the eigenvalues of A is zero. So the equilibrium point is not isolated so that there are infinitely many equilibrium points in its neighborhood. In this case the matrix A has a nontrivial null space of dimension one or two and any vector in the null space is a fixed point of the system. If the dimension of the null space is two, the matrix A will be a zero matrix. Let us first consider $\lambda_1 = 0$ and $\lambda_2 = \lambda \neq 0$ as two eigenvalues of matrix A with $\underset{\sim}{\alpha}$ and β as the corresponding eigenvectors. Then the canonical form of the system (3.4) can be written as $\dot{x} = 0$, $\dot{y} = \lambda y$. Its solution with $x(0) = x_0, y(0) = y_0$ is given by

$$x(t) = x_0, y(t) = y_0 e^{\lambda t}.$$

From the solution we see that the exponential term grows or decays depending on the sign of the eigenvalue λ. The phase portrait for different values of λ is shown in Fig. 3.8.

The first figure shows that all the trajectories converge to the equilibrium subspace for $\lambda < 0$ and diverge away from it for $\lambda > 0$. When $\lambda_1 = \lambda$ and $\lambda_2 = 0$, all points on the y-axis are equilibrium points and are stable/unstable depending upon $\lambda < 0/\lambda > 0$. We now discuss the case when all eigenvalues of the matrix A are zero. The linear system (3.4) then becomes $\dot{x} = 0$, $\dot{y} = 0$ and the solution is $x(t) = x_0, y(t) = y_0$. The solution is independent of time and is depending on the initial conditions. As t increases, $x(t)$ and $y(t)$ remain $x(0)$ and $y(0)$. So every orbit is an equilibrium point and is shown in Fig. 3.9a.

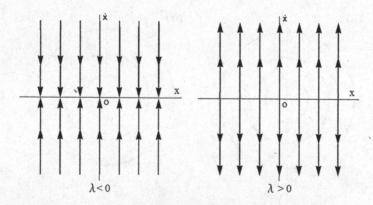

Fig. 3.8 Phase portraits when only one eigenvalue is zero

Fig. 3.9 a Phase portrait
when both eigenvalues are
zero. b A typical phase
portrait when all eigenvalues
are zero

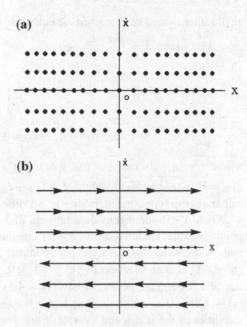

Again, a non-singular coordinate transformation of variables reduces the matrix
A as $A = \begin{pmatrix} 0 & 1 \\ 0 & 0 \end{pmatrix}$ and the system becomes $\dot{x} = y, \dot{y} = 0$ with the solution

$$x(t) = y_0 t + x_0, \quad y(t) = y_0$$

where

$$y(0) = y_0, x(0) = x_0.$$

For given nonzero values of x_0 and y_0 we see that $y(t)$ remains as y_0, while
$x(t) \to -\infty$ for $y_0 > 0$ and $x(t) \to -\infty$ for $y_0 < 0$ as $t \to \infty$. Trajectories are
parallel to the x-axis. The x-axis is the equilibrium subspace. The phase portrait of
the system is shown in Fig. 3.9b.

3.3 Local Stability of Two-Dimensional Linear Systems

We have drawn phase trajectories for four cases in the neighborhood of the equilibrium point of a linear system and named equilibrium points geometrically. Their
stability characters are mentioned. Stability near an equilibrium point means that a
small change of a system at some instant gives only a small change of its behavior

at all future time. The characteristic equation of matrix A associated with the system $\dot{\underset{\sim}{x}} = A\underset{\sim}{x}$, where $A = \begin{pmatrix} a & b \\ c & d \end{pmatrix}$ gives

$$\begin{vmatrix} a - \lambda & b \\ c & d - \lambda \end{vmatrix} = 0$$

or, $\lambda^2 - (a+d)\lambda + (ad - bc) = 0$

or, $\lambda^2 - \tau\lambda + \Delta = 0$

where $\tau = (a+d)$ and $\Delta = (ad - bc)$. Let λ_1, λ_2 be two eigenvalues of A. Then $\lambda_{1,2} = \frac{\tau \pm \sqrt{\tau^2 - 4\Delta}}{2} = \frac{\tau \pm \sqrt{D}}{2}$, where $\tau = (\lambda_1 + \lambda_2)$, $\Delta = \lambda_1\lambda_2$ and $D = (\tau^2 - 4\Delta)$. We can show the types and stabilities of all different fixed points in Fig. 3.10.

When $\Delta < 0$, the eigenvalues are real and have opposite signs. Hence the fixed point is saddle, the left side in $(\Delta - \tau)$ diagram. If $\Delta > 0$, the eigenvalues are either real with same sign (node) or complex (spiral or center). Nodes satisfy $(\tau^2 - 4\Delta) > 0$ and for spirals $(\tau^2 - 4\Delta) < 0$. Some of the equilibrium points, e.g., star and degenerate nodes, satisfy $(\tau^2 - 4\Delta) = 0$. So the parabola represented by $(\tau^2 - 4\Delta) = 0$ in $(\Delta - \tau)$ plane is the borderline between nodes and spirals. The stabilities of the nodes and the spirals are determined by the sign of τ. When $\tau < 0$, both the eigenvalues have negative real parts or both have negative signs and so the fixed point is stable focus or stable node. For unstable spirals/nodes, τ is positive. The fixed point center is neutrally stable and lies on the broadline $\tau = 0$ and the eigenvalues are purely imaginary. If $\Delta = 0$, at least one of the eigenvalues is zero. The origin is not isolated fixed point. There is either a whole line of fixed points or a plane of fixed points. The stabilities of the fixed points depend on the initial conditions of the system as discussed in the earlier section. From Fig. 3.10 it is clear that the fixed points like saddle, node, and spiral are the major fixed points. These fixed points occur in large open region of (Δ, τ) plane. However, the fixed points center, star, degenerate node, non-isolated fixed points, etc. are borderline cases and these fixed points are minor fixed points. But the fixed point center occurs in many

Fig. 3.10 Diagramatic representation of different types of stabilities

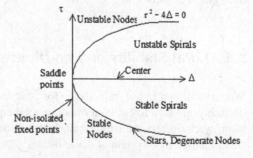

physical systems particularly in mechanical systems where energy is conserved, that is, no dissipation. Finally, we summarize the results in the following way:

Classification of fixed points

(a) node if $\Delta > 0$ and $D \geq 0$,

(b) spiral if $\tau \neq 0$ and $D < 0$

(c) saddle if $\Delta < 0$,

(d) star/degenerate node if $\Delta > 0$ and $D = 0$

(e) center if $\tau = 0$ and $\Delta > 0$,

Stability criteria

(a) stable and attracting if $\tau < 0$ and $\Delta > 0$,

(b) stable if $\tau \leq 0$ and $\Delta > 0$,

(c) unstable if $\tau > 0$ or $\Delta < 0$.

Depending upon the signs of the eigenvalues λ_1, λ_2 of the matrix A, the stability character of the equilibrium points may be stated as follows:

1. Eigenvalues are real and distinct:

   - **Saddle** if λ_1, λ_2 are of opposite signs,
   - **Unstable node** if λ_1, λ_2 are both positive,
   - **Stable node** if λ_1, λ_2 are both negative.

2. Eigenvalues are real and repeated ($\lambda_1 = \lambda_2 = \lambda$, say):

   - **Unstable star node** if $\lambda > 0$.
   - **Stable star node** if $\lambda < 0$.
   - **Unstable improper node** if $\lambda > 0$.
   - **Stable improper node** if $\lambda < 0$.

   First two cases occur when $A = \lambda I$, I being the identity matrix.

3. Complex eigenvalues $(\alpha \pm i\beta), \alpha, \beta \in \mathbb{R}$:

   - **Unstable spiral or focus** if $\alpha > 0$
   - **Stable spiral or focus** if $\alpha < 0$
   - **Neutrally stable center** if $\alpha = 0$

3.4 Linearization and Its Limitations

Consider a two-dimensional autonomous system (3.1) and let (x^*, y^*) be a fixed point of this system so that $P(x^*, y^*) = 0$ and $Q(x^*, y^*) = 0$. Let $\xi = (x - x^*)$, $\eta = (y - y^*)$ denote the components of a small disturbance from the fixed point (x^*, y^*).

We need to derive the differential equations for the perturbation variables ξ and η. We proceed as follows:

$$\dot{\xi} = \dot{x} = P(x^* + \xi, y^* + \eta)$$

$$= P(x^*, y^*) + \xi \frac{\partial P}{\partial x} + \eta \frac{\partial P}{\partial y} + O(\xi^2, \eta^2, \xi\eta) \ \text{[Taylor's series expansion]}$$

$$= \xi \frac{\partial P}{\partial x} + \eta \frac{\partial P}{\partial y} + O(\xi^2, \eta^2, \xi\eta) \quad \text{[Since } P(x^*, y^*) = 0\text{]}$$

where the partial derivates are evaluated at the fixed point (x^*, y^*). Similarly, we obtain $\dot{\eta} = \xi \frac{\partial Q}{\partial x} + \eta \frac{\partial Q}{\partial y} + O(\xi^2, \eta^2, \xi\eta)$.

Hence the perturbation variables (ξ, η) evolve according to the following equation:

$$\begin{pmatrix} \dot{\xi} \\ \dot{\eta} \end{pmatrix} = \begin{pmatrix} \frac{\partial P}{\partial x} & \frac{\partial P}{\partial y} \\ \frac{\partial Q}{\partial x} & \frac{\partial Q}{\partial y} \end{pmatrix}_{(x^*, y^*)} \begin{pmatrix} \xi \\ \eta \end{pmatrix} + \text{quadratic terms.}$$

The matrix $J = \begin{pmatrix} \frac{\partial P}{\partial x} & \frac{\partial P}{\partial y} \\ \frac{\partial Q}{\partial x} & \frac{\partial Q}{\partial y} \end{pmatrix}_{(x^*, y^*)}$ is called the **Jacobian matrix** of the nonlinear system (3.1) evaluated at the fixed point (x^*, y^*). Assuming the quadratic and higher order terms are small and if we neglect them, then we obtain a linearized system for two-dimensional nonlinear autonomous system (3.1). This mathematical technique is known as linearization of a nonlinear system about a fixed point.

3.4.1 Limitations of Linearization

First of all we see what we did in the name of linearization. One should know it clearly. In the linearization process the quadratic and higher order terms are being neglected due to their smallness with respect to the equilibrium point. It is well known from matrix theory that the eigenvalues of a matrix depend continuously on its perturbation quantity. However, some of the eigenvalues remain unchanged their characters under arbitrarily small perturbation, but some of them will change their positions under small perturbation. We can say that if the equilibrium point $x = 0$ of the system $\dot{x} = Ax$ is a node, focus, or saddle types, then they will remain their same characters for sufficiently small perturbations. However, the situation is different if the equilibrium points are center/star node/imperfect nodes, that is, all minor or border line equilibrium points. These will go either a stable focus or an unstable focus under small perturbation of the system. So, the node, focus, and saddle equilibrium points are said to be structurally stable equilibrium points because they maintain the same qualitative behaviors under small perturbation, that is, these equilibrium points have similar behavior both in the nonlinear system and its linearization (see the book of Andronov, Leontovich, Gordon and Maier [3] for proof). On the other hand, the equilibrium points of center/star/imperfect nodes are not structurally stable, in general. More specifically, a **hyperbolic fixed point** (real $(\lambda) \neq 0$) will retain the same stability/instability characters for small perturbations in the neighborhood of the fixed point. We shall now illustrate an example for this limitation. Consider the following two-dimensional system represented by

$$\dot{x} = -y + \mu x (x^2 + y^2)$$
$$\dot{y} = x + \mu y (x^2 + y^2)$$

where μ is a parameter. This is a nonlinear system and the linearized form of the system about the equilibrium point (0, 0) is

$$\left. \begin{array}{l} \dot{x} = -y \\ \dot{y} = x \end{array} \right\}$$

This can also be written in matrix notation as

$$\dot{\underset{\sim}{x}} = A \underset{\sim}{x}, \text{ where } A = \begin{pmatrix} 0 & -1 \\ 1 & 0 \end{pmatrix} \text{ and } \underset{\sim}{x} = \begin{pmatrix} x \\ y \end{pmatrix}.$$

The characteristic equation of A is

$$\begin{vmatrix} -\lambda & -1 \\ 1 & -\lambda \end{vmatrix} = 0 \Rightarrow \lambda^2 + 1 = 0 \Rightarrow \lambda = \pm i$$

Eigenvalues are purely imaginary. Hence, the equilibrium fixed point (0, 0) of the linearized system is a center for all values of the parameter μ and it represents a closed path near the equilibrium point. We find analytical solution of the given system by converting (x, y) into polar coordinates (r, θ). So $x = r \cos \theta, y = r \sin \theta$, where $r^2 = x^2 + y^2$ and $\tan \theta = y/x$. Differentiating above relations with respect to t, we get

$$r\dot{r} = x\dot{x} + y\dot{y} = x\{-y + \mu x(x^2 + y^2)\} + y\{x + \mu y(x^2 + y^2)\}$$
$$= \mu x^2 (x^2 + y^2) + \mu y^2 (x^2 + y^2) = \mu r^4$$

or, $\dot{r} = \mu r^3$ (3.15)

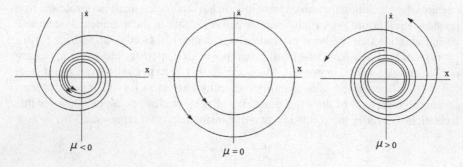

Fig. 3.11 Phase trajectories for different values of the parameter μ

and

$$\dot{\theta} = \frac{1}{\left(\frac{y}{x}\right)^2 + 1} \cdot \frac{x\dot{y} - y\dot{x}}{x^2} = \frac{1}{x^2 + y^2} \left[x\{x + \mu y(x^2 + y^2)\} - y\{-y + \mu x(x^2 + y^2)\} \right]$$

$$= \frac{1}{x^2 + y^2} (x^2 + y^2) = 1$$

or, $\dot{\theta} = 1$ (3.16)

Solutions of (3.15) and (3.16) are obtained as

$$\left. \begin{array}{l} \frac{1}{2r^2} = -\mu t + \frac{1}{2r_0^2} \\ \theta(t) = t + \theta_0 \end{array} \right\}$$ (3.17)

where r_0, θ_0 be the given initial conditions.

If $\mu < 0$, $r(t)$ approaches to 0 monotonically but spirally as $t \to \infty$. Therefore in this case, the origin is a stable spiral. If $\mu = 0$, then $r(t) = r_0$ for all t representing a closed path. The equilibrium point origin is then a center. If $\mu > 0$, then $r(t)$ moves away from 0 spirally as $t \to -\infty$. The equilibrium point origin is an unstable spiral. The phase trajectories are displayed in Fig. 3.11. Hence the linearizion gives the equilibrium point as center whereas the original system has stable or unstable focus.

3.5 Nonlinear Simple Pendulum

A simple pendulum consists of a bob of mass m suspended from a fixed point O by a light string of length L, which is allowed to swing in the vertical plane. In the absence of frictional forces the equation of motion is given by

Fig. 3.12 Sketch of simple pendulum

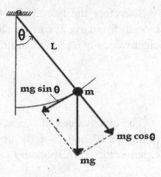

$$(mL)\ddot{\theta} = -mg \sin \theta \qquad (3.18)$$

where θ denotes the angular displacement, measured in anti-clockwise sense (Fig. 3.12).

The equation of motion can be written as

$$\ddot{\theta} = -\left(\frac{g}{L}\right) \sin \theta = -\omega^2 \sin \theta, \omega = \sqrt{(g/L)} > 0 \qquad (3.19)$$

This is a second-order nonlinear autonomous differential equation. We set $\theta = x$ and $\dot{\theta} = y$. Then Eq. (3.19) can be written as

$$\left. \begin{array}{l} \dot{x} = y \\ \dot{y} = -\omega^2 \sin x \end{array} \right\} \qquad (3.20)$$

The equilibrium points of (3.20) are given by $x = n\pi$, $n = 0, \pm 1, \pm 2, \ldots$, and $y = 0$. The system has infinite number of fixed points $(n\pi, 0)$. We now calculate the Jacobian matrix of the given nonlinear system about the equilibrium points. So we have

$$J(x, y) = \begin{bmatrix} \frac{\partial f}{\partial x} & \frac{\partial f}{\partial y} \\ \frac{\partial g}{\partial x} & \frac{\partial g}{\partial y} \end{bmatrix} = \begin{bmatrix} 0 & 1 \\ -\omega^2 \cos x & 0 \end{bmatrix}.$$

The Jacobian matrix $J(x, y)$ at the equilibrium points $(0, 0)$ and $(\pm\pi, 0)$ are $J(0, 0) = \begin{bmatrix} 0 & 1 \\ -\omega^2 & 0 \end{bmatrix}$ and $J(\pm\pi, 0) = \begin{bmatrix} 0 & 1 \\ \omega^2 & 0 \end{bmatrix}$. The eigenvalues of $J(0, 0)$ are the roots of the equation

$$\begin{vmatrix} -\lambda & 1 \\ -\omega^2 & -\lambda \end{vmatrix} = 0 \Rightarrow \lambda^2 + \omega^2 = 0 \Rightarrow \lambda = \pm i\omega \text{ which are purely imaginary.}$$

Therefore the fixed point $(0, 0)$ is a center, and the neighboring orbits are all closed. Similarly it can be shown that the fixed points $(2n\pi, 0)$ are centers. The eigenvalues of $J(\pm\pi, 0)$ are the solutions of the equation

$$\begin{vmatrix} -\lambda & 1 \\ \omega^2 & -\lambda \end{vmatrix} = 0 \Rightarrow \lambda^2 - \omega^2 = 0 \Rightarrow \lambda = \pm\omega.$$

The eigenvalues of $J(\pm\pi, 0)$ are real but opposite in signs. So the equilibrium points $(\pi, 0)$ and $(-\pi, 0)$ are saddle. In general, $((2n+1)\pi, 0)$ are saddles. The eigenvectors corresponding to the eigenvalues $(\pm\omega)$ of $J(\pm\pi, 0)$ are

$$\begin{bmatrix} 0 & 1 \\ \omega^2 & 0 \end{bmatrix} \begin{bmatrix} e_1 \\ e_2 \end{bmatrix} = \pm\omega \begin{bmatrix} e_1 \\ e_2 \end{bmatrix}$$

$$\text{or,} e_2 = \pm\omega e_1 \text{ and } \omega^2 e_1 = \pm\omega e_2$$

The eigenvector for ω is $(1, \omega)^T$ and $(1, -\omega)^T$ for $(-\omega)$. Here $(1, \omega)^T$ is a repelling eigen axis corresponding to the eigenvalue ω, while $(1, -\omega)^T$ is an attracting eigen axis corresponding to $(-\omega)$. The phase diagram for the undamped nonlinear pendulum is shown in Fig. 3.13.

The saddle connection that is the tangent to the eigen axes, connecting two saddle points formed separatrix loops in which the orbits are closed curve encircling the equilibrium point. The closed orbit corresponds to oscillatory motion of the pendulum. Orbit outside the separatrix is unbounded curve corresponding to whirling motion of the pendulum. Connecting the whole of these separatrix loops and the closed curves is agreed with physical behavior of the pendulum problem. Pendulum has two distinct vertical equilibrium states, one is stable and the other is unstable. The points O, A, B represent states of physical equilibrium positions of the pendulum and are called equilibrium points of the system. The points at $(\pm\pi, 0)$ correspond to the unstable equilibrium points in which the pendulum is sliding up

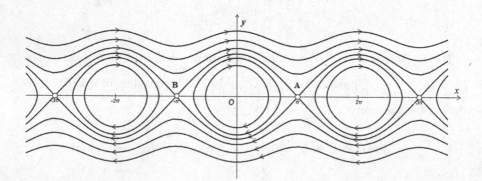

Fig. 3.13 Phase portrait for the undamped pendulum

in vertical positions. The pendulum problem is conservative. The energy function can be calculated easily as follows:

Multiplying the pendulum equation by $\dot{\theta}$ and then integrating, we get

$$\dot{\theta}(\ddot{\theta} + \omega^2 \sin \theta) = 0 \Rightarrow \frac{1}{2}\dot{\theta}^2 - \omega^2 \cos \theta = \text{constant}.$$

The energy function E is given by

$$E = \frac{1}{2}v^2 - \omega^2 \cos \theta, (\because \dot{\theta} = v).$$

It has local minimum at the origin. For small angle approximation it gives the close orbits (ellipses for $\omega \neq 1$) given as

$$(\omega\theta)^2 + v^2 \simeq 2(E + \omega^2)$$

This agrees the results of phase plane analysis. The pendulum problem is also reversible. A system in $\mathbb{R}^2$ is said to be **reversible** if the system is invariant under the transformation $x \to -x$ and $y \to -y$. In other way, the dynamics looks the same in forward or backward times (see the book of Strogatz [4] for more physical examples). What can we get from the flow evolution operator of the pendulum problem? Consider the linear case, so that the equation of motion is $\ddot{\theta} + \omega^2\theta = 0$. The flow evolution operator $\phi(t, \underset{\sim}{\theta})$ is deduced as follows:

The solution of the above equation is $\theta(t) = A \cos \omega t + B \sin \omega t$, where A, B are constants. This implies

$$\dot{\theta}(t) = -A\omega \sin \omega t + B\omega \cos \omega t.$$

Let the initial condition be at $t = 0$, $\theta = \theta_0$ and $\dot{\theta} = \dot{\theta}_0$. Using it in the above two equations, we obtain $A = \theta_0$ and $B = \dot{\theta}_0/\omega$. Therefore, $\theta(t) = \theta_0 \cos \omega t + (\dot{\theta}_0/\omega) \sin \omega t$. Now, setting $x = \theta$ and $y = \dot{\theta}$, we have

$$x(t) = \theta(t) = x_0 \cos \omega t + \frac{y_0}{\omega} \sin \omega t$$
$$\text{and} \quad y(t) = \dot{\theta}(t) = y_0 \cos \omega t - x_0\omega \sin \omega t$$

where $x_0 = x(t = 0) = \theta(t = 0) = \theta_0$ and $y_0 = \dot{\theta}_0$.

Therefore the evolution operator ϕ_t for the simple pendulum is

$$\phi_t(\underset{\sim}{x}) = \phi_t\begin{pmatrix} x \\ y \end{pmatrix} = \begin{pmatrix} x \cos \omega t + \frac{y}{\omega} \sin \omega t \\ y \cos \omega t - x\omega \sin \omega t \end{pmatrix} \text{ for all } \underset{\sim}{x} = \begin{pmatrix} x \\ y \end{pmatrix} \in \mathbb{R}^2 \text{ and } t \in \mathbb{R}.$$

We shall now verify that $\phi_t(\phi_s(\underset{\sim}{x})) = \phi_{t+s}(\underset{\sim}{x})$ for all $\underset{\sim}{x} \in \mathbb{R}^2$ and $t, s \in \mathbb{R}$.

Fig. 3.14 A typical phase
portrait of damped pendulum

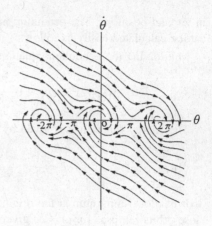

We have for all $\underset{\sim}{x} \in \mathbb{R}^2$, and $t, s \in \mathbb{R}$,

$$
\begin{aligned}
\phi_t(\phi_s(\underset{\sim}{x})) = \phi_t &\begin{pmatrix} x\cos\omega s + \frac{y}{\omega}\sin\omega s \\ y\cos\omega s - x\omega\sin\omega s \end{pmatrix} \\
&= \begin{pmatrix} \{x\cos\omega s + (y/\omega)\sin\omega s\}\cos\omega t + \frac{y\cos\omega s - x\omega\sin\omega s}{\omega}\sin\omega t \\ (y\cos\omega s - x\omega\sin\omega s)\cos\omega t - \{x\cos\omega s + (y/\omega)\sin\omega s\}\omega\sin\omega t \end{pmatrix} \\
&= \begin{pmatrix} x\cos\omega(t+s) + (y/\omega)\sin\omega(t+s) \\ y\cos\omega(t+s) - x\omega\sin(t+s) \end{pmatrix} \\
&= \phi_{t+s}\begin{pmatrix} x \\ y \end{pmatrix} = \phi_{t+s}(\underset{\sim}{x}) = \phi_{s+t}(\underset{\sim}{x}).
\end{aligned}
$$

The flow evolution operator satisfies the property $\phi(t, \underset{\sim}{x}_0) = \underset{\sim}{x}_0$, which gives the equilibrium point of the flow generating by the system. The flow evolution operator forms a commutative dynamical group. Interested readers can try for finding the closed-form solution of nonlinear pendulum problem using elliptic integral (see Jordan and Smith [5]) and also try to find evolution operator for this case.

The damping case is important physically. We now see what happen in adding a small amount of linear damping to the pendulum equation. The equation of motion for linear damping becomes

$$
\ddot{\theta} + \alpha\dot{\theta} + \omega^2\sin\theta = 0
$$

where $\alpha > 0$ is the damping strength. The energy of this dissipative system is calculated easily and obtained as $E = \frac{1}{2}\dot{\theta}^2 - \omega^2\cos\theta$. Therefore, the rate of change of energy is

$$\frac{\mathrm{d}E}{\mathrm{d}t} = \dot{\theta}\ddot{\theta} + \omega^2\dot{\theta}\sin\theta = \dot{\theta}(\ddot{\theta} + \omega^2\sin\theta) = -\alpha\dot{\theta}^2 < 0 \, [\because \alpha > 0].$$

Hence, the energy in the dissipative pendulum decreases monotonically along the path of the pendulum, except at the equilibrium point of the pendulum while the energy of the undamped pendulum (no dissipation) is constant. It can be shown easily that the centers become spirals while the saddles remain the same. A typical sketch of phase portrait is displayed in Fig. 3.14.

One can find the flow evolution operator for this system easily. For simplicity we give the evolution operator $\phi(t, \underset{\sim}{x})$ for the dissipative linear system $\ddot{\theta} + 2\omega\dot{\theta} + \omega^2\theta = 0$. The flow evolution operator is expressed as $\phi(t, \underset{\sim}{x}) = \begin{pmatrix} \{(1 + \omega t)x + ty\}e^{-\omega t} \\ \{-\omega^2 tx + (1 - \omega t)y\}e^{-\omega t} \end{pmatrix}$, which satisfies all flow properties. Moreover, $\phi(t + s, \underset{\sim}{x}) = \phi(s + t, \underset{\sim}{x}), \forall t, s \in \mathbb{R}$. This implies that linear dissipative flow forms a commutative dynamical group. Interested readers can try to establish the properties of damping pendulum problem.

3.6 Linear Oscillators

Linear system in two dimensions admits oscillatory motion and the linear super-position principle is valid. In case of nonlinear system this principle is no longer valid and the frequency of oscillation, in general, amplitude-dependent. When a system has sustained oscillations of finite amplitude, then it is usually referred to as a harmonic oscillator. This is a model for the linear LC circuit or pendulum problem. We can see that the physical mechanism resulting in these oscillations is a periodic exchange (without dissipation) of the energy stored in the system. There are two fundamental problems in linear oscillator. First, we have seen that a small perturbation will destroy the oscillation, and the linear oscillator is not structurally stable. The amplitude of oscillation is dependent on the initial conditions. The two fundamental problems of a linear oscillator can be overcome in nonlinear oscillator. Nonlinear oscillators have the following important properties, viz.,

 (i) the nonlinear oscillator is structurally stable;
(ii) the amplitude of oscillation at steady state is independent of initial conditions or insensitive of initial conditions.

The properties of nonlinear oscillator are very important in practical applications particularly in electrical, electronic devices and mechanical systems. The linear oscillators are predictable and their oscillations change stability characters for the small change of initial conditions. Generally, the oscillations are represented by an inhomogeneous linear second-order differential equation of the following standard form

$$\ddot{x} + \alpha\dot{x} + \omega_0^2 x = f \sin \omega t, \alpha, \omega, f > 0$$

Here $\alpha\dot{x}$ is the linear damping force and f, ω_0, and ω are, respectively, the forcing amplitude, the natural frequency, and angular (forcing) frequency. Let $x \equiv x(t)$ be the displacement of the body from the rest of mass m with the force of inertia $m\ddot{x}$. Linear oscillator may be in different types, viz., (i) free oscillator, (ii) damped oscillator, and (iii) damped and forced oscillators.

When the damping and forcing terms are absent, then the oscillator is called free and executes a harmonic oscillation. The solutions subject to the initial conditions $x(0) = A$, $\dot{x}(0) = 0$ is given by $x(t) = A \cos \omega_0 t$, with period $T = 2\pi/\omega_0$. Note that the period is independent of the amplitude A. The trajectory plot in x versus t (Fig. 3.15a) and the phase portrait in the $x - \dot{x}$ plane are shown in the following figures for the initial conditions $x(0) = 2, \dot{x}(0) = 0$. In Fig. 3.15b the motion is characterized by closed-curves where the phase point $(x, \dot{x})$ moves continuously on the closed curves as time goes on. The different closed curves correspond to the slightly different initial conditions.

The figure indicates that the amplitude of oscillation is constant.

When the damping force is present, the equation is represented by $\ddot{x} + \alpha\dot{x} + \omega_0^2 x = 0$. The solution is obtained as $x(t) = Ae^{m_1 t} + Be^{m_2 t}$, where A and B are arbitrary constants and $m_{1,2} = \frac{1}{2}\left[-\alpha \pm \sqrt{\alpha^2 - 4\omega_0^2}\right]$. The three possibilities may arise:

(i)   Strong damping $(\alpha^2 - 4\omega_0^2) > 0$
(ii) Weak damping $(\alpha^2 - 4\omega_0^2) < 0$
(iii) Critical damping $(\alpha^2 - 4\omega_0^2) = 0$.

We shall discuss these three cases using dynamical system principle and closed-form solutions.

The given second-order system can be written as

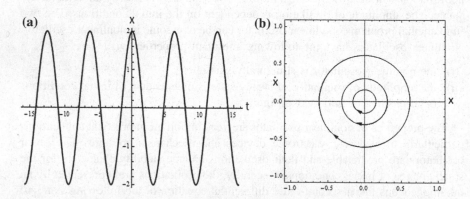

Fig. 3.15 a Flow trajectory x versus t, **b** phase portrait $\dot{x}$ versus x

Fig. 3.16 Typical flow trajectories for strong damping force

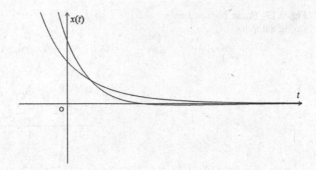

$$\left.\begin{array}{l} \dot{x} = y \\ \dot{y} = -\alpha y - \omega_0^2 x \end{array}\right\}$$ (3.21)

Clearly, $(0, 0)$ is the only fixed point of the system. In matrix notation the system (3.21) can be expressed as

$$\begin{pmatrix} \dot{x} \\ \dot{y} \end{pmatrix} = \begin{pmatrix} 0 & 1 \\ -\omega_0^2 & -\alpha \end{pmatrix} \begin{pmatrix} x \\ y \end{pmatrix}$$ (3.22)

Therefore, $\tau = a + d = -\alpha$, $\Delta = \beta$ and $D = (\tau^2 - 4\Delta) = (\alpha^2 - 4\omega_0^2)$.

(i) Strong Damping (D > 0)

In this case, the eigenvalues, say λ_1, λ_2 are real, distinct, and negative, since $\alpha > 0$. The solution of the system is $x(t) = Ae^{\lambda_1 t} + Be^{\lambda_2 t}$, where A, B are arbitrary constants. Figure 3.16 displays two typical phase trajectories of the system. The phase path of the system is given by

$$x = x(t) = Ae^{\lambda_1 t} + Be^{\lambda_2 t}, y = \dot{x} = \lambda_1 Ae^{\lambda_1 t} + \lambda_2 Be^{\lambda_2 t}$$

The set of points (x, y) can be treated as a parametric representation of the phase path with parameter t, for fixed values of A and B. The phase path is shown in Fig. 3.17.

From this figure it is clear that the equilibrium point origin is a stable node. In the strong damping force all trajectories, starting at some initial condition, terminate at the origin as $t \to \infty$.

(ii) Weak Damping (D < 0)

Eigenvalues for this weak damping case are complex with negative real part. The solution is obtained as

$$x(t) = Ae^{-\frac{1}{2}\alpha t} \cos\left\{\frac{1}{2}\sqrt{-D}t + \epsilon\right\}$$

Fig. 3.17 Phase portrait for strong damping

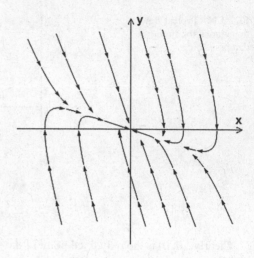

Fig. 3.18 A typical flow trajectory for weak damping

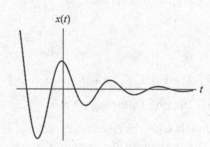

where A and $\in$ are real and arbitrary constants with $A > 0$. The graphical representation of a typical solution is shown in Fig. 3.18 which represents an oscillation with decreasing amplitude $\left(Ae^{-\frac{1}{2}\alpha t}\right)$ and the oscillation decays more rapidly for large values of α. The phase diagram is shown in Fig. 3.19.

The equilibrium point origin is a stable focus and the trajectory spiral moves toward the equilibrium point.

(iii) **Critical Damping (D = 0)**

In the phase plane analysis,, the eigenvalues are equal and negative: $\lambda_1 = \lambda_2 = -(\alpha/2)$, $\alpha > 0$. The equilibrium point is therefore stable. The solution is given by $x(t) = (A + Bt)e^{-\frac{1}{2}\alpha t}$. For $\alpha > 2\omega_0$, the solution is non-oscillatory and exponentially approaches toward the origin. Figure 3.20 depicts the phase diagram. It indicates that all trajectories of the oscillator reach asymptotically to the origin as $t \to \infty$, staring from different initial conditions.

Damped and forced oscillator: The equation of motion for a damped-forced oscillator is expressed as

Fig. 3.19 Phase portrait $A = 1, \alpha = 1, \omega_0 = 1, \in = 0.1$ in $(x - \dot{x})$ plane

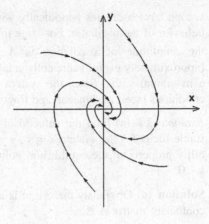

Fig. 3.20 Phase portrait for critical damping

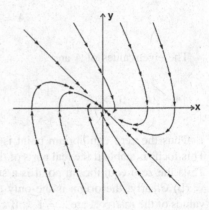

$$\ddot{x} + \alpha \dot{x} + \omega_0^2 x = f \sin \omega t$$

where $\alpha > 0$, ω_0 and ω are the original and forcing frequencies of the oscillator. The equation is linear and its solution is given by

$$x(t) = A_t \omega_0 / c e^{-\alpha t/2} \cos(ct - \beta) + A_p \cos(\omega t - \gamma)$$

where

$$A_p = \frac{f}{(\omega_0^2 - \omega^2)^2 + \alpha^2 \omega^2}, \gamma = \tan^{-1}\left(\frac{\omega^2 - \omega_0^2}{\alpha \omega}\right).$$

The constants A_t and β are determined from initial conditions and c can be obtained from the relation $c = \sqrt{\omega_0^2 - (\alpha^2/4)}$. Note that if $\alpha > 0$, the first term of the solution (independent of f and ω) decreases exponentially fast while the second term oscillates periodically with time. So the first term gives the transient behavior of the oscillator. For large time, the frequency of oscillation is $\omega/2\pi$ and the amplitude of oscillation is A_p. At the resonance value $\sqrt{\omega_0^2 - (\alpha^2/2)}$ (approximately ω_0 for sufficiently small damping) the amplitude of oscillation takes a maximum value. Since the system is linear, its motion is insensitive to initial conditions (see Lakshmanan and Rajasekar [6] for linear and nonlinear oscillators).

Example 3.1 (a) Find the value(s) of $k \in \mathbb{R}$ such that zero equilibrium will be a stable focus for the system $\dot{x} = y$, $\dot{y} = -x + ky$, (b) determine the nature and stability property of the equilibrium point of the system $\dot{x} = x, \dot{y} = ky$ for $k > 0$ and $k < 0$.

Solution (a) Obviously the origin is an equilibrium point of the system. Here the coefficient matrix A is

$$A = \begin{bmatrix} 0 & 1 \\ -1 & k \end{bmatrix}$$

The eigenvalues of A are

$$|A - \lambda I| = 0 \Rightarrow \begin{vmatrix} -\lambda & 1 \\ -1 & k - \lambda \end{vmatrix} = 0 \Rightarrow \lambda^2 - k\lambda + 1 = 0 \Rightarrow \lambda = \frac{k \pm \sqrt{k^2 - 4}}{2}$$

Thus the zero equilibrium point is a focus if $k^2 - 4 < 0$, that is, if $-2 < k < 2$. This focus is stable if the real parts of the eigenvalues are negative. This gives $k < 0$. Thus the zero equilibrium point is a stable focus if $-2 < k < 0$.

(b) Clearly, the origin is the only equilibrium point of the system. The eigenvalues of the matrix A are $\lambda = 1, k$. If $k > 0$ but $k \neq 1$, then the eigenvalues are real, distinct, and positive. So in this case, the origin is an unstable node. If $k < 0$, then the eigenvalues are real and distinct but they are of opposite signs. So in this case, the origin is a saddle point. If $k = 1$, then the eigenvalues are equal and positive in sign. Hence, the equilibrium point $(0, 0)$ is a star node and it is unstable.

Example 3.2 Locate the critical point and find its nature for the system $\dot{x} = x + y$, $\dot{y} = x - y + 1$. Also find the phase path of the system.

Solution For the critical points, we have

$$\dot{x} = 0 \text{ and } \dot{y} = 0.$$

That is, $x + y = 0$ and $x - y + 1 = 0$. Solving, we get $x = -1/2$, $y = 1/2$. So, the critical point of the system is $(-1/2, 1/2)$. We now translate the origin $(0, 0)$ at the critical point $(-1/2, 1/2)$ using the transformation

$$x = \xi - \frac{1}{2} \text{ and } y = \eta + \frac{1}{2}.$$

Under this transformation the equations become

$$\dot{\xi} = \xi - \frac{1}{2} + \eta + \frac{1}{2} = \xi + \eta$$

$$\text{and} \quad \dot{\eta} = \xi - \frac{1}{2} - \eta - \frac{1}{2} + 1 = \xi - \eta$$

That is, $\quad \begin{pmatrix} \dot{\xi} \\ \dot{\eta} \end{pmatrix} = \begin{pmatrix} 1 & 1 \\ 1 & -1 \end{pmatrix} \begin{pmatrix} \xi \\ \eta \end{pmatrix}$

The characteristic equation of the coefficient matrix gives

$$(1 - \lambda)(-1 - \lambda) - 1 = 0$$

$$\text{or,} \quad 1 - \lambda^2 + 1 = 0$$

$$\text{or,} \quad \lambda^2 = 2$$

$$\text{or,} \quad \lambda = \pm\sqrt{2}.$$

So, the characteristic roots are of opposite signs. Thus the critical point $\left(-\frac{1}{2}, \frac{1}{2}\right)$ is a saddle. We now find the phase paths of the system. The differential equation of the phase path is

$$\frac{dy}{dx} = \frac{\dot{y}}{\dot{x}} = \frac{x - y + 1}{x + y}$$

$$\Rightarrow x\,dy + y\,dx = x\,dx - y\,dy + dx$$

$$\Rightarrow d(xy) = x\,dx - y\,dy + dx$$

Integrating, we get

$$xy = \frac{x^2}{2} - \frac{y^2}{2} + x + k$$

$$\text{or,} \quad 2xy = x^2 - y^2 + 2x + c,$$

where c is an arbitrary constant. This is the required equation for the phase path of the system.

Example 3.3 Indicate the nature and stability of the fixed point of the linear system

$$\dot{x} = 10x - y$$
$$\dot{y} = 25x + 2y$$

Solution The system has only one fixed point (0, 0). The system can be written as $\dot{x} = A\underset{\sim}{x}$, where $\underset{\sim}{x} = \begin{pmatrix} x \\ y \end{pmatrix}$ and $A = \begin{pmatrix} 10 & -1 \\ 25 & 2 \end{pmatrix}$. The characteristic equation of A is

$$\begin{vmatrix} 10 - \lambda & -1 \\ 25 & 2 - \lambda \end{vmatrix} = 0$$
$$\Rightarrow \quad \lambda^2 - 12\lambda + 45 = 0$$
$$\Rightarrow \quad \lambda = \frac{12 \pm \sqrt{144 - 170}}{2} = \frac{12 \pm i\sqrt{26}}{2} = 6 \pm i\frac{\sqrt{26}}{2}.$$

Thus, the eigenvalues of A are complex with positive real part. Hence, the fixed point (0, 0) is an unstable focus.

Example 3.4 Find the nature and stability of the fixed points of

$$\dot{x} = -ax + y, \dot{y} = -x - ay$$

for different values of the parameter a.

Solution The only fixed point of the system is (0, 0). The system can be written as $\dot{x} = A\underset{\sim}{x}$, where $\underset{\sim}{x} = \begin{pmatrix} x \\ y \end{pmatrix}$ and $A = \begin{pmatrix} -a & 1 \\ -1 & -a \end{pmatrix}$. The eigenvalues of A are obtained as follows:

$$\begin{vmatrix} -a - \lambda & 1 \\ -1 & -a - \lambda \end{vmatrix} = 0$$
$$\Rightarrow \quad (\lambda + a)^2 + 1 = 0$$
$$\Rightarrow \quad \lambda = -a \pm i.$$

When $a > 0$, the eigenvalues are complex with negative real parts. So, the fixed point (0, 0) is a stable spiral.

If $a < 0$, then the eigenvalues are complex with positive real parts. So, the fixed point (0, 0) is an unstable spiral.

If $a = 0$, then the eigenvalues are purely imaginary and the fixed point is center which is neutrally stable.

Example 3.5 Draw the phase diagrams for linear harmonic undamped oscillators represented by (i) $\ddot{x} + \omega^2 x = 0$, (ii) $\ddot{x} - \omega^2 x = 0$.

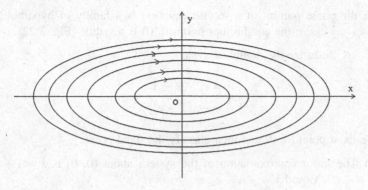

Fig. 3.21 Phase portrait for case (i)

Solution Both the systems can be written as

$$\left. \begin{array}{l} \dot{x} = y \\ \dot{y} = \mp\omega^2 x \end{array} \right\}$$

The only equilibrium point of the systems is $(0, 0)$. We can easily find the phase path of the systems as follows:

$$\frac{\mathrm{d}y}{\mathrm{d}x} = \frac{\dot{y}}{\dot{x}} = \frac{\mp\omega^2 x}{y} = \mp\omega^2 \frac{x}{y}$$

$$\text{or,} \quad y\,\mathrm{d}y \pm \omega^2 x\mathrm{d}x = 0$$

Integrating, we get $y^2 \pm \omega^2 x^2 = c$, c being the integrating constant.

In the first case the phase portrait (Fig. 3.21) is the family of ellipses with center at $(0, 0)$. This implies that all solutions of the system are periodic. The equilibrium point $(0, 0)$ is a center.

Fig. 3.22 Phase portrait for case (ii)

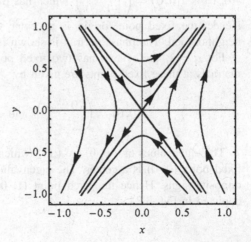

While the phase portrait of a second equation is a family of hyperbolas with asymptotes $y = \pm \omega x$, the equilibrium point $(0, 0)$ is a saddle (Fig. 3.22).

Example 3.6 Linearize the nonlinear system

$$\dot{x} = e^{-x-3y} - 1$$
$$\dot{y} = -x(1 - y^2)$$

about the fixed point $(0, 0)$ and then classify the fixed point.

Solution The linear approximation of the system about $(0, 0)$ is $\dot{x} = 1 + (-x - 3y) - 1 = -x - 3y$ and $\dot{y} = -x$.

We write the system as $\dot{\underset{\sim}{x}} = A\underset{\sim}{x}$ where

$$A = \begin{pmatrix} -1 & -3 \\ -1 & 0 \end{pmatrix}$$

Now, $\tau = \text{tr}(A) = a + d = -1$ $\Delta = \det(A) = (ad - bc) = -3 < 0$ and $D = (\tau^2 - 4\Delta) = 13 > 0$

Hence, the fixed point $(0, 0)$ is a saddle.

Example 3.7 Draw the phase portraits of $\ddot{x} + x = ax^2$ for the three values $a = 0, -1, 1$.

Solution We can write the second-order equation as $\dot{x} = y, \dot{y} = ax^2 - x$. The fixed points are obtained by solving the equations $y = 0$ and $ax^2 - x = 0$. For $a = 0$, the system has only one fixed point at $(0, 0)$, for $a = 1$, it has two fixed points at $(0, 0)$, $(1, 0)$ and for $a = -1$, the fixed points are $(0, 0)$ and $(-1, 0)$. We now discuss the stabilities of these fixed points.

For $a = 0$, the system has only one fixed point at the origin. The Jacobian matrix $J(0, 0)$ is $J(0,0) = \begin{pmatrix} 0 & 1 \\ -1 & 0 \end{pmatrix}$ which has purely imaginary eigenvalues $\lambda = \pm i$. Hence, the fixed point origin is a center which represents closed paths in its neighborhood. The phase portrait is shown in Fig. 3.23.

For $a = 1$, the system has two fixed points $(0, 0)$ and $(1, 0)$. The Jacobian matrices at these fixed points are given by

$$J(0,0) = \begin{pmatrix} 0 & 1 \\ -1 & 0 \end{pmatrix} \text{ and } J(1,0) = \begin{pmatrix} 0 & 1 \\ 1 & 0 \end{pmatrix}.$$

The eigenvalues of $J(0, 0)$ are $(\pm i)$, which are purely imaginary. Therefore, the fixed point $(0, 0)$ is a center. The eigenvalues of $J(1, 0)$ are (± 1), which are of opposite signs. Hence the fixed point $(1, 0)$ is a saddle. The phase diagram is depicted in Fig. 3.24.

Fig. 3.23 Phase portrait for $a = 0$

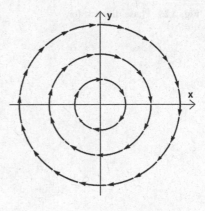

Fig. 3.24 Phase diagram for $a = 1$

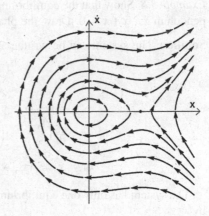

The center and saddles are clearly depicted in the figure.

For $a = -1$ the fixed points of the system are $(0, 0)$ and $(-1, 0)$. Now,

$$J(0,0) = \begin{pmatrix} 0 & 1 \\ -1 & 0 \end{pmatrix} \text{ and } J(-1,0) = \begin{pmatrix} 0 & 1 \\ 1 & 0 \end{pmatrix}$$

The eigenvalues of the matrix $J(0, 0)$ are $(\pm i)$, which are purely imaginary. So, the fixed point $(0, 0)$ is a center. The eigenvalues of $J(-1, 0)$ are (± 1), which are of opposite signs. Hence, the fixed point $(-1, 0)$ is a saddle. The phase diagram is represented in Fig. 3.25.

Fig. 3.25 Phase diagram for $a = -1$

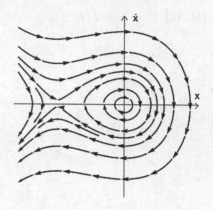

Example 3.8 Show that the equilibrium point is a stable focus for the damped linear pendulum $\ddot{x} + \dot{x} + x = 0$. Draw the phase diagram of the system.

Solution The system can be written as

$$\left.\begin{array}{l} \dot{x} = y \\ \dot{y} = -x - y \end{array}\right\}$$

$$\text{or,} \quad \frac{\mathrm{d}}{\mathrm{d}t}\begin{pmatrix} x \\ y \end{pmatrix} = \begin{pmatrix} 0 & 1 \\ -1 & -1 \end{pmatrix}\begin{pmatrix} x \\ y \end{pmatrix}$$

$$\text{or,} \quad \dot{\underset{\sim}{x}} = A\underset{\sim}{x}, \text{ where } A = \begin{pmatrix} 0 & 1 \\ -1 & -1 \end{pmatrix}$$

The system has only one equilibrium point $(0, 0)$. The eigenvalues of A are given by

$$|A - \lambda I| = 0$$

$$\text{or,} \quad \begin{vmatrix} -\lambda & 1 \\ -1 & -1 - \lambda \end{vmatrix} = 0$$

$$\text{or,} \quad \lambda^2 + \lambda + 1 = 0$$

$$\text{or,} \quad \lambda = \frac{-1 \pm \sqrt{1 - 4}}{2} = -\frac{1}{2} \pm i\frac{\sqrt{3}}{2}$$

Eigenvalues are complex with negative real part. Therefore the equilibrium point origin is a stable focus, that is, all orbits are spirals converging to the origin. The phase diagram is shown in Fig. 3.26.

Fig. 3.26 Phase portrait for the given system

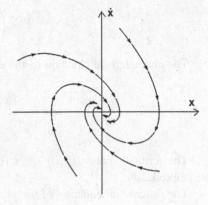

Example 3.9 Obtain all critical points of the system $\dot{x} = \sin y, \dot{y} = \cos x$. Linearize the system about the critical point $\left(\frac{\pi}{2}, 0\right)$. Find the equation of the phase path.

Solution For critical points we set

$$\sin y = 0 \Rightarrow y = n\pi, n = 0, \pm 1, \pm 2, \ldots$$

and $$\cos x = 0 \Rightarrow x = (2m+1)\frac{\pi}{2} = \left(m + \frac{1}{2}\right)\pi, m = 0, \pm 1, \pm 2, \ldots$$

Thus we obtain the critical points

$$\left(\left(m + \frac{1}{2}\right)\pi, n\pi\right), \quad \begin{array}{l} n = 0, \pm 1, \pm 2, \ldots \\ m = 0, \pm 1, \pm 2, \ldots \end{array}$$

Consider the critical point $\left(\frac{\pi}{2}, 0\right)$. We translate the origin $(0, 0)$ to the critical point $\left(\frac{\pi}{2}, 0\right)$. Then we have

$$x = \xi + \frac{\pi}{2} \text{ and } y = \eta + 0.$$

The system becomes

$$\dot{\xi} = \sin(\eta + 0) = \sin \eta = \eta - \frac{\eta^3}{3!} + \cdots$$

$$\dot{\eta} = \cos\left(\xi + \frac{\pi}{2}\right) = -\sin \xi = -\xi + \frac{\xi^3}{3!} + \cdots$$

Neglecting higher order terms, we get the linearized system as $\dot{\xi} = \eta$ and $\dot{\eta} = -\xi$. In matrix notation we write it as

$$\begin{pmatrix} \dot{\xi} \\ \dot{\eta} \end{pmatrix} = \begin{pmatrix} 0 & 1 \\ -1 & 0 \end{pmatrix} \begin{pmatrix} \xi \\ \eta \end{pmatrix}.$$

The characteristic equation of the coefficient matrix is

$$\begin{vmatrix} -\lambda & 1 \\ -1 & -\lambda \end{vmatrix} = 0$$

or, $\quad \lambda^2 + 1 = 0 \Rightarrow \lambda = \pm i$

The critical point $\left(\frac{\pi}{2}, 0\right)$ is a center which is neutrally stable but not asymptotically.

The differential equation of the phase path is

$$\frac{dy}{dx} = \frac{\dot{y}}{\dot{x}} = \frac{\cos x}{\sin y}$$

or, $\quad \sin y \, dy - \cos x \, dx = 0$

or, $\quad \sin x + \cos y = c,$

c being an arbitrary constant. This is the required phase path of the system.

Example 3.10 Classify the equilibrium points of the system

$$\dot{x} = x - y$$
$$\dot{y} = x^2 - 1$$

Solution Let $f(x, y) = x - y$ and $g(x, y) = x^2 - 1$.

For fixed points of the system, we have

$$f(x, y) = 0 \text{ and } g(x, y) = 0$$

Now, $f(x, y) = 0 \Rightarrow x - y = 0 \Rightarrow x = y$

and $g(x, y) = 0 \Rightarrow x^2 - 1 = 0 \Rightarrow x = \pm 1.$ So, $x = y = \pm 1.$

Thus the fixed points of the system are $(1, 1)$ and $(-1, -1)$.

The Jacobian matrix of the system is

$$J(x, y) = \begin{pmatrix} \frac{\partial f}{\partial x} & \frac{\partial f}{\partial y} \\ \frac{\partial g}{\partial x} & \frac{\partial g}{\partial y} \end{pmatrix} = \begin{pmatrix} 1 & -1 \\ 2x & 0 \end{pmatrix}.$$

At the critical point $(1, 1)$,

$$J(1, 1) = \begin{pmatrix} 1 & -1 \\ 2 & 0 \end{pmatrix}$$

whose eigenvalues are $\left(1 \pm i\sqrt{7}\right)/2$, which are complex conjugate with positive real part. Therefore, the critical point $(1, 1)$ is an unstable focus.

Again at the equilibrium point $(-1, -1)$, the Jacobian matrix is

$$J(-1, -1) = \begin{pmatrix} 1 & -1 \\ -1 & 0 \end{pmatrix}$$

which has two real, distinct eigenvalues with opposite signs. Hence the equilibrium point $(-1, -1)$ is saddle.

Example 3.11 Consider a nonlinear system

$$\dot{x} = 1 - (a+1)x + bx^2 y$$
$$\dot{y} = ax - bx^2 y$$

where a and b are positive parameters.

 (i) Show that $(1, a/b)$ is the only critical point of the system,
(ii) linearize the system about this critical point.

Solution Let $f(x, y) = 1 - (a+1)x + bx^2 y$ and $g(x, y) = ax - bx^2 y$
 (i) For critical points, we have $f(x, y) = 0$ and $g(x, y) = 0$
Now,

$$f(x, y) = 0 \Rightarrow 1 - (a+1)x + bx^2 y = 0$$

Again, $g(x, y) = 0 \Rightarrow ax - bx^2 y = 0 \Rightarrow$ either $x = 0$ or $xy = a/b$. But x cannot be zero, which implies $xy = a/b$. From above, we get

$$1 - (a+1)x + ax = 0 \Rightarrow 1 - x = 0 \Rightarrow x = 1.$$

Using this value of x, we get $y = a/b$. This shows that $(1, a/b)$ is the only critical point of the system.
 (ii) We calculate

$$\frac{\partial f}{\partial x} = -(a+1) + 2bxy, \frac{\partial f}{\partial y} = bx^2, \frac{\partial g}{\partial x} = a - 2bxy, \frac{\partial g}{\partial y} = -bx^2$$

At the fixed point $\left(1, \frac{a}{b}\right)$,

$$\frac{\partial f}{\partial x} = -(a+1) + 2a = a - 1, \frac{\partial f}{\partial y} = b, \frac{\partial g}{\partial x} = a - 2a = -a, \frac{\partial g}{\partial y} = -b$$

So, the linearized system about the fixed point $(1, a/b)$ is

$$\begin{pmatrix} \dot{x} \\ \dot{y} \end{pmatrix} = \begin{pmatrix} a-1 & b \\ -a & -b \end{pmatrix} \begin{pmatrix} x \\ y \end{pmatrix}$$

or, $\dot{x} = (a-1)x + by,\ \dot{y} = -ax - by$

Example 3.12 Find all fixed points of the system

$$\dot{x} = x(y-1)$$
$$\dot{y} = 3x - 2y + x^2 - 2y^2$$

Linearize the system about the fixed point (0, 0). Comment about the stability around this fixed point.

Solution Let $P(x,y) = x(y-1)$ and $Q(x,y) = 3x - 2y + x^2 - 2y^2$
For fixed points, we have

$$P(x,y) = 0 \text{ and } Q(x,y) = 0.$$
$$\text{Now}, P(x,y) = 0 \Rightarrow x(y-1) = 0 \Rightarrow x = 0 \text{ or } y = 1$$
$$\text{and}, Q(x,y) = 0 \Rightarrow 3x - 2y + x^2 - 2y^2 = 0$$

When $x = 0$, we get $y^2 + y = 0 \Rightarrow y = 0, -1$. Similarly, when $y = 1$, we get $x = 1, -4$. So, the fixed points of the system are $(0,0), (0,-1), (1,1)$, and $(-4,1)$. At the fixed point $(0,0)$, $\frac{\partial P}{\partial x} = -1, \frac{\partial P}{\partial y} = 0, \frac{\partial Q}{\partial x} = 3, \frac{\partial Q}{\partial y} = -2$. So, the corresponding linearized system about the fixed point $(0,0)$ is

$$\left.\begin{array}{c} \dot{x} = -x \\ \dot{y} = 3x - 2y \end{array}\right\}$$

which can also be written as $\underset{\sim}{\dot{x}} = A\underset{\sim}{x}$, where $A = \begin{pmatrix} -1 & 0 \\ 3 & -2 \end{pmatrix}$. The characteristic equation of A yields

$$\begin{vmatrix} -1-\lambda & 0 \\ 3 & -2-\lambda \end{vmatrix} = 0$$
$$\Rightarrow\ (\lambda+1)(\lambda+2) = 0 \Rightarrow \lambda = -1, -2$$

The eigenvalues of A are $\lambda_1 = -2, \lambda_2 = -1$. Since $\lambda_1 < \lambda_2 < 0$, the fixed point $(0, 0)$ is a node which is asymptotically stable.

3.7 Exercises

1. Locate the equilibrium points and find the equations for phase paths of the following two-dimensional systems:

   (a) $\dot{x} = x+y, \dot{y} = x - y + 1$
   (b) $\dot{x} = y^3, \dot{y} = x^3$
   (c) $\dot{x} = 1 - y, \dot{y} = x^2 - y^2$

2. Find the Jordan forms of the following matrices:

   (i) $\begin{bmatrix} 2 & 1 \\ -2 & 4 \end{bmatrix}$, (ii) $\begin{bmatrix} 3 & 1 \\ -2 & 0 \end{bmatrix}$, (iii) $\begin{bmatrix} 2 & 1 \\ -2 & 4 \end{bmatrix}$, (iv) $\begin{bmatrix} 3 & -4 \\ 1 & -1 \end{bmatrix}$, (v) $\begin{bmatrix} 10 & -1 \\ 25 & 2 \end{bmatrix}$.

3. Find the nature of the critical points of the following linear systems:

   (a) $\dot{x} = -3x+y, \dot{y} = x - 3y$
   (b) $\dot{x} = -x - 2y, \dot{y} = 4x - 5y$
   (c) $\dot{x} = 5x+2y, \dot{y} = -17x - 5y$
   (d) $\dot{x} = -3x+4y, \dot{y} = -2x+3y$
   (e) $\dot{x} = -4x - y, \dot{y} = x - 2y$
   (f) $\dot{x} = 3x - 4y, \dot{y} = x - y$

4. Solve the linear system $\ddot{x} + \frac{1}{9}x = 0$. Find trajectories and sketch some of them.

5. (a) Show that the improper node for the system $\dot{x} = ax + by, \dot{y} = cx + dy$ becomes a star node if $a = d, b = c = 0$; (b) give an example with justification of a critical point which is always stable but neither a positive attractor nor a negative attractor.

6. Obtain the equilibrium points of the following nonlinear systems and classify them according to their linear approximations.

   (a) $\dot{x} = \sin y, \dot{y} = x + x^3$,
   (b) $\dot{x} = -2x - y + 2, \dot{y} = xy$
   (c) $\dot{x} = 1 - xy, \dot{y} = x - y^3$
   (d) $\dot{x} = x(5 - x - 2y), \dot{y} = y(1 - x - y)$
   (e) $\dot{x} = -6y + 2xy - 8, \dot{y} = y^2 - x^2$
   (f) $\ddot{x} + \text{sgn}(x) + x^2 = 0$
   (g) $\ddot{x} + \sin x = 0$
   (h) $\ddot{x} + x^3 = 0$
   (i) $\ddot{x} + x^{1/3} = 0$
   (j) $\ddot{x} + \mu x + x^2 = 0, \mu \in \mathbb{R}$
   (k) $\ddot{x} + \mu x + x^3 = 0, \mu \in \mathbb{R}$

7. Draw the phase path of the following systems:

   (a) $\ddot{x} + \frac{1}{2}\text{sgn}(\dot{x})|x| + x = 0$
   (b) $\ddot{x} + \frac{1}{2}|\dot{x}|\dot{x} + x^3 = 0$

(c) $\ddot{x} + (2x^2 + \dot{x}^2 - 2)\dot{x} + 2x = 0$

(d) $\dot{x} = -x + y - x^2 + xy, \dot{y} = x + y - x^2 - xy.$

8. Obtain three consecutive equilibrium points and their stabilities for the non-linear simple pendulum $\ddot{x} - \omega^2 \sin x = 0$. Hence, draw the phase diagram for different values of ω.

9. Define evolution operator in $\mathbb{R}^2$. Find the evolution operator for the simple harmonic oscillators:

(a) $\ddot{x} + \omega^2 x = 0$

(b) $\ddot{x} - \omega^2 x = 0$

10. (a) Find the evolution operators of the following linear systems and find the intervals of the time t for which each operator is defined. Also, verify that
$\phi_t(\phi_s(\underset{\sim}{x})) = \phi_{t+s}(\underset{\sim}{x})$

   i. $\dot{x} = x, \dot{y} = x + y$
   ii. $\ddot{x} + a\dot{x} + b = 0, a, b > 0.$

(b) Find the evolution operator for the nonlinear pendulum problems
   (i) $\ddot{\theta} + \omega^2 \sin \theta = 0$, (ii) $\ddot{\theta} + \alpha\dot{\theta} + \omega^2 \sin \theta = 0, \alpha, \omega > 0.$

11. Convert the nonlinear system

$$\dot{x} = -y + x\{1 - (x^2 + y^2)\}, \quad \dot{y} = x + y\{1 - (x^2 + y^2)\}$$

into a polar form and then find the flow evolution operator of the system.

12. Consider the linear system

$$\dot{\underset{\sim}{x}} = A\underset{\sim}{x}, \text{ where } \underset{\sim}{x} = \begin{pmatrix} x \\ y \end{pmatrix} \text{ and } A = \begin{pmatrix} a & b \\ c & d \end{pmatrix}$$

where the eigenvalues λ_1, λ_2 of the matrix A are real and satisfy the condition $\lambda_1 < \lambda_2 < 0$. Draw the phase diagram. Also, mention the nature of the critical point.

13. Find the nature of the critical point $(0, 0)$ for the linear system $\dot{\underset{\sim}{x}} = A\underset{\sim}{x}$, where

$$A = \begin{pmatrix} \lambda & 0 \\ 0 & \lambda \end{pmatrix}, \underset{\sim}{x} = \begin{pmatrix} x \\ y \end{pmatrix} \text{ and } \lambda \neq 0, \text{ the eigenvalue of the matrix } A. \text{ Also,}$$

sketch the phase diagrams for $\lambda > 0$ and $\lambda < 0$.

14. Transform the following system of equations

$$\dot{x} = \alpha x + \beta y$$
$$\dot{y} = -\beta x + \alpha y$$

into the polar form and hence solve it, where α, β are real numbers. Draw the phase diagram for different signs of α, β and mention the geometrical interpretations.

15. Find the nature and stability property of the equilibrium points of the system

$$\dot{x} = -ax + y, \dot{y} = -x - ay$$

for different values of the parameter a.

16. Consider the system

$$\dot{x} = x(a + 1 + y) - 2a$$
$$\dot{y} = y(a + 4 - 2x) + a$$

where a is a real parameter.

(i)   Show that $(2, -1)$ is a critical point of the system.
(ii) Show that for some b, the Jacobian matrix for the system at the critical point is $(2, -1)$ is $J(2, -1) = \begin{bmatrix} a & b \\ b & a \end{bmatrix}$.
(iii) For what values of a and b is $(2, -1)$ stable for the linearization?

17. Find all equilibrium points of the system

$$\dot{x} = x(1 - x - ay), \dot{y} = by(x - y)$$

where a, b are positive constants. Determine their stabilities and also sketch the phase portraits.

18. Show that the system $\ddot{\theta} + (g/L) \sin \theta = 0$ is a conservative in $(\theta - \dot{\theta})$ plane and study the stability of its critical points. Also, find α-, ω-limits for the system.

19. Find the solutions of the forced harmonic oscillator problems
(i) $\ddot{x} + x = f \cos \omega t$ (ii) $\ddot{x} + \dot{x} + x = f \cos \omega t$,   (iii) $\ddot{x} + \dot{x} + x = f_1 \cos \omega_i t + f_2 \cos \omega_2 t$ (iv) $\ddot{x} + \dot{x} + \in x \cos \omega t = 0$, $\in \ll 1$, where f, f_1, f_2 are constants and $\omega, \omega_1, \omega_2$ are forcing frequencies.

20. Find all critical points of the system

$$\left. \begin{array}{l} \dot{x} = -x + y - x^2 + xy \\ \dot{y} = x + y - x^2 - xy \end{array} \right\}.$$

Also, sketch a phase portrait that includes all critical points of the system.

21. Find all equilibrium points of the system $\dot{x} = 2x - xy, \dot{y} = 2x^2 - y$ and discuss their qualitative behaviors. Also, draw the phase portrait of the system.

22. Show that the plane autonomous system $\dot{x} = x - y, \dot{y} = 4x^2 + 2y^2 - 6$ has critical points at $(1, 1)$ and $(-1, -1)$. Also, show that they are unstable.

23. Is the equilibrium point $(0, 0)$ of the system of equations

$$\left. \begin{array}{l} \dot{x} = 3x \\ \dot{y} = x + 2y \end{array} \right\}$$

stable? Justify your answer.

24. Consider the system

$$\dot{x} = -x + \alpha y - \beta xy + y^2$$
$$\dot{y} = -(\alpha + \beta)x + \beta x^2 - xy$$

where $\alpha > 0$ and $\beta \neq 0$.

(a) Find all critical points of the system.
(b) Determine the type of each isolated critical point, for all values of the parameters $\alpha(> 0)$ and $\beta(\neq 0)$.
(c) Draw the phase portrait of the system for $\alpha = 1$, $\beta = -2$ and discuss the qualitative behavior of the system.

25. Consider the system $\dot{x} = ax + by, \dot{y} = cx + dy$ with $ad - bc = 0$. Show that the system has infinitely many non-isolated equilibrium points. Also, determine the phase paths of the system.

26. Consider the system $\dot{x} = y, \dot{y} = a$, where the control parameter a can take the values ± 1.

(a) Draw the phase portraits for $a = 1$ as well as for $a = -1$.
(b) Superpose the two phase portraits and develop a strategy for switching the control between ± 1 so that any point in the phase plane can be moved to the origin in finite time.

27. What do you mean by linearization of a nonlinear system? State its limitations. Consider the Brusselator system which represents the mathematical model for a chemical reaction $\dot{x} = a + x^2 y - (1 + b)x, \dot{y} = bx - x^2 y$, where a, b are nonzero parameters. Show that $(a, b/a)$ is the only critical point of the system. Linearize the system about the critical point. Discuss stability behaviors at the critical point when $a = 1$, $b = 2$. Sketch the phase portraits.

28. Consider the system

$$\dot{x} = -x - \frac{y}{\ln \sqrt{x^2 + y^2}}, \dot{y} = -y + \frac{x}{\ln \sqrt{x^2 + y^2}}$$

(a) Linearize the system about the equilibrium point origin of the system and show that the origin is a stable node.
(b) Find the phase portrait of the nonlinear system near the origin and then show that the portrait represents a stable focus.
(c) State reason for this.

29. Define α and ω limit sets in $\mathbb{R}^2$. Find α and ω limits for $\dot{\theta} + \omega^2 \sin\theta = 0$, $\omega > 0$.

30. Prove that the nonlinear equation $2f''' = -ff''$ is transformed to $\dot{x} = x(1 + x + y)$, $\dot{y} = y(2 + x - y)$ using the transformation $x = ff'/f''$, $y = f'^2/ff''$, $t = \log|f'|$. Find the fixed points and draw the phase diagram of the transformed system (this equation is known as Blasius (*Hermann Blasius* 1908) equation for boundary layer flow of an incompressible fluid over a flat plate).

References

1. Arrowsmith, D.K., Place, L.M.: Dynamical Systems: Differential equations, Maps, and Chaotic Behaviour. Chapman and Hall/CRC (1992)
2. Perko, L.: Differential Equations and Dynamical Systems, 3rd edn. Springer, New York (2001)
3. Andronov, A.A., Leontovitch, E.A., Gordon, I.I., Maier, A.G.: Qualitative Theory of Second-Order Dynamical Systems. Willey, New York (1973)
4. Strogatz, S.H.: Nonlinear Dynamics and Chaos with Application to Physics, Biology, Chemistry and Engineering. Perseus Books, L.L.C, Massachusetts (1994)
5. Jordan, D.W., Smith, P.: Non-linear Ordinary Differential Equations. Oxford University Press (2007)
6. Lakshmanan, M., Rajasekar, S.: Nonlinear Dynamics: Integrability. Springer, Chaos and Patterns (2003)
7. Irwin, M.C.: Smooth Dynamical Systems. World Scientific (2001)

Chapter 4
Stability Theory

Stability of solutions is an important qualitative property in linear as well as non-linear systems. The objective of this chapter is to introduce various methods for analyzing stability of a system. In fact, stability of a system plays a crucial role in the dynamics of the system. In the context of differential equations rigorous mathematical definitions are often too restrictive in analyzing the stability of solutions. Different kinds of methods on stability were developed in the theory of differential equations. We begin with the stability analysis of linear systems. Stability theory originates from the classical mechanics, the laws of statics and dynamics. The ideas in mechanics had been enriched by many mathematicians and physicists like *Evangelista Torricelli* (1608–1647), *Christiaan Huygens* (1629–1695), *Joseph-Louis Lagrange* (1736–1813), *Henri Poincaré* (1854–1912), and others. In the beginning of the twentieth century the principles of stability in mechanics were generalized by the Russian mathematician *A.M. Lyapunov* (1857–1918). There are many stability theories in the literature but we will discuss a few of them in this chapter which are practically the most useful.

4.1 Stability of Linear Systems

This section describes the stability analysis of a linear system of homogeneous first-order differential equations. The systems with constant coefficients can be written as

$$\dot{x}_i = \sum_{j=1}^{n} a_{ij} x_j; \quad i = 1, 2, \ldots, n \qquad (4.1)$$

where $a_{ij}(i, j = 1, 2, \ldots, n)$ are constants. In matrix notation, (4.1) can be written as

$$\dot{\underset{\sim}{x}} = A \underset{\sim}{x} \qquad (4.2)$$

© Springer India 2015
G.C. Layek, *An Introduction to Dynamical Systems and Chaos*,
DOI 10.1007/978-81-322-2556-0_4

where A is an $n \times n$ matrix and $\underset{\sim}{x} = (x_1, x_2, \ldots, x_n)^t$ is a column vector. The characteristic equation of (4.2) is $\det(A - \lambda I) = 0$. Depending upon the roots of the characteristic equation the following cases may arise for stability of solutions of (4.2):

(i) If all the roots of the characteristic equation of (4.2) have negative real part, then all solutions of (4.2) are asymptotically stable. Moreover, the solutions tend to the equilibrium point origin as $t \to \infty$;

(ii) If at least one root of the characteristic equation has a positive real part, then all solutions are unstable;

(iii) If the characteristic equation has simple roots, purely imaginary or zero and the other roots exist and have a negative real part, then all solutions of the system are stable, but not asymptotically.

In case of nonhomogeneous linear systems we prove the following theorem for stability.

Theorem 4.1 *The solutions of the nonhomogeneous linear system* $\dot{x}_i = \sum_{j=1}^{n} a_{ij}(t)x_j + b_i(t)$; $i = 1, 2, \ldots, n$ *are all simultaneously either stable or unstable.*

Proof Let $\underset{\sim}{f}(t) = (f_1(t), f_2(t), \ldots, f_n(t))$ be any particular solution of the nonhomogeneous linear system

$$\dot{x}_i = \sum_{j=1}^{n} a_{ij}(t)x_j + b_i(t), \ i = 1, 2, \ldots, n \qquad (4.3)$$

Consider the transformation $y_i(t) = x_i(t) - f_i(t), i = 1, 2, \ldots, n$, which transforms the particular solution $\underset{\sim}{f}(t)$ of (4.3) into a trivial solution. Applying this transformation to (4.3), we get the homogeneous linear system

$$\dot{y}_i = \sum_{j=1}^{n} a_{ij}(t)y_j(t), \ i = 1, 2, \ldots, n \qquad (4.4)$$

Thus any particular solution of (4.3) has the same stability behavior as that of the trivial solution of (4.4). Suppose that the trivial solution of (4.4) is stable. Then by definition of stability, for any $\varepsilon > 0$ there is a $\delta = \delta(\varepsilon) > 0$ such that for every other solution $y_i, i = 1, 2, \ldots, n$ of (4.4),

$$|y_i(t_0) - 0| < \delta \Rightarrow |y_i(t) - 0| < \varepsilon \ \forall t \geq t_0.$$

Substituting $y_i(t) = x_i(t) - f_i(t), i = 1, 2, \ldots, n$, we see that for every solution $x_i(t), i = 1, 2, \ldots, n$ of (4.3),

$$|x_i(t_0) - f_i(t_0)| < \delta \Rightarrow |x_i(t) - f_i(t)| < \varepsilon \ \forall t \geq t_0.$$

This implies that the particular solution $f_i(t), i = 1, 2, \ldots, n$ of (4.3) is stable. One can prove the instability of the particular solution similarly. This completes the proof.

4.2 Methods for Stability Analysis

There does not exist a single method which will suffice for stability analysis of a system. We begin with the Lyapunov stability analysis.

(I) Lyapunov method

First, we shall explain Lyapunov method with respect to equilibrium points of a system. Let $\underset{\sim}{x}^*$ be the equilibrium point of a nonlinear system $\underset{\sim}{x}^* = f(\underset{\sim}{x}), \ \underset{\sim}{x} \in \mathbb{R}^n$.

If any orbit that passes close to the equilibrium point stays close to it for all time, then we say that the equilibrium point $\underset{\sim}{x}^*$ is Lyapunov stable. Mathematically, it is defined as follows:

An equilibrium point $\underset{\sim}{x}^*$ of a system $\dot{\underset{\sim}{x}} = f(\underset{\sim}{x}), \underset{\sim}{x} \in \mathbb{R}^n$ is said to be *Lyapunov stable* if and only if for any $\varepsilon > 0$ there exists a $\delta(\varepsilon) > 0$ such that the orbit $\phi\left(t, \underset{\sim}{x}\right)$ of the system satisfies the following relation:

$$\|\underset{\sim}{x} - \underset{\sim}{x}^*\| < \delta \Rightarrow \|\phi(t, \underset{\sim}{x}) - \underset{\sim}{x}^*\| < \varepsilon, \ \forall t \geq 0.$$

(Starts near $\underset{\sim}{x}^*$) (Stayed nearby orbit)

The equilibrium point $\underset{\sim}{x}^*$ is said to be *asymptotically stable* if

 (i) it is stable, and
(ii) the orbit $\phi(t, \underset{\sim}{x})$ approaches to $\underset{\sim}{x}^*$ as $t \to \infty$.

Thus, for asymptotically stable equilibrium point we can find a $\delta > 0$ such that

$$\|\underset{\sim}{x} - \underset{\sim}{x}^*\| < \delta \Rightarrow \|\phi(t, \underset{\sim}{x}) - \underset{\sim}{x}^*\| \to 0 \text{ as } t \to \infty.$$

For an asymptotic stable equilibrium point $\underset{\sim}{x}^*$, the set $D(\underset{\sim}{x}^*) = \{\underset{\sim}{x} \in \mathbb{R}^n | \lim_{t \to \infty} \|\phi(t, \underset{\sim}{x}) - \underset{\sim}{x}^*\| = 0\}$ is called the domain of asymptotic stability of $\underset{\sim}{x}^*$. If $D = \mathbb{R}^n$, then $\underset{\sim}{x}^*$ is globally stable (asymptotically).

An equilibrium point x^* which satisfies only the condition (ii) of the definition of asymptotic stability is called *quasi-asymptotically stable*. An equilibrium point which is not stable is said to be *unstable*. The diagrammatic representations of Lyapunov, asymptotic, and quasi-asymptotic stabilities about the equilibrium point are shown in Fig. 4.1a–c.

The solution $u(t)$ of a system is said to be **uniformly** stable if there exists a $\delta(\varepsilon) > 0$ for all $\varepsilon > 0$ such that for any other solution $v(t)$, the inequality $|u(t_0) - v(t_0)| < \delta$ implies $|u(t) - v(t)| < \varepsilon$ for all $t \geq t_0$. The solution $u(t)$ is said to be unstable when no such δ exists. Again, a stable solution $u(t)$ is said to be asymptotically stable if $|u(t) - v(t)| \to 0$ as $t \to \infty$. From this stability criterion we see that the Lyapunov stability condition is quite restrictive. The two neighboring solutions remain close to each other at the same time. We now discuss few less restrictive stability methods below.

(II) **Poincaré method**

This stability criterion is related with different time scales, say t' and t. Let Γ' and Γ be two orbits represented by $x(t)$ and $y(t)$, respectively, for all t. The orbit Γ is orbitally stable if for any $\varepsilon > 0$, there exists $\delta(\varepsilon) > 0$ such that if $\|x(0) - y(\tau)\| < \delta$ for some time τ, then there exists $t'(t)$ such that $\|x(t) - y(t')\| < \varepsilon$, $\forall t > 0$. The orbit is said to be asymptotically stable if the orbit Γ' tends toward Γ as $t \to \infty$. This is the most significant test for stability analysis but it is very difficult to establish mathematically.

(III) **Lagrange method**

This is a simple criterion for stability analysis. The solutions of the system $\dot{x} = f(x, t)$ are said to be bounded stable if $\|x(t)\| \leq M < \infty$, $\forall t$. This is also known as bounded stability.

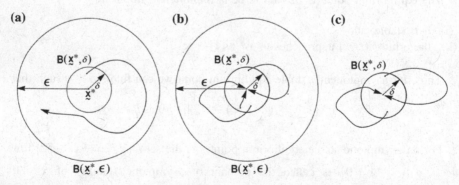

Fig. 4.1 **a** Lyapunov stability, **b** asymptotic stability, and **c** quasi-asymptotic stability of an equilibrium point x^*

(IV) Lyapunov's direct method

The Russian mathematician A.M. Lyapunov generalized the stability conditions which are used in analyzing the stability of a system, in particular the stability in the neighborhood of equilibrium points of a system. This is known as Lyapunov's second method or direct method for stability. He nicely introduced a scalar function, $L(\underset{\sim}{x})$ later called it the Lyapunov function, such that $L(\underset{\sim}{x}^{*}) = 0$ and $L(\underset{\sim}{x}) > 0$ when $\underset{\sim}{x} \neq \underset{\sim}{x}^{*}$ in the neighborhood of $\underset{\sim}{x}^{*}$, the equilibrium point of the system $\dot{\underset{\sim}{x}} = f(\underset{\sim}{x})$. The function $L \equiv L(x_1, x_2, \ldots, x_n)$ is said to be positively (resp. negatively) definite in a domain $D \subset \mathbb{R}^n$ if $L(\underset{\sim}{x}) > 0$ (resp. <0) for all $\underset{\sim}{x} \in D, \underset{\sim}{x} \neq \underset{\sim}{0}$. Similarly, L is called positively (resp. negatively) semi-definite in D if $L(\underset{\sim}{x}) \geq 0$ (resp. ≤ 0) for all $\underset{\sim}{x} \in D$. When the function $L(t, \underset{\sim}{x})$ depends explicitly on time t, these definitions can be redefined as follows:

The function $L(t, \underset{\sim}{x})$ is said to be positively (resp. negatively) definite in D if there exists a function $G(\underset{\sim}{x})$ in D such that $G(\underset{\sim}{x})$ is continuous in D, $G(\underset{\sim}{0}) = 0$ and $0 < G(\underset{\sim}{x}) \leq L(t, \underset{\sim}{x})$ (resp. $L(t, \underset{\sim}{x}) \leq G(\underset{\sim}{x}) < 0$) for all $\underset{\sim}{x} \in D \setminus \{0\}, t \geq t_0$. Similarly, the semi-definite functions can be defined. The total derivative or orbital derivative of L in the direction of the vector field $f(\underset{\sim}{x})$ is defined as

$$\frac{dL}{dt} = \underset{\sim}{f} \cdot \nabla L = f(\underset{\sim}{x}) \frac{\partial L}{\partial \underset{\sim}{x}}.$$

Let $D \subseteq \mathbb{R}^n$ be an open neighborhood of the equilibrium point $\underset{\sim}{x}^{*}$. Then the function $L : D \to \mathbb{R}$, satisfying the following properties:

(i) L is continuously differentiable,

(ii) $L > 0$ for all $\underset{\sim}{x} \in D \setminus \{\underset{\sim}{x}^{*}\}$ and $L\left(\underset{\sim}{x}^{*}\right) = 0$,

is called a Lyapunov function. Moreover, if $\frac{dL}{dt} \leq 0$ in D, then $\underset{\sim}{x}^{*}$ is stable. This condition implies that the point $\underset{\sim}{x}(t)$ moves along a path where $L(\underset{\sim}{x})$ does not increase. Hence, $\underset{\sim}{x}(t)$ will remain close to the point $\underset{\sim}{x}^{*}$ and come to $\underset{\sim}{x}^{*}$ if $\frac{dL}{dt} = 0$. There is no systematic procedure to deduct the Lyapunov function $L(\underset{\sim}{x})$. However, in case of conservative system it L is the energy of the system. In fact, Lyapunov constructed this function on the basis of the principle of energy in mechanics.

Theorem 4.2 (*Lyapunov theorem*) *Suppose that the origin is an equilibrium point of $\dot{\underset{\sim}{x}} = f(\underset{\sim}{x}), \underset{\sim}{x} \in \mathbb{R}^n$ and let $L = L(x_1, x_2, \ldots, x_n)$ be a Lyapunov function in a neighborhood D of the origin. If*

(i) *the orbital derivative $\dot{L} \leq 0$ in D, that is, if $\dot{L}$ is negative semi-definite in D, the origin is stable,*

(ii) *$\dot{L} < 0$ in $D \setminus \{0\}$, that is, if $\dot{L}$ negative definite in D, then the origin is asymptotically stable,*

(iii) *$\dot{L} > 0$, that is, positive definite in D, the origin is unstable.*

For proof see Hartmann [1].

For an application of the theorem we illustrate the stability of pendulum problem.

Simple undamped pendulum: Consider the simple pendulum problem governed by the equation

$$ml\ddot{\theta} = -mg\sin\theta,$$

$$\text{that is, } \ddot{\theta} = -\left(\frac{g}{l}\right)\sin\theta \qquad (4.5)$$

in which a bob of mass m is suspended from a light string of length l, where θ represents the angle between the string and the vertical axis at some instant t, and g is the acceleration due to gravity. With $x = \theta$ and $y = \dot{\theta}$, we can rewrite the Eq. (4.5) as a system of equations

$$\left.\begin{array}{l} \dot{x} = y \\ \dot{y} = -\left(\frac{g}{l}\right)\sin x \end{array}\right\} \qquad (4.6)$$

Consider the function $L(x,y) = \frac{1}{2}ml^2y^2 + mgl(1 - \cos x)$, $(x,y) \in \mathbb{R}^2$, which is simply the total energy of the system. Let $G = \{(x,y) \in \mathbb{R}^2 : -2\pi < x < 2\pi\}$. We see that $L(0,0) = 0$ and $L > 0$ in $G\setminus\{(0,0)\}$. Therefore, L is positive definite in G. We now calculate the derivate $\dot{L}$ of L along the trajectory of (4.6) as

$$\dot{L} = \frac{dL}{dt} = \frac{\partial L}{\partial x}\dot{x} + \frac{\partial L}{\partial y}\dot{y} = [mgl\sin x]y + ml^2y\left[-\left(\frac{g}{l}\right)\sin x\right] = 0$$

Thus, conditions of Theorem 4.2 are satisfied. Hence, the fixed point origin is stable. Note that the origin is not asymptotically stable, since $\dot{L} \equiv 0$.

Damped pendulum: Consider the damped pendulum governed by the equation

$$ml\ddot{\theta} = -mg\sin\theta - \mu l\dot{\theta} \qquad (4.7)$$

which is simply obtained by taking into account the effect of damping force (frictional force) $\mu l\dot{\theta}$, $\mu > 0$ being the coefficient of friction. As previous, with $x = \theta$ and $y = \dot{\theta}$, we can rewrite Eq. (4.7) as

$$\left.\begin{array}{l} \dot{x} = y \\ \dot{y} = -\left(\frac{g}{l}\right)\sin x - \left(\frac{\mu}{m}\right)y \end{array}\right\} \qquad (4.8)$$

The origin $O(0,0)$ is a fixed point of the system. We now determine its stability. As earlier, consider the function

$$L(x,y) = \frac{1}{2}ml^2y^2 + mgl(1 - \cos x), \quad (x,y) \in \mathbb{R}^2.$$

Then L is positive definite in G, as defined earlier. Now calculate $\dot{L}$ along the trajectory of (4.8) as follows:

$$\dot{L} = \frac{dL}{dt} = \frac{\partial L}{\partial x}\dot{x} + \frac{\partial L}{\partial y}\dot{y} = [mgl\sin x]y + ml^2 y\left[-\left(\frac{g}{l}\right)\sin x - \left(\frac{\mu}{m}\right)y\right] = -\mu l^2 y^2.$$

Now, in G we can find points, in particular $(x,y) = (\pi/2, 0)$, such that $\dot{L} = 0$. So, $\dot{L} \leq 0$ in G. Therefore, by Theorem 4.2 the fixed point origin is stable. However, the phase portrait near the origin gives some other picture: the origin is asymptotically stable (see Fig. 3.14). So, we discard this particular choice of L and consider a more general form of L as

$$L(x,y) = \frac{1}{2}ml^2\left[ax^2 + 2bxy + cy^2\right] + mgl(1 - \cos x)$$

We shall now determine the values of a, b, and c for which the origin are asymptotically stable, that is, L is positive definite and $\dot{L}$ is negative definite in some neighborhood of the origin. It can be shown that the first right-hand member in the expression of L is positive definite if and only if $a > 0$, $c > 0$, $ac - b^2 > 0$. The orbital derivative $\dot{L}$ of the Lyapunov function L is given by

$$\dot{L} = \frac{\partial L}{\partial x}\dot{x} + \frac{\partial L}{\partial y}\dot{y} = ml^2\left[ax + by + \left(\frac{g}{l}\right)\sin x\right]y + ml^2(bx + cy)\left[-\left(\frac{g}{l}\right)\sin x - \left(\frac{\mu}{m}\right)y\right]$$

$$= ml^2\left[\left\{a - b\left(\frac{\mu}{m}\right)\right\}xy + \left(\frac{g}{l}\right)(1 - c)y\sin x - \left(\frac{g}{l}\right)bx\sin x + \left\{b - c\left(\frac{\mu}{m}\right)\right\}y^2\right]$$

The right-hand side of the above expression contains two sign indefinite terms, xy and $y\sin x$. We need to discard them in our problem and it leads to the following relations:

$$a - b\left(\frac{\mu}{m}\right) = 0, \quad 1 - c = 0 \Rightarrow b = a\left(\frac{m}{\mu}\right), \quad c = 1.$$

With this choice, $\dot{L}$ takes the form

$$\dot{L} = -ml^2\left[\left(\frac{g}{l}\right)\left(\frac{m}{\mu}\right)ax\sin x + \left\{\left(\frac{\mu}{m}\right) - \left(\frac{m}{\mu}\right)a\right\}y^2\right].$$

To make $\dot{L}$ negative definite, we must have

$$\left(\frac{\mu}{m}\right) - \left(\frac{m}{\mu}\right)a > 0 \Rightarrow 0 < a < \left(\frac{\mu}{m}\right)^2.$$

Then $0 < b < \left(\frac{\mu}{m}\right)$. Now, the term $x \sin x > 0$ if all $x : -\pi < x < \pi$ with $x \neq 0$. Let $N = \{(x,y) \in \mathbb{R}^2 : -\pi < x < \pi\}$. Then L is positive definite and $\dot{L}$ is negative definite in N. Therefore, by Theorem 4.2 the fixed point origin is asymptotically stable, as required.

Theorem 4.3 *Consider a nonautonomous system* $\dot{\underset{\sim}{x}} = f(t, \underset{\sim}{x})$ *with* $f(t, \underset{\sim}{0}) = \underset{\sim}{0}$, $\underset{\sim}{x} \in D \subseteq \mathbb{R}^n$, *and* $t \geq t_0$. *The Lyapunov function* $L(t, \underset{\sim}{x})$ *is defined in a neighborhood of the origin and positively definite for* $t \geq t_0$. *Then*

   (i) *if the orbital derivative is negatively semi-definite, the solution is stable;*
  (ii) *if the orbital derivative is negative definite, the solution is asymptotically stable; and*
 (iii) *if the orbital derivative in positive definite, the solution is unstable.*

Example 4.1 Show that the solution of the autonomous system $\dot{x} = y$, $\dot{y} = -x$ with $x(0) = 0$, $y(0) = 0$ is stable in the sense of Lyapunov.

Solution The solution of the system with $x(0) = x_0$, $y(0) = y_0$ is given as

$$x(t) = x_0 \cos t + y_0 \sin t, \quad y(t) = -x_0 \sin t + y_0 \cos t$$

and the solution subject to the given initial condition is $x(t) = 0$, $y(t) = 0$. Choose an arbitrary real $\varepsilon > 0$. We have to find a $\delta(\varepsilon) > 0$ such that for $|x_0 - 0| < \delta$ and $|y_0 - 0| < \delta$,

$$|x(t) - 0| = |x_0 \cos t + y_0 \sin t| < \varepsilon, \text{ and } |y(t) - 0| = |-x_0 \sin t + y_0 \cos t| < \varepsilon$$

hold for all $t \geq 0$. We see that

$$|x_0 \cos t + y_0 \sin t| \leq |x_0 \cos t| + |y_0 \sin t| \leq |x_0| + |y_0|.$$

Similarly, $|-x_0 \sin t + y_0 \cos t| \leq |x_0| + |y_0|$. Take $\delta = \varepsilon/2$. This gives

$$\text{for } |x_0| < \delta \text{ and } |y_0| < \delta$$
$$\Rightarrow |x_0 \cos t + y_0 \sin t| < \varepsilon/2 + \varepsilon/2 = \varepsilon, \ \forall t \geq 0.$$

Hence, the solution $x(t) = 0$, $y(t) = 0$ is stable in the sense of Lyapunov but the stability is not asymptotic.

Example 4.2 Prove that each solution of the equation $\dot{x} + x = 0$ is asymptotically stable.

Solution The general solution is given by $x(t) = Ae^{-t}$, where A is an arbitrary constant. The solutions $x_1(t)$ and $x_2(t)$ of the equation that satisfy the initial conditions $x_1(t_0) = x_1^0$ and $x_2(t_0) = x_2^0$ are $x_1(t) = x_1^0 e^{-(t-t_0)}$ and $x_2(t) = x_2^0 e^{-(t-t_0)}$, respectively. We see that

$$|x_2(t) - x_1(t)| = |x_2^0 - x_1^0| e^{-(t-t_0)} \rightarrow 0 \text{ as } t \rightarrow \infty.$$

This implies that every solution of the equation is asymptotically stable.

Example 4.3 Prove that all solutions of the system $\dot{x} = \sin^2 x$ are bounded on $(-\infty, +\infty)$ but the solution $x(t) = 0$ is unstable as $t \rightarrow \infty$.

Solution Clearly, $x = n\pi$; $n = 0, \pm 1, \pm 2, \ldots$ are the obvious solutions of the given equation. Other solutions are obtained as

$$\text{cosec}^2 x \, dx = dt \Rightarrow \cot x = \cot x_0 - t \quad [\text{assuming } x(0) = x_0]$$
$$\Rightarrow x = \cot^{-1}(\cot x_0 - t), \ x_0 \neq n\pi.$$

All above solutions are bounded on $(-\infty, +\infty)$. The solution $x(t) = 0$ is unstable as $t \rightarrow \infty$, because for any $x_0 \in (0, \pi)$ we have $\lim_{t \rightarrow \infty} x(t) = \pi$. So, boundedness of solution does not imply that it is stable. Similarly, stability of a solution does not ensure that it is bounded. Thus, bounded and stability of solutions are independent properties of a system.

Example 4.4 Using suitable Lyapunov functions, examine the stabilities for the following systems: (i) $\ddot{x} + x = 0$, (ii) $\dot{x} = x$, $\dot{y} = -y$ at the origin.

Solution (i) The given system can be written as $\dot{x} = y$, $\dot{y} = -x$. The origin is the equilibrium point of the system. We take Lyapunov function as $L(x, y) = x^2 + y^2$, which is positive definite in the neighborhood of the origin and $L(0, 0) = 0$. The orbital derivative of L is given by

$$\frac{dL}{dt} = \frac{\partial L}{\partial x} \dot{x} + \frac{\partial L}{\partial y} \dot{y} = 2xy - 2xy = 0.$$

Hence, $\dfrac{dL}{dt}$ is semi-negative definite. So, the system is stable at $(0, 0)$. The phase paths of the system are obtained as

$$\frac{dy}{dx} = -\frac{x}{y} \Rightarrow x^2 + y^2 = k^2, \quad k \neq 0.$$

which represent concentric circles with center at the origin. Hence, the system is not asymptotically stable at the origin.

(ii) We take Lyapunov function $L(x,y) = x^2 - y^2$, in the neighborhood of origin, which is positive definite in arbitrarily close to $(0, 0)$ ($L > 0$ along the straight line $y = 0$) and $L(0,0) = 0$. The orbital derivative of L is

$$\frac{dL}{dt} = \frac{\partial L}{\partial x}\dot{x} + \frac{\partial L}{\partial y}\dot{y} = 2(x^2 + y^2) > 0.$$

So the equilibrium point origin is unstable. In this case, the origin is a saddle point. The path of the system is $xy = k$, k being an arbitrary constant, which is a rectangular hyperbola.

Example 4.5 Examine different stability criteria satisfied by the linear harmonic oscillator $\ddot{x} + x = 0$.

Solution The harmonic oscillator can be written as a system of differential equations as

$$\dot{x} = y, \dot{y} = -x.$$

The solution of the system is given by

$$x(t) = A\cos t + B\sin t, y(t) = -A\sin t + B\cos t,$$

where A and B are constants. Let $\varepsilon > 0$ be given. Assume $u(t) = A_1\cos t + B_1\sin t$ and $v(t) = A_2\cos t + B_2\sin t$ are two solutions of the equation, where $A_1^2 + A_2^2 \neq 0$ and $B_1^2 + B_2^2 \neq 0$. Then we get

$$\begin{aligned}|u(t) - v(t)| &= |(A_1 - A_2)\cos t + (B_1 - B_2)\sin t| \\ &\leq |A_1 - A_2||\cos t| + |B_1 - B_2||\sin t| \\ &\leq |A_1 - A_2| + |B_1 - B_2| < \varepsilon\end{aligned}$$

if $|A_1 - A_2| < \varepsilon/2$ and $|B_1 - B_2| < \varepsilon/2$. Take $\delta = \varepsilon/2$. Then, $\delta = \delta(\varepsilon) > 0$ and $|u(0) - v(0)| \leq |A_1 - A_2| < \delta$. Thus, the solution is uniformly stable.

Now, take $L(x,y) = x^2 + y^2$ as Lyapunov function, which is the energy of the harmonic oscillator. Then

$$\frac{dL}{dt} = 2x\dot{x} + 3y\dot{y} = 2xy - 2xy = 0.$$

Hence, the origin of the harmonic oscillator is stable in the sense of Lyapunov but it is not asymptotically stable. This can be shown easily that the system is orbitally stable in the sense of Poincaré but not asymptotically stable.

The solutions of the system $\dot{x} = y, \dot{y} = -x$ are given by

$$x(t) = A\cos t + B\sin t, \ y(t) = -A\sin t + B\cos t,$$

where $A, B \in \mathbb{R}$ are arbitrary constants. Then

$$\|\underset{\sim}{x}(t)\| = \sqrt{A^2 + B^2} < \infty \text{ for all } t.$$

Hence, the solutions of the harmonic oscillator have bounded stability in the sense of Lagrange.

Example 4.6 Show that the system $\dot{x} = -y(x^2 + y^2)^{1/2}, \dot{y} = x(x^2 + y^2)^{1/2}$ is orbitally stable but not Lyapunov stable.

Solution Convert the system into the polar coordinates (r, θ) using $x = r\cos\theta, y = r\sin\theta$. In (r, θ) coordinates, the system becomes

$$\dot{r} = 0, \dot{\theta} = r,$$

which has the solution

$$r = r_0, \theta = r_0 t + \theta_0,$$

where $r_0 = r(0)$ and $\theta_0 = \theta(0)$, the initial condition of the system. Therefore, the solution of the original system is given by

$$x(t) = r_0 \cos(r_0 t + \theta_0), y(t) = r_0 \sin(r_0 t + \theta_0).$$

This shows that the amplitude and frequency of the solutions depend upon r_0. Hence, the system is orbitally stable. The solutions represent concentric circles with center at the origin. Consider two neighboring points $(r_0, 0)$ and $(r_0 + \varepsilon, 0)$ on the concentric circles as two initial solutions, where ε is very small. After some time t, these two points move to $(r_0, r_0 t)$ and $(r_0, (r_0 + \varepsilon)t)$. This yields the angle difference between the solutions as

$$\Delta\theta = (r_0 + \varepsilon)t - r_0 t = \varepsilon t.$$

Hence, when $t = (2n + 1)\pi/\varepsilon, \Delta\theta = (2n + 1)\pi$, that is, the solutions are diametrically opposite to one another and in this case, the distance between them is $2r_0 + \varepsilon$. So, there always exists a time at which the solutions move further away from each other. Hence, the solution is not stable in the sense of Lyapunov.

Example 4.7 Investigate the stability of the system

$$\frac{dx}{dt} = -(x - 2y)(1 - x^2 - 3y^2)$$

$$\frac{dy}{dt} = -(y + x)(1 - x^2 - 3y^2)$$

at the fixed point origin.

Solution Take the Lyapunov function $L(x,y) = x^2 + 2y^2$. Then L is positive definite in the neighborhood of $(0, 0)$ and $L(0,0) = 0$. The orbital derivative is calculated as

$$\frac{dL}{dt} = \frac{\partial L}{\partial x}\dot{x} + \frac{\partial L}{\partial y}\dot{y} = 2x\left[-(x-2y)(1-x^2-3y^2)\right] + 4y\left[-(y+x)(1-x^2-3y^2)\right]$$

$$= -2(x^2+2y^2)(1-x^2-3y^2) < 0, \text{ in the neighbourhood of}(0,0)$$

and is equal to zero only when $x = y = 0$. So, $\dot{L}$ is negative definite, and hence the fixed point origin is asymptotically stable.

Example 4.8 Using a suitable Lyapunov function shows that the origin is an asymptotically stable equilibrium point of the system

$$\dot{x} = -2y + yz - x^3$$
$$\dot{y} = x - xz - y^3$$
$$\dot{z} = xy - z^3$$

Solution Obviously, $(0, 0, 0)$ is the equilibrium point of the system. We take $L(x,y,z) = x^2 + 2y^2 + z^2$ as a Lyapunov function for which we can test the stability of the equilibrium point origin. The orbital derivative of L is given by

$$\dot{L} = 2x\dot{x} + 4y\dot{y} + 2z\dot{z}$$

$$= 2x(-2y + yz - x^3) + 4y(x - xz - y^3) + 2z(xy - z^3)$$

$$= -4xy + 2xyz - 2x^4 + 4xy - 4xyz - 4y^4 + 2xyz - 2z^4$$

$$= -(2x^4 + 4y^4 + 2z^4) < 0, \text{ and } \dot{L} = 0 \text{ only at } (0,0,0).$$

This implies that $\dot{L}$ is negative definite for $(x,y,z) \neq (0,0,0)$. Hence, by Lyapunov theorem on stability, the origin is an asymptotically stable.

4.3   Stability of Linearized Systems

Let us consider a nonlinear system represented as

$$\dot{x} = f(x); x \in \mathbb{R}^n. \tag{4.9}$$

Without loss of generality we assume that $x = 0$ is an equilibrium point of the system. So when $\| x \| \ll 1$, we can expand $f(x)$ in the form of a Taylor series in a neighborhood of $\underset{\sim}{x} = \underset{\sim}{0}$. Neglecting second- and higher order terms. we get a linear system as

$$\dot{\underset{\sim}{x}} = A\underset{\sim}{x} \tag{4.10}$$

where $A = J(0)$, the Jacobian of the system evaluated at the origin. The linear system (4.10) is known as the linearization of the nonlinear system (4.9). An equilibrium point of a system is hyperbolic if the corresponding Jacobian matrix evaluated at the point has eigenvalues with nonzero real part. If not, then it is said to be non-hyperbolic. The flows in the neighborhood of hyperbolic fixed point retain the character under sufficiently small perturbation. On the other hand, non-hyperbolic fixed points and corresponding flows are easily changed under small perturbation. The non-hyperbolic fixed points are weak, whereas hyperbolic fixed points are robust in the context of flows. The phase portrait near a hyperbolic fixed point of a nonlinear system is topologically equivalent to the phase portrait of the corresponding linear system. This means that there is a homeomorphism which maps the local phase portrait onto the other preserving directions of trajectories. A homeomorphism is a continuous map with a continuous inverse. The flow near a hyperbolic fixed point is structurally stable. A phase portrait is said to be structurally stable if its topology does not change under an arbitrarily small perturbation to the vector field of the system. For example, the phase portrait of a saddle point (hyperbolic type) is structurally stable, whereas the center (non-hyperbolic type) is not structurally stable. By adding a small amount of damping force to the undamped pendulum equation makes, the center becomes a spiral. For hyperbolic fixed points and their flows we discuss some important theorems.

Theorem 4.4 (Hartman–Grobman) Let $\underset{\sim}{x} = \underset{\sim}{0}$ *be a hyperbolic equilibrium point of the nonlinear system* (4.9) *with* $f \in C^1$ *(continuously differentiable of order one). Then the stability type of the equilibrium point origin for the nonlinear system is same as that of the linear system* $\dot{\underset{\sim}{x}} = A\underset{\sim}{x}$, *which is the linearization of* (4.9) *in the neighborhood of* $\|\underset{\sim}{x}\| \ll 1$. *Also, there exists a homeomorphism* $H(\underset{\sim}{x})$ *which maps the orbits of the nonlinear system* (4.9) *onto the orbits of the corresponding linear system in the neighborhood of the origin.*

The Hartman–Grobman theorem gives a very important result in the local qualitative theory of a dynamical system. This theorem shows that near a hyperbolic-type equilibrium point, the nonlinear system has the same qualitative behavior (locally) as the corresponding linearized system. Also, one can find the local solution of the nonlinear system through homeomorphism.

Example 4.9 Using Hartman–Grobman theorem discuss the local stability of the equilibrium point for the system $\dot{x} = x - y^2, \dot{y} = -y$. Also, find the homeomorphic mapping.

Solution Clearly, the origin is the only equilibrium point of the system. At the origin, the Jacobian matrix of the nonlinear system is given by

$$J = \begin{pmatrix} 1 & 0 \\ 0 & -1 \end{pmatrix}.$$

The matrix has nonzero real eigenvalues $1, -1$. The origin is of hyperbolic type and it is a saddle. According to Hartman–Grobman theorem, it is a saddle-type equilibrium point of the given nonlinear system. We shall now find the homeomorphic mapping H. The solutions of the nonlinear system and the corresponding linearized system with the initial conditions $x(0) = x_0, y(0) = y_0$ are given by

$$x(t) = x_0 e^t + \frac{y_0^2}{3}(e^{-2t} - e^t), y(t) = y_0 e^{-t}$$

and $x(t) = x_0 e^t, y(t) = y_0 e^{-t},$

respectively. Therefore, the flow of the nonlinear system is

$$\varphi_t(x, y) = \begin{pmatrix} xe^t + \frac{y^2}{3}(e^{-2t} - e^t) \\ ye^{-t} \end{pmatrix}.$$

and the flow of the linear system is

$$e^{At} = \begin{pmatrix} e^t & 0 \\ 0 & e^{-t} \end{pmatrix}.$$

Now consider the map

$$H(x, y) = \begin{pmatrix} x - \frac{y^2}{3} \\ y \end{pmatrix}.$$

Clearly, H is continuous, and $H^{-1}(x, y) = (x + \frac{y^2}{3})$ exists and also continuous,

that is, the mapping H is a homeomorphism. Now, for all $(x, y) \in \mathbb{R}^2$ and for all $t \geq 0$, we see that

$$H(\varphi_t(x,y)) = H\begin{pmatrix} xe^t + \dfrac{y^2}{3}(e^{-2t} - e^t) \\ ye^{-t} \end{pmatrix}$$

$$= \begin{pmatrix} xe^t + \dfrac{y^2}{3}(e^{-2t} - e^t) - \dfrac{y^2}{3}e^{-2t} \\ ye^{-t} \end{pmatrix}$$

$$= \begin{pmatrix} xe^t - \dfrac{y^2}{3}e^t \\ ye^{-t} \end{pmatrix}$$

$$= \begin{pmatrix} e^t & 0 \\ 0 & e^{-t} \end{pmatrix}\begin{pmatrix} x - \dfrac{y^2}{3} \\ y \end{pmatrix}$$

$$= e^{At}H(x,y).$$

Therefore, $H \circ \varphi_t = e^{At} \circ H \ \forall t \geq 0$. This relation shows that the two flows are connected by the mapping H.

4.4 Topological Equivalence and Conjugacy

Two autonomous systems are said to be topologically equivalent in a neighborhood of the origin if there exists a homeomorphism $H : U \to V$, where U and V are two open sets containing the origin, such that the trajectories of nonlinear system (4.9) in U are mapped onto the trajectories of the corresponding linear system (4.10) in V and preserve their orientation by time in the sense that if a trajectory is directed from $\underset{\sim}{x}_1$ to $\underset{\sim}{x}_2$ in U, then its image is directed from $H(\underset{\sim}{x}_1)$ to $H(\underset{\sim}{x}_2)$ in V. If the homeomorphism H preserves the parameterization by time, then the systems (4.9) and (4.10) are said to be topologically conjugate in a neighborhood of the origin.

The following theorem is very useful for topologically equivalent of two linear systems.

Theorem 4.5 *Two linear systems $\dot{\underset{\sim}{x}} = A\underset{\sim}{x}$ and $\dot{\underset{\sim}{y}} = B\underset{\sim}{y}$, whose all eigenvalues have nonzero real parts, are topologically equivalent if and only if the number of eigenvalues with positive (and corresponding negative) real parts are the same for both the systems (see Arnold [2]).*

Example 4.10 Show that the systems $\dot{\underset{\sim}{x}} = A\underset{\sim}{x}$ and $\dot{\underset{\sim}{y}} = B\underset{\sim}{y}$ where $A = \begin{pmatrix} -2 & -5 \\ -5 & -2 \end{pmatrix}$ and $B = \begin{pmatrix} 3 & 0 \\ 0 & -7 \end{pmatrix}$ are topologically conjugate.

Solution Consider the map $H(\underset{\sim}{x}) = C\underset{\sim}{x}$, where

$$C = \frac{1}{\sqrt{2}} \begin{pmatrix} 1 & -1 \\ 1 & 1 \end{pmatrix}.$$

Clearly, the matrix C is invertible with the inverse

$$C^{-1} = \frac{1}{\sqrt{2}} \begin{pmatrix} 1 & 1 \\ -1 & 1 \end{pmatrix}.$$

Also, it is easy to verify that $B = CAC^{-1}$, that is, A and B are similar matrices. Then

$$e^{Bt} = Ce^{At}C^{-1}.$$

Let $\underset{\sim}{y} = H(\underset{\sim}{x}) = C\underset{\sim}{x}$. Then $\underset{\sim}{x} = C^{-1}\underset{\sim}{y}$ and

$$\dot{\underset{\sim}{y}} = C\dot{\underset{\sim}{x}} = CA\underset{\sim}{x} = CAC^{-1}\underset{\sim}{y} = B\underset{\sim}{y}.$$

Let $\underset{\sim}{x}(t) = e^{At}\underset{\sim}{x}_0$ be the solution of the system $\dot{\underset{\sim}{x}} = A\underset{\sim}{x}$ with the initial condition $\underset{\sim}{x}(0) = \underset{\sim}{x}_0$. Then $\underset{\sim}{y}(t) = C\underset{\sim}{x}(t) = Ce^{At}\underset{\sim}{x}_0 = e^{Bt}C\underset{\sim}{x}_0$. This shows that if $\underset{\sim}{x}(t) = e^{At}\underset{\sim}{x}_0$ is a solution of the first system through $\underset{\sim}{x}_0$, then $\underset{\sim}{y}(t) = e^{Bt}C\underset{\sim}{x}_0$ is a solution of the second system through $C\underset{\sim}{x}_0$. Thus the mapping H maps the trajectories of the first system onto the trajectories of the second and since $Ce^{At} = e^{Bt}C$, and H also preserves the parameterization. The map H is a homeomorphism. Therefore, the given two systems are topologically conjugate. Note that the map $H(\underset{\sim}{x}) = C\underset{\sim}{x}$ is simply a rotation through $45°$ as shown in Fig. 4.2.

4.5 Linear Subspaces

The dynamics of a system may be restricted to manifolds which are embedded in the phase space. We give very formal definition of manifold below.

Manifold: The concept of manifold is very important in dynamical system, especially in stability theory, bifurcation, etc. A manifold in the n-dimensional Euclidean space $\mathbb{R}^n$ is defined as an $m(m \leq n)$-dimensional continuous region embedded in $\mathbb{R}^n$ and is represented by equations, say $f_j(\underset{\sim}{x}) = 0, j = 1, 2, \ldots, n - m$ in $\underset{\sim}{x} = (x_1, x_2, \ldots, x_n) \in \mathbb{R}^n$. In other words, an n-dimensional topological

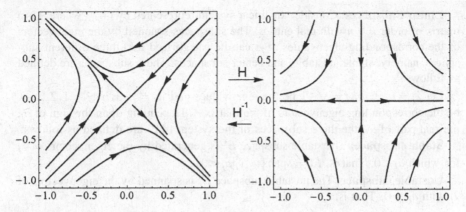

Fig. 4.2 Topologically equivalent flows

manifold is a Housdroff space (topological space such that any two distinct points possess distinct neighborhoods) such that every point has an open neighborhood N_i which is homeomorphic to an open set of E^n. If the functions $f_j(\underset{\sim}{x})$ are differentiable, then the manifold is called a differentiable manifold. More specifically, let M be a differentiable manifold. We may consider simply an open set of an Euclidean space, or a sphere or a torous as examples. A function on a differential manifold M is a diffeomorphism iff it is smooth, invertible, and its inverse is also smooth. On the other hand, an endomorphism of M is a smooth function from M to itself. A curve is an example of a one-dimensional manifold, and a surface is a two-dimensional manifold (see the book Tu [3] for details on manifolds). Our next target is to find the manifolds for some dynamical systems. First, consider the simple linear harmonic oscillator represented by the equation $m\ddot{x} = -kx$. With $\dot{x} = y$ we have the system

$$\dot{x} = y, \dot{y} = -\left(\frac{k}{m}\right)x.$$

This is a conservative system and its phase space is the two-dimensional Euclidean plane $\mathbb{R}^2$. It is easy to show that the Hamiltonian of the harmonic oscillator is constant and is given by $H(x,y) = \frac{1}{2}my^2 + \frac{1}{2}kx^2 = $ constant. The Hamiltonian represents a one-dimensional differential manifold in $\mathbb{R}^2$ and all solutions of the system lie on this manifold. The manifold is a system of ellipses in the phase plane. All these ellipses are topologically equivalent to the unit circle $S = \{(x,y) \in \mathbb{R}^2 : x^2 + y^2 = 1\}$ under the homeomorphism $h(x,y) = \left(\frac{x}{\sqrt{2H/k}}, \frac{y}{\sqrt{2H/m}}\right)$. Since $\frac{dH}{dt} = 0$ for all $(x,y) \in \mathbb{R}^2$, the Hamiltonian H is an integral of motion (this notion will discuss in later chapter) and the manifold H is also known as an *integral manifold*. All these manifolds for different values of constants are topologically equivalent to the unit circle S.

Linear Subspaces: Consider the linear system represented by (4.2), where A is a matrix of order $n \times n$ with real entries. The subspaces spanned by the eigenvectors of the corresponding eigenvalues of A can be categorized into three different subspaces, namely, stable, unstable, and center subspaces. These subspaces are defined as follows:

Let $\lambda_j = a_j \pm ib_j; i = \sqrt{-1}$ be the eigenvalues and $w_j = u_j \pm iv_j; j = 1, 2, \ldots, k$ be the corresponding eigenvectors of the matrix A. Depending upon the sign of a_j, the real part of λ_j, the three subspaces of the system (4.2) are defined as follows:

Stable subspace: The stable subspace E^s is generated by the eigenvectors of λ_j for which $a_j < 0$. That is, $E^s = \text{span}\{u_j, v_j | a_j < 0\}$.

Unstable subspace: The unstable subspace E^u is spanned by the eigenvectors of λ_j with $a_j > 0$. That is,

$$E^u = \text{span}\{u_j, v_j | a_j > 0\}.$$

Center subspace The center subspace occurs when the eigenvalues are purely imaginary. It is defined as $E^c = \text{span}\{u_j, v_j | a_j = 0\}$.

Example 4.11 Find the linear subspaces for the system $\dot{\underset{\sim}{x}} = A\underset{\sim}{x}$ with $\underset{\sim}{x}(0) = \underset{\sim}{x}_0$, where

$$A = \begin{pmatrix} -3 & 0 & 0 \\ 0 & 3 & -2 \\ 0 & 1 & 1 \end{pmatrix}.$$

Solution The characteristic equation of A has the roots $\lambda = -3, 2 \pm i$. So, the fixed points are hyperbolic type. The eigenvector corresponding to $\lambda_1 = -3$ is $(1, 0, 0)^t$, and that for $\lambda_2 = 2 + i$ is

$$w_2 = \begin{pmatrix} 0 \\ 1+i \\ 1 \end{pmatrix} = \begin{pmatrix} 0 \\ 1 \\ 1 \end{pmatrix} + i \begin{pmatrix} 0 \\ 1 \\ 0 \end{pmatrix} = u_2 + iv_2, \text{where } u_2 = \begin{pmatrix} 0 \\ 1 \\ 1 \end{pmatrix} \text{ and } v_2$$

$$= \begin{pmatrix} 0 \\ 1 \\ 0 \end{pmatrix}.$$

Therefore, the stable and unstable subspaces are given by

$$E^s = \text{span}\{u_j, v_j | a_j < 0\}$$
$$= \text{span}\{(1, 0, 0)^t\} = x\text{-axis in the phase space}$$

and

$$E^u = \text{span}\{u_j, v_j | a_j > 0\} = \text{span}[(0, 1, 0)^t, (0, 1, 1)^t] = yz\text{-plane}.$$

There is no center subspace E^C, since no eigenvalue is purely imaginary.

Example 4.12 Find all linear subspaces of the system $\dot{\underset{\sim}{x}} = A\underset{\sim}{x}$, where

$$A = \begin{pmatrix} 0 & 0 & 1 \\ 0 & 1 & 0 \\ -1 & 0 & 0 \end{pmatrix}.$$

Solution Clearly, the origin is the unique equilibrium point of the system. The eigenvalues of the matrix A are $1, \pm i$. It can be easily obtained that the eigenvectors

corresponding to $\lambda_1 = 1$ is $w_1 = \begin{pmatrix} 0 \\ 1 \\ 0 \end{pmatrix}$ and that for $\lambda_2 = i$ is

$$w_2 = \begin{pmatrix} 1 \\ 0 \\ i \end{pmatrix} = \begin{pmatrix} 1 \\ 0 \\ 0 \end{pmatrix} + i \begin{pmatrix} 0 \\ 0 \\ 1 \end{pmatrix} = u_2 + iv_2, \text{ where } u_2 = \begin{pmatrix} 1 \\ 0 \\ 0 \end{pmatrix} \text{ and } v_2 = \begin{pmatrix} 0 \\ 0 \\ 1 \end{pmatrix}.$$

Since the system has positive and purely imaginary eigenvalues, it has unstable and center subspaces, given by

$$E^u = \text{span}\{u_j, v_j | a_j > 0\} = \text{span}[(0, 1, 0)^t] = y\text{-axis in the phase space}$$

and

$$E^c = \text{span}\{u_j, v_j | a_j = 0\} = \text{span}[(1, 0, 0)^t, (0, 0, 1)^t]$$
$$= xz\text{-plane in the phase space}$$

These two subspaces are presented in Fig. 4.3. Note that the system has no stable subspace, since it has no negative eigenvalue.

Theorem 4.6 *Consider a system* $\dot{\underset{\sim}{x}} = A\underset{\sim}{x}$, *where A is a $n \times n$ matrix with real entries. Then phase space $\mathbb{R}^n$ can be decomposed as*

$$\mathbb{R}^n = E^u \oplus E^s \oplus E^c$$

where E^u, E^s, and E^c are the unstable, stable, and of the system, respectively. Furthermore, these subspaces are invariant with respect to the flow.

Fig. 4.3 Unstable and center subspaces of the given system

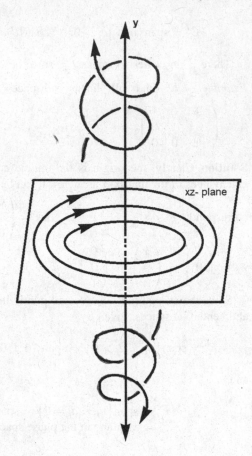

4.6 Hyperbolicity and Its Persistence

The flow in the neighborhood of hyperbolic fixed point has some special characteristic features. The special flow characteristic around the hyperbolic fixed point is called hyperbolicity. There are two important theorems, namely (i) Hartman–Grobman theorem and (ii) Stable manifold theorem for hyperbolic fixed points. The first theorem proves that there exists a continuous invertible map in some neighborhood of the hyperbolic fixed point which maps the nonlinear flow to the linear flow preserving the sense of time and the second theorem implies that the local structure of hyperbolic fixed points of nonlinear flows is the same as the linear flows in terms of the existence and transversality of local stable and unstable manifolds. We now define the local stable and unstable manifolds as follows:

Let U be some neighborhood of a hyperbolic fixed point $\underset{\sim}{x}^*$. The local stable manifold, denoted by $W_{loc}^s(\underset{\sim}{x}^*)$, is defined as

$$W_{loc}^{s}(\underset{\sim}{x}^{*}) = \left\{ \underset{\sim}{x} \in U | \phi_{t}(\underset{\sim}{x}) \to \underset{\sim}{x}^{*} \text{ as } t \to \infty, \phi_{t}(\underset{\sim}{x}) \in U \; \forall t \geq 0 \right\}.$$

Similarly, the local unstable manifold is defined as

$$W_{loc}^{u}(\underset{\sim}{x}^{*}) = \left\{ \underset{\sim}{x} \in U | \phi_{t}(\underset{\sim}{x}) \to \underset{\sim}{x}^{*} \text{ as } t \to -\infty, \phi_{t}(\underset{\sim}{x}) \in U \; \forall t \leq 0 \right\}.$$

The stable manifold theorem states that these manifolds exist and have the same dimension as the stable and unstable manifolds of the corresponding linear system $\dot{\underset{\sim}{x}} = A\underset{\sim}{x}$, if $\underset{\sim}{x}^{*}$ is a hyperbolic equilibrium point, and that they are tangential to the manifolds of linear system at $\underset{\sim}{x}^{*}$. This notion is known as hyperbolicity of a system.

Hyperbolic flow: If all the eigenvalues of the $n \times n$ matrix A are nonzero, then the flow $e^{At} : \mathbb{R}^{n} \to \mathbb{R}^{n}$ is called a hyperbolic flow, and the linear system $\dot{\underset{\sim}{x}} = A\underset{\sim}{x}$ is then called a hyperbolic linear system.

Invariant manifold: An invariant set $D \subset \mathbb{R}^{n}$ is said to be a $C^{r}(r \geq 1)$ invariant manifold if the set D has a structure of a C^{r} differentiable manifold. Similarly, the positively and negatively invariant manifolds are defined. In other words, a subspace $D \subseteq \mathbb{R}^{n}$ is said to be invariant if any flow starting in this subspace will remain within it for all future time.

The linear subspaces E^{s}, E^{u}, and E^{c} are all invariant subspaces of the linear system $\dot{\underset{\sim}{x}} = A\underset{\sim}{x}$ with respect to the flow e^{At}.

Theorem 4.7 (*Stable manifold theorem*) *Let* $\underset{\sim}{x}^{*} = 0$ *be a hyperbolic equilibrium point of the system* $\dot{\underset{\sim}{x}} = \underset{\sim}{f}(\underset{\sim}{x}), \underset{\sim}{x} \in C^{1}$, *and* E^{s} *and* E^{u} *be the stable an unstable manifolds of the corresponding linear system* $\dot{\underset{\sim}{x}} = A\underset{\sim}{x}$. *Then there exists local stable and unstable manifolds* $W_{loc}^{s}(0)$ *and* $W_{loc}^{u}(0)$ *of the nonlinear system with the same dimension as that of* E^{s} *and* E^{u}, *respectively. These manifolds are tangential to* E^{s} *and* E^{u}, *respectively, at the origin and are smooth as the function* $\underset{\sim}{f}$.

Let $\underset{\sim}{x}_{0}$ be a hyperbolic fixed point of the nonlinear system. Then $\underset{\sim}{x}_{0}$ is called a sink if all the eigenvalues of the linear system have strictly negative real parts, and a source if all the eigenvalues have strictly positive real parts. Otherwise, $\underset{\sim}{x}_{0}$ is a saddle. A sketch of stable and unstable manifolds is given in Fig. 4.4.

Example 4.13 Find the local stable and unstable manifolds of the system $\dot{x} = x - y^{2}, \dot{y} = -y$.

Solution The system has the unique equilibrium point at the origin, (0, 0). Also, the origin is a saddle equilibrium point of the corresponding linearized system $\dot{x} = x, \dot{y} = -y$ with the invariant linear stable and unstable subspaces as

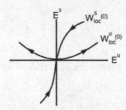

Fig. 4.4 Stable and unstable manifolds at the origin

$$E^s(0,0) = \{(x,y) : x = 0\} \text{ and } E^u(0,0) = \{(x,y) : y = 0\},$$

respectively. Therefore by stable manifold theorem, the system has local stable and unstable manifolds

$$W_{loc}^s(0,0) = \left\{(x,y) : x = S(y), \frac{\partial S}{\partial y}(0) = 0\right\} \text{ and}$$

$$W_{loc}^u(0,0) = \left\{(x,y) : y = U(x), \frac{\partial U}{\partial x}(0) = 0\right\},$$

respectively. We now find these manifolds.

Stable manifold: For the local stable manifold, we expand $S(y)$ as a power series in the neighborhood of the origin as follows:

$$S(y) = \sum_{i \geq 0} s_i y^i = s_0 + s_1 y + s_2 y^2 + s_3 y^3 + \cdots.$$

Since at $y = 0, S = 0$ and $\frac{\partial S}{\partial x} = 0$, we have $s_0 = s_1 = 0$. Therefore,

$$x = S(y) = \sum_{i \geq 2} s_i y^i = s_2 y^2 + s_3 y^3 + s_4 y^4 + s_5 y^5 + \cdots\cdots.$$

Now,

$$\dot{x} = x - y^2 = \left(s_2 y^2 + s_3 y^3 + s_4 y^4 + s_5 y^5 + \cdots\right) - y^2$$
$$= (s_2 - 1)y^2 + s_3 y^3 + s_4 y^4 + s_5 y^5 + \cdots$$

Again,

$$x = S(y) \Rightarrow \dot{x} = \frac{\partial S}{\partial y}\dot{y} = (2s_2 y + 3s_3 y^2 + 4s_4 y^3 + 5s_5 y^4 + \cdots)(-y)$$
$$= -(2s_2 y^2 + 3s_3 y^3 + 4s_4 y^4 + 5s_5 y^5 + \cdots)$$

Therefore, we have

$$(s_2 - 1)y^2 + s_3y^3 + s_4y^4 + s_5y^5 + \cdots = -(2s_2y^2 + 3s_3y^3 + 4s_4y^4 + 5s_5y^5 + \cdots).$$

Equating the coefficients of like powers of y from both sides of the above relation, we get

$$s_2 = 1/3, \quad s_3 = s_4 = \cdots = 0.$$

Therefore, $x = \frac{y^2}{3}$, and hence, the local stable manifold of the nonlinear system in the neighborhood of the equilibrium point origin is

$$W_{loc}^s(0,0) = \left\{ (x,y) : x = \frac{y^2}{3} \right\}.$$

Unstable manifold: For the local unstable manifold we expand $U(x)$ as

$$U(x) = \sum_{i \geq 0} u_i x^i = u_0 + u_1 x + u_2 x^2 + u_3 x^3 + \cdots.$$

As previous, $u_0 = u_1 = 0$. Therefore,

$$y = U(x) = \sum_{i \geq 2} u_i x^i = u_2 x^2 + u_3 x^3 + u_4 x^4 + u_5 x^5 + \cdots.$$

Now,

$$\dot{y} = -y = -(u_2 x^2 + u_3 x^3 + u_4 x^4 + u_5 x^5 + \cdots).$$

But

$$\dot{y} = \frac{\partial U}{\partial x} \dot{x} = (2u_2 x + 3u_3 x^2 + 4u_4 x^3 + 5u_5 x^4 + \cdots)(x - y^2)$$

$$= (2u_2 x + 3u_3 x^2 + 4u_4 x^3 + 5u_5 x^4 + \cdots)\left\{ x - (u_2 x^2 + u_3 x^3 + u_4 x^4 + u_5 x^5 + \cdots)^2 \right\}$$

Therefore, we must have

$$(u_2 x^2 + u_3 x^3 + u_4 x^4 + u_5 x^5 + \cdots) = (2u_2 x + 3u_3 x^2 + 4u_4 x^3 + 5u_5 x^4 + \cdots)$$
$$\left\{ (u_2 x^2 + u_3 x^3 + u_4 x^4 + u_5 x^5 + \cdots)^2 - x \right\}$$

Equating the coefficients of like powers of x from both sides of the above relation, we get

Fig. 4.5 Local stable and
unstable manifolds at the
equilibrium point origin

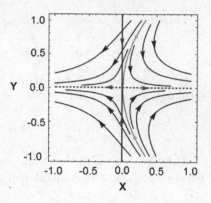

$$u_2 = u_3 = u_4 = \cdots = 0.$$

Therefore, $y = U(x) = 0$. Hence, the local unstable manifold of the nonlinear system in the neighborhood of the origin is (Fig. 4.5)

$$W_{loc}^u(0,0) = \{(x,y) : y = 0\} = E^u(0,0).$$

4.6.1 Persistence of Hyperbolic Fixed Points

In the previous section, we have seen important features of hyperbolic fixed points that near a hyperbolic fixed point the nonlinear and its corresponding linear systems have the same qualitative features locally. In this section, we study another important feature that hyperbolic equilibrium points persist their character under sufficiently small perturbation. Let the origin be a hyperbolic fixed point of the linear system $\dot{\underset{\sim}{x}} = \underset{\sim}{f}(\underset{\sim}{x}); x \in \mathbb{R}^n$. Consider the perturbed system

$$\dot{\underset{\sim}{x}} = \underset{\sim}{f}(\underset{\sim}{x}) + \varepsilon \underset{\sim}{g}(\underset{\sim}{x}) \tag{4.11}$$

where $\underset{\sim}{g}$ is a smooth vector field defined in $\mathbb{R}^n$ and ε is a sufficiently small perturbation quantity. The fixed points of (4.11) are given by

$$\underset{\sim}{f}(\underset{\sim}{x}) + \varepsilon \underset{\sim}{g}(\underset{\sim}{x}) = 0.$$

Expanding in Taylor series about $\underset{\sim}{x} = \underset{\sim}{0}$ and using $\underset{\sim}{f}(\underset{\sim}{0}) = \underset{\sim}{0}$, we get

$$Df(\underset{\sim}{0})\underset{\sim}{x} + \varepsilon\left[\underset{\sim}{g}(\underset{\sim}{0}) + Dg(\underset{\sim}{0})\underset{\sim}{x}\right] + O\left(|\underset{\sim}{x}|^2\right) = 0$$

$$\Rightarrow \left[Df(\underset{\sim}{0}) + \varepsilon Dg(\underset{\sim}{0})\right]\underset{\sim}{x} + \varepsilon\underset{\sim}{g}(\underset{\sim}{0}) + O\left(|\underset{\sim}{x}|^2\right) = 0$$

Since the origin is hyperbolic, the eigenvalues of $Df(\underset{\sim}{0})$ are nonzero and so the eigenvalues of $\left[Df(\underset{\sim}{0}) + \varepsilon Dg(\underset{\sim}{0})\right]$ are nonzero for sufficiently small ε. Hence, $\det\left[Df(\underset{\sim}{0}) + \varepsilon Dg(\underset{\sim}{0})\right] \neq 0$, that is, $\left[Df(\underset{\sim}{0}) + \varepsilon Dg(\underset{\sim}{0})\right]^{-1}$ exists. Therefore the fixed points of (4.11) are given by

$$\underset{\sim}{x}^* = \varepsilon\left[Df(\underset{\sim}{0}) + \varepsilon Dg(\underset{\sim}{0})\right]^{-1}\underset{\sim}{g}(\underset{\sim}{0}) + O\left(|\underset{\sim}{x}|^2\right).$$

We now determine whether the point is hyperbolic or not. Since ε is small, we can find a neighborhood of $\varepsilon = 0$ in which the eigenvalues of $\left[Df(\underset{\sim}{x}) + \varepsilon Dg(\underset{\sim}{x})\right]$ have nonzero real part for sufficiently small $\underset{\sim}{x}$. So, for sufficiently small ε, the eigenvalues of the perturbed equation do not change. So, the equilibrium points retain their character, that is, they are of hyperbolic type. This proves that the character of hyperbolic fixed point remains unchanged when the system undergoes small perturbation.

Theorem 4.8 (*Center manifold theorem*) *Consider a nonlinear system* $\dot{\underset{\sim}{x}} = \underset{\sim}{f}(\underset{\sim}{x})$ *where* $\underset{\sim}{f} \in C^r(E), r \geq 1, E$ *being an open subset of* $\mathbb{R}^n$ *containing a non-hyperbolic fixed point, say* $\underset{\sim}{x}^* = 0$ *of the system. Suppose that the Jacobian matrix,* $J = D\underset{\sim}{f}(0)$, *of the system at the origin has j eigenvalues with positive real parts, k eigenvalues with negative real parts, and* $m(= n - j - k)$ *eigenvalues with zero real parts. Then there exists a j-dimensional* C^r*-class unstable manifold* $W^u(0)$*, a k-dimensional* C^r*-class stable manifold* $W^s(0)$*, and an m-dimensional* C^r*-class center manifold* $W^c(0)$ *tangent to subspaces* E^u, E^s, E^c *of the corresponding linear system* $\dot{\underset{\sim}{x}} = A\underset{\sim}{x}$ *at the origin, respectively. Furthermore, these manifolds are invariant under the flow* ϕ_t *of the nonlinear system. The manifolds* $W^s(0)$ *and* $W^u(0)$ *are unique but the local center manifold* $W^c(0)$ *is not unique.*

Example 4.14 Find the manifolds of the system $\dot{x} = x, \dot{y} = y^2$.

Solution The system has a non-hyperbolic fixed point at the origin. The unstable subspace $E^u(0,0)$ of the linearized system at the origin is the x-axis and the center subspace is the y-axis. No stable subspace occurs for this system. Using the technique of power series expansion, discussed in Example 4.13, we see that the unstable manifold at the origin is the x-axis and its center manifold is the y-axis, that is, the line $x = 0$. However, there are other center manifolds of the system. From the equations, we have

$$\frac{dy}{dx} = \frac{y^2}{x},$$

which have the solution $x = ke^{-1/y}$ for $y \neq 0$. Thus, the center manifold of the origin is

$$W_{loc}^c(0,0) = \left\{ (x,y) \in \mathbb{R}^2 : x = ke^{-1/y} \text{ for } y > 0, x = 0 \text{ for } y \leq 0 \right\}.$$

It represents a one-parameter (k) family of center manifolds of the origin. Note that if we use the technique of power series expansion for the center manifold, we only get $x = 0$ as the center manifold. This example also shows that the center manifold is not unique.

4.7 Basin of Attraction and Basin Boundary

Let $\underset{\sim}{x}^*$ be an attracting fixed point of the linear system (4.2). We define the basin of attraction in some neighborhood of $\underset{\sim}{x}^*$ subject to some initial condition $\underset{\sim}{x}(0) = \underset{\sim 0}{x}$ to be the set of points such that $\underset{\sim}{x}(t) \to \underset{\sim}{x}^*$ as $t \to \infty$. The boundary of this attracting set is called the basin boundary, also known as separatrix, separating the stable and unstable regions.

We discuss the basin of attraction and basin boundary with the help of the model for two interacting species. The well-known Lotka–Volterra model is considered which exhibits the basin of attraction and basin boundary for some situations. Consider the Lotka–Volterra model represented by the system of equations as

$$\dot{x} = x(3 - x - 2y), \dot{y} = y(3 - 2x - y)$$

where $x(t)$ and $y(t)$ are the populations the two interacting species, say rabbits and sheep, respectively, and $x, y \geq 0$. We shall first find the fixed points of the system, which can be obtained by solving the equations

$$x(3 - x - 2y) = 0 \text{ and } y(3 - 2x - y) = 0.$$

Solving we get four fixed points, $(0,0), (0,3), (3,0), (1,1)$. The Jacobian matrix of the system is given by

$$J(x,y) = \begin{pmatrix} 3 - 2x - 2y & -2x \\ -2y & 3 - 2x - 2y \end{pmatrix}.$$

At the fixed point $(0, 0)$, $J(0,0) = \begin{pmatrix} 3 & 0 \\ 0 & 3 \end{pmatrix}$. The eigenvalues of $J(0,0)$ are $3, 3$, which are positive. So, the fixed point origin is an unstable node. All trajectories leave the origin parallel to the eigenvector $(0, 1)^T$ for $\lambda = 3$ which spans the y-axis. The phase portrait near the origin is shown in Fig. 4.6.

At the fixed point $(0, 3)$, $J(0,3) = \begin{pmatrix} -3 & 0 \\ -6 & -3 \end{pmatrix}$, which has the eigenvalues $-3, -3$. So, the fixed point $(0, 3)$ is a stable node. Trajectories approach along the eigen direction with the eigenvalue $\lambda = -3$ spanning the eigenvector $(0, 1)^T$. The phase portrait near the fixed point $(0, 3)$ which is a stable node looks like as presented in Fig. 4.7.

At $(3, 0)$, we have $J(3,0) = \begin{pmatrix} -3 & -6 \\ 0 & -3 \end{pmatrix}$. The eigenvalues of $J(3,0)$ are $-3, -3$. So, as previous the fixed point $(3, 0)$ is also a stable node. Trajectories approach along the eigen direction with the eigenvalue $\lambda = -3$ spanning the eigenvector $(1, 0)^T$. The phase portrait near the fixed point $(3, 0)$ is depicted in Fig. 4.8.

Fig. 4.6 Local phase portrait near the fixed point origin

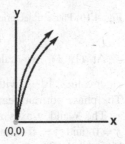

Fig. 4.7 Local phase portrait near the fixed point $(0, 3)$

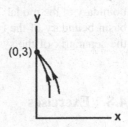

Fig. 4.8 Local phase portrait near the fixed point $(3, 0)$

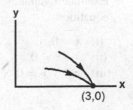

Fig. 4.9 Local phase portrait
near the fixed point (1, 1)

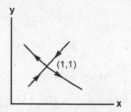

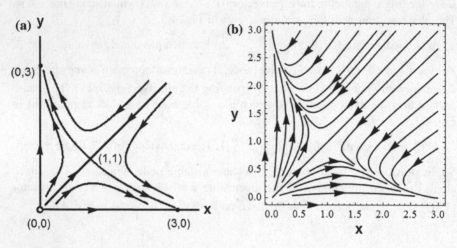

Fig. 4.10 Phase trajectories of the given system

At (1, 1), we calculate $J(1,1) = \begin{pmatrix} -1 & -2 \\ -2 & -1 \end{pmatrix}$, which gives two distinct eigenvalues, 1, −3, with opposite signs. Therefore, the fixed point (1, 1) is a saddle. The phase portrait near (1, 1) is shown in Fig. 4.9.

The x and y axes represent the straight line trajectories because $\dot{x} = 0$ when $x = 0$ and $\dot{y} = 0$ when $y = 0$. All trajectories of the system are presented in Fig. 4.10. This figure also clearly depicts the attracting points and the basin boundary of the model. The attracting points of the system are (3, 0) and (0, 3). The basin boundary of the two attracting points is the straight line $y = x$, which is also the separatrix of the system.

4.8   Exercises

1. Examine Lyapunov, Poincaré, and Lagrange stability criteria for the following equations:

   (i) $\dot{x} = 0$,
   (ii) $\dot{x} + x = 0$,
   (iii) $\dot{x} = y, \dot{y} = 0$.

2. Find the general solution of the nonlinear oscillator $\dot{x} = -y(x^2+y^2)^{\frac{1}{2}}$, $\dot{y} = x(x^2+y^2)^{\frac{1}{2}}$. Also, examine whether it is Lyapunov or orbitally stable.

3. Define Lyapunov function and Lyanupov stability. Examine the stability in the Lyapunov sense for the following equations:

   (i) $\dot{x}+x = 2$, $x(0) = 1$,
   (ii) $\dot{x}-x = 2$, $x(0) = -1$,
   (iii) $\dot{x} = 5$, $x(0) = 0$

4. Using a suitable Lyapunov function, prove that the system $\dot{x} = -x+4y$, $\dot{y} = -x - y^3$ has no closed orbits.

5. Examine asymptotic stability through the construction of suitable Lyapunov function L for the system $\dot{x} = 2y(z-1), \dot{y} = -x(z-1), \dot{z} = xy$.

6. Using suitable Lyapunov functions examine the stability at the equilibrium point origin for the following systems:

   (i) $\dot{x} = y+x^3$, $\dot{y} = x - y^3$
   (ii) $\dot{x} = y - xg(x, y)$, $\dot{y} = -x - yg(x, y)$ where the function $g(x, y)$ can be expanded in a convergent power series with $g(0, 0) = 0$,
   (iii) $\dot{x} = 2xy+x^3$, $\dot{y} = x^2 - y^5$,
   (iv) $\dot{x} = y - x^3$, $\dot{y} = -x - y^3$

7. Investigate the stability of the system $\dot{x} = -5y - 2x^3$, $\dot{y} = 5x - 3y^3$ at $(0, 0)$ using Lyapunov direct method.

8. State Hartman–Grobmann theorem and discuss its significance. Using theorem describe the local stability behavior near equilibrium points of the following nonlinear systems (i) $\dot{x} = y^2 - x+2, \dot{y} = x^2 - y^2$. (ii) $\dot{x} = -y$, $\dot{y} = x - x^5$. Also, draw the phase portrait.

9. Find the stable, unstable, and center subspaces for the linear system $\dot{\underset{\sim}{x}} = A\underset{\sim}{x}$

   when the matrix A is given by

   (i) $A = \begin{pmatrix} 2 & 0 \\ 0 & -1 \end{pmatrix}$

   (ii) $A = \begin{pmatrix} 1 & 2 \\ 4 & -1 \end{pmatrix}$,

   (iii) $A = \begin{pmatrix} 3 & 5 \\ -1 & 1 \end{pmatrix}$,

   (iv) $A = \begin{pmatrix} 2 & 1 \\ 1 & 0 \end{pmatrix}$,

   (v) $A = \begin{pmatrix} -1 & 1 \\ 1 & -1 \end{pmatrix}$,

   (vi) $A = \begin{pmatrix} 0 & 1 \\ -1 & 0 \end{pmatrix}$,.

$$\text{(vii)} \quad A = \begin{pmatrix} 10 & -1 & 0 \\ 25 & 2 & 0 \\ 0 & 0 & -3 \end{pmatrix}$$

$$\text{(viii)} \quad A = \begin{pmatrix} 0 & -2 & 0 \\ 1 & 2 & 0 \\ 0 & 0 & -2 \end{pmatrix}$$

10. Obtain the local stable and unstable manifolds for the system $\dot{x} = -x$, $\dot{y} = y + x^2$ and give a rough sketch of the manifolds.

11. Obtain the stable and unstable manifolds for the system
    $\dot{x} = -x + \sigma + \frac{x^2}{y}, \dot{y} = -y + x^2$, where σ is a parameter.

12. Find the fixed point and investigate their stability for the system
    $\dot{x} = x(3 - 2x - y)$
    $\dot{y} = y(2 - x - y)$
    Also, draw the basin of attraction and basin boundary.

13. Find the basin of attraction and basin boundary for the following systems:

    (i) $\dot{x} = x(1 - x - 2y), \dot{y} = y(1 - 2x - y)$
    (ii) $\dot{x} = x(1 - x - 2y), \dot{y} = y(1 - 3x - y)$
    (iii) $\dot{x} = x(1 - x - 3y), \dot{y} = y(1 - 2x - y)$
    (iv) $\dot{x} = x(1 - x - 5y), \dot{y} = 2y(1 - 3x - y)$
    (v) $\dot{x} = x(3 - x - 2y), \dot{y} = y(2 - x - y)$.

References

1. Hartmann, P.: Ordinary Differential Equations. Wiley, New York (1964)
2. Arnold, V.I.: Mathematical Methods of Classical Mechanics. Springer, New York (1984)
3. Tu, L.W.: An Introduction to Manifold. Springer (2011)
4. Krasnov, M.L.: Ordinary Differential Equations. MIR Publication, Moscow, English translation (1987)
5. Jackson, E.A.: Perspectives of Nonlinear Dynamics, vol. 1. Cambridge University Press (1989)
6. Glendinning, P.: Stability, instability and chaos: an introduction to the theory of nonlinear differential equations. Cambridge University Press (1994)
7. Jordan, D.W., Smith, P.: Non-linear Ordinary Differential Equations. Oxford University Press (2007)
8. Perko, L.: Differential Equations and Dynamical Systems, 3rd edn. Springer, New York (2001)

Chapter 5
Oscillations

It is well known that some important properties of nonlinear equations can be determined through qualitative analysis. The general theory and solution methods for linear equations are highly developed in mathematics, whereas a very little is known about nonlinear equations. Linearization of a nonlinear system does not provide the actual solution behaviors of the original nonlinear system. Nonlinear systems have interesting solution features. It is a general curiosity to know in what conditions an equation has periodic or bounded solutions. Systems may have solutions in which the neighboring trajectories are closed, known as limit cycles. What are the conditions for the existence of such limiting solutions? In what conditions does a system have unique limit cycle? These were some questions both in theoretical and engineering interest at the beginning of twentieth century. This chapter deals with oscillatory solutions in linear and nonlinear equations, their mathematical foundations, properties, and some applications.

5.1 Oscillatory Solutions

In our everyday life, we encounter many systems either in engineering devices or in natural phenomena, some of which may exhibit oscillatory motion and some are not. The undamped pendulum is such a device that executes an oscillatory motion. Oscillatory motions have wide range of applications. It is in fact that no system in macroscopic world is a simple oscillator because of damping force, however small present in the system. Of course, in some cases these forces are so small that we can neglect them in respect to time scale and treat the system as a simple oscillator. Oscillation and periodicity are closely related to each other. Both linear and nonlinear systems may exhibit oscillation, but qualitatively different. Linear oscillators have interesting properties. To explore these properties, we begin with second order linear systems. Note that the first-order linear system cannot have oscillatory solution.

© Springer India 2015
G.C. Layek, *An Introduction to Dynamical Systems and Chaos*,
DOI 10.1007/978-81-322-2556-0_5

Consider a second-order linear homogeneous differential equation represented as

$$\ddot{x} + a(t)\dot{x} + b(t)x = 0 \qquad (5.1)$$

where $a(t)$ and $b(t)$ are real-valued functions of the real variable t, and are continuous on an interval $I \subset \mathbb{R}$, that is, $a(t), b(t) \in C(I)$.

A solution $x(t)$ of the Eq. (5.1) is said to be **oscillating** on I, if it vanishes there at least two times, that is, if $x(t)$ has at least two zeros on I. Otherwise, it is called non-oscillating on I. For example, consider a linear equation

$$\ddot{x} - m^2 x = 0, \quad x, m \in \mathbb{R}.$$

Its general solution is given by

$$x(t) = \begin{cases} ae^{mt} + be^{-mt}, \text{if } m \neq 0 \\ a + bt, \text{if } m = 0 \end{cases}$$

which is non-oscillating in $\mathbb{R}$. On the other hand, the general solution

$$x(t) = a\cos(mt) + b\sin(mt) = A\sin(mt + \delta)$$

of the equation $\ddot{x} + m^2 x = 0$, $m \in \mathbb{R}$, $m \neq 0$ is oscillating, where $A = \sqrt{(a^2 + b^2)}$ and $\delta = \tan^{-1}(a/b)$ are the amplitude and the initial phase of the solution, respectively. All solutions of this equation are oscillating with period of oscillation $(2\pi/m)$. The distance between two successive zeros is (π/m). The above two equations give a good illustration of the existence/nonexistence oscillatory solutions for the general second-order linear equation

$$\ddot{x} + p(t)x = 0, \quad p \in C(I) \qquad (5.2)$$

The Eq. (5.2) can be derived from (5.1) by applying the transformation

$$x(t) = z(t)\exp\left(-\frac{1}{12}\int_{t_0}^{t} a(\tau)d\tau\right), \quad t_0, t \in I$$

with $p(t) = -\frac{a^2(t)}{4} - \frac{\dot{a}(t)}{2} + b(t)$. The transformation preserves the zeros of the solutions of the equations. We now derive condition(s) for which the Eq. (5.2) has oscillating and/or non-oscillating solutions. We first assume that $p(t) = $ constant. If $p > 0$, every solution

$$x(t) = a\cos(\sqrt{p}t) + b\sin(\sqrt{p}t) = A\sin(\sqrt{p}t + \delta)$$

of (5.2) has infinite number of zeros and the distance between two successive zeros is $\left(\pi/\sqrt{p}\right)$. So the solution is oscillating. On the other hand, if $p \leq 0$, we cannot find any nonzero solution of (5.2) that vanish at more than one point. The solution is called non-oscillating.

5.2 Theorems on Linear Oscillatory Systems

Theorem 5.1 (On non-oscillating solutions) *If $p(t) \leq 0$ for all $t \in I$, then all nontrivial solutions of (5.2) are non-oscillating on I.*

Proof If possible, let a solution $X(t) \not\equiv 0$ of (5.2) has at least two zeros on I and let $t_0, t_1 (t_0 < t_1)$ be two of them. We also assume that the function $X(t)$ has no zeros in the interval (t_0, t_1). Since $X(t)$ is continuous and have no zeros in (t_0, t_1), it has same sign (positive or negative) in (t_0, t_1) .

Without loss of generality, we assume that $X(t) > 0$ in (t_0, t_1). Then from (Fig. 5.1), it follows that $X(t)$ is maximum at some point, say $c \in (t_0, t_1)$ and consequently, $\ddot{X}(t) < 0$ in some neighborhood of c. Now, if $p(t) \leq 0$ on I, then from (5.2) it follows that $\ddot{X}(t) \geq 0$ on I, that is, $\ddot{X}(t) \geq 0$ in the neighborhood of c. This gives a contradiction. So, our assumption is wrong, and hence the solution $X(t)$ of (5.2) cannot have two or more than two zeros on I. Consequently, $X(t)$ is non-oscillating on I. Since $X(t)$ is arbitrary, every nontrivial solutions of (5.2) are non-oscillating on I. This completes the proof.

Corollary 5.1 *If $p(t) > 0$ on I, all nontrivial solutions of (5.2) are oscillating on I.*

Lemma 5.1 *Zeros of any nontrivial solution of Eq. (5.1) in I are simple and isolated.*

Corollary 5.2 *Any nontrivial solution of (5.1) has a finite number of zeros on any compact interval I.*

Theorem 5.2 (Sturm's separation theorem) *Let t_0, t_1 be two successive zeros of a nontrivial solution $x_1(t)$ of the Eq. (5.1) and $x_2(t)$ be another linearly independent solution of that equation. Then there exists exactly one zero of $x_2(t)$ between t_0 and t_1, that is, the zeros of two linearly independent solutions separate each other.*

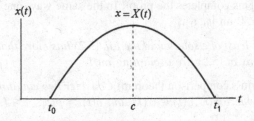

Fig. 5.1 Graphical representation of solution $X(t)$

Proof Without loss of generality, we assume that $t_0 < t_1$. If possible, let $x_2(t)$ has no zeros in the interval (t_0, t_1). Also, since $x_1(t)$ and $x_2(t)$ are linearly independent, and $x_1(t)$ has two zeros t_0 and t_1 in I, $x_2(t)$ does not vanish at $t = t_0, t_1$. Then the Wronskian

$$W(x_1, x_2; t) = \begin{vmatrix} x_1(t) & x_2(t) \\ \dot{x}_1(t) & \dot{x}_2(t) \end{vmatrix} = x_1(t)\dot{x}_2(t) - x_2(t)\dot{x}_1(t) \tag{5.3}$$

does not vanish on $[t_0, t_1]$. We assume that $W(x_1, x_2; t) > 0$ on $[t_0, t_1]$. Dividing both sides of (5.3) by $x_2^2(t)$, we get

$$\frac{W(x_1, x_2; t)}{x_2^2(t)} = \frac{x_1(t)\dot{x}_2(t) - x_2(t)\dot{x}_1(t)}{x_2^2(t)} = -\frac{d}{dt}\left(\frac{x_1(t)}{x_2(t)}\right) \tag{5.4}$$

Integrating (5.4) with respect to t from t_0 to t_1,

$$\int_{t_0}^{t_1} \frac{W(x_1, x_2; t)}{x_2^2(t)} dt = -\int_{t_0}^{t_1} d\left(\frac{x_1(t)}{x_2(t)}\right)$$

$$= -\left(\frac{x_1(t)}{x_2(t)}\right)\Bigg|_{t_0}^{t_1}$$

$$= \frac{x_1(t_0)}{x_2(t_0)} - \frac{x_1(t_1)}{x_2(t_1)}$$

$$= 0.$$

(Since $x_2(t)$ does not have zeros at t_0, t_1, that is, $x_2(t_0) \neq 0$ and $x_2(t_1) \neq 0$.) This gives a contradiction, because

$$\frac{W(x_1, x_2; t)}{x_2^2(t)} > 0 \ \forall t \in [t_0, t_1].$$

So, our assumption is wrong, and hence $x_2(t)$ has at least one zero in the interval (t_0, t_1). To prove the uniqueness, let t_2, t_3 be two distinct zeros of $x_2(t)$ in (t_0, t_1) with $t_2 < t_3$. That is, $x_2(t_2) = x_2(t_3) = 0$, where $t_0 < t_2 < t_3 < t_1$. Since $x_1(t)$ and $x_2(t)$ are linearly independent, $x_1(t)$ must have at least one zero in (t_2, t_3), that is, in (t_0, t_1). This is a contradiction, which ensures that $x_2(t)$ has exactly one zero between t_0 and t_1. This completes the proof. In the same way, one can also prove it when $W(x_1, x_2; t) < 0$ on $[t_0, t_1]$.

Corollary 5.3 *If at least one solution of the Eq. (5.2) has more than two zeros on I, then all the solutions of (5.2) are oscillating on I.*

Theorem 5.3 (Sturm's comparison theorem) *Consider two equations $\ddot{y} + p(t)y = 0$ and $\ddot{z} + q(t)z = 0$ where $p(t), q(t) \in C(I)$ and $q(t) \geq p(t), \ t \in I$. Let a pair t_0, t_1*

$(t_0 < t_1)$ *of successive zeros of a nontrivial solution* $y(t)$ *be such that there exists* $t \in (t_0, t_1)$ *such that* $q(t) > p(t)$. *Then any nontrivial solution* $z(t)$ *has at least one zero between* t_0 *and* t_1.

Proof Let $y(t)$ be a solution of the first equation such that $y(t_0) = y(t_1) = 0$ and $y(t) > 0$ in $t \in (t_0, t_1)$. If possible, let there exist a solution $z(t)$ such that $z(t) > 0 \ \forall t \in (t_0, t_1)$. Note that if there exists a solution $z(t) < 0$ we may consider the solution $-z(t)$ instead of $z(t)$. Multiplying the first equation by $z(t)$ and the second by $y(t)$ and then subtracting, we get

$$\ddot{y}(t)z(t) - \ddot{z}(t)y(t) = (q(t) - p(t))y(t)z(t)$$

$$\Rightarrow \frac{d}{dt}(\dot{y}(t)z(t) - \dot{z}(t)y(t)) = (q(t) - p(t))y(t)z(t)$$

Integrating this w.r.to t from t_0 to t_1 and using $y(t_0) = y(t_1) = 0$, we get

$$\dot{y}(t_1)z(t_1) - \dot{z}(t_0)y(t_0) = \int_{t_0}^{t_1} (q(t) - p(t))y(t)z(t)dt \qquad (5.5)$$

The right-hand side of (5.5) is positive, since $y(t), z(t)$ are positive on (t_0, t_1) and $q(t) \geq p(t) \ \forall t \in (t_0, t_1)$. But the left-hand side is nonpositive, because $\dot{y}(t_0) > 0$, $\dot{y}(t_1) < 0$ and $z(t_1) \geq 0$. So, we arrive at a contradiction. This completes the proof.

Sturm's comparison theorem has a great importance on determining the distance between two successive zeros of any nontrivial solution of (5.2). Let us consider three equations $\ddot{x} + q(t)x = 0$, $\ddot{y} + my = 0$ and $\ddot{z} + Mz = 0$, where $q(t) > 0$ for all t, $m = \min_{t \in [t_0, t_1]} q(t)$, and $M = \max_{t \in [t_0, t_1]} q(t)$. We also assume that $M > m$ so that $q(t)$ is not constant over the interval. Applying Sturm's comparison theorem for first two equations, we see that the distance between two successive zeros of any solution of $\ddot{x} + q(t)x = 0$ is not greater than $(\pi / \sqrt{m})$. Similarly, taking the first and the third equations and then applying Sturm's comparison theorem, we see that the distance between two successive zeros of $\ddot{x} + q(t)x = 0$ is not smaller than $(\pi / \sqrt{M})$. If $\lim_{t \to \infty} q(t) = q > 0$, then any solution of the equation $\ddot{x} + q(t)x = 0$ is infinitely oscillating, and the distance between two successive zeros tends to $(\pi / \sqrt{q})$. From this discussion, we have the following theorem.

Theorem 5.4 (Estimate of distance between two successive zeros of solutions of the Eq. (5.2)) *Let the inequality*

$$0 < m^2 \leq p(t) \leq M^2$$

be true on $[t_0, t_1] \subset I$. *Then the distance d between two successive zeros of any nontrivial solution of (5.2) is estimated as*

$$\frac{\pi}{M} \le d \le \frac{\pi}{m}.$$

We illustrate two examples as follows:

(I) Estimate the distance between two successive zeros of the equation

$$\ddot{x} + 2t^2\dot{x} + t(2t^3 + 3)x = 0$$

on $[1, 2]$.

Solution Consider the transformation $x(t) = z(t)e^{-t^3/3}$. It transforms the given equation into the equation

$$\ddot{z} + (t^4 + t)z = 0.$$

Comparing this with (5.2), we get

$$p(t) = t^4 + t.$$

Since $1 \le t \le 2$, $0 < 2 \le p(t) \le 18$. Therefore by Theorem 5.4, we estimated as

$$\frac{\pi}{3\sqrt{2}} \le d \le \frac{\pi}{\sqrt{2}}.$$

(II) Transform the Bessel's equation of order n into the form $\ddot{x} + p(t)x = 0$. Show that if $0 \le n < 1/2$ the distance between two successive zeros of $x(t)$ is less than π and tends to π when the number of oscillating solutions increases.

Solution The Bessel's equation of order n is given by

$$t^2\ddot{y} + t\dot{y} + (t^2 - n^2)y = 0, \ t > 0.$$

Taking $y(t) = x(t)t^{-1/2}$, we obtain the transformed equation as

$$\ddot{x} + \left(1 - \frac{n^2 - 1/4}{t^2}\right)x = 0.$$

Therefore, we have

$$p(t) = 1 - \frac{n^2 - 1/4}{t^2}.$$

Now, if $0 \leq n < 1/2$, then $p(t) > 1$ and the distance between two successive zeros of $x(t)$ is

$$d < \frac{\pi}{1} = \pi.$$

If the number of zeros increases, for a sufficiently large t, $p(t)$ can be made arbitrarily close to 1, and so, the distance between two successive zeros of $x(t)$ will tend to π. In general, the Bessel's equation of order n the expression $\left(1 - \frac{n^2 - 1/4}{t^2}\right)$ can be made arbitrarily to unite a sufficiently large t. Therefore, for sufficiently large values of t, the distance between two successive zeros of the solutions of Bessel's equation is arbitrarily close to π.

We now give a lower estimate of the distance between two successive zeros of the Eq. (5.1) without reducing it into the Eq. (5.2).

Theorem 5.5 (de la Vallée Poussian) *Let the coefficients $a(t)$ and $b(t)$ of the Eq. (5.1) be such that*

$$|a(t)| \leq M_1, \quad |b(t)| \leq M_2, \, t \in I.$$

Then the distance d between two successive zeros of any nontrivial solution of (5.1) satisfies

$$d \geq \frac{\sqrt{4M_1^2 + 8M_2} - 2M_1}{M_2}.$$

See the book of Tricomi [1].

Remark 5.1 The lower estimate of the distance d between two successive zeros of the Eq. (5.1) can be determined using one of the followings:

(1) Use directly the statement of Theorem 5.5;
(2) First, apply the transformation given earlier to reduce (5.1) into (5.2) and then use the left part of the inequality of the Theorem 5.4.

In general, we cannot say which one is a better estimation for the distance between two successive zeros. In some cases, Theorem 5.5 gives better estimation. Let us consider two examples for this purpose.

(I) Consider the equation

$$\ddot{x} + 2t^2\dot{x} + t(2t^3 + 3) = 0, \quad t \in [1, 2].$$

(1) Applying Theorem 5.4

$$p(t) = t^4 + t, \, 1 \leq t \leq 2 \Rightarrow 2 \leq p(t) \leq 18 \Rightarrow d \geq \frac{\pi}{3\sqrt{2}}$$

(2) Applying Theorem 5.5

$$|a(t)| = |2t^2| \le 8 = M_1, \ |b(t)| = |t(2t^3 + 3)| \le 38 = M_2.$$

Therefore, the distance d is given by

$$d \ge \frac{\sqrt{4 \times 8^2 + 8 \times 38} - 2 \times 8}{38} \approx 0.2.$$

Therefore, Theorem 5.4 gives a better result for approximate distance between successive zeros.

We now give our attention on nonlinear oscillating systems.

5.3 Nonlinear Oscillatory Systems

Linear oscillators follow the superposition principle and the frequency of oscillation is independent of the amplitude and the initial conditions. In contrast, nonlinear oscillators do not follow the superposition principle and their frequencies of oscillations depend on both the amplitude and the initial conditions. Nonlinear oscillations have characteristic features such as resonance, jump or hysteresis, limit cycles, noisy output, etc. Some of these features are useful in communication engineering, electrical circuits, cyclic action of heart, cardiovascular flow, neural systems, biological and chemical reactions, dynamic interactions of various species, etc. Relaxation oscillations are special type of periodic phenomena. This oscillation is characterized by intervals of time in which very little change takes place, followed by short intervals of time in which significant change occurs. Relaxation oscillations occur in many branches of physics, engineering sciences, economy, and geophysics. In mathematical biology one finds applications in the heartbeat rhyme, respiratory movements of lungs, and other cyclic phenomena. We illustrate four physical problems where nonlinear oscillations are occurred.

(i) *Simple pendulum*: The simplest nonlinear oscillating system is the undamped pendulum. The equation of motion of a simple pendulum is given by $\ddot{\theta} + (g/L)\sin\theta = 0$, where g and L are the acceleration due to gravity and length of the light inextensible string, respectively. Due to the presence of the nonlinear term $\sin\theta$, the equation is nonlinear. For small angle approximations, that is, $\sin\theta = \theta - \frac{\theta^3}{6} + O(\theta^5)$, the equation can be written as $\ddot{\theta} + (g/L)\theta - (g/6L)\theta^3 = 0$. This is a good approximation even for angles as large as $\pi/4$. The original system has equilibrium points at $(n\pi, 0), n \in \mathbb{Z}$. The equilibrium solutions $(\pi, 0)$ and $(-\pi, 0)$ are unstable whereas the equilibrium solution $(0, 0)$ is stable. Consider a periodic solution with the initial condition

$\theta(0) = a, \dot{\theta}(0) = 0$ where $0 < a < \pi$. We now calculate the period of the periodic solution. The equation $\ddot{\theta} + (g/L) \sin \theta = 0$ has the first integral

$$\frac{1}{2}\dot{\theta}^2 - \left(\frac{g}{L}\right)\sin\theta = -\frac{g}{L}\cos a$$

$$\Rightarrow \frac{d\theta}{dt} = \pm[2(g/L)(\cos\theta - \cos a)]^{1/2}$$

satisfying the above initial condition. The period T of the periodic solutions is then given by

$$T = 4\int_0^a \frac{d\theta}{[2(g/L)(\cos\theta - \cos a)]^{1/2}}.$$

The period T can be expressed in terms of Jacobian elliptic functions. So, the period T depends nontrivially on the initial condition. On the other hand, the linear pendulum has the constant period $T = 2\pi\sqrt{g/L}$.

(ii) *Nonlinear electric circuits*: As we know that electric circuits may be analyzed by applying Kirchoff's laws. For a simple electric circuit, the voltage drop across the inductance is $L\frac{dI}{dt}$, L being the inductance. But for an iron-core inductance coil, the voltage drop can be expressed as $d\phi/dt$, where ϕ is the magnetic flux. The voltage drop across the capacitor is Q/C, where Q is the charge on the capacitor and C the capacitance. The current in the circuit is followed by $I = \frac{dQ}{dt}$. The equation of the current in the circuit for an iron-core inductance coil connected parallel to a charged condenser may be expressed by equation

$$\frac{d\phi}{dt} + \frac{Q}{C} = 0 \Rightarrow \frac{d^2\phi}{dt^2} + \frac{I}{C} = 0.$$

For an elementary circuit, there is a linear relationship between the current and the flux, that is, $I = \phi/L$. It is known that for an iron-core inductance the relationship is $I = A\phi - B\phi^3$, where A and B are positive constants, for small values of magnetic flux. This gives the equation of the current in the circuit as

$$\frac{d^2\phi}{dt^2} + \left(\frac{A}{C}\right)\phi - \left(\frac{B}{C}\right)\phi^3 = 0.$$

It is a nonlinear second-order equation which may exhibit oscillatory solutions for certain values of the parameters A, B and C (see Mickens [2], Lakshmanan and Rajasekar [3]).

(iii) *Brusselator chemical reactions*: It is a widely used model for chemical reactions invented by Prigogine and Lefever (1968). The following set of chemical reactions were considered

$$\left.\begin{aligned} A &\to X \\ B+X &\to D+Y \\ Y+2X &\to 3X \\ X &\to E \end{aligned}\right\}$$

The net effect of the above set of reactions is to convert the two reactants A and B into the products D and E. If the initial concentrations of A and B are large then the maximum rate concentrations of X and Y are expressed by the following equations

$$\left.\begin{aligned} \frac{dx}{dt} &= a - (1+b)x + x^2 y \\ \frac{dy}{dt} &= bx - x^2 y \end{aligned}\right\}$$

The constants a and b are proportional to the concentrations of chemical reactants A and B and the dimensionless variables x and y are proportional to the chemical reactants X and Y, respectively. The nonlinear equations may have a stable limit cycle for certain values the parameters a and b. This system exhibits oscillatory changes in the concentrations of X and Y depending upon the values of a and b.

(iv) *Glycolysis*: It is a fundamental biochemical reaction in which living cells get energy by breaking down glucose. This biochemical process may give rise to oscillations in the concentrations of various intermediate chemical reactants. However, the period of oscillations is of the order of minutes. The oscillatory pattern of reactions and the time period of oscillations are very crucial in reaching the final product. A set of equations was derived by Sel'kov (1968) for this biochemical reaction. The biochemical reaction process equations may be expressed by

$$\left.\begin{aligned} \dot{x} &= -x + ay + x^2 y \\ \dot{y} &= b - ay - x^2 y \end{aligned}\right\}$$

where x and y are proportional to the concentrations of adenosine diphosphate (ADP) and fructose-6-phospate (F6P). The positive constants a and b are kinetic parameters. The same nonlinear term $(x^2 y)$ is present in both the equations with opposite signs. A stable limit cycle may exist for certain relations of a and b. The existence of the stable limit cycle for the glycolysis mechanism indicates that the biochemical reaction reaches its desire product finally.

5.4 Periodic Solutions

The existence of periodic solutions is of great importance in dynamical systems. Consider a nonlinear autonomous system represented by

$$\left.\begin{array}{l} \dot{x} = f(x, y) \\ \dot{y} = g(x, y) \end{array}\right\} \tag{5.6}$$

where the functions $f(x, y)$ and $g(x, y)$ are continuous and have continuous first-order partial derivatives throughout the xy-plane. Global properties of phase paths are those which describe their behaviors over large region of the phase plane. The main problem of the global theory is to examine the existence of closed paths of the system (5.6). Close path solutions correspond to periodic solutions of a system. A solution $(x(t), y(t))$ of (5.6) is said to be periodic if there exists a real number $T > 0$ such that

$$x(t + T) = x(t) \text{ and } y(t + T) = y(t) \forall t.$$

The least value of T for which this relation is satisfied is called the period (prime period) of the periodic solution. Note that if a solution of (5.6) is periodic of period T, then it is periodic of period nT for every $n \in \mathbb{N}$. The periodic solutions represent a closed path which is traversed once as t increases from t_0 to $(t_0 + T)$ for any t_0. Conversely, if $C = [x(t), y(t)]$ is a closed path of the system, then $(x(t), y(t))$ is a periodic solution. There are some systems which have no closed paths and so they have no periodic solutions. We now discuss the existence/nonexistence criteria for closed paths.

5.4.1 Gradient Systems

An autonomous system $\dot{\underset{\sim}{x}} = \underset{\sim}{f}(\underset{\sim}{x})$ in $\mathbb{R}^n$ is said to be a gradient system if there exists a single valued, continuously differentiable scalar function $V = V(\underset{\sim}{x})$ such that $\dot{\underset{\sim}{x}} = -\nabla V$. The function V is known as the potential function of the system in compare with the energy in a mechanical system. In terms of components, the equation can be written as

$$\dot{x}_i = -\frac{\partial V}{\partial x_i}; \ i = 1, 2, \ldots, n.$$

Every one-dimensional system can be expressed as a gradient system. Consider a two-dimensional system (5.6). This will represent a gradient system if there exists a potential function $V = V(x, y)$ such that

$$\dot{x} = -\frac{\partial V}{\partial x}, \dot{y} = -\frac{\partial V}{\partial y}$$

$$\text{that is,} \quad f = -\frac{\partial V}{\partial x}, g = -\frac{\partial V}{\partial y}.$$

Differentiating partially the first equation by y and the second by x and then subtracting, we get

$$\frac{\partial f}{\partial y} - \frac{\partial g}{\partial x} = 0 \Rightarrow \frac{\partial f}{\partial y} = \frac{\partial g}{\partial x}.$$

This is the condition for which a two-dimensional system is expressed as a gradient system.

Theorem 5.6 *Gradient systems cannot have closed orbits.*

Proof If possible, let there be a closed orbit C in a gradient system in $\mathbb{R}^n$. Then there exists a potential function V such that $\dot{x} = -\nabla V$. Consider a change ΔV of the potential function V in one circuit. Let T be the time of one complete rotation along the closed orbit C. Since V is a single valued scalar function, we have $\Delta V = 0$. Again using the definition, we get

$$\Delta V = \int_0^T \frac{dV}{dt} dt = \int_0^T (\nabla V \cdot \dot{x}) dt \left[\because \frac{dV}{dt} = \frac{\partial V}{\partial x_1} \dot{x}_1 + \frac{\partial V}{\partial x_2} \dot{x}_2 + \cdots + \frac{\partial V}{\partial x_n} \dot{x}_n = \nabla V \cdot \dot{x} \right]$$

$$= -\int_0^T (\dot{x} \cdot \dot{x}) dt$$

$$= -\int_0^T \|\dot{x}\|^2 dt,$$

where $\|\dot{x}\|$ is the norm of x in $\mathbb{R}^n$. This is a contradiction. So, our assumption is wrong. Hence there are no closed orbits in a gradient system. This completes the proof.

We give few examples as follows:

(I) Consider the two-dimensional system $\dot{x} = 2xy + y^3$, $\dot{y} = x^2 + 3xy^2 - 2y$. Here we take $f(x, y) = 2xy + y^3$ and $g(x, y) = x^2 + 3xy^2 - 2y$. Now, calculate the derivatives as

$$\frac{\partial f}{\partial y} = 2x + 3y^2, \quad \frac{\partial g}{\partial x} = 2x + 3y^2.$$

Since $\frac{\partial f}{\partial y} = \frac{\partial g}{\partial x}$, the system is a gradient system, and so it has no closed path. The given system does not exhibit periodic solution. We now determine the potential function $V = V(x, y)$ for this system. By definition, we get

$$\frac{\partial V}{\partial x} = -f = -2xy - y^3, \quad \frac{\partial V}{\partial y} = -g = -x^2 - 3xy^2 + 2y.$$

From the first relation,

$$\frac{\partial V}{\partial x} = -2xy - y^3 \Rightarrow V = -x^2y - xy^3 + h(y)$$

where $h(y)$ is a function of y only. Differentiating this relation partially with respect to y and then using the value of $\frac{\partial V}{\partial y}$, we get

$$-x^2 - 3xy^2 + 2y = -x^2 - 3xy^2 + \frac{dh(y)}{dy}$$

$$\Rightarrow \frac{dh}{dy} = 2y$$

$$\Rightarrow h(y) = y^2 \text{ [Neglecting the constant of integration.]}$$

Therefore, the potential function of the system is $V(x, y) = -x^2y - xy^3 + y^2$.

(II) Let $V : \mathbb{R}^n \to \mathbb{R}$ be the potential of a gradient system $\underset{\sim}{x} = \underset{\sim}{f}(\underset{\sim}{x})$, $\underset{\sim}{x} \in \mathbb{R}^n$. Show that $\dot{V}(\underset{\sim}{x}) \leq 0$ $\forall \underset{\sim}{x}$ and $\dot{V}(\underset{\sim}{x}) = 0$ if and only if $\underset{\sim}{x}$ is an equilibrium point.

Solution Using the chain rule of differentiation, we have

$$\dot{V}(\underset{\sim}{x}) = \sum_{i=1}^{n} \frac{\partial V}{\partial x_i} \dot{x}_i$$

$$= \nabla V \cdot \underset{\sim}{x}$$

$$= \nabla V \cdot (-\nabla V) \quad [\text{Since } \underset{\sim}{x} = -\nabla V]$$

$$= -|\nabla V|^2 \leq 0.$$

Now, $\dot{V} = 0$ if and only if $\nabla V = 0$, that is, if and only if $\underset{\sim}{x} = 0$. Hence $\underset{\sim}{x}$ is an equilibrium point.

5.4.2   *Poincaré Theorem*

Given below is an important theorem of Poincaré on the existence of a closed path of a two-dimensional system.

Theorem 5.7 *A closed path of a two-dimensional system* (5.6) *necessarily surrounds at least one equilibrium point of the system.*

Proof If possible, let there be a closed path C of the system (5.6) that does not surround any equilibrium point of the system. Let A be the region (area) bounded by C. Then $f^2 + g^2 \neq 0$ in A. Let θ be the angle between the tangent to the closed path C and the x-axis. Then

$$\oint_C d\theta = 2\pi \tag{5.7}$$

But $\tan\theta = \dfrac{dy}{dx} = \dfrac{g}{f}$. On differentiation, we get

$$\sec^2\theta \, d\theta = \frac{f dg - g df}{f^2}$$

$$\Rightarrow \left(1 + \frac{g^2}{f^2}\right) d\theta = \frac{f dg - g df}{f^2}$$

$$\Rightarrow d\theta = \frac{f dg - g df}{f^2 + g^2}$$

Substituting this value in (5.7),

$$\oint_C \left(\frac{f dg - g df}{f^2 + g^2}\right) = 2\pi$$

$$\Rightarrow \oint_C \left\{ \left(\frac{f}{f^2 + g^2}\right) dg - \left(\frac{g}{f^2 + g^2}\right) df \right\} = 2\pi$$

Using Green's theorem in plane, we have

$$\iint_A \left\{ \frac{\partial}{\partial f}\left(\frac{f}{f^2 + g^2}\right) + \frac{\partial}{\partial g}\left(\frac{g}{f^2 + g^2}\right) \right\} df dg = \oint_C \left\{ \left(\frac{f}{f^2 + g^2}\right) dg - \left(\frac{g}{f^2 + g^2}\right) df \right\} \tag{5.8}$$

$$= 2\pi$$

But

$$\frac{\partial}{\partial f}\left(\frac{f}{f^2+g^2}\right)+\frac{\partial}{\partial g}\left(\frac{g}{f^2+g^2}\right)=\frac{g^2-f^2}{f^2+g^2}+\frac{f^2-g^2}{f^2+g^2}=0.$$

So, finally we get $0=2\pi$, which is a contradiction. Hence a closed path of the system (5.6) must surround at least one equilibrium point of the system. This completes the proof.

This theorem also implies that a system without any equilibrium point in a given region cannot have closed paths in that region.

5.4.3 Bendixson's Negative Criterion

Bendixson's negative criterion gives an easiest way of ruling out the existence of periodic orbit for a system in $\mathbb{R}^2$. The theorem is as follows.

Theorem 5.8 *There are no closed paths in a simply connected region of the phase plane of the system (5.6) on which $\left(\frac{\partial f}{\partial x}+\frac{\partial g}{\partial y}\right)$ is not identically zero and is of one sign.*

Proof Let D be a simply connected region of the system (5.6) in which $\left(\frac{\partial f}{\partial x}+\frac{\partial g}{\partial y}\right)$ is of one sign. If possible, let C be a closed curve in D and A be the area of the region. Then by divergence theorem, we have

$$\iint_A \left(\frac{\partial f}{\partial x}+\frac{\partial g}{\partial y}\right)\mathrm{dxdy} = \oint_C (f,g)\cdot \underset{\sim}{n}\,\mathrm{d}l$$

$$= 0\,[\because (f,g)\perp \underset{\sim}{n}\,].$$

where $\mathrm{d}l$ is an undirected line element of the path C. This is a contradiction, since $\left(\frac{\partial f}{\partial x}+\frac{\partial g}{\partial y}\right)$ is of one sign, that is, either positive or negative, and hence the integral cannot be zero. Therefore, C cannot be a closed path of the system. This completes the proof.

We now illustrate the theorem through some examples.

Example 1 Show that the equation $\ddot{x}+f(x)\dot{x}+g(x)=0$ cannot have periodic solutions whose phase path lies in a region where f is of one sign.

Solution Let D be the region where f is of one sign. Given equation can be written as

$$\left.\begin{array}{l} \dot{x} = y = F(x, y) \\ \dot{y} = -yf(x) - g(x) = G(x, y) \end{array}\right\}$$

Therefore,

$$\frac{\partial F}{\partial x} + \frac{\partial G}{\partial y} = 0 - f(x) = -f(x).$$

This shows that $\left(\frac{\partial F}{\partial x} + \frac{\partial G}{\partial y} \right)$ is of one sign in D, since f is of one sign in D. Hence by Bendixson's negative criterion, there is no closed path of the system in D. Therefore, the given equation cannot have periodic solution in the region where f is of one sign.

Example II Consider the system $\dot{x} = p(y) + x^n$, $\dot{y} = q(x)$, where $p, q \in C^1$ and $n \in \mathbb{N}$. Derive a sufficient condition for n so that the system has no periodic solution.

Solution Let $f(x, y) = p(y) + x^n$ and $g(x, y) = q(x)$. Then

$$\frac{\partial f}{\partial x} + \frac{\partial g}{\partial y} = \frac{\partial}{\partial x}(p(y) + x^n) + \frac{\partial}{\partial y}q(x) = nx^{n-1}.$$

This expression is of one sign if n is odd. So, by Bendixson's negative criterion the given system has no periodic solutions, if n is odd. This is the required sufficient condition for n.

Example III Show that the system

$$\dot{x} = -y + x(x^2 + y^2 - 1), \quad \dot{y} = x + y(x^2 + y^2 - 1)$$

has no closed orbits inside the circle with center at $(0, 0)$ and radius $\frac{1}{\sqrt{2}}$.

Solution Let $f(x, y) = -y + x(x^2 + y^2 - 1)$ and $g(x, y) = x + y(x^2 + y^2 - 1)$. Now

$$\frac{\partial f}{\partial x} + \frac{\partial g}{\partial y} = 3x^2 + y^2 - 1 + x^2 + 3y^2 - 1 = 4\left(x^2 + y^2 - \frac{1}{2}\right)$$

Clearly, $\left(\frac{\partial f}{\partial x} + \frac{\partial g}{\partial y} \right)$ is of one sign inside the circle $x^2 + y^2 = \frac{1}{2}$, which is a simply connected region in $\mathbb{R}^2$. Hence by Bendixson's negative criterion, the given system has no closed orbits inside the circle with center at $(0, 0)$ and radius $\frac{1}{\sqrt{2}}$. We also see that $\left(\frac{\partial f}{\partial x} + \frac{\partial g}{\partial y} \right)$ is of one sign outside the circle $x^2 + y^2 = \frac{1}{2}$. But it is not simply connected. So, we cannot apply Bendixson's criterion on this region.

5.4.4 Dulac's Criterion

There are some systems for which Bendixson's negative criterion fails in ruling out the existence of closed paths. For example, consider the system $\dot{x} = y$, $\dot{y} = x - y + y^2$. Using Bendixson's negative criterion, one cannot be ascertained the ruling out of the closed path of the system. However, the generalization of this criterion may give the ruling out of the closed path and this is due to Dulac, who made the generalization of Bendixon's negative criterion. The criterion is given below.

Theorem 5.9 *Consider the system* $\dot{x} = f(x, y)$, $\dot{y} = g(x, y)$, *where* $f(x, y)$, $g(x, y)$ *are continuously differentiable functions in a simply connected region D of* $\mathbb{R}^2$. *If there exists a real-valued continuously differentiable function* $\rho = \rho(x, y)$ *in D such that*

$$\frac{\partial(\rho f)}{\partial x} + \frac{\partial(\rho g)}{\partial y}$$

is of one sign throughout D, then the system has no closed orbits (periodic solutions) lying entirely in D.

Proof If possible, let there be a closed orbit C in the region D on which $\left(\frac{\partial(\rho f)}{\partial x} + \frac{\partial(\rho g)}{\partial y}\right)$ is of one sign. Let A be the area bounded by C. Then by divergence theorem, we have

$$\iint_A \left(\frac{\partial(\rho f)}{\partial x} + \frac{\partial(\rho g)}{\partial y}\right) \mathrm{d}x\mathrm{d}y = \oint_C (\rho f, \rho g) \cdot \underset{\sim}{n} \, \mathrm{d}l = \oint_C \rho(f, g) \cdot \underset{\sim}{n} \, \mathrm{d}l$$

where $\underset{\sim}{n}$ is the unit outward drawn normal to the closed orbit C and $\mathrm{d}l$ is an elementary line element along C (Fig. 5.2).

Since (f, g) is perpendicular to $\underset{\sim}{n}$, we have

$$\oint_C \rho(f, g) \cdot \underset{\sim}{n} \, \mathrm{d}l = 0.$$

So, $\iint_A \left(\frac{\partial(\rho f)}{\partial x} + \frac{\partial(\rho g)}{\partial y}\right) \mathrm{d}x\mathrm{d}y = 0$. This yields a contradiction, since $\left(\frac{\partial(\rho f)}{\partial x} + \frac{\partial(\rho g)}{\partial y}\right)$ is of one sign. Hence no such closed orbit C can exist.

Fig. 5.2 Sketch of the domains

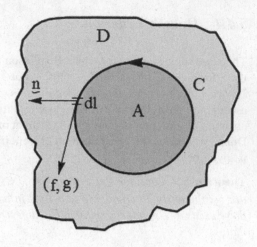

Remarks

(i)   The function $\rho = \rho(x, y)$ is called the weight function.

(ii) Bendixson's negative criterion is a particular case of Dulac's criterion with $\rho = 1$.

(iii) The main difficulty of Dulac's criterion is to choose the weight function ρ. There is no specific rule for choosing this function.

Example Show that the system $\dot{x} = x(\alpha - ax - by)$, $\dot{y} = y(\beta - cx - \mathrm{d}y)$ where $a, d > 0$ has no closed orbits in the positive quadrant in $\mathbb{R}^2$.

Solution Let $D = \{(x, y) \in \mathbb{R}^2 : x, y > 0\}$. Clearly, D is a simply connected region in $\mathbb{R}^2$. Let $f(x, y) = x(\alpha - ax - by)$, $g(x, y) = y(\beta - cx - \mathrm{d}y)$. Consider the weight function $\rho(x, y) = \frac{1}{xy}$. Then

$$\frac{\partial(\rho f)}{\partial x} + \frac{\partial(\rho g)}{\partial y} = \frac{\partial}{\partial x}\left(\frac{\alpha - ax - by}{y}\right) + \frac{\partial}{\partial y}\left(\frac{\beta - cx - \mathrm{d}y}{x}\right)$$

$$= -\left(\frac{a}{y} + \frac{d}{x}\right) < 0 \ \forall (x, y) \in D.$$

Again, the functions f, g and ρ are continuously differentiable in D. Hence by Dulac's criterion, the system has no closed orbits in the positive quadrant $x, y > 0$ of $\mathbb{R}^2$.

5.5 Limit Cycles

A limit cycle (cycle in limiting sense) is an isolated closed path such that the neighboring paths (or trajectories) are not closed. The trajectories approach to the closed path or move away from it spirally. This is a nonlinear phenomenon and occurs in many physical systems such as the path of a satellite, biochemical processes, predator–prey model, nonlinear electric circuits, economical growth model, ecology, beating of heart, self-excited vibrations in bridges and airplane wings, daily rhythms in human body temperature, hormone secretion, etc. Linear systems cannot support limit cycles. There are basically three types of limit cycles, namely stable, unstable, and semistable limit cycles. A limit cycle is said to be stable (or attracting) if it attracts all neighboring trajectories. If the neighboring trajectories are repelled from the limit cycle, then it is called an unstable (or repelling) limit cycle. A semistableLimit cycle:semistable limit cycle is one which attracts trajectories from one side and repels from the other. These three types of limit cycles are shown in Fig. 5.3.

Scientifically, the stable limit cycles are very important.

5.5.1 Poincaré–Bendixson Theorem

So far we have discussed theorems that describe the procedure for the nonexistence of periodic orbit of a system in some region in the phase plane. It is extremely difficult to prove the existence of a limit cycle or periodic solution for a nonlinear system of $n \geq 3$ variables. The Poincaré–Bendixson theorem permits us to prove the existence of at least one periodic orbit of a system in $\mathbb{R}^2$ under certain conditions The main objective of this theorem is to find an 'annular region' that does not contain any equilibrium point of the system, in which one can find at least one periodic orbit. The proof of the theorem is quite complicated because of the topological concepts involved in it.

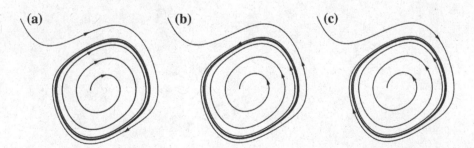

Fig. 5.3 **a** Stable, **b** unstable, **c** semistable limit cycles

Theorem 5.10 *Suppose that*

(i)   *R is a closed, bounded subset of the phase plane.*

(ii) $\dot{\underset{\sim}{x}} = \underset{\sim}{f}\,(\underset{\sim}{x})$ *is a continuously differentiable vector field on an open set containing R.*

(iii) *R does not contain any fixed points of the system.*
      There exists a trajectory C of the system that lies in R for some time t_0, say, and remains in R for all future time $t \geq t_0$.
      Then C is either itself a closed orbit or it spirals towards a closed orbit as time $t \to \infty$. In either case, the system has a closed orbit in R.

For explaining the theorem, we consider a region R containing two dashed curves together with the ring-shaped region between them as depicted in Fig. 5.4. Every path C through a boundary point at $t = t_0$ must enter in R and can leave it. The theorem asserts that C must spiral toward a closed path C_0. The closed path C_0 must surround a fixed point, say P and the region R must exclude all fixed points of the system.

The Poincaré–Bendixson theorem is quite satisfying from the theoretical point of view. But in general, it is rather difficult to apply. We give an example that shows how to use the theorem to prove the existence of at least one periodic orbit of the system.

Example Consider the system $\dot{x} = x - y - x(x^2 + 2y^2)$, $\dot{y} = x + y - y(x^2 + 2y^2)$. The origin $(0,0)$ is a fixed of the system. Other fixed points must satisfy $\dot{x} = 0$, $\dot{y} = 0$. These give

$$x = y + x(x^2 + 2y^2), \; x = -y + y(x^2 + 2y^2).$$

A sketch of these two curves shows that they cannot intersect except at the origin. So, the origin is the only fixed point of the system. We now covert the

Fig. 5.4 Sketch of the annular region

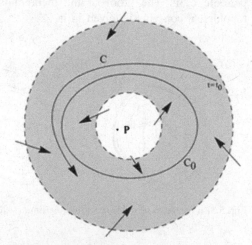

system into polar coordinates (r, θ) using the relations $x = r\cos\theta$, $y = r\sin\theta$ where $r^2 = x^2 + y^2$ and $\tan\theta = y/x$. Differentiating the expression $r^2 = x^2 + y^2$ with respect to t, we have

$$\begin{aligned}
r\dot{r} &= x\dot{x} + y\dot{y} \\
&= x[x - y - x(x^2 + 2y^2)] + y[x + y - y(x^2 + 2y^2)] \\
&= x^2 + y^2 - (x^2 + y^2)(x^2 + 2y^2) \\
&= r^2 - r^2(r^2 + r^2\sin^2\theta) \\
&= r^2 - (1 + \sin^2\theta)r^4 \\
\Rightarrow \dot{r} &= r - (1 + \sin^2\theta)r^3.
\end{aligned}$$

Similarly, differentiating $\tan\theta = y/x$ with respect to t, we get $\dot{\theta} = 1$.

We see that $\dot{r} > 0$ for all θ if $(r - 2r^3) > 0$, that is, if $r^2 < 1/2$, that is, if $r < 1/\sqrt{2}$, and $\dot{r} < 0$ for all θ if $(r - r^3) < 0$, that is, if $r^2 > 1$, that is, if $r > 1$. We now define an annular region R on which we can apply the Poincaré–Bendixson theorem. Consider the annular region

$$R = \left\{ (r, \theta) : \frac{1}{\sqrt{2}} \le r \le 1 \right\}.$$

Since the origin is the only fixed point of the system, the region R does not contain any fixed point of the system. Again, since $\dot{r} > 0$ for $r < 1/\sqrt{2}$ and $\dot{r} < 0$ for $r > 1$, all trajectories in R will remain in R for all future time. Hence, by Poincaré–Bendixson theorem, there exists at least one periodic orbit of the system in the annular region R.

Again, consider another system below

$$\dot{x} = -y + x(x^2 + y^2)\sin\frac{1}{\sqrt{x^2 + y^2}}$$

$$\dot{y} = x + y(x^2 + y^2)\sin\frac{1}{\sqrt{x^2 + y^2}}.$$

The origin is an equilibrium point of the system. In polar coordinates, the system can be transformed as

$$\dot{r} = r^3\sin\frac{1}{r}, \quad \dot{\theta} = 1.$$

This system has limit cycles Γ_n lying on the circle $r = 1/(n\pi)$. These limit cycles concentrated at the origin for large value of n, that is, the distance between the limit cycle Γ_n and the equilibrium point origin decreases as n increases, and finally the distance becomes zero as $n \to \infty$. Among these limit cycles, the limit cycles Γ_{2n} are stable while the other are unstable. The existence of finite numbers of limit cycles is of great importance physically. The following theorem gives the criterion for a system to have finite number of limit cycles.

Theorem 5.11 (Dulac) *In any bounded region of the plane, a planar analytic system $\dot{x} = f(x)$ with $f(x)$ analytic in $\mathbb{R}^2$ has at most a finite number of limit cycles. In other words, any polynomial system has at most a finite number of limit cycles in $\mathbb{R}^2$.*

Theorem 5.12 (Poincaré) *A planar analytic system (5.6) cannot have an infinite number of limit cycles that accumulate on a cycle of (5.6).*

5.5.2 Liénard System

Consider the system

$$\left.\begin{array}{l} \dot{x} = y - F(x) \\ \dot{y} = -g(x) \end{array}\right\} \tag{5.9}$$

This system can also be written as a second-order differential equation of the form

$$\ddot{x} + f(x)\dot{x} + g(x) = 0 \tag{5.10}$$

where $f(x) = F'(x)$. This equation is popularly known as *Liénard equation*, according to the the French Physicist A. *Liénard*, who invented this equation in the year 1928 in connection with nonlinear electrical circuit. The Liénard equation is a generalization of the *van der Pol* (Dutch Electrical Engineer) oscillator in connection with diode circuit as

$$\ddot{x} + \mu(x^2 - 1)\dot{x} + x = 0,$$

$\mu \geq 0$ is the parameter. The Liénard equation can also be interpreted as the motion of a unit mass subject to a nonlinear damping force $(-f(x)\dot{x})$ and a nonlinear restoring force $(-g(x))$. Under certain conditions on the functions F and g, Liénard proved the existence and uniqueness of a stable limit cycle of the system.

Theorem 5.13 (Liénard Theorem) *Suppose two functions $f(x)$ and $g(x)$ satisfy the following conditions:*

 (i) *$f(x)$ and $g(x)$ are continuously differentiable for all x,*
 (ii) *$g(-x) = -g(x)$ $\forall x$, that is, $g(x)$ is an odd function,*
 (iii) *$g(x) > 0$ for $x > 0$,*
 (iv) *$f(-x) = f(x)$ $\forall x$, that is, $f(x)$ is an even function,*

(v) *The odd function,* $F(x) = \int_0^x f(u)du$, *has exactly one positive zero at* $x = \alpha$,
say, $F(x)$ *is negative for* $0 < x < \alpha$, *and* $F(x)$ *is positive and nondecreasing
for* $x > \alpha$ *and* $F(x) \to \infty$ *as* $x \to \infty$.

*Then the Liénard equation (5.10) has a unique stable limit cycle surrounding the
origin of the phase plane.*

The theorem can also be stated as follows.

Under the assumptions that $F, g \in C^1(\mathbb{R})$, F *and* g *are odd functions of* x,
$xg(x) > 0$ *for* $x \neq 0$, $F(0) = 0$, $F'(0) < 0$, F *has single positive zero at* $x = \alpha$, *and
* F *increases monotonically to infinity for* $x \geq \alpha$ *as* $x \to \infty$, *it follows that the
Liénard system (5.9) has a unique stable limit cycle.*

The Liénard system is a very special type of equation that has a unique stable
limit cycle. The following discussions are given for the existence of finite numbers
of limit cycles for some special class of equations. In 1958, the Chinese mathematician Zhang Zhifen proved a theorem for the existence and the number of limit
cycles for a system. The theorem is given below.

Theorem 5.14 (Zhang theorem-I) *Under the assumptions that* $a < 0 < b$,
$F, g \in C^1(a, b)$, $xg(x) > 0$ *for* $x \neq 0$, $G(x) \to \infty$ *as* $x \to a$ *if* $a = -\infty$ *and
* $G(x) \to \infty$ *as* $x \to b$ *if* $b = \infty$, $f(x)/g(x)$ *is monotone increasing on* $(a, 0) \cap (0, b)$
and is not constant in any neighborhood of $x = 0$, *it follows that the system (5.9)
has at most one limit cycle in the region* $a < x < b$ *and if it exists it is stable, where*

$$G(x) = \int_0^x g(u)du.$$

Again in 1981, he proved another theorem relating the number of limit cycles of
the Liénard type systems.

Theorem 5.15 (Zhang theorem-II) *Under the assumptions that* $g(x) = x$,
$F \in C^1(\mathbb{R})$, $f(x)$ *is an even function with exactly two positive zeros* $a_1, a_2 (a_1 < a_2)$
with $F(a_1) > 0$ *and* $F(a_2) < 0$, *and* $f(x)$ *is monotone increasing for* $x > a_2$, *it
follows that the system (5.9) has at most two limit cycles.*

If $g(x) = x$ and $F(x)$ is a polynomial, then one can ascertain the number of limit
cycles of a system from the following theorem.

Theorem 5.16 (Lins, de Melo and Pugh) *The system (5.9) with* $g(x) = x$,
$F(x) = a_1 x + a_2 x^2 + a_3 x^3$, *and* $a_1 a_3 < 0$ *has exactly one limit cycle. It is stable if
* $a_1 < 0$ *and unstable if* $a_1 > 0$.

Remark The Russian mathematician Rychkov proved that the system (5.9) with
$g(x) = x$ and $F(x) = a_1 x + a_3 x^3 + a_5 x^5$ has at most two limit cycles.

See Perko [4] for detail discussions.

The great mathematician David Hilbert presented 23 outstanding mathematical
problems to the Second International Congress of Mathematics in 1900 and the

16th Hilbert Problem was to determine the maximum number of limit cycles H_n, of an n th degree polynomial system

$$\dot{x} = \sum_{i+j=0}^{n} a_{ij} x^i y^j$$

$$\dot{y} = \sum_{i+j=0}^{n} b_{ij} x^i y^j$$

For given $(a, b) \in \mathbb{R}^{(n+1)(n+2)}$, the number of limit cycles $H_n(a, b)$ of the above system must be finite. This can be ascertained from Dulac's theorem. So even for a nonlinear system in $\mathbb{R}^2$, it is difficult to determine the number of limit cycles. In the year 1962, Russian mathematician N.V. Bautin proved that any quadratic system has at most three limit cycles. However, in 1979 the Chinese mathematicians S.L. Shi, L.S. Chen, and M.S. Wang established that a quadratic system has four limit cycles. It had been proved by Y.X. Chin in 1984. A cubic system can have at least 11 limit cycles. Thus the determination of the number of limit cycles is extremely difficult for a system in general.

5.5.3 van der Pol Oscillator

We now discuss van der Pol equation which is a special type nonlinear oscillator. This type of oscillator is appeared in electrical circuits, diode valve, etc. The van der Pol equation is given by

$$\ddot{x} + \mu(x^2 - 1)\dot{x} + x = 0, \ \mu > 0.$$

Comparing with the Liénard equation (5.10), we get $f(x) = \mu(x^2 - 1)$ and $g(x) = x$. Clearly, the conditions (i) to (iv) of the Liénard theorem are satisfied. We only check the condition (v). So we have

$$F(x) = \int_0^x f(u)du = \int_0^x \mu(u^2 - 1)du$$

$$= \mu\left[\frac{u^3}{3} - u\right]_0^x = \mu\left(\frac{x^3}{3} - x\right)$$

$$\Rightarrow F(x) = \frac{1}{3}\mu x(x^2 - 3).$$

The function $F(x)$ is odd. $F(x)$ has exactly one positive zero at $x = \sqrt{3}$. $F(x)$ is negative for $0 < x < \sqrt{3}$, and it is positive and nondecreasing for $x > \sqrt{3}$ and

$F(x) \to \infty$ as $x \to \infty$. So, the condition (v) is satisfied for $x = \sqrt{3}$. Hence, the van der Pol equation has a unique stable limit cycle, provided $\mu > 0$. The graphical representation of the limit cycle and the solution $x(t)$ are displayed below taking initial point $x(0) = 0.5$, $\dot{x}(0) = 0$ and $\mu = 1.5$ and 0.5, respectively.

The behavior of solutions for large values of the parameter μ can be understood from the figures. The graph of t versus x for $\mu = 1.5$ (cf. Fig. 5.5a) is characterized by fast changes of the position x near certain values of time t. The van der Pol equation can be expressed as $\ddot{x} + \mu\dot{x}(x^2 - 1) = \frac{d}{dx}\left(\dot{x} + \mu(\frac{1}{3}x^3 - x)\right)$. Now, we put $f(x) = -x + x^3/3$ and $\mu y = \dot{x} + \mu f(x)$. So, the van der Pol equation becomes

$$\dot{x} = \mu(y - f(x))$$
$$\dot{y} = -x/\mu$$

Therefore,

$$\frac{dy}{dx} = \frac{\dot{y}}{\dot{x}} = -\frac{x}{\mu^2(y - f(x))} \Rightarrow (y - f(x))\frac{dy}{dx} = -\frac{x}{\mu^2}.$$

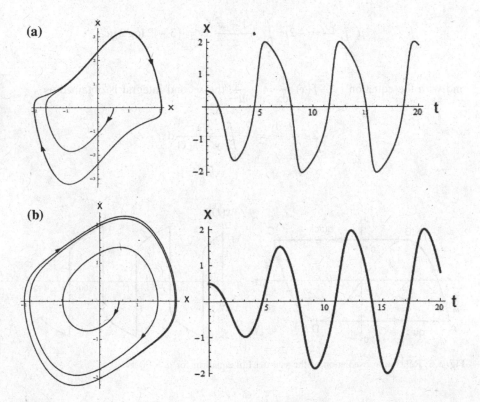

(a)

(b)

Fig. 5.5 **a** van der Pol oscillator for $\mu = 1.5$. **b** van der Pol oscillator for $\mu = 0.5$

For $\mu \gg 1$, the right-hand side of the above equation is small and the orbits can be described by the equation

$$(y - f(x))\frac{dy}{dx} = 0 \Rightarrow \text{either } y = f(x) \text{ or } y = \text{constant.}$$

A sketch of the function $y = f(x) = -x + x^3/3$ indicates that the variable $x(t)$ changes quickly in the outside of the curve, whereas the variable $y(t)$ is changing very slow. Using the Poincaré-Bendixson theorem in the annular region $R = \{(x, y) : -2 \leq x \leq 2, -1 \leq y \leq 1\} \backslash \{(x, y) : -1 \leq x \leq 1, -0.5 \leq y \leq 0.5\}$, which does not contain the equilibrium point origin of the system, the limit cycle must be located in a neighborhood of the curve $y = f(x)$, as shown in Fig. 5.6. Now, the relaxation period T is given by

$$T = -\mu \int_{ABCDA} \frac{dy}{x} = -2\mu \int_{AB} \frac{dy}{x} - 2\mu \int_{BC} \frac{dy}{x}.$$

where the first integral corresponds with the slow motion and so it gives the largest contribution to the period T. With $y = f(x) = -x + x^3/3$, we see that

$$-2\mu \int_{AB} \frac{dy}{x} = -2\mu \int_{-2}^{-1} \frac{-1 + x^2}{x} dx = (3 - 2\log 2)\mu$$

and with the equation $(y - f(x))\frac{dy}{dx} = -\frac{x}{\mu^2}$, the second integral is obtained as

$$-2\mu \int_{BC} \frac{dy}{x} = \frac{2}{\mu} \int_{BC} \frac{x}{y - f(x)} dx.$$

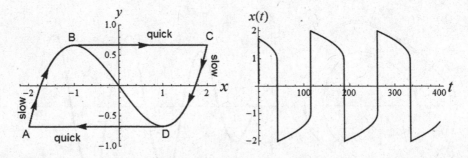

Fig. 5.6 Relaxation oscillation of the van der Pol equation for $\mu = 90$

For $\mu \gg 1$, we get approximately the equation $(y - f(x))\dfrac{dy}{dx} = 0$. The orders of

$(y - f(x))$ and $\dfrac{dy}{dx}$ depend upon the parameter μ and they are inversely proportional
to each other. One can obtain the orders of x and y. From Fig. 5.6, it is easily found
that for $\mu \gg 1$, $y \sim O(\mu^{2/3})$, and so $y - f(x) \sim O(\mu^{-2/3})$. Therefore, the integral
$-2\mu \displaystyle\int_{BC} \dfrac{dy}{x}$ must be of order $O(\mu^{-1/3})$ for $\mu \gg 1$. Hence the period T of the
periodic solution of the van der Pol equation is given by

$$T = (3 - 2\log 2)\mu + O(\mu^{-1/3}) \text{ when } \mu \gg 1.$$

Relaxation oscillation is a periodic phenomenon in which a slow build-up is
followed by a fast discharge. One may think that there are two time scales, viz., the
fast and slow scales that operate sequentially.

We now discuss another consequence of nonlinear oscillations. Weakly nonlinear oscillations have been observed in many physical systems. Let us consider
the van der Pol equation $\ddot{x} + \varepsilon(x^2 - 1)\dot{x} + x = 0$ and the Duffing equation
$\ddot{x} + x + \varepsilon x^3 = 0$ where $\varepsilon \ll 1$. The phase diagrams for the two nonlinear oscillators
with the initial condition $x(0) = a, \dot{x}(0) = 0$ are shown in Fig. 5.7.

The phase trajectories for the van der Pol oscillator are slowly building up and
the trajectories take many cycles to grow amplitude. The trajectories finally reach to
circular limit cycle of radius 2 as shown in Fig. 5.7a. In case of Duffing oscillator,
the phase trajectories form a closed path and the frequency of oscillation is
depending on ε and a. The Duffing equation is conservative. It has a nonlinear

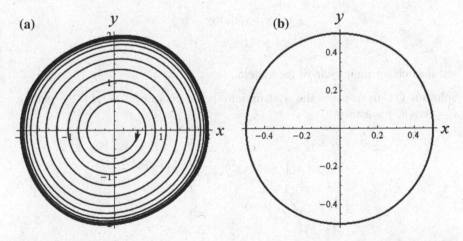

Fig. 5.7 Limit cycles of the **a** van der Pol oscillator and **b** duffing oscillator for $\varepsilon = 0.1, a = 0.5$

center at the origin. For sufficiently small ε, all orbits close to the origin are periodic with no change in amplitude in long term. The mathematical theories such as regular perturbation and method of averaging are well-established and can be found in Gluckenheimer and Holmes [5], Grimshaw [6], Verhulst [7], and Strogatz [8]. We present few examples for limit cycles as follows:

Example 5.1 Show that the equation $\ddot{x} + \mu(x^4 - 1)\dot{x} + x = 0$ has a unique stable limit cycle if $\mu > 0$.

Solution Comparing the given equation with the Liénard equation (5.9), we get $f(x) = \mu(x^4 - 1)$, $g(x) = x$. Let $\mu > 0$. Clearly, the conditions (i) to (iv) of the Liénard theorem are satisfied. We now check the condition (v). We have

$$F(x) = \int_0^x f(u)du = \int_0^x \mu(u^4 - 1)du$$

$$= \mu\left[\frac{u^5}{5} - u\right]_0^x = \mu\left(\frac{x^5}{5} - x\right)$$

$$\Rightarrow F(x) = \frac{1}{5}\mu x(x^4 - 5) = \frac{1}{5}\mu x(x^2 + \sqrt{5})(x^2 - \sqrt{5}).$$

The function $F(x)$ is odd. $F(x)$ has exactly one positive zero at $x = \sqrt[4]{5}$, is negative for $0 < x < \sqrt[4]{5}$, is positive and nondecreasing for $x > \sqrt[4]{5}$ and $F(x) \to \infty$ as $x \to \infty$. So, the condition (v) is satisfied for $x = \sqrt[4]{5}$. Thus the given equation has a unique stable limit cycle if $\mu > 0$.

Example 5.2 Find analytical solution of the following system

$$\dot{x} = -y + x(1 - x^2 - y^2)$$
$$\dot{y} = x + y(1 - x^2 - y^2)$$

and then obtain limit cycle of the system.

Solution Let us convert the system into polar coordinates (r, θ) by putting $x = r\cos\theta$, $y = r\sin\theta$,

$$r\dot{r} = x\dot{x} + y\dot{y}$$
$$= x[-y + x(1 - x^2 - y^2)] + y[x + y(1 - x^2 - y^2)]$$
$$= (x^2 + y^2)(1 - x^2 - y^2)$$
$$= r^2(1 - r^2)$$
$$\Rightarrow \dot{r} = r(1 - r^2)$$

Similarly, differentiating $\tan\theta = y/x$ with respect to t, we get

$$\sec^2(\theta)\dot\theta = \frac{x\dot y - y\dot x}{x^2}$$

$$\Rightarrow \left(1 + \frac{y^2}{x^2}\right)\dot\theta = \frac{x[x + y(1 - x^2 - y^2)] - y[-y + x(1 - x^2 - y^2)]}{x^2}$$

$$\Rightarrow (x^2 + y^2)\dot\theta = x^2 + y^2$$

$$\Rightarrow \dot\theta = 1$$

So, the system becomes

$$\left.\begin{array}{l} \dot r = r(1 - r^2) \\ \dot\theta = 1 \end{array}\right\} \tag{5.11}$$

We now solve this system. We have

$$\dot r = \frac{dr}{dt} = r(1 - r^2) \Rightarrow \left(\frac{1}{r} + \frac{r}{1 - r^2}\right) dr = dt$$

Integrating, we get

$$\log r - \frac{1}{2}\log(1 - r^2) = t - \frac{1}{2}\log c \Rightarrow \frac{1}{r^2} = 1 + ce^{-2t}$$

where c is an arbitrary constant. Similarly

$$\dot\theta = \frac{d\theta}{dt} = 1 \Rightarrow \theta(t) = t + \theta_0$$

where $\theta_0 = \theta(t = 0)$. Therefore, the solution of the system (5.11) is

$$\left.\begin{array}{l} r = \frac{1}{\sqrt{1 + ce^{-2t}}} \\ \theta = t + \theta_0 \end{array}\right\}$$

Hence, the corresponding general solution of the original system is given by

$$\left.\begin{array}{l} x(t) = \frac{\cos(t + \theta_0)}{\sqrt{1 + ce^{-2t}}} \\ y(t) = \frac{\sin(t + \theta_0)}{\sqrt{1 + ce^{-2t}}} \end{array}\right\}$$

Now, if $c = 0$, we have the solutions $r = 1$, $\theta = t + \theta_0$. This represents the closed path $x^2 + y^2 = 1$ in anticlockwise direction (since θ increases as t increases).

If $c < 0$, it is clear that $r > 1$ and $r \to 1$ as $t \to \infty$. Again, if $c > 0$, we see that $r < 1$ and $r \to 1$ as $t \to \infty$. This shows that there exists a single closed path ($r = 1$) and all other paths approach spirally from the outside or inside as $t \to \infty$. The following figure is drawn for different values of c (Fig. 5.8).

The important point is that the closed orbit or isolated closed orbit of the system solely depends on some parameters.

From Fig. 5.9, we see that all solutions of the equation tend to the periodic solution (limit cycle) $S = \{(x,y) : x^2 + y^2 = 1\}$.

Example 5.3 Show that the system

$$\dot{x} = y + \frac{x}{\sqrt{x^2 + y^2}}\{1 - (x^2 + y^2)\}, \; \dot{y} = -x + \frac{y}{\sqrt{x^2 + y^2}}\{1 - (x^2 + y^2)\}$$

has a stable limit cycle.

Fig. 5.8 Solution curves for
different values of c

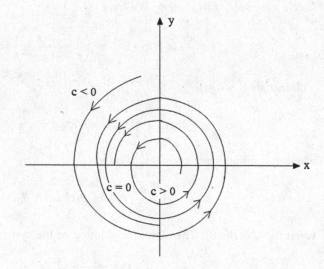

Fig. 5.9 Sketch of the limit
cycle of the system

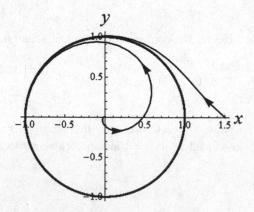

Solution Let us convert the system into polar coordinate (r, θ) using $x = r\cos\theta$, $y = r\sin\theta$. Then

$$r^2 = x^2 + y^2, \quad \tan\theta = \frac{y}{x}$$

Then the system can be written as

$$\dot{x} = y + \frac{x}{r}(1 - r^2), \quad \dot{y} = -x + \frac{y}{r}(1 - r^2)$$

Differentiating $r^2 = x^2 + y^2$ with respect to t, we have

$$
\begin{aligned}
r\dot{r} &= x\dot{x} + y\dot{y} \\
&= x\left[y + \frac{x}{r}(1 - r^2)\right] + y\left[-x + \frac{y}{r}(1 - r^2)\right] \\
&= \frac{(x^2 + y^2)}{r}(1 - r^2) \\
&= r(1 - r^2) \\
\Rightarrow \dot{r} &= 1 - r^2
\end{aligned}
$$

Similarly, differentiating $\tan\theta = y/x$ with respect to t, we get

$$
\begin{aligned}
\sec^2(\theta)\dot{\theta} &= \frac{x\dot{y} - y\dot{x}}{x^2} \\
\Rightarrow \left(1 + \frac{y^2}{x^2}\right)\dot{\theta} &= \frac{x\left[-x + \frac{y}{r}(1 - r^2)\right] - y\left[y + \frac{x}{r}(1 - r^2)\right]}{x^2} \\
\Rightarrow (x^2 + y^2)\dot{\theta} &= -(x^2 + y^2) \\
\Rightarrow \dot{\theta} &= -1
\end{aligned}
$$

So, the system becomes

$$\left.\begin{aligned} \dot{r} &= 1 - r^2 \\ \dot{\theta} &= -1 \end{aligned}\right\}$$

We now solve this system. We have

$$\frac{dr}{dt} = \dot{r} = 1 - r^2 \Rightarrow \left(\frac{1}{1+r} + \frac{1}{1-r}\right)dr = 2dt$$

Integrating, we get

$$\log(1+r) - \log(1-r) = 2t + \log A$$

$$\Rightarrow \log\left(\frac{1+r}{1-r}\right) = 2t + \log A$$

$$\Rightarrow \frac{1+r}{1-r} = Ae^{2t}$$

$$\Rightarrow r = \frac{Ae^{2t} - 1}{Ae^{2t} + 1}$$

where $A = \frac{1+r_0}{1-r_0}$, $r_0 \neq 1$ being the initial condition.

Now, $r \to 1$ as $t \to \infty$ and the limit cycle in this case is a circle of unit radius.

If $r_0 > 1$, the spiral goes itself into the circle $r = 1$ from outside in clockwise direction (as $\dot\theta = -1$) and if $r_0 < 1$, it goes itself onto the unit circle $r = 1$ from the inside in the same direction. Therefore, the limit cycle is stable.

Example 5.4 Find the limit cycle of the system

$$\dot{x} = (x^2 + y^2 - 1)x - y\sqrt{x^2 + y^2}, \; \dot{y} = (x^2 + y^2 - 1)y + x\sqrt{x^2 + y^2}$$

and investigate its stability.

Solution Let us convert the system into polar coordinates (r, θ) putting

$$x = r\cos\theta, \; y = r\sin\theta$$

where $r^2 = x^2 + y^2$ and $\tan\theta = y/x$. Then the system can be written as

$$\dot{x} = (r^2 - 1)x - yr, \; \dot{y} = (r^2 - 1)y + xr.$$

Differentiating $r^2 = x^2 + y^2$ with respect to t

$$r\dot{r} = x\dot{x} + y\dot{y}$$
$$= x[(r^2 - 1)x - yr] + y[(r^2 - 1)y + xr]$$
$$= (x^2 + y^2)(r^2 - 1) = r^2(r^2 - 1)$$
$$\Rightarrow \dot{r} = r(r^2 - 1)$$

Differentiating $\tan\theta = y/x$ with respect to t

$$\sec^2(\theta)\dot\theta = \frac{x\dot{y} - y\dot{x}}{x^2}$$

$$\Rightarrow \left(1 + \frac{y^2}{x^2}\right)\dot\theta = \frac{x[(r^2 - 1)y + xr] - y[(r^2 - 1)x - yr]}{x^2}$$

$$\Rightarrow (x^2 + y^2)\dot\theta = (x^2 + y^2)r$$

$$\Rightarrow \dot\theta = r$$

Therefore, the given system reduces to

$$\left.\begin{array}{l} \dot{r} = r(r^2 - 1) \\ \dot{\theta} = r \end{array}\right\}$$

Now,

$$\frac{\mathrm{d}r}{\mathrm{d}t} = \dot{r} = r(r^2 - 1) \Rightarrow \left(\frac{r}{r^2 - 1} - \frac{l}{r}\right) \mathrm{d}r = \mathrm{d}t$$

Integrating, we have

$$\log(r^2 - 1) - 2 \log r = 2t + \log c \Rightarrow r = \frac{1}{\sqrt{1 - ce^{2t}}}$$

where $c = (r_0^2 - 1)/r_0^2$, $r_0 \neq 0$ being the initial condition. This gives an unstable limit cycle at $r = 1$ as presented in Fig. 5.10.

Example 5.5 Show that the system

$$\dot{x} = -y + x\left(\sqrt{x^2 + y^2} - 1\right)\left(2 - \sqrt{x^2 + y^2}\right),$$

$$\dot{y} = x + y\left(\sqrt{x^2 + y^2} - 1\right)\left(2 - \sqrt{x^2 + y^2}\right)$$

has exactly two limit cycles.

Fig. 5.10 Graphical representation of the unstable limit cycle for the given system

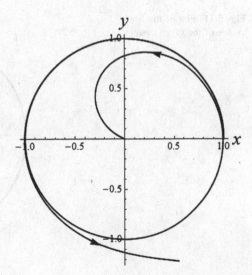

Solution Setting $x = r\cos\theta, y = r\sin\theta$, the given system can be transformed into polar coordinates (r, θ) as

$$\dot{r} = r(r-1)(2-r), \quad \dot{\theta} = 1.$$

It has limit cycles when $r(r-1)(2-r) = 0$, that is, $r = 0, 1, 2$. But the point $r = 0$ corresponds to the unique stable equilibrium point of the given system. So the system has exactly two limit cycles at $r = 1$ and $r = 2$. The limit cycle at $r = 1$ is unstable, while the limit cycle at $r = 2$ is stable. A computer generated plot of the limit cycles is presented in the Fig. 5.11.

Example 5.6 Show that the system

$$\dot{x} = -y + x(x^2 + y^2)\sin\left(\log\left(\sqrt{x^2 + y^2}\right)\right),$$

$$\dot{y} = x + y(x^2 + y^2)\sin\left(\log\left(\sqrt{x^2 + y^2}\right)\right)$$

has infinite number of periodic orbits.

Solution The given system can be transformed into polar coordinates as follows:

$$\dot{r} = r^3 \sin(\log r), \quad \dot{\theta} = 1.$$

It has periodic orbits when $\sin(\log r) = 0$, that is, $\log r = \pm n\pi$. Hence there is an infinite sequence of isolated periodic orbits with period $e^{-n\pi}, n = 1, 2, 3, \ldots$.

Fig. 5.11 Plot of the limit cycles of the given system

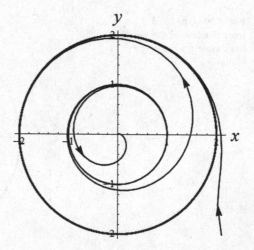

5.6 Applications

In this section, we discuss two dynamical processes viz., the biochemical reaction of living cells and the interaction dynamics of prey and predator populations. Both the dynamical processes have limit cycles under the fulfillment of certain conditions among parameters of the model equations.

5.6.1 Glycolysis

Every animate object requires energy to perform their daily activities. This energy is produced through respiration. We all know that our daily meals contain hues amount of Glucose ($C_6H_{12}O_6$, $12H_2O$). Glycolysis is a process that breaks down the glucose and produce Pyruvic acid and two mole ATP. Through Craves cycle, the Pyruvic acid is finally converted to energy, which is stored in cells. So, glycolysis is a fundamental biochemical reaction in which living cells get energy by breaking down glucose. It was observed that the process may give rise to oscillations in the concentrations of various intermediate chemical reactants. A set of equations for this oscillatory motion was derived by Sel'kov (1968). In dimensionless form, the equations are expressed by

$$\left. \begin{array}{l} \dot{x} = -x + ay + x^2y \\ \dot{y} = b - ay - x^2y \end{array} \right\}$$

where x and y are proportional to the concentrations of adenosine diphosphate (ADP) and fructose-6-phospate (F6P). The positive constants a and b are kinetic parameters of the glycolysis process. The same nonlinear term is present in both equations with opposite signs. The system has a unique equilibrium point

$$(x^*, y^*) = \left(b, \frac{b}{a+b^2} \right).$$

At this equilibrium point, the Jacobian matrix is

$$J = \begin{pmatrix} \frac{b^2-a}{a+b^2} & a+b^2 \\ -\frac{2b^2}{a+b^2} & -(a+b^2) \end{pmatrix}.$$

Therefore, $\Delta = \det(J) = (a+b^2) > 0$ and the sum of diagonal elements of the Jacobian matrix, $\tau = -\frac{b^4 + (2a-1)b^2 + a(1+a)}{a+b^2}$. So, the fixed point is stable when $\tau < 0$, and unstable when $\tau > 0$. A bifurcation may occur when $\tau = 0$, that is, when $b^2 = \frac{(1-2a) \pm \sqrt{(1-8a)}}{2}$. A plot of (a, b) plane representing a parabola is presented in Fig. 5.12.

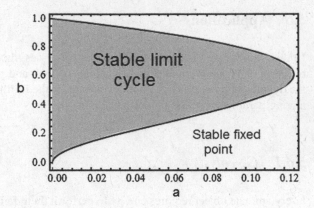

Fig. 5.12 Representation of the stable fixed point and limit cycles in the (a, b)-plane

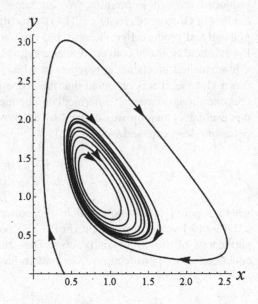

Fig. 5.13 Limit cycle of the Glycolysis problem for $a = 0.07$ and $b = 0.8$

The figure shows that the region corresponding to $\tau > 0$ is bounded. The unique stable limit cycle has been obtained for particular values of a and b satisfying above relation of a and b. A typical limit cycle for $a = 0.07$ and $b = 0.8$ is shown in Fig. 5.13.

5.6.2   *Predator-Prey Model*

Volterra and Lotka were formulated a model to describe the interaction of two species, the prey and the predator. The dynamical equation for predator–prey model are given by

$$\left. \begin{array}{l} \dot{x} = x(\alpha - \beta y) \\ \dot{y} = y(\delta x - \gamma) \end{array} \right\} \tag{5.12}$$

where α, β, γ, and δ are all positive constants. $x(t)$, $y(t)$ and t represent the population density of the prey, the population density of the predator and time, respectively. Here α represents the growth rate of the prey in the absence of intersection with the predators, whereas γ represents the death rate of the predators in the absence of interaction with the prey and β, δ are the interaction parameters and are all constants (for simple model). Note that the survival of the predators completely depends on the population of the prey. If $x(0) = 0$ then $y(t) = y(0) \exp(-\gamma t)$ and $\lim_{t \to \infty} y(t) = 0$. We now study the model under dynamical system's approach. First we calculate the fixed points of the system, which are found by solving the equations $\dot{x} = \dot{y} = 0$. This gives the following fixed points of the system:

$$x^* = 0, y^* = 0; \text{ and } x^* = \frac{\gamma}{\delta}, y^* = \frac{\alpha}{\beta}.$$

The Jacobian of the linearization of the model Eq. (5.12) is obtained as

$$J(x, y) = \begin{pmatrix} \alpha - \beta y & -\beta x \\ \delta y & \delta x - \gamma \end{pmatrix}.$$

It is easy to show that the matrix $J(0,0)$, at the fixed point origin, has the eigenvalues α and $(-\gamma)$, which are opposite in sign, and therefore the fixed point origin is a saddle point. Now, at the fixed point $\left(\frac{\gamma}{\delta}, \frac{\alpha}{\beta}\right)$ the Jacobian matrix is calculated as follows:

$$J = \begin{pmatrix} 0 & -\frac{\beta \gamma}{\delta} \\ \frac{\alpha \delta}{\beta} & 0 \end{pmatrix}$$

which yields the purely imaginary eigenvalues $\left(\pm i\sqrt{\alpha \gamma}\right)$. Therefore, fixed point $\left(\frac{\gamma}{\delta}, \frac{\alpha}{\beta}\right)$ always forms a center which gives a closed path in the neighborhood of the fixed point. This fixed point is nonhyperbolic type, so the stability cannot be determined from the linearised system. The phase path of the system can be obtained easily as

$$\frac{dy}{dx} = \frac{y(\delta x - \gamma)}{x(\alpha - \beta y)},$$

which gives the solution curves $f(x, y) \equiv x^\gamma y^\alpha e^{-(\delta x + \beta y)} = k$, where k is a constant. For different values of k we will get different solution curves. The solution curves can also be written as $f(x, y) \equiv g(x)h(y) = k$, where $g(x) = x^\gamma e^{-\delta x}$ and $h(y) = y^\alpha e^{-\beta y}$. Linearisation of the system gives periodic solutions in a neighborhood of the equilibrium point. Using the solution $f(x, y)$, this result can be verified easily for the nonlinear original system. Note that each of the functions $g(x)$ and $h(y)$ have the same form; they are positive $x, y \in (0, \infty)$. Also, they attain a single maximum in this interval. It is easy to verify that the functions $g(x)$ and $h(y)$ attain their maximums at the points $x = \frac{\gamma}{\delta}$ and $y = \frac{\alpha}{\beta}$, respectively. Therefore, the function $f(x, y)$ has its maximum at the fixed point $\left(\frac{\gamma}{\delta}, \frac{\alpha}{\beta}\right)$ and the trajectories surrounding this point are closed curves, since the point $\left(\frac{\gamma}{\delta}, \frac{\alpha}{\beta}\right)$ is a center. Figure 5.14 depicts the function $g(x)$ and $h(y)$ for some typical values of the parameters involved in the system.

The graphical representations of two solutions curves are explained as follows. The number of preys is increasing when there is less number of predators. The prey population approaches its maximum value at α/β. Thereafter, suddenly the number of predators increases explosively at the cost of the preys. As the number of preys decreases the number of predators has to decrease. These features clearly depict in the figure.

The phase portrait of the system is presented in Fig. 5.15. This shows close curves in the neighborhood of the equilibrium point.

The equations of Volterra and Lotka are too simple model to represent the populations of prey predator in real situations. However, the equations model a crude approximation for two species living together with interactions.

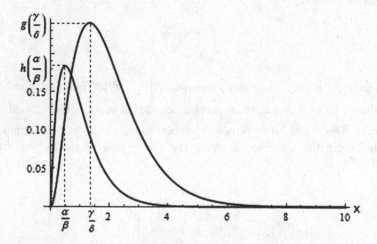

Fig. 5.14 Graphs of $g(x)$ and $h(x)$ for some typical values of α, β, γ, and δ

Fig. 5.15 Phase portrait of
the system (5.12) for
$\alpha = 1.5, \beta = 1, \gamma = 2, \delta = 1$

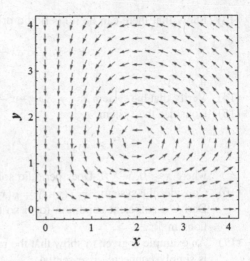

5.7 Exercises

(1) What do you mean by oscillation of a system? Give examples of two systems in which one is oscillating and the other is non-oscillating.

(2) Prove that one-dimensional system cannot oscillate.

(3) Prove that zeros of two linearly independent solutions of the equation $\ddot{x} + p(t)x = 0,\; p(t) \in C(I)$ separate one another.

(4) Prove that if at least one solution of the equation $\ddot{x} + p(t)x = 0$ has more than two zeros on an interval I, then all solutions of this equation are oscillating on I.

(5) Find the relation between $\alpha > 0$ and $k \in \mathbb{N}$, sufficient for any solution of $\ddot{x} + (\alpha \sin t)x = 0$ to be non-oscillating on $[0, 2k\pi]$.

(6) Find the relation between $A > 0$ and $n \in \mathbb{N}$, sufficient for any solution of $\ddot{x} + (e^{-At} \cos t)x = 0$ to be non-oscillating on $[0, 2n\pi]$.

(7) For which $a > 0$ and $n \in \mathbb{N}$ all solutions of the equation $\ddot{x} + \alpha x e^{-at} = 0$ oscillate on $[0, 2n\pi]$, where α is a nonzero real.

(8) Discuss nonlinear oscillation of a system with example. Give few examples where nonlinear oscillations are needed for practical applications.

(9) Consider the system $\dot{x} = \sin y,\; \dot{y} = x \cos y$. Verify that the system is a gradient system. Also find the potential function. Show that the system has no closed orbits.

(10) Show that the system $\dot{x} = 2xy + y^3,\; \dot{y} = x^2 + 3xy^2 - 2y$ has no closed orbits.

(11) Determine the critical points and characterize these points for the system $\ddot{x} - \mu \dot{x} - (1 - \mu)(2 - \mu)x = 0, \mu \in \mathbb{R}$. Sketch the flow in $x - \dot{x}$ phase plane.

(12) Show that the system $\dot{x} = x(2 - 3x - y),\; \dot{y} = y(4x - 4x^2 - 3)$ has no closed orbits in the positive quadrant in $\mathbb{R}^2$.

(13) Show that there can be no periodic orbit of the system

$$\dot{x} = y, \ \dot{y} = ax + by - x^2 y - x^3$$

if $b < 0$.

(14) Verify that the system $\dot{x} = y$, $\dot{y} = -x + y + x^2 + y^2$ has no periodic solutions.

(15) Consider the system

$$\dot{x} = y(1 + x - y^2), \ \dot{y} = x(1 + y - x^2)$$

where $x \geq 0$, $y \geq 0$. Does periodic solution exist?

(16) Consider the system $\dot{x} = f(y)$, $\dot{y} = g(x) + y^{n+1}$, where $f, g \in C^1$ and $n \in \mathbb{N}$. Derive a sufficient condition for n so that the system has no periodic solutions in $\mathbb{R}^2$.

(17) An example is given to show that the requirement that in Dulac's criterion D is simply connected is essential.

(18) Show that the system $\dot{x} = \alpha x - ax^2 + bxy$, $\dot{y} = \beta y - cy^2 + dxy$ with $a, c > 0$ has no periodic orbits in the positive quadrant of $\mathbb{R}^2$.

(19) Consider the system $\dot{x} = -y + x^2 - xy$, $\dot{y} = x + xy$. Show that Bendixson's negative criterion will not give the existence of closed orbits and but Dulac's criterion will be useful for taking the function $\rho(x,y) = -(1+y)^{-3}(1+x)^{-1}$.

(20) Show that the following systems have no periodic solutions
  (a) $\dot{x} = y + x^3$, $\dot{y} = x + y + y^3$.
  (b) $\dot{x} = y$, $\dot{y} = x^9 + (1 + x^4)y$.
  (c) $\dot{x} = \sin \dot{x} \sin y + y^3$, $\dot{y} = \cos x \cos y + x^3 + y$.
  (d) $\dot{x} = x(1 + y^2) + y$, $\dot{y} = y(1 + x^2)$.
  (e) $\dot{x} = x(y^2 + 1) + y$, $\dot{y} = x^2 y + x$.
  (f) $\dot{x} = -x + y^2$, $\dot{y} = -y^3 + x^2$.
  (g) $\dot{x} = y^2$, $\dot{y} = -y^3 + x^2$.
  (h) $\dot{x} = x^2 - y - 1$, $\dot{y} = xy - 2y$.
  (i) $\dot{x} = -xe^y + y^2$, $\dot{y} = x - x^2 y$.
  (j) $\dot{x} = x + y + x^3 - y^3$, $\dot{y} = -x + 2y - x^2 y + y^3$.
  (k) $\dot{x} = x(y - 1)$, $\dot{y} = x + y - 2y^2$.

(21) Show that the system

$$\dot{x} = -y + x(x^2 + y^2 - 2x - 3), \ \dot{y} = x + y(x^2 + y^2 - 2x - 3)$$

has no closed orbits inside the circle with center $\left(\frac{3}{4}, 0\right)$ and radius $\frac{\sqrt{33}}{4}$.

(22) Show that the system $\dot{x} = x + y^3 - xy^2$, $\dot{y} = 3y + x^3 - x^2 y$ has no periodic orbit in the region $\{(x,y) \in \mathbb{R} : x^2 + y^2 \leq 4\}$.

(23) Consider the system $\dot{x} = y - x^3 + \mu x$, $\dot{y} = -x$. For what values of the parameter μ does a periodic solution exist? Describe what happens as $\mu \to 0$?

(24) Consider the system $\dot{x} = f(x,y)$, $\dot{y} = g(x,y)$, where $f, g \in C^1(D)$, D being a simply connected region in $\mathbb{R}^2$. If $\frac{\partial g}{\partial x} = \frac{\partial f}{\partial y}$ in D, show that the system has no closed orbit in D. Hence show that the system $\dot{x} = 2xy + y$, $\dot{y} = x + x^2 - y^2$ has no closed orbits.

(25) Define 'limit cycle' and 'periodic solution'. Give few physical phenomena where limit cycles form. Find the limit cycles and investigate their stabilities for the following systems:

(a) $\dot{x} = (x^2 + y^2 - 1)x - y\sqrt{x^2 + y^2}$, $\dot{y} = (x^2 + y^2 - 1)y + x\sqrt{x^2 + y^2}$
(b) $\dot{x} = -y + x(x^2 + y^2 - 1)$, $\dot{y} = x + y(x^2 + y^2 - 1)$
(c) $\dot{x} = -y + x(\sqrt{x^2 + y^2} - 1)^2$, $\dot{y} = x + y(\sqrt{x^2 + y^2} - 1)^2$
(d) $\dot{x} = y + x(x^2 + y^2)^{-1/2}$, $\dot{y} = -x + y(x^2 + y^2)^{-1/2}$
(e) $\dot{x} = y + x(1 - x^-y^2)$, $\dot{y} = -x + y(1 - x^2 - y^2)$

(26) Show that the system

$$\dot{x} = x - y\left(x^2 + \frac{3}{2}y^2\right), \quad \dot{y} = x + y - y\left(x^2 + \frac{1}{2}y^2\right)$$

has a periodic solution.

(27) Show that the system

$$\dot{x} = x - y - x(x^2 + 5y^2), \quad \dot{y} = x + y - y(x^2 + y^2)$$

has a limit cycle in some 'trapping region' to be determined.

(28) State Poincaré–Bendixson theorem. Prove that the system given by

$$\dot{r} = r(1 - r^2) + \mu r \cos\theta, \quad \dot{\theta} = 1$$

has a closed orbit for $\mu = 0$.

(29) Show the system $\dot{x} = -y + x(1 - x^2 - y^2)$, $\dot{y} = x + y(1 - x^2 - y^2)$ has at least one periodic solution.

(30) Consider the system $\dot{x} = y + ax(1 - 2b - r^2)$, $\dot{y} = -x + ay(1 - r^2)$, where $r^2 = x^2 + y^2$ and a and b are two constants with $0 \le b \le 1/2$ and $0 < a \le 1$. Prove the system has at least one periodic orbit and if there are several periodic orbits, then they have the same period $P(b, a)$. Also prove that if $b = 0$, then there is exactly one periodic orbit of the system.

(31) Consider the system

$$\dot{x} = -y + x(x^4 + y^4 - 3x^2 + 2x^2y^2 - 3y^2 + 1),$$
$$\dot{y} = x + y(x^4 + y^4 - 3x^2 + 2x^2y^2 - 3y^2 + 1)$$

(a) Determine the stability of the fixed points of the system.
(b) Convert the system into polar coordinates, using $x = r\cos\theta$ and $y = r\sin\theta$.
(c) Use Poincaré–Bendixson theorem to show that the system has a limit cycle in an annular region to be determined.

(32) Consider the system

$$\dot{x} = -y + x(1 - 2x^2 - 3y^2), \ \dot{y} = x + y(1 - 2x^2 - 3y^2).$$

(a) Find all fixed points of the system and define their stability.
(b) Transfer the system into polar coordinates (r, θ).
(c) Find a trapping region $R(a,b) = \{(r,\theta) : a \le r \le b\}$ and then use Poincaré–Bendixson theorem to prove that the system has a limit cycle in the region R.

(33) Show that the system given by

$$\dot{r} = r(1 - r^2) + \mu r \cos\theta$$
$$\dot{\theta} = 1$$

has a closed orbit in the annular region $\sqrt{1-\mu} < r < \sqrt{1+\mu}$ for all $\mu \ll 1$.

(34) Consider the system $\dot{x} = x - y - x^3/3$, $\dot{y} = -x$. Show that the system has a limit cycle in some annular region to be determined.

(35) Show that the system represented by $\ddot{x} + \mu(x^2 - 1)\dot{x} + \tanh x = 0$, for $\mu > 0$, has exactly one periodic solution and classify its stability.

(36) Show that the equation $\ddot{x} + x = \mu(1 - \dot{x}^2)x$, $\mu > 0$ has a unique periodic solution and classify its stability.

(37) Show that the equation

$$\ddot{x} + \frac{x^3 - x}{x^2 + 1}\dot{x} + x = 0$$

has exactly one stable limit cycle.

(38) Show that the system $\dot{x} = 2x + y + x^3$, $\dot{y} = 3x - y + y^3$ has no limit cycles and hence no periodic solutions, using Bendixson's negative criterion or otherwise, while the system $\dot{x} = -y$, $\dot{y} = x$ has periodic solutions.

(39) Show that the system $\dot{r} = r(r^2 - 2r\cos\theta - 3), \dot{\theta} = 1$ contains one or more limit cycles in the annulus $1 < r < 3$.

(40) Show that the Rayleigh equation $\ddot{x} + x - \mu(1 - \dot{x}^2)x = 0, \mu > 0$ has a unique periodic solution.

(41) Show that the system $\dot{x} = y + x(x^2 + y^2 - 1)$, $\dot{y} = -x + y(x^2 + y^2 - 1)$ has an unstable limit cycle.

(42) Show that the system

$$\dot{x} = y + x\sqrt{(x^2+y^2)}(x^2+y^2-1)^2, \ \dot{y} = -x + y\sqrt{(x^2+y^2)}(x^2+y^2-1)^2$$

has a semistable limit cycle.

(43) Find the limit cycle of the system $\dot{z} = -iz + \alpha(1 - |z|^2)$, $z \in \mathbb{C}$, where α is a real parameter. Also investigate its stability.

(44) Find the radius of the limit cycle and its period approximately for the equation represented by $\ddot{x} + \epsilon(x^2-1)\dot{x} + x - \epsilon x^3 = 0$, $0 < \epsilon < 1$.

(45) Find the phase path of the Lotka–Volterra competating model equations $\dot{x} = \alpha x - \beta xy$, $\dot{y} = \beta xy - \gamma y$ with $x, y \geq 0$ and positive parameters α, β, γ. Also, show that the solution represents periodic solutions in the neighborhood of the critical point $(\gamma/\beta, \alpha/\beta)$.

(46) Consider the following Lotka–Volterra model taking into account the saturation affect caused by a large number of prey in the prey–predator populations $\dot{x} = \alpha x - \beta xy/(1 + \delta x)$, $\dot{y} = \beta xy/(1 + \delta x) - \gamma y$ with $x, y \geq 0$ and $\alpha, \beta, \gamma, \delta > 0$. Determine the critical points and sketch the flow in the linearized system. Explain the saturation affect of the prey for the cases $\delta \to 0$ and $\delta \to \infty$.

(47) Discuss the solution behaviors of the two populations $x(t)$ and $y(t)$ in the Lotka–Volterra model $\dot{x} = \alpha x - \beta xy$, $\dot{y} = \beta xy - \gamma y$ when the birth rate α of a prey is much smaller than the death rate γ of the predator, that is, $\alpha/\gamma = \epsilon_1$ a small quantity.

(48) Find the creeping velocity of the system with large friction $\ddot{x} + \mu \dot{x} + f(x) = 0$, $\mu \gg 0$. Also, show that for a creeping motion the solution $x(t)$ follows approximately $\dot{x} = -\frac{1}{\mu}f(x)$ which is called gradient flow.

(49) Find the period of oscillation $T(\varepsilon)$ for the Duffing oscillator $\ddot{x} + x + \varepsilon x^3 = 0$ with $x(0) = a, \dot{x}(0) = 0$, where $0 < \varepsilon \ll 1$.

(50) Obtain the frequency of small oscillation (amplitude $\ll 1$) of the equation $\ddot{x} + x - x^3/6 = 0$.

References

1. Tricomi, F.: Differential Equations. Blackie and Son, London (1961)
2. Mickens. R.E.: Oscillations in Planar Dynamic Systems. World Scientific (1996)
3. Lakshmanan, M., Rajasekar, S.: Nonlinear Dynamics: Integrability. Springer, Chaos and Patterns (2003)
4. Perko, L.: Differential Equations and Dynamical Systems, 3rd edn. Springer, New York (2001)
5. Gluckenheimer, J., Holmes, P.: Nonlinear Oscillations, Dynamical Systems, and Bifurcations of Vector Fields. Springer (1983)
6. Grimshaw, R.: Nonlinear Ordinary Differential Equations. CRC Press (1993)

7. Verhulst, F.: Nonlinear Differential Equations and Dynamical Systems, 2nd edn. Springer (1996)
8. Strogatz, S.H.: Nonlinear Dynamics and Chaos with Application to Physics, Biology, Chemistry and Engineering. Perseus Books, L.L.C, Massachusetts (1994)
9. Ya Adrianova, L.: Introduction to Linear Systems of Differential Equations. American Mathematical Society (1995)
10. Krasnov, M.L.: Ordinary Differential Equations. MIR Publication, Moscow (1987). (English translation)
11. Demidovick, B.P.: Lectures on Mathematical Theory of Stability. Nauka, Moscow (1967)

Chapter 6
Theory of Bifurcations

Bifurcation means a structural change in the orbit of a system. The bifurcation of a system had been first reported by the French mathematician *Henri Poincaré* in his work. The study of bifurcation is concerned with how the structural change occurs when the parameter(s) are changing. The structural change and the transition behavior of a system are the central part of dynamical evolution. The point at which bifurcation occurs is known as the bifurcation point. The behavior of fixed point and the nature of trajectories may change dramatically at bifurcation points. The characters of attractor and repellor are altered, in general when bifurcation occurs. The diagram of the parameter values versus the fixed points of the system is known as the bifurcation diagram. This chapter deals with important bifurcations of one and two-dimensional systems, their mathematical theories, and some physical applications.

6.1 Bifurcations

The dynamics of a continuous system $\dot{\underset{\sim}{x}} = f(\underset{\sim}{x}, \mu)$ depends on the parameter $\mu \in \mathbb{R}$. It is often found that as μ crosses a critical value, the properties of dynamical evolution, e.g., its stability, fixed points, periodicity etc. may change. Moreover, a completely new orbit may be created. Basically, a structurally unstable system is termed as bifurcation. The bifurcation diagram is very useful in understanding the dynamical behavior of a system. Bifurcations associated with a single parameter are called codimension-1 bifurcations. On the other hand, bifurcations connected with two parameters are known as codiemension-2 bifurcations. These bifurcations give many interesting dynamics and have a wide range of applications in biological and physical sciences. Various bifurcations and their theories are the integral part of nonlinear systems. We discuss some important bifurcations in one- and two-dimensional systems in the following sections.

© Springer India 2015
G.C. Layek, *An Introduction to Dynamical Systems and Chaos*,
DOI 10.1007/978-81-322-2556-0_6

6.2   Bifurcations in One-Dimensional Systems

The dynamics of a vector field on the real line of a system is very restricted as we have seen in the preceding chapters. However, the dynamics in one-dimensional systems depending upon parameters is interesting and have wide applications in science and engineering. Consider a one-dimensional continuous system

$$\dot{x}(t) = f(x, \mu); \quad x, \mu \in \mathbb{R} \tag{6.1}$$

depending on the parameter μ, where $f : \mathbb{R} \times \mathbb{R} \to \mathbb{R}$ is a smooth function of x and μ. The equilibrium points of (6.1) are the solutions of the equation

$$f(x, \mu) = 0. \tag{6.2}$$

The Eq. (6.2) clearly indicates that all the equilibrium points of the system (6.1) depend on the parameter μ, and they may change their stabilities as μ varies. Thus, bifurcations of a one-dimensional system are associated with the stabilities of its equilibrium points. Such bifurcations are known as local bifurcations as they occur in the neighborhood of the equilibrium points. Such types of bifurcations are occurred in the population growth model, outbreak insect population model, chemical kinetics model, bulking of a beam, etc. In the following subsections, three important bifurcations, namely the saddle-node, pitchfork, and transcritical bifurcations are discussed in depth for one-dimensional systems.

6.2.1   Saddle-Node Bifurcation

Consider the one-dimensional system

$$\dot{x}(t) = f(x, \mu) = \mu + x^2; \quad x \in \mathbb{R} \tag{6.3}$$

with μ as the parameter. Equilibrium points of (6.3) are obtained as

$$f(x, \mu) = 0 \Rightarrow \mu + x^2 = 0 \Rightarrow x^2 = -\mu \tag{6.4}$$

Depending upon the sign of the parameter μ, we have three possibilities. When $\mu < 0$, the system has two fixed points, $x_{1,2}^* = \pm\sqrt{-\mu}$. They merge at $x^* = 0$ when $\mu = 0$ and disappear when $\mu > 0$. We shall now analyze the system's behavior under flow consideration in the real line. The system $\dot{x} = f(x, \mu)$ represents a vector field $f(x, \mu)$ on the real line and gives the velocity vector $\dot{x}$ at each position x of the flow. As we discussed earlier, arrows point to the right direction if $\dot{x} > 0$ and to the left if $\dot{x} < 0$. So, the flow is to the right direction when $\dot{x} > 0$ and to the left when $\dot{x} < 0$. At the points where $\dot{x} = 0$, there are no flows and such points are called fixed

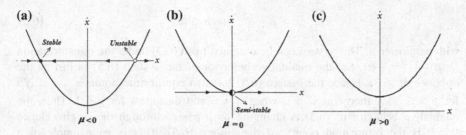

Fig. 6.1 Phase portraits for **a** $\mu < 0$, **b** $\mu = 0$, and **c** $\mu > 0$

points or equilibrium points of the system (6.3). The graph of the vector filed $f(x, \mu)$ in the $x - \dot{x}$ plane represents a parabola, as shown in Fig. 6.1.

When $\mu < 0$, there are two fixed points of the system and are shown in Fig. 6.1a. According to the flow imagination, the figure indicates that the fixed point at $x = \sqrt{-\mu}$ is unstable, whereas the fixed point at $x = -\sqrt{-\mu}$ is stable. From the figure, we also see that when μ approaches to zero from blow, the parabola moves up and the two fixed points move toward each other and they merge at $x = 0$ when $\mu = 0$. There are no fixed points of the system for $\mu > 0$, as shown in Fig. 6.1c. This is a very simple system but its dynamics is highly interesting. The bifurcation in the dynamics occurred at $\mu = 0$, since the vector fields for $\mu < 0$ and $\mu > 0$ are qualitatively different. The diagram of the parameter μ versus the fixed point x^* is known as the bifurcation diagram of the system and the point $\mu = 0$ is called the bifurcation point or the turning point of the trajectory of the system. The bifurcation diagram is shown in Fig. 6.2.

This is an example of a saddle-node bifurcation even though the system is one-dimensional. Actually, it is a subcritical saddle-node bifurcation, since the fixed points exist for values of the parameter below the bifurcation point $\mu = 0$. Consider another simple one-dimensional system

Fig. 6.2 Saddle-node bifurcation diagram for the one-dimensional system (6.3)

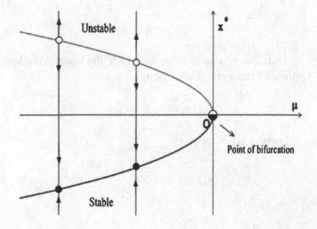

$$\dot{x} = \mu - x^2; \quad x, \mu \in \mathbb{R} \tag{6.5}$$

with parameter μ. This system can be obtained from (6.3) under the transformation $(x, \mu) \mapsto (-x, -\mu)$. So, the qualitative behavior of the system (6.5) is just as the opposite of (6.3). Hence the system (6.5) has two equilibrium points $x^*_{1,2} = \pm\sqrt{\mu}$ for $\mu > 0$, they merge at $x^* = 0$ when $\mu = 0$ and disappear for $\mu < 0$. Thus, the qualitative behavior of (6.5) is changing as μ passes through the origin. Hence $\mu = 0$ is the bifurcation point of the system (6.5). This is an example of a supercritical saddle-node bifurcation, since the equilibrium points exist for values of μ above the bifurcation point $\mu = 0$. The name 'saddle-node bifurcation' is not properly given because the actual bifurcation that occurred in this one-dimensional system is inconsistent with the name "saddle-node." The name is coined in comparison to the bifurcation pattern in two-dimensional systems in which a saddle and a node coincide and then disappear as the parameter exceeds the critical value. The saddle-node bifurcation in a one-dimensional system is connected with appearance and disappearance (vice versa) of the fixed points of the system as the parameter exceeds the critical value.

6.2.2   Pitchfork Bifurcation

We now discuss pitchfork bifurcation in a one-dimensional system which appears when the system has symmetry between left and right directions. In such a system, the fixed points tend to appear and disappear in symmetrical pair. For example, consider the one-dimensional system

$$\dot{x}(t) = f(x, \mu) = \mu x - x^3; \quad x, \mu \in \mathbb{R} \tag{6.6}$$

Replacing x by $-x$ in (6.6), we get

$$-\dot{x} = -\mu x + x^3 = -(\mu x - x^3)$$
$$\Rightarrow \dot{x} = \mu x - x^3$$

Thus the system is invariant under the transformation $x \mapsto -x$. The equilibrium points of the system are obtained as

$$f(x, \mu) = 0 \Rightarrow \mu x - x^3 = 0 \Rightarrow x = 0, \pm\sqrt{\mu}.$$

For $f(x, \mu) = \mu x - x^3$,

$$\frac{\partial f}{\partial x}(x, \mu) = \mu - 3x^2, \frac{\partial f}{\partial x}(0, \mu) = \mu, \frac{\partial f}{\partial x}(\pm\sqrt{\mu}, \mu) = -2\mu.$$

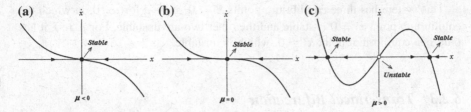

Fig. 6.3 Phase diagram for **a** $\mu < 0$, **b** $\mu = 0$, and **c** $\mu > 0$

When $\mu = 0$, the system has only one equilibrium point $x^* = 0$ and it is a equilibrium point in nature, since $\frac{\partial f}{\partial x}(0,0) = 0$. For $\mu > 0$, three equilibrium points occur at $x^* = 0, \pm\sqrt{\mu}$, in which the equilibrium point origin ($x^* = 0$) is a source (unstable) and the other two equilibrium points are sink (stable). For $\mu < 0$, the system has only one stable equilibrium point at the origin. The phase diagram in the $x - \dot{x}$ plane is depicted in Fig. 6.3.

From the diagram we see that when μ increases from negative to zero, the equilibrium point origin is still stable but much more weakly, because of its non-hyperbolic nature. When $\mu > 0$, the origin becomes unstableequilibrium point and two new stable equilibrium points appear on either side of the origin located at $x = -\sqrt{\mu}$ and $x = \sqrt{\mu}$. The bifurcation diagram of the system is shown in Fig. 6.4. From the pitchfork-shape bifurcation diagram, the name 'pitchfork' becomes clear. But it is basically a pitchfork trifurcation of the system. The bifurcation for this vector field is called a supercritical pitchfork bifurcation, in which a stable equilibrium bifurcates into two stable equilibria. Transforming (x, μ) into $(-x, -\mu)$, we can directly obtain another pitchfork bifurcation, the subcritical pitchfork bifurcation, described by the system

$$\dot{x}(t) = \mu x + x^3. \tag{6.7}$$

Fig. 6.4 Pitchfork bifurcation diagram for the one-dimensional system (6.6)

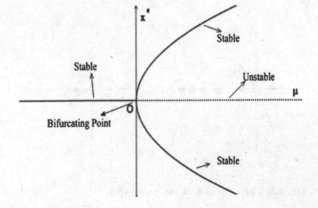

This system has three equilibrium points $x^* = 0, \pm\sqrt{-\mu}$ for $\mu < 0$, in which the equilibrium point $x^* = 0$ is stable and the other two are unstable. For $\mu > 0$, it has only one equilibrium point $x^* = 0$, which is unstable.

6.2.3 Transcritical Bifurcation

There are many parameter-dependent physical systems for which an equilibrium point must exist for all values of a parameter of the system and can never disappear. But it may change its stability character as the parameter varies. The transcritical bifurcation is one such type of bifurcation in which the stability characters of the fixed points are changed for varying values of the parameters. Consider the one-dimensional system

$$\dot{x} = f(x, \mu) = \mu x - x^2; \quad x \in \mathbb{R} \tag{6.8}$$

with $\mu \in \mathbb{R}$ as the parameter. The equilibrium points of this system are obtained as

$$f(x, \mu) = 0 \Rightarrow \mu x - x^2 = 0 \Rightarrow x = 0, \mu.$$

Thus the system has two equilibrium points $x^* = 0, \mu$. We calculate

$$\frac{\partial f}{\partial x}(x, \mu) = \mu - 2x, \text{ so, } \frac{\partial f}{\partial x}(0, \mu) = \mu, \frac{\partial f}{\partial x}(\mu, \mu) = -\mu.$$

This shows that for $\mu = 0$ the system has only one equilibrium point at $x^* = 0$, which is nonhyperbolicequilibrium points. For $\mu \neq 0$, it has two distinct equilibrium points $x^* = 0, \mu$, in which the equilibrium point origin is a source (unstable) for $\mu > 0$ and it is a sink (stable) for $\mu < 0$. The other equilibrium point $x^* = \mu$ is unstable if $\mu < 0$ and stable for $\mu > 0$. The phase diagrams for the above three cases are shown in Fig. 6.5.

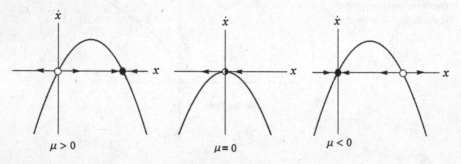

Fig. 6.5 Phase portraits of the system (6.8)

Fig. 6.6 Bifurcation diagram
of the system (6.8)

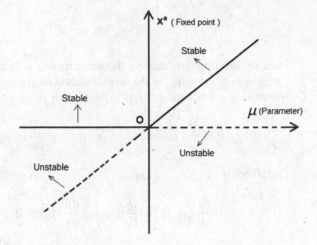

This type of bifurcation is known as transcritical bifurcation. In this bifurcation, an exchange of stabilities has taken place between the two fixed points of the system. The bifurcation diagram is presented in Fig. 6.6.

6.3   Bifurcations in One-Dimensional Systems: A General Theory

So far we have discussed bifurcations based on the flow of vector fields. We now derive a general mathematical theory for bifurcations in one-dimensional systems. Consider a general one-dimensional system

$$\dot{x}(t) = f(x, \mu); \quad x, \mu \in \mathbb{R} \tag{6.9}$$

where $f : \mathbb{R} \times \mathbb{R} \to \mathbb{R}$ is a smooth function. If μ_0 be the bifurcation point and x_0 be the corresponding equilibrium point of the system, then x_0 is nonhyperbolic if

$$\frac{\partial f}{\partial x}(x_0, \mu_0) = 0. \tag{6.10}$$

We first establish the condition for the saddle-node bifurcation.

6.3.1   Saddle-Node Bifurcation

We assume that

$$\frac{\partial f}{\partial \mu}(x_0, \mu_0) \neq 0 \tag{6.11}$$

Then by the implicit function theorem, there exists a unique smooth function $\mu = \mu(x)$ with $\mu(x_0) = \mu_0$, in the neighborhood of (x_0, μ_0) such that $f(x, \mu(x)) = 0$. Differentiating the equation $f(x, \mu(x)) = 0$ with respect to x, we have

$$0 = \frac{df}{dx}(x, \mu(x)) = \frac{\partial f}{\partial x}(x, \mu(x)) + \frac{\partial f}{\partial \mu}(x, \mu(x)) \frac{d\mu}{dx}(x) \tag{6.12}$$

Therefore, at (x_0, μ_0), we get

$$\frac{\partial f}{\partial x}(x_0, \mu_0) + \frac{\partial f}{\partial \mu}(x_0, \mu_0) \frac{d\mu}{dx}(x_0) = 0$$

$$\Rightarrow \frac{d\mu}{dx}(x_0) = -\frac{\frac{\partial f}{\partial x}(x_0, \mu_0)}{\frac{\partial f}{\partial \mu}(x_0, \mu_0)} \tag{6.13}$$

$$\Rightarrow \frac{d\mu}{dx}(x_0) = 0$$

Again, differentiating the equation $f(x, \mu(x)) = 0$ with respect to x twice, we get

$$0 = \frac{d^2 f}{dx^2}(x, \mu(x)) = \frac{d}{dx}\left(\frac{\partial f}{\partial x}(x, \mu(x)) + \frac{\partial f}{\partial \mu}(x, \mu(x))\frac{d\mu}{dx}(x)\right)$$

$$= \frac{\partial^2 f}{\partial x^2}(x, \mu(x)) + 2\frac{\partial^2 f}{\partial x \partial \mu}(x, \mu(x))\frac{d\mu}{dx}(x) + \frac{\partial^2 f}{\partial^2 \mu}(x, \mu(x))\left(\frac{d\mu}{dx}(x)\right)^2$$

$$+ \frac{\partial f}{\partial \mu}(x, \mu(x))\frac{d^2\mu}{dx^2}(x). \tag{6.14}$$

Therefore, at (x_0, μ_0),

$$\frac{\partial^2 f}{\partial x^2}(x_0, \mu_0) + \frac{\partial f}{\partial \mu}(x_0, \mu_0)\frac{d^2\mu}{dx^2}(x_0) = 0. \tag{6.15}$$

Now, recall the saddle-node bifurcation diagram of the system (6.3). In this diagram, the unique curve $x = \mu^2$ of fixed points, passing through $(0, 0)$, lies entirely on only one side of the bifurcation point $\mu = 0$. This will be possible only if

$$\frac{d^2\mu}{dx^2} \neq 0$$

in the neighborhood of $(x, \mu) = (0, 0)$. Compared to this, we take

$$\frac{d^2\mu}{dx^2}(x_0) \neq 0.$$

Therefore from (6.15), we see that

$$\frac{\partial^2 f}{\partial x^2}(x_0, \mu_0) \neq 0. \tag{6.16}$$

We state the result in the following theorem.

Theorem 6.1 (Saddle-node bifurcation) *Suppose the system* $\dot{x}(t) = f(x, \mu)$, $x, \mu \in \mathbb{R}$ *has an equilibrium point* $x = x_0$ *at* $\mu = \mu_0$ *satisfying the conditions*

$$f(x_0, \mu_0) = 0, \frac{\partial f}{\partial x}(x_0, \mu_0) = 0.$$

If

$$\frac{\partial f}{\partial \mu}(x_0, \mu_0) \neq 0, \frac{\partial^2 f}{\partial x^2}(x_0, \mu_0) \neq 0,$$

then the system has a saddle-node bifurcation at (x_0, μ_0).

Similarly, one can easily derive the conditions under which the system $\dot{x}(t) = f(x, \mu)$, $x, \mu \in \mathbb{R}$ possess transcritical and pitchfork bifurcations. In this book, we only state the following theorem for these two bifurcations. See Wiggins [1] for proof.

Theorem 6.2 (Transcritical and pitchfork bifurcations) *Suppose the system* $\dot{x}(t) = f(x, \mu)$, $x, \mu \in \mathbb{R}$ *has an equilibrium point* $x = x_0$ *at* $\mu = \mu_0$ *satisfying the conditions*

$$f(x_0, \mu_0) = 0, \frac{\partial f}{\partial x}(x_0, \mu_0) = 0.$$

(i) *If*

$$\frac{\partial f}{\partial \mu}(x_0, \mu_0) = 0, \frac{\partial^2 f}{\partial x^2}(x_0, \mu_0) \neq 0 \text{ and } \frac{\partial^2 f}{\partial x \partial \mu}(x_0, \mu_0) \neq 0,$$

then the system has a transcritical bifurcation at (x_0, μ_0).

(ii) *If*

$$\frac{\partial f}{\partial \mu}(x_0, \mu_0) = 0, \ \frac{\partial^2 f}{\partial x^2}(x_0, \mu_0) = 0, \ \frac{\partial^2 f}{\partial x \partial \mu}(x_0, \mu_0) \neq 0 \ and \ \frac{\partial^3 f}{\partial x^3}(x_0, \mu_0) \neq 0,$$

then the system has a pitchfork bifurcation at (x_0, μ_0).

We now derive the normal forms of these bifurcations in one-dimensional systems. By normal form of a system we mean the most simplified mathematical form from which one can easily understand the type of bifurcations occurred in the system.

(a) Normal form of saddle-node bifurcation:

Suppose the system (6.9) has an equilibrium point at $x = x_0$ for $\mu = \mu_0$ for which all the saddle-node bifurcation conditions are satisfied, that is,

$$f(x_0, \mu_0) = 0, \frac{\partial f}{\partial x}(x_0, \mu_0) = 0, \frac{\partial f}{\partial \mu}(x_0, \mu_0) \neq 0 \ and \ \frac{\partial^2 f}{\partial x^2}(x_0, \mu_0) \neq 0, \qquad (6.17)$$

Expanding $f(x, \mu)$ in a Taylor series in the neighborhood of (x_0, μ_0), we have

$$\begin{aligned}
\dot{x} &= f(x, \mu) \\
&= f(x_0, \mu_0) + (x - x_0)\frac{\partial f}{\partial x}(x_0, \mu_0) + (\mu - \mu_0)\frac{\partial f}{\partial \mu}(x_0, \mu_0) + \frac{1}{2!}(x - x_0)^2 \frac{\partial^2 f}{\partial x^2}(x_0, \mu_0) \\
&\quad + (x - x_0)(\mu - \mu_0)\frac{\partial^2 f}{\partial x \partial \mu}(x_0, \mu_0) + \frac{1}{2!}(\mu - \mu_0)^2 \frac{\partial^2 f}{\partial \mu^2}(x_0, \mu_0) + \cdots \\
&= (\mu - \mu_0)\frac{\partial f}{\partial \mu}(x_0, \mu_0) + \frac{1}{2!}(x - x_0)^2 \frac{\partial^2 f}{\partial x^2}(x_0, \mu_0) + \ldots \\
&= \alpha(\mu - \mu_0) + \beta(x - x_0)^2 + \cdots \qquad\qquad (6.18)
\end{aligned}$$

where $\alpha = \frac{\partial f}{\partial \mu}(x_0, \mu_0)$ and $\beta = \frac{1}{2!}\frac{\partial^2 f}{\partial x^2}(x_0, \mu_0)$ are nonzero real. The Eq. (6.18) refers to as the normal form of the saddle-node bifurcation. This is a great advantage for determining the bifurcation which a system undergoes.

(b) Normal form of transcritical bifurcation:

Suppose that the system (6.9) has an equilibrium point $x = x_0$ at $\mu = \mu_0$ for which the transcritical bifurcation conditions are satisfied as given in Theorem 6.2(i). Using the Taylor series expansion of $f(x, \mu)$ in the neighborhood of (x_0, μ_0), we have

$$\dot{x} = f(x, \mu)$$

$$= f(x_0, \mu_0) + (x - x_0)\frac{\partial f}{\partial x}(x_0, \mu_0) + (\mu - \mu_0)\frac{\partial f}{\partial \mu}(x_0, \mu_0) + \frac{1}{2!}(x - x_0)^2\frac{\partial^2 f}{\partial x^2}(x_0, \mu_0)$$

$$+ (x - x_0)(\mu - \mu_0)\frac{\partial^2 f}{\partial x \partial \mu}(x_0, \mu_0) + \frac{1}{2!}(\mu - \mu_0)^2\frac{\partial^2 f}{\partial \mu^2}(x_0, \mu_0) + \cdots$$

$$= (x - x_0)(\mu - \mu_0)\frac{\partial^2 f}{\partial x \partial \mu}(x_0, \mu_0) + \frac{1}{2!}(x - x_0)^2\frac{\partial^2 f}{\partial x^2}(x_0, \mu_0) + \cdots$$

$$= \alpha(x - x_0)(\mu - \mu_0) + \beta(x - x_0)^2 + \cdots$$

$$(6.19)$$

where $\alpha = \frac{\partial^2 f}{\partial x \partial \mu}(x_0, \mu_0)$ and $\beta = \frac{1}{2!}\frac{\partial^2 f}{\partial x^2}(x_0, \mu_0)$ are nonzero real. The Eq. (6.19) refers to the normal form of the transcritical bifurcation.

(c) Normal form of pitchfork bifurcation:

Suppose that the system (6.9) has an equilibrium point $x = x_0$ at $\mu = \mu_0$, satisfying all the pitchfork bifurcation conditions given in Theorem 6.2(ii). We now expand the function $f(x, \mu)$ in the neighborhood of (x_0, μ_0) in Taylor series expansion as presented below.

$$\dot{x} = f(x, \mu)$$

$$= f(x_0, \mu_0) + (x - x_0)\frac{\partial f}{\partial x}(x_0, \mu_0) + (\mu - \mu_0)\frac{\partial f}{\partial \mu}(x_0, \mu_0) + \frac{1}{2}(x - x_0)^2\frac{\partial^2 f}{\partial x^2}(x_0, \mu_0)$$

$$+ (x - x_0)(\mu - \mu_0)\frac{\partial^2 f}{\partial x \partial \mu}(x_0, \mu_0) + \frac{1}{2}(\mu - \mu_0)^2\frac{\partial^2 f}{\partial \mu^2}(x_0, \mu_0)$$

$$+ \frac{1}{6}(x - x_0)^3\frac{\partial^3 f}{\partial x^3}(x_0, \mu_0) + \frac{1}{2}(x - x_0)^2(\mu - \mu_0)\frac{\partial^3 f}{\partial x^2 \partial \mu}(x_0, \mu_0)$$

$$+ \frac{1}{2}(x - x_0)(\mu - \mu_0)^2\frac{\partial^3 f}{\partial x \partial \mu^2}(x_0, \mu_0) + \frac{1}{6}(\mu - \mu_0)^2\frac{\partial^2 f}{\partial \mu^2}(x_0, \mu_0) + \cdots$$

$$= (x - x_0)(\mu - \mu_0)\frac{\partial^2 f}{\partial x \partial \mu}(x_0, \mu_0) + \frac{1}{6}(x - x_0)^3\frac{\partial^3 f}{\partial x^3}(x_0, \mu_0) + \cdots$$

$$= \alpha(x - x_0)(\mu - \mu_0) + \beta(x - x_0)^3 + \cdots \tag{6.20}$$

where $\alpha = \frac{\partial^2 f}{\partial x \partial \mu}(x_0, \mu_0)$ and $\beta = \frac{1}{6}\frac{\partial^3 f}{\partial x^3}(x_0, \mu_0)$ are nonzero real. This is the normal form of the pitchfork bifurcation.

Example 6.1 Show that the system

$$\dot{x} = x(1 - x^2) - a(1 - e^{-bx})$$

undergoes a transcritical bifurcation at $x = 0$ when the parameters a and b satisfy a certain relation to be determined.

Solution Clearly, $x = 0$ is an equilibrium point of the given system for all values of the parameters a and b. This indicates that the system will exhibit a transcritital bifurcation. For small x, the expansion of e^{-bx} gives

$$e^{-bx} = 1 - bx + \frac{b^2 x^2}{2!} - O(x^3).$$

Therefore,

$$\dot{x} = x(1 - x^2) - a(1 - e^{-bx})$$

$$= x(1 - x^2) - a(bx - \frac{b^2 x^2}{2!} + O(x^3))$$

$$= (1 - ab)x + \frac{1}{2}ab^2 x^2. \quad \text{[Neglecting cube and higher powers of } x]$$

For transcritical bifurcation at $x = 0$, $1 - ab = 0$, that is, $ab = 1$. Hence the system undergoes a transcritical bifurcation at $x = 0$ when $ab = 1$.

Example 6.2 Describe the bifurcation of the system $\dot{x} = x^3 - 5x^2 - (\mu - 8) x + \mu - 4$.

Solution Let $f(x, \mu) = x^3 - 5x^2 - (\mu - 8)x + \mu - 4$. The equilibrium points are given by

$$x^3 - 5x^2 - (\mu - 8)x + \mu - 4 = 0$$

$$\Rightarrow (x - 1)(x^2 - 4x - \mu + 4) = 0$$

Clearly, $x = 1$ is a fixed point of the system for all values of μ. The other two fixed points are $x_{\pm} = 2 \pm \sqrt{\mu}$, which are real and distinct for $\mu > 0$. They coincide with the fixed point $x = 2$ for $\mu = 0$ and vanish when $\mu < 0$. Therefore, the system has a saddle-node bifurcation at $x = 2$ with $\mu = 0$ as the bifurcation point. We can also verify this using Theorem 6.1. Take $x_0 = 2$ and $\mu_0 = 0$. Now, calculate

$$\frac{\partial f}{\partial x}(x, \mu) = 3x^2 - 10x + 8 - \mu, \frac{\partial f}{\partial \mu}(x, \mu) = -x + 1, \frac{\partial^2 f}{\partial x^2}(x, \mu) = 6x - 10.$$

We see that

$$f(x_0, \mu_0) = 0, \frac{\partial f}{\partial x}(x_0, \mu_0) = 0, \frac{\partial f}{\partial \mu}(x_0, \mu_0) = -1 \neq 0, \frac{\partial^2 f}{\partial x^2}(x_0, \mu_0) = 2 \neq 0.$$

Therefore, by Theorem 6.1 the system has a saddle-node bifurcation at (x_0, μ_0), where $x_0 = 2$ and $\mu_0 = 0$.

Example 6.3 Consider the system $\dot{x} = x^3 - \mu$; $x, \mu \in \mathbb{R}$. Does the system has any bifurcation in neighborhood of its fixed points? Justify.

Solution Here $f(x, \mu) = x^3 - \mu$. The fixed points are given by

$$f(x, \mu) = 0 \Rightarrow x^3 - \mu = 0 \Rightarrow x = \mu^{1/3}.$$

Calculate

$$\frac{\partial f}{\partial x}(x, \mu) = 3x^2, \text{ so } \frac{\partial f}{\partial x}(\mu^{1/3}, \mu) = 3\mu^{2/3}.$$

The qualitative behavior of the system does not change with the variation of the parameter μ. So, bifurcation does not occur in the neighborhood of its fixed points.

6.4 Imperfect Bifurcation

Consider the system represented by the Eq. (6.3). Suppose this system exhibits a saddle-node bifurcation at the point $(x, \mu) = (0, 0)$. If we add a quantity $\varepsilon \in \mathbb{R}$ in this equation and then apply Theorem 6.1, we see that the system also has a saddle-node bifurcation at $(x, \mu) \stackrel{.}{=} (0, -\varepsilon)$. Thus an addition of the term ε in (6.3) does not change its bifurcation character. In similar way, addition of the term εx in the Eq. (6.3) will not produce any new bifurcation pattern, provided that the parameter $\mu \neq 0$. This bifurcation is structurally stable. The other two bifurcations, mentioned earlier, are not structurally stable. They can alter under arbitrarily small perturbations and produce new bifurcations. These bifurcations are called imperfect bifurcations and the parameter (perturbation quantity) is known as the imperfection parameter. For example, consider the system

$$\dot{x}(t) = \varepsilon + \mu x - x^2 \tag{6.21}$$

where $\varepsilon, \mu \in \mathbb{R}$ are parameters. If $\varepsilon = 0$, it reduces to the system (6.8) and so, it has a transcritical bifurcation. We shall now analyze the system for $\varepsilon \neq 0$. The equilibrium points of (6.21) are the solutions of the equation

$$\dot{x} = 0 \Rightarrow \varepsilon + \mu x - x^2 = 0$$

$$\Rightarrow x = \frac{\mu \pm \sqrt{\mu^2 + 4\varepsilon}}{2}.$$

If $\varepsilon < 0$, then (6.21) has two distinct equilibrium points

$$x_{\pm}^* = \frac{\mu \pm \sqrt{\mu^2 + 4\varepsilon}}{2}$$

when the parameter μ lies in $(-\infty, -2\sqrt{-\varepsilon}) \cup (2\sqrt{-\varepsilon}, \infty)$ in which x^*_+ is stable and x^*_- is unstable. These two equilibrium points merge at $x^* = \mu/2$ when $\mu = \pm 2\sqrt{-\varepsilon}$ and disappear when μ lies in $(-2\sqrt{-\varepsilon}, 2\sqrt{-\varepsilon})$. Thus, for $\varepsilon < 0$, the tran-scritical bifurcation for $\varepsilon = 0$ perturbs into two saddle-node bifurcations at $(-\sqrt{-\varepsilon}, -2\sqrt{-\varepsilon})$ and $(\sqrt{-\varepsilon}, 2\sqrt{-\varepsilon})$ with bifurcation points $\mu = -2\sqrt{-\varepsilon}$ and $\mu = 2\sqrt{-\varepsilon}$, respectively.

Again, if $\varepsilon > 0$, then $(\mu^2 + 4\varepsilon) > 0$ for all μ. Therefore, in this case, the system has two distinct (nonintersecting) solution curves, one is stable and the other is unstable, and so no bifurcations will appear as μ varies. In conclusion, the addition of small quantity in a system will change the bifurcation character when the bifurcation pattern is not structurally stable.

6.5 Bifurcations in Two-Dimensional Systems

Dynamics of two-dimension systems are vast and their qualitative behaviors are determined by the nature of equilibrium points, periodic orbits, limit cycles, etc. The parameters and their critical values for bifurcations are highly associated with system's evolution and have physical significances. The critical parameter value is a deciding factor for a system undergoing bifurcation solutions. We shall now for-mulate a simple problem where the critical value for qualitative change in the system can be obtained very easily. Consider a circular tube suspended by a string attached to its highest point and carrying a heavy mass m, which is rotating with an angular velocity ω about the vertical axis (see Fig. 6.7).

The angular motion of the mass m is determined by the following equation without taking into account the damping force,

$$ma\ddot{\theta} = ma\omega^2 \sin\theta \cos\theta - mg \sin\theta$$

$$\text{or,} \quad \ddot{\theta} = \left(\omega^2 \cos\theta - \frac{g}{a}\right) \sin\theta,$$

where a is the radius of the circular tube and g is the acceleration due to gravity and $\mu = \omega^2 a/g$. The right-hand side of the above equation may be denoted by

$$f(\theta, \mu) = \frac{g}{a}(\mu \cos\theta - 1)\sin\theta.$$

The equilibrium positions are given by $f(\theta, \mu) = 0$. Thus, there exist two positions of equilibrium and are given by

$\sin\theta = 0 \Rightarrow \theta = 0, \pi, -\pi$ according to the problem and $\cos\theta = \frac{1}{\mu} = \frac{g}{a\omega^2}$.

If $\omega^2 < g/a$, that is, if $\mu < 1$, then $\cos\theta > 1$ and so $\theta = 0$ is the only position of equilibrium of the system, and it is stable for small θ,

Fig. 6.7 Sketch of a rotating circular tube carrying a mass m

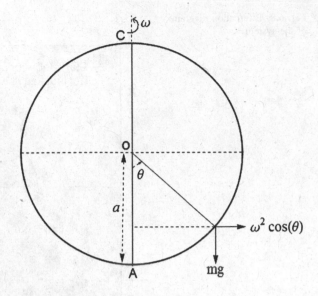

$$\frac{d^2\theta}{dt^2} = -\frac{g}{a}(1 - \mu)\theta.$$

Any small displacement, say $\theta = \theta_0$ with $\dot{\theta} = 0$ will result in small oscillations about the lowest point $\theta = 0$.

As ω increases beyond the critical value $\omega_{cr} > \sqrt{\frac{g}{a}}$, the equilibrium at $\theta = 0$ loses its stability and a new position of equilibrium $\theta = \cos^{-1}(g/a\omega^2)$ is created. This is a position of stable equilibrium. Thus, we see that a bifurcation occurs when the angular velocity ω crosses the critical value $\omega_{cr} = \sqrt{\frac{g}{a}}$, that is, $\mu = 1$. The bifurcation diagram is presented in Fig. 6.8.

This simple example illustrates how bifurcation occurs and how the behavior of the system alters before and after the bifurcating point. In the following subsections, we shall discuss few common bifurcations that frequently occur in two-dimensional systems, viz., (i) saddle-node bifurcation, (ii) transcritical bifurcation, (iii) pitchfork bifurcation, (iv) Hopf bifurcation, and (v) homoclinic and heteroclinic bifurcations.

6.5.1 Saddle-Node Bifurcation

Consider a parameter-dependent two-dimensional system

$$\dot{x} = \mu - x^2, \dot{y} = -y; \mu \in \mathbb{R} \tag{6.22}$$

The fixed points of the system are the solutions of the equations

Fig. 6.8 Bifurcation diagram
of the system

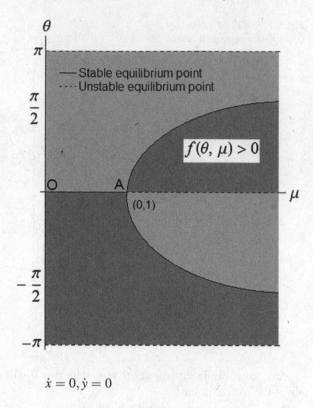

$$\dot{x} = 0, \dot{y} = 0$$

which yield

$$\mu - x^2 = 0, y = 0.$$

For $\mu > 0$, the Eq. (6.22) has two distinct fixed points at $(\sqrt{\mu}, 0)$ and $(-\sqrt{\mu}, 0)$. These two fixed points merge at the origin $(0, 0)$ when $\mu = 0$ and they vanish when $\mu < 0$. This is a same feature as we have seen in one-dimensional saddle-node bifurcation. We shall now determine the stabilities of the fixed points. This needs to evaluate the Jacobian matrix of the system for local stability behavior and is given by

$$J(x, y) = \begin{pmatrix} -2x & 0 \\ 0 & -1 \end{pmatrix}.$$

We first consider the case $\mu > 0$. Here the system has two fixed points $(\sqrt{\mu}, 0)$ and $(-\sqrt{\mu}, 0)$. The Jacobian

$$J(\sqrt{\mu}, 0) = \begin{pmatrix} -2\sqrt{\mu} & 0 \\ 0 & -1 \end{pmatrix}$$

at the fixed point $(\sqrt{\mu}, 0)$ has two eigenvalues $(-2\sqrt{\mu})$ and (-1), which are real and negative. Hence the fixed point $(\sqrt{\mu}, 0)$ is a stable node. Similarly, calculating the eigenvalues of $J(-\sqrt{\mu}, 0)$ we can show that the fixed point $(-\sqrt{\mu}, 0)$ is saddle.

Consider the second case, $\mu = 0$. In this case, the system has a single fixed point $(0, 0)$. The Jacobian matrix at $(0, 0)$ is

$$J(0,0) = \begin{pmatrix} 0 & 0 \\ 0 & -1 \end{pmatrix}$$

with eigenvalues $0, (-1)$. This indicates that the fixed point $(0, 0)$ is semi-stable. For $\mu < 0$, the system has no fixed points.

Thus we see that the system (6.22) has two fixed points, one is a stable node and the other is a saddle point, when $\mu > 0$. As μ decreases, the saddle and the stable node approach each other. They collide at $\mu = 0$ and disappear when $\mu < 0$. The phase portraits are shown for different values of the parameter in Fig. 6.9.

From the phase diagram we see that when the parameter is positive, no matter how small, all trajectories in the region $\{(x, y) : x > -\sqrt{\mu}\}$ reach steadily at the stable node origin of the system. As soon as μ crosses the origin, an exchange of stability takes place and this clearly indicates in the phase portrait of the system. When μ is negative, all trajectories eventually escape to infinity. This type of bifurcation is known as saddle-node bifurcation. The name "saddle-node" is because its basic mechanism is the collision of two fixed points, viz., a saddle and a node of the system. Here $\mu = 0$ is the bifurcation point. The bifurcation diagram is same as that for the one-dimensional system.

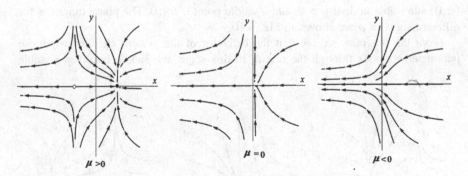

Fig. 6.9 Phase portrait of the system for different values of the parameter μ

6.5.2 Transcritical Bifurcation

Consider a two-dimensional parametric system expressed by

$$\dot{x} = \mu x - x^2, \dot{y} = -y; \qquad (6.23)$$

with parameter $\mu \in \mathbb{R}$. This system has always two distinct fixed points $(0,0)$ and $(\mu, 0)$ for $\mu \neq 0$. For $\mu = 0$, these two fixed points merge at $(0, 0)$. This is why this bifurcation is called as transcritical bifurcation. The Jacobian matrix of the system (6.23) is given by

$$J(x, y) = \begin{pmatrix} \frac{\partial f}{\partial x} & \frac{\partial f}{\partial y} \\ \frac{\partial g}{\partial x} & \frac{\partial g}{\partial y} \end{pmatrix} = \begin{pmatrix} \mu - 2x & 0 \\ 0 & -1 \end{pmatrix}$$

At the point $(0, 0)$,

$$J(0,0) = \begin{pmatrix} \mu & 0 \\ 0 & -1 \end{pmatrix}$$

which has eigenvalues μ and (-1). Therefore, the fixed point $(0,0)$ of the system (6.23) is a stable node if $\mu < 0$ and it is a saddle point if $\mu > 0$. For $\mu = 0$, the fixed point is semi-stable. Again, at $(\mu, 0)$,

$$J(\mu, 0) = \begin{pmatrix} -\mu & 0 \\ 0 & -1 \end{pmatrix}.$$

The eigenvalues of $J(\mu, 0)$ are $(-\mu)$ and (-1), showing that the fixed point $(\mu, 0)$ is a stable node if $\mu > 0$, and a saddle point if $\mu < 0$. The phase diagrams for different signs of μ are shown in Fig. 6.10.

From the diagram, we see that the behavior of the system changes when the parameter μ passes through the origin. In this stage, the saddle becomes a stable

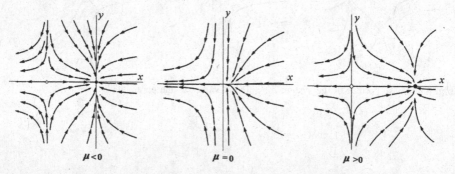

Fig. 6.10 Phase portrait of the system for different values of the bifurcation parameter μ

node and the stable node becomes a saddle. That is, when μ passes through the origin from left, the fixed point origin changes to a saddle from a stable node and the fixed point $(\mu, 0)$ changes from a saddle to a stable node. This type of bifurcation is known as transcritical bifurcation. Here $\mu = 0$ is the bifurcation point. The feature is same as in one-dimensional system where no fixed points are disappeared.

6.5.3 Pitchfork Bifurcation

There are two types of pitchfork bifurcations, namely supercritical and subcritical pitchfork bifurcations. In the present section, we deal with these two bifurcations scenario, first the supercritical pitchfork bifurcation and then the subcritical pitchfork bifurcation will be illustrated.

Consider a two-dimensional system represented by

$$\dot{x} = \mu x - x^3, \dot{y} = -y; \tag{6.24}$$

where $\mu \in \mathbb{R}$ is the parameter. For $\mu < 0$, the system (6.24) has only one equilibrium point at the origin. The Jacobian matrix at this fixed point is given by

$$J(0,0) = \begin{pmatrix} \mu & 0 \\ 0 & -1 \end{pmatrix}.$$

The eigenvalues of $J(0,0)$ are μ, (-1), showing that the fixed point origin is a stable node. For $\mu > 0$, the system has three fixed points $(0, 0)$, $(\sqrt{\mu}, 0)$, and $(-\sqrt{\mu}, 0)$. The Jacobian matrix of (6.24) calculated at these fixed points are given by

$$J(0,0) = \begin{pmatrix} \mu & 0 \\ 0 & -1 \end{pmatrix}, J(\sqrt{\mu}, 0) = \begin{pmatrix} -2\mu & 0 \\ 0 & -1 \end{pmatrix}, J(-\sqrt{\mu}, 0) = \begin{pmatrix} -2\mu & 0 \\ 0 & -1 \end{pmatrix}.$$

The eigenvalues of $J(0,0)$ are μ, (-1), which are opposite in signs. So, the equilibrium point $(0,0)$ is a saddle for $\mu > 0$. Clearly, the eigenvalues of Jacobian matrix show that the other two fixed points are stable nodes. The phase diagrams for different values of the bifurcation parameter μ are presented in Fig. 6.11.

The diagram shows that as soon as the parameter μ crosses the bifurcation point origin, the fixed point origin bifurcates into a saddle point from a stable node. In this situation, it also gives birth to two stable nodes at the points $(\sqrt{\mu}, 0)$ and $(-\sqrt{\mu}, 0)$. The amplitudes of the newly created stable nodes grow with the parameter. This type of bifurcation is known as supercritical pitchfork bifurcation. We shall now discuss the subcritical pitchfork bifurcation. Consider a parameter-dependent two-dimensional system represented by

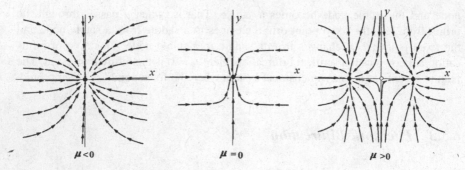

Fig. 6.11 Phase portrait for different parameter values of the system

$$\dot{x} = \mu x + x^3, \dot{y} = -y; \qquad (6.25)$$

with the parameter $\mu \in \mathbb{R}$. When $\mu < 0$, the system (6.25) has three distinct fixed points, namely $(0,0)$, $(\sqrt{\mu}, 0)$, and $(-\sqrt{\mu}, 0)$. The Jacobians of the system evaluated at these fixed points are given by

$$J(0,0) = \begin{pmatrix} \mu & 0 \\ 0 & -1 \end{pmatrix}, J(\sqrt{\mu}, 0) = \begin{pmatrix} 4\mu & 0 \\ 0 & -1 \end{pmatrix}, J(-\sqrt{\mu}, 0) = \begin{pmatrix} 4\mu & 0 \\ 0 & -1 \end{pmatrix}.$$

The eigenvalues of $J(0,0)$ are μ, (-1), which are of same sign. Thus, the fixed point origin is a stable node for $\mu < 0$. Similarly, calculating the eigenvalues of the other two Jacobian matrices of the system one can see that the fixed points $(\pm\sqrt{\mu}, 0)$ are saddle points. For $\mu > 0$, the system has a single fixed point at the origin, which is saddle. If we draw the phase portrait of the system, then we can see that as soon as the parameter crosses the bifurcation point $\mu = 0$, the stable node at the origin coincides with the saddles and then bifurcates into a saddle. This type of bifurcation is known as subcritical pitchfork bifurcation.

6.5.4 Hopf Bifurcation

So far we have discussed bifurcations of systems with real eigenvalues, either positive or negative, of the corresponding Jacobian matrix evaluated at the fixed points of the corresponding system. We shall now discuss a very interesting periodic bifurcation phenomenon for a two-dimensional system where the eigenvalues are complex. This type of bifurcating phenomenon in two-dimensional or higher dimensional systems was studied by the German Scientist *Eberhard Hopf* (1902–1983) and it was named Hopf bifurcation due to the recognition of his work. This type of bifurcation was also recognized by Henri Poincaré and later by A.D. Andronov in 1930. Hopf bifurcation occurs when a stable equilibrium point losses its stability and gives birth to a limit cycle and vice versa. There are two types of Hopf bifurcations, viz., supercritical and subcritical Hopf bifurcations. When stable limit cycles are created for an unstable equilibrium point, then the bifurcation is called a supercritical Hopf bifurcation. In engineering applications point of view, this type of bifurcation is also termed as soft or safe bifurcation because the amplitude of the limit cycles build up gradually as the parameter varies from the bifurcation point. On the other hand, when an unstable limit cycle is created for a stable equilibrium point, then the bifurcation is called a subcritical Hopf bifurcation. It is also known as a hard bifurcation. In case of subcritical Hopf bifurcation, a steady state solution could become unstable as parameter varies and the nonzero solutions could tend to infinity. We shall now illustrate the supercritical and subcritical Hopf bifurcations below.

6.5.4.1   Supercritical Hopf Bifurcation

Consider a two-dimensional system with parameter $\mu \in \mathbb{R}$,

$$\dot{x} = \mu x - y - x(x^2 + y^2), \dot{y} = x + \mu y - y(x^2 + y^2). \qquad (6.26)$$

The system has a unique fixed point at the origin. In polar coordinates, the system can be written as

$$\dot{r} = \mu r - r^3, \dot{\theta} = 1,$$

which are decoupled, and so easy to analyze. The phase portraits for $\mu < 0$ and $\mu > 0$ are shown in Fig. 6.12.

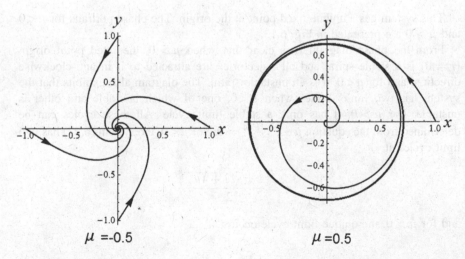

$\mu = -0.5$ $\mu = 0.5$

Fig. 6.12 Phase portraits of the system for $\mu = -0.5$ and $\mu = 0.5$, respectively

When $\mu < 0$, the fixed point origin ($r = 0$) is a stable spiral and all trajectories are attracted to it in anti-clockwise direction. For $\mu = 0$, the origin is still a stable spiral, though very weak. For $\mu > 0$, the origin is an unstable spiral, and in this case, there is a stable limit cycle at $r = \sqrt{\mu}$. We are now interested to see how the eigenvalues behave when the parameter is varying. The Jacobian matrix at the fixed point origin is calculated as

$$J(0,0) = \begin{pmatrix} \mu & -1 \\ 1 & \mu \end{pmatrix}$$

which has the eigenvalues ($\mu \pm i$). Thus origin is a stable spiral when $\mu < 0$ and an unstable spiral when $\mu > 0$. Therefore as expected the eigenvalues cross the imaginary axis from left to right as the parameter changes from negative to positive values. Thus we see that a supercritical Hopf bifurcation occurs when a stable spiral changes into an unstable spiral surrounded by a limit cycle.

6.5.4.2   Subcritical Hopf Bifurcation

Consider a two-dimensional system represented by

$$\begin{aligned}
\dot{x} &= \mu x - y + x(x^2 + y^2) - x(x^2 + y^2)^2 \\
\dot{y} &= x + \mu y + y(x^2 + y^2) - y(x^2 + y^2)^2;
\end{aligned} \tag{6.27}$$

where $\mu \in \mathbb{R}$ is the parameter. In polar coordinates, the system can be transformed as

$$\dot{r} = \mu r + r^3 - r^5, \dot{\theta} = 1.$$

This system has a unique fixed point at the origin. The phase portraits for $\mu < 0$ and $\mu > 0$ are presented in Fig. 6.13.

From the phase diagram it is clear that when $\mu > 0$, the fixed point origin ($r = 0$) is a stable spiral and all trajectories are attracted to it in anti-clockwise direction, and for $\mu < 0$, it is an unstable spiral. The diagram also exhibits that the system has two limits cycles when $\mu < 0$, one of which is stable and other is unstable. For $\mu > 0$, it has only a stable limit cycle. All these cycles can be determined from the equation $\mu + r^2 - r^4 = 0$. For $\mu < 0$, the system (6.27) has two limit cycles at

$$r^2 = \frac{1 \pm \sqrt{1 + 4\mu}}{2},$$

and for $\mu > 0$, the unique limit cycle occurs at

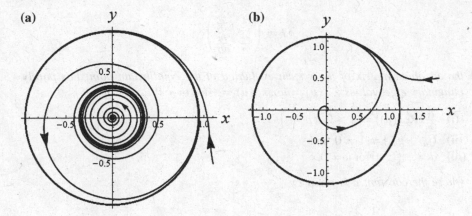

Fig. 6.13   Phase portraits of the system for **a** $\mu = -0.1$ and **b** $\mu = 0.5$

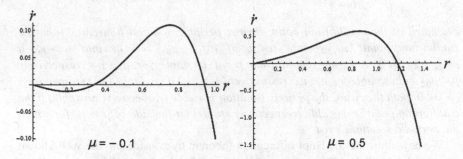

Fig. 6.14   A sketch of $\dot{r}$ versus $(\mu r + r^3 - r^5)$ for $\mu = -0.1$, and 0.5

$$r^2 = \frac{1 + \sqrt{1 + 4\mu}}{2}.$$

A sketch of $\dot{r}$ versus $(\mu r + r^3 - r^5)$ for two different values of μ is shown in Fig. 6.14.

From this figure it is clear that when $\mu < 0$, the limit cycle at $r^2 = \frac{1 + \sqrt{1 + 4\mu}}{2}$ is stable, while the limit cycle at $r^2 = \frac{1 - \sqrt{1 + 4\mu}}{2}$ is unstable, and for $\mu > 0$, the limit cycle at $r^2 = \frac{1 + \sqrt{1 + 4\mu}}{2}$ is stable.

Theorem 6.3 (Hopf bifurcation) *Let* (x_0, y_0) *be an equilibrium point of a planer autonomous system*

$$\dot{x} = f(x, y, \mu), \quad \dot{y} = g(x, y, \mu)$$

depending on some parameter $\mu \in \mathbb{R}$, *and let*

$$J = \begin{pmatrix} \frac{\partial f}{\partial x} & \frac{\partial f}{\partial y} \\ \frac{\partial g}{\partial x} & \frac{\partial g}{\partial y} \end{pmatrix},$$

the Jacobian matrix of the system evaluated at the equilibrium point has purely imaginary eigenvalues $\lambda_+(\mu) = i\omega, \lambda_-(\mu) = -i\omega; \omega \neq 0$ *at* $\mu = \mu_0$. *If*

(i) $\frac{d}{d\mu}(\text{Re}\lambda(\mu)) = \frac{1}{2} > 0$ at $\mu = 0$,

(ii) $(f_{\mu x} + g_{\mu y}) = 1 \neq 0$, and

(iii) $a = -\frac{\mu}{8} \neq 0$ for $\mu \neq 0$.

where the constant a is given by

$$a = \frac{1}{16}\left(f_{xxx} + g_{xxy} + f_{xyy} + g_{yyy}\right)$$

$$+ \frac{1}{16\omega}\left\{f_{xy}\left(f_{xx} + f_{yy}\right) - g_{xy}\left(g_{xx} + g_{yy}\right) - f_{xx}g_{xx} + f_{yy}g_{yy}\right\},$$

evaluated at the equilibrium point, then a periodic solution bifurcates from the equilibrium point (x_0, y_0) *into* $\mu < \mu_0$ *if* $a(f_{\mu x} + g_{\mu y}) > 0$ *or into* $\mu > \mu_0$ *if* $a(f_{\mu x} + g_{\mu y}) < 0$. *Also, the equilibrium point is stable for* $\mu > \mu_0$ *(respectively* $\mu < \mu_0$) *and unstable for* $\mu < \mu_0$ *(respectively* $\mu > \mu_0$) *if* $(f_{\mu x} + g_{\mu y}) < 0$ *(respectively* > 0). *In both the cases, the periodic solution is stable (respectively unstable) if the equilibrium point is unstable (respectively stable) on the side of* $\mu = \mu_0$ *for which the periodic solutions exist.*

We now illustrate the Hopf bifurcation theorem by considering the well-known van der Pol oscillator. The equation for van der Pol is given by $\ddot{x} + \mu(x^2 - 1)\dot{x} + x = 0, \mu \geq 0$. Setting $\dot{x} = y$, the equation can be written as

$$\left. \begin{array}{l} \dot{x} = y = f(x, y, \mu) \\ \dot{y} = -x + \mu(1 - x^2)y = g(x, y, \mu) \end{array} \right\}$$

The system has the equilibrium point $(x_0, y_0) = (0, 0)$, and the corresponding Jacobian matrix at (0, 0) has the eigenvalues

$$\lambda_{\pm}(\mu) = \frac{\mu \pm \sqrt{\mu^2 - 4}}{2},$$

which are complex for $0 \leq \mu < 2$ and real for $\mu \geq 2$. For $\mu = 0$, the eigenvalues are purely imaginary: $\lambda_+(\mu) = i, \lambda_-(\mu) = -i$. Now, at the equilibrium point

(i) $\frac{d}{d\mu}(\text{Re}\lambda(\mu)) = \frac{1}{2} > 0$ at $\mu = 0$,

(ii) $(f_{\mu x} + g_{\mu y}) = 1 \neq 0$, and

(iii) $a = -\frac{\mu}{8} \neq 0$ for $\mu \neq 0$.

Also, at this point, we see that $a(f_{\mu x} + g_{\mu y}) = -\frac{\mu}{8} < 0$ for $\mu > 0$. So, by Theorem 6.3, the system has a periodic solution (limit cycle) for $\mu > 0$. The stability of the limit cycle depends on the sign of $(f_{\mu x} + g_{\mu y})$, which is positive (equal to 1) at $(0, 0)$. Hence for $\mu > 0$ the equilibrium point origin must be unstable and the limit cycle must be stable.

6.5.5 Homoclinic and Heteroclinic Bifurcations

A separatrix is a phase path which separates distinct regions in the phase plane. This could be the paths which enter from a saddle point or a limit cycle or a path joining two equilibrium points of a system. There are paths which have special interest in context of dynamical systems. When a phase path joins an equilibrium point to itself, the path is a special form of separatrix known as a homoclinic path. On the other hand, any phase path which joins an equilibrium point to another is called heteroclinic path. In $\mathbb{R}^2$ homoclinic paths are associated with saddles while heteroclinic paths are connected with hyperbolic equilibrium points such as saddle-saddle, node-saddle, and spiral-saddle. Phase paths which join one saddle to itself or two distinct saddles are known as saddle connections. The graphical representations of homoclinic and heteroclinic paths are shown in Fig. 6.15.

For an example, consider the equation $\ddot{x} + x^5 - x = 0$. With $\dot{x} = y$, it can be written as

$$\dot{x} = y$$
$$\dot{y} = x - x^5$$

The equilibrium points are given by $(0, 0)$ and $(\pm 1, 0)$. The equilibrium point origin is a saddle and the other two are centers. Homoclinic paths can be formed through the origin. The equation of the phase paths is given by

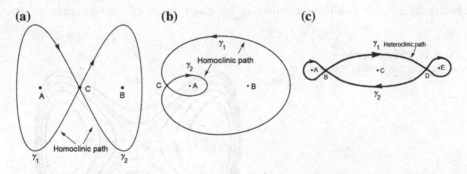

Fig. 6.15 Graphical representations of homoclinic and heteroclinic paths

$$\frac{dy}{dx} = \frac{\dot{y}}{\dot{x}} = \frac{x - x^5}{y},$$

which on integration gives

$$y^2 = x^2 - \frac{x^6}{3} + C,$$

where C is some constant. For homoclinic paths, the curve must pass through the saddle point origin and so we have $C = 0$. Therefore,

$$y^2 = x^2 - \frac{x^6}{3}.$$

This equation represents two homoclinic paths, one in the interval $0 \le x \le 3^{1/4}$ and the other in $-3^{1/4} \le x \le 0$. Figure 6.16 represents the phase diagram of the system in which the trajectory (red marked curve) represents the two homoclinic paths through the origin.

We now study the system with respect to some perturbations. Consider the above equation with a damping term $\varepsilon \dot{x}$, that is, consider the equation $\ddot{x} + \varepsilon \dot{x} + x^5 - x = 0$, where ε is a small perturbation quantity. Setting $\dot{x} = y$, we can re-write the equation as

$$\dot{x} = y$$
$$\dot{y} = -\varepsilon y + x - x^5$$

which has the equilibrium points $(0, 0), (\pm 1, 0)$, same as the unperturbed equation. The Jacobians at the equilibrium points indicate that the origin is always a saddle point, and the other two equilibrium points, $(\pm 1, 0)$, are stable spirals if $\varepsilon > 0$ and unstable spirals if $\varepsilon < 0$. The phase diagrams for $\varepsilon <, =, > 0$ are presented in Fig. 6.17, from which we see that as ε passes through the origin from left a heteroclinic spiral-saddle connection is appeared for $\varepsilon < 0$ and changes to a

Fig. 6.16 Phase trajectories and homoclinic paths of the system

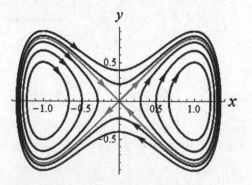

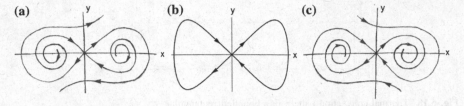

Fig. 6.17 Phase diagrams of the system for **a** $\varepsilon < 0$, **b** $\varepsilon = 0$, and **c** $\varepsilon > 0$

homoclinic saddle connection for $\varepsilon = 0$ and then to a heteroclinic saddle-spiral connection for $\varepsilon > 0$. So, the system has undergone a bifurcation in the neighborhood of $\varepsilon = 0$, the bifurcation point of the system. This type of bifurcation is known as homoclinic bifurcation (see Jordan and Smith [2] for details).

6.6 Lorenz System and Its Properties

The Massachusetts Institute of Technology (M.I.T.) Meteorologist *Edward Norton Lorenz* (1917–2008) in the year 1963 had derived a three-dimensional system from a drastically simplified model of convection rolls in atmospheric flow. The simplified model may be written in normalized form as follows:

$$\left.\begin{array}{l} \dot{x} = \sigma(y - x) \\ \dot{y} = rx - y - xz \\ \dot{z} = xy - bz \end{array}\right\} \tag{6.28}$$

where $\sigma, r, b > 0$ are all parameters. The system (6.28) has two simple nonlinear terms xz and xy in the second and third equations, respectively. Lorenz discovered that this simple looking deterministic system could have extremely erratic or complicated dynamics over a wide range of parameter values σ, r, and b. The solutions oscillate irregularly in a bounded phase space. When he plotted the trajectories in three dimensions, he discovered a new concept in the theory of dynamical system. Moreover, unlike stable fixed points or limit cycle, the strange attractor appeared in the phase space is not a point neither a curve nor a surface. It is a fractal with fractional dimension between 2 and 3. We shall study this simple looking system thoroughly below.

Consider a fluid layer of depth h, confined between two very long, stress-free, rigid and isothermal, horizontal plates in which the lower plate has a temperature T_0 and the upper plate has a temperature T_1 with $T_0 > T_1$. Let $\Delta T = T_0 - T_1$ be the temperature difference between the plates. As long as the control parameter the temperature difference ΔT is small, the fluid layer remains static and so it is stable. As ΔT crosses a critical value, this static fluid layer becomes unstable and as a result a convection roll appears in the fluid layer. This phenomenon is known as thermal

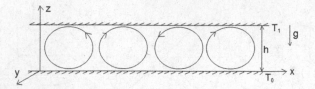

Fig. 6.18 Thermal convection pattern in a bounded rectangular

Fig. 6.19 Phase trajectory of
the Lorenz system in the *xz*-
plane

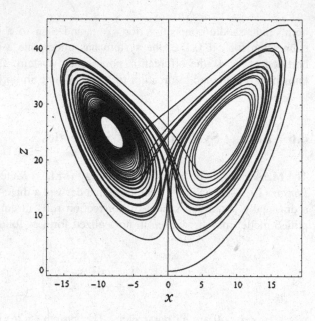

convection. We take the *x*-axis in the horizontal direction and the *z*-axis in the
vertical direction. From the symmetry of the problem, all flow variables are independent of the *y*-coordinate and the velocity of the fluid in the *y*-direction is zero.

Under Boussinesq approximations (the effects of temperature is considered only
for body force term in the equation of motion), the governing equations of motions
for incompressible fluid flows viz., the continuity equation, momentum equations,
and thermal convection may be written for usual notations as (Batchelor [3],
Chandrasekhar [4]) (Fig. 6.18).

$$\nabla \cdot \underset{\sim}{V} = 0 \qquad (6.29)$$

$$\frac{\partial \underset{\sim}{V}}{\partial t} + (\underset{\sim}{V} \cdot \nabla)\underset{\sim}{V} = -\frac{1}{\rho_0}\nabla p + \nu\nabla^2 \underset{\sim}{V} - \frac{\rho}{\rho_0}g\hat{z} \qquad (6.30)$$

$$\frac{\partial T}{\partial t} + (\underset{\sim}{V} \cdot \nabla)T = \kappa \nabla^2 T \tag{6.31}$$

where $\rho = \rho(T)$ is the fluid density at temperature T as given by

$$\rho(T) = \rho_0\{1 - \alpha(T - T_0)\} \tag{6.32}$$

$\rho_0 = \rho(T_0)$ is the fluid density at the reference temperature T_0, $v(=\mu/\rho)$ the kinematic viscosity of the fluid, μ being the coefficient of dynamic fluid viscosity, α the coefficient of thermal expansion, κ the coefficient of thermal diffusion, g the acceleration of gravity acting in the downward direction, $\hat{z}$ the unit vector along the z-axis, $\underset{\sim}{v} = (u, 0, w)$ is the fluid velocity at some instant t in the convectional motion, and $T = T(x, z, t)$ is the temperature of the fluid at that time.

The boundary conditions are prescribed as follows:

$$T = T_0 \text{ at } z = 0 \text{ and } T = T_1 \text{ at } z = h.$$

Consider the perturbed quantities (when convection starts) T', ρ' and p' defined as

$$T = T_b(z) + T'(x, z, t), \quad p = p_b(z) + p'(x, z, t) \text{ and } \rho = \rho_b(z) + \rho'(x, z, t).$$

where $T_b(z) = T_0 - (T_0 - T_1)\frac{z}{h}$ is the temperature at the steady state, $\rho_b(z) = \rho_0\{1 - \alpha(T_b(z) - T_{01})\}$ is the corresponding fluid density, and $p_b(z)$ is the corresponding pressure given by $dp_b/dz = -g\rho_b(z)$, which is obtained in the conduction state and by putting $\underset{\sim}{v} = 0$ in the equation of motion (6.30).

Substituting these in the Eqs. (6.30)–(6.32), we get

$$\frac{\partial \underset{\sim}{V}}{\partial t} + (\underset{\sim}{V} \cdot \nabla)\underset{\sim}{V} = -\frac{1}{\rho_0}\nabla p' + v\nabla^2 \underset{\sim}{V} - \frac{\rho'}{\rho_0}g\hat{z} \tag{6.33}$$

$$\frac{\partial T'}{\partial t} + (\underset{\sim}{V} \cdot \nabla)T' - \frac{(T_0 - T_1)}{h} = \kappa \nabla^2 T' \tag{6.34}$$

$$\rho' = -\rho_0 \alpha T' \tag{6.35}$$

Using (6.35) into (6.33), we have

$$\frac{\partial \underset{\sim}{V}}{\partial t} + (\underset{\sim}{V} \cdot \nabla)\underset{\sim}{V} = -\frac{1}{\rho_0}\nabla p' + v\nabla^2 \underset{\sim}{V} - \alpha g T'\hat{z}. \tag{6.36}$$

The boundary conditions become

$$T' = 0 \text{ at } z = 0, h \tag{6.37}$$

Consider the dimensionless quantities

$$x^* = \frac{x}{h}, z^* = \frac{z}{h}, \underset{\sim}{V}^* = \frac{h}{k}\underset{\sim}{V}, t^* = \frac{k}{h^2}t, p^* = \frac{h^2}{k^2}p', \theta^* = \frac{T'}{T_0 - T_1}$$

where θ^* represents the temperature deviation. Then Eqs. (6.29), (6.36), and (6.34), respectively, become (omitting the asterisk (*) for the dimensionless quantities)

$$\nabla \cdot \underset{\sim}{V} = 0 \tag{6.38}$$

$$\frac{\partial \underset{\sim}{V}}{\partial t} + (\underset{\sim}{V} \cdot \nabla)\underset{\sim}{V} = -\frac{1}{\rho_0}\nabla p + \sigma\nabla^2\underset{\sim}{V} - \sigma R\theta\hat{z} \tag{6.39}$$

$$\frac{\partial \theta}{\partial t} + (\underset{\sim}{V} \cdot \nabla)\theta = \nabla^2\theta \tag{6.40}$$

where $\sigma = \nu/\kappa$ is the Prandtl number measuring the ratio of fluid kinematic viscosity and the thermal diffusivity and $R = \alpha g(T_0 - T_1)h^3/\nu\kappa$ is the Rayleigh number characterizing basically the ratio of temperature gradient and the product of the kinematic fluid viscosity and the thermal diffusivity. Again, the boundary conditions become

$$\theta = 0 \text{ at } z = 0, 1. \tag{6.41}$$

This is known as Rayleigh–Benard convection in the literature. Let $\psi = \psi(x, z, t)$ be the steam function which is a scalar function representing a curve in the fluid medium in which tangent at each point gives velocity vector and satisfying the following relations for two-dimensional flow consideration

$$u = -\frac{\partial \psi}{\partial z}, w = \frac{\partial \psi}{\partial x}. \tag{6.42}$$

Then the continuity equation (6.38) is automatically satisfied. Also, in this case, the vorticity vector has only one nonzero component ω in the y-direction expressed by

$$\omega = \frac{\partial u}{\partial z} - \frac{\partial w}{\partial x} = -\frac{\partial^2\psi}{\partial z^2} - \frac{\partial^2\psi}{\partial x^2} = -\nabla^2\psi. \tag{6.43}$$

Taking curl of the Eq. (6.39) and then projecting the modified equation in the y-direction, we have

$$\frac{\partial \omega}{\partial t} + (\underset{\sim}{V} \cdot \nabla)\omega = \sigma \nabla^2 \omega - \sigma R \frac{\partial \theta}{\partial x} \tag{6.44}$$

But,

$$\begin{aligned}
(\underset{\sim}{V} \cdot \nabla)\omega &= u \frac{\partial \omega}{\partial x} + w \frac{\partial \omega}{\partial z} \\
&= -\frac{\partial \psi}{\partial z} \frac{\partial \omega}{\partial x} + \frac{\partial \psi}{\partial x} \frac{\partial \omega}{\partial z} \\
&= \frac{\partial(\omega, \psi)}{\partial(x, z)} \\
&= J(\omega, \psi)
\end{aligned}$$

Similarly, $(\underset{\sim}{V} \cdot \nabla)\theta = J(\theta, \psi)$. Therefore, the Eqs. (6.40) and (6.44), respectively, reduce to

$$\frac{\partial \theta}{\partial t} + J(\theta, \psi) - w = \nabla^2 \theta \tag{6.45}$$

$$\frac{\partial \omega}{\partial t} + J(\omega, \psi) = \sigma \nabla^2 \omega - \sigma R \frac{\partial \theta}{\partial x} \tag{6.46}$$

where

$$\omega = -\nabla^2 \psi. \tag{6.47}$$

We assumed that the boundaries $z = 0, 1$ are stress-free, an idealized boundary conditions. So we have other boundary conditions as given by

$$\psi = \frac{\partial^2 \psi}{\partial z^2} = 0 \text{ at } z = 0, 1. \tag{6.48}$$

We shall now convert the above set PDEs (6.45) and (6.46) into ODEs using Galerkin expansion of ψ and θ. Let the Galerkin expansions of ψ and θ satisfying the boundary conditions be

$$\psi(x, z, t) = A(t) \sin(\pi z) \sin(kx). \tag{6.49}$$

$$\theta(x, z, t) = B(t) \sin(\pi z) \cos(kx) - C(t) \sin(2\pi z). \tag{6.50}$$

where k is the wave number and $A(t)$, $B(t)$, and $C(t)$ are some functions of time t. Then

$$\omega = -\nabla^2 \psi = (\pi^2 + k^2)\psi, \nabla^2 \omega = -\nabla^2 \psi = -(\pi^2 + k^2)^2 \psi \text{ and } J(\omega, \psi) = 0.$$

Therefore, the Eq. (6.46) gives

$$(\pi^2 + k^2)\frac{dA}{dt}\sin(\pi z)\sin(kx) = kR\sigma B(t)\sin(\pi z)\sin(kx)$$
$$- \sigma(\pi^2 + k^2)^2 A(t)\sin(\pi z)\sin(kx).$$

This is true for all values of x and z. Therefore, we must have

$$(\pi^2 + k^2)\frac{dA}{dt} = kR\sigma B(t) - \sigma(\pi^2 + k^2)^2 A(t)$$

$$\Rightarrow \frac{dA}{dt} = \frac{kR\sigma}{(\pi^2 + k^2)}B(t) - \sigma(\pi^2 + k^2)A(t) \qquad (6.51)$$

Similarly, from the Eq. (6.45), we get

$$\frac{dB}{dt} = kA(t) - (\pi^2 + k^2)B(t) - \pi kA(t)C(t) \qquad (6.52)$$

$$\frac{dC}{dt} = \frac{\pi k}{2}A(t)B(t) - 4\pi^2 C(t) \qquad (6.53)$$

Rescale the variables t, $A(t)$, $B(t)$, and $C(t)$ as follows:

$$\tau = (\pi^2 + k^2)t, \ X(\tau) = \frac{k/k_c}{2 + (k/k_c)^2}A(t),$$

$$Y(\tau) = \frac{(k/k_c)^2 R}{k_c^3 \left\{2 + (k/k_c)^2\right\}^3}B(t) \text{ and } Z(\tau) = \frac{\sqrt{2}(k/k_c)^2 R}{k_c^3 \left\{2 + (k/k_c)^2\right\}^3}C(t),$$

where $k_c = \frac{\pi}{\sqrt{2}}$ is the wave number corresponding to the convection threshold. Substituting these in the Eqs. (6.51)–(6.53), we finally obtain the Lorenz equations as

$$\left.\begin{array}{l} \frac{dx}{d\tau} = \sigma(Y - X) \\ \frac{dy}{d\tau} = rX - Y - XZ \\ \frac{dz}{d\tau} = XY - bZ \end{array}\right\} \qquad (6.54)$$

where $r = R/R_c$ is known as the reduced Rayleigh number, $b = 8\big/(2 + (k/k_c)^2)$, and $R_c = (\pi^2 + k^2)^3/k^2$. Using the wave number corresponding to the convection threshold, that is, using $k = k_c$, we get $R_c = 27\pi^4/4$ and $b = 8/3$. The system (6.54) is an autonomous system of dimension three. The system, although looks very simple, is very complicated to solve analytically, because the system represents a set of nonlinear equations in $\mathbb{R}^3$ with the nonlinear terms XZ and XY in the

Fig. 6.20 Time solution (x, t) of the Lorenz system

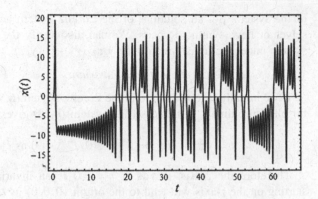

second and third equations, respectively. Lorenz studied the system for the parameter values $\sigma = 10$, $b = 8/3$ and $r = 28$. Numerical solutions (Runge–Kutta integration) of the (6.54) for these values are obtained and shown graphically below. It clearly indicates from the figure that the solutions never settle down to any simple periodic orbit or limit cycle, that is, they never repeat; it is an aperiodic motion wandering in a random manner (see the time series solution graphically, Fig. 6.20). With these parameters values and the initial conditions $(X, Y, Z) = (0, 1, 0)$, the trajectory of solutions of the Lorenz system in the phase plane (XZ-plane) looks like a butterfly (Fig. 6.19).

From the above orbit diagram for Lorenz system, the following qualitative features can be drawn:

(a) The orbit is not closed;
(b) The orbit diagram or the set of trajectories do not depict a transition stage but a well-organized regular structure;
(c) The orbit describes a number of loops on the left and on the right without any regularity in the number of loops and the loops on both sides are in opposite directions of rotations;
(d) The number of loops on the left and on the right depends in a very sensitive way on the infinitesimal change of initial conditions. Transient solution does not exhibit any periodic pattern.
(e) This is an attracting set with a dimension greater than two and was named "strange attractor" by Ruelle and Takens.

We now derive some characteristics of the Lorenz system below.

6.6.1 Properties of Lorenz System

(i) *Lorenz system is symmetric with respect to x and y axes.*

If we replace (x, y, z) by $(-x, -y, z)$ in (6.54), it remains invariant. Thus, if $(x(t), y(t), z(t))$ is a solution of the Lorenz system, then $(-x(t), -y(t), z(t))$ is also a solution of the system. So, all solutions of the Lorenz system are either symmetric them-selves or have symmetric pairs. We can also verify the symmetry of the Lorenz system under the transformation $(x, y, z) \mapsto (-x, -y, z)$.

(ii) *z-axis of the Lorenz system is invariant.*

If we initially take $x = y = 0$ in the Lorenz system (6.54), we see that $x = y = 0$ for all future time t. In this case, the system (6.28) gives

$$\dot{z} = -bz \Rightarrow z(t) = z(0)e^{-bt} \to 0 \text{ as } t \to \infty.$$

Therefore the z-axis, that is, $x = y = 0$ is an invariant set and all solutions starting on the z-axis will tend to the origin $(0, 0, 0)$ as $t \to \infty$.

(iii) *Lorenz system is dissipative in nature.*

In the dissipative system, the volume occupied in the phase space decreases as the system evolves in time. Let $V(t)$ be an arbitrary volume enclosed by a closed surface $S(t)$ in the phase space and let $S(t)$ changes to $S(t + dt)$ in the time interval dt. Let $\hat{n}$ be the outward drawn unit normal to the surface S. f is the velocity of any point, then the dot product $(f \cdot \hat{n})$ is the outward normal component of velocity.

Therefore, in time dt, a small elementary area dA sweeps out a volume $(\underset{\sim}{f} \cdot \hat{n}) dA dt$.

Therefore, $V(t + dt) = V(t) + $ (volume swept out by small area of surface which is integrated over all such elementary areas).

Hence we get

$$V(t + dt) = V(t) + \iint\limits_{S} (\underset{\sim}{f} \cdot \hat{n}) dA dt$$

$$\Rightarrow \frac{V(t + dt) - V(t)}{dt} = \iint\limits_{S} (\underset{\sim}{f} \cdot \hat{n}) dA = \int\limits_{V} (\nabla \cdot f) dV \text{ [Divergence Theorem]}$$

$$\Rightarrow \dot{V}(t) = \frac{dV}{dt} = \int\limits_{V} (\nabla \cdot \underset{\sim}{f}) dV$$

$$(6.55)$$

So for the Lorenz system, we have

$$\nabla \cdot \underset{\sim}{f} = \frac{\partial}{\partial x}(\sigma(y - x)) + \frac{\partial}{\partial y}(rx - y - xz) + \frac{\partial}{\partial z}(xy - bz)$$

$$= -(\sigma + 1 + b).$$

Therefore, from (6.55)

$$\dot{V} = \int_V -(\sigma + 1 + b)dV = -(\sigma + 1 + b)V$$

which gives the solution $V(t) = V(0)e^{-(1+\sigma+b)t}$, $V(0)$ being the initial volume. This implies that the volumes in the phase space decreases (shrink) exponentially fast and finally reaches an attracting set of zero volume. Hence, Lorenz system is dissipative in nature.

(iv) *Lorenz system shows a pitchfork bifurcation at origin when $r \to 1$.*

The fixed points of the Lorenz system are obtained by solving the equations

$$\sigma(y - x) = 0, \quad rx - y - xz = 0, \quad xy - bz = 0.$$

These give

$$x = y = z = 0 \text{ and } x = y = \pm\sqrt{b(r-1)}, z = (r-1).$$

Clearly, the origin $(0, 0, 0)$ is a fixed point for all values of the parameters. The system has another two fixed points for $r > 1$, which are given by

$$x^* = y^* = \pm\sqrt{b(r-1)}, z^* = (r-1).$$

Lorenz called these fixed points as

$$c^+ = \left(\sqrt{b(r-1)}, \sqrt{b(r-1)}, (r-1)\right) \text{ and}$$
$$c^- = \left(-\sqrt{b(r-1)}, -\sqrt{b(r-1)}, (r-1)\right).$$

Clearly, these two fixed points are symmetric in x and y coordinates. As $r \to 1$, they coincide with the fixed point origin, which gives a pitchfork bifurcation of the system. The fixed point origin is the bifurcating point. It is impossible for the Lorenz system to have either repelling fixed points or repelling closed orbits.

(v) *Linear stability analysis of the Lorenz system about the fixed point origin.*

The linearized form of the Lorenz system about the fixed point origin is given by

$$\dot{x} = \sigma(y - x)$$
$$\dot{y} = rx - y$$
$$\dot{z} = -bz$$

Now, the z-equation is decoupled so it gives

$$\dot{z} = -bz \Rightarrow z(t) = z(0)e^{-bt} \rightarrow 0 \text{ as } t \rightarrow \infty.$$

The other two equations can be written as

$$\begin{pmatrix} \dot{x} \\ \dot{y} \end{pmatrix} = \begin{pmatrix} -\sigma & \sigma \\ r & 1 \end{pmatrix} \begin{pmatrix} x \\ y \end{pmatrix}.$$

Hence sum of the diagonal elements of the matrix, $\tau = -\sigma - 1 = -(\sigma+1) < 0$ and its determinant $\Delta = (-\sigma)(-1) - \sigma r = \sigma(1-r)$. If $r > 1$, then $\Delta < 0$ and so the fixed point origin is a saddle. Since the system is three dimensional, a new type of saddle is created. This saddle has one outgoing and two incoming directions. If $r < 1$, then $\Delta > 0$ and all directions are incoming and the fixed point origin is a sink (stable node).

(vi) *The fixed point origin of the Lorenz system is globally stable for* $0 < r < 1$.

Let us consider the Lyapunov function for the Lorenz system as

$$V(x, y, z) = \frac{x^2}{\sigma} + y^2 + z^2.$$

Then the directional derivative or orbital derivative is given by

$$\dot{V} = \frac{2x\dot{x}}{\sigma} + 2y\dot{y} + 2z\dot{z}$$

$$\Rightarrow \frac{\dot{V}}{2} = \frac{x\dot{x}}{\sigma} + y\dot{y} + z\dot{z}$$

$$= x(y - x) + y(rx - y - xz) + z(xy - bz)$$

$$= -x^2 + (1 + r)xy - y^2 - bz^2$$

$$= -\left[x^2 - 2x\left(\frac{1+r}{2}\right)xy + \left(\frac{1+r}{2}\right)^2 y^2 \right] + \left(\frac{1+r}{2}\right)^2 y^2 - y^2 - bz^2$$

$$= -\left[x - \left(\frac{1+r}{2}\right)y \right]^2 - \left[1 - \left(\frac{1+r}{2}\right)^2 \right] y^2 - bz^2.$$

Thus we see that $\dot{V} < 0$ if $r < 1$ for all $(x, y, z) \neq (0, 0, 0)$ and $\dot{V} = 0$ iff $(x, y, z) = (0, 0, 0)$. Therefore, according to Lyapunov stability theorem, the fixed point origin of the Lorenz system is globally stable if the parameter $r < 1$.

(vii) *Linear stability at the fixed points* $c^{\pm}$.

Eigenvalues of the Jacobian matrix at the critical points $c^{\pm}$ of the Lorenz system satisfy the equation

$$\lambda^3 + (\sigma + b + 1)\lambda^2 + b(\sigma + r)\lambda + 2\sigma b(r - 1) = 0.$$

For $1 < r < r_H$, the three roots of the above cubic equation have all negative real parts, where

$$r_H = \frac{\sigma(3 + b + \sigma)}{\sigma - b - 1}.$$

If $r = r_H$, two of the eigenvalues are purely imaginary, and so Hopf bifurcation occurs. This bifurcation turns out to be subcritical for $r < r_H$, where two unstable periodic solutions exist for two critical values of fixed points. At $r = r_H$ these periodic solutions disappeared. For $r > r_H$ each of the two critical points have one negative real eigenvalue and two eigenvalues with positive real part, gives the unstable solution.

(viii) *Boundedness of solutions in the Lorenz system*:

There is a solid ellipsoid E given by

$$rx^2 + \sigma y^2 + \sigma(z - 2r)^2 \le c < \infty$$

such that all solutions of the Lorenz system enter E within finite time and therefore remain in E.

To prove it we take

$$\varphi(x, y, z) \equiv rx^2 + \sigma y^2 + \sigma(z - 2r)^2 = c.$$

We shall show that there exists $c = c_{cr}$ such that for all $c > c_{cr}$ the trajectory is directed toward to the ellipsoid E at any point on the E. We have

$$\hat{n}_\varphi \cdot \dot{\underset{\sim}{r}} = \frac{\nabla\varphi}{|\nabla\varphi|} \cdot \dot{\underset{\sim}{r}} = \frac{1}{|\nabla\varphi|}\left[\frac{\partial\varphi}{\partial x}\dot{x} + \frac{\partial\varphi}{\partial y}\dot{y} + \frac{\partial\varphi}{\partial z}\dot{z}\right]$$

$$= \frac{1}{|\nabla\varphi|}[2rx\sigma(y - x) + 2\sigma y(rx - y - xz) + 2\sigma(z - 2r)(xy - bz)]$$

$$= -\frac{2\sigma}{|\nabla\varphi|}\left[rx^2 + y^2 + b(z - r)^2 - br^2\right] < 0$$

So, the trajectory is directed inward to E if (x, y, z) lies inside of the ellipsoid

$$D \equiv rx^2 + y^2 + b(z - r)^2 = br^2.$$

Now, for the ellipsoid D we have

$$\frac{x^2}{\left(\sqrt{br}\right)^2} + \frac{y^2}{\left(r\sqrt{b}\right)^2} + \frac{(z-r)^2}{r^2} = 1,$$

whose center is $(0, 0, r)$ and the length of the semi-axes are $\sqrt{br}$, $r\sqrt{b}$, r respectively. Similarly, the center of the ellipsoid $E \equiv rx^2 + \sigma y^2 + \sigma(z - 2r)^2 = c$ is $(0, 0, 2r)$ and the length of semi-axes are $\sqrt{c/r}$, $\sqrt{c/\sigma}$, and $\sqrt{c/\sigma}$, respectively. Since the x and y coordinates of the centers of both the ellipsoids are 0, 0 respectively, the extent of the ellipsoid E in the x and y directions exceed the extent of the ellipsoid D in the same direction if

$$\sqrt{c/r} > \sqrt{br}; \quad \sqrt{c/\sigma} > r\sqrt{b}$$

that is, $c > br^2; c > b\sigma r^2$.

Next, along the z-axis the ellipsoid D is contained

$$0 = r - r < z < r + r = 2r$$

while for the ellipsoid E,

$$2r - \sqrt{c/\sigma} < z < 2r + \sqrt{c/\sigma}.$$

But $2r < 2r + \sqrt{c/\sigma}$ for all c. So, the lowest point $(0, 0, 0)$ of the ellipsoid D lies above the lowest point $\left(0, \, 0, \, 2r - \sqrt{c/\sigma}\right)$ of the ellipsoid E if

$$2r - \sqrt{c/\sigma} < 0, \text{that is}, c > 4r^2\sigma.$$

Let $c = c_{cr} = \max\{br^2, b\sigma r^2, 4r^2\sigma\}$. Then the ellipsoid D lies entirely within the ellipsoid E_{cr}. Hence for any point (x, y, z) exterior to D, the trajectory is directed inward E. All such trajectories must enter E_{cr} after some finite time and remain inside as $\hat{n}_\varphi \cdot \dot{\underset{\sim}{r}}$ can never be positive (Fig. 6.21).

Fig. 6.21 Graphical representation of bounded solutions of Lorenz system

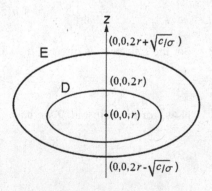

6.7 Applications

In this section, we discuss few important mathematical models which can be analyzed using dynamical principles.

6.7.1 Interacting Species Model

When two or more species (large in numbers) compete for a limited common food supply, naturally one will inhibit others growth. This interaction between the species can be studied mathematically by models, first time introduced by Alfred J. Lotka and Viot Volterra. Consider two species, say rabbits and sheep in a grassy field with limited grass supply. Let $x(t)$ and $y(t)$ be the normalized populations of the rabbits and the sheep, respectively. Then their growth rates $\dot{x}/x$ and $\dot{y}/y$ will decreasing functions of x and y, respectively. Assuming that they decrease linearly, we have the following equations

$$\left. \begin{array}{l} \dot{x} = x(a - bx - cy) \\ \dot{y} = y(d - ex - fy) \end{array} \right\} \tag{6.56}$$

where a, b, c, d, e, and f are all positive constants, and $x(t) \geq 0, y(t) \geq 0$. We now nondimensionalize these equations by setting

$$x'(\tau) = \frac{b}{a}x, y'(\tau) = \frac{f}{d}y, \tau = at, \alpha = \frac{cd}{af}, \beta = \frac{ae}{bd}, k = \frac{d}{a}.$$

Then from (6.56), we have (after removing the dashed)

$$\left. \begin{array}{l} \dot{x} = x(1 - x - \alpha y) \\ \dot{y} = ky(1 - \beta x - y) \end{array} \right\} \tag{6.57}$$

Let $f(x, y) = x(1 - x - \alpha y)$ and $\dot{y} = ky(1 - \beta x - y)$. We now find all the fixed points of the system (6.57), which are the solutions of the equations $x(1 - x - \alpha y) = 0$ and $ky(1 - \beta x - y) = 0$ simultaneously. Solving these two equations, we obtain the following four fixed points:

$$x^* = 0, y^* = 0;$$
$$x^* = 0, y^* = 1;$$
$$x^* = 1, y^* = 0;$$
$$x^* = \frac{1 - \alpha}{1 - \alpha\beta}, y^* = \frac{1 - \beta}{1 - \alpha\beta}.$$

The last fixed point is relevant with our discussion if $\alpha\beta \neq 1$ with $x^* \geq 0$ and $y^* \geq 0$. Stability of these fixed points requires the Jacobian matrix of the system, given by

$$J(x,y) = \begin{pmatrix} 1 - 2x - \alpha y & -\alpha x \\ -k\beta y & k - k\beta x - 2ky \end{pmatrix}.$$

At the fixed point $x^* = y^* = 0$, $J(0,0) = \begin{pmatrix} 1 & 0 \\ 0 & k \end{pmatrix}$. The eigenvalues of $J(0,0)$ are $\lambda_1 = 1, \lambda_2 = k$, which are positive. This implies that the fixed point $(0,0)$ is unstable. At the fixed point $(x^*, y^*) = (0, 1)$, we have $J(0,1) = \begin{pmatrix} 1 - \alpha & 0 \\ -k\beta & -k \end{pmatrix}$, which has eigenvalues $\lambda_1 = (1 - \alpha)$ and $\lambda_2 = -k$. So, the fixed point $(0,1)$ is stable if $\alpha > 1$ and unstable if $\alpha < 1$. For the third fixed point, namely $(x^*, y^*) = (1,0)$, we have $J(1,0) = \begin{pmatrix} -1 & -\alpha \\ 0 & k(1-\beta) \end{pmatrix}$, which gives the eigenvalues $\lambda_1 = -1$ and $\lambda_2 = k(1 - \beta)$. So, the fixed point $(1,0)$ is stable if $\beta > 1$ and unstable if $\beta < 1$. We now consider the last fixed point $(x^*, y^*) = \left(\frac{1-\alpha}{1-\alpha\beta}, \frac{1-\beta}{1-\alpha\beta} \right)$. The Jacobian matrix at this fixed point is calculated as $J(x^*, y^*) = \frac{1}{(1-\alpha\beta)} \begin{pmatrix} \alpha - 1 & \alpha(\alpha - 1) \\ k\beta(\beta - 1) & k(\beta - 1) \end{pmatrix}$, which has the eigenvalues

$$\lambda_\pm = \frac{(\alpha - 1) + k(\beta - 1) \pm \sqrt{\{(\alpha - 1) + k(\beta - 1)\}^2 - 4k(1 - \alpha\beta)(\alpha - 1)(\beta - 1)}}{2(1 - \alpha\beta)}.$$

Depending on the values of the parameters k, α, and β, the eigenvalues are either real or complex conjugate. We now determine the sign(s) of the eigenvalues for different values of the parameters. Note that all these parameters are positive. We now have the following four possible cases:

(i) $\alpha < 1, \beta < 1$, (ii) $\alpha < 1, \beta > 1$, (iii) $\alpha > 1, \beta < 1$, and (iv) $\alpha > 1, \beta > 1$.

Note that

$$\lambda_\pm = \frac{(\alpha - 1) + k(\beta - 1) \pm \sqrt{\{(\alpha - 1) - k(\beta - 1)\}^2 + 4k\alpha\beta(\alpha - 1)(\beta - 1)}}{2(1 - \alpha\beta)}$$

$$\lambda_+ + \lambda_- = \frac{(\alpha - 1) + k(\beta - 1)}{(1 - \alpha\beta)}$$

and $\lambda_+ \lambda_- = \frac{k(\alpha - 1)(\beta - 1)}{(1 - \alpha\beta)}$

Case I.

Let $\alpha < 1$ and $\beta < 1$. Then $\alpha\beta < 1$. So, in this case the system has four fixed points as mentioned earlier. The fixed points $(0, 1)$ and $(1, 0)$ are unstable. We now see that

Fig. 6.22 Phase portrait for
$\alpha = 0.5$, $\beta = 0.5$ and $k = 1$

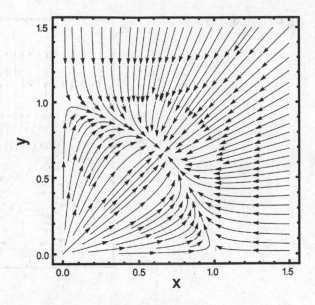

$\lambda_- < 0$ and $\lambda_+ \lambda_- > 0$. Therefore, $\lambda_+ < 0$. Thus, the fixed point $(((1-\alpha)/(1-\alpha\beta)), (1-\beta)/(1-\alpha\beta))$ is stable, and all trajectories approach to this fixed point (see Fig. 6.22).

Case II.
Let $\alpha < 1$ and $\beta > 1$. Then $(1 - \alpha\beta)$ cannot be of same sign. So, in this case the system has only three fixed points $(0,0), (1,0),$ and $(0,1)$, among which the fixed point $(1,0)$ is stable, and $(0,1)$ is unstable. All trajectories approach to the stable fixed point $(1,0)$ (see Fig. 6.23).

Case III.
Let $\alpha > 1$ and $\beta < 1$. Then $(1 - \alpha\beta)$ cannot be of same sign. Therefore, as previous, the system has only three fixed points, $(0,0)$, $(1,0)$ and $(0,1)$. The fixed point $(1,0)$ is unstable, while the fixed point $(0,1)$ is stable. All trajectories approach to the stable fixed point $(0,1)$ as time goes on (see Fig. 6.24).

Case IV.
Let $\alpha > 1$ and $\beta > 1$. Then $\alpha\beta > 1$. Therefore, the system contains all the four fixed points mentioned previously. The fixed points $(1,0)$ and $(0,1)$ are stable. In this case we also see that $\lambda_+ < 0$ and $\lambda_+ \lambda_- < 0$. Therefore, we must have $\lambda_- > 0$. So, the eigenvalues are of opposite signs. Therefore, the fixed point $(((1-\alpha)/(1-\alpha\beta)), (1-\beta)/(1-\alpha\beta))$ is a saddle point, which is unstable. The phase portrait is shown in the Fig. 6.25.

The figure shows that the fixed points $(1,0)$ and $(0,1)$ are sink. It also shows that some trajectories starting from the unstable fixed point $(0,0)$ must go to the stable

Fig. 6.23 Phase portrait for
$\alpha = 0.5$, $\beta = 2$ and $k = 1$

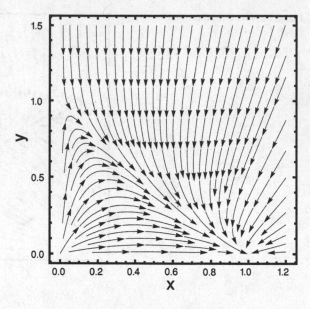

Fig. 6.24 Phase portrait for
$\alpha = 2$, $\beta = 0.5$ and $k = 1$

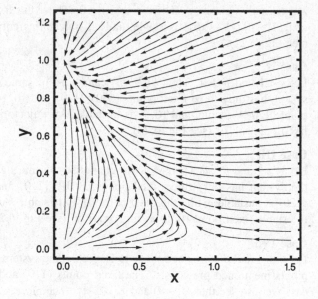

node $(1, 0)$ on the x-axis and other must go to the stable node $(0, 1)$ on the y-axis. There are also some trajectories that will go to the saddle point $(((1 - \alpha)/ (1 - \alpha\beta), (1 - \beta)/(1 - \alpha\beta))$. These trajectories are a part of the stable incoming trajectory of the saddle. This incoming trajectory divides the whole domain (positive quadrant of the xy plane) into two nonoverlapping domains with different qualitative features. This separating stable incoming trajectory is called separatrix of the system which indicates in Fig. 6.25.

Fig. 6.25 Phase portrait for $\alpha = 2$, $\beta = 2$ and $k = 1$

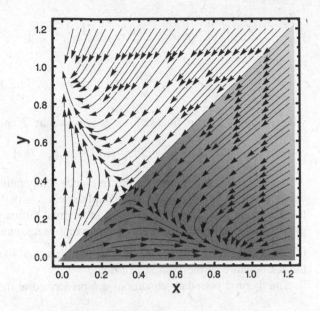

6.7.2 Convection of Couple-Stress Fluid Layer

In this section we study the convectional motion of couple-stress fluid layer heated underneath. When we mix some additives with a fluid, the forces present in the fluid and in the additives oppose one another. This opposition produces a couple-force, which in result generates a couple-stress in the fluid medium. These types of fluids are known as couple-stress fluids. Some examples are flowing blood, synovial fluid in the joints, the mixture of sugar in water, etc. Couple-stress fluids play a significant role in medical as well as in engineering sciences. In general, couple-stress fluids are non-Newtonian fluids, that is, they do not satisfy Newton's law of viscosity (stress is linearly proportional to the rate of strain). We are interested to study the convection of couple-stress fluid layer. The convection of Newtonian horizontal fluid layer had been studied extensively, see Chandrasekar [4], Drazin and Reid [5]. Consider the convection of a rectangular layer of couple-stress fluid of depth h, confined between two stress-free boundaries. The fluid layer is heated from below. We take the x-axis along the lower boundary, and the z-axis vertically upward. The lower surface is held at a temperature T_0 and the upper is at T_1 with $T_0 > T_1$. Let $\Delta T = T_0 - T_1$ be the temperature difference between the boundaries. The governing equations of motion of an incompressible couple-stress fluid in the absence of body couple are given by

$$\nabla \cdot \underset{\sim}{V} = 0 \tag{6.58}$$

$$\frac{\partial \underset{\sim}{V}}{\partial t} + (\underset{\sim}{V} \cdot \nabla)\underset{\sim}{V} = -\frac{1}{\rho_0}\nabla p + \frac{1}{\rho_0}(\mu - \mu_c\nabla^2)\nabla^2\underset{\sim}{V} - \frac{\rho}{\rho_0}g\hat{z} \qquad (6.59)$$

$$\frac{\partial T}{\partial t} + (\underset{\sim}{V} \cdot \nabla)T = \kappa\nabla^2 T \qquad (6.60)$$

where $\rho = \rho(T)$ is the fluid density at temperature T and is expressed by

$$\rho(T) = \rho_0\{1 - \alpha(T - T_0)\} \qquad (6.61)$$

$\rho_0 = \rho(T_0)$ is the density at the reference temperature T_0, μ is the dynamical coefficient of viscosity, μ_c is the couple-stress viscosity, α is the coefficient of thermal expansion, κ is the coefficient of thermal diffusion, g is the acceleration of gravity in the downward direction, $\hat{z}$ in the unit vector along z-axis, $\underset{\sim}{V} = (u, 0, w)$ is the fluid velocity at some instant t in the convectional motion, and $T = T(x, y, t)$ is the temperature of the fluid at that time.

The thermal boundary conditions are prescribed at the boundary surfaces as

$$T = T_0 \text{ at } z = 0 \text{ and } T = T_1 \text{ at } z = h.$$

Consider the perturbed quantities T', ρ' and p' defined by

$$T = T_b(z) + T'(x, z, t), \ p = p_b(z) + p'(x, z, t) \text{ and } \rho = \rho_b(z) + \rho'(x, z, t)$$

where $T_b(z) = T_0 - (T_0 - T_1)\frac{z}{h}$ is the temperature at the steady state, $\rho_b(z) = \rho_0\{1 - \alpha(T_b(z) - T_{01})\}$ is the corresponding fluid density and $p_b(z)$ is the corresponding pressure given by $dp_b/dz = -g\rho_b(z)$, which is obtained by putting $\underset{\sim}{V} = 0$ in the Eq. (6.59). Substituting these in Eqs. (6.58)–(6.60), we get

$$\frac{\partial \underset{\sim}{V}}{\partial t} + (\underset{\sim}{V} \cdot \nabla)\underset{\sim}{V} = -\frac{1}{\rho_0}\nabla p' + \frac{1}{\rho_0}(\mu - \mu_c\nabla^2)\nabla^2\underset{\sim}{V} - \frac{\rho'}{\rho_0}g\hat{z} \qquad (6.62)$$

$$\frac{\partial T'}{\partial t} + (\underset{\sim}{V} \cdot \nabla)T' - \frac{(T_0 - T_1)}{h}w = \kappa\nabla^2 T' \qquad (6.63)$$

$$\rho' = -\rho_0\alpha T' \qquad (6.64)$$

Using (6.64) in (6.62) we get the following equation

$$\frac{\partial \underset{\sim}{V}}{\partial t} + (\underset{\sim}{V} \cdot \nabla)\underset{\sim}{V} = -\frac{1}{\rho_0}\nabla p' + \frac{1}{\rho_0}(\mu - \mu_c\nabla^2)\nabla^2\underset{\sim}{V} - \alpha g T'\hat{z} \qquad (6.65)$$

The boundary conditions become

$$T' = 0 \text{ at } z = 0, h. \tag{6.66}$$

Consider the dimensionless quantities

$$x^* = \frac{x}{h}, z^* = \frac{z}{h}, V^* = \frac{h}{k}V, t^* = \frac{k}{h^2}t, p^* = \frac{1}{\rho_0}\frac{h^2}{k^2}p', \theta^* = \frac{T'}{T_0 - T_1}$$

where θ^* represents the temperature deviation. Then the Eqs. (6.58), (6.62), and (6.63), respectively, become (omitting the asterisk (*) for the dimensionless quantities)

$$\nabla \cdot V = 0 \tag{6.67}$$

$$\frac{\partial V}{\partial t} + (V \cdot \nabla)V = -\nabla p + \sigma(1 - C\nabla^2)\nabla^2 V - \sigma R\theta\hat{z} \tag{6.68}$$

$$\frac{\partial \theta}{\partial t} + (V \cdot \nabla)\theta - w = \nabla^2\theta \tag{6.69}$$

where $\sigma = v/\kappa = \mu/\kappa\rho_0$ is the Prandtl number, $R = \alpha g(T_0 - T_1)h^3/v\kappa$ is the Rayleigh number, and $C = \mu_c/\mu h^2$ is the couple-stress parameter. Actually, the parameter C is the ratio of the coefficient couple-stress fluid viscosity and the coefficient of fluid viscosity. The boundary conditions are represented by

$$\theta = 0 \text{ at } z = 0, 1. \tag{6.70}$$

Let $\psi = \psi(x, z, t)$ be the steam function satisfying the relation as

$$u = -\frac{\partial \psi}{\partial z}, w = \frac{\partial \psi}{\partial x}. \tag{6.71}$$

Then the continuity Eq. (6.67) is automatically satisfied and the vorticity vector has only one nonzero component ω in the y-direction given by

$$\omega = \frac{\partial u}{\partial z} - \frac{\partial w}{\partial x} = -\frac{\partial^2 \psi}{\partial z^2} - \frac{\partial^2 \psi}{\partial x^2} = -\nabla^2\psi. \tag{6.72}$$

Taking curl of the Eq. (6.68) and then projecting the modified equation in the y-direction, we have

$$\frac{\partial \omega}{\partial t} + (V \cdot \nabla)\omega = \sigma(1 - C\nabla^2)\nabla^2\omega - \sigma R\frac{\partial \theta}{\partial x}. \tag{6.73}$$

But we have a relation

$$(\underset{\sim}{V} \cdot \nabla)\omega = u\frac{\partial \omega}{\partial x} + w\frac{\partial \omega}{\partial z}$$

$$= -\frac{\partial \psi}{\partial z}\frac{\partial \omega}{\partial x} + \frac{\partial \psi}{\partial x}\frac{\partial \omega}{\partial z}$$

$$= \frac{\partial(\omega, \psi)}{\partial(x, z)}$$

$$= J(\omega, \psi)$$

Similarly, $(\underset{\sim}{V} \cdot \nabla)\theta = J(\theta, \psi)$. Therefore, the equations for heat transfer and the equation of motion, respectively, reduce to

$$\frac{\partial \theta}{\partial t} + J(\theta, \psi) - w = \nabla^2 \theta \qquad (6.74)$$

$$\frac{\partial \omega}{\partial t} + J(\omega, \psi) = \sigma(1 - C\nabla^2)\nabla^2\omega - \sigma R\frac{\partial \theta}{\partial x} \qquad (6.75)$$

where

$$\omega = -\nabla^2 \psi \qquad (6.76)$$

Since the boundaries $z = 0, 1$ are stress-free, so

$$\psi = \frac{\partial^2 \psi}{\partial z^2} = 0 \text{ at } z = 0, 1. \qquad (6.77)$$

Let

$$\psi(x, z, t) = A(t)\sin(\pi z)\sin(kx). \qquad (6.78)$$

$$\theta(x, z, t) = B(t)\sin(\pi z)\cos(kx) - D(t)\sin(2\pi z). \qquad (6.79)$$

be the Galerkin expansions of ψ and θ satisfying the above boundary conditions. Then

$$\omega = -\nabla^2 \psi = (\pi^2 + k^2)\psi, \nabla^2 \omega = -(\pi^2 + k^2)^2 \psi \text{ and } J(\omega, \psi) = 0.$$

Therefore, the equation of motion becomes

$$(\pi^2 + k^2)\frac{dA}{dt}\sin(\pi z)\sin(kx) = kR\sigma B(t)\sin(\pi z)\sin(kx)$$

$$- \sigma(1 + C(\pi^2 + k^2))(\pi^2 + k^2)^2 A(t)\sin(\pi z)\sin(kx).$$

which is true for all x and z. Therefore,

$$(\pi^2 + k^2)\frac{dA}{dt} = kR\sigma B(t) - \sigma(1 + C(\pi^2 + k^2))(\pi^2 + k^2)^2 A(t)$$

$$\Rightarrow \frac{dA}{dt} = \frac{kR\sigma}{(\pi^2 + k^2)}B(t) - \sigma(1 + C(\pi^2 + k^2))(\pi^2 + k^2)A(t) \quad (6.80)$$

Similarly, the heat transfer equation reduces to

$$\frac{dB}{dt} = kA(t) - (\pi^2 + k^2)B(t) - \pi kA(t)D(t) \quad (6.81)$$

$$\frac{dD}{dt} = \frac{\pi k}{2}A(t)B(t) - 4\pi^2 D(t) \quad (6.82)$$

Rescaling the variables t, $A(t)$, $B(t)$, and $D(t)$ as

$$\tau = (\pi^2 + k^2)t, \; X(\tau) = \frac{k/k_c}{2 + (k/k_c)^2}A(t),$$

$$Y(\tau) = \frac{(k/k_c)^2 R}{k_c^3\{2 + (k/k_c)^2\}^3}B(t) \text{ and } Z(\tau) = \frac{\sqrt{2}(k/k_c)^2 R}{k_c^3\{2 + (k/k_c)^2\}^3}C(t),$$

where $k_c = \frac{\pi}{\sqrt{2}}$ is the wave number corresponding to the convection threshold, we finally obtain the Lorenz-like system at the convection threshold

$$\left.\begin{array}{l}\frac{dx}{d\tau} = \sigma\{Y - (1 + C_1)X\} \\ \frac{dy}{d\tau} = rX - Y - XZ \\ \frac{dz}{d\tau} = XY - bZ\end{array}\right\} \quad (6.83)$$

where $r = R/R_c$, $R_c = (\pi^2 + k^2)^3/k^2$, $b = 8/(2 + (k/k_c)^2)$, and $C_1 = \frac{\pi^2\{2 + (k/k_c)^2\}}{2}C$. Using the wave number corresponding to the convection threshold we get $R_c = 27\pi^4/4$ and $b = 8/3$. Note that if $C = 0$, then it gives the Lorenz system Newtonian incompressible fluid.

We shall now illustrate the role of couple-stress parameter in the evolution of the system. This system is also dissipative in nature and the rate of dissipation depends upon the Couple-stress parameter C. The fixed points of the system (6.83) are given by

$$(x^*, y^*, z^*) = (0, 0, 0), \left(\pm\sqrt{\frac{b(r - 1 - C_1)}{1 + C_1}}, \pm\sqrt{(1 + C_1)b(r - 1 - C_1)}, r - 1 - C_1\right),$$

in which the equilibrium point $(0, 0, 0)$ exists for all values of the parameter r whether the other two fixed points $c^{\pm} = \left(\pm\sqrt{\frac{b(r-1-C_1)}{1+C_1}}, \pm\sqrt{(1 + C_1)b(r - 1 - C_1)},\right.$

$r - 1 - C_1)$ exist when $r \geq (1 + C_1)$ and these two equilibrium points correspond to the convective solutions of the system. They coincide with the origin when $r = (1 + C_1)$. As in the previous case, one can easily verify that the equilibrium point origin is stable for $0 < r < r_P = (1 + C_1)$ and unstable when $r > r_P$. On the other side,

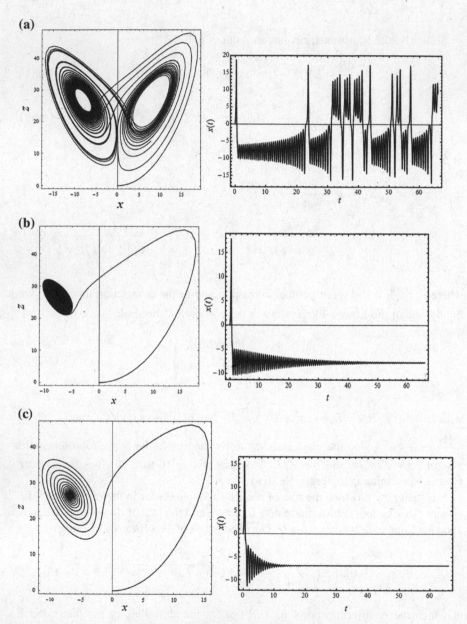

Fig. 6.26 Phase trajectories and transient behaviors of the system for **a** $C_1 = 0.1$, **b** $C_1 = 0.2$, **c** $C_1 = 0.5$, **d** $C_1 = 1.0$, **e** $C_1 = 5.0$, with $\sigma = 10$, $b = 8/3$ and $r = 28$

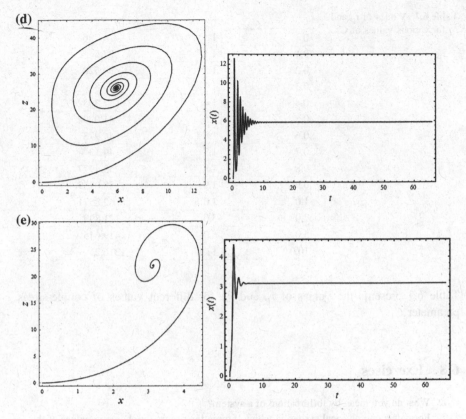

Fig. 6.26 (continued)

the equilibrium points $c^\pm$ are stable for $(1+C_1)<r<r_H$ and unstable for $r>r_H$, where $r_H = \frac{\sigma(1+C_1)^2\{3+b+\sigma(1+C_1)\}}{\sigma(1+C_1)-b-1}$. So the critical values of r are at $r=r_P$ for stable fixed points and $r=r_H$ for unstable fixed points, and the system exhibits pitchfork bifurcation first and then Hopf bifurcation. This is same as Lorenz system but the contribution of C_1 in dynamics is important. Figure 6.26 depicts the phase trajectories and transient behaviors of the system for different values of the parameter C_1. From the figure we see that the convective motion is delayed with increasing values of C_1 for a fixed value of r.

With the increase of couple-stress parameter the butterfly structure of the phase trajectories and the aperiodic transient behaviors are significantly reduced. This implies that chaotic motion generated by the Lorenz system for the above values is suppressed with increasing values of C_1. So, the couple-stress parameter of the fluid inhibits the convection in the horizontal fluid layer heated underneath. From the figure it clearly indicates that the flow trajectory approaches an attractor toward the fixed points. The attractor moves to the fixed point c^- at the value $C_1 = 0.2$. Further increase values of C_1 the attractor approaches toward the fixed point c^+.

Table 6.1 Values of r_p and	C_1	r_p	r_H
r_H for various values of C_1	0	1	24.7368
	0.1	1.1	27.5
	0.2	1.2	30.528
	0.3	1.3	33.8
	0.4	1.4	37.3032
	0.5	1.5	41.0294
	0.6	1.6	44.973
	0.7	1.7	49.13
	0.8	1.8	53.4977
	0.9	1.9	58.0739
	1.0	2.0	62.8571
	2.0	3.0	121.899
	5.0	6.0	419.645
	10.0	11.0	1316.21

Table 6.1 presents the values of r_p and r_H for different values of couple-stress parameter C_1.

6.8 Exercises

1. What do you mean by 'bifurcation' of a system?
2. Formulate one physical system in which bifurcation occurs for changing values of the parameter. Draw also bifurcation diagram.
3. Find the critical value of μ in which bifurcation occurs for the following systems:

    (i) $\dot{x} = \mu x + x^2$, (ii) $\dot{x} = 1 + \mu x + x^2$, (iii) $\dot{x} = \mu x + x^3$, (iv) $\dot{x} = \mu x - x^3$, (v)
    $\dot{x} = x^2 - \mu$, $\mu \in \mathbb{R}$, (v) $\dot{x} = \mu x + x^2 - x^3 + x^4$, (vi) $\dot{x} = -\mu_1 x - \mu_2 x^2$.

4. Determine the bifurcation point (μ_0, x_0) for the system $\dot{x} = \mu x - cx^2, (c \neq 0)$. Sketch the phase portraits for $\mu < \mu_0$ and $\mu > \mu_0$.
5. Determine the bifurcation points for the system $\dot{x} = \mu - A\cos(\pi x)$. Sketch the flow in the (x, μ) plane.
6. Draw the bifurcation diagram for the following systems when the parameter $\mu \in \mathbb{R}$ varies:

    (i) $\dot{r} = r(\mu + r), \dot{\theta} = -1$
    (ii) $\dot{r} = r(\mu - r)(\mu - r^2), \dot{\theta} = -1$

7. Discuss Hopf bifurcation for the system $\dot{r} = r(\mu - r^2), \dot{\theta} = -1; \mu \in \mathbb{R}$.
8. Plot phase portraits and also sketch the bifurcation diagrams for the following systems

    (i) $\dot{x} = x, \dot{y} = \mu - y^4, \mu \in \mathbb{R}$

(ii) $\dot{r} = r(\mu - r)(\mu + r), \dot{\theta} = 1; \mu \in \mathbb{R}$

(iii) $\dot{r} = \mu r(r - \mu)^2, \dot{\theta} = 1; \mu \in \mathbb{R}$.

9. A biochemical reaction model for certain chemical reactions, known as Brusselator model had been modeled. The model is described mathematically given by

$$\dot{x} = a - (b+1)x + x^2 y$$
$$\dot{y} = bx - x^2 y$$

where $a, b > 0$ and x and y are concentrations ($x, y \geq 0$). Discuss bifurcation of the model. In what criterion the Brusselator model admits Hopf bifurcation.

10. What do you mean by 'separatrix' of a system? What does it signify physically ? Find the fixed points of the systems and plot the separatrix and also the basin of attractions of the stable fixed points for the following systems:

(i) $\dot{x} = x(4 - x - y), \dot{y} = y(3 - x - y),$

(ii) $\dot{x} = x(3 - x - 2y), \dot{y} = y(2 - x - y),$

(iii) $\dot{x} = x(4 - x - 2y), \dot{y} = y(3 - x - y),$

(iv) $\dot{x} = x(4 - x - 2y), \dot{y} = y(3 - 2x - y)$.

11. Plot the phase portrait for the following system

$$\dot{x} = x\left(1 - \frac{x}{2} - y\right), \ \dot{y} = y\left(x - 1 - \frac{y}{2}\right).$$

Indicate the stable and unstable manifolds of the fixed points.

12. Consider the Holling-Tanner model for predator-prey interaction

$$\dot{x} = x\left(1 - \frac{x}{7}\right) - \frac{6xy}{(7 + 7x)}, \dot{y} = 0.2y\left(1 - \frac{Ny}{x}\right),$$

where N is a constant with $x(t) \neq 0$ and $y(t)$ representing the populations of prey and predators respectively. Sketch the phase portraits for (i) $N = 2.5$, and (ii) $N = 0.5$.

13. An age-dependent population can be modeled by the differential equations

$$\dot{x} = y + x(a - bx), \dot{y} = y(c + (a - bx)),$$

where x is the population, y is the birth rate, and a, b, and c are all positive constants. Find the equilibrium points of the systems and determine the long-term solution. Also draw the phase portrait.

14. Consider a simple model equations which are used in laser physics as $\dot{x} = xy - x, \dot{y} = a + by - xy$. Determine the fixed points and draw the phase portraits around the fixed points. Draw the bifurcation diagram when the parameters a and b vary.

References

1. Wiggins, S.: Introduction to Applied Nonlinear Dynamical Systems and Chaos, 2nd edn. Springer (2003)
2. Jordan, D.W., Smith, P.: Non-linear Ordinary Differential Equations. Oxford University Press (2007)
3. Batchelor, G.K.: An Introduction to Fluid Dynamics. Cambridge University Press (2000)
4. Chandrasekhar, S.: Hydrodynamic and Hydromagnetic Stability. Oxford University Press (1961)
5. Drazin, P.G., Reid, W.H.: Hydrodynamic Stability. Cambridge University Press (2004)
6. Strogatz, S.H.: Nonlinear Dynamics and Chaos with Application to Physics, Biology, Chemistry and Engineering. Perseus Books L.L.C, Massachusetts (1994)
7. Dingjun, L., Xian, W., Deming, Z., Maoan, H.: Bifurcation Theory and Methods of Dynamical Systems. World Scientific
8. Gluckenheimer, J., Holmes, P.: Nonlinear Oscillations, Dynamical Systems, and Bifurcations of Vector Fields. Springer (1983)
9. Hale, J., Kocak, H.: Dynamics and Bifurcations. Springer, New York (1991)
10. Marsden, J.E., McCracken, M.: The Hopf Bifurcation and its Applications. Springer, New York (1976)
11. Murray, J.D.: Mathematical Biology: I. An introduction, 3rd edn. Springer (2002)
12. Lorenz, E.N.: Deterministic non-periodic flow. J. Atmos. Sci. **20**, 130–141 (1963)
13. Vadasz, P.: Subcritical transitions to chaos and hysteresis in a fluid layer heated from below. Int. J. Heat Mass Trans. **43**, 705–724 (2000)
14. Stokes, V.K.: Couple stresses in fluids. Phys. Fluids **9**, 1709–1715 (1966)

Chapter 7
Hamiltonian Systems

It is well known fact that Newton's equation of deterministic motion correctly describes the motion of a particle or a system of particles in an inertial frame. In Newtonian set up there is no chance for unpredictable nature of motion. On the other hand, sometimes the particle may be restricted in its motion so that it is forced to follow a specified path or some forces may act on the particles to keep them on the surface. Thus it is out of question to treat such cases using the Newtonian formalism. Besides, if the forces of constraints acting on a system are unknown to us in advance then the Newton's equation of motion remains undefined. To get over such situation, Lagrangian mechanics, introduced by renowned Italian mathematician Joseph Louis Lagrange in 1788, provides a technique of two kinds. In the first kind of Lagrange's formulation, Newton's equation of motion is solved by evaluating the forces of constraints using the constraint relations. But it is a tedious procedure. Moreover, Newton's equation of motion is applicable in an inertial frame only. The forces of constraints operating on a dynamical system regulate some of these coordinates to vary independently this means that all the coordinates which describe the configuration of a dynamical system moving under the forces of constraints may not necessarily be independent. Consequently, the resulting equations of motion are not independent. So, to describe the configuration of the dynamical system and also to obtain a general equation of motion valid in any coordinate system, a set of independent coordinates is required. This gives a general equation of motion which is known as Lagrange's equation of motion. It is valid in any coordinate system, and the knowledge of constraint forces is not necessary for its derivation instead the knowledge of work, energy and principle of virtual work are needed. Thus, Lagrangian's method can provide a much fresher way of solving some physical systems compared to Newtonian mechanics, in particular for the system moving under some constraints. Thus Lagrangian mechanics is a reformulation of classical mechanics in terms of arbitrary coordinates.

In Lagrangian mechanics, the Lagrange's equation of motion is a second order differential equation, where the Lagrangian variables are the generalized coordinates and generalized velocities with time t as parameter. Apart from Lagrangian formulation there is another formulation in terms of Hamiltonian function. The corresponding dynamics is called Hamiltonian dynamics named after the famous Scottish mathematician Sir William Rowan Hamilton (1805–1865). Hamilton

© Springer India 2015
G.C. Layek, *An Introduction to Dynamical Systems and Chaos*,
DOI 10.1007/978-81-322-2556-0_7

originated this formulation of classical mechanics in 1833 which is applicable to a holonomic system described by a set of generalized coordinates. Hamiltonian mechanics is founded on the basis of Lagrangian formulation where the basic variables are the generalized coordinates and the generalized momenta. This reformulation provides a deeper understanding of the equations of motion of a dynamical system compared to the Lagrangian formulation and makes possible to write the equation of motion in a very stylish, yet simple way. The main beneficial thing about this formulation is that rather than providing a more convenient way of solving a particular problem, Hamiltonian mechanics gives a deeper understanding of the general structure of classical mechanics. Also, it makes clear its relationship with the quantum mechanics and other related areas of science. In this chapter we shall learn the basics of Lagrangian and Hamiltonian mechanics, and also Hamiltonian flows in the phase space, symplectic transformations and Hamilton–Jacobi equation.

7.1   Generalized Coordinates

The position of a point in space is generally specified by its position vector with respect to a fixed set of coordinate system or by the help of three Cartesian coordinates (x, y, z) of that point. Generally, the positions of N points are determined by N vectors or by $3N$ Cartesian coordinates. But the position of a system can be determined not only by using Cartesian coordinates, but there also exists alternative coordinates systems or alternative parameters by which one can determine the position of a system completely at any time t. These coordinates are called the generalized coordinates. Therefore generalized coordinates are the independent coordinates which completely specify or describe the configuration of a dynamical system at any given time. Now if we consider a set of quantities say, $q_1, q_2, \ldots, q_n$, defining the position of a dynamical system as generalized coordinates of the system then the set of their first order derivatives $\dot{q}_1, \dot{q}_2, \ldots, \dot{q}_n$ are called as generalized velocities. Any set of parameters which gives the representation of the configuration of a dynamical system without any ambiguity can serve the purpose of generalized coordinates. So one can use angles, axes, moments or any set of parameters as the generalized coordinates. But one should be careful while making choice of generalized coordinates as it is totally dependent on skill. Correct choice of generalized coordinates; make the problem look easy while the problem becomes difficult to handle for a wrong choice of the generalized coordinates. Some examples of generalized coordinates are as follows:

(i)   For a simple pendulum of length l, the generalized coordinate is the angular displacement θ from the vertical.

(ii) For a spherical pendulum of fixed length l, the generalized coordinates are θ, ϕ; θ, ϕ being the spherical polar coordinates.

(iii) Consider a rod lying on a plane surface. The generalized coordinates are x, y, θ, where (x, y) are the coordinates of one end of the rod and θ is the angle between x-axis and the rod.

(iv) Consider a lamina lying in a plane. For this case, the generalized coordinates are x, y, θ, where (x, y) are the coordinates of the centroid and θ is the angle made by a line fixed in the plane.

So far we have defined the generalized coordinate and the generalized velocities. The generalized momentum is the product of mass and the generalized velocity. If p_j is the generalized momentum of the jth particle whose mass is m_j and generalized velocity is $\dot{q}_j$ then $p_j = m_j \dot{q}_j$.

7.1.1 Configuration and Phase Spaces

The configuration of a dynamical system is described instantaneously by the generalized coordinates. The n-generalized coordinates $q_1, q_2, \ldots, q_n$ correspond to a particular point in the n-dimensional space. The n-dimensional space spanned by these n-generalized coordinates of a dynamical system is called the configuration space of that system. The state of the system changes with time and the system point traces out a curve in moving through the configuration space. The curve traces out by the system point is known as trajectory or the path of the motion of the dynamical system. On the other hand phase space is generally a $2n$-dimensional space spanned by n generalized coordinates and n generalized momenta where the qualitative behavior of a dynamical system is represented geometrically. A $2n$-dimensional space spanned by n generalized coordinates and n generalized momenta of a dynamical system is called the phase space of that dynamical system. At any instant of time a point in a phase space is called the phase point. As the dynamical system evolves with time the phase point moves through the phase space thereby traced a path, known as phase curve. When one additional dimension in terms of time t is added to the phase space then the phase space is a $(2n + 1)$-dimensional space which is called as state space.

For instance, Hamiltonian system which does not depend on time t explicitly is a $2n$-dimensional phase space. The axes of a Hamiltonian system give the values of generalized coordinate q and generalized momentum p. Hamiltonian of such systems are conserved quantities and gives the energy of the system. The trajectories of Hamiltonian system therefore can go only to those regions of phase space where the energy of the system remains same as to the initial point of the trajectory. The trajectories of a Hamiltonian system are thus confined to a $2n - 1$-dimensional constant energy surface.

7.2 Classification of Systems

A dynamical system is said to be holonomic if it is possible to give arbitrary and independent variations to the generalized coordinates without breaching the constraint relations. Otherwise it is called nonholonomic system. In mechanics the constraint is very important and can be found in Sommerfeld [1], Goldstein [2] and Arnold [3].

More specifically, a system is said to be holonomic if it contains only the holonomic constraints. If there is any nonholonomic constraint then the system is called nonholonomic. For instance, if $q_1, q_2, \ldots, q_n$ be the n generalized coordinates of a dynamical system then for a holonomic system it is possible to change q_r to $(q_r + \delta q_r)$ without changing the other coordinates.

Again consider two particles of masses m_1 and m_2 connected by a string of length l moving in space. If $\vec{r}_1$ and $\vec{r}_2$ are the position vectors of masses at time t then clearly we have $|\vec{r}_2 - \vec{r}_1| \leq l$ or $l^2 - (\vec{r}_2 - \vec{r}_1)^2 \geq 0$. In this case the system is a holonomic system with unilateral constraint. If the conditions of the constraints are expressed by means of non-integrable relations of the following form $a_m \delta t + \sum_{j=1}^{n} a_{jm} \delta q_j = 0$ for $m = 1, 2, \ldots, k(< n)$ where a's are functions of generalized coordinates then the system is called a non-holonomic dynamical system.

Example 7.1 Examine whether the motion of a vertical wheel on a horizontal plane is holonomic or non holonomic.

Solution Consider the motion of a vertical wheel of radius a rolling on a perfectly rough horizontal plane specified by the coordinate axes O_x, O_y. The contact point P traces out some curve C on the xy-plane. If θ be the angle of rotation of the wheel when the contact point P has travelled a distance s (measured from P_o) along the curve then $s = a\theta$ (assuming that the wheel rolls without sliding). Now, $\delta s = a \, \delta \theta$ (Fig. 7.1).

Fig. 7.1 Motion of a vertical wheel on a horizontal plane

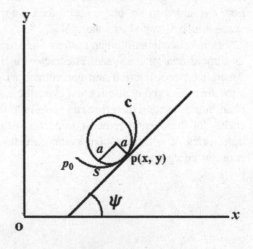

If the coordinates of P are (x, y) and if the tangent at P makes an angle ψ with O_x then

$$\delta x = \cos \psi \, \delta s = a \cos \psi \, \delta \theta$$

and

$$\delta y = \sin \psi \, \delta s = a \sin \psi \, \delta \theta$$

The coordinates (x, y, θ, ψ) form a nonholonomic system.

7.2.1 Degrees of Freedom

The degrees of freedom of a system are the minimum number of generalized coordinates needed to describe the configuration of a system or to specify the exact position of an object of that system. In other words, the minimum number of independent parameters necessary to describe the configuration of the dynamical system at any time is called the degrees of freedom.

We shall now give some examples of degrees of freedom which will help in understanding the idea more clearly.

Example 7.2 If a system is made up of N particles, we need $3N$ coordinates to specify the positions of all the particles of the system. If a system of N particles are subjected to C constraints (i.e. if some of the particles are connected by C relations), there will be (3N-C) number of independent coordinates only. So, the number of degrees of freedom is (3N-C).

Example 7.3 If a point mass is constrained to move in a plane (two dimensions) the number of spatial coordinates necessary to describe its motion is two. So the degrees of freedom in this case are two.

Example 7.4 Consider a particle moving on a surface $x^2 + y^2 + z^2 = a^2$. In this case the degrees of freedom is 2, though the degrees of freedom in 3-dimensional Cartesian coordinate system is 3. Again if the particle is inside the sphere (i.e., $x^2 + y^2 + z^2 - a^2 < 0$), then the degrees of freedom is 3.

Example 7.5 Consider a system of three free objects. The system has 9 degrees of freedom. If by imposing some constraints the free spaces between the objects are fixed, then the number of degrees of freedom of the system will be $9 - 3 = 6$. These six degrees of freedom can be chosen in any way. For example, the three coordinates of the centre of mass with the 3 angles of their inclinations to a fixed frame of reference.

The set of coordinates used to describe a system can be selected freely, keeping in mind that, the number of coordinates minus the number of constraints must give the number of degrees of freedom for that system.

7.2.1.1 Some Important Features of the Degrees of Freedom of a System

1. The number of degrees of freedom is independent of the choice of coordinate system.
2. The number of coordinates and number of constraints do not have to be the same for all possible choices.
3. There are freedoms of choices of origin, coordinate system.

7.3 Basic Problem with the Constraints

The most fundamental problem associated with the forces of constraints is that they are unknown beforehand. So, in the absence of knowledge of the total force acting on the system, it is impossible to solve Newton's equation of motion which is a relation between the total force and the acquired acceleration. The total force is the sum of the externally applied force and the force of constraints. Let us try to overcome this situation.

Consider the motion of a particle of mass m under the velocity dependent (nonholonomic) constraint

$$g\left(\vec{r},\dot{\vec{r}},t\right) = 0 \tag{7.1}$$

Let $\vec{f}^{(a)}$ and $\vec{f}$ be the externally applied forces and constraint forces respectively acting on the particle. So, the total force acting on the particle is given by $\vec{F} = \vec{f}^{(a)} + \vec{f}$. The Newton's equation of motion therefore becomes

$$m\ddot{\vec{r}} = \vec{F} = \vec{f}^{(a)} + \vec{f} \tag{7.2}$$

The numbers of equations are four whereas the numbers of unknowns are six. Therefore, the problem does not possess unique solution. To obtain a unique solution one needs additional constraint relations. The search for additional relations gives rise to Lagranges's equations of motion of the first kind. We shall now give the derivation of Lagrange's equation of first kind.

7.3.1 *Lagrange Equation of Motion of First Kind*

Consider a holonomic, bilateral constraint given by

$$g(\vec{r}, t) = 0 \qquad (7.3)$$

Differentiating Eq. (7.3) with respect to time t we get,

$$\frac{\partial g}{\partial t} + \frac{\partial g}{\partial \vec{r}} \cdot \dot{\vec{r}} = 0 \qquad (7.4)$$

Again differentiating Eq. (7.4) with respect to time 't' we get,

$$\frac{\partial^2 g}{\partial t^2} + \frac{\partial^2 g}{\partial \vec{r} \partial t} \cdot \dot{\vec{r}} + \frac{\mathrm{d}}{\mathrm{d}t}\left(\frac{\partial g}{\partial \vec{r}}\right) \cdot \dot{\vec{r}} + \frac{\partial g}{\partial \vec{r}} \cdot \ddot{\vec{r}} = 0 \qquad (7.5)$$

The above constraint relation on the total acceleration $(\ddot{\vec{r}})$ is therefore directly affected by the vector $\frac{\partial g}{\partial \vec{r}}$. Only the component of acceleration (hence the force) parallel to the vector $\frac{\partial g}{\partial \vec{r}}$ enters the above constraint relation due to scalar nature of the product $\frac{\partial g}{\partial \vec{r}} \cdot \ddot{\vec{r}}$. In other words, $\vec{f}$ must be parallel to $\frac{\partial g}{\partial \vec{r}}$, that is

$$\vec{f} = \lambda \frac{\partial g}{\partial \vec{r}} \qquad (7.6)$$

where λ is a scalar.

Let us consider a nonholonomic, bilateral constraint of the form $g\left(\vec{r}, \dot{\vec{r}}, t\right) = 0$ and taking time derivative we get, $\frac{\partial g}{\partial t} + \frac{\partial g}{\partial \vec{r}} \cdot \dot{\vec{r}} + \frac{\partial g}{\partial \dot{\vec{r}}} \cdot \ddot{\vec{r}} = 0$.

Arguing as above, one can get,

$$\vec{f} = \lambda \frac{\partial g}{\partial \dot{\vec{r}}}. \qquad (7.7)$$

Since, $g(\vec{r}, t) = 0$ or, $g\left(\vec{r}, \dot{\vec{r}}, t\right) = 0$ is given, hence $\vec{f}$ is known except for λ.

Now there are four unknowns and four independent equations which can give simultaneous solution for $\vec{r}$. Hence $\vec{f}$ can be uniquely specified along with λ. Newton's equations of motion now take the following form:

For holonomic one particle system

$$m\ddot{\vec{r}} - \vec{f}^{(a)} - \lambda \frac{\partial g(\vec{r}, t)}{\partial \vec{r}} = 0, \qquad (7.8)$$

and for nonholonomic system

$$m\ddot{\vec{r}} - \vec{f}^{(a)} - \lambda \frac{\partial g\left(\vec{r}, \dot{\vec{r}}, t\right)}{\partial \vec{r}} = 0. \qquad (7.9)$$

The Eq. (7.8) or (7.9) is sometimes called Lagrange's equation of motion of the first kind and λ is called Lagrange's multiplier.

This can easily be generalized for the motion of a system of N particles having k bilateral constraints, viz.

$$g_i(\vec{r}, t) = 0, \ i = 1, 2, \ldots, k \quad \text{(for holonomic system)}$$

and

$$g_i\left(\vec{r}, \dot{\vec{r}}, t\right) = 0, \ i = 1, 2, \ldots, k \quad \text{(for nonholonomic system)}.$$

Thus Lagrange's equations of motion of the first kind for the jth particle having mass m_j become

$$m_j\ddot{\vec{r}}_j - \vec{f}_j^{(a)} - \sum_{i=1}^{k} \lambda_i \frac{\partial g_i(\vec{r}_j, t)}{\partial \vec{r}_j} = 0, \ j = 1, 2, \ldots, N \quad \text{(for holonomic system)}$$

$$(7.10)$$

$$m_j\ddot{\vec{r}}_j - \vec{f}_j^{(a)} - \sum_{i=1}^{k} \lambda_i \frac{\partial g_i\left(\vec{r}_j, \dot{\vec{r}}_j, t\right)}{\partial \dot{\vec{r}}_j} = 0, \ j = 1, 2, \ldots, N \quad \text{(for nonholonomic system)}$$

$$(7.11)$$

$\vec{f}_j^{(a)}$ being the total externally applied force on the jth particle of the system.

These vector equations for holonomic and nonholonomic systems can be applied to the systems containing scleronomic or rheonomic bilateral constraints forms. The total number of scalar equations is $3N + k$ ($3N$ equations for motion and k number of constraints). The total number of unknowns are $3N + k$ ($3N$ components for $\vec{r}$ and k number of unknown λ). Since these equations are coupled, so obtaining solutions of these equations become rather complicated. So, Lagrange's equations of motion of the first kind are of little help and find a few applications in practice. But if solved then the solution provides the complete description of the dynamical problems of diversified nature.

Let us now show that the order of differentiation is immaterial in Lagrange's equation of motion.

Suppose that the dynamical system be comprised of N particles of masses $m_i (i = 1, 2, \ldots, N)$. Let $\vec{r}_i$ be the position vector of the ith particle having mass m_i. The position of the system at time t is specified by n generalized coordinates denoted by $q_1, q_2, \ldots, q_n$. Then each $\vec{r}_i$ is a function of $q_1, q_2, \ldots, q_n$ and time t, that is $\vec{r}_i = \vec{r}_i(q_1, q_2, \ldots, q_n)$.

Time derivative of the generalized coordinate q_i is called the generalized velocity of the ith particle. The velocity of the ith particle is given by

$$\dot{\vec{r}} = \sum_{j=1}^{n} \frac{\partial \vec{r}_i}{\partial q_j} \dot{q}_j + \frac{\partial \vec{r}_i}{\partial t}$$

Differentiating again this equation with respect to generalized coordinate 'q_j' we have,

$$\frac{\partial \dot{\vec{r}}_i}{\partial \dot{q}_j} = \frac{\partial \vec{r}_i}{\partial q_j} \text{ for } j = 1, 2, \ldots, n$$

Again,

$$\frac{\mathrm{d}}{\mathrm{d}t}\left(\frac{\partial \vec{r}_i}{\partial q_j}\right) = \frac{\partial \dot{\vec{r}}_i}{\partial q_j}$$

This proves that the order of differentiation with respect to 't' and 'q_j' are immaterial.

7.3.2 Lagrange Equation of Motion of Second Kind

Let the system contains N particles of masses $m_i (i = 1, 2, \ldots, N)$. The position of the system at time t is specified by n generalized coordinates $q_1, q_2, \ldots, q_n$. If $\vec{r}_i$ be the position vector of the ith mass then

$$r_i = \vec{r}_i(q_1, q_2, \ldots, q_n)(i = 1, 2, \ldots, N)$$

From the generalized D'Alembert's principle we have

$$\sum_{i=1}^{N} \left(\vec{F}_i - m_i \ddot{\vec{r}}_i\right).\delta \vec{r}_i = 0 \tag{7.12}$$

where $\vec{F}_i$'s being the external forces acting on the system and $\delta \vec{r}_i$'s are the small instantaneous virtual displacements consistent with the constraints.

From Eq. (7.12) we have,

$$\sum_{i=1}^{N} m_i \ddot{\vec{r}}_i.\delta \vec{r}_i = \sum_{i=1}^{N} \vec{F}_i.\delta \vec{r}_i \tag{7.13}$$

Since $\vec{r}_i = \vec{r}_i(q_1, q_2, \ldots, q_n)$, $\delta\vec{r}_i = \sum_{e=1}^{n} \frac{\partial \vec{r}_i}{\partial q_e} \delta q_e$.

Then $\delta w = \sum_{i=1}^{N} \vec{F}_i . \delta\vec{r}_i = \sum_{i=1}^{N} \vec{F}_i . \left(\sum_{e=1}^{n} \frac{\partial \vec{r}_i}{\partial q_e} \right) \delta q_e = \sum_{e=1}^{n} \left(\sum_{i=1}^{N} \vec{F}_i . \frac{\partial \vec{r}_i}{\partial q_e} \right) \delta q_e = \sum_{e=1}^{n} Q_e \delta q_e$

where $Q_e = \frac{\partial w}{\partial q_e}$ being the generalized force associated with the generalized coordinates q_e, $(e = 1, 2, \ldots, n)$.

Now,

$$
\begin{aligned}
\sum_{i=1}^{N} m_i \ddot{\vec{r}}_i . \delta\vec{r}_i &= \sum_{i=1}^{N} m_i \ddot{\vec{r}}_i . \left(\sum_{e=1}^{n} \frac{\partial \vec{r}_i}{\partial q_e} \delta q_e \right) = \sum_{e=1}^{n} \left(\sum_{i=1}^{N} m_i \ddot{\vec{r}}_i . \frac{\partial \vec{r}_i}{\partial q_e} \right) \delta q_e \\
&= \sum_{e=1}^{n} \left[\frac{\mathrm{d}}{\mathrm{d}t} \left(\sum_{i=1}^{N} m_i \dot{\vec{r}}_i . \frac{\partial \vec{r}_i}{\partial q_e} \right) - \sum_{i=1}^{N} m_i \dot{\vec{r}}_i . \frac{\mathrm{d}}{\mathrm{d}t} \left(\frac{\partial \vec{r}_i}{\partial q_e} \right) \right] \delta q_e \\
&= \sum_{e=1}^{n} \left[\frac{\mathrm{d}}{\mathrm{d}t} \left(\sum_{i=1}^{N} m_i \dot{\vec{r}}_i . \frac{\partial \vec{r}_i}{\partial q_e} \right) - \sum_{i=1}^{N} m_i \dot{\vec{r}}_i . \frac{\partial}{\partial q_e} \left(\frac{\mathrm{d}\vec{r}_i}{\mathrm{d}t} \right) \right] \delta q_e \\
&= \sum_{e=1}^{n} \left[\frac{\mathrm{d}}{\mathrm{d}t} \left(\sum_{i=1}^{N} m_i \dot{\vec{r}}_i . \frac{\partial \vec{r}_i}{\partial q_e} \right) - \sum_{i=1}^{N} m_i \dot{\vec{r}}_i . \frac{\partial \dot{\vec{r}}_i}{\partial q_e} \right] \delta q_e
\end{aligned}
$$

T = Kinetic Energy of the system $= \frac{1}{2} \sum_{i=1}^{N} m_i \dot{\vec{r}}_i^2$

Therefore, $\frac{\partial T}{\partial \dot{q}_e} = \sum_{i=1}^{n} m_i \dot{\vec{r}}_i . \frac{\partial \dot{\vec{r}}_i}{\partial \dot{q}_e} = \sum_{i=1}^{n} m_i \dot{\vec{r}}_i . \frac{\partial \vec{r}_i}{\partial q_e}$ (since, $\frac{\partial \dot{\vec{r}}_i}{\partial \dot{q}_e} = \frac{\partial \vec{r}_i}{\partial q_e}$).

$$
\frac{\partial T}{\partial q_e} = \sum_{i=1}^{N} m_i \dot{\vec{r}}_i . \frac{\partial \dot{\vec{r}}_i}{\partial q_e}
$$

Thus, $\sum_{i=1}^{N} m_i \ddot{\vec{r}}_i . \delta\vec{r}_i = \sum_{e=1}^{n} \left[\frac{\mathrm{d}}{\mathrm{d}t} \left(\frac{\partial T}{\partial \dot{q}_e} \right) - \frac{\partial T}{\partial q_e} \right] \delta q_e$.

Substituting the above expression into Eq. (7.13) and transferring all the terms in one side we have,

$$
\sum_{e=1}^{n} \left[Q_e - \left\{ \frac{\mathrm{d}}{\mathrm{d}t} \left(\frac{\partial T}{\partial \dot{q}_e} \right) - \frac{\partial T}{\partial q_e} \right\} \right] \delta q_e = 0 \qquad (7.14)
$$

Case (i) System with n degrees of freedom

In this case the coordinates are free coordinates and can be varied arbitrarily. So, the coefficients of each δq_e must vanish separately, giving

$$Q_e - \left\{ \frac{\mathrm{d}}{\mathrm{d}t} \left(\frac{\partial T}{\partial \dot{q}_e} \right) - \frac{\partial T}{\partial q_e} \right\} = 0 \quad \text{for} \quad e = 1, 2, \ldots, n$$

or,

$$\frac{\mathrm{d}}{\mathrm{d}t} \left(\frac{\partial T}{\partial \dot{q}_e} \right) - \frac{\partial T}{\partial q_e} = Q_e \quad \text{for} \quad e = 1, 2, \ldots, n \tag{7.15}$$

These equations are second order differential equations and are called as the Lagrange's equations of motion of a dynamical system with n degrees of freedom.

If in addition, the system is conservative then $Q_e = -\frac{\partial V}{\partial q_e}$ where $V = V(q_e, t)$ is the potential function.

Substituting the value for Q_e in Eq. (7.15) we have,

$$\frac{\mathrm{d}}{\mathrm{d}t} \left(\frac{\partial T}{\partial \dot{q}_e} \right) - \frac{\partial T}{\partial q_e} = -\frac{\partial V}{\partial q_e}$$

If we assume that V is independent of the generalized velocity $\dot{q}_e$, then we can write the above equation as $\frac{\mathrm{d}}{\mathrm{d}t} \left\{ \frac{\partial (T-V)}{\partial \dot{q}_e} \right\} - \frac{\partial (T-V)}{\partial q_e} = 0$.

If we set $L = T - V$, known as Lagrangian of the system then Lagrange's equations of motion can be written as

$$\frac{\mathrm{d}}{\mathrm{d}t} \left(\frac{\partial L}{\partial \dot{q}_e} \right) - \frac{\partial L}{\partial q_e} = 0, e = 1, 2, \ldots, n$$

Note that if the system contains some forces derivable from a potential function and some other forces not derivable from a potential function then the Lagrange's equation of motion can be written as

$$\frac{\mathrm{d}}{\mathrm{d}t} \left(\frac{\partial L}{\partial \dot{q}_e} \right) - \frac{\partial L}{\partial q_e} = Q'_e, e = 1, 2, \ldots, n \tag{7.16}$$

where all the potential forces have been included in the Lagrangian L and the non-potential forces are given by Q'_e.

Case (ii) Holonomic dynamical system with k bilateral constraints

For a holonomic dynamical system with k bilateral constraints, the generalized coordinates are connected by k independent relations of the following form:

$$f_j(q_1, q_2, \ldots, q_n, t) = 0, j = 1, 2, \ldots, k (k < n) \tag{7.17}$$

Let us now consider a virtual change of the system at time t consistent with the constraints in which the coordinates $q_1, q_2, \ldots, q_n$ are changed to

$q_1 + \delta q_1, q_2 + \delta q_2, \ldots, q_n + \delta q_n$. Therefore, from Eq. (7.17) we have, $f_j(q_1 + \delta q_1, q_2 + \delta q_2, \ldots, q_n + \delta q_n, t) = 0$, this can be expanded in a series like

$$f_j(q_1, q_2, \ldots, q_n, t) + \sum_{e=1}^{n} \frac{\partial f_j}{\partial q_e} \delta q_e + O(\delta q_e)^2 = 0$$

Since changes δq_e are small we have

$$\sum_{e=1}^{n} \frac{\partial f_j}{\partial q_e} \delta q_e = 0 \quad \text{for} \quad j = 1, 2, \ldots, k(k < n) \tag{7.18}$$

It is evident from (7.18) that the changes $\delta q_1, \delta q_2, \ldots, \delta q_k$ are not independent. We now introduce k arbitrary parameters $\lambda_1, \lambda_2, \ldots, \lambda_k$. We now multiply the Eq. (7.18) by these k parameters and sum up to obtain $\sum_{j=1}^{k} \lambda_j \sum_{e=1}^{n} \frac{\partial f_j}{\partial q_e} \delta q_e = 0$ or,

$$\sum_{e=1}^{n} \sum_{j=1}^{k} \left(\lambda_j \frac{\partial f_j}{\partial q_e} \right) \delta q_e = 0 \tag{7.19}$$

Adding Eq. (7.19) to the Eq. (7.14), we get

$$\sum_{e=1}^{n} \left[Q_e - \left\{ \frac{\mathrm{d}}{\mathrm{d}t} \left(\frac{\partial T}{\partial \dot{q}_e} \right) - \frac{\partial T}{\partial q_e} \right\} + \sum_{j=1}^{k} \lambda_j \frac{\partial f_j}{\partial q_e} \right] \delta q_e = 0 \tag{7.20}$$

Now choose the parameters $\lambda_1, \lambda_2, \ldots, \lambda_k$ in such a way that the coefficients of $\delta q_1, \delta q_2, \ldots, \delta q_k$ vanish separately. This gives

$$Q_e - \left\{ \frac{\mathrm{d}}{\mathrm{d}t} \left(\frac{\partial T}{\partial \dot{q}_e} \right) - \frac{\partial T}{\partial q_e} \right\} + \sum_{j=1}^{k} \lambda_j \frac{\partial f_j}{\partial q_e} = 0 \quad \text{for} \quad e = 1, 2, \ldots, k\,(<n)$$

or,

$$\frac{\mathrm{d}}{\mathrm{d}t} \left(\frac{\partial T}{\partial \dot{q}_e} \right) - \frac{\partial T}{\partial q_e} = Q_e + \sum_{j=1}^{k} \lambda_j \frac{\partial f_j}{\partial q_e} \text{for } e = 1, 2, \ldots, k(<n) \tag{7.21}$$

Now Eq. (7.19) takes the following form

$$\sum_{e=k+1}^{n} \left[Q_e - \left\{ \frac{\mathrm{d}}{\mathrm{d}t} \left(\frac{\partial T}{\partial \dot{q}_e} \right) - \frac{\partial T}{\partial q_e} \right\} + \sum_{j=1}^{k} \lambda_j \frac{\partial f_j}{\partial q_e} \right] \delta q_e = 0 \tag{7.22}$$

Since the variations $\delta q_{k+1}, \delta q_{k+2}, \ldots, \delta q_n$ are arbitrary and independent we must have

$$Q_e - \left\{ \frac{\mathrm{d}}{\mathrm{d}t}\left(\frac{\partial T}{\partial \dot{q}_e} \right) - \frac{\partial T}{\partial q_e} \right\} + \sum_{j=1}^{k} \lambda_j \frac{\partial f_j}{\partial q_e} = 0 \quad \text{for} \quad e = k+1, k+2, \ldots, n$$

or,

$$\frac{\mathrm{d}}{\mathrm{d}t}\left(\frac{\partial T}{\partial \dot{q}_e} \right) - \frac{\partial T}{\partial q_e} = Q_e + \sum_{j=1}^{k} \lambda_j \frac{\partial f_j}{\partial q_e} \quad \text{for} \quad e = k+1, k+2, \ldots, n. \tag{7.23}$$

The Eqs. (7.20) and (7.22) together give

$$\frac{\mathrm{d}}{\mathrm{d}t}\left(\frac{\partial T}{\partial \dot{q}_e} \right) - \frac{\partial T}{\partial q_e} = Q_e + \sum_{j=1}^{k} \lambda_j \frac{\partial f_j}{\partial q_e} \quad \text{for} \quad e = 1, 2, \ldots, k, k+1, k+2, \ldots, n$$

$$\tag{7.24}$$

These are the Lagrange's equations of motion for a holonomic dynamical system with k bilateral constraints.

These equations have $(n + k)$ unknown quantities $q_1, q_2, \ldots, q_n; \lambda_1, \lambda_2, \ldots, \lambda_k$. In order to solve the n number of equations given by (7.24) we have to supply k equations of constraints.

Case (iii) Nonholonomic dynamical system

In this case the changes $\delta q_1, \delta q_2, \ldots, \delta q_k$ are connected by k-nonintegrable relations of the following form

$$a_j \delta t + \sum_{e=1}^{n} a_{je} \delta q_e = 0 \quad \text{for} \quad j = 1, 2, \ldots, k (< n) \tag{7.25}$$

where a_j's are the functions of the coordinates.

For virtual changes at time t,

$$\sum_{e=1}^{n} a_{je} \delta q_e = 0 \quad \text{for } j = 1, 2, \ldots, k (< n) \tag{7.26}$$

From Eq. (7.25) it is clear that the changes $\delta q_1, \delta q_2, \ldots, \delta q_k$ are not independent. We now multiply Eq. (7.25) by k arbitrary parameters $\lambda_j (j = 1, 2, \ldots, k)$ and sum up to obtain

$$\sum_{j=1}^{k} \lambda_j \sum_{e=1}^{n} a_{je} \delta q_e = 0 \quad \text{or,} \quad \sum_{e=1}^{n} \left(\sum_{j=1}^{k} \lambda_j a_{je} \right) \delta q_e = 0 \tag{7.27}$$

Adding Eq. (7.27) with the (7.14), we get

$$\sum_{e=1}^{n} \left[Q_e - \left\{ \frac{\mathrm{d}}{\mathrm{d}t} \left(\frac{\partial T}{\partial \dot{q}_e} \right) - \frac{\partial T}{\partial q_e} \right\} + \sum_{j=1}^{k} \lambda_j a_{je} \right] \delta q_e = 0 \qquad (7.28)$$

Let us choose $\lambda_1, \lambda_2, \ldots, \lambda_k$ such that the coefficients $\delta q_1, \delta q_2, \ldots, \delta q_k$ vanish separately.

This gives $Q_e - \left\{ \frac{\mathrm{d}}{\mathrm{d}t} \left(\frac{\partial T}{\partial \dot{q}_e} \right) - \frac{\partial T}{\partial q_e} \right\} + \sum_{j=1}^{k} \lambda_j a_{je} = 0$ for $e = 1, 2, \ldots, n$

$$\text{or,} \quad \frac{\mathrm{d}}{\mathrm{d}t} \left(\frac{\partial T}{\partial \dot{q}_e} \right) - \frac{\partial T}{\partial q_e} = Q_e + \sum_{j=1}^{k} \lambda_j a_{je} \text{ for } e = 1, 2, \ldots, n \qquad (7.29)$$

Equations (7.29) are the Lagrange's equations of motion for nonholonomic dynamical system.

The equations of constraints are added in the modified form:

$$a_j + \sum_{e=1}^{n} a_{je} \dot{q}_e = 0 \quad \text{for} \quad j = 1, 2, \ldots, k(<n) \qquad (7.30)$$

If $Q_e = -\frac{\partial V}{\partial q_e}$ and $L = T - V$ then we have,

$$\frac{\mathrm{d}}{\mathrm{d}t} \left(\frac{\partial L}{\partial \dot{q}_e} \right) - \frac{\partial L}{\partial q_e} = \sum_{j=1}^{k} \lambda_j a_{je}$$

7.3.2.1   Physical Significance of λ's

Let us suppose that we remove the constraints of the system and instead of constraints let us apply external forces Q'_e in such a manner so as to keep the motion of the system unchanged. Clearly, the extra applied forces must be equal to the forces of constraints. Then under the influence of these forces Q'_e the equations of motion are $\frac{\mathrm{d}}{\mathrm{d}t} \left(\frac{\partial L}{\partial \dot{q}_e} \right) - \frac{\partial L}{\partial q_e} = Q'_e$ for $e = 1, 2, \ldots, n$.

But this must be identical with $\frac{\mathrm{d}}{\mathrm{d}t} \left(\frac{\partial L}{\partial \dot{q}_e} \right) - \frac{\partial L}{\partial q_e} = \sum_{j=1}^{k} \lambda_j a_{je}$ for $e = 1, 2, \ldots, n$.

Hence one can identify $\sum_{j=1}^{k} \lambda_j a_{je}$ with Q'_e, the generalized forces of constraints.

7.3.2.2   Cyclic Coordinates (Ignorable Coordinates)

If a coordinate is explicitly absent in the Lagrangian function L of a dynamical system then the coordinate is called a cyclic or ignorable coordinate. Thus if q_k is a cyclic coordinate, then the form of the Lagrangian is $L \equiv L(\dot{q}_k, t)$ and $\frac{\partial L}{\partial q_k} = 0$. The cyclic co-ordinate is very important in deriving Hamilton's equations of motion.

Example 7.6 Find the Lagrange's equation of motion of a simple pendulum

Solution The generalized coordinate of a simple pendulum of length l is the angle variable θ. The velocity of the ball is $l\dot{\theta}$ where l is the length of the string of the pendulum. A simple pendulum oscillating in a vertical plane constitutes a conservative holonomic dynamical system with one degree of freedom.

Here, Kinetic Energy $= T = \frac{1}{2} m l^2 \dot{\theta}^2$, m is the mass of the ball.

Potential Energy $= V = mgh = mgl(1 - \cos \theta)$.

Therefore, Lagrangian of the system $= L = T - V = \frac{1}{2} m l^2 \dot{\theta}^2 - mgl(1 - \cos \theta)$.

Lagrange's equation of motion is $\frac{d}{dt}\left(\frac{\partial L}{\partial \dot{\theta}}\right) - \frac{\partial L}{\partial \theta} = 0$ or, $m l^2 \ddot{\theta} + mgl \sin \theta = 0$

or, $m l^2 \ddot{\theta} = -mgl \sin \theta$ or, $\ddot{\theta} = -\frac{g}{l} \sin \theta \simeq -\frac{g}{l}\theta$ (if the amplitude of oscillation is small then θ is small and so $\sin \theta \simeq \theta$).

Time period is given by $2\pi\sqrt{\frac{l}{g}}$, g is the acceleration due to gravity.

Example 7.7 For a dynamical system Lagrangian is given by $L = \frac{1}{2}(\dot{x}^2 + \dot{y}^2 + \dot{z}^2) - V(x, y, z) + A\dot{x} + B\dot{y} + C\dot{z}$ where A, B, C are functions of (x, y, z). Show that Lagrange's equations of motion are $\ddot{x} + \dot{y}\left(\frac{\partial A}{\partial y} - \frac{\partial B}{\partial x}\right) + \dot{z}\left(\frac{\partial A}{\partial z} - \frac{\partial C}{\partial x}\right) + \frac{\partial V}{\partial x} = 0$ and similar ones.

Solution The generalized coordinates for the given dynamical system are x, y, z.

Now the Lagrange's equation of motion corresponding to x-coordinate is

$$\frac{d}{dt}\left(\frac{\partial L}{\partial \dot{x}}\right) - \frac{\partial L}{\partial x} = 0$$

or, $\frac{d}{dt}(\dot{x} + A) - \left[-\frac{\partial V}{\partial x} + \frac{\partial A}{\partial x}\dot{x} + \frac{\partial B}{\partial x}\dot{y} + \frac{\partial C}{\partial x}\dot{z}\right] = 0$

or, $\ddot{x} + \frac{dA}{dt} + \frac{\partial V}{\partial x} - \frac{\partial A}{\partial x}\dot{x} - \frac{\partial B}{\partial x}\dot{y} - \frac{\partial C}{\partial x}\dot{z} = 0$

or, $\ddot{x} + \dot{y}\left(\frac{\partial A}{\partial y} - \frac{\partial B}{\partial x}\right) + \dot{z}\left(\frac{\partial A}{\partial z} - \frac{\partial C}{\partial x}\right) + \frac{\partial V}{\partial x} = 0$ since $\frac{dA}{dt} = \frac{\partial A}{\partial x}\dot{x} + \frac{\partial A}{\partial y}\dot{y} + \frac{\partial A}{\partial z}\dot{z}$.

In an analogous way one can obtain the other two Lagrange's equations of motion corresponding to y and z coordinates.

Example 7.8 Obtain the Lagrangian and also the Lagrange's equation of motion of a harmonic oscillator.

Solution A harmonic oscillator consists of a single particle of mass m moving in a straight line which can be taken as x-axis (see Fig. 7.2). The particle is attracted

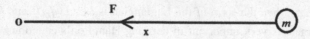

Fig. 7.2 Sketch of a harmonic oscillator

towards the origin by a force which varies proportionally with the distance of the particle from the origin.

Then the Kinetic Energy of the harmonic oscillator is given by $T = \frac{1}{2}m\dot{x}^2$ whereas the Potential Energy is given by $V = \frac{1}{2}kx^2$, k being a constant.

Then the Lagrangian of the motion is $L = T - V = \frac{1}{2}m\dot{x}^2 - \frac{1}{2}kx^2$.

Lagrange's equation of motion is $\frac{d}{dt}\left(\frac{\partial L}{\partial \dot{x}}\right) - \frac{\partial L}{\partial x} = 0$ or, $m\ddot{x} + kx = 0$.

Example 7.9 Find the Lagrangian of motion of a particle of unit mass moving in a central force field under a force that varies inversely as the square of the distance from the centre O (Fig. 7.3).

Solution Here r, θ are the generalized coordinates. Let V be the potential.

Then, $-\frac{dV}{dr} = F = -\frac{\mu}{r^2}$ which on integration gives $V = -\frac{\mu}{r} + \text{Constant}$ $T = $ Kinetic Energy of the particle $= \frac{1}{2}\left(\dot{r}^2 + r^2\dot{\theta}^2\right)$.

The Lagrangian of the motion is $L = T - V = \frac{1}{2}\left(\dot{r}^2 + r^2\dot{\theta}^2\right) + \frac{\mu}{r} - \text{Constant}$.

Now, Lagrange's equation of motion is given by $\frac{d}{dt}\left(\frac{\partial L}{\partial \dot{q}_e}\right) - \frac{\partial L}{\partial q_e} = 0, e = 1, 2$

Here, $q_1 = r, q_2 = \theta$.

Lagrange's equation of motion corresponding to r-coordinate gives $\frac{d}{dt}\left(\frac{\partial L}{\partial \dot{r}}\right) - \frac{\partial L}{\partial r} = 0$ or, $\frac{d}{dt}(\dot{r}) - r\dot{\theta}^2 + \frac{\mu}{r^2} = 0$ or, $\ddot{r} - r\dot{\theta}^2 + \frac{\mu}{r^2} = 0$ and $\frac{d}{dt}\left(\frac{\partial L}{\partial \dot{\theta}}\right) - \frac{\partial L}{\partial \theta} = 0$ or,

$\frac{d}{dt}\left(r^2\dot{\theta}\right) = 0$ which gives $r^2\dot{\theta} = \text{constant}$

Fig. 7.3 Motion under a central force field

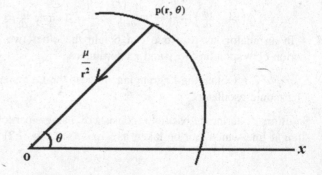

Theorem 7.1 *For a scleronomous, conservative dynamical system of n degrees of freedom the quantity $\left(\sum_{k=1}^{n} \dot{q}_k \frac{\partial L}{\partial \dot{q}_k} - L\right)$ is a constant of motion where $q_1, q_2, \ldots, q_n$ are the n generalized coordinates and L is the Lagrangian of the system.*

Proof Differentiating the quantity $\sum_{k=1}^{n} \dot{q}_k \frac{\partial L}{\partial \dot{q}_k} - L$ with respect to t, we get

$$\frac{d}{dt}\left(\sum_{k=1}^{n} \dot{q}_k \frac{\partial L}{\partial \dot{q}_k} - L\right) = \sum_{k=1}^{n} \ddot{q}_k \frac{\partial L}{\partial \dot{q}_k} + \sum_{k=1}^{n} \dot{q}_k \frac{d}{dt}\left(\frac{\partial L}{\partial \dot{q}_k}\right) - \frac{dL}{dt}$$

Now, $L = L(q_k, \dot{q}_k)$

Therefore, $\frac{dL}{dt} = \sum_{k=1}^{n} \frac{\partial L}{\partial q_k} \dot{q}_k + \sum_{k=1}^{n} \frac{\partial L}{\partial \dot{q}_k} \ddot{q}_k$

Thus, $\frac{d}{dt}\left(\sum_{k=1}^{n} \dot{q}_k \frac{\partial L}{\partial \dot{q}_k} - L\right) = \sum_{k=1}^{n} \ddot{q}_k \frac{\partial L}{\partial \dot{q}_k} + \sum_{k=1}^{n} \dot{q}_k \frac{d}{dt}\left(\frac{\partial L}{\partial \dot{q}_k}\right) - \sum_{k=1}^{n} \frac{\partial L}{\partial q_k} \dot{q}_k - \sum_{k=1}^{n} \frac{\partial L}{\partial \dot{q}_k} \ddot{q}_k$

$$= \sum_{k=1}^{n} \left[\frac{d}{dt}\left(\frac{\partial L}{\partial \dot{q}_k}\right) - \frac{\partial L}{\partial q_k}\right] \dot{q}_k = 0,$$

since by Lagrange's equations of motion $\frac{d}{dt}\left(\frac{\partial L}{\partial \dot{q}_k}\right) - \frac{\partial L}{\partial q_k} = 0$ for $k = 1, 2, \ldots, n$.

Therefore, $\left(\sum_{k=1}^{n} \dot{q}_k \frac{\partial L}{\partial \dot{q}_k} - L\right) = $ constant $= E$ (say)

Thus when the Lagrangian is independent of time, Lagrange's equations possess the integral of motion, known as the energy integral of the system.

Theorem 7.2 *For a scleronomous, conservative dynamical system of n degrees of freedom the total energy $E = T + V$ is constant*

Proof We know that $L = T - V = T(q_k, \dot{q}_k) - V(q_k)$

Now,

$$E = \sum_{k=1}^{n} \dot{q}_k \frac{\partial(T - V)}{\partial \dot{q}_k} - (T - V) = \sum_{k=1}^{n} \dot{q}_k \frac{\partial T}{\partial \dot{q}_k} - T + V$$

$$= 2T - T + V = T + V$$

since, $\frac{\partial V}{\partial \dot{q}_k} = 0$ in a velocity independent potential field and since in a scleronomous system, Kinetic Energy T is a homogeneous quadratic function of generalized velocities $\dot{q}_k(k = 1, 2, \ldots, n)$ i.e. $2T = \sum_{k=1}^{n} \dot{q}_k \frac{\partial T}{\partial \dot{q}_k}$ (by Euler's formula for homogeneous function). In such a system the total energy is conserved.

Theorem 7.3 (Law of conservation of generalized momentum) *The generalised momentum corresponding to a cyclic coordinate of a system is an integral of motion or constant of motion.*

Proof Lagrange's equation of motion corresponding to the coordinate q_k is given by

$$\frac{d}{dt}\left(\frac{\partial L}{\partial \dot{q}_k}\right) - \frac{\partial L}{\partial q_k} = 0, L \text{ being the Lagrangian}$$

If q_k is cyclic coordinate then $\frac{\partial L}{\partial q_k} = 0$. Thus from the above equation, we have

$$\frac{d}{dt}\left(\frac{\partial L}{\partial \dot{q}_k}\right) = 0 \text{ i.e. } \frac{\partial L}{\partial \dot{q}_k} = \text{constant or, } p_k = \text{constant} \tag{7.31}$$

where $p_k = \frac{\partial L}{\partial \dot{q}_k}$ is the generalized momentum corresponding to the generalized coordinate q_k.

Therefore, the generalized momentum conjugate to a cyclic coordinate is an integral of motion. So, when the Lagrangian is time independent there is an energy integral for the system. In case of cyclic co-ordinate there is also an integral for the system, known as a momentum integral.

Theorem 7.4 *Let q_j be a cyclic coordinate such that δq_j corresponds to a rotation of the system of particles around some axis, then the angular momentum of the system is conserved.*

Proof Let the system be conservative. Then the potential energy V depends on position only. It is well known that kinetic energy T depends on translational velocities which are not affected by rotation. As the position coordinate q_j is affected by rotation δq_j, the kinetic energy T does not depend on the coordinate q_j.
So,

$$\frac{\partial V}{\partial \dot{q}_j} = 0 \quad \text{and} \quad \frac{\partial T}{\partial q_j} = 0 \tag{7.32}$$

Now, Lagrange's equation of motion corresponding to the generalized coordinate q_j can be written as

$$\frac{d}{dt}\left(\frac{\partial T}{\partial \dot{q}_j}\right) - \frac{\partial T}{\partial q_j} = -\frac{\partial V}{\partial q_j} \tag{7.33}$$

Using (7.32) in (7.33) we have, $\frac{d}{dt}\left\{\frac{\partial(T-V)}{\partial \dot{q}_j}\right\} = -\frac{\partial V}{\partial q_j}$ or, $\frac{d}{dt}\left(\frac{\partial L}{\partial \dot{q}_j}\right) = -\frac{\partial V}{\partial q_j}$
or, $\frac{d}{dt}(p_j) = -\frac{\partial V}{\partial q_j}$ i.e.

$$\dot{p}_j = Q_j \tag{7.34}$$

where $Q_j = -\frac{\partial V}{\partial q_j}$ is the generalized force corresponding to the generalized coordinate q_j.

As q_j is a rotational coordinate so p_j is the component of the total angular momentum along the axis of rotation.

The magnitude of change in the position vector $\vec{r}_i$ due to the change in rotational coordinate q_j is $|\delta \vec{r}_i| = \vec{r}_i \sin \theta \, \delta q_j$ which gives $\left| \frac{\partial \vec{r}_i}{\partial q_j} \right| = \vec{r}_i \sin \theta$

Now, $\frac{\partial \vec{r}_i}{\partial q_j}$ is perpendicular to both $\vec{r}_i$ and $\vec{n}$ where $\vec{n}$ is the unit vector along the axis of rotation. Therefore, $\frac{\partial \vec{r}_i}{\partial q_j} = \vec{n} \times \vec{r}_i$.

Now, $p_j = \frac{\partial T}{\partial \dot{q}_j} = \sum_i m_i \vec{v}_i \cdot \frac{\partial \vec{r}_i}{\partial q_j} = \sum_i m_i \vec{v}_i \cdot \vec{n} \times \vec{r}_i = \vec{n} \cdot \sum_i \vec{L}_i = \vec{n} \cdot \vec{L}$

where $\vec{L} = \sum_i \vec{L}_i = \sum_i m_i \vec{v}_i \times \vec{r}_i$ is the total angular momentum along the axis of rotation.

It is known that if q_j is cyclic then the generalized momentum p_j is constant.

Hence, one can find that if the rotational coordinate is cyclic, the component of total angular momentum along the axis of rotation remains constant.

Corollary 7.1 *If the rotational coordinate is cyclic, the component of the applied torque along the axis of rotation vanishes.*

Proof Generalised force Q_j is given by

$$Q_j = \sum_i \vec{F}_i \cdot \frac{\partial \vec{r}_i}{\partial q_j},$$

$\vec{F}_i$ being the force acting on the ith particle of the system and $\vec{r}_i$ is the position vector of the ith particle.

Again, $Q_j = \sum_i \vec{F}_i \cdot \vec{n} \times \vec{r}_i = \vec{n} \cdot \sum_i \vec{F}_i \times \vec{r}_i = \vec{n} \cdot \sum_i \vec{N}_i = \vec{n} \cdot \vec{N}, \quad \vec{N} = \sum_i \vec{N}_i = \sum_i \vec{F}_i \times \vec{r}_i$ being the total torque acting on the system.

From Eq. (7.34) we have $Q_j = 0$ since p_j is constant.

Thus if the rotational coordinate is cyclic, the component of the applied torque along the axis of rotation vanishes.

Example 7.10 The Lagrangian of a particle of mass m moving in a central force field is given by (in polar coordinates)

$$L = \frac{1}{2} m \left(\dot{r}^2 + r^2 \dot{\theta}^2 \right) - V(r).$$

Discuss its motion.

Solution Clearly, r, θ are the generalized coordinates and since θ is not present in the Lagrangian L, it is the cyclic coordinate.

Therefore,

$$p_\theta = \frac{\partial L}{\partial \dot{\theta}} = mr^2 \dot{\theta} = \text{constant}$$

which indicates that angular momentum of the particle is a constant of motion.

Lagrange's equation of motion corresponding to the r-coordinate is
$\frac{d}{dt}\left(\frac{\partial L}{\partial \dot{r}}\right) - \frac{\partial L}{\partial r} = 0$ or, $m\ddot{r} - mr\dot{\theta}^2 + \frac{\partial V}{\partial r} = 0$ or, $m\ddot{r} - \frac{p_\theta^2}{mr^3} + \frac{\partial V}{\partial r} = 0$.

As p_θ is constant, so we have to deal with only one degree of freedom.

7.3.2.3 Routh's Process for the Ignoration of Coordinates

Consider a dynamical system with n degrees of freedom specified by n generalized coordinates $q_1, q_2, \ldots, q_n$. Let the system has j-cyclic coordinates $q_1, q_2, \ldots, q_j$ ($j < n$). We shall show that the dynamical system has $(n\text{-}j)$ degrees of freedom. Clearly the generalized momenta against the cyclic coordinates would be

$$p_k = \frac{\partial L}{\partial \dot{q}_k} = \text{constant} = \beta_k \quad \text{for} \quad k = 1, 2, \ldots, j \qquad (7.35)$$

Let us define a new function R as

$$R = L - \sum_{k=1}^{j} \beta_k \dot{q}_k, L \qquad (7.36)$$

being the Lagrangian of the system

With the help of (7.36), R can be expressed as a function of $q_{j+1}, q_{j+2}, \ldots, q_n; \dot{q}_{j+1}, \dot{q}_{j+2}, \ldots, \dot{q}_n; \beta_1, \beta_2, \ldots, \beta_j; t$

that is, $R = R(q_{j+1}, q_{j+2}, \ldots, q_n; \dot{q}_{j+1}, \dot{q}_{j+2}, \ldots, \dot{q}_n; \beta_1, \beta_2, \ldots, \beta_j; t)$.

The function R is called the Routhian function.

Taking a virtual change of R we get from Eq. (7.36),

$$\delta R = \delta L - \sum_{k=1}^{j} \delta\beta_k \dot{q}_k - \sum_{k=1}^{j} \beta_k \delta\dot{q}_k. \qquad (7.37)$$

Also, $L = L(q_{j+1}, q_{j+2}, \ldots, q_n; \dot{q}_1, \dot{q}_2, \ldots, \dot{q}_n; t)$.

$$\therefore \delta L = \sum_{k=j+1}^{n} \frac{\partial L}{\partial q_k} \delta q_k + \sum_{k=1}^{n} \frac{\partial L}{\partial \dot{q}_k} \delta\dot{q}_k$$

$$= \sum_{k=j+1}^{n} \frac{\partial L}{\partial q_k} \delta q_k + \sum_{k=1}^{j} \frac{\partial L}{\partial \dot{q}_k} \delta\dot{q}_k + \sum_{k=j+1}^{n} \frac{\partial L}{\partial \dot{q}_k} \delta\dot{q}_k. \qquad (7.38)$$

Substituting (7.38) in the Eq. (7.37) we get

$$\sum_{k=j+1}^{n} \frac{\partial R}{\partial q_k}\delta q_k + \sum_{k=j+1}^{n} \frac{\partial R}{\partial \dot{q}_k}\delta \dot{q}_k + \sum_{k=1}^{j} \frac{\partial R}{\partial \beta_k}\delta \beta_k$$

$$= \sum_{k=j+1}^{n} \frac{\partial L}{\partial q_k}\delta q_k + \sum_{k=1}^{j} \frac{\partial L}{\partial \dot{q}_k}\delta \dot{q}_k + \sum_{k=j+1}^{n} \frac{\partial L}{\partial \dot{q}_k}\delta \dot{q}_k - \sum_{k=1}^{j} \delta \beta_k \dot{q}_k - \sum_{k=1}^{j} \beta_k \delta \dot{q}_k \quad (7.39)$$

$$= \sum_{k=j+1}^{n} \frac{\partial L}{\partial q_k}\delta q_k + \sum_{k=j+1}^{n} \frac{\partial L}{\partial \dot{q}_k}\delta \dot{q}_k - \sum_{k=1}^{j} \delta \beta_k \dot{q}_k,$$

since by Eq. (7.35), $\frac{\partial L}{\partial \dot{q}_k} = \beta_k$ for $k = 1, 2, \ldots, j$.

Now Eq. (7.39) involves changes δq_k, $\delta \dot{q}_k (k = j+1, j+2, \ldots, n)$ and $\delta \beta_k (k = 1, 2, \ldots, j)$ which are arbitrary and independent.

This leads to the following equations

$$\frac{\partial R}{\partial q_k} = \frac{\partial L}{\partial q_k}, \frac{\partial R}{\partial \dot{q}_k} = \frac{\partial L}{\partial \dot{q}_k} \quad \text{for} \quad k = j+1, j+2, \ldots, n \qquad (7.40)$$

And

$$-\frac{\partial R}{\partial \beta_k} = \dot{q}_k \quad \text{for} \quad k = 1, 2, \ldots, j. \qquad (7.41)$$

By Lagrange's equation of motion we have

$$\frac{\mathrm{d}}{\mathrm{d}t}\left(\frac{\partial L}{\partial \dot{q}_k}\right) - \frac{\partial L}{\partial q_k} = 0 \quad \text{for} \quad k = j+1, j+2, \ldots, n.$$

Using (7.40) we have,

$$\frac{\mathrm{d}}{\mathrm{d}t}\left(\frac{\partial R}{\partial \dot{q}_k}\right) - \frac{\partial R}{\partial q_k} = 0 \quad \text{for} \quad k = j+1, j+2, \ldots, n.$$

From (7.41) we have

$$q_k = -\int \frac{\partial R}{\partial \beta_k}\,dt + \text{constant} \quad \text{for} \quad k = 1, 2, \ldots, j \qquad (7.42)$$

This shows that R behaves as the Lagrangian L of a new dynamical system having $(n\text{-}j)$ degrees of freedom.

Example 7.11 Use the method of ignorable coordinates to reduce the degrees of freedom of a spherical pendulum to one.

Solution A spherical pendulum is a simple pendulum of length l. The bob of the pendulum moves on the surface of a sphere of radius equal to the length of the pendulum.

Let (l, θ, ϕ) be the position of the bob at time t in spherical polar coordinate system. Let h be the height of the bob from the horizontal plane.

$$\text{Potential energy} = V = mgh = mg\{l - l\cos(\pi - \theta)\} = mgl(1 + \cos\theta)$$

$$\text{Kinetic energy} = T = \frac{1}{2}m\left\{\left(l\dot{\theta}\right)^2 + \left(l\sin\theta\dot{\phi}\right)^2\right\} = \frac{1}{2}ml^2\left(\dot{\theta}^2 + \sin^2\theta\dot{\phi}^2\right).$$

Now, Lagrangian of the system $= L = T - V = \frac{1}{2}ml^2\left(\dot{\theta}^2 + \sin^2\theta\dot{\phi}^2\right) - mgl(1 + \cos\theta)$.

Clearly, ϕ is a cyclic coordinate.

$$\therefore \frac{\partial L}{\partial \dot{\phi}} = p_\phi = \text{constant} = \beta_\phi \text{ which gives } ml^2\sin^2\theta\dot{\phi} = \beta_\phi \text{ or, } \dot{\phi} = \frac{\beta_\phi}{ml^2\sin^2\theta}.$$

Now, Routhian function is given by

$$R = L - \sum_{k=1}^{j}\beta_k\dot{q}_k = L - \beta_\phi\dot{\phi} = \frac{1}{2}ml^2\left(\dot{\theta}^2 + \sin^2\theta\dot{\phi}^2\right) - mgl(1 + \cos\theta) - \beta_\phi\dot{\phi}$$

$$= \frac{1}{2}ml^2\left(\dot{\theta}^2 + \frac{\beta_\phi^2}{m^2l^4\sin^2\theta}\right) - mgl(1 + \cos\theta) - \frac{\beta_\phi^2}{ml^2\sin^2\theta}$$

$$= \frac{1}{2}ml^2\dot{\theta}^2 - \frac{1}{2}\frac{\beta_\phi^2}{ml^2\sin^2\theta} - mgl(1 + \cos\theta)$$

$$= R(\theta, \dot{\phi}, \beta_\phi)$$

The equation of motion is $\frac{d}{dt}\left(\frac{\partial R}{\partial \dot{\theta}}\right) - \frac{\partial R}{\partial \theta} = 0,$,

$$\text{or, } ml^2\ddot{\theta} - \frac{\beta_\phi^2\cos\theta}{ml^2\sin^3\theta} - mgl\sin\theta = 0.$$

The above equation can be interpreted as representing a system with single degree of freedom.

Example 7.12 In a dynamical system the kinetic energy and the potential energy are given by $T = \frac{1}{2}\frac{\dot{q}_1^2}{a + bq_2^2} + \frac{1}{2}\dot{q}_2^2, V = c + dq_2^2$ where a, b, c, d are constants.

Determine $q_1(t)$ and $q_2(t)$ by Routh's process for ignorable coordinates.

Solution Here, Lagrangian is given by

$$L = \frac{1}{2}\frac{\dot{q}_1^2}{a + bq_2^2} + \frac{1}{2}\dot{q}_2^2 - c - dq_2^2.$$

Clearly, q_1 is a cyclic coordinate and so, $\frac{\partial L}{\partial q_1} = 0$.

Generalized momentum corresponding to the coordinate q_1 is $p_1 = \frac{\partial L}{\partial \dot{q}_1} =$ constant $= \beta_1$(say) i.e. $\frac{\dot{q}_1}{a + bq_2^2} = \beta_1$. Now the Routhian function is given by

$$
\begin{aligned}
R = L - \beta_1 \dot{q}_1 &= \frac{1}{2} \frac{\dot{q}_1^2}{a + bq_2^2} + \frac{1}{2}\dot{q}_2^2 - c - dq_2^2 - \beta_1 \dot{q}_1 \\
&= \frac{1}{2}\frac{\beta_1^2 (a + bq_2^2)^2}{(a + bq_2^2)} + \frac{1}{2}\dot{q}_2^2 - c - dq_2^2 - \beta_1^2(a + bq_2^2) \\
&= -\frac{1}{2}\beta_1^2(a + bq_2^2) + \frac{1}{2}\dot{q}_2^2 - c - dq_2^2.
\end{aligned}
$$

The equation of motion for the coordinate q_2 is

$$
\frac{\mathrm{d}}{\mathrm{d}t}\left(\frac{\partial R}{\partial \dot{q}_2}\right) - \frac{\partial R}{2} = 0, \text{ or, } \ddot{q}_2 + \beta_1^2 b q_2 + 2dq_2 = 0, \text{ or, } \ddot{q}_2 + A^2 q_2 = 0
$$

where $A^2 = \beta_1^2 b + 2d$.

Thus we have, $q_2 = B \sin(At + \in)$ where B and $\in$ are arbitrary constants.

Again, we have $\frac{\dot{q}_1}{a + bq_2^2} = \beta_1$ or, $\dot{q}_1 = \beta_1(a + bq_2^2) = \beta_1\{a + bB^2 \sin^2(At + \in)\}$.

Upon integration the above relation, we have

$$
\begin{aligned}
q_1 &= \beta_1 at + \frac{1}{2}\beta_1 bB^2 \int \{1 - \cos 2(At + \in)\}dt \\
&= \beta_1 at + \frac{1}{2}\beta_1 bB^2 t - \frac{1}{4A}\beta_1 bB^2 \sin 2(At + \in) + C
\end{aligned}
$$

where C is an arbitrary constant.

7.4 Hamilton Principle

Sir W.R. Hamilton gave his famous principle of least action, also known as Hamilton's principle of least action in 1834 which states that

The variation of the integral $\int_1^2 Ldt$ between the actual path and any neighboring virtual path of a dynamical system moving from one configuration to another, coterminous in both space and time with the actual path, is zero. That is,

$$\delta \int\limits_{1}^{2} L(q, \dot{q}, t) dt = 0,$$

L being the Lagrangian of the dynamical system, $q, \dot{q}$ being the generalized coordinate and generalized velocity, respectively and δ denotes the variation of the integral.

7.5 Noether Theorem

Conservation laws of a dynamical system in classical mechanics where time t is one of the independent variable and other variables are spatial variables is the first integral of motion or constant of motion of the system. Conservation laws reduces the degrees of freedom of a system, thus makes the system simple to integrate. But, deducing conservation laws of a system is not an easy task. One can notice that the idea of conservation laws and symmetry with respect to group of transformations are interrelated for instance, the translational symmetry provides the conservation of linear momentum, rotational symmetry provides the conservation of angular momentum etc.

German mathematician Emalie Emmy Noether in 1918 gave a general theory of conservation laws and symmetry transformations recognizing the importance of the relation between the symmetry and conservation laws. Actually, Noether gave two theorems in this regard. Herein we only give the statement of the first theorem; we shall discuss other in the later chapter.

Theorem 7.5 (Noether's First theorem) *Every conservation law gives rise to a one-parameter symmetry group of transformation and vice versa.*

Let us now explain this theorem

Consider that L be the Lagrangian of a system in a coordinate system $(q, \dot{q}, t)$ and L' be the Lagrangian in the coordinate system $(q', \dot{q}', t)$ obtained under the coordinate transformations $q' = q'(q, \dot{q}, t), \dot{q}' = \dot{q}'(q, \dot{q}, t)$. Then this transformation of coordinates is said to be a symmetry transformation of the Lagrangian if

$$L'(q', \dot{q}', t) = L(q, \dot{q}, t).$$

Noether's theorem states that if the coordinates of a Lagrangian of a system has a set of continuous symmetry transformations $\bar{t} = \mathcal{P}(t, \in)$ where $\mathcal{P}(\in = 0) = t$ and ϵ, $\bar{q}_k = Q_k(q_k, \in)$ where $Q_k(\in = 0) = q_k$, ϵ is a continuous parameter then the functional $I = \int_{x_1}^{x_2} L(q_k, \dot{q}_k, t) dt$ with arbitrary end points x_1 and x_2 is an invariant with a set of quantities that remain conserved along the trajectories of the system, given by:

$$\sum_{k=1}^{n} p_k \frac{dQ_k}{d\epsilon}\Big|_{\{\epsilon=0\}} - \left(\sum p_k \dot{q}_k - L(q_k, \dot{q}_k, t)\right)\frac{d\mathcal{P}}{d\epsilon}\Big|_{\{\epsilon=0}\} = \text{constant}$$

where $p_k = \frac{\partial L}{\partial \dot{q}_k}$ is the momentum conjugate to generalized coordinate q_k.

Example 7.13 Show that if the functional $I = \int L(q, \dot{q}, t)dt$ is invariant with respect to the homogeneity in time t then the total energy of a system is conserved.

Solution Homogeneity of time t means that the physical laws governing a system do not change for any arbitrary shift in the origin of time. For such a system the Lagrangian L is independent of time t. For homogeneity in time t, one have the transformations $\bar{t} = t + \in, \bar{q} = q$. Therefore $\mathcal{P} = t + \in$ and $Q = q$. Hence the conserved quantity is $p\dot{q} - L(q, \dot{q})$, which is the total energy of the system.

Example 7.14 Show that if the functional $I = \int L(x, y, z, \dot{x}, \dot{y}, \dot{z}, t)dt$ is invariant with respect to the homogeneity of space then the total linear momentum of the system is conserved.

Solution Homogeneity of space means that the space possesses the same property everywhere. For homogeneity in x, one has the transformations $\bar{t} = t, \bar{x} = x + \in$, $\bar{y} = y, \bar{z} = z$. Therefore $\mathcal{P} = t$ and $Q = (x + \in, y, z)$. Hence the conserved quantity is $p_x = \text{constant}$. Similarly, for homogeneity in y gives $p_y = \text{constant}$ and for homogeneity in z gives $p_z = \text{constant}$. Accordingly the total linear momentum $p = p_x \hat{i} + p_y \hat{j} + p_z \hat{k}$ is conserved.

Example 7.15 Show that if the functional $I = \int L(x, y, z, \dot{x}, \dot{y}, \dot{z}, t)dt$ is invariant with respect to the isotropy of space then the total angular momentum of the system is conserved.

Solution Isotropy of space means that an arbitrary rotation of a system about an axis does not change the system. For isotropy of space about z-axis, one has the transformations $\bar{t} = t, \bar{x} = x \cos\theta + y \sin\theta$,
$\bar{y} = y \cos\theta - x \sin\theta, \bar{z} = z$. Therefore $\mathcal{P} = t$ and $Q(\theta) = (x \cos\theta + y \sin\theta, y \cos\theta - x \sin\theta, z)$.

Now $\frac{\partial Q}{\partial \theta}\big|_{\theta=0} = (y, -x, 0)$. Hence the conserved quantity is $yp_x - xp_y = \text{constant}$. Similarly for isotropy about x-axis, $yp_z - zp_y = \text{constant}$ and for isotropy about y-axis, $xp_z - zp_x = \text{constant}$. Thus the components of angular momentum are conserved quantity under rotation of space. Accordingly the total Angular momentum $H = q_k \times p_k$ is conserved.

Remarks Conservation theorem of generalized momentum is a particular case of Noether's theorem.

7.6   Legendre Dual Transformations

Legendre dual transformation as the name suggests is a transformation that transforms functions on a vector space to functions on the dual space. Legendre transformation is a standard technique for generating a new pair of independent variables (x, z) from an initial pair (x, y). The transformation is completely invertible i.e. applying the transformation twice one gets the initial pair of variables (x, y).

If $L = L(q_k, \dot{q}_k, t)$ is the Lagrangian of a dynamical system where q_k is the generalized coordinate, $\dot{q}_k$ is the generalized velocity then the generalized momenta are given by $p_k = \frac{\partial L}{\partial \dot{q}_k}, k = 1, 2, \ldots, n$. If one wants to eliminate $\dot{q}_k$ in terms of p_k then this elimination considers L as a function of $q_k, \frac{\partial L}{\partial \dot{q}_k}$ and time. The transformation which does this (mathematically) is known as Legendre transformation.

Theorem 7.6 *Let a function $F(x_1, x_2, \ldots, x_n)$ depending explicitly on the n independent variables $x_1, x_2, \ldots, x_n$ be transformed to another function $G(y_1, y_2, \ldots, y_n)$, which is expressed explicitly in terms of the new set of n independent variables $y_1, y_2, \ldots, y_n$. These new variables are connected by the old variables by a given set of relations*

$$y_i = \frac{\partial F}{\partial x_i}, i = 1, 2, \ldots, n \tag{7.43}$$

and the form of G is given by

$$G(y_1, y_2, \ldots, y_n) = \sum_{i=1}^{n} x_i y_i - F(x_1, x_2, \ldots, x_n) \tag{7.44}$$

Then the variables $x_1, x_2, \ldots, x_n$ satisfy the dual transformations viz. the relations:

$$x_i = \frac{\partial G}{\partial y_i}, i = 1, 2, \ldots, n \tag{7.45}$$

and

$$F(x_1, x_2, \ldots, x_n) = \sum_{i=1}^{n} x_i y_i - G(y_1, y_2, \ldots, y_n) \tag{7.46}$$

The transformations (7.46) between the two set of variables given by the Eqs. (7.43) and (7.45) are known as Legendre dual transformations.

Proof It is given that,

$$G(y_1, y_2, \ldots, y_n) = \sum_{i=1}^{n} x_i y_i - F(x_1, x_2, \ldots, x_n)$$

$$\therefore \delta G = \sum_{i=1}^{n} \frac{\partial G}{\partial y_i} \delta y_i.$$

Again, $\delta G = \sum_{i=1}^{n} x_i \delta y_i + \sum_{i=1}^{n} y_i \delta x_i - \sum_{i=1}^{n} \frac{\partial F}{\partial x_i} \delta x_i$.

Thus we have,

$$\sum_{i=1}^{n} \frac{\partial G}{\partial y_i} \delta y_i = \sum_{i=1}^{n} x_i \delta y_i + \sum_{i=1}^{n} y_i \delta x_i - \sum_{i=1}^{n} \frac{\partial F}{\partial x_i} \delta x_i$$

$$= \sum_{i=1}^{n} x_i \delta y_i + \sum_{i=1}^{n} \left(y_i - \frac{\partial F}{\partial x_i} \right) \delta x_i.$$

$$\therefore x_i = \frac{\partial G}{\partial y_i}, i = 1, 2, \ldots, n$$

since it is given that, $y_i = \frac{\partial F}{\partial x_i}, i = 1, 2, \ldots, n$ and δy_i's are arbitrary (as all y_i's are independent).

This proves the duality of the transformations.

Relation (7.46) can simply be obtained by rearranging terms of the relation (7.44). Moreover, starting from the relation (7.45) one can easily prove the relation (7.43), exactly in the same fashion.

7.7 Hamilton Equations of Motion

In Lagrangian mechanics, $q_k, \dot{q}_k$ are treated as the Lagrangian variables with time t as a parameter where q_k are the n-generalized coordinates and $\dot{q}_k$ are the n-generalized velocities ($k = 1, 2, \ldots, n$). In Hamiltonian mechanics, the basic variables are the generalized coordinates q_k and the generalized momenta $p_k (k = 1, 2, \ldots, n)$ defined by the equations

$$p_k = \frac{\partial L}{\partial \dot{q}_k}, k = 1, 2, \ldots, n \tag{7.47}$$

where L is the Lagrangian of the system. With the aid of the Legendre dual transformation the Lagrangian of a system can be converted into a new function H, known as Hamiltonian function, by the relation

$$H = \sum_{k=1}^{n} \dot{q}_k p_k - L \tag{7.48}$$

where $L = L(q_k, \dot{q}_k, t)$.

With the help of (7.47), $\dot{q}_k$'s can be eliminated from the right hand side of (7.48) which can then be expressed in terms of (q_k, p_k, t) so that $H = H(q_k, p_k, t)$.

Theorem 7.7 *The system of n second order differential equation known as Lagranges's equation of motion are equivalent to the 2n first order differential equation known as Hamilton's equation of motion given by*

$$\dot{q}_k = \frac{\partial H}{\partial p_k}, \ \dot{p}_k = -\frac{\partial H}{\partial q_k}$$

where H is the Legender dual transformation of the Lagrangian function L given by

$$H = \sum p_i \dot{q}_i - L(q_i, \dot{q}_i, t)$$

Proof The Legender dual transformation of the Lagrangian $L(q_k, \dot{q}_k, t)$ is given by

$$H = \sum p_i \dot{q}_i - L(q_i, \dot{q}_i, t)$$

Let us consider a virtual change of H at time t. Then,

$$\delta H = \sum_{k=1}^{n} \delta \dot{q}_k p_k + \sum_{k=1}^{n} \dot{q}_k \delta p_k - \delta L(q_k, \dot{q}_k, t) \tag{7.49}$$

Now, $H = H(q_k, p_k, t)$.

$$\therefore \delta H = \sum_{k=1}^{n} \frac{\partial H}{\partial q_k} \delta q_k + \sum_{k=1}^{n} \frac{\partial H}{\partial p_k} \delta p_k + \frac{\partial H}{\partial t} \delta t \tag{7.50}$$

Again,

$$\delta L = \sum_{k=1}^{n} \frac{\partial L}{\partial q_k} \delta q_k + \sum_{k=1}^{n} \frac{\partial L}{\partial \dot{q}_k} \delta \dot{q}_k + \frac{\partial L}{\partial t} \delta t \tag{7.51}$$

Substituting (7.50) and (7.51) in (7.49) we have,

$$\sum_{k=1}^{n} \frac{\partial H}{\partial q_k} \delta q_k + \sum_{k=1}^{n} \frac{\partial H}{\partial p_k} \delta p_k + \frac{\partial H}{\partial t} \delta t$$

$$= \sum_{k=1}^{n} \delta \dot{q}_k p_k + \sum_{k=1}^{n} \dot{q}_k \delta p_k - \sum_{k=1}^{n} \frac{\partial L}{\partial q_k} \delta q_k - \sum_{k=1}^{n} \frac{\partial L}{\partial \dot{q}_k} \delta \dot{q}_k - \frac{\partial L}{\partial t} \delta t \qquad (7.52)$$

$$= \sum_{k=1}^{n} \dot{q}_k \delta p_k - \sum_{k=1}^{n} \dot{p}_k \delta q_k - \frac{\partial L}{\partial t} \delta t,$$

as $p_k = \frac{\partial L}{\partial \dot{q}_k}, k = 1, 2, \ldots, n$ so Lagrange's equations of motion give

$$\frac{d}{dt}\left(\frac{\partial L}{\partial \dot{q}_k} \right) - \frac{\partial L}{\partial q_k} = 0 \text{ i.e. } \dot{p}_k = \frac{\partial L}{\partial q_k} \text{ for } k = 1, 2, \ldots, n.$$

Equation (7.52) is true for all arbitrary variations. Hence we get,

$$\frac{\partial H}{\partial q_k} = -\dot{p}_k, \frac{\partial H}{\partial p_k} = \dot{q}_k \quad \text{for} \quad k = 1, 2, \ldots, n$$

and

$$\frac{\partial H}{\partial t} = -\frac{\partial L}{\partial t}.$$

Thus we have

$$\dot{q}_k = \frac{\partial H}{\partial p_k}, \; \dot{p}_k = -\frac{\partial H}{\partial q_k} \quad \text{for } k = 1, 2, \ldots, n \qquad (7.53)$$

These equations are called Hamilton's equations of motion of a dynamical system.

Theorem 7.8 *If the Lagrangian does not explicitly depend on time, Hamiltonian is also independent of time and Hamiltonian is a constant of motion or first integral of the system.*

Proof We have Hamiltonian H as defined by $H = \sum_{k=1}^{n} \dot{q}_k p_k - L$, where L is the Lagrangian of the system. Now, if $H = H(q_k, p_k, t)$ then using Hamilton's equations of motion and $\frac{\partial H}{\partial t} = -\frac{\partial L}{\partial t}$, we have

$$\frac{dH}{dt} = \sum_{i=1}^{n} \frac{\partial H}{\partial q_i} \dot{q}_i + \sum_{i=1}^{n} \frac{\partial H}{\partial p_i} \dot{p}_i + \frac{\partial H}{\partial t} = -\sum_{i=1}^{n} \dot{p}_i \dot{q}_i + \sum_{i=1}^{n} \dot{q}_i \dot{p}_i - \frac{\partial L}{\partial t},$$

If in addition the Lagrangian does not explicitly depend on time that is, $\frac{\partial L}{\partial t} = 0$ then $\frac{\partial H}{\partial t} = 0$.

Hence the Hamiltonian is also independent of time t.

Moreover, using the above result we have $\frac{dH}{dt} = 0$, that is $H = $ constant. Thus, Hamiltonian is a constant of motion.

Theorem 7.9 *For a scleronomous dynamical system in a velocity independent potential field the total energy (sum of the Kinetic Energy and Potential Energy) remains conserved.*

Proof If the Hamiltonian H is explicitly independent of time t then $H = H(q_k, p_k)$, $k = 1, 2, \ldots, n$ where p_k's are the generalized momenta corresponding to the generalized coordinates q_k of the system.

Now,

$$\frac{dH}{dt} = \sum_{k=1}^{n} \frac{\partial H}{\partial q_k} \delta q_k + \sum_{k=1}^{n} \frac{\partial H}{\partial p_k} \delta p_k \tag{7.54}$$

By Hamilton's equations of motion we have,

$$\dot{q}_k = \frac{\partial H}{\partial p_k}, \ \dot{p}_k = -\frac{\partial H}{\partial q_k} \quad \text{for } k = 1, 2, \ldots, n.$$

Substituting these in (7.54) we get,

$$\frac{dH}{dt} = 0 \text{ which gives } H = \text{constant}. \tag{7.55}$$

Moreover, in a scleronomous system, the kinetic Energy T is a homogeneous quadratic function of generalized velocities. Therefore, by Euler's theorem on homogeneous function we have,

$$2T = \sum_{k=1}^{n} \dot{q}_k \frac{\partial T}{\partial \dot{q}_k} \tag{7.56}$$

We have,

$$H = \sum_{k=1}^{n} \dot{q}_k p_k - L = \sum_{k=1}^{n} \dot{q}_k \frac{\partial L}{\partial \dot{q}_k} - (T - V) = \sum_{k=1}^{n} \dot{q}_k \frac{\partial (T - V)}{\partial \dot{q}_k} - T + V$$

Since in a velocity independent potential field, $\frac{\partial V}{\partial \dot{q}_k} = 0$,

$\therefore H = \sum_{k=1}^{n} \dot{q}_k \frac{\partial T}{\partial \dot{q}_k} - T + V = 2T - T + V = T + V$. [using (7.56)]

Therefore, in a scleronomous dynamical system in a velocity independent potential field the total energy (sum of the Kinetic Energy and Potential Energy) remains conserved.

Theorem 7.10 *If all the coordinates of a dynamical system are cyclic they can be determined by integration. In particular, if the system is scleronomous, the coordinates are all linear functions of time t.*

Proof It is known that a coordinate q_k is said to be cyclic if it is explicitly absent from the Hamiltonian.

Since $H = \sum_{k=1}^{n} \dot{q}_k p_k - L$, so q_k's are cyclic coordinates.

As all the coordinates are cyclic, so $H = H(p_k, t)$ where p_k's are the n-generalized momenta of the system.

Hamilton's equations of motion give $\dot{q}_k = \frac{\partial H}{\partial p_k} = f_k(p_k, t)$, $k = 1, 2, \ldots, n$.

Since q_k's are the cyclic coordinates, $p_k = \text{constant} = \beta_k(\text{say})$, $k = 1, 2, \ldots, n$.

Therefore, $\dot{q}_k = f_k(\beta_k, t)$, this on integration gives

$$q_k = \int f_k(\beta_k, t) dt + \text{constant}, k = 1, 2, \ldots, n.$$

So, the coordinates can be determined by integration.

For a scleronomous system, $\dot{q}_k = f_k(p_k)$, which on integration gives

$$q_k = \int f_k(\beta_k) dt + \text{constant} = f_k(\beta_k) \int dt + \text{constant} = t f_k(\beta_k) + \text{constant}, \quad \text{for } k$$
$$= 1, 2, \ldots, n.$$

So, the coordinates are linear functions of time t.

Example 7.16 Obtain the equation of motion of a simple pendulum using Hamiltonian function.

Solution Here, θ is the generalized coordinate. The velocity of the ball is $l\dot{\theta}$ where l is the length of the string of the pendulum. A simple pendulum oscillating in a vertical plane constitutes a conservative holonomic dynamical system.

Here, Kinetic Energy $= T = \frac{1}{2} m l^2 \dot{\theta}^2$, m is the mass of the ball.

Potential Energy $= V = mgh = mgl(1 - \cos \theta)$.

Therefore, Lagrangian of the system $= L = T - V = \frac{1}{2} m l^2 \dot{\theta}^2 - mgl(1 - \cos \theta)$.

Since the system is scleronomous and conservative,

$$H = T + V = \frac{1}{2} m l^2 \dot{\theta}^2 + mgl(1 - \cos \theta)$$

Generalized momentum $p_\theta = \frac{\partial L}{\partial \dot{\theta}} = m l^2 \dot{\theta}$ or, $\dot{\theta} = \frac{p_\theta}{m l^2}$.

$$\therefore H = \frac{1}{2} m l^2 \left(\frac{p_\theta}{m l^2} \right)^2 + mgl(1 - \cos \theta) = \frac{1}{2} \frac{p_\theta^2}{m l^2} + mgl(1 - \cos \theta) = H(\theta, p_\theta).$$

Hamilton's equations of motion are given by

$$\dot{\theta} = \frac{\partial H}{\partial p_\theta} = \frac{p_\theta}{m l^2} \text{ i.e. } p_\theta = m l^2 \dot{\theta} \quad \text{and} \quad \dot{p}_\theta = -\frac{\partial H}{\partial \theta} = -mgl \sin \theta$$

$$\therefore ml^2\ddot{\theta} = -mgl \sin\theta \text{ or, } \ddot{\theta} = -\frac{g}{l}\sin\theta \simeq -\frac{g}{l}\theta \text{ (when } \theta \text{ is very small)}.$$

Time period is given by $2\pi\sqrt{\frac{l}{g}}$, g is the acceleration due to gravity.

Example 7.17 For a dynamical system Hamiltonian H is given by

$$H = q_1p_1 - q_2p_2 - aq_1^2 + bq_2^2$$

Find q_1, q_2, p_1, p_2 in terms of time t. Hence find the corresponding equation of motion of the dynamical system.

Solution Hamilton's equations of motion give

$$\dot{q}_1 = \frac{\partial H}{\partial p_1} = q_1 \text{ which on integration gives } q_1 = Ce^t, \ C = \text{constant}$$

$$\dot{p}_1 = -\frac{\partial H}{\partial q_1} = -p_1 + 2aq_1 \text{ or, } \dot{p}_1 + p_1 = 2aCe^t.$$

Multiplying both sides by the integrating factor e^t we get,

$$\frac{\mathrm{d}}{\mathrm{d}t}(p_1e^t) = 2aCe^{2t}$$

Integrating this we get, $p_1 = aCe^t + De^{-t}, D = \text{constant}$

$$\dot{q}_2 = \frac{\partial H}{\partial p_2} = -q_2 \text{ or, } q_2 = Ae^{-t}, A = \text{constant}$$

$$\dot{p}_2 = -\frac{\partial H}{\partial q_2} = p_2 - 2bq_2 = p_2 - 2Abe^{-t} \text{ or, } \dot{p}_2 - p_2 - 2Abe^{-t} = 0$$

Multiplying both sides by the integrating factor e^{-t} we get,
$\frac{\mathrm{d}}{\mathrm{d}t}(p_2e^{-t}) = -2Abe^{-2t}$ which on integration gives $p_2 = Abe^{-t} + Be^t, B = \text{constant}$.
Therefore the required equation of motion is

$$\dot{p}_1 + p_1 - 2aCe^t = 0$$
$$\dot{p}_2 - p_2 - 2Abe^{-t} = 0$$

Example 7.18 Hamiltonian of a dynamical system is given by $H = \frac{1}{2}\sum_{i=1}^{3}\left(p_i^2 + \mu^2 q_i^2\right)$ where p_i, q_i are the n generalized momenta and generalized coordinates, μ is a constant. Show that $F_1 = q_2p_3 - q_3p_2$ and $F_2 = \mu q_1 \cos\mu t - p_1 \sin\mu t$ are constants of motion.

Solution Now, $H = \frac{1}{2}\left(p_1^2 + p_2^2 + p_3^2 + \mu^2 q_1^2 + \mu^2 q_2^2 + \mu^2 q_3^2\right).$

$$F_1 = q_2 p_3 - q_3 p_2 = F_1(q_2, q_3, p_2, p_3)$$

$$\therefore \frac{dF_1}{dt} = \frac{\partial F_1}{\partial q_2}\dot{q}_2 + \frac{\partial F_1}{\partial q_3}\dot{q}_3 + \frac{\partial F_1}{\partial p_2}\dot{p}_2 + \frac{\partial F_1}{\partial p_3}\dot{p}_3$$

$$= \frac{\partial F_1}{\partial q_2}\frac{\partial H}{\partial p_2} + \frac{\partial F_1}{\partial q_3}\frac{\partial H}{\partial p_3} - \frac{\partial F_1}{\partial p_2}\frac{\partial H}{\partial q_2} - \frac{\partial F_1}{\partial p_3}\frac{\partial H}{\partial q_3}$$

(using Hamilton's equations of motion)

$$= p_3 p_2 - p_2 p_3 + q_3 \mu^2 q_2 - q_2 \mu^2 q_3 = 0$$

So, $F_1 =$ constant.

Again, we have

$$\frac{dF_2}{dt} = \frac{\partial F_2}{\partial q_1}\dot{q}_1 + \frac{\partial F_2}{\partial p_1}\dot{p}_1 + \frac{\partial F_2}{\partial t} = \frac{\partial F_2}{\partial q_1}\frac{\partial H}{\partial p_1} - \frac{\partial F_2}{\partial p_1}\frac{\partial H}{\partial q_1} + \frac{\partial F_2}{\partial t}$$

(using Hamilton's equations of motion)

$$= p_1 \mu \cos \mu t + q_1 \mu^2 \sin \mu t - q_1 \mu^2 \sin \mu t - p_1 \mu \cos \mu t = 0$$

So, $F_2 =$ constant.

7.7.1 Differences Between Lagrangian and Hamiltonian of a Dynamical System

1. Lagrangian $L(q, \dot{q}, t)$ is described always in the configuration space of appropriate dimension and the configuration space is described only by the total number of generalized coordinates of the system whereas Hamiltonian $H(q, p, t)$ is described in the phase space of requisite dimension set by the equal number of generalized coordinates and generalized momenta.
2. Lagrangian mechanics provides a second order differential equation of motion corresponding to each degree of freedom of the system whereas Hamiltonian mechanics provides us two first order differential equations of motion corresponding to each degree of freedom.
3. Hamilton's equations of motion can only be derived for holonomic dynamical systems whereas Lagrange's equations of motion can be derived for holonomic as well as nonholonomic dynamical systems.

So far an introduction of basic concepts of Hamiltonian function has been given in the context of classical mechanics. We shall now give the concept with respect to the flow of a dynamical system.

7.8 Hamiltonian Flows

We have knowledge that solutions of a system of differential equations are geometrically curves which depict a flow in the space $\mathbb{R}^n$. The system of differential equation represents a vector field where the direction of the tangent at a given point of the solution curve is given by the first order differential equation written in a solved form. Herein we shall discuss the dynamical properties of the flow generated by Hamiltonian vector fields in the phase space.

A system of differential equation on $\mathbb{R}^{2n}$ corresponding to n degrees of freedom of a system is said to be Hamiltonian flows or systems if it can be expressed as

$$\dot{q}_i = \frac{\partial H}{\partial p_i}, \ \dot{p}_i = -\frac{\partial H}{\partial q_i}, \ i = 1, 2, \ldots, n \qquad (7.57)$$

where $H(\underset{\sim}{x}) = H(q_1, q_2, \ldots, q_n, p_1, p_2, \ldots, p_n)$ is a twice continuously differentiable function. The dimension of the phase space is $2n$ for the n generalized coordinates q_i and n generalized momenta p_i, $i = 1, 2, \ldots, n$. The function H is called Hamiltonian of the system as defined earlier. Furthermore, Hamiltonian functions are commutative. The Hamiltonian vector field is derived from the Hamiltonian function H denoted by $X_H(x)$ and is given by $X_H(x) \equiv \left(\frac{\partial H}{\partial p}, -\frac{\partial H}{\partial q} \right)$. From Liouville theorem (see Chap. 1) it has been established that the flow induced by a time independent Hamiltonian is volume preserving. The flow takes place on bounded energy manifolds and all orbits will return after some time to a neighbourhood of the starting point. This is a consequence of the recurrence theorem by Poincarè stated below.

Theorem 7.11 (Poincarè Recurrence Theorem) *Let f be a bijective, continuous, volume preserving mapping of a bounded domain $D \subset \mathbb{R}^n$ into itself. Then each neighbourhood U of each point in D contains a point x which returns to U after repeated applications of the mapping $f^{(n)}x \in U$ for some $n \in \mathbb{N}$.*

It follows that $H(x)$ is a first integral of $\dot{x} = f(x)$ if and only if $\frac{dH}{dt} = \nabla H \cdot f(x) = 0$ $\forall x \in \mathbb{R}^n$, $H(x)$ is identically constant on any open subset of $\mathbb{R}^n$. Furthermore, if there is a first integral, that is, a Hamiltonian function then the orbits of the system are contained in the one-parameter family of level curves $H(x) = k$.

Given a system $\dot{x} = f(x), x \in \mathbb{R}^n$, a set $S \subseteq \mathbb{R}^n$ which is the union of whole orbits of the system, is called an invariant set for the system. For example, for a Hamiltonian system in $\mathbb{R}^2$ the level sets $H(x_1, x_2) = k$ are invariant sets, since H is constant along any orbit. More generally, a function $H : \mathbb{R}^n \to \mathbb{R}$ of class C^1, is called a first integral of the system $\dot{x} = f(x)$ if H is constant on every orbit, $\frac{d}{dt} H(x, t) = \nabla H(x(t)) \cdot f(x(t)) = 0$ $\forall t$.

In general, the first integral of a dynamical system is defined as follows.

First integral of a system A continuously differentiable function $f : D \to \mathbb{R}$, $D \subseteq \mathbb{R}^n$ is said to be a first integral of the system $\dot{x} = F(x), x = (x_1, x_2, \ldots, x_n) \in X \subseteq \mathbb{R}^n$ on the region $D \subseteq X$ if

$$\mathcal{D}_t f(x(t)) = 0, \ x \in X \subseteq \mathbb{R}^2$$

where $\mathcal{D}_t f = \frac{\partial f}{\partial x} \dot{x} = \frac{\partial f}{\partial x_1} \dot{x}_1 + \frac{\partial f}{\partial x_2} \dot{x}_2 + \cdots + \frac{\partial f}{\partial x_n} \dot{x}_n$, $x = (x_1, x_2, \ldots, x_n)$ and $\mathcal{D}_t$ is called the orbital derivative.

The first integral $f(x(t))$ is constant for any solution $x(t)$ of the system i.e. $f(x(t)) = C$(constant) and is therefore also called as constant of motion. The first integral of a system when exists is not unique for if $f(x)$ is a first integral of a system then $f(x) + C$ and $Cf(x)$, where $C \in \mathbb{R}$ are also the first integral of the system. The first integral of a system, as the name suggests is obtained by integrating just once the system $\dot{x} = F(x)$.

The level curve of the first integral of a system is denoted by L_c and is defined as $L_c = \{x | f(x) = C\}$. The first integral of motion f is constant on every trajectory of the system. Hence every trajectory of a system is a member of some level curve of f. Each level curve is therefore a union of trajectories of the system. The level set which contains family of trajectories of the system is called an integral manifold.

A system is said to be conservative if it has a first integral of motion on the whole plane i.e. $D = \mathbb{R}^n$.

Theorem 7.12 *If $f \in C^1(E)$, where E is an open, simply connected subset of $\mathbb{R}^2$, then the system $\dot{x} = f(x)$ is a Hamiltonian system on E if and only if $\nabla \cdot f(x) = 0 \quad \forall x \in E$.*

Proof Hamiltonian system is given by $\dot{q} = \frac{\partial H}{\partial p}, \dot{p} = -\frac{\partial H}{\partial q}$, therefore

$$\nabla \cdot \left(\frac{\partial H}{\partial p}, -\frac{\partial H}{\partial q} \right) = \frac{\partial^2 H}{\partial q \partial p} - \frac{\partial^2 H}{\partial p \partial q} = 0 \quad \text{for} (q, p) \in E$$

since $f(x) \equiv (f_1(q,p), f_2(q,p)) = \left(\frac{\partial H}{\partial p}, -\frac{\partial H}{\partial q} \right)$ is a continuously differentiable function.

Hence, $\nabla \cdot f(x) = 0 \quad \forall x \in E$ for a Hamiltonian system, where $f(x) \equiv (f_1(q,p), f_2(q,p)) = \left(\frac{\partial H}{\partial p}, -\frac{\partial H}{\partial q} \right)$.

For the converse part suppose that $\dot{x} = f_1(x,y), \dot{y} = f_2(x,y)$ where $f_1, f_2 \in C^1(E)$ and $\nabla \cdot f(x) = 0 \quad \forall x \in E$ holds. Therefore

$$\nabla \cdot (f_1(x,y), f_2(x,y)) = 0$$
$$\frac{\partial f_1}{\partial x} + \frac{\partial f_2}{\partial y} = 0$$
$$\frac{\partial f_1}{\partial x} = -\frac{\partial f_2}{\partial y}$$

Clearly, the system is Hamiltonian.

Theorem 7.13 *The flow defined by a Hamiltonian system with one degree of freedom is area preserving.*

Proof The rate of change of area of a system $\dot{x} = f(x), x = (x, y), f = (f_1, f_2)$ is given by

$$\frac{1}{A}\frac{dA}{dt} = \frac{\partial f_1}{\partial x} + \frac{\partial f_2}{\partial y}$$

Now for a Hamiltonian system $\frac{\partial f_1}{\partial x} = -\frac{\partial f_2}{\partial y}$ (since $\nabla \cdot f(x) = 0$) hence

$$\frac{1}{A}\frac{dA}{dt} = 0 \text{ or, } A = \text{constant}$$

Thus the area of the flow generated by the Hamiltonian system is preserved.

7.8.1 Integrable and Non-Integrable Systems

The most crucial aspect of any system is its integrability. An integrable system is that system whose solution curves and therefore the geometry of the flow in its embedding space can be determined with certainty. At this juncture, question may arise that when the Hamiltonian system will be integrable. The answer to this question is given as follows:

When the degrees of freedom of a Hamiltonian system are equal to the constants of motion of the system then the Hamiltonian system is said to be integrable. For instance, all Hamiltonian system with only one degree of freedom for which the Hamiltonian function H is analytic are integrable. Moreover, all Hamiltonian systems for which Hamilton's equations of motion are linear in generalized coordinates and momenta are integrable. Furthermore, if it is possible to separate the nonlinear Hamilton's equations of motion into decoupled one degree freedom systems, then the system is integrable.

On the other hand, if the degrees of freedom are more than the constants of motion of the system then the system is called non-integrable.

We have seen in our earlier discussions that if a coordinate is explicitly absent in a system then the corresponding momentum is a constant of motion. But for a Hamiltonian system it is not always the momenta which are constants of motion but there are constants of motion which are expressed in terms of the generalized coordinates and generalized momenta.

Generally the trajectory of an integrable system with n-degrees of freedom moves on an n-dimensional surface of a torus which lies in the $2n$-dimensional phase space. Now if the motion of the trajectory is constrained by some constants of motion then the dimension of the surface in which it stays in the phase space would get reduced. For instance, the trajectories of a Hamiltonian system which has k numbers of constants of motion, lie on a $(2n - k)$-dimensional surface in the phase space. Since the motion of the trajectories is always restricted to these surfaces so these surfaces are called as invariant tori. Nevertheless, systems with more than one degree of freedom may not be integrable. If a Hamiltonian system with two degrees of freedom is integrable then there exist exactly two constants of motion. The trajectories of the system therefore would move on the two dimensional surface $(2 \times 2 - 2 = 2)$ of a torus lying in a four dimensional phase space. The trajectories of the system are periodic or quasi-periodic, and do not show any chaotic behavior. Now if the system is slightly nonintegrable due to some perturbation then the constants of motion of the system is no longer constant except the energy of the system. Since the energy of the Hamiltonian system is conserved even now so, the trajectories of the system are constrained to move on a three dimensional surface $(2 \times 2 - 1 = 3)$ in the four dimensional phase space. The trajectories of this three dimensional motion of the system are no longer periodic and displays chaotic motion. When the amount of non-integrability get increased then the trajectories of the system move off the tori and the tori are destroyed. The trajectories of the system therefore can move throughout the phase space without any restriction. Note that for non-integrability, a system must have at least two degrees of freedom which implies a phase space of at least four dimensions. The phase portraits of such system are very difficult to obtain, (see Hilborn [13]).

Theorem 7.14 *For any Hamiltonian system, the Hamiltonian $\left(x, p\right)$ is a conserved quantity or first integral of the system.*

Solution Let us consider a Hamiltonian system with x as generalized coordinate and p as generalized momentum. Now, Hamilton's equation of motion is given by

$$
\left.
\begin{aligned}
\dot{x} &= \frac{\partial H}{\partial p} \\[2mm]
\dot{p} &= -\frac{\partial H}{\partial x}
\end{aligned}
\right\}
\tag{7.58}
$$

We can write $\dfrac{dH}{dt} = \dfrac{\partial H}{\partial x}\dfrac{dx}{dt} + \dfrac{\partial H}{\partial p}\dfrac{dp}{dt} = \dot{x}\dfrac{\partial H}{\partial x} + \dot{p}\dfrac{\partial H}{\partial p} = \dfrac{\partial H}{\partial p}\dfrac{\partial H}{\partial x} - \dfrac{\partial H}{\partial x}\dfrac{\partial H}{\partial p} = 0$

Hence, the Hamiltonian $H\left(x, p\right)$ = constant of motion or integral of motion along the trajectories of the system.

Example 7.19 Show that the system $\dot{x} = -y, \dot{y} = x$ is conservative.

Solution The given system can be written as

$$\frac{dy}{dx} = -\frac{x}{y}, y \neq 0$$

The solution of this equation is $x^2 + y^2 = c$, $y \neq 0$, where c is a positive constant. However

$$\frac{d}{dt}f(\mathbf{x}) = \frac{\partial f}{\partial x}\dot{x} + \frac{\partial f}{\partial y}\dot{y} = -2xy + 2xy = 0 \quad \forall (x, y) \in \mathbb{R}^2$$

Hence the given system is a conservative system.

Example 7.20 Find out whether the system $\dot{x} = x$, $\dot{y} = y$ is a conservative system or not.

Solution The given system can be written as

$$\frac{dy}{dx} = \frac{y}{x}, \ x \neq 0$$

The solution of this equation is $y = cx$, $x \neq 0$, where c is a positive constant. However the first integral of the system $f(x, y) = \frac{y}{x}$, $x \neq 0$ satisfies

$$\frac{d}{dt}f(\mathbf{x}) = \frac{\partial f}{\partial x}\dot{x} + \frac{\partial f}{\partial y}\dot{y} = -y/x + y/x = 0 \quad \forall\, (x, y)\backslash(0, y) \in \mathbb{R}^2$$

Hence the given system is not a conservative system.

Example 7.21 Find a first integral of the system

$$\dot{x} = xy, \ \dot{y} = \ln x, x > 1$$

in the region indicated. Hence, sketch the phase portrait.

Solution The given system of equation can be written as a single system

$$\frac{dy}{dx} = \frac{\ln x}{xy}$$

Separation of variables yields

$$ydy = \frac{\ln x}{x}dx$$

$$\therefore y^2 = (\ln x)^2 + \text{constant}$$

This is the first integral of the system. Since the first integral $f(x, y) = y^2 - (\ln x)^2$ of the given system satisfies

$$\frac{d}{dt}f(\mathbf{x}) = \frac{\partial f}{\partial x}\dot{x} + \frac{\partial f}{\partial y}\dot{y} = -2y \ln x + 2y \ln x = 0 \quad \forall (x, y)\backslash(0, y) \in \mathbb{R}^2$$

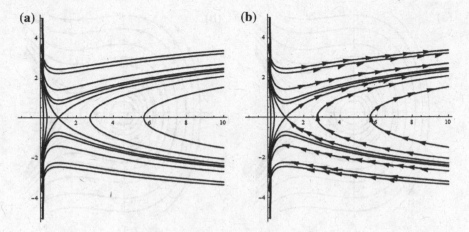

Fig. 7.4 a Level curves and **b** Phase portrait of the first integral

hence the system is not conservative. The sketch the phase portrait of the given system in the $x - \dot{x}$ plane is shown in Fig. 7.4. The direction of the trajectories is given by the direction of $\dot{x}$. Now $\dot{x} > 0$ when $y > 0$ and $\dot{x} < 0$ for $y < 0$. Hence the direction of the trajectories are from left to right in the upper half-plane and from right to left in the lower half plane. The level curves give the shape of the trajectories without the directions.

Example 7.22 Find a Hamiltonian H for a moving particle along a straight line, given by the equation of motion

$$\ddot{x} = -x + \beta x^2, \beta > 0, x \text{ is the displacement}$$

Sketch the Level curve of the Hamiltonian H in the phase plane.

Solution The given nonlinear equation is converted into the following system of equations

$$\dot{x} = y$$
$$\dot{y} = -x + \beta x^2$$

If we consider the variable x as generalized coordinate and y as genaralised momentum then we have

$$\frac{\partial H}{\partial y} = y$$

$$\frac{\partial H}{\partial x} = x - \beta x^2$$

$$\therefore \frac{dH}{dt} = \dot{x}\frac{\partial H}{\partial x} + \dot{y}\frac{\partial H}{\partial y} = \dot{x}(x - \beta x^2) + y\dot{y} = 0$$

and the Hamiltonian of the system is given by $H(x,y) = \frac{x^2}{2} - \frac{\beta}{3}x^3 + \frac{y^2}{2}$.

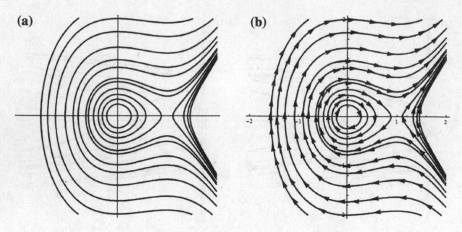

(a)   **(b)**

Fig. 7.5 Level curve and phase portrait of $H(x,y) = \frac{x^2}{2} - \frac{\beta}{3}x^3 + \frac{y^2}{2}$ for $\beta = 1$

Now we sketch the phase portrait of the given system in the $x - \dot{x}$ plane shown in Fig. 7.5. The direction of the trajectories is given by the direction of $\dot{x}$. Now $\dot{x} > 0$ when $y > 0$ and $\dot{x} < 0$ for $y < 0$. Hence the direction of the trajectories are from left to right in the upper half-plane and from right to left in the lower half plane.

Example 7.23 Find a Hamiltonian H for the undamped pendulum, given by the equation of motion

$$\ddot{x} + \sin x = 0$$

Sketch the level curves of the Hamiltonian H in the phase plane.

Solution The given equation of pendulum can be written as the following system of equation

$$\dot{x} = y$$
$$\dot{y} = -\sin x$$

If we consider the variable x as the generalized coordinate and y as the generalized velocity then we will have

$$\frac{\partial H}{\partial y} = y$$
$$\frac{\partial H}{\partial x} = \sin x$$

Fig. 7.6 Level curve of undamped pendulum

Hence $\frac{dH}{dt} = \dot{x}\frac{\partial H}{\partial x} + \dot{y}\frac{\partial H}{\partial y} = \dot{x}\sin x + \dot{y}y = 0$

Whence Hamiltonian $H = -\cos x + \frac{y^2}{2}$. The level curves are shown in Fig. 7.6.

Example 7.24 Find the Hamiltonian for the system

$$\dot{x} = y$$
$$\dot{y} = -x + x^3$$

Also, sketch the Hamiltonian of the system in the phase plane.

Solution If x is the generalized coordinate of the system and y the generalized velocity, then clearly,

$$\frac{\partial H}{\partial x} = x - x^3 \quad \text{and} \quad \frac{\partial H}{\partial y} = y$$

Hence $\frac{dH}{dt} = \dot{x}\frac{\partial H}{\partial x} + \dot{y}\frac{\partial H}{\partial y} = \dot{x}(x - x^3) + \dot{y}y = 0$. Therefore the Hamiltonian of the system is

$$H = \frac{x^2}{2} - \frac{x^4}{4} + \frac{y^2}{2}$$

Now we sketch the phase portrait of the given system in the $x - \dot{x}$ plane shown in Fig. 7.7. The direction of the trajectories is given by the direction of $\dot{x}$. Now $\dot{x} > 0$ when $y > 0$ and $\dot{x} < 0$ for $y < 0$. Hence the direction of the trajectories are from left to right in the upper half-plane and from right to left in the lower half plane.

Example 7.25 Find a conserved quantity for the system $\ddot{x} = a - e^x$ and also sketch the phase portrait for $a < 0, = 0, \quad$ and $\quad > 0$.

Solution The given system can be written as the following system of equation

$$\dot{x} = p$$
$$\dot{p} = a - e^x$$

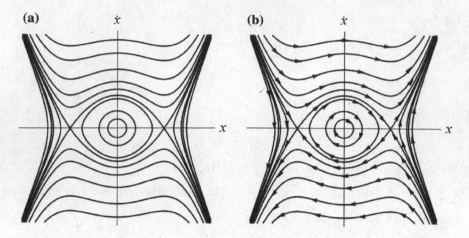

(a) $\dot{x}$ **(b)** $\dot{x}$

Fig. 7.7 a Level curves of $H(x,y) = \frac{x^2}{2} - \frac{x^4}{4} + \frac{y^2}{2}$. **b** Phase portrait of the system $\dot{x} = y,\ \dot{y} = -x + x^3$

For this system $\frac{\partial H}{\partial p} = p$ and $\frac{\partial H}{\partial x} = e^x - a$. Therefore, the Hamiltonian H is given by

$$dH = (e^x - a)dx + pdp$$

$$H = e^x - ax + \frac{p^2}{2}$$

The given system of equation is a Hamiltonian system and the Hamiltonian H is the conserved quantity of the system. When $a = 0$ then the given system has no fixed points and for $a = -1$ the system has complex fixed points $((2n+1)i\pi, 0)$ while for $a = 1$ the only real fixed point is $(0,0)$ which is a center and the trajectories in the phase plane are closed curves about the fixed point origin. The sketch the phase portrait of the given system in the $x - \dot{x}$ plane shown in Fig. 7.8.

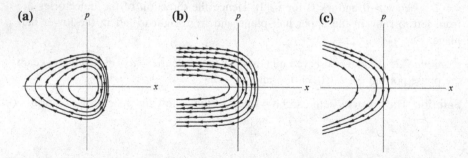

(a) p **(b)** p **(c)** p

Fig. 7.8 Phase portrait of the Hamiltonian $H = e^x - ax + \frac{p^2}{2}$ for $a > 0, a = 0$ and $a < 0$, **a** $a = 1$, **b** $a = 0$, **c** $a = -1$

The direction of the trajectories is given by the direction of $\dot{x}$. Now $\dot{x} > 0$ when $p > 0$ and $\dot{x} < 0$ for $p < 0$. Hence the direction of the trajectories are from left to right in the upper half-plane and from right to left in the lower half plane.

Example 7.26 Show that the Hamiltonian H is $\frac{p^2}{2m} + \frac{\mu x^2}{2}$ for a simple harmonic oscillator of mass m and spring constant μ. Also, show that H is the total energy.

Solution The equation of motion of a simple linear harmonic oscillator with spring constant μ is given by

$$m\ddot{x} + \mu x = 0$$

This equation can be written as a system of equation given below

$$\dot{x} = \frac{p}{m}$$

$$\dot{p} = -\mu x$$

Now, $\frac{\partial H}{\partial p} = \frac{p}{m}$ and $\frac{\partial H}{\partial x} = \mu x$. Hence the Hamiltonian $H(x,p)$ is given by

$$dH = \frac{\partial H}{\partial x} dx + \frac{\partial H}{\partial p} dp$$

$$H = \mu x dx + \frac{p}{m} dp = \frac{\mu x^2}{2} + \frac{p^2}{2m}$$

Since the given system is a Hamiltonian system, the conserved quantity is the Hamiltonian function H. The only fixed point of the system is the origin $(0,0)$ obtained by solving $\frac{\partial H}{\partial p} = \frac{p}{m} = 0$ and $\frac{\partial H}{\partial x} = \mu x = 0$. The trajectories of the system are ellipse. The sketch of the phase portrait is given in the Fig. 7.9. In the upper half plane $x > 0$, $\dot{x} > 0$ hence the direction of the trajectories is from left to right and in

Fig. 7.9 Phase portrait of simple harmonic oscillator

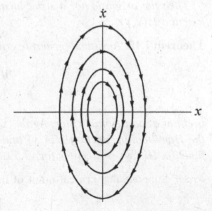

the lower half plane $(x < 0)$, $\dot{x} < 0$ hence the direction of the trajectories is from right to left.

Now the total energy of the system = Kinetic energy of the system + potential energy of the system.

Kinetic energy $= \frac{1}{2}m\dot{x}^2$ and potential energy $= -\int -\mu x \, dx = \frac{\mu x^2}{2}$ (since the system is conservative and the force acting on the system is $-\mu x$). Therefore the total energy of the system is $\frac{m\dot{x}^2}{2} + \frac{\mu x^2}{2} = \frac{p^2}{2m} + \frac{\mu x^2}{2}$ (Since $\dot{x} = \frac{p}{m}$). Hence, the Hamiltonian H is equal to the total energy of the system.

7.8.2 Critical Points of Hamiltonian Systems

We have learnt the qualitative analysis of a nonlinear dynamical system in Chap. 3 by evaluating the fixed points of the system and various behaviors in its neighbourhood. The fixed points of a conservative Hamiltonian system $\dot{x} = H_y, \dot{y} = -H_x$ are given by

$$\frac{\partial H}{\partial x} = 0, \frac{\partial H}{\partial y} = 0.$$

The Hamiltonian H is a conserved quantity along any phase path of the system, and gives the shapes of the trajectories of the flow generated by the Hamiltonian vector field X_H in the phase plane of the system. And the fixed points of the system give the local dynamics in its neighbourhood. Moreover, if the Hamiltonian H of a system is known than one can directly yield the fixed points of the system.

Lemma 7.1 *If the origin is a focus of the Hamiltonian system*

$$\dot{x} = \frac{\partial H}{\partial y}, \dot{y} = -\frac{\partial H}{\partial x}$$

Then the origin is not a strict local maximum or minimum of the Hamiltonian function $H(x, y)$.

Theorem 7.15 *Any nondegenerate critical point of an analytic Hamiltonian system*

$$\dot{x} = \frac{\partial H}{\partial y}, \dot{y} = -\frac{\partial H}{\partial x} \tag{7.59}$$

is either a saddle or a center; Again (x_0, y_0) is a saddle for (7.59) iff it is a saddle of the Hamiltonian function $H(x, y)$ and a strict local maximum or minimum of the function $H(x, y)$ is a center for (7.59).

Proof Suppose that critical point of the Hamiltonian system

$$\dot{x} = H_y(x,y), \dot{y} = -H_x(x,y)$$

is at origin. Therefore $H_x(0,0) = 0$, $H_y(0,0) = 0$. The linearized system of the Hamiltonian system at the origin is given by

$$\dot{x} = Ax \tag{7.60}$$

where $A = \begin{pmatrix} H_{yx}(0,0) & H_{yy}(0,0) \\ -H_{xx}(0,0) & -H_{xy}(0,0) \end{pmatrix}$.

Trace of $A = 0$ and the det $A = H_{xx}(0,0)H_{yy}(0,0) - H_{xy}^2(0,0)$. The critical point at the origin is a saddle of the function $H(x, y)$ if and only if it is saddle of the Hamiltonian system. Now origin is a saddle of the Hamiltonian system if and only if it is a saddle of the linearized system (7.60) i.e. det $A < 0$. Also if tr$A = 0$ and det $A > 0$ then the origin is a center for the system (7.60) and then the origin is either a center or focus for the Hamiltonian system. Now if the nondegenerate critical point $(0, 0)$ is a strict local maximum or minimum of the Hamiltonian $H(x, y)$ then det $A > 0$ and then according to the above lemma the origin is not a focus for the Hamiltonian system (7.59) i.e. the origin is a center for the Hamiltonian system (7.59).

Example 7.27 Hamiltonian H for the undamped pendulum, is given by $H = 1 - \cos x + \frac{y^2}{2}$. Calculate its fixed points and sketch the phase portrait of the Hamiltonian H.

Solution The fixed points of the system is given by

$$\frac{\partial H}{\partial x} = \sin x = 0,$$

$$\frac{\partial H}{\partial y} = y = 0$$

Hence the fixed points are $(n\pi, 0)$, $n \in \mathbb{Z}$.

We shall now analyze the trajectories in the neighbourhood of $(0, 0)$. In this case the linearized system is

$$\dot{x} = y$$

$$\dot{y} = -x + \frac{x^3}{3!} - \frac{x^5}{5!} + \cdots$$

Neglecting the higher order terms, the linearized system is obtained as

$$\dot{x} = y$$
$$\dot{y} = -x$$

In matrix form it can be written as

$$\begin{pmatrix} \dot{x} \\ \dot{y} \end{pmatrix} = \begin{pmatrix} 0 & 1 \\ -1 & 0 \end{pmatrix} \begin{pmatrix} x \\ y \end{pmatrix}$$

The eigen values of the system are complex with real part zero. Hence the origin $(0,0)$ is the center of the system. About the fixed point $(-\pi,0)$ the Hamiltonian equation of motion can be linearized by introducing $(x,y) \rightarrow (x-\pi,y)$. We obtain

$$\dot{x} = y$$
$$\dot{y} = -\sin(x-\pi) = \sin(\pi-x)$$
$$= \sin x = x - \frac{x^3}{3!} + \cdots$$

Neglecting higher order terms we get the linearized system as

$$\dot{x} = y$$
$$\dot{y} = x$$

In matrix form the above system can be written as $\begin{pmatrix} \dot{x} \\ \dot{y} \end{pmatrix} = \begin{pmatrix} 0 & 1 \\ 1 & 0 \end{pmatrix} \begin{pmatrix} x \\ y \end{pmatrix}$

In this case the system has real distinct eigen values of opposite signs. Hence the fixed point $(-\pi,0)$ is a saddle of the system.

About the fixed point $(\pi,0)$ the Hamiltonian equation of motion can be linearized by introducing $(x,y) \rightarrow (x+\pi,y)$. We obtain

$$\dot{x} = y$$
$$\dot{y} = -\sin(\pi+x) = \sin x$$
$$= x - \frac{x^3}{3!} + \cdots$$

In this case also the linearized system is (Fig. 7.10)

$$\dot{x} = y$$
$$\dot{y} = x$$

Hence the fixed point $(\pi,0)$ is a saddle of the system.

Now we will see the nature of the fixed points $(2\pi,0)$. Here introducing $(x,y) \rightarrow (x+2\pi,y)$ we have the

$$\dot{x} = y$$
$$\dot{y} = -\sin(2\pi+x) = -\sin x$$
$$= -x + \frac{x^3}{3!} + \cdots$$

Fig. 7.10 Phase portrait of undamped pendulum

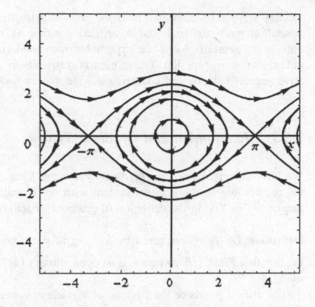

Neglecting the higher order terms we have the linearized system

$$\dot{x} = y$$
$$\dot{y} = -x$$

In matrix form the above system can be written as

$$\begin{pmatrix} \dot{x} \\ \dot{y} \end{pmatrix} = \begin{pmatrix} 0 & 1 \\ -1 & 0 \end{pmatrix} \begin{pmatrix} x \\ y \end{pmatrix}$$

which gives complex eigen values with real part zero, hence $(2\pi, 0)$ is a center similarly the fixed point $(-2\pi, 0)$ is also a center. Hence, we can conclude that the fixed points $(2n\pi, 0), n \in \mathbb{Z}$ are centers and the fixed points $((2n+1)\pi, 0), n \in \mathbb{Z}$ are saddles. The phase portrait of the given system is given in Fig. 7.10. The given system is a Hamiltonian system hence the system is a conservative system with no dissipation and the total energy of the system is given by the Hamiltonian H, which is the conserved quantity. The separatrices of the system divide the phase space into two types of qualitatively different behaviors of the trajectories. Each trajectory corresponds to a particular value of the energy H. The trajectories inside the sepratrices around the fixed points $(2n\pi, 0), n \in \mathbb{Z}$ have small values of the energy and are nearly circles which describes the usual to and fro (Oscillatory) motion of the pendulum about the equilibrium points. The sepratrices which connects the saddles at $(\pm\pi, 0)$ corresponds to the motion with total energy $H = 2$ and the pendulum tends to the unstable vertical position as $t \to \pm\infty$. The trajectories outside the sepratrix loops are hyperbolas with total energy $H > 2$ and the pendulum swing over the top. The motion corresponds to clockwise motion for negative angular velocity and counterclockwise motion for positive angular velocity.

7.8.3 Hamiltonian and Gradient Systems

We have already defined the gradient systems in Chap. 4. Herein we will discuss the relationship of the gradient system with the Hamiltonian system. For convenience we are giving the definition of gradient system once again as follows

Definition 7.6 A system given by $\dot{x} = -\operatorname{grad} F(x)$, $\operatorname{grad} F = \left(\frac{\partial F}{\partial x_1}, \ldots, \frac{\partial F}{\partial x_n}\right)^T$ and the function $F \in C^2(E)$, where E is an open subset of $\mathbb{R}^n$ is called a gradient system on E.

The critical points or fixed points of a gradient system is given by the function $F(x)$ where $\operatorname{grad} F(x) = 0$. The points for which $\operatorname{grad} F(x) \neq 0$ are called regular points of the function $F(x)$. At regular points of $F(x)$ the vector $\operatorname{grad} F(x)$ is perpendicular to the level surface $F(x) = \text{constant}$ through the regular point.

We know that any system orthogonal to a two dimensional system $\dot{x} = f(x, y), \dot{y} = g(x, y)$ is given as $\dot{x} = g(x, y), \dot{y} = -f(x, y)$. The critical points of these two systems are same. Moreover, the centre of a planar system corresponds to the nodes of its orthogonal system. Also the saddle and foci of the planar system are the saddles and foci of its orthogonal system. At regular points the trajectories of the planar system and its orthogonal system are orthogonal to each other. Again if the planar system is a Hamiltonian system then the system orthogonal to this system is a gradient system and conversely. It follows that there is an interesting relationship between the gradient system and the Hamiltonian system. The following theorem gives the relationship between the Hamiltonian and gradient system.

Theorem 7.16 *The planar system given by* $\dot{x} = f(x, y), \dot{y} = g(x, y)$ *is a Hamiltonian system if and only if the system orthogonal to it, given by* $\dot{x} = g(x, y), \dot{y} = -f(x, y)$ *is a gradient system.*

Proof Suppose that the system

$$\begin{aligned} \dot{x} &= f(x, y) \\ \dot{y} &= g(x, y) \end{aligned} \tag{7.61}$$

is a Hamiltonian system, therefore $\nabla \cdot (f, g) = 0$

i.e.$\frac{\partial f}{\partial x} + \frac{\partial g}{\partial y} = 0$. So there exist a function say H for which we can write $f = \frac{\partial H}{\partial y}$ and $g = -\frac{\partial H}{\partial x}$. Now the system orthogonal to this system is

$$\dot{x} = g(x, y)$$
$$\dot{y} = -f(x, y)$$ (7.62)

which can be written as

$$\dot{x} = -\frac{\partial H}{\partial x}, \dot{y} = -\frac{\partial H}{\partial y}$$

or $\dot{x} = -\text{grad}\, H(x)$, $x = (x, y)$ where $\text{grad}\, H = \left(\frac{\partial H}{\partial x}, \frac{\partial H}{\partial y}\right)$. This is by definition a gradient system. Conversely suppose that the system (7.61) orthogonal to (7.60) is a gradient system therefore there exists a function say H such that we can write $g = -\frac{\partial H}{\partial x}$ and $f = \frac{\partial H}{\partial y}$. For this the system (7.60) can be written as

$$\dot{x} = \frac{\partial H}{\partial y}, \dot{y} = -\frac{\partial H}{\partial x}$$

For which it can be easily checked that $\nabla \cdot \left(\frac{\partial H}{\partial y}, -\frac{\partial H}{\partial x}\right) = 0$. Hence the system (7.61) is a Hamiltonian system when the system (7.62) is a gradient system.

Note that the trajectories of the gradient system (7.62) cross the surface $H(x, y) = $ constant orthogonally.

For the Hamiltonian system in higher dimensional spaces say for n degrees of freedom is given by

$$\left.\begin{array}{l} \dot{x} = \dfrac{\partial H}{\partial y} \\[2mm] \dot{y} = -\dfrac{\partial H}{\partial x} \end{array}\right\}$$ (7.63)

Then the system orthogonal to (7.63)

$$\left.\begin{array}{l} \dot{x} = -\dfrac{\partial H}{\partial x} \\[2mm] \dot{y} = -\dfrac{\partial H}{\partial y} \end{array}\right\}$$ (7.64)

is a gradient system in $\mathbb{R}^{2n}$.

Example 7.28 For the Hamiltonian function $H(x, y) = y \sin x$, sketch the phase portraits of the Hamiltonian system and its gradient system.

Fig. 7.11 Phase portrait of
the Hamiltonian and its
gradient system

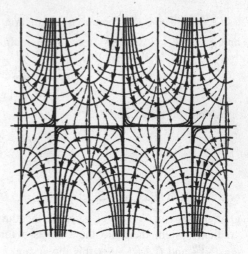

Solution For $H(x, y) = y \sin x$, the Hamiltonian system is given by

$$\dot{x} = \sin x$$
$$\dot{y} = -y \cos x \tag{7.65}$$

Therefore its gradient system is given by

$$\dot{x} = -y \cos x$$
$$\dot{y} = -\sin x$$

The critical points of the Hamiltonian system and the gradient system are at $(n\pi, 0), n \in \mathbb{Z}$. The phase portrait of the Hamiltonian and Gradient system is shown in Fig. 7.11.

7.9 Symplectic Transformations

The symplectic or canonical transformation is an important coordinate transformation from $\mathbb{R}^{2n}$ to $\mathbb{R}^{2n}$ as it preserves the flow generated by the Hamiltonian vector field X_H of the Hamiltonian system. Under symplectic transformation, Hamilton equation of motion is form invariant. Also, the Hamilton H in terms of new coordinate, obtained under this transformation is such that the new system becomes comparatively easier as it reveals all cyclic coordinates and conserved quantities. We will first define the symplectic form.

7.9.1 Symplectic Forms

Let $x = (q, p)$, then the Hamiltonian system can be written as $\dot{x} = JDH(x)$, where
$J = \begin{pmatrix} O & I \\ -I & O \end{pmatrix}$, O, I denotes the $n \times n$ null and unit matrices respectively and
$DH(\dot{x}) = \left(\frac{\partial H}{\partial q}, \frac{\partial H}{\partial p} \right)$.

A bilinear form Ω is said to be symplectic on the phase space $\mathbb{R}^{2n}$ if it is a skew-symmetric and nondegenerate that is, the matrix representation of the bilinear form is non-singular and skew-symmetric. A vector space is said to be a symplectic vector space if it is furnished with a symplectic form. A symplectic form for the vector space, particularly for the phase space $\mathbb{R}^{2n}$ is given by

$$\Omega(u, v) = \langle u, Jv \rangle, u, v \in \mathbb{R}^{2n}$$

where $\langle \cdot, \cdot \rangle$ is the standard Euclidean inner product on $\mathbb{R}^{2n}$ and J is the nonsingular, skew-symmetric matrix defined above. This particular symplectic structure on $\mathbb{R}^{2n}$ is called the Canonical symplectic form.

7.9.2 Symplectic Transformation

An $r(\geq 1)$ times continuously differentiable diffeomorphism $\phi : \mathbb{R}^{2n} \to \mathbb{R}^{2n}$ is said to be a Symplectic or Canonical transformation if, $\Omega(u, v) = \Omega(D\phi(x)u, D\phi(x)v)$ $\forall x, u, v \in \mathbb{R}^{2n}$.

7.9.3 Derivation of Hamilton's Equations from Symplectic Form

Hamilton's equation of motion in phase space can be derived from the symplectic form. For this consider the phase space $\mathbb{R}^{2n}$ for the Hamiltonian vector field $X_H(x) \equiv \left(\frac{\partial H}{\partial p}, -\frac{\partial H}{\partial q} \right)$ obtained from the Hamiltonian function H. The symplectic structure for $X_H(x)$ is expressed by

$$\Omega(X_H(x), v) = \langle DH(x), v \rangle, x \in \mathcal{U} \subset \mathbb{R}^{2n}, v \in \mathbb{R}^{2n} \qquad (7.66)$$

Now if $X_H(x) = (\dot{q}, \dot{p})$ is an arbitrary vector field on some subset $\mathcal{U} \subset \mathbb{R}^{2n}$ with $DH = \left(\frac{\partial H}{\partial q}, \frac{\partial H}{\partial p} \right)$ then the above equation becomes

$$\Omega((\dot{q},\dot{p}),\upsilon) = \langle(\dot{q},\dot{p}),J\upsilon\rangle = \left\langle \left(\frac{\partial H}{\partial q},\frac{\partial H}{\partial p}\right),\upsilon\right\rangle \tag{7.67}$$

Now, $J = \begin{pmatrix} 0 & I \\ -I & 0 \end{pmatrix} = -J^T$, therefore we get

$$\langle(\dot{q},\dot{p}),J\upsilon\rangle = \langle -J(\dot{q},\dot{p}),\upsilon\rangle = \langle(-\dot{p},\dot{q}),\upsilon\rangle \tag{7.68}$$

Substituting (7.68) into (7.67), we get

$$\langle(-\dot{p},\dot{q}),\upsilon\rangle = \left\langle \left(\frac{\partial H}{\partial q},\frac{\partial H}{\partial p}\right),\upsilon\right\rangle$$

For the symplectic form, the inner product must be nondegenerate. Using linearity principle of symplectic form and for fixed $\upsilon \in \mathbb{R}^{2n}$, the above form can be written as

$$\left\langle (-\dot{p},\dot{q}) - \left(\frac{\partial H}{\partial q},\frac{\partial H}{\partial p}\right),\upsilon\right\rangle = 0$$

This relation holds for all υ. By non degeneracy of the symplectic form we have

$$(-\dot{p},\dot{q}) - \left(\frac{\partial H}{\partial q},\frac{\partial H}{\partial p}\right) = 0$$

$$\Rightarrow \dot{p} = \frac{\partial H}{\partial q}, \dot{q} = -\frac{\partial H}{\partial p}.$$

Hence, the Hamiltonian canonical equations are established.

Example 7.29 Show that a transformation $\phi : \mathbb{R}^{2n} \to \mathbb{R}^{2n}$ is symplectic with respect to the canonical symplectic form if $D\phi(x)^T JD\phi(x) = J$ $\quad \forall x,u,\upsilon \in \mathbb{R}^{2n}$.

Solution We have for symplectic transformation $\Omega(u,\upsilon) = \Omega(D\phi(x)u,D\phi(x)\upsilon)$ $\forall x,u,\upsilon \in \mathbb{R}^{2n}$. Now for a canonical symplectic form this relation can be written as

$$\langle u,J\upsilon\rangle = \langle D\phi(x)u,JD\phi(x)\upsilon\rangle = \langle u,(D\phi(x))^T JD\phi(x)\upsilon\rangle$$

Since this holds for all $u,\upsilon \in \mathbb{R}^{2n}$, therefore $(D\phi(x))^T JD\phi(x) = J$.

Theorem 7.17. *The flow generated by the Hamiltonian vector field X_H defined on some open convex set $\mathcal{U} \in \mathbb{R}^{2n}$ is a one parameter family of symplectic (canonical) transformation and conversely if the flow generated by a vector field comprise of symplectic transformation for each t, then the vector field is a Hamiltonian vector field.*

For proof see the book of Stephen Wiggins [11].

7.10 Poisson Brackets

Poisson bracket is a connection between a pair of dynamical variables for any holonomic system which remains invariant under any symplectic transformation. This relation is helpful in testing whether a phase space transformation is symplectic or not. Also, using Poisson bracket new integrals of motion can be constructed from those already known. Poisson bracket of the variables of the Hamiltonian system in the phase space $\mathbb{R}^{2n}$ is given as follows:

The Poisson bracket of any two $C^r, r \geq 2$ (r times continuously differentiable functions) functions F, G of $x \in \mathbb{R}^{2n}$ such that $F, G : \mathcal{U} \subset \mathbb{R}^{2n} \to \mathbb{R}$ is a function defined by the following notation:

$$\{F, G\} = \Omega(X_F, X_G) = \langle X_F, J X_G \rangle \tag{7.69}$$

From (7.69), we can write

$$\{F, G\} = \Omega(X_F, X_G) = \langle X_F, J X_G \rangle = \langle J^T X_F, X_G \rangle = \langle -J X_F, X_G \rangle = -\langle J X_F, X_G \rangle$$
$$= -\langle X_G, J X_F \rangle = -\{G, F\}$$

This implies that the Poisson bracket is anti-symmetric.

If $F \equiv H$, the Hamiltonian vector fields $X_H(x)$ obtained from the Hamilton function H is given by

$$X_H(x) = \left(\frac{\partial H}{\partial p}, -\frac{\partial H}{\partial q} \right), \text{ and similarly } X_G(x) = \left(\frac{\partial G}{\partial p}, -\frac{\partial G}{\partial q} \right),$$

The Poisson bracket of H and G is therefore given as

$$\{H, G\} = \sum_{i=1}^{n} \frac{\partial H}{\partial q_i} \frac{\partial G}{\partial p_i} - \frac{\partial H}{\partial p_i} \frac{\partial G}{\partial q_i},$$

this implies that

$$\{F, F\} = 0. \tag{7.70}$$

The Hamilton's equation of motion $\dot{q}_i = \frac{\partial H}{\partial p_i}$ and $\dot{p}_i = -\frac{\partial H}{\partial q_i}$ can be written in terms of Poisson bracket. For this, the rate of change of any function F along the trajectories generated by the Hamiltonian vector field X_H is given by

$$\frac{dF}{dt} = \sum_{i=1}^{n} \left(\frac{\partial F}{\partial q_i} \dot{q}_i + \frac{\partial F}{\partial p_i} \dot{p}_i \right) = \sum_{i=1}^{n} \left(\frac{\partial F}{\partial q_i} \frac{\partial H}{\partial p_i} - \frac{\partial F}{\partial p_i} \frac{\partial H}{\partial q_i} \right) = \{F, H\} \tag{7.71}$$

So the above relation is an alternative way to write Hamilton's equation of motion for all function $F : \mathcal{U} \to \mathbb{R}$. Again the Hamilton's equation of motion

$\dot{q} = \frac{\partial H}{\partial p}, \dot{p} = -\frac{\partial H}{\partial q}$ can be easily established from the relation $\frac{dF}{dt} = \{F, H\}$. For that consider the time derivative of the function F, given as $\frac{dF}{dt} = \sum_{i=1}^{n} \frac{\partial F}{\partial q_i} \dot{q}_i + \frac{\partial F}{\partial p_i} \dot{p}_i$, then subtracting the latter relation from the former, we have

$$\{F, H\} - \sum_{i=1}^{n} \frac{\partial F}{\partial q_i} \dot{q}_i + \frac{\partial F}{\partial p_i} \dot{p}_i = 0$$

Now since the Poisson bracket of F, H is $\sum_{i=1}^{n} \frac{\partial F}{\partial q_i} \frac{\partial H}{\partial p_i} - \frac{\partial F}{\partial p_i} \frac{\partial H}{\partial q_i}$, we have

$$\sum_{i=1}^{n} \left(\frac{\partial H}{\partial p_i} - \dot{q}_i \right) \frac{\partial F}{\partial q_i} - \left(\frac{\partial H}{\partial q_i} + \dot{p}_i \right) \frac{\partial F}{\partial p_i} = 0$$

which gives $\dot{q}_i = \frac{\partial H}{\partial p_i}$ and $\dot{p}_i = -\frac{\partial H}{\partial q_i}$.

The relation (7.70) and (7.71) yields the following proposition.

Proposition 7.1 *The Hamiltonian $H(q, p)$ is constant along trajectories of the Hamiltonian vector field X_H.*

In other words, any function F which satisfies $\dot{F} = \{F, H\} = 0$ is an integral or constant of motion with respect to the flow generated by the Hamiltonian vector field X_H. We have already established this result without using Poisson bracket that has a fine geometric structure and can be obtained as follows:

We have $\{F, H\} = \langle X_F, J X_H \rangle = -\langle J X_F, X_H \rangle = 0$

Now, $J X_F = J \begin{pmatrix} \frac{\partial F}{\partial P} \\ -\frac{\partial F}{\partial q} \end{pmatrix} = \begin{pmatrix} O & I \\ -I & O \end{pmatrix} \begin{pmatrix} \frac{\partial F}{\partial P} \\ -\frac{\partial F}{\partial q} \end{pmatrix} = -\begin{pmatrix} \frac{\partial F}{\partial q} \\ \frac{\partial F}{\partial p} \end{pmatrix}$. Thus the vector

$-\left(\frac{\partial F}{\partial q}, \frac{\partial F}{\partial p} \right)$ is the vector perpendicular to the surface due to the level set of the integral of motion F, at each point at which it is evaluated. Thus geometrically the vector field is tangent to the surface given by the level set of the integral (see Wiggins [11]).

The Poisson bracket of any two functions F and G satisfies the following properties

(i) $\{F, G\}_{p,q} = -\{G, F\}_{p,q} = \{G, F\}_{q,p}$.
(ii) $\{F, C\} = 0$, if C is a constant.
(iii) $\{F_1 + F_2, G\} = \{F_1, G\} + \{F_2, G\}$.
(iv) $\{F_1 F_2, G\} = F_1 \{F_2, G\} + \{F_1, G\} F_2$.
(v)   For any three functions F, G, H, Poisson bracket satisfies the Jacobi identity

$$\{F, \{G, H\}\} + \{G, \{H, F\}\} + \{H, \{F, G\}\} = 0$$

Furthermore, the Poisson bracket of any two functions F and G remains invariant under a symplectic transformation that is, $\{F, G\}_{p,q} = \{F, G\}_{P,Q}$, where P, Q are obtained from p, q under symplectic transformation. Again the Poisson bracket of the variables P, Q obtained from p, q, given by $\{P_i, Q_j\} = \delta_{ij}$, $\{P_i, P_j\} = 0 = \{Q_i, Q_j\}$ are useful for testing the canonicality of any phase space transformation.

7.11   Hamilton–Jacobi Equation

Here we have another formulation of motion of a system that is, Hamilton–Jacobi equation. At this point one may ask why to formulate any other method when we already have Hamilton's canonical equations of motions. The answer is that Hamilton's equations of motions are system of first order differential equation which though looks simple is usually harder to solve. Thus Hamilton's formulation though provides us a simpler formulation over the Lagrangian by giving liberty in choosing the generalized coordinate and momenta, but the difficulty in solving the problem remains same. Hamilton–Jacobi equation on the other hand is a single partial differential equation. Even though, it is also not easy to solve this equation, but can be solved when variables are separable. Hamilton–Jacobi equation was formulated by Carl Gustov Jacob Jacobi (1804–1851) which is useful in particular for solving conservative periodic systems. This theory is regarded as the most complex, yet significant and strong approach for solving problems in classical mechanics. By using Hamilton–Jacobi equation all hidden constants of motion in spite of having complicated form can be found out. We shall now give the formulation of Hamilton–Jacobi equation as follows:

Suppose that $\bar{q}(t)$ is the extremal of the action integral $\int_{t_0}^{t} L(q, \dot{q}, t)\mathrm{d}t$ with $\bar{q}(t_0) = q_0$ and $\bar{q}(t) = q$, where q_0 and t_0 are fixed. Mathematically this can be written as

$$A(q, t) = \int_{t_0}^{t} L(\bar{q}, \dot{\bar{q}}, t)dt \qquad (7.72)$$

For an infinitesimal change h, its differential is given by

$$dA(q, t)h = \int_{t_0}^{t} (L(q_i + h, \dot{q}_i + \dot{h}, t) - L(q_i, \dot{q}_i, t))dt + o(h^2)$$

$$= \int_{t_0}^{t} \sum_{i=1}^{N} \left(\frac{\partial L}{\partial q_i} - \frac{d}{dt}\frac{\partial L}{\partial \dot{q}_i} \right) h_i dt + \sum_{i=1}^{N} \frac{\partial L}{\partial \dot{q}_i} h_i \Big|_{t_0}^{t} \quad \text{(using integration by parts)}$$

this gives $dA(q, t)h = \sum_{i=1}^{N} \frac{\partial L}{\partial \dot{q}_i} h_i = \sum_{i=1}^{N} p_i h_i$ (by using $p_i = \frac{\partial L}{\partial \dot{q}_i}$), the first term vanishes since $\bar{q}(t)$ is an extremal of the action integral and $\bar{q}(t_0) = q_0$ is fixed for each extremal. Since the variation is taken only for q in computing the differential of A one obtains the relation $p_i = \frac{\partial A}{\partial q_i}$. Again from the action integral (7.72) we have $\frac{dA}{dt} = L$.

Therefore

$$\frac{dA}{dt} = \sum_{i=1}^{N} \frac{\partial A}{\partial q_i} \dot{q}_i + \frac{\partial A}{\partial t} = L \Rightarrow \sum_{i=1}^{N} p_i \dot{q}_i + \frac{\partial A}{\partial t} = L \Rightarrow L - \sum_{i=1}^{N} p_i \dot{q}_i = \frac{\partial A}{\partial t}$$

which can be rewritten as

$$H(q_i, p_i, t) = \frac{\partial A}{\partial t} \text{ or, } \frac{\partial A}{\partial t} - H\left(q_i, \frac{\partial A}{\partial q_i}, t\right) = 0 \qquad (7.73)$$

The above equation is a first order partial differential equation for the function $A(q, t)$ containing $(n + 1)$ partial derivatives, known as Hamilton–Jacobi equation. The trajectories of the Hamilton's canonical equation can be obtained from the solutions of the Hamilton–Jacobi equation which follows from the following Jacobi theorem, proved by Jacobi in the year 1845.

Theorem 7.18 *If the Hamilton Jacobi equation given by (7.73) admits a complete integral $A = f(q_1, q_2, \ldots, q_N, t; c_1, c_2, \ldots, c_N) + \alpha$ then the equations $\frac{\partial f}{\partial \alpha_i} = \beta_i, \frac{\partial f}{\partial q_i} = p_i$ with the 2N arbitrary constant $a_i, c_i, \alpha_i, \beta_i$, respectively gives the 2N parameter family of solutions of Hamilton's equations $\dot{q}_i = \frac{\partial H}{\partial p_i}, \dot{p}_i = -\frac{\partial H}{\partial q_i} i = 1, \ldots, N$.*

Proof For a complete integral of the Hamilton Jacobi equation we must have $\det\left(\frac{\partial^2 f}{\partial q \partial \alpha}\right) \neq 0$ and therefore one can get the solution of the N equations $\frac{\partial f}{\partial \alpha_i} = \beta_i$ for q_i as a function of t and the $2N$ constants α_i, β_i. Substituting these functions in $\frac{\partial f}{\partial q_i} = p_i$ one will obtain p_i as a function of t and the $2N$ constants α_i, β_i. Now in order to obtain the proof of the theorem differentiate $\frac{\partial f}{\partial \alpha_i} = \beta_i$ with respect to t, which gives

$$\frac{\partial^2 f}{\partial t \partial \alpha_i} + \sum_{j=1}^{N} \frac{\partial^2 f}{\partial q_j \partial \alpha_i} \frac{dq_j}{dt} = 0 \qquad (7.74)$$

Now differentiate $\frac{\partial f}{\partial t} + H(q, \frac{\partial f}{\partial q}, t) = 0$ with respect to α_i which gives

$$\frac{\partial^2 f}{\partial t \partial \alpha_i} + \sum_{j=1}^{N} \frac{\partial H}{\partial p_j} \frac{\partial^2 f}{\partial q_j \partial \alpha_i} = 0 \qquad (7.75)$$

Now using $\det\left(\frac{\partial^2 f}{\partial q \partial \alpha}\right) \neq 0$ and subtracting (7.75) from (7.74), one will obtain

$$\sum_{j=1}^{N} \left(\frac{dq_j}{dt} - \frac{\partial H}{\partial p_j}\right) \frac{\partial^2 f}{\partial q_j \partial \alpha_i} = 0 \Rightarrow \dot{q}_j = \frac{\partial H}{\partial p_j} \quad j = 1, 2, \ldots$$

Similarly, differentiating $\frac{\partial f}{\partial q_i} = p_i$, with respect to t, we have

$$\frac{dp_i}{dt} = \frac{\partial^2 f}{\partial t \partial q_i} + \sum_{j=1}^{N} \frac{\partial^2 f}{\partial q_j \partial q_i} \frac{dq_j}{dt} \qquad (7.76)$$

Again differentiating $\frac{\partial f}{\partial t} + H(q, \frac{\partial f}{\partial q}, t) = 0$ with respect to q_i, we will have

$$\frac{\partial^2 f}{\partial t \partial q_i} + \sum_{j=1}^{N} \frac{\partial H}{\partial p_j} \frac{\partial^2 f}{\partial q_j \partial q_i} + \frac{\partial H}{\partial q_i} = 0 \qquad (7.77)$$

Now substituting $\dot{q}_j = \frac{\partial H}{\partial p_j}$ in (7.76) and then using (7.76) in (7.77), we will have

$$\dot{p}_i = -\frac{\partial H}{\partial q_i} \quad i = 1, 2, \ldots, N$$

Let us elaborate this with the following examples.

Example 7.30 Find the trajectory of the motion of a free particle using Hamilton–Jacobi equations.

Solution The Hamiltonian of a free particle is $H = \frac{1}{2}m(\dot{q})^2$ and the momentum $p = m\dot{q}$, therefore $H = \frac{p^2}{2m}$. Again $p = \frac{\partial A}{\partial q}$. Therefore the Hamilton–Jacobi equation becomes $\frac{1}{2m}\left(\frac{\partial A}{\partial q}\right)^2 + \frac{\partial A}{\partial t} = 0$.

Solving this equation one can get, $A(q, E, t) = \sqrt{2mE}q - Et$ where E is the non-additive constant.

This is a complete integral.

Now, using the solution of the Hamilton–Jacobi equation, the solution of the Hamilton's equation is obtained as follows

$$\frac{\partial A}{\partial \alpha} = \beta \Rightarrow \beta = \sqrt{\frac{m}{2E}}q - t \Rightarrow q = \sqrt{\frac{2E}{m}}(t + \beta)$$
$$\text{and} \, p = \frac{\partial A}{\partial q} = \sqrt{2mE}$$

The constant β can be obtained when the initial condition for the trajectories are prescribed.

Example 7.31 Find the trajectories of the simple harmonic oscillator using Hamilton–Jacobi equation.

Solution For harmonic oscillator, the Hamiltonian function $H = \frac{1}{2m}(p^2 + m^2\omega^2 q^2)$.

Now, $p = \frac{\partial A}{\partial q}$ gives the Hamilton–Jacobi equation $\frac{1}{2m}\left[\left(\frac{\partial A}{\partial q}\right)^2 + m^2\omega^2 q^2\right] +$
$\frac{\partial A}{\partial t} = 0$

Let us try to solve this using method of separation of variables. Suppose $A(q; \alpha; t) = W(q; \alpha) - \alpha t$, α is a non-additive constant.

We have from Hamilton-Jacobi equation, $\frac{1}{2m}\left[\left(\frac{\partial W}{\partial q}\right)^2 + m^2\omega^2 q^2\right] = \alpha \Rightarrow$
$W = \sqrt{2m\alpha}\int\sqrt{1 - \frac{m^2\omega^2 q^2}{2\alpha}}dq$.

Thus the solution of the Hamilton–Jacobi equation is obtained as

$$A(q; \alpha; t) = \sqrt{2m\alpha}\int\sqrt{1 - \frac{m^2\omega^2 q^2}{2\alpha}}dq - \alpha t$$

Now using the above solution of Hamilton–Jacobi equation, the solution of Hamilton's canonical equation can be obtained as follows $\beta = \frac{\partial A}{\partial \alpha}$
gives $\beta = \sqrt{\frac{m}{2\alpha}}\int\frac{dq}{\sqrt{1-\frac{m\omega^2 q^2}{2\alpha}}} - t$ which on integration gives $t + \beta = \frac{1}{\omega}\sin^{-1}q\sqrt{\frac{m\omega^2}{2\alpha}}$.
Thus we have,

$$q = \sqrt{\frac{2\alpha}{m\omega^2}}\sin\,\omega(t + \beta), p = \frac{\partial S}{\partial q} = \frac{\partial W}{\partial q} = \sqrt{2m\alpha - m^2\omega^2 q^2}$$
$$= \sqrt{2m\alpha}\cos\,\omega(t + \beta)$$

Remarks Consider the time independent Hamilton–Jacobi equation given by

$$H\left(q, \frac{\partial G_2}{\partial q}\right) + \frac{\partial G_2}{\partial t} = 0 \tag{7.78}$$

To solve this let us use the method of separation of variables and take

$$G_2(q, t) = W(q) + T(t) \tag{7.79}$$

Thus we have,

$$H\left(q, \frac{\partial W}{\partial q}\right) = E, \frac{\partial T}{\partial t} = -E \tag{7.80}$$

The first Eq. (7.80) is known as time independent Hamilton–Jacobi equation and E represents the total constant energy. We then write

$$G_2 = W(q_1, q_2, \ldots, q_n; \alpha_1, \alpha_2, \ldots, \alpha_n) - E(\alpha_1, \alpha_2, \ldots, \alpha_n)t$$

with the canonical transformation

$$p = \frac{\partial W(q; \alpha)}{\partial q}, \beta = \frac{\partial W(q; \alpha)}{\partial \alpha} - \frac{\partial E(\alpha)}{\partial \alpha} t. \tag{7.81}$$

This may be interpreted as a transformation from $(q, p) \rightarrow (Q, P)$ where $Q = \beta + \frac{\partial E}{\partial \alpha} t, P = \alpha$. The transformed Hamiltonian is $E(P)$ since $E = E(\alpha_1, \alpha_2, \ldots, \alpha_n) = E(\alpha) = E(P)$.

7.12 Exercises

1. Prove that for a scleronomous system the kinetic energy T is a homogeneous quadratic function of $\dot{q}_k, k = 1, 2, \ldots, n$ and hence $2T = \sum_{k=1}^{n} \dot{q}_k \frac{\partial T}{\partial \dot{q}_k}$

2. Find the Lagrangian and Lagrange's equation of motion for a simple harmonic oscillator of mass m and spring constant k freely moving in a plane. Show that this system gives two independent integrals.

3. Find the energy integral for the simple pendulum and hence show that it is constant on trajectories.

4. Formulate Lagrange's equations of motion for the double pendulum.

5. Find Hamiltonian and Hamilton's equations of motion for a particle of mass m moving in the xy-plane under the influence of a central force depending on the distance from the origin.

6. Find the Hamiltonian function $H(x, y)$ for the system $\dot{x} = y, \dot{y} = -\mu(e^{-x} - e^{-2x})$.

7. Prove that node and focus types fixed points cannot exists for Hamiltonian systems.

8. Prove that $\frac{dH}{dt} = \frac{\partial H}{\partial t}$ where H is the Hamiltonian.

9. State and prove the theorem for conservation of linear momentum.

10. Prove the theorem for conservation of angular momentum.

11. State and prove the theorem for conservation of energy.

12. A particle of mass m moves in a plane. Find Hamilton's equations of motion.

13. Find the Hamilton's equations of motion for a compound pendulum oscillating in a vertical plane about a fixed horizontal axis.

14. A bead is sliding on a uniformly rotating wire in a force-free space. Obtain the equations of motion in terms of Hamiltonian.

15. Construct the Hamiltonian and find the Hamilton's equations of motion of a coplanar double pendulum placed in a uniform gravitational field.

16. A particle of mass m is attracted to a fixed point O by an inverse square law of force $F_r = -\frac{\mu}{r^2}$ where $\mu(> 0)$ is a constant. Using Hamiltonian, obtain the equations of motion of the particle.

17. Use Hamiltonian to obtain the Hamilton's equations of motion of a projectile in space.

18. The Hamiltonian of a dynamical system is given by $H = qp^2 - qp + bp$ where b is a constant. Obtain the equations of motion.

19. Using cylindrical coordinates obtain the Hamilton's equations of motion for a particle of mass m moving inside the frictionless cone $x^2 + y^2 = z^2 \tan^2 \alpha$.

20. If the kinetic energy $T = \frac{1}{2}mr^2$ and the potential energy $V = \frac{1}{r}\left(1 + \frac{r^2}{c^2}\right)$, find the Hamiltonian.

    Determine whether (i) $H = T + V$ and (ii) $\frac{dH}{dt} = 0$.

21. For the Hamiltonian $H = q_1 p_1 - q_2 p_2 - a q_1^2 - b q_2^2$, solve the Hamilton's equations of motion and prove that $\frac{p_2 - b q_2}{q_1} = \text{constant}$ and $q_1 q_2 = \text{constant}$, a,b are constants and q_1, q_2, p_1, p_2 are generalized coordinates.

22. The Lagrangian of a system of one degree of freedom can be written as $L = \frac{m}{2}\left(\dot{q}^2 \sin \omega t + q\dot{q}\omega \sin 2\omega t + q^2 \omega^2\right)$. Determine the corresponding Hamiltonian. Is it conserved?

23. Define first integral of a system. When a system is conservative? What can you say about the dynamics of conservative system?

24. Find the first integrals of the following system and also check whether the systems are conservative or not
    (i) $\dot{x} = y$, $\dot{y} = x^2 + 1$ (ii) $\dot{x} = x(y+1)$, $\dot{y} = -y(x+1)$ (iii) $\dot{x} = -(1-y)x$, $\dot{y} = -(1-x)y$
    (iv) $\dot{x} = -x^3$, $\dot{y} = -x^2 y$ (v) $\dot{x} = -x + 2xy^2$, $\dot{y} = -x^2 y^3$

25. Find the Hamiltonian H of the following system
    (i) $\dot{x} = 4p$, $\dot{p} = -2x$ (ii) $\dot{x} = -2p$, $\dot{p} = -2x$ (iii) $\dot{x} = \sin x$, $\dot{p} = -p\cos x$ (iv) $\dot{x} = -2p + 4$, $\dot{p} = 2 - 2x$
    (v) $\dot{x} = -2x - 2y - 2$, $\dot{p} = -2x + 2y - 2$
    Also sketch the phase portraits and level curves.

26. Show that the system
    $$\dot{x} = a_{11}x + a_{12}y + Ax^2 - 2Bxy + Cy^2$$
    $$\dot{y} = a_{21}x + a_{11}y + Dx^2 - 2Axy + By^2$$
    is a Hamiltonian system with one degree of freedom.

27. For each of the following Hamiltonian functions sketch the phase portraits for the Hamiltonian system and the gradient system. Also draw both the phase portraits on the same phase plane.
    (i) $H(x,y) = x^2 + 2y^2$ (ii) $H(x,y) = x^2 - y^2$ (iii) $H(x,y) = \frac{\lambda x^2}{2} + \frac{y^2}{2}$ (iv) $H(x,y) = \frac{x^2}{2} - \frac{x^4}{4} + \frac{y^2}{2}$
    (iv) $H(x,y) = \frac{x^2}{2} - \frac{\beta}{3}x^3 + \frac{y^2}{2}$

28. Prove that the transformation $U : \mathbb{R}^2 \to \mathbb{R}^2$ is symplectic iff it is both area and orientation preserving.

29. Prove that the matrix S satisfies the properties (i) $s^T = -s = s^{-1}$ (ii) $\det(s) = 1$.

30. For a symplectic differentiation $h : \mathbb{R}^{2n} \to \mathbb{R}^{2n}$, prove that
    (i) the mapping h preserves volumes in $\mathbb{R}^{2n}$,
    (ii) h^{-1} is symplectic.
    (iii) $\left[Dh(x)\right]^{-1} = S^T \left[Dh(x)\right]^T S$

31. Prove that the composition of two symplectic transformation is a symplectic transformation.

32. Show that two dimensional volume preserving vector fields are Hamiltonian.

33. Prove that the Poisson bracket is anti-symmetric.

34. Show that Poisson bracket satisfies the Jacobi identity $\{F,\{G,H\}\} + \{G,\{H,F\}\} + \{H,\{F,G\}\} = 0$.

35. Using Poisson bracket show that if time t is explicitly absent in the Hamiltonian of a system, then the energy of the system is conserved.

References

1. Sommerfeld, A.: Mechanics. Academic Press, New York (1952)
2. Goldstein, H.: *Classical Mechanics*. Narosa Publishing Home, New Delhi (1980)
3. Arnold, V.I.: Mathematical Methods of Classical Mechanics. Springer, New York (1984)
4. Takawale, R.G., Puranik, P. S.: Introduction to Classical Mechanics, Tata Mc-Graw Hill, New Delhi
5. Marion, Thomtron: Classical Dynamics of Particles and Systems, Third edn. Horoloma Book Jovanovich College Publisher
6. Taylor, J.R.: Classical mechanics. University Science Books, USA (2005)
7. Synge, J.L., Graffith, B.A.: Principles of Mechanics. McGraw-Hill, New York (1960)
8. Shapiro, J.A.: Classical Mechanics (2003)
9. Bhatia, V.B.: Classical Mechanics: With Introduction to Nonlinear Oscillations and Chaos. Narosa Publishing House, New Delhi (1997)
10. Landau, L.D., Lifshitz, E.M.: Mechanics (Course of Theoretical Physics, vol. 1)
11. Wiggins, S.: Introduction to Applied Nonlinear Dynamical Systems and Chaos, 2nd edn. Springer, Berlin (2003)
12. Arrowsmith, D.K., Place, L.M.: Dynamical Systems: Differential equations, maps and Chaotic behavior. Chapman and Hall/CRC, London (1992)
13. Hilborn, R.C.: Chaos and Nonlinear Dynamics: An introduction for Scientists and Engineers. Oxford University Press, Oxford (2000)

Chapter 8
Symmetry Analysis

We have learnt the qualitative analysis of nonlinear systems in previous chapters. Symmetry is an inherent property of natural phenomena as well as man-made devices. Naturally, the concept of symmetry is exploited to study the linear as well as nonlinear problems. As discussed earlier analytical solution of nonlinear equations are difficult to determine except few special types of equations. Symmetry analysis of a nonlinear system reduces it to a simpler system, sometimes to a linear system exhibiting similar properties to the nonlinear system. By analysing the simpler system one can tell the behavior of the nonlinear system. The theory of symmetry analysis was discovered by Norwegian mathematician Marius Sophus Lie (1842–1899). Lie theory is a systematic procedure for analysing nonlinear systems. The main advantage of Lie theory is that it comprise of symmetry transformations which form a group. The group property ensures the identification of all the symmetry transformations of a system under which the system remains invariant. In addition, the transformations are continuous as a result if the parameters are slightly changed; the new transformation is again a symmetry transformation of the system. Symmetry of a nonlinear system determines its solvability; moreover, it describes the symmetries intrinsic in the physical system represented by the nonlinear system, and consequently assists in understanding the nature of complex physical phenomena. It is therefore worthwhile to study the operations or transformations on coordinates under which the laws of a physical system remain unchanged. The symmetry and its mathematical foundations are great findings in the nineteenth century's science. The Russian mathematician I.M. Yaglom wrote in his biography [1] of Sophus Lie and Felix Klein: "It is my firm belief that of all of the general scientific ideas which arose in the 19th century and were inherited by our century, none contributed so much to the intellectual atmosphere of our time as the idea of symmetry."

Keeping in mind the immense importance of symmetry, particularly in analysing nonlinear systems we devote this chapter on basic idea of group of transformations, Lie group of transformations, some theorems on Lie symmetry, its invariance principle and its algorithm, and symmetry analysis of some important systems.

© Springer India 2015
G.C. Layek, *An Introduction to Dynamical Systems and Chaos*,
DOI 10.1007/978-81-322-2556-0_8

8.1   Symmetry

Symmetry has been a wide spread interest of philosophers, artists, architect, mathematicians, and physicists since centuries. Symmetry is an inherent property of an object and is a subject of enormous significance. Natural and man-made objects are symmetric in shape for beauty and longevity. We can see many natural objects in our surroundings that exhibit symmetry. From the tiny snowflakes to sunflowers and honeycomb to cauliflower or broccoli in our earth to the Milky Way galaxy outside earth everything possesses some kind of symmetry. Many animals including human beings have bilateral symmetry. Broccoli or cauliflower gives a good example of symmetry where each part has the same geometric shape as the whole one. So, symmetrical objects are those that preserve some features of their origins when grow up in size. In other words, object which when subjected to a certain change, the shape remains invariant are called symmetrical object.

In geometry circles of different radii are symmetrical as the ratio of circle circumference to the diameter is always a constant denoted by π irrespective of the diameter or magnitude of circumference. Equilateral triangles of different edge sizes are symmetrical with respect to its angle which always measures 60°. Similarly, pentagons of different sizes are symmetrical each having an angle measure of 72°. In general, polygons of different sizes are symmetrical with equal angle measures. These are few examples we have given but one will get many of them in geometry or natural objects. We shall discuss invariance property of an object in which its geometrical form remains invariant under certain group of transformations. In school geometry we learn transformations of geometric figures from one position to another like translation, reflection, rotation, and dilation and their combinations. These transformations with geometrical figures are described below.

(i) **Translational invariance** This is an important principle in geometry as well as physical sciences. When shape, size, magnitude, and orientation of a geometric object remain invariant under the change of location of the object then the transformation is known as translation and this is defined by $(x, y) \rightarrow (x + a, y)$.

The (Fig. 8.1) clearly shows the change of position of the triangle $\triangle ABC$ 3 units along the axis keeping its area and structure remains same.

(ii) **Rotational invariance** When an object undergoes angular displacement (say θ), that is orientation of the object keeping its shape and size unaltered. This transformation is known as rotation and the transformation law is given by $(x, y) \rightarrow (x \cos \theta + y \sin \theta, y \cos \theta - x \sin \theta)$.

The snowflake has six arms with identical pattern in each arm (Fig. 8.2). Rotating the snowflake anticlockwise by 30°, 60°, 90°, 120° (Fig. 8.3) we see that 60° and 120° rotations of the snowflake give similar structure to the original snowflake. In addition, if we do not put any identification marks of each arm, the snowflakes will be indistinguishable from the original one. So there exists exactly six rotation transformations 60°, 120°, 180°, 240°, 300°, 360° under which the snowflake shown in Fig. 8.2 remains invariant.

Fig. 8.1 Translation of
triangle ABC 3 units along
x-axis

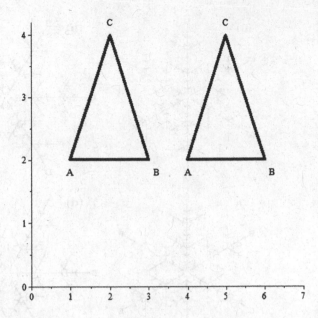

Fig. 8.2 Snowflake

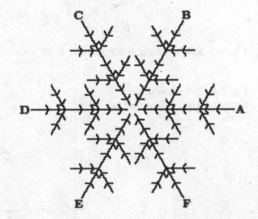

(iii) **Reflectional invariance** Reflection is a transformation of a point, line, or
geometric figure that results in a mirror image of the original. Mathematically,
this can be expressed as $(x, y) \rightarrow (x, -y)$.
Thus the triangles (Fig. 8.4) are reflectional invariance.

(iv) **Dilation invariance** Dilation (also known as scaling transformation or similarity transformation) is a transformation which differs from the above three in
the sense that though it can keep the shape of an object invariant, fails to keep
its size that is either it is expanded or contracted. Mathematically, this transformation is defined by $(x, y) \rightarrow (ax, by)$ when $a = b$, dilation is uniform and
when $a \neq b$, then the transformation is nonuniform (Fig. 8.5).

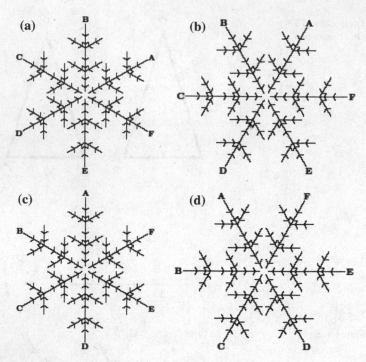

(a) **(b)**

(c) **(d)**

Fig. 8.3 a 30° rotation. **b** 60° rotation. **c** 90° rotation. **d** 120° rotation

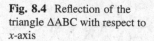

Fig. 8.4 Reflection of the triangle ΔABC with respect to x-axis

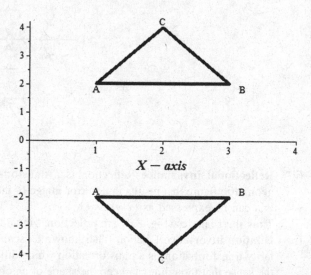

Thus, the three conventional transformations translation, rotation, and reflection do not alter the area of any geometric figure (congruent). But the dilation transformation either reduces or increases the area and thus the object is known as

Fig. 8.5 Dilation of the triangle $\triangle ABC$ by 3 units

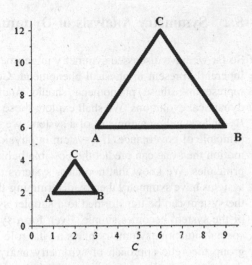

similar object. The above transformations are simple transformations and one can easily obtain them from transformations of coordinates. One frequently come across with transformations which are the combination of these transformations but apart from the above transformations and their combinations, an object may also admit new kind of transformations which leave it invariant and may not have simple geometrical representation like the conventional transformations. These new kind of transformations have been used and studied extensively in differential geometry using abstract notion of groups of transformations developed by Eliè-Joseph Cartan (1869–1951) on manifolds which are open sets and are the generalization of Euclidean space to higher dimensional spaces. In differential geometry, geometry is treated with the help of differential calculus and vector spaces. The differential geometry is not the scope of this book and our main concern is the theory of symmetry group of transformation. The main concept of invariant transformations was brought in by the mathematician Marius Sophus Lie of Norway in the late nineteenth century. He however originated the idea in relation to solving differential equations using a general approach. His approach since then is known as Lie group analysis, which satisfies the group properties of algebra as well as continuity and differentiability properties of calculus. The approach of Lie is interesting as it determines the transformations which keep the properties of a geometric object or a physical system invariant. Also, all known and unknown invariant transformations can be ascertained by this single approach of Lie. It thus helps in recognizing the transformations under which the property of a geometric object remains invariant. Hence for determining these new kinds of symmetry transformations Lie group theory has considerable importance. Herein we shall focus our attention to study the groups of transformations, symmetries, and their properties applicable to dynamical systems only.

8.2   Symmetry Analysis of Dynamical Systems

So far we have discussed symmetry in geometry. Symmetry is a natural entity. It is inherently present in physical phenomena. Consequently, the governing equations representing these phenomena should maintain symmetry in course of their dynamical evolutions. We shall explore these symmetries in the following sections. By understanding symmetry of a system, we may understand the mechanism of its principle of governance. If a system in physical space follows symmetry during its motion then one can predict the possible behavior of the system and its underlying principles. We know that nonlinear systems are harder to solve. If these nonlinear systems have symmetry then by determining the possible symmetries of the system, the system can be transformed to a simpler system. Consequently, further analysis of the system becomes simple. Even for a system governing turbulent flow under special situation, symmetry plays a vital role in its dynamical evolution. Recently, group theoretic approach of symmetry analysis has been developed in turbulent flows. It is found that some quantities in turbulent flows are invariant (see Oberlack [2]).

Determining symmetry solutions of a dynamical system are not new and are used since ancient times either by intuition or using dimension analysis. At that time there was not a systematic procedure for finding symmetries of a system either ordinary or partial differential equations. Lie group analysis is an extraordinary method which unifies all the ad hoc methods developed mainly in the seventeenth and eighteenth centuries used for integrating ordinary differential equations and provides us a nice way to solve ordinary differential equation without using any ad hoc or guessing methods and any prior knowledge of its physical nature. Besides, Lie group analysis can give us solutions which may give new physical insights of the problem which were earlier unnoticed. This new theory of Lie renders the classification of the ordinary differential equations in terms of their symmetry groups which aids in recognizing the full set of equations which can be integrated or simplified to an equation of lower order.

In spite of the elegant theory of Lie groups to differential equation, the concept did not get its significance and fades in the popularity of combination of Lie group theory and differential geometry for more than half a century. The primary idea was to associate a uniquely defined geometric structure of a certain form with every differential equation. It was the American mathematician Garret Birkhoff [3] who for the first time applied Lie group theory to the differential equation of fluid mechanics and thereafter the work of Russian mathematician L.V. Ovsyannikov [4–7] arouse a broad interest in the subject among applied mathematicians and originate the modern applied group analysis. Since Ovsyannikov Lie group theory is much more developed by Bluman and Cole [8, 9], S. Kumei [10, 11], P.J. Olver [12–14], N. H. Ibragimov [15, 16], and others. Nowadays, Lie group theory is itself an area of wide research interest and is in continuous use to determine the possible symmetries of a problem with the aid of symbolic software packages viz., Maple, Mathematica, Macsyma, Reduce, etc.

8.3   Group of Transformations

Sophus Lie discovered unified transformations which satisfy the group properties with respect to a binary law of composition and a continuous parameter. This is known as continuous group of transformations. We shall begin with by defining group, subgroup, etc., as follows.

Group: A set G of elements $(a, b, c, \ldots)$ with a law of composition denoted by o between the elements of G is called group if the elements satisfy the following properties:

  (i) $\forall a, b \in G, aob \in G.$ (**Closure Property**)
 (ii) $\forall a, b, c \in G, ao(boc) = (aob)oc$ (**Associative Property**)
(iii) $\forall a \in G$ such that $aoe = a = eoa$, then e is called the identity element of $G.$ (**Identity Property**)
 (iv) $\forall a, b \in G$ such that $aob = e$, b is called the inverse of a (**Inverse Property**)

If $aob = boa \, \forall a, b \in G$ then the set G with respect to the composition o is called an **abelian** group.

Subgroup: If G is a group and H be any subset of $G(H \subset G)$, then H is called as a subgroup of G if H exhibits all the properties of group G with respect to the group composition law of G.

Continuous group: A group G is said to be continuous, if the elements of the group depends continuously upon the group parameters. This means that a small change in the parameter determining a particular element of the group can change the transformations continuously.

Finite Continuous group: A group G is said to be a finite continuous group if it contains a finite number of parameters and the elements of the group depends continuously on these parameters. A continuous group containing $r(<\infty)$ finite parameters is called an r-parameter group.

Infinite continuous group: A continuous group G is said to be an infinite continuous group if it contains an arbitrary function.

Group of transformations: A set $G = \{T_i : X \to Y, X, Y \subseteq \mathbb{R}^n\}$ of transformations is said to form a group of transformations if the identity T_0, the inverse transformations $T_i^{-1} \in G$, and the composition of any two transformations $T_i, T_j \in G$ also belongs to G.

Lie Group of transformations: The transformation $\bar{x} = \mathcal{X}(x, \epsilon), \mathcal{X} : D \times S \to D, D \subseteq \mathbb{R}^n$ is a Lie group of transformations if the transformation $\bar{x}$ satisfies following properties:

  (i) ϵ is a continuous parameter, that is $\epsilon \in S \subset \mathbb{R}$.
 (ii) The set S of parameters with the law of composition $\phi(\epsilon, \delta)$ form an abelian group with the identity e, $\phi(\epsilon, \delta)$ is an analytic function of $\epsilon, \delta, \forall \epsilon, \delta \in S$.
(iii) $\mathcal{X}(x; e) = x, \forall x \in D$
 (iv) The transformations are invertible, i.e., $\mathcal{X}(\bar{x}; \epsilon^{-1}) = x;$

(v) $\mathcal{X}(\mathcal{X}(x; \epsilon); \delta) = \mathcal{X}(x; \phi(\epsilon, \dot{\delta})), \forall x \in D, \forall \epsilon, \delta \in S.$
(vi) $\mathcal{X}$ is C^∞ with respect to x in D and an analytic function of ϵ in S.

Note that Lie groups of transformations deal with C^∞ functions. A real-valued function $f : U \to R$ is said to be C^k at $p \in U$, k being a positive integer if its partial derivatives $\frac{\partial^j f}{\partial x^{i_1} x^{i_2} \dots x^{i_j}}$, of all orders $j \le k$ exist and are continuous at p. The function $f : U \to R$ is C^∞ at p if it is C^k for all $k \ge 0$.

Geometrically, the Lie group of transformation means that any point $x \in \mathbb{R}^n$ is transformed by the group of transformation into the point $\bar{x} = \mathcal{X}(x, \epsilon)$. The locus of the point is a continuous curve and by the group property every point of continuous curve is transformed into the point of the same continuous curve. The continuous group which involves arbitrary functions is called infinite continuous group. Lie groups are basically differentiable manifolds. Manifolds are locally open subsets of a Euclidean space $\mathbb{R}^n$ in which coordinates can be changed arbitrarily. The manifolds are the generalization of curves and surfaces to higher dimensional spaces. Thus, a one-dimensional manifold is a curve and a two-dimensional manifold is a surface. By differentiable manifolds we mean that there exists a bijective infinitely differentiable map ϕ from an open subset U of $\mathbb{R}^n$ to an open subset V of $\mathbb{R}^n$ with infinitely differentiable inverse. Symmetry groups are just the collection of transformations on such open subsets satisfying certain elementary group axioms which allows one to compose successive symmetries.

One Parameter Continuous Group of Transformation
A set S of all such transformations $T_\epsilon : \bar{x} = \phi(x, \epsilon)$ which depends upon a single parameter ϵ, ϵ takes on all values in an interval $\mathcal{U} \subset \mathbb{R}$ is said to form a one parameter group of transformation if $T_\epsilon = I$ for a unique value $\epsilon = \epsilon_0 \in \mathcal{U}$. Also, for each $\epsilon \in \mathcal{U}$ there exists a unique $\epsilon^{-1} \in \mathcal{U}$ such that $T_\epsilon^{-1} = T_{\epsilon^{-1}} \in S$ and $T_{\epsilon_1} T_{\epsilon_2} = T_{\epsilon_3} \in S$, where $\epsilon_1, \epsilon_2, \epsilon_3 \in \mathcal{U}$ and $\epsilon_3 = \phi(\epsilon_1, \epsilon_2)$ is a continuous function. In other words, T_ϵ is a single-valued transformation $\forall \epsilon \in \mathcal{U} \subset \mathbb{R}$. For instance, the translation group: $\bar{x} = x + \epsilon, \bar{y} = y$ is a one parameter group of transformation. Also the rotation group: $\bar{x} = x \cos \epsilon + y \sin \epsilon, \bar{y} = y \cos \epsilon - x \sin \epsilon$ is a one parameter group of transformation. Other examples of one parameter group of transformations in a plane are dilation transformation, Lorentz transformation, Galilean transformation, projective transformation, etc.

Local and Global Transformation Groups
Sophus Lie defines the group of transformation as local groups which are not the full Lie groups and valid only for group elements close to the identity element. Local transformation groups act only locally which means that the group of transformations may be defined neither for all elements of the group nor for all points on the manifold. In contrast, if the transformations are defined for all elements of the group and for all points on the manifold then the group is a global group of transformations. For instance, rotation group, projective group, etc. are local transformation group whereas translation, Galilean, Lorentz group of transformations, etc. are global group of transformations.

Example 8.1 Show that the translational group $\bar{x} = x + \epsilon$, $\bar{y} = y$ is a Lie group.

Solution The given group is $\mathcal{X}((x,y), \epsilon) = (x + \epsilon, y)$. When $\epsilon = 0, \mathcal{X}((x,y), 0) = (x, y)$, the identity transformation. Also $\mathcal{X}((x + \epsilon, y), -\epsilon) = (x + \epsilon - \epsilon, y) = (x, y)$ gives the inverse transformation. Again, $\mathcal{X}(\mathcal{X}((x,y), \epsilon), \delta) = \mathcal{X}((x + \epsilon, y), \delta) = (x + \epsilon + \delta, y) = \mathcal{X}((x,y), \epsilon + \delta)$, where the group composition $\phi(\epsilon, \delta) = \epsilon + \delta$ is commutative. Hence, the given translational group is a Lie group. Translation group is an example of global Lie group.

Example 8.2 Show that the uniform dilation group $\bar{x} = e^\epsilon x$, $\bar{y} = e^\epsilon y$ is a Lie group.

Solution The given group is $\mathcal{X}((x,y), \epsilon) = (e^\epsilon x, e^\epsilon y)$. When $\epsilon = 0, \mathcal{X}((x,y), 0) = (x, y)$, the identity transformation. Also $\mathcal{X}((e^\epsilon x, e^\epsilon y), -\epsilon) = (e^{-\epsilon} e^\epsilon x, e^{-\epsilon} e^\epsilon y) = (x, y)$ gives the inverse transformation. Again

$$\mathcal{X}(\mathcal{X}((x,y), \epsilon), \delta) = \mathcal{X}((e^\epsilon x, e^\epsilon y), \delta)$$
$$= (e^\delta e^\epsilon x, e^\delta e^\epsilon y) = (e^{\delta + \epsilon} x, e^{\delta + \epsilon} y)$$
$$= \mathcal{X}((x,y), \delta + \epsilon),$$

where the group composition $\phi(\epsilon, \delta) = \delta + \epsilon$ is commutative. Hence, the given dilation group is a Lie group.

Example 8.3 Show that the composition of translational and reflection group represented by $\bar{x} = -x + \epsilon$, $\bar{y} = y$ is not a Lie group.

Solution The given group is $\mathcal{X}((x,y), \epsilon) = (-x + \epsilon, y)$. When $\epsilon = 0, \mathcal{X}((x,y), 0) = (-x, y)$, hence there is no identity transformation of the group of transformation. Again, $\mathcal{X}(\mathcal{X}((x,y), \epsilon), \delta) = \mathcal{X}((-x + \epsilon, y), \delta) = (x - \epsilon + \delta, y)$, here a change of sign occurs in x. Hence, the given group is not a Lie group.

Example 8.4 Show that the hyperbolic group of transformation $\bar{x} = x + \epsilon$, $\bar{y} = \frac{xy}{x + \epsilon}$ is a Lie group.

Solution The given group is $\mathcal{X}((x,y), \epsilon) = (x + \epsilon, \frac{xy}{x + \epsilon})$. When $\epsilon = 0, \mathcal{X}((x,y), 0) = (x, y)$, the identity transformation. Also, $\mathcal{X}((x + \epsilon, \frac{xy}{x + \epsilon}), -\epsilon) = (x + \epsilon - \epsilon, \frac{xy}{x + \epsilon - \epsilon}) = (x, y)$ gives the inverse transformation. Again, $\mathcal{X}(\mathcal{X}(\epsilon, (x,y)), \delta) = \mathcal{X}((x + \epsilon, \frac{xy}{x + \epsilon}), \delta) = (x + \epsilon + \delta, \frac{(x + \epsilon)\frac{xy}{x + \epsilon}}{x + \epsilon + \delta}) = (x + \epsilon + \delta, \frac{xy}{x + \epsilon + \delta}) = \mathcal{X}((x,y), \epsilon + \delta)$, where the group composition $\phi(\epsilon, \delta) = \epsilon + \delta$ is commutative. Hence, the given hyperbolic group is a Lie group.

Example 8.5 Determine whether the rotation transformation $\bar{x} = x + \epsilon y$, $\bar{y} = y - \epsilon x$ is a Lie group of transformation.

Solution The given group is $\mathcal{X}((x,y),\epsilon) = (x+\epsilon y, y-\epsilon x)$. When $\epsilon = 0$, $\mathcal{X}((x,y),0) = (x, y)$, the identity transformation. Also,

$$\mathcal{X}((x+\epsilon y, y-\epsilon x), -\epsilon) = (x+\epsilon y - \epsilon(y-\epsilon x), y-\epsilon x+\epsilon(x+\epsilon y))$$
$$= ((1+\epsilon^2)x, (1+\epsilon^2)y)$$

Again,

$$\mathcal{X}(\mathcal{X}(\epsilon, (x,y)), \delta) = \mathcal{X}((x+\epsilon y, y-\epsilon x), \delta)$$
$$= (x+\epsilon y + \delta(y-\epsilon x), y-\epsilon x - \delta(x+\epsilon y))$$
$$= ((1-\epsilon\delta)x + (\epsilon+\delta)y, (1-\epsilon\delta)y - (\epsilon+\delta)x)$$
$$= \mathcal{X}((1-\epsilon\delta)x, (1-\epsilon\delta)y), \epsilon+\delta)$$

Hence, the given group is not a Lie group globally and is a local group near the identity $\epsilon = 0$.

Example 8.6 Show that the one parameter projective transformation $\bar{x} = \frac{x}{1-\epsilon x}, \bar{y} = \frac{y}{1-\epsilon x}$ is a local Lie group of transformation.

Solution The given group is $\mathcal{X}((x,y),\epsilon) = \left(\frac{x}{1-\epsilon x}, \frac{y}{1-\epsilon x}\right)$.
When $\epsilon = 0, \mathcal{X}((x,y),0) = (x, y)$, the identity transformation.
Also,

$$\mathcal{X}\left(\left(\frac{x}{1-\epsilon x}, \frac{y}{1-\epsilon x}\right), -\epsilon\right) = \left(\frac{\frac{x}{1-\epsilon x}}{1+\epsilon\frac{x}{1-\epsilon x}}, \frac{\frac{y}{1-\epsilon x}}{1+\epsilon\frac{x}{1-\epsilon x}}\right) = (x, y)$$

Again

$$\mathcal{X}(\mathcal{X}(\epsilon, (x,y)), \delta) = \mathcal{X}\left(\left(\frac{x}{1-\epsilon x}, \frac{y}{1-\epsilon x}\right), \delta\right) = \left(\frac{\frac{x}{1-\epsilon x}}{1-\delta\cdot\frac{x}{1-\epsilon x}}, \frac{\frac{y}{1-\epsilon x}}{1-\delta\cdot\frac{x}{1-\epsilon x}}\right)$$
$$= \left(\frac{x}{1-(\epsilon+\delta)x}, \frac{y}{1-(\epsilon+\delta)x}\right) = \mathcal{X}((x,y), \epsilon+\delta)$$

where the group composition $\phi(\epsilon,\delta) = \epsilon+\delta$ is commutative. This leads to the conclusion that the given group is a Lie group globally. However, this is not correct as the conclusion does not hold if $1-\epsilon x$ or $1-(\epsilon+\delta)x = 0$. So for the given group to become a Lie group, the parameter range must be $0 \leq \epsilon < \frac{1}{x}$. One may think that the given group will be Lie group by taking intervals excluding these points but one should not omit the possibility that the composition of any two elements may give the prohibited value. However, if one takes an interval very close to the identity 0 then the composition of any two elements belongs to $[0, \frac{1}{x})$. However, a further composition may give the prohibited value. So, it is not possible to solve the problem by only fixing a small interval, as the locus of any point x has a singularity when $\epsilon = \frac{1}{x}, x \neq 0$. Thus, the given projective group of transformation is a local group of transformation where the composition is defined for the transformation sufficiently close to the identity 0.

8.3.1 Symmetry Group of Transformations

We shall now define the symmetry group of transformations which is the fundamental discovery of Lie group theory. Lie's method on group of transformations is the unified theory to capture all symmetry groups of transformations either linear or nonlinear systems. Naturally, it is inquisitive that when a group of transformation is called a symmetry group of transformations. Well, the answer is given as follows

A group of transformation $G : \bar{x} = f(x, y, \epsilon), \bar{y} = g(x, y, \epsilon)$ with $\bar{x} = x$ and $\bar{y} = y$ when $\epsilon = 0$ is said to be a symmetry group of transformation of a system of differential equation

$$F_\alpha(x, y, y', \ldots, y^{(n)}) = 0, \alpha = 1, 2, \ldots \tag{8.1}$$

If this system remains invariant under G and the solution of the system maps to another solution of the system. The solution is known as invariant solution of the system under G. Mathematically, this can be expressed as follows:

$$F_\alpha(x, y, y', \ldots, y^n) = 0 \Leftrightarrow F_\alpha(\bar{x}, \bar{y}, \bar{y}', \ldots, \bar{y}^n) = 0 \tag{8.2}$$

and

$$y(x) = \bar{y}(\bar{x}) \tag{8.3}$$

Thus under a symmetry group of transformations a differential or algebraic systems read the same in terms of the transformed variables as was in terms of the original variables and the integral (solution) curves of the system transforms into a solution curve of the same system. Therefore, the solution curves preserve themselves under G. Naturally, one would look for those integral curves which remain invariant under G, known as invariant solution curve. We have already discussed that Lie groups are differentiable manifolds (see Olver [17]). One can also define manifold $M \subset X \times U \subset J^k$ as the locus of the solution curves $y = f(x)$ of the differential equation $\mathcal{F}(x, y, y^{(1)}, \ldots, y^{(k)}) = 0$, where $J^k = X \times U^k$ is a total space called the jet space whose coordinates represents the independent variables, dependent variables, and the derivatives of the dependent variables up to order k of the kth-order differential equation $\mathcal{F}(x, y, y^{(1)}, \ldots, y^{(k)}) = 0$ and is the prolongation of the basic space $X \times U$ whose coordinates are the independent and the dependent variables of $\mathcal{F}(x, u, u^{(1)}, \ldots, u^{(k)}) = 0$. If $y = f(x)$ lies in the manifold M then the kth prolongation of $f(x)$ lies in the jet space J^k. Thus a kth order differential equation $\mathcal{F}(x, u, u^{(1)}, \ldots, u^{(k)}) = 0$ implies a manifold $M \subset J^k$ and vice versa. A group of transformation $\chi(x, \epsilon)$ transforms a manifold $M \subset J^k$ (jet bundle) into the manifold

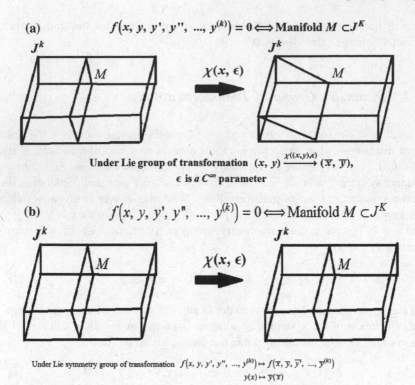

(a) $\quad f(x, y, y', y'', ..., y^{(k)}) = 0 \Longleftrightarrow$ Manifold $M \subset J^K$

Under **Lie group of transformation** $(x, y) \xrightarrow{\chi((x,y),\epsilon)} (\overline{x}, \overline{y})$,
ϵ is a C^∞ parameter

(b) $\quad f(x, y, y', y'', ..., y^{(k)}) = 0 \Longleftrightarrow$ Manifold $M \subset J^K$

Under Lie symmetry group of transformation $f(x, y, y', y'', ..., y^{(k)}) \mapsto f(\overline{x}, \overline{y}, \overline{y'}, ..., \overline{y}^{(k)})$
$y(x) \mapsto \overline{y}(\overline{x})$

Fig. 8.6 a Transformation group mapping a manifold **M** into the manifold **M′. b** Symmetry group maps a manifold **M** into the same manifold **M**

$M' \subset J^k$ (shown in the Fig. 8.6a) and this becomes a symmetry transformation when $\chi(x, \epsilon)$ transforms the manifold M into itself (shown in Fig. 8.6b).

8.3.2 Infinitesimal Transformations

The most important aspect of Lie theory is that the transformations are replaced by their infinitesimal forms. According to Lie the largest symmetry group G can be formed simply by finding the infinitesimal transformations of a one parameter Lie group of transformations

$$\overline{x}^i = \mathcal{X}^i(x^i, \epsilon)$$

ϵ is a continuous parameter with $\overline{x}^i = x^i$ when $\epsilon = 0$

Expanding the left-hand side of the above transformation by Taylor series expansion in the parameter ϵ in a neighborhood of $\epsilon = 0$, we have

$$\bar{x}^i = x^i + \epsilon \frac{\partial \mathcal{X}^i}{\partial \epsilon}\bigg|_{\epsilon=0} + \frac{\epsilon^2}{2}\left(\frac{\partial^2 \mathcal{X}^i}{\partial \epsilon^2}\bigg|_{\epsilon=0}\right)$$

$$\bar{x}^i = x^i + \epsilon \xi^i(x) + O(\epsilon^2).$$

(8.4)

where the quantities $\xi^i(x) = \frac{\partial \mathcal{X}^i}{\partial \epsilon}\big|_{\epsilon=0}$ are smooth functions and are called the infinitesimals of the group. The transformations $\bar{x}^i = x^i + \epsilon \xi^i(x), x = (x^1, x^2, \ldots, x^n)$ are called the infinitesimal transformation of the Lie group of transformation.

Lie group of point transformation When the infinitesimals ξ^i of the Lie group of continuous transformations depend upon the dependent and independent variables of a system of equation then the transformations are called as Lie group of point transformations.

8.3.3 Infinitesimal Generator

The infinitesimals of a Lie group of transformations are smooth functions and form the generator of the group of transformations as in usual group structure. The infinitesimal generator denoted by $X(x)$ is in fact a tangent vector assigned by a vector field X to a curve say C passing through a point x in a manifold M. The infinitesimal generator (or operator) of the group G is represented by the following linear differential operator

$$X(x) = \xi(x) \cdot \nabla = \sum_{i=1}^n \xi_i(x) \frac{\partial}{\partial x_i}, \ x = (x_i, i = 1 \ldots n),$$

(8.5)

where $\xi(x)$ and $\nabla = \left(\frac{\partial}{\partial x_1}, \frac{\partial}{\partial x_2}, \frac{\partial}{\partial x_3}, \ldots, \frac{\partial}{\partial x_n}\right)$ are the infinitesimals of G and the gradient operator, respectively.

The space of all tangent vectors to all possible curves passing through a given point x in the manifold M is called as tangent space to M at x denoted by $T_x M$ and the collection of all tangent spaces $T_x M$ for all points $x \in M$ is called the tangent bundle of M, i.e., $TM = \bigcup_{x \in M} T_x M$

When the infinitesimal generator X operates on any differentiable function $F(x) = F(x_1, x_2, \ldots, x_n)$ then $XF(x)$ is given by

$$XF(x) = \xi(x).\nabla F(x) = \sum_{i=1}^n \xi_i(x) \frac{\partial F(x)}{\partial x_i}$$

8.3.4 Extended Infinitesimal Operator

The infinitesimal generator X can be extended or prolonged to all derivatives involved in the equation $F(x,y,y',y'',\ldots,y^{(n)})=0$ and is obtained by finding the infinitesimal transformations of all the higher order derivatives (neglecting second and higher order terms)

The infinitesimal transformation of y' is given by

$$\bar{y}' = \frac{d\bar{y}}{d\bar{x}} = \frac{D(y+\epsilon\eta(x,y))}{D(x+\epsilon\xi(x,y))} = \frac{\left(\frac{\partial}{\partial x}+y'\frac{\partial}{\partial y}+y''\frac{\partial}{\partial y'}+\cdots\right)(y+\epsilon\eta(x,y))}{\left(\frac{\partial}{\partial x}+y'\frac{\partial}{\partial y}+y''\frac{\partial}{\partial y'}+\cdots\right)(x+\epsilon\xi(x,y))}$$

$$= \frac{\epsilon\eta_x(x,y)+y'+\epsilon y'\eta_y(x,y)}{1+\epsilon(\xi_x(x,y)+y'\xi_y(x,y))}$$

$$= y' + \epsilon(\eta_x(x,y)+y'\eta_y(x,y))(1-\epsilon(\xi_x(x,y)+y'\xi_y(x,y)))$$

$$= y' + \epsilon(\eta_x(x,y)+y'\eta_y(x,y)-y'(\xi_x(x,y)+y'\xi_y(x,y)))$$

$$= y' + \epsilon\left(\eta_x(x,y)+y'(\eta_y(x,y)-\xi_x(x,y))-(y')^2\xi_y(x,y)\right)$$

$$= y' + \epsilon\eta^1$$

where

$$\eta^1 = \eta_x(x,y)+y'(\eta_y(x,y)-\xi_x(x,y))-(y')^2\xi_y(x,y) = D(\eta(x,y))-y'D(\xi(x,y))$$

The infinitesimal generator is therefore prolonged once and can be written as

$$X^{[1]} = \xi(x,y)\frac{\partial}{\partial x}+\eta(x,y)\frac{\partial}{\partial y}+\eta^1(x,y,y')\frac{\partial}{\partial y'} = X+\eta^1(x,y,y')\frac{\partial}{\partial y'}$$

acting in the space (x,y,y'). Similarly, one can evaluate the transformation of higher order derivatives of the dependent variables and construct the higher prolongations of X. D is the total differential operator; for a partial differential equation involving more than one independent variables the total differential operator is given by

$$D_i = \frac{\partial}{\partial x^i}+y_i^j\frac{\partial}{\partial y_i}+y_{ii_1}^j\frac{\partial}{\partial y_{i_1}^j}+\cdots+y_{ii_1i_2\cdots i_n}^j\frac{\partial}{\partial y_{i_1i_2\cdots i_n}^j}$$

where $x^i=(x^1,x^2,\cdots,x^n)$ are the independent variables and $y^j=(y^1,y^2,\ldots)$ are the dependent variable and $y_{i_1i_2\cdots i_n}^j$ denotes the nth derivative of the dependent variable y^j with respect to all the independent variables

The extended generator is then expressed by

$$X^{[k]} = \xi^i \frac{\partial}{\partial x} + \eta^j \frac{\partial}{\partial y^j} + \zeta_{i_1}^j \frac{\partial}{\partial y_i^j} + \zeta_{i_1 i_2}^j \frac{\partial}{\partial y_{i_1 i_2}^j} + \ldots + \zeta_{i_1 i_2 \ldots i_k}^j \frac{\partial}{\partial y_{i_1 i_2 \ldots i_k}^j} + \ldots$$

$$= X + \zeta_{i_1}^\beta \frac{\partial}{\partial y_i^\beta} + \zeta_{i_1 i_2}^\beta \frac{\partial}{\partial y_{i_1 i_2}^\beta} + \ldots + \zeta_{i_1 i_2 \ldots i_k}^\beta \frac{\partial}{\partial y_{i_1 i_2 \ldots i_k}} + \ldots \tag{8.6}$$

where the coefficients $\zeta_{i_1 i_2 \ldots i_k}^j = D_{i_1} \ldots D_{i_k}(\eta^j - \xi^i y_i^j) + \xi^i y_{i i_1 \ldots i_k}^j, k = 1, 2, \ldots$

8.3.5 Invariance Principle

The main idea in Lie group of transformations is to find the transformations which keep a given system of equations invariant. The key concept in the Lie group theory is the infinitesimal generator of a symmetry group and the infinitesimal criterion. This is a vector field on the underlying manifold or subset of $\mathbb{R}^n$ in which the flow coincides with the one parameter group it generates over $\mathbb{R}^n$. Now we shall give the criterion which ensures the invariance of a given system of equations. The invariance criterion also finds out the infinitesimals of the transformation which then determines the generator of the system of equations.

A system of differential equations given by

$$F_\alpha(x, y, y^{(1)}, y^{(2)}, \ldots, y^{(n)}) = 0 \tag{8.7}$$

is said to be invariant under the Lie group of transformations with an infinitesimal generator X if and only if

$$XF_\alpha|_{F_\alpha=0} = 0, \quad \alpha = 1, 2, \ldots \tag{8.8}$$

where the infinitesimal generator X is prolonged to all derivatives involved in Eq. (8.7). Equation (8.8) is known as defining equation of the system which ascertains all the infinitesimal symmetries of (8.7). Lie's fundamental discovery was that the complicated nonlinear conditions of invariance for a system under the group of transformations can be replaced by equivalent, simple, and linear conditions involving the generator of the group. The above invariance conditions give a system of linear equations, known as determining equations. The left-hand side of (8.8) contains the independent variables x, dependent variables y, the derivatives $y^{(1)}, y^{(2)}, \ldots$, and the unknown infinitesimal coefficients ξ^i and η^j and their derivatives and should be satisfied with respect to all the variables involved. Consequently the Eq. (8.8) splits into an overdetermined system (since it contains more equations than unknowns) of partial differential equations which are generally linear and are easy to solve in many cases by hand. Thus, the closed form of the system can be explicitly determined. Several symbolic manipulation computer programs are now available for this symmetry group calculations for instance the mathematica packages Intro-to-symmetry.m developed by Brian J. Cantwell (see Cantwell [18]), yalie. m, etc. The main advantage of Lie group we have discussed so far herein is that one need not apply any guess or assumption to calculate the symmetry transformation of a system. Lie's theory calculates the symmetry groups of a problem in an algorithmic way. The algorithmic approach is however lengthy for systems involving many variables, when one uses hand calculation. Nowadays several symbolic software have been developed that can be used for the calculations of transformations and symmetries, see the packages. A step-by-step procedure for calculating symmetry groups of differential equation known as Lie algorithm is given below.

Algorithm to calculate Lie Symmetry

(a) Determine the number of dependent variables and independent variables contained in the differential equation

$$\mathcal{F}(x, u, u^{(1)}, \cdots, u^{(k)}) = 0 \tag{8.9}$$

where $x = (x_i, i = 1, 2, \ldots, n)$ are the independent variables, $u = (u_j, j = 1, 2, \ldots, n)$ are the dependent variables, $u^{(k)}$ is the kth derivative of u with respect to the independent variables.

(b) Write down the infinitesimal generator corresponding to the differential Eq. (8.9)

(c) Apply the extended infinitesimal generator to the differential Eq. (8.9)

(d) Write the constraint relation $\mathcal{F} = 0$ in a solved form where the highest order linear derivative is written in terms of the others and then replace the corresponding term in the invariance condition.

(e) Isolate the terms containing derivatives of all orders of dependent variable with respect to the independent variable and the various product terms and those free of these terms, and set coefficients of each term equal to zero.

(f) Obtain the infinitesimal coefficients by solving the overdetermined system thus obtained which are generally linear and can be easily calculated by hand or using symbolic software (Mathematica, Maple, Macsyma, Reduce, etc.).

We shall now give an example which illustrate the use of Lie algorithm for calculating symmetry groups and hence the generator of a second-order nonlinear differential equation.

Consider a second-order ordinary nonlinear differential equation $y'' + \left(\frac{1}{x}\right)y' - e^y = 0$ with the dependent variable y and independent variable x. Let us suppose that it admits the generator $X = \xi \frac{\partial}{\partial x} + \eta \frac{\partial}{\partial y}$. Since it is a second-order equation, therefore the generator is twice extended and is given by $X^{[2]} = \xi(x,y)\frac{\partial}{\partial x} + \eta(x,y)\frac{\partial}{\partial y} + \eta^1(x,y)\frac{\partial}{\partial y'} + \eta^2(x,y)\frac{\partial}{\partial y''}$. The infinitesimal criterion $X^{[2]}F\big|_{F=0} = 0$ gives

$$\frac{-1}{x^2}\xi(x,y)y' - \eta(x,y)e^y + \frac{1}{x}\eta^1 + \eta^2 = 0.$$

where $\eta^1 = \eta_x + (\eta_y - \xi_x)y' - \xi_y y'^2$
and

$$\eta^2 = \eta_{xx} + (2\eta_{xy} - \xi_{xx})y_x + (\eta_{yy} - 2\xi_{xy})y_x^2 - \xi_{yy}y_x^3 + (\eta_y - 2\xi_x - 3\xi_y y_x)y_{xx}$$

According to Lie algorithm substitution of the above quantities and replacing $y'' = -\left(\frac{1}{x}\right)y' + e^y$ in the above equation, we have

$$-\eta(x,y)e^y + \frac{1}{x}\left(\eta_x + (\eta_y - \xi_x)y' - \frac{1}{x^2}\xi(x,y)y' - \xi_y y'^2\right) + \eta_{xx} + (2\eta_{xy} - \xi_{xx})y'$$

$$+ (\eta_{yy} - 2\xi_{xy})(y')^2 - \xi_{yy}(y')^3 + (\eta_y - 2\xi_x - 3\xi_y y')\left(-\left(\frac{1}{x}\right)y' + e^y\right) = 0$$

After rearranging the terms in the above equation, the equation can be written as

$$-\eta(x,y)e^y + (\eta_y - 2\xi_x)e^y + \eta_{xx} + \frac{\eta_x}{x} + \left(\frac{\xi_x}{x} + 2\eta_{xy} - \xi_{xx} - \frac{\xi(x,y)}{x^2} - 3\xi_y e^y\right)y'$$

$$+ \left(\frac{2\xi_y}{x} + (\eta_{yy} - 2\xi_{xy})\right)(y')^2 - \xi_{yy}(y')^3 = 0$$

The above equation contains y', y'^2, y'^3 and terms free of these variables. The invariance condition will be satisfied if each term is individually zero, so that we get the following overdetermined system of equations as

$$(-\eta(x,y) + \eta_y - 2\xi_x)e^y + \eta_{xx} + \frac{\eta_x}{x} = 0$$

$$\frac{\xi_x}{x} - \frac{\xi}{x^2} + 2\eta_{xy} - \xi_{xx} - 3\xi_y e^y = 0$$

$$\frac{2\xi_y}{x} + (\eta_{yy} - 2\xi_{xy}) = 0$$

$$-\xi_{yy} = 0$$

It follows from $\xi_{yy} = 0$, $\xi = a(x)y + b(x)$ and then from $\frac{2\xi_y}{x} + \eta_{yy} - 2\xi_{xy} = 0$, $\eta_{yy} = 2\left(a'(x) - \frac{a(x)}{x}\right)$ which gives $\eta = \left(a'(x) - \frac{a(x)}{x}\right)y^2 + c(x)y + d(x)$. Since ξ and η are polynomials in y and the determining equation involves e^y, therefore $\xi_y = 0$ and $\eta_y - 2\xi_x - \eta = 0$ which gives $a(x) = 0$ and $c(x) - 2b'(x) - c(x)y - d(x) = 0$ which is satisfied if $c(x) = 0, 2b'(x) + d(x) = 0$. Finally, we will have $\xi = b(x)$ and $\eta = -2b'(x)$. Now, from $\frac{\xi_x}{x} - \frac{\xi}{x^2} + 2\eta_{xy} - \xi_{xx} - 3\xi_y e^y = 0$, we have $b''(x) - \frac{b'(x)}{x} + \frac{b(x)}{x^2} = 0$, which gives $b(x) = x(c_1 + c_2 \ln x)$. Thus the infinitesimals are $\xi(x,y) = x(c_1 + c_2 \ln x)$ and $\eta(x,y) = -2(c_1 + c_2 \ln x) - 2c_2 x$ where c_1, c_2 are arbitrary constants. Thus, the given nonlinear ordinary differential equation admits a two parameter (c_1, c_2) symmetry group generated by the infinitesimal generator given by

$$X_1 = x \frac{\partial}{\partial x} - 2 \frac{\partial}{\partial y}, X_2 = x \ln x \frac{\partial}{\partial x} - 2(1 + \ln x) \frac{\partial}{\partial y}$$

which gives the invariants as $\psi = y + 2\ln x$ for the generator X_1 and $\psi = y + 2\ln x + 2\ln(\ln x)$ for the generator X_2, the invariants are found by solving the characteristic form of the generators. We will see later in this chapter that the solution of the given equation can be obtained using the infinitesimals (ξ, η).

8.4 Canonical Parameters

The group parameter is said to be canonical if the composition law of the parameters reduces to the addition that is $\phi(\epsilon, \delta) = \epsilon + \delta$. By changing the parameter ϵ appropriately the composition law in any local group of transformation can be transformed one into another. The composition law $\phi(\epsilon, \delta)$ can be written as addition $\epsilon + \delta$ by changing the parameter ϵ to a canonical form. When the parameter ϵ becomes canonical the following property of Lie group of transformation holds

$$\mathcal{X}(\mathcal{X}(x; \epsilon); \delta) = \mathcal{X}(x; \phi(\epsilon, \delta))$$

and is written as

$$\mathcal{X}(\mathcal{X}(x; \epsilon); \delta) = \mathcal{X}(x; \epsilon + \delta) \tag{8.10}$$

Also, $\mathcal{X}(x; 0) = x$ and $\mathcal{X}(\bar{x}; -\epsilon) = x$ are hold good.

8.4.1 Some Theorems, Lemmas, and Definitions

We shall give few important theorems for Lie group of transformations which are useful in finding symmetries, laws of transformations, etc. The following lemma is useful for proving theorems. When the group parameter of local one parameter group of transformation becomes canonical then the transformation law satisfies (8.10) and can also be written as $\mathcal{X}(\mathcal{X}(x; \epsilon); \phi(\epsilon^{-1}, \epsilon + \Delta\epsilon)) = \mathcal{X}(x; \epsilon + \Delta\epsilon)$ for $\delta = \Delta\epsilon$. This relation is proved by the following lemma

Lemma 8.1

$$\mathcal{X}(x; \epsilon + \Delta\epsilon) = \mathcal{X}(\mathcal{X}(x; \epsilon); \phi(\epsilon^{-1}, \epsilon + \Delta\epsilon)) \tag{8.11}$$

Proof

$\mathcal{X}(\mathcal{X}(x; \epsilon); \phi(\epsilon^{-1}, \epsilon + \Delta\epsilon) = \mathcal{X}(x; \phi(\epsilon, \phi(\epsilon^{-1}, \epsilon + \Delta\epsilon)))$ (follows by the property, $\mathcal{X}(\mathcal{X}(x, \epsilon); \delta) = \mathcal{X}(x; \phi(\epsilon, \delta))$)

$\qquad\qquad = \mathcal{X}(x; \phi(\phi(\epsilon, \epsilon^{-1}), \epsilon + \Delta\epsilon))$ (Associative property of group composition)

$\qquad\qquad = \mathcal{X}(x; \phi(e, \epsilon + \Delta\epsilon))$ (e is the identity element)

$\qquad\qquad = \mathcal{X}(x; \epsilon + \Delta\epsilon)$ (by the identity property) $\qquad\qquad\qquad\square$

Theorem 8.1 (First fundamental theorem of Lie) *The infinitesimal form of the local one parameter group of transformation* $G : \bar{x} = \mathcal{X}(x; \epsilon), \epsilon$ *is a canonical parameter, with the infinitesimal generator* $X = \xi(x)\frac{\partial}{\partial x}$ *can be constructed by solving the Lie equations*

$$\frac{d\bar{x}}{d\epsilon} = \xi(\bar{x}), \quad \text{with the initial conditions } \bar{x}|_{\epsilon=0} = x.$$

Proof A local one parameter group $\bar{x} = \mathcal{X}(x, \epsilon)$, where the parameter ϵ is canonical can be represented as follows

$$\mathcal{X}(x; \epsilon + \Delta\epsilon) = \mathcal{X}(\mathcal{X}(x; \epsilon); \Delta\epsilon) \tag{8.12}$$

Now expanding both sides of the relation (8.12) in $\Delta\epsilon$ about $\Delta\epsilon = 0$, we obtain

$$\mathcal{X}(x; \epsilon) + \frac{\partial\mathcal{X}(x; \epsilon)}{\partial\epsilon}\Delta\epsilon + O((\Delta\epsilon)^2) = \mathcal{X}(\mathcal{X}(x; \epsilon); \Delta\epsilon)$$

$$\bar{x} + \frac{\partial\bar{x}}{\partial\epsilon}\Delta\epsilon + O((\Delta\epsilon)^2) = \mathcal{X}(x; \epsilon) + \frac{\partial\mathcal{X}(\bar{x}; \Delta\epsilon)}{\partial\Delta\epsilon}\bigg|_{\Delta\epsilon=0}\Delta\epsilon + O((\Delta\epsilon)^2)$$

$$= \bar{x} + \frac{\partial\mathcal{X}(\bar{x}; \Delta\epsilon)}{\partial\Delta\epsilon}\bigg|_{\Delta\epsilon=0}\Delta\epsilon + O((\Delta\epsilon)^2)$$

$$\frac{\partial\bar{x}}{\partial\epsilon}\Delta\epsilon + O((\Delta\epsilon)^2) = \xi(\bar{x})\Delta\epsilon + O((\Delta\epsilon)^2)$$

$$\tag{8.13}$$

Equating both sides of the above relation, we have

$$\frac{\partial\bar{x}}{\partial\epsilon} = \xi(\bar{x}) \tag{8.14}$$

with $\bar{x} = x$ at $\epsilon = 0$. Since $\xi(x)$ are continuous and so being smooth functions, it follows from the existence and uniqueness theorem for a first-order differential equation that there exists a unique solution of (8.14). $\square$

Theorem 8.2 *For any arbitrary composition law, $\gamma = \phi(\epsilon, \delta)$, so that γ is thrice continuously differentiable, there exists a canonical parameter $\tilde{\tau}$, given by*

$$\tilde{\tau}(\epsilon) = \int\limits_0^\epsilon \Gamma(\epsilon')d\epsilon'$$

where $\Gamma(\epsilon) = \frac{\partial\phi(\epsilon,\delta)}{\partial\epsilon}\Big|_{(\epsilon,\delta)=(\epsilon,\epsilon^{-1})}$ and $\Gamma(0) = 1$.

We shall now give an example of transforming a parameter to canonical parameter.

Consider the one parameter nonuniform scaling group $\bar{x} = \epsilon x$, $\bar{y} = \epsilon^2 y, 0 < \epsilon < \infty$ for which identity parameter $e = 1$, inverse parameter is $\epsilon^{-1} = \frac{1}{\epsilon}$ and the law of composition of the parameters is $\phi(\epsilon, \delta) = \epsilon\delta$. Therefore, $\Gamma(\epsilon) = \frac{\partial\phi(\epsilon,\delta)}{\partial\epsilon}\Big|_{(\epsilon,\delta)=(\epsilon,\epsilon^{-1})} = \epsilon^{-1}$

Hence, the canonical parameter $\tilde{\tau}(\epsilon) = \int_0^\epsilon \Gamma(\epsilon')d\epsilon' = \int_1^\epsilon \frac{1}{\epsilon'}d\epsilon' = \log\epsilon$ or $\epsilon = e^\tau$. The scaling group becomes $\bar{x} = e^\tau x$, $\bar{y} = e^{2\tau}y$ and the group composition law becomes $\phi(\tau_1, \tau_2) = \tau_1 + \tau_2, -\infty < \tau < \infty$.

Theorem 8.3 *The Lie group of transformations G can also be constructed by constructing the exponential map*

$$\bar{x}^i = e^{\epsilon X}x^i, \quad i = 1 \cdots n$$

where

$$e^{\epsilon X} = 1 + \frac{\epsilon}{1!}X + \frac{\epsilon^2}{2!}X^2 + \cdots + \frac{\epsilon^k}{k!}X^k + \cdots$$

In other words, the one parameter Lie group of transformations G is equivalent to

$$\bar{x} = e^{\epsilon X}x = \left(\sum_{k=0}^\infty \frac{\epsilon^k}{k!}X^k\right)x \qquad X^k = XX^{(k-1)}, k = 1, 2, \ldots X^0 = 1$$

Proof The infinitesimal generators X of a group of transformations G is given by

$$X(x) = \sum_{i=1}^n \xi_i(x)\frac{\partial}{\partial x_i} \tag{8.15}$$

In transformed variables $\bar{x} = \mathcal{X}(x; \epsilon)$, the infinitesimal generator can be written as

$$X(\bar{x}) = \sum_{i=1}^{n} \xi_i(\bar{x}) \frac{\partial}{\partial \bar{x}_i} \qquad (8.16)$$

Expanding $\bar{x} = X(x; \epsilon)$ in ϵ about $\epsilon = 0$, we have

$$\bar{x} = x + \epsilon \left(\frac{\partial \mathcal{X}(x; \epsilon)}{\partial \epsilon} \bigg|_{\epsilon=0} \right) + \cdots = \sum_{m=0}^{\infty} \frac{\epsilon^m}{m!} \left(\frac{\partial^m \mathcal{X}(x; \epsilon)}{\partial \epsilon^m} \bigg|_{\epsilon=0} \right) = \sum_{m=0}^{\infty} \frac{\epsilon^m}{m!} \left(\frac{\partial^m \bar{x}}{\partial \epsilon^m} \bigg|_{\epsilon=0} \right)$$

$$(8.17)$$

Let $F(x)$ be any differentiable function then

$$\frac{\mathrm{d}}{\mathrm{d}\epsilon} F(\bar{x}) = \frac{\partial F(\bar{x})}{\partial \bar{x}} \frac{\partial \bar{x}}{\partial \epsilon} = \sum_{i=1}^{n} \xi_i(\bar{x}) \frac{\partial F(\bar{x})}{\partial \bar{x}_i} = X(\bar{x}) F(\bar{x})$$

Hence, when $F(\bar{x}) = \bar{x}$ then

$$\frac{\mathrm{d}\bar{x}}{\mathrm{d}\epsilon} = X(\bar{x})\bar{x}$$

$$\frac{\mathrm{d}^2\bar{x}}{\mathrm{d}\epsilon^2} = \frac{\mathrm{d}}{\mathrm{d}\epsilon}\left(\frac{\mathrm{d}\bar{x}}{\mathrm{d}\epsilon}\right) = X^2(\bar{x})\bar{x} \qquad (8.18)$$

Proceeding in this way, one will obtain

$$\frac{\mathrm{d}^m\bar{x}}{\mathrm{d}\epsilon^m} = X^m(\bar{x})\bar{x}, m = 1, 2, \ldots$$

$$\frac{\mathrm{d}^m\bar{x}}{\mathrm{d}\epsilon^m}\bigg|_{\epsilon=0} = X^m(x)x, m = 1, 2, \ldots$$

Hence from (8.18),

$$\bar{x} = \sum_{m=0}^{\infty} \frac{\epsilon^m}{m!}\left(\frac{\partial^m\bar{x}}{\partial\epsilon^m}\bigg|_{\epsilon=0}\right) = \sum_{m=0}^{\infty} \frac{\epsilon^m}{m!}X^m(x)x = e^{\epsilon X}x \qquad \square$$

Corollary 8.1 *For any differentiable function $F(x)$, $F(\bar{x}) = e^{\epsilon X}F(x)$,*

Proof We have

$$F(\bar{x}) = F(\mathcal{X}(x; \epsilon)) = F(x) + \epsilon\left(\frac{\partial F(\mathcal{X}(x; \epsilon))}{\partial \epsilon}\bigg|_{\epsilon=0}\right) + \cdots$$

$$= \sum_{m=0}^{\infty} \frac{\epsilon^m}{m!}\left(\frac{\partial^m F(\bar{x})}{\partial \epsilon^m}\bigg|_{\epsilon=0}\right) = \sum_{m=0}^{\infty} \frac{\epsilon^m}{m!}X^m F(x) = e^{\epsilon X}F(x)$$

Hence, $F(\bar{x}) = e^{\epsilon X}F(x)$.

Note that if the group of transformation is known then it is possible to construct the corresponding generator by finding the infinitesimals. Here, we shall give some examples in this regard.   □

Example 8.7 Determine the one parameter group of transformation for the following infinitesimal generator

$$X = xy \frac{\partial}{\partial x} + y^2 \frac{\partial}{\partial y}$$

using both Lie equation and exponential map.

Solution The Lie equations corresponding to the above infinitesimal generator are

$$\frac{d\bar{x}}{d\epsilon} = \bar{x}\bar{y}, \quad \frac{d\bar{y}}{d\epsilon} = \bar{y}^2$$

with initial condition $\bar{x} = x$, $\bar{y} = y$ at $\epsilon = 0$
upon integration,

$$\bar{x} = \frac{-c_2}{c_1 + \epsilon}, \quad \bar{y} = \frac{-1}{\epsilon + c_1}$$

Applying the initial conditions one obtain $x = \frac{-c_2}{c_1}$ and $y = \frac{-1}{c_1}$. Therefore, the corresponding group of transformation is $\bar{x} = \frac{x}{1-\epsilon y}$, $\bar{y} = \frac{y}{1-\epsilon y}$.

Let us obtain this group of transformation using the exponential map $\bar{x} = e^{\epsilon X} x$, $\bar{y} = e^{\epsilon X} y$.   Now $X(x) = xy, X^2(x) = X(X(x)) = X(xy) = 2xy^2$, $X^3(x) = X(X^2(x)) = 3!xy^3, \cdots, X^k(x) = X(X^{k-1}(x)) = k!xy^k$. In the same way $X(y) = y^2, X^2(y) = 2!y^3, X^3(y) = 3!y^4, \ldots, X^k(y) = k!y^{k+1}$

Substitution of the above expression in the exponential map gives

$$\bar{x} = \left(1 + \epsilon X + \frac{\epsilon^2 X^2}{2!} + \frac{\epsilon^3 X^3}{3!} + \cdots \right) x = x + \epsilon xy + \frac{\epsilon^2 2xy^2}{2!} + \frac{\epsilon^3 3!xy^3}{3!} + \cdots$$

$$= (1 + \epsilon y + \epsilon^2 y^2 + \epsilon^3 y^3 + \cdots) x$$

$$\therefore \bar{x} = \frac{x}{1 - \epsilon y}$$

Similarly

$$\bar{y} = \left(1 + \epsilon X + \frac{\epsilon^2 X^2}{2!} + \frac{\epsilon^3 X^3}{3!} + \cdots \right) y = y + \epsilon y^2 + \frac{\epsilon^2 2!y^3}{2!} + \frac{\epsilon^3 3!y^4}{3!} + \cdots$$

$$= (1 + \epsilon y + \epsilon^2 y^2 + \epsilon^3 y^3 + \cdots) y$$

$$\therefore \bar{y} = \frac{y}{1 - \epsilon y}$$

Example 8.8 Find the one parameter group of transformation for the following infinitesimal generator using Lie equation and exponential map

$$X = y \frac{\partial}{\partial x} - x \frac{\partial}{\partial y}$$

Solution

(i) The Lie equation can be written as

$$\frac{d\bar{x}}{d\epsilon} = \bar{y}, \quad \frac{d\bar{y}}{d\epsilon} = -\bar{x}$$

The above equation is solved by Euler's method, which gives the solution as

$$\bar{x} = C_1 \cos \epsilon + C_2 \sin \epsilon, \quad \bar{y} = C_2 \cos \epsilon - C_1 \sin \epsilon$$

using $\bar{x} = x$, $\bar{y} = y$ at $\epsilon = 0$, we have $C_1 = x$ and $C_2 = y$, therefore the rotation group of transformation is given by

$$\bar{x} = x \cos \epsilon + y \sin \epsilon, \quad \bar{y} = y \cos \epsilon - x \sin \epsilon$$

(ii) Use of exponential map
The exponential map are given as

$$\bar{x} = e^{\epsilon X}(x), \bar{y} = e^{\epsilon X}(y)$$

Now we have $X(x) = y$, $X^2(x) = X(X(x)) = X(y) = -x$, $X^3(x) = X(-x) = -y$, $X^4(x) = X(-y) = x$ and so on. Likewise, $X(y) = -x$, $X^2(y) = X(-x) = -y$, $X^3(y) = X(-y) = x$, $X^4(y) = y$ and so on. Substituting these expressions in the exponential map gives,

$$\bar{x} = \left(1 + \epsilon X + \frac{(\epsilon X)^2}{2!} + \cdots\right)x = x + \epsilon X(x) + \frac{(\epsilon)^2}{2!} X^2(x) + \frac{\epsilon^3}{3!} X^3(x) + \frac{\epsilon^4}{4!} X^4(x) + \cdots$$

$$= x + \epsilon y - \frac{(\epsilon)^2}{2!} x - \frac{\epsilon^3}{3!} y + \frac{\epsilon^4}{4!} x = \left(x - \frac{(\epsilon)^2}{2!} x + \frac{\epsilon^4}{4!} x + \cdots\right) + \left(\epsilon y - \frac{\epsilon^3}{3!} y + \cdots\right)$$

$$= x \cos \epsilon + y \sin \epsilon$$

and

$$\bar{y} = \left(1 + \epsilon X + \frac{(\epsilon X)^2}{2!} + \cdots\right)y = y + \epsilon X(y) + \frac{(\epsilon)^2}{2!} X^2(y) + \frac{\epsilon^3}{3!} X^3(y) + \frac{\epsilon^4}{4!} X^4(y) + \cdots$$

$$= y - \epsilon x - \frac{(\epsilon)^2}{2!} y + \frac{\epsilon^3}{3!} x + \frac{\epsilon^4}{4!} y = \left(y - \frac{(\epsilon)^2}{2!} y + \frac{\epsilon^4}{4!} y + \cdots\right) - \left(\epsilon x - \frac{\epsilon^3}{3!} x + \cdots\right)$$

$$= y \cos \epsilon - x \sin \epsilon$$

Hence, one obtain the rotation group of transformations $\bar{x} = x \cos \epsilon + y \sin \epsilon$ and $\bar{y} = y \cos \epsilon - x \sin \epsilon$

Example 8.9 Consider the one parameter group of transformation $\phi : \mathbb{R} \times \mathbb{R}^2$ defined by $\phi(t, p) = (xe^{2t}, ye^{3t}), p = (x, y) \in \mathbb{R}^2$, find the infinitesimals generator.

Solution The infinitesimal generator can be found out by evaluating the lie equations

$$\xi(x, y) = \frac{d\bar{x}}{dt}\bigg|_{t=0} = 2\bar{x}e^{2t}\big|_{t=0} \quad \text{with} \quad \bar{x} = x \text{ at } t = 0 \text{ and}$$

$$\eta(x, y) = \frac{d\bar{y}}{dt}\bigg|_{t=0} = 3\bar{y}e^{3t}\big|_{t=0} \text{ with } \bar{y} = y \text{ at } t = 0$$

Hence, $\xi(x, y) = 2x$ and $\eta(x, y) = 3y$ which gives the generator $X = 2x\frac{\partial}{\partial x} + 3y\frac{\partial}{\partial y}$

Some definitions: Any function $F(x) \in C^\infty$ is said to be invariant under the Lie group of transformations $\bar{x} = \mathcal{X}(x; \epsilon)$, ϵ is a continuous parameter if and only if under this transformation $F(\bar{x}) = F(x)$. We shall now give the definition of invariant points, invariant surface, invariant curves, and invariant solution.

Invariant point A point x is said to be an invariant point for a one parameter Lie group of transformations if and only if under a one parameter Lie group of transformation $\bar{x} = \mathcal{X}(x; \epsilon)$, the relation $\bar{x} = x$ holds.

Invariant surface A surface $F(x) = 0$ is said to be an invariant surface for a Lie group of transformation if and only if under a one parameter Lie group of transformation $\bar{x} = \mathcal{X}(x; \epsilon)$, $F(\bar{x}) = 0$ when $F(x) = 0$.

Invariant curve A curve $F(x, y)$ for a one parameter Lie group of transformations $\bar{x} = \mathcal{X}(x; \epsilon), \bar{y} = \mathcal{Y}(y; \epsilon)$ is said to be an invariant curve if and only if $F(\bar{x}, \bar{y}) = 0$ when $F(x, y) = 0$ under the action of this transformation.

Invariant solution A solution $y = \psi(x)$ of an nth-order ordinary differential equation $y^{(n)} = f(x, y, y', \ldots, y^{(n-1)})$ that is invariant under the infinitesimal generator $X = \xi(x, y)\frac{\partial}{\partial x} + \eta(x, y)\frac{\partial}{\partial y}$ is said to be an invariant solution if and only if $y = \psi(x)$ is an invariant curve of the infinitesimal generator $X = \xi(x, y)\frac{\partial}{\partial x} + \eta(x, y)\frac{\partial}{\partial y}$ that is, if $\xi(x, y)\psi'(x) - \eta(x, y) = 0$.

We shall now give some theorems related to the conditions under which a given function is invariant.

Theorem 8.4 *Any function $F(x) \in C^\infty$ is an invariant under a group of continuous transformations G with the infinitesimal generator X if and only if the following relation is satisfied by $F(x)$*

$$XF(x) = 0.$$

Proof Consider the infinitely differentiable function $F(x)$ and the group of transformation $G : \bar{x} = \mathcal{X}(x; \epsilon)$, which satisfies $F(\bar{x}) = F(x)$ then

$$F(\bar{x}) = e^{\epsilon X} F(x) = \sum_{m=0}^{\infty} \frac{\epsilon^m}{m!} X^m F(x)$$

$$= \left(1 + \epsilon X + \frac{(\epsilon X)^2}{2!} + \cdots\right) F(x)$$

$$F(x) = F(x) + \epsilon X F(x) + \frac{\epsilon^2}{2!} X^2 F(x) + \cdots$$

Equating both sides we have, $XF(x) = 0, X^2 F(x) = 0$, and so on.

Again, let $F(x)$ be any continuously differentiable function satisfying $XF(x) = 0$. Then obviously $X^2 F(x) = X(XF(x)) = 0$, similarly $X^n F(x) = 0, n = 3, 4, \ldots$. Then from the following relation

$$F(\bar{x}) = e^{\epsilon X} F(x) = \sum_{m=0}^{\infty} \frac{\epsilon^m}{m!} X^m F(x) = F(x) + \epsilon X F(x) + \frac{\epsilon^2}{2!} X^2 F(x) + \cdots$$

We have $F(\bar{x}) = F(x)$. $\qquad\qquad\square$

Theorem 8.5 *Any function* $F(x) \in C^{\infty}$ *for a Lie group of transformation* $\bar{x} = \mathcal{X}(x; \epsilon)$ *satisfy the relation*

$$XF(x) = 1$$

if and only if the following identity holds:

$$F(\bar{x}) = F(x) + \epsilon$$

Proof First, suppose that $F(x)$ satisfies $F(\bar{x}) = F(x) + \epsilon$, then

$$F(\bar{x}) = F(x) + \epsilon X F(x) + \frac{\epsilon^2}{2!} X^2 F(x) + \cdots$$

$$F(x) + \epsilon = F(x) + \epsilon X F(x) + \frac{\epsilon^2}{2!} X^2 F(x) + \cdots$$

equating both sides, we get $XF(x) = 1$.

For the converse, suppose that $XF(x) = 1$ holds, then obviously $X^n F(x) = 0, n = 2, 3, \ldots$ then

$$\bar{x} = F(x) + \epsilon X F(x) + \frac{\epsilon^2}{2!} X^2 F(x) + \cdots$$

$$F(\bar{x}) = F(x) + \epsilon \qquad\qquad\square$$

Theorem 8.6 *A surface $F(x) = 0$ is an invariant surface under a Lie group of transformation $\bar{x} = \mathcal{X}(x; \epsilon)$ if and only if $XF(x) = 0$ when $F(x) = 0$.*

Theorem 8.7 *A curve $F(x, y) = 0$ where $F(x, y) = y - f(x)$, is an invariant curve under a Lie group of transformation $\bar{x} = \mathcal{X}(x; \epsilon)$ and $\bar{y} = \mathcal{Y}(y; \epsilon)$ with the infinitesimals ξ and η if and only if*

$$XF(x, y) = \xi(x, y)\frac{\partial F(x, y)}{\partial x} + \eta(x, y)\frac{\partial F(x, y)}{\partial y}$$

$$= -\xi(x, y)\frac{\partial f(x)}{\partial x} + \eta(x, y) = 0$$

when $F(x, y) = y - f(x) = 0$.

Theorem 8.8 *A point x is an invariant point under a Lie group of transformation $\bar{x} = \mathcal{X}(x; \epsilon)$ if and only if the corresponding infinitesimal $\xi(x) = 0$.*

Proof Suppose that the point x is invariant then $\bar{x} = x$. Now expanding the right-hand side of the transformation $\bar{x} = \mathcal{X}(x; \epsilon)$ in $\Delta\epsilon$ about $\epsilon = 0$ we have

$$\bar{x} = x + \Delta\epsilon\left(\left.\frac{\partial \mathcal{X}(x; \epsilon)}{\partial \epsilon}\right|_{\epsilon=0}\right) + \cdots$$

$$x = x + \left(\left.\frac{\partial \mathcal{X}(x; \epsilon)}{\partial \epsilon}\right|_{\epsilon=0}\right)\Delta\epsilon + \cdots$$

(8.19)

which gives $\left.\frac{\partial \mathcal{X}(x;\epsilon)}{\partial \epsilon}\right|_{\epsilon=0} = 0$, hence $\xi(x) = 0$

Conversely, suppose that $\xi(x) = 0$ then $\left.\frac{\partial \mathcal{X}(x;\epsilon)}{\partial \epsilon}\right|_{\epsilon=0} = 0$, then from (8.19) we have $\bar{x} = x$. $\qquad\qquad\qquad\qquad\qquad\qquad\qquad\qquad\qquad\qquad\qquad\qquad\qquad\square$

8.5 Lie Group Theoretic Method for First-Order ODEs

So far we get acquainted with the basics of Lie group of transformations and some important theorems. We now concentrate on solving ordinary differential equations using Lie invariance principle. Consider a first-order ordinary differential equation of the form

$$\frac{dy}{dx} = f(x, y)$$

(8.20)

The generator X of one parameter Lie group of transformation is given by

$$X = \xi(x,y)\frac{\partial}{\partial x} + \eta(x,y)\frac{\partial}{\partial y}$$

When invariance criterion (8.8) is applied to the first order Eq. (8.20), the corresponding determining equation contains only the infinitesimals (ξ, η) and their partial derivatives with respect to x and y, dependent variable y, and independent variable x and does not contain y' or any higher order derivatives of y, and hence cannot be split into an overdetermined system of equations. In this case the symmetry group contains an arbitrary function. Thus, admits an infinite-dimensional continuous group. The solutions (ξ, η) of the determining equation are often found out by inspection. When it is possible to express ξ, η by $\frac{\eta(x,y)}{\xi(x,y)} = f(x,y)$, the process of reducing (8.20) to quadrature is identical to solving (8.20). In this case the group is trivial (as its characteristic form represents the equation itself), and is always admitted by (8.20), but cannot be neglected as it also gives a symmetry solution. For this case the infinitesimals (ξ, η) is of no use as they do not provide any information of the behavior of solutions. However, for $\frac{\eta(x,y)}{\xi(x,y)} \neq f(x,y)$, the groups are nontrivial. In this case, Lie groups are useful for determining the solution of Eq. (8.20). A first-order differential equation can be integrated by two means, viz. (i) determination of an integrating factor, and (ii) use of canonical coordinates. The following theorem is relevant for determining integrating factor of first-order differential equation.

Theorem 8.9 *A first-order ordinary differential equation written in the differential form* $M(x,y)\mathrm{d}x + N(x,y)\mathrm{d}y = 0$ *having a one parameter group* G *with an infinitesimal generator* $X = \xi(x,y)\frac{\partial}{\partial x} + \eta(x,y)\frac{\partial}{\partial y}$ *has an integrating factor* $\mu = \frac{1}{\xi M + \eta N}$ *provided that* $\xi M + \eta N \neq 0$.

For proof see the book Blumen and Kumei [19].

We shall define the canonical variables and integration technique which are useful for finding general solution of first-order differential equation.

If a change of variables occur from (x, y) to (t, u) such that the Lie group of transformation $\bar{x} = f(x, y, \epsilon)$, $\bar{y} = g(x, y, \epsilon)$ reduces to the group of translations $\bar{t} = t + \epsilon$, $\bar{u} = u$ with the generator $X = \frac{\partial}{\partial t}$, then the changed variables (t, u) are known as canonical variables. Now consider a first-order differential equation $y' = f(x, y)$ admitting a one parameter Lie group with group operator

$$X = \xi\frac{\partial}{\partial x} + \eta\frac{\partial}{\partial y} \tag{8.21}$$

Then the canonical coordinates $r = R[x, y]$ and $\theta = \Theta[x, y]$ are obtained by solving the equations $XR = 0$ and $X\Theta = 1$. The original variables are then substituted by these canonical variables in the original ordinary differential equation. The new equation obtained in this way is an invariant under the translation group

$\bar{r} = r, \bar{\theta} = \theta + s$ and hence does not depend on θ. The ordinary differential equation in terms of the canonical coordinates is then given by

$$\frac{d\theta}{dr} = \frac{\Theta_x + \Theta_y \frac{dy}{dx}}{R_x + R_y \frac{dy}{dx}} = H(r) \tag{8.22}$$

as a result, the general solution of a first-order ordinary differential equation is given by

$$\theta = \int\limits^{R(x,y)} H(\tilde{r})\,d\tilde{r} + C, \; C = \text{constant}. \tag{8.23}$$

For a differential equation of order more than one admitting one parameter group, the canonical variable method reduces the order of the differential equation by one.

Now, we shall give some examples which illustrate how the canonical variables are obtained when the symmetry group of transformations are known.

Example 8.10 Determine, the canonical variables corresponding to the following one parameter Lie group of transformations (i) Rotation group: $\tilde{x} = x \cos \epsilon + y \sin \epsilon, \tilde{y} = y \cos \epsilon - x \sin \epsilon$ and   (ii) Lorentz group: $\tilde{x} = x \cosh \epsilon + y \sinh \epsilon, \; \tilde{y} = y \cosh \epsilon + x \sinh \epsilon$

Solution

(i) Rotation group

$$\tilde{x} = x \cos \epsilon + y \sin \epsilon,$$
$$\tilde{y} = y \cos \epsilon - x \sin \epsilon$$

The infinitesimal generator is $X = y\frac{\partial}{\partial x} - x\frac{\partial}{\partial y}$. Let (t,u) be the canonical coordinates which is obtained by solving the equations $Xt = 0$ and $Xu = 1$. The characteristic differential equation corresponding to $Xt = y\frac{\partial t}{\partial x} - x\frac{\partial t}{\partial y} = 0$ is given by

$$\frac{dy}{dx} = \frac{-x}{y}$$

which gives $t = x^2 + y^2$. Similarly $Xu = y\frac{\partial u}{\partial x} - x\frac{\partial u}{\partial y} = 1$ corresponds to

$$\frac{dx}{y} = \frac{dy}{-x} = \frac{du}{1}$$

which gives $u = \tan^{-1}\frac{y}{x}$. Hence the canonical variables are $(t,u) = (x^2 + y^2, \tan^{-1}\frac{y}{x})$.

(i) Lorentz group

This group gives Lorentz transformation as

$$\tilde{x} = x \cosh \epsilon + y \sinh \epsilon,$$

$$\tilde{y} = y \cosh \epsilon + x \sinh \epsilon$$

The infinitesimal generator associated with Lorentz transformation is given by $X = y\frac{\partial}{\partial x} + x\frac{\partial}{\partial y}$. The canonical variables (t, u) satisfy the equations $Xt = 0, Xu = 1$. The equation corresponding to $Xt = y\frac{\partial t}{\partial x} + x\frac{\partial t}{\partial y} = 0$ is

$$\frac{dx}{y} = \frac{dy}{x} = \frac{dt}{0}$$

This gives $t = \text{constant} = y^2 - x^2$. Again, the characteristic equation corresponding to $Xu = y\frac{\partial u}{\partial x} + x\frac{\partial u}{\partial y} = 1$ is given by

$$\frac{dx}{y} = \frac{dy}{x} = \frac{du}{1}$$

Upon solving the above equation, one would obtain $u = \log(x+y)$. Hence the canonical variables are $(t, u) = (y^2 - x^2, \log(x+y))$.

Example 8.11 Find out the group of transformation for the well-known first-order nonlinear ordinary differential equation, Riccati equation

$$\mathcal{F} = \frac{dy}{dx} + y^2 - \frac{2}{x^2} = 0 \tag{8.24}$$

Also, obtain the solution of the equation.

Solution Riccati equation is a nonlinear first-order differential equation of the general form $y' = p(x) + q(x)y + r(x)y^2, p(x), r(x) \neq 0$ which has nonlinearity in the quadratic term of the dependent variable. Ricatti equation is named after Riccati, J. Francesco (1676-1754) who investigated the equation $y' = ay^2 + bx^m$ in 1974. The general equation was studied by J.d'Alembert in 1763 and proved by French mathematician Liouville in 1841 as one of the simplest differential equation that cannot be integrated by quadratures. But by suitable change of the variables this equation can be reduced to a linear second-order differential equation. The given equation is a particular form of the Riccati equation where $p(x) = \frac{-2}{x^2}, q(x) = 0$ and $r(x) = 1$. Here, we will look for the symmetry transformation of the variables (x, y)

of this equation using Lie symmetry approach, which will reduce it to a linear equation, and hence solvable by quadratures.

The infinitesimal generator for the given equation is obtained as

$$X = \xi \frac{\partial}{\partial x} + \eta \frac{\partial}{\partial y}$$

The determining equation is given by

$$X^{[1]} \mathcal{F} = 0 \text{ whenever } \mathcal{F} = 0 \tag{8.25}$$

where $X^{[1]}$ is the first prolongation of X, given by

$$X^{[1]} = \xi \frac{\partial}{\partial x} + \eta \frac{\partial}{\partial y} + \eta^{(1)} \frac{\partial}{\partial y'}$$

where

$$\eta^{(1)} = D(\eta(x,y)) - y'D(\xi(x,y)) = \eta_x + y'\eta_y - y'(\xi_x + y'\xi_y)$$
$$= \eta_x + (\eta_y - \xi_x)y' - \xi_y y'^2$$

where D is the total differential operator given by $D = \frac{\partial}{\partial x} + y'\frac{\partial}{\partial y} + y''\frac{\partial}{\partial y'} + \cdots$

Hence from (8.25), $\xi\left(\frac{4}{x^3}\right) + \eta(2y) + \eta_x + (\eta_y - \xi_x)y' - \xi_y y'^2 = 0$

Replacing y' by $\frac{2}{x^2} - y^2$ in the above equation gives

$$\left(\frac{4}{x^3}\right)\xi + 2\eta y + \eta_x + (\eta_y - \xi_x)\left(\frac{2}{x^2} - y^2\right) - \xi_y\left(\frac{2}{x^2} - y^2\right)^2 = 0 \tag{8.26}$$

This is the determining equation of the infinitesimals of (8.25) but one can notice that (8.26) does not contain y' so (8.26) cannot be split into an overdetermined system of equations. However, one can obtain the infinitesimal coefficients as

$$\xi(x,y) = -cx, \text{ and } \eta(x,y) = cy$$

Thus, the Riccatti equation admits a one parameter dilation group spanned by the generator

$$X = -x\frac{\partial}{\partial x} + y\frac{\partial}{\partial y} \tag{8.27}$$

The group of transformation is obtained by solving the Lie equations as $\bar{x} = xe^{-\epsilon}$ and $\bar{y} = ye^{\epsilon}$. Thus from the above example we can conclude that the method of determining equation is not efficient for a first-order ordinary differential equation or a system of first-order ordinary differential equation.

Let us now solve the Riccati equation by the following two procedures:

(i) **Integrating factor method**

The Riccati equation admits the one parameter group with infinitesimal generator

$$X = -x\frac{\partial}{\partial x} + y\frac{\partial}{\partial y}$$

The integrating factor is obtained as

$$\mu = \frac{1}{-(y^2 - 2/x^2)\cdot x + y} = \frac{x}{xy + 2 - x^2y^2}$$

Multiplying both sides of the Riccati equation by this integrating factor, we get

$$\frac{xdy + (xy^2 - 2/x)dx}{xy + 2 - x^2y^2} = \frac{xdy + ydx}{xy + 2 - x^2y^2} - \frac{dx}{x} = \frac{d(xy)}{2 + xy - (xy)^2} - \frac{dx}{x} = 0 \qquad (8.28)$$

upon integration, we obtain

$$x^3\frac{xy - 2}{xy + 1} = C \text{ or } y = \frac{2x^3 + C}{x(x^3 - C)}, C = \text{ const.}$$

(ii) **Canonical coordinates method**

The canonical variables (r, θ) are obtained by solving the following equations

$$X(r) = -x\frac{\partial r}{\partial x} + y\frac{\partial r}{\partial y} = 0 \qquad (8.29)$$

which gives, $r(x, y) = xy$ and

$$X(\theta) = -x\frac{\partial \theta}{\partial x} + y\frac{\partial \theta}{\partial y} = 1 \qquad (8.30)$$

This gives the canonical variable $\theta(x, y) = -\log x$.

The Riccati equation in terms of canonical variables is then given by

$$\frac{d\theta}{dr} = \frac{\theta_x + \theta_y \cdot y'}{r_x + r_y y'} = \frac{-1/x}{y + x(y^2 - 2/x^2)} = \frac{-1}{xy + x^2y^2 - 2} = \frac{-1}{r + r^2 - 2}$$

On solving $d\theta = \frac{-dr}{r+r^2-2}$

One would obtain, $\theta = \frac{1}{3}\log\left|\frac{r+2}{r-1}\right|$ or $\log\left|\frac{r+2}{r-1}\right| - 3\theta = $ const.

In terms of original variables (x, y), one would obtain the solution as

$$y = \frac{2x^3 + C}{x(x^3 - C)}, \quad C = \text{Constant}$$

This is a general solution of the Riccatti equation under group theoretic approach.

Example 8.12 Find out the group of point transformation for the first-order ordinary differential equation

$$\frac{dy}{dx} = \frac{y}{x} + \frac{y^3}{x^4} \tag{8.31}$$

Solution The infinitesimal generator in this case is given by

$$X = \xi\frac{\partial}{\partial x} + \eta\frac{\partial}{\partial y}$$

The determining equation is obtained as follows:

$$X^{[1]}\mathcal{F} = 0 \text{ whenever } \mathcal{F} = 0. \tag{8.32}$$

where $X^{[1]}$ is the first prolongation of X, given by

$$X^{[1]} = \xi\frac{\partial}{\partial x} + \eta\frac{\partial}{\partial y} + \eta^{(1)}\frac{\partial}{\partial y'}$$

and

$$\begin{aligned}
\eta^{(1)} &= D(\eta(x,y)) - y'D(\xi(x,y)) \\
&= \eta_x + y'\eta_y - y'(\xi_x + y'\xi_y) \\
&= \eta_x + (\eta_y - \xi_x)y' - \xi_y y'^2
\end{aligned}$$

Hence from (8.32) and replacing y' by $y/x + y^3/x^4$, we have the determining equation as

$$\frac{-\eta}{x} + \frac{3y^2}{x^4}\eta - \frac{y}{x^2}\xi - \frac{4y^3\xi}{x^5} - \frac{y}{x}\eta_y - \frac{y^3}{x^4}\eta_y + \frac{y^2}{x^2}\xi_y + \frac{2y^4}{x^5}\xi_y - \frac{y^6}{x^8}\xi_y + \eta_x - \frac{y}{x}\xi_x - \frac{y^3}{x^4}\xi_x = 0$$

The determining equation is satisfied by $\xi(x,y) = x^2$ and $\eta(x,y) = xy$. The corresponding infinitesimal generator is given by

$$X = x^2 \frac{\partial}{\partial x} + xy \frac{\partial}{\partial y} \qquad (8.33)$$

The solution of the equation is given by the following two procedures

(i) **Integrating factor method of solution**

The given differential equation can be written in the differential form as

$$x^4 \, dy = (yx^2 + y^3) \, dx \text{ or, } (yx^3 + y^3) \, dx - x^4 \, dy = 0 \qquad (8.34)$$

Here $M(x,y) = yx^3 + y^3$ and $N(x,y) = -x^4$. Hence the integrating factor is

$$\mu = \frac{1}{M\xi + N\eta} = \frac{1}{x^2(yx^3 + y^3) + xy(-x^4)} = \frac{1}{x^2 y^3}$$

Multiplying, both sides of Eq. (8.34) by the integrating factor μ, gives

$$\left(\frac{x}{y^2} + \frac{1}{x^2}\right) dx - \frac{x^2}{y^3} \, dy = 0 \Rightarrow \frac{1}{x^2} \, dx + \frac{x}{y} \, d\left(\frac{x}{y}\right) = 0$$

which on integration gives the solution as $y = \pm x \sqrt{\frac{x}{cx+2}}$, where c is an arbitrary integration constant.

(ii) **Canonical variable method of solution**

The canonical variables r and s are obtained by evaluating $Xr = 0$ and $Xs = 1$ corresponding to the infinitesimal generator (8.33) and is given as

$$r = \frac{y}{x}, \, s = \frac{-1}{x}$$

The equation in terms of canonical variables is given by

$$\frac{ds}{dr} = \frac{s_x + s_y \cdot y'}{r_x + r_y y'} = \frac{x^3}{y^3} = \frac{1}{r^3}$$

on integration, we have $s = -\frac{1}{2r^2} + \text{constant}$, in terms of original variables the solution is $y^2 = \frac{x^3}{cx+2}$. Hence, $y = \pm x \sqrt{\frac{x}{cx+2}}$, where c is an arbitrary integration constant.

8.6 Multiparameters Groups

So far, we have dealt with various one parameter groups. But when one deals with higher order ordinary differential equations or partial differential equations, one often obtains multiparameter groups. The multiparameter group is generally the combination of various one parameter groups. Each parameter of the multiparameter group corresponds to a one parameter group. Hence for each one parameter group one will obtain an infinitesimal generator. The infinitesimal generator of a multiparameter group is the linear combination of the infinitesimal generator of each of the different one parameter continuous groups. The infinitesimal generator of each of the one parameter Lie group of transformations of a multiparameter group forms a vector space, known as Lie algebra with respect to a commutator called Lie bracket. Lie algebra is useful in determining additional symmetries of a system from known symmetries. We will now discuss briefly the Lie algebra and its properties in the following section.

8.6.1 Lie Algebra

The infinitesimal generators $X = \zeta^i \frac{\partial}{\partial x^i}$ of multiparameter Lie group of transformations constitute a linear closed space, known as Lie algebra and is denoted by the operator $\mathcal{L}$. The commutator Lie bracket of infinitesimal generators is defined as

$$[X^p, X^q] = X^p(X^q) - X^q(X^p) \tag{8.35}$$

The generators satisfy the following properties of Lie algebra with respect to the commutator (8.35).

(i) The commutator of any two infinitesimal generators of an r-parameter Lie group is also an infinitesimal generator of $\mathcal{L}$ (Lie algebra) and can be expressed as

$$[X^p, X^q] = c_{pq}^k X^k$$

where c_{pq}^k are structure constants which satisfy the relations $c_{pq}^k = -c_{pq}^k$.

(ii) The group operators satisfy the associative rule with respect to addition

$$X^a + (X^b + X^c) = (X^a + X^b) + X^c$$

(iii) The commutator of the group operators satisfies the bilinear property:

$$[\alpha X^p + \beta X^q, X^r] = \alpha[X^p, X^r] + \beta[X^q, X^r]$$

where α, β are real constants.

(iv) The commutator of the group operators are skew-symmetric: $[X^p, X^q] = -[X^q, X^p]$

(v) The commutator of the group operators satisfies the Jacobi identity:

$$[X^p, [X^q, X^r]] + [X^q, [X^r, X^p]] + [X^r, [X^p, X^q]] = 0$$

The property of skew-symmetry and Jacobi identity imposes further restriction on structure constants that is

$$c_{pq}^k = -c_{qp}^k$$
$$c_{pn}^m c_{qr}^n + c_{rn}^m c_{pq}^n + c_{qn}^m c_{rp}^n = 0$$

Proof For the first equality, we have from (iv) $[X^p, X^q] = -[X^q, X^p]$, again we have from (i), $[X^p, X^q] = c_{pq}^k X^k$ from where it follows that $c_{pq}^k X^k = -c_{qp}^k X^k$ and hence $c_{pq}^k = -c_{qp}^k$.

For the second equality we have from (i) $[X^p, X^q] = c_{pq}^k X^k$,

Now $[X^p, [X^q, X^r]] = [X^p, c_{qr}^n X^n] = c_{qr}^n [X^p, X^n] = c_{qr}^n c_{pn}^m X^m$,

Similarly $[X^q, [X^r, X^p]] = [X^q, c_{rp}^n X^n] = c_{rp}^n [X^q, X^n] = c_{rp}^n c_{qn}^m X^m$

and $[X^r, [X^p, X^q]] = [X^r, c_{pq}^n X^n] = c_{pq}^n [X^r, X^n] = c_{pq}^n c_{rn}^m X^m$. We have from Jacobi identity (v),

$$[X^p, [X^q, X^r]] + [X^q, [X^r, X^p]] + [X^r, [X^p, X^q]] = 0$$

Which gives $c_{qr}^n c_{pn}^m X^m + c_{rp}^n c_{qn}^m X^m + c_{pq}^n c_{rn}^m X^m = 0$, i.e., $c_{qr}^n c_{pn}^m + c_{rp}^n c_{qn}^m + c_{pq}^n c_{rn}^m = 0$. □

Result If $f, g \in C^\infty$, then $[fX, gY] = fg[X, Y] + f(Xg)Y - g(Yf)X$.

Proof Let $h \in C^\infty$, then we have

$$
\begin{aligned}
[fX, gY]h &= fX((gY)h) - gY((fX)h) \\
&= fX(gYh) - gY(fXh) \\
&= f(Xg)(Yh) + fgX(Yh) - g(Yf)(Xh) - gfY(Xh) \\
&= f(Xg)Yh + fg(X(Yh) - gfY(Xh) - g(Yf)(Xh) \\
&= fg(XY - YX)h + f(Xg)Yh - g(Yf)Xh \\
&= (fg[X, Y] + f(Xg)Y - g(Yf)X)h \\
[fX, gY] &= fg[X, Y] + f(Xg)Y - g(Yf)X
\end{aligned}
$$

  □

Theorem 8.10 *Lie algebra $\mathcal{L}$ is abelian if $[X, Y] = 0 \; \forall X, Y \in \mathcal{L}$.*

Proof The Lie bracket for X and Y is given by $[X, Y] = XY - YX$. Suppose that the operators commute, i.e., $XY = YX$ then $[X, Y] = 0$. For the converse, suppose that $[X, Y] = 0$ then $XY - YX = 0$ which gives $XY = YX$. $\square$

Theorem 8.11 *If r linearly independent operators $X_1, X_2, \ldots, X_r$ of the form $\xi_i \frac{\partial}{\partial x_i} + \eta_j \frac{\partial}{\partial y_j}$ spans an r-dimensional Lie algebra $\mathcal{L}_r$, then the n times extended generator span a Lie algebra with the same structure constants as the algebra $\mathcal{L}_r$.*

Theorem 8.12 *The set of all solutions of any determining equation forms a Lie algebra.*

Proof The equation for determining the infinitesimals ξ_i, η_j is given by $XF|_{F=0} = 0$. This equation splits into an overdetermined system of linear homogeneous partial differential equations for the unknowns ξ_i and η_j. These infinitesimals form the infinitesimal generators X_i's. These generators then form a vector space. Now in order to prove the given theorem we have to prove that the vector space formed by the infinitesimal generator obtained by solving the determining equation is closed with respect to the Lie bracket $[X_i, X_j]$. That is if X_i, X_j satisfy the determining equation then their commutator also satisfies the determining equation. Now if F and XF are differential functions then the determining equation for the infinitesimals can also be written as

$$XF = \phi F \tag{8.36}$$

where ϕ is a differential function of finite order that is locally analytic function of a finite number of variables $x, y, y', y'', \ldots$ and belongs to the universal vector space $\mathcal{A}$ of all differential functions of finite order. Let X_1, X_2 are the generators constructed by the infinitesimal obtained by solving (8.36). Therefore,

$$X_1 F = \phi F, \; X_2 F = \varphi F$$

Now, let $X = [X_1, X_2] = X_1 X_2 - X_2 X_1$ then we calculate XF as below:

$$
\begin{aligned}
XF &= (X_1 X_2 - X_2 X_1)F = X_1(X_2 F) - X_2(X_1 F) = X_1(\varphi F) - X_2(\phi F) \\
&= X_1(\varphi)F + \varphi X_1 F - X_2(\phi)F - \phi X_2 F \\
&= X_1(\varphi)F - X_2(\phi)F + \varphi X_1 F - \phi X_2 F = (X_1(\varphi) - X_2(\phi) + \varphi X_1 - \phi X_2)F = \psi F
\end{aligned}
$$

$\square$

The coefficient $\psi = X_1(\varphi) - X_2(\phi) + \varphi X_1 - \phi X_2$ is a differential function of finite order since $\phi, \varphi \in \mathcal{A}$. This proves the theorem.

Theorem 8.13 *A second-order differential equation can admit at most an eight-dimensional Lie algebra. The maximal dimension 8 is attained when the differential equation is either linear or can be linearized by a change of variables.*

Proof Consider the second-order differential equation $y'' = f(x, y, y')$ where f is an arbitrary differential function of the variables x, y, y'. The generator which keeps the equation invariant is $X = \xi(x, y)\frac{\partial}{\partial x} + \eta(x, y)\frac{\partial}{\partial y}$. Now applying the infinitesimal criterion $X^{[2]}(y'' - f(x, y, y' = 0))\big|_{y''-f(x,y,y'=0}= 0$, one will obtain

$$\eta^2 - \eta^1 \frac{\partial f}{\partial y'} - \xi \frac{\partial f}{\partial x} - \eta \frac{\partial f}{\partial y}\bigg|_{y''-f(x,y,y')=0} = 0$$

Substituting the values of η^2, η^1, the above equation become

$$\eta_{xx} + (2\eta_{xy} - \xi_{xx})y' + (\eta_{yy} - 2\xi_{xy})y'^2 - \xi_{yy}y'^3 -_x -\eta f_y + (\eta_y - 2\xi_x$$
$$- 3y'\xi_y)f - (\eta_x + (\eta_y - \xi_x)y' - \xi_y y'^2)f_{y'} = 0 \qquad \square$$

Since f is a differential function, it is expandable in a Taylor series with respect to all its variables, therefore, the function f and its partial derivatives $f_x, f_y, f_{y'}$ can be replaced in the above equation by expanding in a power series with respect to y'. The coefficients are dependent upon x and y. Since y' is an independent variable therefore the coefficient of each power of y' must be zero this means that the series must be identically vanishes which yields

$$\eta_{xx} = \phi_1, 2\eta_{xy} - \xi_{xx} = \phi_2, \eta_{yy} - 2\xi_{xy} = \phi_3, -\xi_{yy} = \phi_4 \qquad (8.37)$$

where ϕ_1, ϕ_2, ϕ_3, ϕ_4 are linear homogeneous function of $\xi, \eta, \xi_x, \xi_y, \eta_x, \eta_y$ whose coefficients are known functions of x and y. Now using the Eq. (8.37) and setting the values of $\xi, \eta, \xi_x, \xi_y, \eta_x, \eta_y, \xi_{xx}, \xi_{xy}$ at any point (x_0, y_0) one can easily calculate all the second and higher derivatives of ξ, η at (x_0, y_0). In this way, one can construct a power series representation of ξ, η near (x_0, y_0). Thus, the infinitesimals are dependent upon eight arbitrary constants $\xi, \eta, \xi_x, \xi_y, \eta_x, \eta_y, \xi_{xx}, \xi_{xy}$ at any point (x_0, y_0). Now if the coefficients of the higher order derivatives are also included then it is obvious that the number of arbitrary constants would get reduced. Hence any second-order differential equation can admit at most eight-dimensional group and hence an eight dimensional Lie algebra.

We shall later see in this chapter that the second-order linear differential equation $y'' = 0$ admits an eight-dimensional group.

Lie algebra has a very important relation with the Lie group as all the information about Lie group can be obtained from the corresponding Lie algebra. The structure of the Lie algebra $\mathcal{L}$ of the group G can be completely determined by its structure constants. Lie algebra can be conveniently summarized by setting up a commutator table of the generators whose entry at the position (i, j) is $[X^i, X^j]$ and due to the anti symmetric property of the commutator, the entry at the (i, i)th position that is at the main diagonal of the table is zero. If all the entries of the table are zero then the corresponding Lie algebra is abelian. Construction of Lie algebra is useful in determining whether a given differential equation is completely solvable by quadratures or not. Lie algebra is also useful in determining the order to which a differential equation can be reduced or simplified. Lie algebra is useful in determining additional symmetries that an equation admits.

Example 8.13 Prove that

$$[X, fY] = f[X, Y] + (Xf)Y, \forall f \in C^\infty, \forall X, Y \in \mathcal{L}$$

Solution We know that for any $f, g \in C^\infty$, Lie bracket of any two infinitesimal generators $X, Y \in \mathcal{L}$ satisfies the relation $[fX, gY] = fg[X, Y] + f(Xg)Y - g(Yf)X$.
Hence

$$[X, fY] = f[X, Y] + (Xf)Y - f(Y(1))X = f[X, Y] + (Xf)Y$$

8.6.2 Subalgebra and Ideal

A subspace $\mathcal{K} \subset \mathcal{L}$ is said to be subalgebra of the Lie algebra $\mathcal{L}$ if the commutator of any two infinitesimal generators $X, Y \in \mathcal{K}$ also belongs to $\mathcal{K}$. The subalgebra $\mathcal{K}$ is called an ideal of $\mathcal{L}$ if for $X \in \mathcal{K}$ and $Y \in \mathcal{L}$, the commutator $[X, Y] \in \mathcal{K}$.

Consider the commutator Table 8.1 of the Lie algebra $\mathcal{L}_6$ formed by a six parameter group. The Lie algebra $\mathcal{L}_6$ contains a number of subalgebras for instance the operators X_1, X_2, X_3 is a subalgebra of $\mathcal{L}_6$. Similarly the operators X_2, X_3, X_5, X_6 forms a subalgebra of $\mathcal{L}_6$ (shown in the Table 8.2). One can obtain several other

Table 8.1 Commutator table

	X_1	X_2	X_3	X_4	X_5	X_6
X_1	0	X_1	0	0	X_4	0
X_2	$-X_1$	0	$-X_3$	0	X_5	0
X_3	0	X_3	0	$-X_1$	X_6-X_2	$-X_3$
X_4	0	0	X_1	0	0	X_4
X_5	$-X_4$	$-X_5$	$-(X_6-X_2)$	0	0	X_5
X_6	0	0	X_3	$-X_4$	$-X_5$	0

Table 8.2 Commutator table

	X_2	X_3	X_5	X_6
X_2	0	$-X_3$	X_5	0
X_3	X_3	0	$X_6 - X_2$	$-X_3$
X_4	$-X_5$	$-(X_6 - X_2)$	0	X_5
X_5	0	X_3	$-X_5$	0

Table 8.3 Commutator table

	X_1	X_2	X_3
X_1	0	X_1	0
X_2	$-X_1$	0	$-X_3$
X_3	0	X_3	0

subalgebras of $\mathcal{L}_6$ from commutator Table 8.1. But the algebra shown in Table 8.2 is not an ideal subalgebra of $\mathcal{L}_6$ as $X_1 \in \mathcal{L}_6$ and X_2 belongs to the subalgebra formed by X_2, X_3, X_5, X_6 say $\mathcal{L}_s$ but $[X_1, X_2] = X_1 \notin \mathcal{L}_s$. Again the Lie algebra formed by X_1, X_2, X_3 shown in commutator Table 8.3 is also not an ideal subalgebra of $\mathcal{L}_6$ but if we consider the algebra formed by X_1, X_2, X_3 then we can see that X_1, X_3 forms an ideal subalgebra of the algebra formed by X_1, X_2, X_3 as $X_2 \in \mathcal{L}(X_1, X_2, X_3)$ and $X_1 \in \mathcal{L}(X_1, X_3)$ and $[X_1, X_2] = X_1 \in \mathcal{L}(X_1, X_3)$ shown in Table 8.3.

8.6.3 Solvable Lie Algebra

A Lie algebra $\mathcal{L}_r$ corresponding to an r-parameters Lie group of transformations is called a solvable Lie algebra if $\exists$ a sequence of sub algebras such that

$$\mathcal{L}_0 \subset \mathcal{L}_1 \subset \mathcal{L}_2 \subset \cdot \mathcal{L}_{r-1} \subset \mathcal{L}_r$$

where each term of the sequence is an ideal Lie subalgebra of its preceding term. $\mathcal{L}_0$ is the null ideal with no operators. We will later see in this chapter that the concept of solvable Lie algebra is vital in reducing the order of a higher order ordinary differential equation and in determining its solvability under symmetry group of transformations.

Theorem 8.14 *Any two-dimensional Lie algebra is a solvable Lie algebra.*

Proof Let us consider a two-dimensional Lie algebra spanned by the generators X_1 and X_2. Their commutation say X is given by $X = [X_1, X_2] = C_1 X_1 + C_2 X_2$ also belongs to the Lie algebra spanned by these operators. Let $X = C_1 X_1 + C_2 X_2$ and $Y = D_1 X_1 + D_2 X_2$ belongs to the Lie algebra spanned by generators X_1 and X_2. Then their commutation is given by

$$\begin{aligned}
[X, Y] = XY - YX &= (C_1 X_1 + C_2 X_2)(D_1 X_1 + D_2 X_2) \\
&\quad - (D_1 X_1 + D_2 X_2)(C_1 X_1 + C_2 X_2) \\
&= (C_1 D_2 - D_1 C_2)(X_1 X_2 - X_2 X_1) = (C_1 D_2 - D_1 C_2)X \qquad \square
\end{aligned}$$

This shows that there exists a chain $X^0 \subset X \subset \mathcal{L}(X, Y)$ where each subset is an ideal Lie subalgebra. Hence any two-dimensional Lie algebra is a solvable Lie algebra.

Theorem 8.15 *Any abelian Lie algebra is a solvable Lie algebra.*

Proof Since in the commutator table representation of abelian Lie algebra, the entry at each position of the table is zero, it is obvious from the definition of solvable Lie algebra that abelian Lie algebra is a solvable Lie algebra. $\square$

Example 8.14 Determine whether the operators X_1, X_2, X_3 form a solvable Lie algebra which has the following commutator table representation (Table 8.4).

Solution It is noticeable here that all the operators appear in the table. Now X_1, X_2 do not form a Lie subalgebra since $[X_1, X_2] = X_3 \notin \mathcal{L}(X_1, X_2)$. We have $\mathcal{L}(X_1, X_3)$ and $\mathcal{L}(X_2, X_3)$ as Lie subalgebras. $X_1 \subset \mathcal{L}(X_1, X_3) \subset \mathcal{L}(X_1, X_2, X_3)$. Now $X_2 \in \mathcal{L}(X_1, X_2, X_3)$ and $X_3 \in \mathcal{L}(X_1, X_3)$ but $[X_2, X_3] = X_2 \notin \mathcal{L}(X_1, X_3)$ hence $\mathcal{L}(X_1, X_3)$ is not an ideal Lie subalgebra of $\mathcal{L}(X_1, X_2, X_3)$. Again $X_2 \subset X_2, X_3 \subset X_1, X_2, X_3$, $X_1 \in \mathcal{L}(X_1, X_2, X_3)$ and $X_3 \in \mathcal{L}(X_2, X_3)$ but $[X_1, X_3] = -X_1 \notin \mathcal{L}(X_2, X_3)$. Hence the Lie algebra $\mathcal{L}(X_1, X_2, X_3)$ is not a solvable Lie algebra.

Example 8.15 Show that the Lie algebra spanned by the infinitesimal generators $X_1 = x\frac{\partial}{\partial x}, X_2 = t\frac{\partial}{\partial t} - 2y\frac{\partial}{\partial y}, X_3 = \frac{\partial}{\partial t}$ with the commutator table representation (Table 8.5).

is a solvable Lie algebra.

Solution Here only X_3 appear in the commutator table. Also, $X_3 \subset \mathcal{L}(X_2, X_3) \subset \mathcal{L}(X_1, X_2, X_3)$, where each subalgebra is an ideal subalgebra. Hence the given infinitesimal generator forms a solvable Lie algebra.

Table 8.4 Commutator table

	X_1	X_2	X_3
X_1	0	X_3	X_1
X_2	X_3	0	X_2
X_3	$-X_1$	$-X_2$	0

Table 8.5 Commutator table

	X_1	X_2	X_3
X_1	0	0	0
X_2	0	0	$-X_3$
X_3	0	X_3	0

8.7 Group Method for Second-Order ODEs

We have discussed the use of integration method and canonical variables method for first-order differential equations. In canonical variables method the equation is reduced to a quadrature and the solution can be obtained. But for a second-order differential equation if it admits a two-dimensional Lie algebra, and then also it is possible to reduce the equation to a quadrature by the change of variables to canonical variables. We shall give an important theorem relevant to second-order differential equation.

Theorem 8.16 *Any two-dimensional Lie algebra $\mathcal{L}_2$ spanned by the infinitesimal generators X_1 and X_2 having infinitesimals (ξ_1, η_1) and (ξ_2, η_2), respectively can be transformed by a proper choice of canonical variables (t, u), to one of the following four standard canonical forms:*

(1) *If $[X_1, X_2] = 0$, $\xi_1\eta_2 - \eta_1\xi_2 \neq 0$. In this case the canonical variables can be obtained by solving*

$$X_1(t) = 1, X_2(t) = 0; X_1(u) = 0, X_2(u) = 1 \tag{8.38}$$

and the operators of $\mathcal{L}_2$ transform to the canonical form $X_1 = \frac{\partial}{\partial t}$, $X_2 = \frac{\partial}{\partial u}$.

(2) *If $[X_1, X_2] = 0$, $\xi_1\eta_2 - \eta_1\xi_2 = 0$. In this case, the canonical variables can be obtained by solving*

$$X_1(t) = 0, X_2(t) = 0; X_1(u) = 1, X_2(u) = t \tag{8.39}$$

and the operators of $\mathcal{L}_2$ transform to the canonical form $X_1 = \frac{\partial}{\partial u}$, $X_2 = t\frac{\partial}{\partial u}$.

(3) *If $[X_1, X_2] = X_1$, $\xi_1\eta_2 - \eta_1\xi_2 \neq 0$. In this case, the canonical variables can be obtained by solving*

$$X_1(t) = 0, X_2(t) = t; X_1(u) = 1, X_2(u) = u \tag{8.40}$$

and the operators of $\mathcal{L}_2$ transform to the canonical form $X_1 = \frac{\partial}{\partial u}$, $X_2 = t\frac{\partial}{\partial t} + u\frac{\partial}{\partial u}$.

(4) *If $[X_1, X_2] = X_1$, $\xi_1\eta_2 - \eta_1\xi_2 = 0$. In this case, the canonical variables can be obtained by solving*

$$X_1(t) = 0, X_2(t) = 0; X_1(u) = 1, X_2(u) = u \tag{8.41}$$

and the operators of $\mathcal{L}_2$ transform to the canonical form $X_1 = \frac{\partial}{\partial u}$, $X_2 = u\frac{\partial}{\partial u}$.

When a second-order ordinary differential equation admits a multiparameter group, say r then any two-dimensional Lie algebra of the r-dimensional Lie algebra can be reduced to one of the above standard forms.

For the proof of this theorem and further details see Ibragimov [20].

8.8   Method of Differential Invariant

The order of a higher order ordinary differential equation can also be reduced by
another convenient method, known as differential invariant method.

The pth-order ordinary differential equation

$$F = F(x, y, y_x, \ldots, y_{px}) = 0 \tag{8.42}$$

is invariant under the p times prolonged infinitesimal generator $X^{[p]}$ if and only if

$$X^{[p]} F = 0 \tag{8.43}$$

where $X^{[p]}$ is given by

$$X^{[p]} = \xi \frac{\partial}{\partial x} + \eta \frac{\partial}{\partial y} + \eta_1 \frac{\partial}{\partial y_x} + \eta_2 \frac{\partial}{\partial y_{xx}} + \cdots + \eta_p \frac{\partial}{\partial y_{px}} \tag{8.44}$$

The quantities $(\xi, \eta, ..)$ are infinitesimals.
The characteristic equation can be formed as

$$\frac{dx}{\xi} = \frac{dy}{\eta} = \frac{dy_x}{\eta_1} = \cdots = \frac{dy_{px}}{\eta_p} \tag{8.45}$$

By solving the first two equalities, one will get first two invariants of the differential Eq. (8.42), the higher order invariants can easily be generated from these
first two without solving the other equalities of (8.45). For instance if $\psi^1(x, y)$ and
$\psi^2(x, y, y_x)$ are the invariants of the once extended group (ξ, η, η_1), then $d\psi^2/d\psi^1$ is
an invariant of the twice extended group $(\xi, \eta, \eta_1, \eta_2)$. Similarly,
$d(d\psi^2)/d(\psi^1)^2, d(d(d\psi^2))/d(\psi^1)^3, \ldots, d^p\psi^2/d(\psi^1)^p$ are invariants known as differential invariants of the pth extended group. In this process of determining higher
order invariants from those already obtained, one would obtain a $(p\text{-}1)$th-order
differential invariant which can be expressed in terms of the first invariant $\psi^1(x, y)$,
second invariant $\psi^2(x, y, y_x)$ and the successive invariant derivative of $\psi^2(x, y, y_x)$
up to order $(p - 2)$. The problem of solving pth-order equation under the one
parameter group (ξ, η) is thus reduced to finding solution of the $(p - 1)$th-order
differential invariant equation and a quadrature. This method of reducing the order
of an ordinary differential equation by means of differential invariants is known as
differential invariant method for reduction. It is also possible to reduce successively
the order of an ordinary differential equation admitting multiparameter groups and
even to completely solve it by means of quadratures provided that the Lie algebra
formed by these multiparameter groups is solvable. A two parameter group is

always solvable. As a result it is always possible to completely solve a second-order equation admitting two parameter group by means of quadratures and to reduce the order by two for a higher order equation. But this may not be true for a group with more than two parameter until and unless its Lie algebra is solvable.

Now we shall give some examples for determining symmetry groups for second-order differential equations and their solutions using either canonical variables method or differential invariant method.

Example 8.16 Find the group of transformations of the most simple second-order ordinary differential equation $y_{xx} = 0$.

Solution The infinitesimal generator in this case is obtained as $X = \xi \frac{\partial}{\partial x} + \eta \frac{\partial}{\partial y}$. The determining equation is given by $X^{[2]}\mathcal{F}\big|_{\mathcal{F}=0} = 0$ where $X^{[2]}$ is the second prolongation of X expressed by

$$X^{[2]} = \xi \frac{\partial}{\partial x} + \eta \frac{\partial}{\partial y} + \zeta_1 \frac{\partial}{\partial y_x} + \zeta_2 \frac{\partial}{\partial y_{xx}}$$

Here $\mathcal{F} = y_{xx}$, therefore from above, we get

$$X^{[2]}(y_{xx})\big|_{(y_{xx}=0)} \equiv \zeta_2\big|_{(y_{xx}=0)} = 0 \tag{8.46}$$

Now,

$$\zeta_1 = D(\eta) - y_x D(\xi)$$
$$= \frac{\partial}{\partial x}(\eta(x,y)) + y_x \frac{\partial}{\partial x}(\eta(x,y)) - y_x(\frac{\partial}{\partial x}(\xi(x,y)) + y_x \frac{\partial}{\partial x}(\xi(x,y))) \tag{8.47}$$
$$= \eta_x + y_x \eta_y - y_x(\xi_x + y_x \xi_y)$$
$$= \eta_x + (\eta_y - \xi_x)y_x + \xi_y y_x^2$$

and

$$\zeta_2 = D(\zeta_1) - y_{xx}D(\xi)$$
$$= \eta_{xx} + (2\eta_{xy} - \xi_{xx})y_x + (\eta_{yy} - 2\xi_{xy})y_x^2 - \xi_{yy}y_x^3 + (\eta_y - 2\xi_x - 3\xi_y y_x)y_{xx} \tag{8.48}$$

Using (8.48), Eq. (8.46) can be written as

$$\eta_{xx} + (2\eta_{xy} - \xi_{xx})y_x + (\eta_{yy} - 2\xi_{xy})y_x^2 - \xi_{yy}y_x^3 \tag{8.49}$$

The invariance condition will be satisfied if each term in (8.49) is individually zero, so that we get the following overdetermined system of equations

$$\eta_{xx} = 0$$
$$2\eta_{xy} - \xi_{xx} = 0$$
$$\xi_{yy} = 0$$
$$\eta_{yy} - 2\xi_{xy} = 0$$

From the first and third relation one will have $\xi = yf(x) + g(x)$ and $\eta = xf(y) + g(y)$ again differentiating once the second relation with respect to x and the fourth relation with respect to y, one will have $\xi_{xxx} = 0$ and $\eta_{yyy} = 0$. This gives $f(x) = a_1x^2 + b_1x + c_1$ and $g(x) = a_2x^2 + b_2x + c_2$, $f(y) = a_3y^2 + b_3y + c_3$ and $g(x) = a_4y^2 + b_4y + c_4$. Again putting these values in the relation second and fourth gives $a_1 = 0 = a_3, a_4 = 2b_1, a_2 = 2b_3$ which gives $f(x) = b_1x + c_1$ and $g(x) = 2b_3x^2 + b_2x + c_2, f(y) = b_3y + c_3$ and $g(x) = 2b_1y^2 + b_4y + c_4$. Redefining the constants, the infinitesimals ξ and η can be written as

$$\xi = c_1 + c_2x + c_3y + c_7x^2 + c_8xy$$
$$\eta = c_4 + c_5x + c_6y + c_7xy + c_8y^2$$

with eight arbitrary constants. Therefore, the given equation admits an eight-dimensional group with the infinitesimal generators

$$X_1 = \frac{\partial}{\partial x}, X_2 = x\frac{\partial}{\partial x}, X_3 = y\frac{\partial}{\partial x}, X_4 = \frac{\partial}{\partial y}, X_5 = x\frac{\partial}{\partial y},$$
$$X_6 = y\frac{\partial}{\partial y}, X_7 = x^2\frac{\partial}{\partial x} + xy\frac{\partial}{\partial y}, X_8 = xy\frac{\partial}{\partial x} + y^2\frac{\partial}{\partial y} \quad (8.50)$$

The corresponding commutator table is given in Table 8.6.

Since the given equation is of second order one needs only a two-dimensional solvable Lie algebra in order to get quadrature. The infinitesimal generators of the

Table 8.6 Commutator table

	X_1	X_2	X_3	X_4	X_5	X_6	X_7	X_8
X_1	0	X_1	0	0	X_4	0	$2X_2 + X_6$	X_3
X_2	$-X_1$	0	$-X_3$	0	X_5	0	X_7	0
X_3	0	X_3	0	$-X_1$	$X_6 - X_2$	$-X_3$	X_8	0
X_4	0	0	X_1	0	0	X_4	X_5	$X_2 + X_6$
X_5	$-X_4$	$-X_5$	$-(X_6-X_2)$	0	0	X_5	0	X_7
X_6	0	0	X_3	$-X_4$	$-X_5$	0	0	X_8
X_7	$-(2X_2+X_6)$	$-X_7$	$-X_8$	$-X_5$	0	0	0	0
X_8	$-X_3$	0	0	$-(X_2+X_6)$	$-X_7$	$-X_8$	0	0

given equation form an eight-dimensional Lie algebra. We know that every two dimensional Lie algebra is solvable. Hence any two-dimensional Lie subalgebra of this eight-dimensional algebra solves the given equation.

Example 8.17 Find the point transformation of the following linear equation for damped harmonic oscillator

$$F = m\frac{d^2x}{dt^2} + b\frac{dx}{dt} + kx = 0.$$

Solution The above equation is a linear second-order differential equation with constant coefficients whose solution can be readily obtained by known method as

$$x = Ce^{\frac{-b}{2m}t}\cos\left(\frac{\sqrt{b^2 - 4km}}{2m}t + \text{constant}\right).$$

Let us now see the solution obtained using group theoretic approach. The infinitesimal generator is given by $X = \xi\frac{\partial}{\partial t} + \eta\frac{\partial}{\partial x}$. Since the equation is of second order, the above generator is twice extended and is given by

$$X^{[2]} = \xi\frac{\partial}{\partial t} + \eta\frac{\partial}{\partial x} + \eta^1\frac{\partial}{\partial x'} + \eta^2\frac{\partial}{\partial x''}$$

Therefore, the determining equation $X^{[2]}F = 0$ gives $\eta k + \eta^1 b + \eta^2 m = 0$. Now substituting the coefficients η^1 and η^2 we will have

$$\eta k + b(\eta_t + (\eta_x - \xi_t)x' - \xi_x(x')^2\eta^2 + m(\eta_{tt} + (2\eta_{tx} - \xi_{tt})x'$$
$$+ (\eta_{xx} - 2\xi_{tx})(x')^2 - \xi_{xx}(x')^3 + (\eta_x - 2\xi_t - 3\xi_x x')x'') = 0$$

Rearranging the terms we get

$$\eta k + b\eta_t + (b\eta_x - b\xi_t + 2m\eta_{tx} - m\xi_{tt})x' + m\eta_{tt} + (m\eta_{xx} - 2m\xi_{tx} - b\xi_x\eta^2)(x')^2$$
$$- m\xi_{xx}(x')^3 + m(\eta_x - 2\xi_t - 3\xi_x x')x'' = 0$$

Replacing x'' by $\frac{-b}{m}x' - \frac{k}{m}x$ in the above equation we have the following overdetermined system of linear partial differential equation

$$-m\xi_{xx} = 0,$$
$$2b\xi_x + m\eta_{xx} - 2m\xi_{tx} = 0,$$
$$k\eta - kx\eta_x + b\eta_t + 2kx\xi_t + m\eta_{tt} = 0,$$
$$3kx\xi_x + b\xi_t + 2m\eta_{tx} - m\xi_{tt} = 0$$

On solving the above overdetermined system of linear partial differential equation, we will have the infinitesimals as $\xi = a_1$ and $\eta = a_2 x$. The corresponding infinitesimal generators are

$$X_1 = \frac{\partial}{\partial t}; \; X_2 = x \frac{\partial}{\partial x}$$

We will now solve the given equation using differential invariant method where we will see that the given differential equation reduces to quadrature.

We have $X_1 = \frac{\partial}{\partial t}$ and $X_2 = x \frac{\partial}{\partial x}$. We will start with $X_2 = x \frac{\partial}{\partial x}$. The characteristic equation corresponding to this generator is given by

$$\frac{dt}{0} = \frac{dx}{x} = \frac{dx_t}{x_t} = \frac{dx_{tt}}{x_{tt}}$$

Now the first two invariants are obtained by solving the first relation and second and third relations, respectively as $t = \eta$(say) and $\frac{x_t}{x} = f(\eta)$(say). Using these invariants the first-order differential invariant is obtained as

$$\frac{df}{d\eta} = \frac{\frac{\partial f}{\partial t} dt + \frac{\partial f}{\partial x} dx + \frac{\partial f}{\partial x_t} dx_t}{\frac{\partial \eta}{\partial t} dt + \frac{\partial \eta}{\partial x} dx} = \frac{\frac{-x_t}{x^2} dx + \frac{1}{x} dx_t}{dt} = -(x_t)^2 + \frac{1}{x} x_{tt} = -f^2 - \frac{b}{m} f - \frac{k}{m}$$

$$(8.51)$$

We will now see that the differential invariant equation (8.51) is an invariant under the generator X_1 which corresponds to the transformations $\bar{t} = t + \epsilon$ and $\bar{x} = x$. Without evaluating the symmetry transformation of (8.51) we can obtain the transformations $\bar{\eta} = \eta + \epsilon$ and $\bar{f} = \frac{\bar{x}_t}{\bar{x}} = \frac{x_t}{x} = f$ using X_1 whence

$$\frac{d\bar{f}}{d\bar{\eta}} + (\bar{f})^2 + \frac{b}{m} \bar{f} + \frac{k}{m} = \frac{df}{d\eta} + f^2 + \frac{b}{m} f + \frac{k}{m} = 0$$

Now integrating (8.51) one would obtain the solution of the Eq. (8.51) as

$$f(\eta) = -\frac{\sqrt{4km - b^2}}{2m} \tan\left(\frac{\sqrt{4km - b^2}}{2m}\right)(\eta + C_1) - \frac{b}{2m}$$

$$\frac{1}{x}\frac{dx}{dt} = -\frac{\sqrt{4km - b^2}}{2m} \tan\left(\frac{\sqrt{4km - b^2}}{2m}\right)(t + C_1) - \frac{b}{2m}$$

This upon integration gives

$$\log x = \log \cos \frac{\sqrt{4km - b^2}}{2m}(t + C_1) - \frac{b}{2m} t + \log(\text{constant}).$$

Hence $x = C_2 e^{\frac{-b}{2m}t} \cos\left(\frac{\sqrt{4km-b^2}}{2m}t + C_1\right)$ is the required solution, where C_1, C_2 being arbitrary constants of integration.

Example 8.18 Find the infinitesimal generator of the exponential oscillator

$$F = \frac{d^2 y}{dx^2} + e^y = 0 \qquad (8.52)$$

and then find the solution using differential invariant method.

Solution The infinitesimal generator is $X = \xi \frac{\partial}{\partial t} + \eta \frac{\partial}{\partial x}$. The determining equation is $X^{[2]} F\big|_{F=0} = 0$, where $X^{[2]}$ the twice extended operator is given by

$$X^{[2]} = \xi \frac{\partial}{\partial x} + \eta \frac{\partial}{\partial y} + \eta^2 \frac{\partial}{\partial y''}$$

Therefore, $X^{[2]}F = 0$ gives $\eta^2 + \eta e^x = 0$. Putting the value of η^2 gives

$$\eta_{xx} + (2\eta_{xy} - \xi_{xx})y' + (\eta_{yy} - 2\xi_{xy})(y')^2 - \xi_{yy}(y')^3 + (\eta_y - 2\xi_x - 3\xi_y y')y'' + \eta e^y$$
$$= 0$$

Now replacing $y'' = -e^y$, we have

$$\eta_{xx} + (2\eta_{xy} - \xi_{xx})y' + (\eta_{yy} - 2\xi_{xy})(y')^2 - \xi_{yy}(y')^3 - (\eta_y - 2\xi_x - 3\xi_y y')e^y + \eta e^y$$
$$= 0$$

On rearranging the terms of the above equation, we have

$$\eta_{xx} - (\eta_y - 2\xi_x - \eta)e^y + (2\eta_{xy} - \xi_{xx} - 3\xi_y e^y)y' + (\eta_{yy} - 2\xi_{xy})(y')^2 - \xi_{yy}(y')^3 = 0$$

Now setting coefficient of each of y', y'', y''' and the term free of these equal to zero, we obtain the following overdetermined system of linear partial differential equations in ξ and η.

$$\begin{aligned} \xi_{yy} &= 0 \\ \eta_{yy} - 2\xi_{xy} &= 0 \\ 2\eta_{xy} - \xi_{xx} - 3\xi_y e^y &= 0 \\ \eta_{xx} - (\eta_y - 2\xi_x - \eta)e^y &= 0 \end{aligned} \qquad (8.53)$$

It follows from $\xi_{yy} = 0$, $\xi = a(x)y + b(x)$ and then from $\eta_{yy} - 2\xi_{xy} = 0$, $\eta = a'(x)y^2 + c(x)y + d(x)$. Since ξ and η are polynomials in y and the determining equation involves e^y therefore $\xi_y = 0$ and $\eta_y - 2\xi_x - \eta = 0$ which gives $\xi = b(x)$ and $\eta = -2b'(x)$. Now from $2\eta_{xy} - \xi_{xx} = 0$, we have $b''(x) = 0$, $b(x) = c_1 + c_2 x$.

Table 8.7 Commutator table

	X_1	X_2
X_1	0	X_1
X_2	$-X_1$	0

The solution of the above overdetermined system of partial differential equation is

$$\xi = c_1 + c_2 x, \ \eta = -2c_2$$

Thus the equation of exponential oscillator admits two parameter Lie group spanned by the generators

$$X_1 = \frac{\partial}{\partial x} \ \text{and} \ X_2 = x\frac{\partial}{\partial x} - 2\frac{\partial}{\partial y}$$

We can easily see that being a two-dimensional Lie algebra, the Lie algebra shown in the Table 8.7 is a solvable Lie algebra with structure constant 1.

We now solve the given equation using differential invariant method.

We have $X_1 = \frac{\partial}{\partial x}$, the characteristic equation corresponds to

$$\frac{dx}{1} = \frac{dy}{0} = \frac{dy_x}{0} = \frac{dy_{xx}}{0}$$

The first two invariants are $y = \eta$ and $y_x = f$. This gives the first differential invariant as

$$\frac{df}{d\eta} = \frac{\frac{\partial f}{\partial x}dx + \frac{\partial f}{\partial y}dy + \frac{\partial f}{\partial y_x}dy_x}{\frac{\partial \eta}{\partial x}dx + \frac{\partial \eta}{\partial y}dy} = \frac{y_{xx}}{y_x}$$

Now substituting y_{xx} by $-e^y$ and y_x by f, we obtain the following first-order differential invariant equation.

$$\frac{df}{d\eta} = \frac{-e^{\eta}}{f}$$

On integration we have $\frac{f^2}{2} = -e^{\eta} + c_1$ or $f = \sqrt{-2e^{\eta} + c_1}$, where c_1 is an integration constant.

Therefore, we have $\frac{\partial y}{\partial x} = \sqrt{-2e^y + c_1}$, which upon integration gives, $\frac{-2}{\sqrt{c_1}} \tanh^{-1}\left(\frac{\sqrt{-2e^y + c_1}}{\sqrt{c_1}}\right) = x + c_2, c_2$ is an integration constant and hence the solution can be written as

$$y = \log\left(\frac{c_1}{2} - \frac{c_1}{2}\tanh^2\left(\frac{\sqrt{c_1}(x+c_2)}{2}\right)\right)$$

We have used the generator X_1 and obtained the solution, we will now see that the invariant transformations of the differential invariant equation can be obtain from the generator X_2. The infinitesimal group of transformations corresponding to X_2 is $\bar{x} = xe^{\epsilon}$ and $\bar{y} = y - 2\epsilon$ which gives $\bar{f} = \frac{d\bar{y}}{d\bar{x}} = e^{-\epsilon}\frac{dy}{dx} = e^{-\epsilon}f$ and $\bar{\eta} = \bar{y} = y - 2\epsilon = \eta - 2\epsilon$. Now

$$\frac{D\bar{f}}{D\bar{\eta}} + \frac{e^{\bar{\eta}}}{\bar{f}} = 0 \Rightarrow e^{-\epsilon}\left(\frac{Df}{D\eta} + \frac{e^{\eta}}{f}\right) = 0$$

Hence, $\frac{D\bar{f}}{D\bar{\eta}} + \frac{e^{\eta}}{\bar{f}} = \frac{Df}{D\eta} + \frac{e^{\eta}}{f} = 0$. That is the differential invariant equation is form invariant under X_2. Hence, the differential invariant equation is invariant under the infinitesimal generator X_2.

Example 8.19 Find the infinitesimal generators of the following nonlinear second-order ordinary differential equation

$$F(x, y, y', y'') = \frac{d^2y}{dx^2} - \frac{1}{y^2}\frac{dy}{dx} + \frac{1}{xy} = 0 \tag{8.54}$$

Solution The infinitesimal generator of the given equation will be $X = \xi\frac{\partial}{\partial x} + \eta\frac{\partial}{\partial y}$. The determining equation is

$$X^{[2]}F = 0 \tag{8.55}$$

where $X^{[2]}$, the twice extended operator is given by $X^{[2]} = \xi\frac{\partial}{\partial x} + \eta\frac{\partial}{\partial y} + \eta^1\frac{\partial}{\partial y'} + \eta^2\frac{\partial}{\partial y''}$

where $\eta^1 = D(\eta(x,y)) - y'D(\xi(x,y)) = \eta_x + (\eta_y - \xi_x)y' - \xi_y(y')^2$

$\eta^2 = D(\eta^1) - y''D(\xi)$

$\qquad = \eta_{xx} + (2\eta_{xy} - \xi_{xx})y' + (\eta_{yy} - 2\xi_{xy})(y')^2 - \xi_{yy}(y')^3 + (\eta_y - 2\xi_x - 3\xi_y y')y''$

Therefore, $X^{[2]}F = 0$ gives $\xi\frac{\partial F}{\partial x} + \eta\frac{\partial F}{\partial y} + \eta^1\frac{\partial F}{\partial y'} + \eta^2\frac{\partial F}{\partial y''} = 0$
which yields

$$\xi\left(\frac{-1}{x^2 y}\right) + \eta\left(\frac{-1}{xy^2} + \frac{2y'}{y^3}\right) + (\eta_x + (\eta_y - \xi_x)y' - \xi_y(y')^2)\left(\frac{-1}{y^2}\right) + (\eta_{xx} + (2\eta_{xy} - \xi_{xx})y'$$

$$+ (\eta_{yy} - 2\xi_{xy})(y')^2 - \xi_{yy}(y')^3 + (\eta_y - 2\xi_x - 3\xi_y y')y'' = 0$$

Replacing y'' by $\frac{y'}{y^2} - \frac{1}{xy}$ and then collecting the coefficients of y', y'^2, y'^3 and the terms free of these in the above equation, one would obtain

$$
\left(\frac{-\xi}{x^2 y} - \frac{\eta}{xy^2} - \frac{\eta_x}{y^2} - \frac{\eta_{xy}}{xy} + \frac{2\xi_x}{xy} + \eta_{xx} \right)
$$
$$
- \xi_{yy}(y')^3 + \left(\frac{2\eta}{y} - \frac{\xi_x}{y^2} + (2\eta_{xy} - \xi_{xx}) \right) y' + \left(\eta_{yy} - 2\xi_{xy} - \frac{3\xi_y}{y^2} \right)(y')^2
$$
$$
= 0
$$

Setting the coefficients equal to zero the determining equation split into an overdetermined system of equation given below as

$$
\xi_{yy} = 0 \tag{8.56}
$$

$$
\eta_{yy} - 2\xi_{xy} - \frac{3\xi_y}{y^2} = 0 \tag{8.57}
$$

$$
\frac{2\eta}{y} - \frac{\xi_x}{y^2} + (2\eta_{xy} - \xi_{xx}) = 0 \tag{8.58}
$$

$$
\frac{-\xi}{x^2 y} - \frac{\eta}{xy^2} - \frac{\eta_x}{y^2} - \frac{\eta_{xy}}{xy} + \frac{2\xi_x}{xy} + \eta_{xx} = 0 \tag{8.59}
$$

From (8.56) we obtain,

$$
\xi(x, y) = f_1(x)y + f_2(x) \tag{8.60}
$$

Here $f_1(x)$ and $f_2(x)$ are arbitrary function of x. From (8.57) we will have

$$
\eta_{yy} - 2(f_1(x))' - \frac{3f_1(x)}{y^2} = 0
$$
$$
\eta_y = 2(f_1(x))'y - \frac{3f_1(x)}{y} + a_1(x)
$$
$$
\eta = (f_1(x))'y^2 - 3f_1(x) \log y + a_1(x)y + a_2(x)
$$

$(f_1(x))'$ denotes the differentiation of the function f_1 with respect to x, and a_1, a_2 are arbitrary function of x. Substituting the above values of $\xi(x,y)$ and $\eta(x,y)$ in Eqs. (8.58) and (8.59), one would get the infinitesimals as

$$
\xi = 2c_1 x + c_2 x^2 \quad \text{and} \quad \eta = c_1 y + c_2 xy
$$

Table 8.8 Commutator table

	X_1	X_2
X_1	0	X_2
X_2	$-X_2$	0

Thus Eq. (8.54) admits the two-dimensional Lie algebra spanned by

$$X_1 = 2x\frac{\partial}{\partial x} + y\frac{\partial}{\partial y} \text{ and } X_2 = x^2\frac{\partial}{\partial x} + xy\frac{\partial}{\partial y}$$

Its Lie commutator is summarized in the Table 8.8.

Commutator table shows that the generators X_1 and X_2 span a two-dimensional Lie algebra hence a solvable Lie algebra with structure constant 1. In order to reduce the given equation to one of the standard forms in Theorem 10, we will take X_1 as X_2 and vice versa. We will see that $[X_1, X_2] = -2X_1$, $\xi_1\eta_2 - \xi_2\eta_1 = -x^2y \neq 0$. Hence X_1 and X_2 span an algebra corresponding to the type 3 standard form in Theorem 8.16. In order to completely satisfy the condition of type 3, multiply X_2 by $-1/2$ and then one will obtain

$$X_1 = x^2\frac{\partial}{\partial x} + xy\frac{\partial}{\partial y}, X_2 = -x\frac{\partial}{\partial x} - \frac{y}{2}\frac{\partial}{\partial y}$$

So, the canonical variables in this case is obtained by solving

$$X_1(t) = 0, X_2(t) = t; X_1(u) = 1, X_2(u) = u$$

which gives the canonical variables as $t = \frac{y}{x}$ and $u = \frac{-1}{x}$. Replacing the original variables by the new canonical variables the given equation becomes

$$\frac{d^2u}{dt^2} + \frac{1}{t^2}\left(\frac{du}{dt}\right)^2 = 0$$

Integrating once, we have $\frac{du}{dt} = \frac{t}{at-1}$, a is the constant of integration.
Integrating again we have the solution as

$$u(t) = \frac{t}{a} + \frac{1}{a^2}\log|at - 1| + b$$

Here b is constant of integration. In terms of original variables the solution is given by

$$\frac{-1}{x} = \frac{y}{ax} + \frac{1}{a^2}\log\left|a\frac{y}{x} - 1\right| + b$$

$$\text{or, } ay + a^2bx + x\log\left|a\frac{y}{x} - 1\right| + a = 0$$

If $a = 0$, then $\frac{du}{dt} = -t$, which on integration gives $u = \frac{-t^2}{2} + c$, c is the arbitrary integration constant. In terms of original variable $y = \pm\sqrt{2cx^2 + 2x}$. The invariant solution of the equation is given by $y = cx$ which is trivial and can be easily obtained. This solution of the given equation is calculated in the following section.

Similarly, one can calculate the infinitesimals for third and higher order differential equations. However, for higher order differential equation it is usually very cumbersome to split the determining equations into an overdetermined system of equation, so it is advised to use the symbolic software programs for this purpose.

8.8.1 Invariant Solution

Consider an nth-order ordinary differential equation $y^{(n)} = F(x, y, y', y'', \ldots, y^{(n-1)})$ admitting a one parameter Lie group of transformation with infinitesimal generator

$$X = \xi(x, y)\frac{\partial}{\partial x} + \eta(x, y)\frac{\partial}{\partial y} \tag{8.61}$$

We have already learnt that the solution given by $y = f(x)$ is an invariant solution of the given differential equation if and only if it satisfies the given differential equation that is, if it solves the given differential equation and is an invariant curve of (8.61). Mathematically, this can be expressed as

$$f^{(n)} = F(x, f, f', f'', \ldots, f^{(n-1)})$$

and

$$\left(\xi(x, y)\frac{\partial}{\partial x} + \eta(x, y)\frac{\partial}{\partial y}\right)(y - f(x)) = 0$$

or,

$$-\xi(x, y)\frac{\partial f(x)}{\partial x} + \eta(x, y) = 0.$$

Now one can find invariant solutions by two ways (see Bluman and Kumei [19] for details). One way is to first find the solution $\phi(x, y, C) = 0$ of $y' = \frac{\eta(x,y)}{\xi(x,y)}$, and then find the arbitrary constant C by substituting this solution into the original equation. But it is not necessary to solve $y' = \frac{\eta(x,y)}{\xi(x,y)}$ to find the invariant solutions of the given differential equation. One can also obtain invariant solution by solving an algebraic expression. The statement of the theorem to find invariant solution of an nth-order ordinary differential equation using algebraic expression is given below

Theorem 8.17 *When an nth-order ordinary differential equation given by*

$$y^{(n)} = F(x, y, y', y'', \ldots, y^{(n-1)}) \tag{8.62}$$

admits a one parameter group G with the infinitesimal generator $X = \xi \frac{\partial}{\partial x} + \eta \frac{\partial}{\partial y}$ such that the infinitesimal $\xi(x, y) \neq 0$ then an invariant solution of the differential equation can be obtained by solving the algebraic expression given by

$$\mathcal{Q}(x, y) = y_n - F(x, y, y', \ldots, y^{(n-1)}) = 0 \tag{8.63}$$

where

$$y_k = Y^{k-1}\psi, k = 1, 2, \ldots, n; \ \psi = \frac{\eta(x, y)}{\xi(x, y)} \ \text{and} \ Y = \frac{\partial}{\partial x} + \psi \frac{\partial}{\partial y}.$$

For the algebraic expression $\mathcal{Q}(x, y) = y_n - F(x, y, y', \ldots, y^{(n-1)}) = 0$, three cases arises

(i)   When $\mathcal{Q}(x, y) = 0$ defines no curves in the xy-plane then the differential Eq. (8.62) has no invariant solution under $X = \xi \frac{\partial}{\partial x} + \eta \frac{\partial}{\partial y}$

(ii) When $\mathcal{Q}(x, y) = 0$ is satisfied for all x, y in the xy-plane then any solution of the differential equation $y' = \frac{\eta(x,y)}{\xi(x,y)}$ is an invariant solution under $X = \xi \frac{\partial}{\partial x} + \eta \frac{\partial}{\partial y}$

(iii) When $\mathcal{Q}(x, y) = 0$ defines curves in the xy-plane then any curve satisfying $\mathcal{Q}(x, y) = 0$ is an invariant solution under group and conversely any invariant solution of (8.62) under (8.61) must satisfy $\mathcal{Q}(x, y) = 0$

Example 8.20 Determine the invariant solution of

$$F(x, y, y', y'') = \frac{d^2y}{dx^2} - \frac{1}{y^2}\frac{dy}{dx} + \frac{1}{xy} = 0$$

Solution The given differential equation is invariant under the following one parameter groups

$$X_1 = 2x\frac{\partial}{\partial x} + y\frac{\partial}{\partial y} \ \text{and} \ X_2 = x^2\frac{\partial}{\partial x} + xy\frac{\partial}{\partial y}$$

Here

$$Y = \frac{\partial}{\partial x} + \left(\frac{y}{2x}\right)\frac{\partial}{\partial y}, \psi = \frac{y}{2x}; y_1 = \psi, \ y_2 = Y\psi = \frac{\partial}{\partial x}\left(\frac{y}{2x}\right) + \left(\frac{y}{2x}\right)\frac{\partial}{\partial y}\left(\frac{y}{2x}\right)$$

$$= -\frac{y}{2x^2} + \frac{y}{4x^2} = \frac{-y}{4x^2}$$

Now the algebraic expression is given by

$$
\begin{aligned}
Q(x,y) = y^{(n)} - F(x,y,y',\ldots,y^{(n-1)}) &= 0 \\
= y'' - F(x,y,y') &= 0 \\
= Y\psi - F(x,y,\psi) = -\frac{y}{4x^2} + \frac{1}{2xy} &= 0
\end{aligned}
$$

$$\therefore Q(x,y) = -y^2 + 2x = 0$$

Hence the invariant solutions are $y = \pm\sqrt{2x}$. Similarly for $X_2 = x^2\frac{\partial}{\partial x} + xy\frac{\partial}{\partial y}$

$$
Y = \frac{\partial}{\partial x} + \left(\frac{y}{x}\right)\frac{\partial}{\partial y}, \psi = \frac{y}{x};\ y_1 = y' = \psi,
$$

$$
y_2 = y'' = Y\psi = \frac{\partial}{\partial x}\left(\frac{y}{x}\right) + \left(\frac{y}{x}\right)\frac{\partial}{\partial y}\left(\frac{y}{x}\right) = -\frac{y}{x^2} + \frac{y}{x^2} = 0.
$$

Now the algebraic expression is given by

$$
\begin{aligned}
Q(x,y) = y^{(n)} - F(x,y,y',\ldots,y^{(n-1)}) &= 0 \\
= y'' - F(x,y,y') &= 0 \\
= Y\psi - F(x,y,\psi) = 0 - \frac{1}{xy} + \frac{1}{xy} &= 0
\end{aligned}
$$

Hence any solution of $y' = \frac{\eta(x,y)}{\xi(x,y)}$ is an invariant solution of the given differential equation under the invariance of the generator X_2. Now the solution of $y' = \frac{\eta(x,y)}{\xi(x,y)} = \frac{y}{x}$ is $y = cx$. Hence $y = cx$ is an invariant solution of the given differential equation for any arbitrary c.

Example 8.21 Find the separatrices of the equation $\frac{dy}{dx} = \frac{x}{y}$ which is invariant under the infinitesimal generator $X = x\frac{\partial}{\partial x} + y\frac{\partial}{\partial y}$.

Solution Separatrices are the curves which separate two qualitatively different trajectories. Since separatrices are the invariant solutions (see Bluman and Kumei [19]) therefore the separatrix solution $y = \psi(x)$ must satisfy $\xi(x,y)\psi'(x) + \eta(x,y) = 0$,i.e., $y'(x) = \psi'(x) = -y/x$ or $y(x) = cx$. Substituting this solution in the given differential equation one will obtain $c = \pm 1$. Hence the separatrices of the given equation are $y = x$ and $y = -x$ with the general solution $y^2(x) = x^2 + c^2$. The solution curves are shown in the Fig. 8.7.

Fig. 8.7 Integral curves of $y' = x/y$

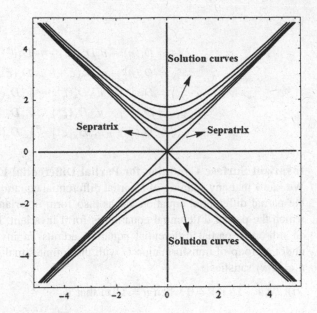

8.9 Group Method for PDEs

Consider the partial differential equation

$$F(x, u, u^{(n)}) = 0 \tag{8.64}$$

where $x = (x_1, x_2, \ldots, x_n)$ are the independent variables, $u = (u_1, u_2, \ldots, u_n)$ are the dependent variable, $u^{(n)}$ is the nth-order partial derivative of u with respect to the independent variables x.

In this case the infinitesimal generator is

$$X = \xi^i \frac{\partial}{\partial x_i} + \eta^i \frac{\partial}{\partial u_i} \quad i = 1, \ldots, n$$

and the nth extended generator is

$$X^{[n]} = \xi^i \frac{\partial}{\partial x_i} + \eta^i \frac{\partial}{\partial u_i} + \zeta_i \frac{\partial}{\partial u_i^{(1)}} + \zeta_{i_1 i_2} \frac{\partial}{\partial u_{i_1 i_2}^{(2)}} + \ldots + \zeta_{i_1 i_2 i_3 \ldots i_n} \frac{\partial}{\partial u_{i_1 i_2 \ldots i_n}^{(n)}} \tag{8.65}$$

For example if $F(x, y, u, u_x, u_y, u_{xx}, u_{xy}, u_{yy}) = 0$ then (8.65) become

$$X^{[2]} = \xi^x \frac{\partial}{\partial x} + \xi^y \frac{\partial}{\partial y} + \eta \frac{\partial}{\partial u} + \zeta_x \frac{\partial}{\partial u_x} + \zeta_y \frac{\partial}{\partial u_y} + \zeta_{xx} \frac{\partial}{\partial u_{xx}} + \zeta_{xy} \frac{\partial}{\partial u_{xy}} + \zeta_{yy} \frac{\partial}{\partial u_{yy}}$$

where

$$\zeta_x = D_x(\eta) - u_x D_x(\xi^x) - u_y D_x(\xi^y)$$
$$\zeta_y = D_y(\eta) - u_x D_y(\xi^x) - u_y D_y(\xi^y)$$
$$\zeta_{xx} = D_x(\zeta_x) - u_{xx} D_x(\xi^x) - u_{xy} D_x(\xi^y)$$
$$\zeta_{xy} = D_x(\zeta_y) - u_{xx} D_x(\xi^x) - u_{xy} D_x(\xi^y)$$
$$\zeta_{yy} = D_y(\zeta_y) - u_{xy} D_y(\xi^x) - u_{yy} D_y(\xi^y)$$

Invariant Surface Condition for Partial Differential Equation

We know that any solution of a partial differential equation is invariant if it satisfies the partial differential equation and is also form invariant under the generator for which the partial differential equation is form invariant. Mathematically speaking, an nth-order partial differential equation admits an invariant solution $u = \Omega(x)$ under a group of transformation G with the infinitesimal generator X if and only if $u = \Omega(x)$ satisfies

(i) $X(u - \Omega(x)) = 0$ when $u = \Omega(x)$ that is,

$$\xi_i(x, \Omega(x)) \frac{\partial \Omega}{\partial x_i} = \eta(x, \Omega(x)) \qquad (8.66)$$

where η is the infinitesimal coefficient corresponding to the dependent variable u. The Eq. (8.66) is called the invariant surface condition.

(ii) The partial differential equation $u^{(k)} = F(x, u, u^{(1)}, \ldots, u^{(k)}) = 0$, where $u^{(k)} = u_{i_1 i_2 \ldots i_j} = \frac{\partial^j \Omega(x)}{\partial x_{i_1} \partial x_{i_2} \ldots \partial x_{i_j}}$.

The symmetry group of a partial differential equation reduces it to an ordinary differential equation also known as similarity equation changing the independent variables to a single similarity variable. However, there is no known procedure for the solution of partial differential equation except when the similarity equation is solvable to quadratures. We will now illustrate group theoretic solution of some partial differential equations both for linear and nonlinear equations.

Example 8.22 Find the point transformation of the one-dimensional heat conducting equation

$$\frac{\partial u}{\partial t} = \frac{\partial^2 u}{\partial x^2}$$

Also, obtain the solutions by taking different combination of the point transformations.

Solution The heat conducting equation is a linear parabolic equation. We have studied the solution of heat equation in graduate texts using separation of variables under prescribed boundary conditions. There, the specification of initial as well as boundary conditions was necessary to obtain a unique solution of the heat equation. The main contribution of Lie symmetry analysis is better understood here as it admits new solution in addition to the solutions obtained by using separation of variables. We will be able to calculate all the invariant solutions of the heat equation just by evaluating its symmetry groups provided that the similarity equation thus obtained could be reduced to quadrature.

The heat equation involves the space coordinate x and time variable t as independent variables and the temperature u as dependent variable. The infinitesimal generator for the heat equation is therefore can be written as

$$X = \xi^1 \frac{\partial}{\partial x} + \xi^2 \frac{\partial}{\partial t} + \eta^1 \frac{\partial}{\partial u} \tag{8.67}$$

Since the heat equation is of order two, the infinitesimal generator can be twice extended.

The second prolongation of X is given by

$$X^{[2]} = X + \zeta^1 \frac{\partial}{\partial u_x} + \zeta^2 \frac{\partial}{\partial u_t} + \zeta^{11} \frac{\partial}{\partial u_{xx}} \tag{8.68}$$

The determining equation is

$$X^{[2]} \mathcal{F} = 0 \text{ whenever } \mathcal{F} = u_t - u_{xx} = 0 \tag{8.69}$$

This gives

$$\zeta^2 - \zeta^{11} = 0 \tag{8.70}$$

$$\zeta^2 = \eta_t + u_t \eta_u - u_x \xi_t^1 - u_x u_t \xi_u^1 - u_t \xi_t^2 - (u_t)^2 \xi_u^2$$

$$\zeta^{11} = D_x(\zeta^1) - u_{xx} D_x(\xi^1) - u_{xt} D_x(\xi^2)$$

$$= D_x(\eta_x + u_x \eta_u - u_x \xi_x^1 - (u_x)^2 \xi_u^1 - u_t \xi_x^2 - u_x u_t \xi_u^2)$$

$$= \eta_{xx} + 2u_x \eta_{xu} + u_{xx} \eta_u + (u_x)^2 \eta_{uu} - 2u_{xx} \xi_x^1 - u_x \xi_{xx}^1 - 2(u_x)^2 \xi_{xu}^1 - 3u_x u_{xx} \xi_u^1$$

$$- (u_x)^3 \xi_{uu}^1 - 2u_{xy} \xi_x^2 - u_t \xi_{xx}^2 - 2u_x u_t \xi_{xu}^2 - (u_t u_{xx} + 2u_x u_{xt}) \xi_u^2 - (u_x)^2 u_t \xi_{uu}^2$$

Substituting ζ^2, ζ^{11} and Setting $u_{xx} = u_t$ in (8.70), we have

$$
\begin{aligned}
&\eta_t + u_t \eta_u - u_x \xi_t^1 - u_x u_t \xi_u^1 - u_t \xi_t^2 - (u_t)^2 \xi_u^2 - \eta_{xx} \\
&- 2u_x \eta_{xu} - u_t \eta_u - (u_x)^2 \eta_{uu} + 2u_t \xi_x^1 + u_x \xi_{xx}^1 \\
&+ 2(u_x)^2 \xi_{xu}^1 + 3u_x u_t \xi_u^1 + (u_x)^3 \xi_{uu}^1 + 2u_{xt} \xi_x^2 + u_t \xi_{xx}^2 \\
&+ 2u_x u_t \xi_{xu}^2 + (u_t u_t + 2u_x u_{xt}) \xi_u^2 + (u_x)^2 u_t \xi_{uu}^2 = 0
\end{aligned}
\tag{8.71}
$$

Now isolating the terms containing u_x, u_t, u_{xt} and those free of these variables, we have

$$
\begin{aligned}
&(\eta_t - \eta_{xx}) + u_t(-\xi_t^2 + \xi_{xx}^2 + \xi_x^1) + u_x u_t(2\xi_u^1 + 2\xi_{xu}^2) \\
&+ (u_x)^2(2\xi_{xu}^1 - \eta_{uu})u_x(-2\eta_{xu} + \xi_{xx}^1 - \xi_t^1) \\
&+ (u_x)^3\xi_{uu}^1 + 2u_{xt}\xi_x^2 - 2u_x u_{xt}\xi_u^2 + (u_x)^2 u_t \xi_{uu}^2 = 0
\end{aligned} \tag{8.72}
$$

Setting each term equal to zero, we have the following nine overdetermined system of equations

$$
\eta_t - \eta_{xx} = 0,\ -\xi_t^2 + \xi_{xx}^2 + \xi_x^1 = 0,\ 2\xi_u^1 + 2\xi_{xu}^2 = 0,\ -2\eta_{xu} + \xi_{xx}^1 - \xi_t^1 = 0
$$
$$
2\xi_{xu}^1 - \eta_{uu} = 0,\ \xi_{uu}^1 = 0,\ 2\xi_x^2 = 0,\ -2\xi_u^2 = 0,\ \xi_{uu}^2 = 0
$$

Upon solving these equations one would obtain the infinitesimals as

$$
\xi^1 = c_1 + \frac{c_3}{2}x - 2tc_5 - 2txc_6
$$
$$
\xi^2 = c_2 + c_3 t + (-2t^2)c_6
$$
$$
\eta = c_4 u + c_5 ux + c_6\left(tu + \frac{ux^2}{2}\right)
$$

So heat equation admits six-dimensional Lie algebra spanned by the operators

$$
X_1 = \frac{\partial}{\partial x},\ X_2 = \frac{\partial}{\partial t},\ X_3 = \frac{x}{2}\frac{\partial}{\partial x} + t\frac{\partial}{\partial t},\ X_4 = u\frac{\partial}{\partial u},\ X_5 = -2t\frac{\partial}{\partial x} + ux\frac{\partial}{\partial u},
$$
$$
X_6 = -2tx\frac{\partial}{\partial x} - 2t^2\frac{\partial}{\partial t} + \left(tu + \frac{ux^2}{2}\right)\frac{\partial}{\partial u} \tag{8.73}
$$

The corresponding commutator table is shown in Table 8.9.

From the above commutator table one can see that X_1 and X_5 generate X_4, X_1, and X_6 generate X_5, X_2, and X_5 generate X_1. Also knowing X_4, X_2, and X_6 generates X_3. Thus, all the six infinitesimal groups can be constructed by X_1, X_2, X_6. We can

Table 8.9 Commutator table of heat equation

	X_1	X_2	X_3	X_4	X_5	X_6
X_1	0	0	$\frac{1}{2}X_1$	0	X_4	X_5
X_2	0	0	X_2	0	$-2X_1$	$-4X_3 + X_4$
X_3	$\frac{-1}{2}X_1$	$-X_2$	0	0	$\frac{1}{2}X_5$	X_6
X_4	0	0	0	0	0	0
X_5	$-X_4$	$2X_1$	$\frac{-1}{2}X_5$	0	0	0
X_6	$-X_5$	$4X_3 - X_4$	$-X_6$	0	0	0

also see that the heat equation is invariant only under the above groups and no new symmetry group can be generated.

Now the heat equation is invariant under the six parameter symmetry group. Thus, the heat equation remains form invariant under the transformations as
(i) $\bar{t} = t, \bar{x} = x + c_1, \bar{u} = u$ (ii) $\bar{t} = t + c_2, \bar{x} = x, \bar{u} = u$ (iii) $\bar{t} = te^{c_3}, \bar{x} = xe^{\frac{c_3}{2}}, \bar{u} = u$
(iv) $\bar{t} = t, \bar{x} = x, \bar{u} = e^{c_4}u$ (v) $\bar{t} = t, \bar{x} = x - 2tc_5, \bar{u} = ue^{xc_5 - 2t(c_5)^2}$ (vi) $\bar{t} = \frac{t}{1 + 2tc_6}$,
$\bar{x} = xe^{\frac{-2t^2 c_6}{1 + 2tc_6}}, \bar{u} = ue^{(\bar{t} + \bar{x}^2/2)c_6}$.

Thus if $u = f(x, t)$ is a solution of the heat equation then $u = f(x + c_1, t)$ is again a solution of the heat equation. Likewise, $u = f(x, t + c_2)$, $u = f(xe^{\frac{c_3}{2}}, te^{c_3})$,
$u = e^{-c_4}f(x, t)$, $u = e^{-xc_5 + 2t(c_5)^2}f(x - 2tc_5, t)$, $u = e^{-(\bar{t} + \bar{x}^2/2)c_6}f\left(xe^{\frac{-2t^2 c_6}{1 + 2tc_6}}, \frac{t}{1 + 2tc_6}\right)$

are also a solution of the heat equation.

So there exist a number of possibilities of symmetry solution. Let us see the possible symmetry solution of the heat equation

(i) Heat equation admits translation group in time t and dilation group in space x. The infinitesimal form of these two groups are $\bar{x} = x, \bar{t} = t + c_2$ and $\bar{u} = e^{c_4}u$. One can show that the heat equation is form invariant under these groups. Combining these two groups one will obtain the generator as

$$X = c_2 \frac{\partial}{\partial t} + c_4 u \frac{\partial}{\partial u}$$

The characteristic equation can be written as

$$\frac{dx}{0} = \frac{dt}{c_2} = \frac{du}{c_4 u}$$

Upon solving the above characteristic equation the invariants are obtained as $x = \tau$ (say) and $u = e^{\frac{c_4}{c_2}t}f(\tau)$, f is an arbitrary function of τ.

Now substituting $u_t = ce^{ct}f(\tau)$, $u_{xx} = e^{ct}f''(\tau)$ in the heat equation the following similarity equation is obtained by

$$f''(\tau) - cf(\tau) = 0$$

This equation will give three solutions as follows:

(a) When $c = 0$, $f(\tau) = \alpha\tau + \beta$, α, β are arbitrary constants hence $u(x, t) = \alpha x + \beta$.

(b) When $c = \mu^2$, $f(\tau) = \alpha e^{\mu\tau} + \beta e^{-\mu\tau}$ which gives $u(x, t) = e^{\mu^2 t}(\alpha e^{\mu x} + \beta e^{-\mu x})$

(c) When $c = -\mu^2,$ $f(\tau) = \alpha \cos \mu\tau + \beta \sin \mu\tau$ which   gives
$u(x,t) = e^{-\mu^2 t}(\alpha \cos \mu x + \beta \sin \mu x)$
α, β are arbitrary constants

(ii) Heat equation admits two translation groups, one in space x and the other in
time t and their combinations give the following infinitesimal generator

$$c_1 \frac{\partial}{\partial x} + c_2 \frac{\partial}{\partial t}$$

The corresponding characteristic equation is given by

$$\frac{dx}{c_1} = \frac{dt}{c_2} = \frac{du}{0}$$

The invariants are given by

$$x - \frac{c_1}{c_2}t = x - ct = \tau(\text{say}) \text{ and } u = f(\tau)$$

Substituting $u = f(\tau), u_t = -cf'(\tau), u_{xx} = f''(\tau)$ the heat equation become

$$f''(\tau) + cf'(\tau) = 0$$

The general solution is readily obtained as

$$f(\tau) = \alpha e^{-c\tau} + \beta \text{ or, } u(x,t) = \alpha e^{-c(x-ct)} + \beta$$

where α, β are arbitrary constants.

(iii) Heat equation admits the following scaling groups

$$\frac{x}{2} \frac{\partial}{\partial x} + t \frac{\partial}{\partial t} \text{ and } u \frac{\partial}{\partial u}$$

The combination of these groups is given by

$$\frac{c_3 x}{2} \frac{\partial}{\partial x} + c_3 t \frac{\partial}{\partial t} + c_4 u \frac{\partial}{\partial u}$$

The corresponding characteristic equation is given by

$$\frac{2dx}{c_3 x} = \frac{dt}{c_3 t} = \frac{du}{c_4 u} \tag{8.74}$$

On solving these equations, the following invariants are obtained

$$\frac{x}{t^{1/2}} = \tau \quad \text{and} \quad ut^{\frac{-c_4}{c_3}} = f(\tau) \quad \text{or} \quad ut^{-c} = f(\tau) \quad \text{where } c = \frac{c_4}{c_3}$$

Substituting $u = t^c f(\tau), u_{xx} = t^{c-1} f''(\tau)$ and $u_t = t^{c-1}(cf(\tau) - \frac{1}{2}\tau f'(\tau))$, the heat equation become

$$f''(\tau) + \left(\frac{1}{2}\tau\right)f'(\tau) - cf(\tau) = 0$$

$$f(\tau) = e^{-\frac{\tau^2}{4}}\left(c_1 \text{Hermite}H\left[-1-2c,\frac{\tau}{2}\right] + c_2 \text{Hypergeometric1}F1\left[\frac{1}{2}(1+2c),\frac{1}{2},\frac{\tau^2}{4}\right]\right)$$

Hence

$$u(x,t) = t^c e^{\frac{-x^2}{4t}}\left(c_1 \text{Hermite}H\left[-1-2c,\frac{x}{2\sqrt{2}}\right] + c_2 \text{Hypergeometric1}F1\left[\frac{1}{2}(1+2c),\frac{1}{2},\frac{x^2}{4t}\right]\right).$$

(iv) Heat equation admits the group

$$-2tx\frac{\partial}{\partial x} - 2t^2\frac{\partial}{\partial t} + \left(t + \frac{x^2}{2}\right)u\frac{\partial}{\partial u}$$

The corresponding characteristic equation can be set up as

$$\frac{dx}{-2tx} = \frac{dt}{-2t^2} = \frac{du}{\left(t + \frac{x^2}{2}\right)u}$$

The invariants are given by $\frac{x}{t} = \tau$ say and $u\sqrt{t}e^{\frac{x^2}{4t}} = f(\tau)$
Now substituting

$$u = \frac{e^{-\frac{x^2}{4t}}}{\sqrt{t}}f(\tau), \quad u_t = \frac{-1}{2t^{3/2}}e^{\frac{-x^2}{4t}}f(\tau) + \frac{x^2}{4t^{5/2}}e^{\frac{-x^2}{4t}}f(\tau) + \frac{-x}{t^{5/2}}e^{\frac{-x^2}{4t}}f'(\tau)$$

$$u_{xx} = \frac{-e^{\frac{-x^2}{4t}}}{2t^{3/2}}f(\tau) + \frac{x^2 e^{\frac{-x^2}{4t}}f(\tau)}{4t^{5/2}} - \frac{x}{t^{5/2}}e^{\frac{-x^2}{4t}}f'(\tau) + \frac{e^{\frac{-x^2}{4t}}f''(\tau)}{t^{5/2}}$$

then the heat equation reduces to

$$f''(\tau) = 0$$

which gives $f(\tau) = \alpha\tau + \beta, \alpha, \beta$ are arbitrary constant. Hence $u(x,t) = \frac{1}{\sqrt{t}}e^{\frac{-x^2}{4t}}\left(\frac{\alpha x}{t} + \beta\right).$

(v)   Heat equation admits the group $X_5 = -2t\frac{\partial}{\partial x} + ux\frac{\partial}{\partial u}$.

One can show that the heat equation is form invariant under these groups. The characteristic equation can be written as

$$\frac{dx}{-2t} = \frac{dt}{0} = \frac{du}{ux}$$

Upon solving the above characteristic equation the invariants are obtained as $\tau = t$ (say) and $u = e^{\frac{-x^2}{4t}}f(\tau)$, f is an arbitrary function of τ.

Now substituting

$$u_t = \frac{x^2}{4t^2}e^{\frac{-x^2}{4t}}f(\tau) + e^{\frac{-x^2}{4t}}f'(\tau), u_{xx} = \frac{-1}{2t}e^{\frac{-x^2}{4t}}f(\tau) + \frac{x^2}{4t^2}e^{\frac{-x^2}{4t}}f(\tau)$$

in the heat equation the following similarity equation is obtained

$$f'(\tau) + \frac{f(\tau)}{2\tau} = 0$$

which gives $f(\tau) = \frac{c}{\sqrt{\tau}}$, where c is an arbitrary constant. Hence $u(x,t) = \frac{c}{\sqrt{t}}e^{\frac{-x^2}{4t}}$.

This completes the analysis of linear heat conducting equation.

Example 8.23 Find the point transformation of the equation

$$\frac{\partial u}{\partial x}\frac{\partial^2 u}{\partial x^2} + \frac{\partial^2 u}{\partial y^2} = 0$$

Solution The given equation is a nonlinear partial differential equation where x, t are the independent variables and u is the dependent variable. The infinitesimal generator for this partial differential equation can be written as

$$X = \xi^x\frac{\partial}{\partial x} + \xi^y\frac{\partial}{\partial y} + \eta\frac{\partial}{\partial u}$$

Since the given equation is of order two the above infinitesimal generator is twice extended.

The extended generator is given by

$$X^{[2]} = \xi^x\frac{\partial}{\partial x} + \xi^y\frac{\partial}{\partial y} + \eta\frac{\partial}{\partial u} + \zeta_x\frac{\partial}{\partial u_x} + \zeta_{xx}\frac{\partial}{\partial u_{xx}} + \zeta_{yy}\frac{\partial}{\partial u_{yy}}$$

Now applying the operator $X^{[2]}$ to the given equation, we have

$$u_{xx}\zeta_x + u_x\zeta_{xx} + \zeta_{yy} = 0$$

which gives

$$
\begin{aligned}
u_{xx}(\eta_x &+ u_x\eta_u - u_x\xi_x^x - (u_x)^2\xi_u^x - u_y\xi_x^y \\
&- u_xu_y\xi_u^y) + u_x(\eta_{xx} + 2u_x\eta_{xu} + u_{xx}\eta_u + (u_x)^2\eta_{uu} - 2u_{xx}\xi_x^x \\
&- u_x\xi_{xx}^x - 2(u_x)^2\xi_{xu}^x - 3u_xu_{xx}\xi_u^x - (u_x)^3\xi_{uu}^x - 2u_{xy}\xi_x^y \\
&- u_y\xi_{xx}^y - 2u_xu_y\xi_{xu}^2 - (u_yu_{xx} + 2u_xu_{xy})\xi_u^y - (u_x)^2u_y\xi_{uu}^y) + \eta_{yy} \\
&+ 2u_y\eta_{yu} + u_{yy}\eta_u + (u_y)^2\eta_{uu} - 2u_{yy}\xi_y^y - u_y\xi_{yy}^y - 2(u_y)^2\xi_{yu}^y \\
&- 3u_yu_{yy}\xi_u^y - (u_y)^3\xi_{uu}^y - 2u_{xy}\xi_y^x - u_x\xi_{yy}^x - 2u_xu_y\xi_{yu}^x \\
&- (u_xu_{yy} + 2u_yu_{xy})\xi_u^x - u_x(u_y)^2\xi_{uu}^x = 0
\end{aligned}
$$

Grouping the coefficients of like terms we have

$$
\begin{aligned}
u_{xx}(\eta_x &+ u_x\eta_u - u_x\xi_x^x - (u_x)^2\xi_u^x - u_y\xi_x^y - u_xu_y\xi_u^y) + u_x(\eta_{xx} + 2u_x\eta_{xu} + u_{xx}\eta_u + (u_x)^2\eta_{uu} \\
&- 2u_{xx}\xi_x^x - u_x\xi_{xx}^x - 2(u_x)^2\xi_{xu}^x - 3u_xu_{xx}\xi_u^x - (u_x)^3\xi_{uu}^x - 2u_{xy}\xi_x^y - u_y\xi_{xx}^y - 2u_xu_y\xi_{xu}^2
\end{aligned}
$$

or,

$$
\begin{aligned}
&- (u_yu_{xx} + 2u_xu_{xy})\xi_u^y - (u_x)^2u_y\xi_{uu}^y) + \eta_{yy} + 2u_y\eta_{yu} + u_{yy}(\eta_u - 2\xi_y^y - 3u_y\xi_u^y - u_x\xi_u^x) \\
&+ (u_y)^2\eta_{uu} - u_y\xi_{yy}^y - 2(u_y)^2\xi_{yu}^y - (u_y)^3\xi_{uu}^y - 2u_{xy}\xi_y^x - u_x\xi_{yy}^x - 2u_xu_y\xi_{yu}^x - 2u_yu_{xy}\xi_u^x - u_x(u_y)^2\xi_{uu}^x = 0
\end{aligned}
$$

Replacing u_{yy} by $-u_xu_{xx}$, we have

$$
\begin{aligned}
u_{xx}(\eta_x) &+ u_xu_{xx}(\eta_u - 3\xi_x^x + 2\xi_y^y) - u_yu_{xx}\xi_x^y \\
&+ u_x(\eta_{xx} - \xi_{yy}^x) + (u_x)^2(2\eta_{xu} - \xi_{xx}^x) \\
&+ (u_x)^3(\eta_{uu} - 2\xi_{xu}^x) - 3(u_x)^2u_{xx}\xi_u^x - (u_x)^4\xi_{uu}^x - 2u_xu_{xy}\xi_x^y \\
&+ u_y(-\xi_{xx}^y + 2\eta_{yu} - \xi_{yy}^y) - 2u_xu_y(\xi_{xu}^y + \xi_{yu}^x) \\
&- 2(u_x)^2u_{xy}\xi_u^y - (u_x)^3u_y\xi_{uu}^y \\
&+ \eta_{yy} + u_yu_xu_{xx}\xi_u^y + (u_y)^2(\eta_{uu} - 2\xi_{yu}^y) \\
&- (u_y)^3\xi_{uu}^y - 2u_{xy}\xi_y^x - 2u_yu_{xy}\xi_u^x - u_x(u_y)^2\xi_{uu}^x = 0
\end{aligned}
$$

Now equating the coefficient of each of $u_x, u_{xy}, u_{xx}, u_{yy}$ and the product of these terms equal to zero, we will obtain the following overdetermined system of linear partial differential equations in ξ, η

$$\eta_x = 0, \ \eta_u - 3\xi_x^x + 2\xi_y^y = 0, \ \xi_x^y = 0, \ 2\eta_{xu} - \xi_{xx}^x = 0, \ \eta_{uu} - 2\xi_{xu}^x = 0$$

$$\xi_u^x = 0, \ -\xi_{xx}^y + 2\eta_{yu} - \xi_{yy}^y = 0, \ \xi_{xu}^y + \xi_{yu}^x = 0, \ \xi_u^y = 0, \ \xi_{uu}^y = 0$$

$$\eta_{yy} = 0, \ \eta_{uu} - 2\xi_{yu}^y = 0, \ \xi_y^x = 0$$

Since ξ^x is a function of x only, we have from the relation $2\eta_{xu} - \xi_{xx}^x = 0$, $\xi^x = c_1 x + c_2$. Again ξ^y is also a function of y only then from the relation $-\xi_{xx}^y + 2\eta_{yu} - \xi_{yy}^y = 0$, we have $-3\xi_{yy}^y = 0$ which gives $\xi^y = c_3 y + c_4$. Using the values of ξ^x and ξ^y in the relation $\eta_u - 3\xi_x^x + 2\xi_y^y = 0$ gives $\eta_u = 3c_1 - 2c_3$ which gives $\eta = (3c_1 - 2c_3)u + f(y)$. From $\eta_{yy} = 0, \frac{d^2}{dy^2}(f(y)) = 0$ which gives $f(y) = c_5 y + c_6$.

Thus the infinitesimals are

$$\xi^x = c_1 x + c_2, \ \xi^y = c_3 y + c_4 \text{ and } \eta = (3c_1 - 2c_3)u + c_5 y + c_6$$

Hence the given equation is invariant under six parameter group generated by the infinitesimal generators

$$X_1 = x\frac{\partial}{\partial x} + 3u\frac{\partial}{\partial u}, \ X_2 = \frac{\partial}{\partial x}, \ X_3 = y\frac{\partial}{\partial y} - 2u\frac{\partial}{\partial u}, \ X_4 = \frac{\partial}{\partial y}, \ X_5 = y\frac{\partial}{\partial u}, \ X_6 = \frac{\partial}{\partial u}$$

The corresponding commutator table is given in the Table 8.10.

From the above commutator table one can see that X_4 and X_5 generates X_6. Hence, the six parameter symmetry group $\mathcal{L}_6$ can be constructed by knowing only X_1, X_2, X_3, X_4, X_5. Also, since $[X_4, X_5] = X_6$, therefore $\mathcal{L}(X_4, X_5)$ is not a subalgebra and become a Lie subalgebra when it also contains X_6. Thus $\mathcal{L}(X_4, X_5, X_6)$ is a subalgebra of $\mathcal{L}_6$. Likewise, looking at the commutator table the various Lie subalgebra of $\mathcal{L}_6$ can be easily determined.

Table 8.10 Commutator table

	X_1	X_2	X_3	X_4	X_5	X_6
X_1	0	$-X_2$	0	0	$-3X_5$	$-3X_6$
X_2	X_2	0	0	0	0	0
X_3	0	0	0	$-X_4$	$3X_5$	$2X_6$
X_4	0	0	X_4	0	X_6	0
X_5	$3X_5$	0	$-3X_5$	$-X_6$	0	0
X_6	$3X_6$	0	$-2X_6$	0	0	0

Now combining the different generators we will obtain different solutions. Here we are giving only two such combinations

(i) Combining X_2, X_5, X_6 we have the generator

$$c_2 \frac{\partial}{\partial x} + (c_5 y + c_6) \frac{\partial}{\partial u}$$

The corresponding characteristic equation is

$$\frac{dx}{c_2} = \frac{dy}{0} = \frac{du}{(c_5 y + c_6)}$$

Solving the above characteristic equation we have the invariants as $y = \tau$(constant) and $u - \frac{(c_5 y - c_6)}{c_2} = f(\tau)$. Substituting these invariants in the given equation, we will obtain

$$\frac{d^2 f}{d\tau^2} = 0$$

Which gives $f(\tau) = a\tau + b$, a, b are arbitrary constants. In terms of original variables the solution is given as $u = \frac{(c_5 y - c_6)}{c_2} + f(\tau) = \left(a + \frac{c_5}{c_2}\right) y + b - \frac{c_6}{c_2}$.

(ii) Combining X_4, X_5, X_6 we have the generator

$$c_4 \frac{\partial}{\partial y} + (c_5 y + c_6) \frac{\partial}{\partial u}$$

The corresponding characteristic equation is

$$\frac{dx}{0} = \frac{dy}{c_4} = \frac{du}{(c_5 y + c_6)}$$

Solving the above equations we will obtain the invariants as $x = $ constant $= \tau$(say)   and $u - \left(\frac{c_5}{2c_4} y^2 + \frac{c_6}{c_4} y\right) = f(\tau)$($f$ is an arbitrary function of τ). Substituting these invariants in the given equation we have

$$f'(\tau) f''(\tau) + \frac{c_5}{c_4} = 0$$

One time integration gives $\left(\frac{df}{d\tau}\right)^2 + 2\frac{c_5}{c_4}\tau = $ constant $= a$(say), further integration gives $f(\tau) = -\frac{c_4}{3c_5}\sqrt{\left(a - \frac{2c_5}{c_4}\tau\right)} + b$. In terms of original variables the solution is given by $u = \left(\frac{c_5}{2c_4} y^2 + \frac{c_6}{c_4} y\right) - \frac{c_4}{3c_5}\sqrt{\left(a - \frac{2c_5}{c_4} x\right)} + b$.

Example 8.24 Find the point transformation of the Burger equation

$$u_t - u_{xx} - uu_x = 0 \tag{8.75}$$

Solution J.M. Burger formulated the Burger equation in 1939 in order to know the physics of governing turbulent motion. The Burger's equation is a nonlinear parabolic type equation where the nonlinear convection uu_x balances the dissipation u_{xx}. Burger's equation is similar to the nonlinear heat conduction equation. This equation has two independent variables x and t and one dependent variable u. Therefore, the infinitesimal generator of this equation has the following form

$$X = \xi^1 \frac{\partial}{\partial x} + \xi^2 \frac{\partial}{\partial t} + \eta \frac{\partial}{\partial u} \tag{8.76}$$

Since the governing equation is of second order, the infinitesimal generator (8.76) can be twice extended or prolonged. The twice extended generator can be written as

$$X^{[2]} = \xi^1(x,t,u) \frac{\partial}{\partial x} + \xi^2(x,t,u) \frac{\partial}{\partial t} + \eta(x,t,u) \frac{\partial}{\partial u} + \zeta^1 \frac{\partial}{\partial u_x} + \zeta^2 \frac{\partial}{\partial u_t} + \zeta^{11} \frac{\partial}{\partial u_{xx}} \tag{8.77}$$

This operator when applied to the given equation gives the determining equation as

$$-\eta u_x - \zeta^1 u + \zeta^2 - \zeta^{11} = 0 \tag{8.78}$$

The coefficients are given as

$$\zeta^1 = \eta_x + u_x(\eta_u - \xi_x^1) - (u_x)^2 \xi_u^1 - u_t \xi_x^2 - u_x u_t \xi_u^2$$

$$\zeta^2 = \eta_t + u_t \eta_u - u_x \xi_t^1 - (u_t)^2 \xi_u^2 - u_t \xi_t^2 - u_x u_t \xi_u^1$$

and

$$\zeta^{11} = \eta_{xx} + u_x(2\eta_{xu} - \xi_{xx}^1) + u_{xx}(\eta_u - 2\xi_x^1) + (u_x)^2(\eta_{uu} - 2\xi_{xu}^1)$$
$$- 3u_x u_{xx} \xi_u^1 - (u_x)^3 \xi_{uu}^1 - 2u_{xt} \xi_x^2 - u_t \xi_{xx}^2 - 2u_x u_t \xi_{xu}^2 - u_t u_{xx} \xi_u^2$$
$$- 2u_x u_{xt} \xi_u^2 - (u_x)^2 u_t \xi_{uu}^2$$

Substituting the above infinitesimal coefficient in the determining equation and then setting u_{xx} as $u_t - uu_x$ we have

$$- \eta u_x - u(\eta_x + u_x(\eta_u - \xi_x^1) - (u_x)^2 \xi_u^1 - u_t \xi_x^2 - u_x u_t \xi_u^2) + \eta_t$$

$$+ u_t \eta_u - u_x \xi_t^1 - (u_t)^2 \xi_u^2 - u_t \xi_t^2 - u_x u_t \xi_u^1 - \eta_{xx} - u_x(2\eta_{xu} - \xi_{xx}^1) + \dot{u} u_x(\eta_u - 2\xi_x^1)$$

$$- u_t(\eta_u - 2\xi_x^1) - (u_x)^2(\eta_{uu} - 2\xi_{xu}^1) - 3u(u_x)^2 \xi_u^1 + 3u_x u_t \xi_u^1 + (u_x)^3 \xi_{uu}^1 + 2u_{xt} \xi_x^2 + u_t \xi_{xx}^2$$

$$+ 2u_x u_t \xi_{xu}^2 + (u_t)^2 \xi_u^2 - u u_t u_x \xi_u^2 + 2u_x u_{xt} \xi_u^2 + (u_x)^2 u_t \xi_{uu}^2 = 0$$

Rearranging the terms of the above equation, we have

$$- u_x(\eta + 2\eta_{xu} - \xi_{xx}^1 + \xi_t^1 + u\xi_x^1) + (\eta_t - \eta_{xx}$$

$$- u\eta_x) + u_t(\xi_{xx}^2 - \xi_t^2 + 2\xi_x^1 + u\xi_x^2) - (u_x)^2(\eta_{uu} - 2\xi_{xu}^1 + 2u\xi_u^1)$$

$$+ u_x u_t(2\xi_u^1 + 2\xi_{xu}^2) + (u_x)^3 \xi_{uu}^1 + 2u_{xt} \xi_x^2 + 2u_x u_{xt} \xi_u^2 + (u_x)^2 u_t \xi_{uu}^2 = 0$$

Splitting the above equation we obtain the following overdetermined equation

$$\xi_u^2 = 0, \; \xi_{uu}^1 = 0, \; \xi_{uu}^2 = 0, \; \xi_x^2 = 0, \; -2u\xi_u^1 - \eta_{uu} + 2\xi_{xu}^1 = 0$$

$$\xi_u^1 + \xi_{xu}^2 = 0, \eta_t - u\eta_x - \eta_{xx} = 0, \; -\eta - \xi_t^1 - u\xi_x^1 - 2\eta_{xu} + \xi_{xx}^1 = 0$$

$$-\xi_t^2 + 2\xi_x^1 + u\xi_x^2 + \xi_{xx}^2 = 0$$

Solving the equations the infinitesimals are obtained as

$$\xi^1(x, t, u) = c_1 - c_3 t - c_4 x - c_5 xt$$

$$\xi^2(x, t, u) = -2c_4 t - c_5 t^2 + c_2 \tag{8.79}$$

$$\eta(x, t, u) = c_3 + c_5 x + (c_4 + c_5 t)u$$

Thus, Burger equation admits five-dimensional Lie algebra spanned by the generators.

$$X_1 = \frac{\partial}{\partial x}, \; X_2 = \frac{\partial}{\partial t}, \; X_3 = -t\frac{\partial}{\partial x} + \frac{\partial}{\partial u},$$

$$X_4 = -x\frac{\partial}{\partial x} - 2t\frac{\partial}{\partial t} + u\frac{\partial}{\partial u}, \; X_5 = -xt\frac{\partial}{\partial x} - t^2\frac{\partial}{\partial t} + (x + tu)\frac{\partial}{\partial u}$$

From the commutator table (Table 8.11) we can see that X_2 and X_3 generates X_1. Also X_1 and X_5 generate X_3 and X_2 and X_5 generate X_4. Thus knowing X_1, X_2, and

Table 8.11 Commutator table of Burger's equation

	X_1	X_2	X_3	X_4	X_5
X_1	0	0	0	$-X_1$	X_3
X_2	0	0	$-X_1$	$-2X_2$	X_4
X_3	0	X_1	0	X_3	0
X_4	X_1	$2X_2$	$-X_3$	0	$-2X_5$
X_5	$-X_3$	$-X_4$	0	$2X_5$	0

X_5 the five-dimensional Lie algebra of the Burger equation can be constructed. This suggests that X_1, X_5 without X_3 and X_2, X_5 without X_4 cannot be a Lie subalgebra of the Lie group $\mathcal{L}_5$. Hence $\mathcal{L}(X_1, X_3, X_5)$ and $\mathcal{L}(X_2, X_4, X_5)$ is a Lie subalgebra of $\mathcal{L}_5$. Again each one-dimensional Lie subgroup is a Lie subalgebra of $\mathcal{L}_5$.

Let us now see few of the possible similarity solution of the Burger equation

(i) For translation group $X_1 = \frac{\partial}{\partial x}$, we have $u = f(t)$, which gives $f'(t) = 0$. Hence $u = $ constant is the trivial solution.

(ii) For translation group $X_2 = \frac{\partial}{\partial t}$, we have $u = f(x)$, which gives

$$\frac{d^2 f}{dx^2} + f(x)\frac{df}{dx} = 0 \tag{8.80}$$

Integrating once we obtain

$$\frac{df}{dx} + \frac{f^2}{2} = \text{constant} = a(\text{say}) \tag{8.81}$$

When $a = 0$, the Eq. (8.81) upon integration gives $f(x) = \frac{2}{x - a_1}$ or $u(x,t) = \frac{2}{x - a_1}$, where a_1 is the integration constant.
When $a = \mu^2 (\mu \neq 0)$, then

$$f(x) = \sqrt{2}\mu \tanh\left(\frac{1}{2}\left(\sqrt{2}x v - 2\sqrt{2}v C_1\right)\right)$$

$$u(x,t) = \sqrt{2}\mu \tanh\left(\frac{1}{2}\left(\sqrt{2}x v - 2\sqrt{2}v C_1\right)\right).$$

When $a = -\mu^2 (\mu \neq 0)$, then

$$f(x) = \sqrt{2}\mu \tan\left(\frac{1}{2}\left(-\sqrt{2}x \mu + 2\sqrt{2}\mu C_1\right)\right)$$

(iii) Combining X_1 and X_2 we get $c_1\frac{\partial}{\partial x} + c_2\frac{\partial}{\partial t}$. The characteristic equation is therefore given by

$$\frac{dx}{c_1} = \frac{dt}{c_2} = \frac{du}{0}$$

This gives the invariants as $x - \frac{c_1}{c_2}t = x - ct = \tau$ (say) and $u = \text{constant} = f(\tau)$ (say). Substituting into the Burger equation gives the similarity equation as

$$f''(\tau) + f(\tau)f'(\tau) + cf'(\tau) = 0$$

Integrating once one would obtain $f'(\tau) + \frac{1}{2}(f(\tau))^2 + cf(\tau) = a(\text{say})$

Which upon integration gives $f(\tau) = \sqrt{c^2 + 2a}\tanh\left(b - \frac{\sqrt{c^2+2a}}{2}\tau\right) + c$ or $f(\tau) = \sqrt{c^2 + 2a}\coth\left(b + \frac{\sqrt{c^2+2a}}{2}\right)$ according as $|f(\tau) + c| < |\sqrt{c^2 + 2a}|$ or $|f(\tau) + c| > |\sqrt{c^2 + 2a}|$.

(iv) For the generator $X_3 = -t\frac{\partial}{\partial x} + \frac{\partial}{\partial u}$, we have the invariants $t = \tau$(constant) and $u = \frac{-x}{\tau} + f(\tau)$, then we have

$$\frac{df}{d\tau} + \frac{f(\tau)}{\tau} = 0 \tag{8.82}$$

Upon integration the above equation gives $f(\tau) = \frac{c}{\tau}$, c is an arbitrary integration constant. In terms of original variables we have $u(x,t) = \frac{-x+c}{t}$.

For the generator $X_4 = -x\frac{\partial}{\partial x} - 2t\frac{\partial}{\partial t} + u\frac{\partial}{\partial u}$ the characteristic equations are $\frac{dx}{-x} = \frac{dt}{-2t} = \frac{du}{u}$ which gives the invariants as $\frac{x}{\sqrt{t}} = \tau$(say) and $u = \frac{f(\tau)}{\sqrt{t}}$. Now substituting u, $u_t = \frac{-1}{2t^{3/2}}f(\tau) - \frac{1}{2t^{3/2}}\tau f'(\tau)$, $u_x = \frac{1}{t}f'(\tau)$, and $u_{xx} = \frac{1}{t^{3/2}}f''(\tau)$ in the Burger's equation, one will obtain the similarity equation as

$$\frac{-1}{2}f(\tau) - \frac{1}{2}\tau f'(\tau) = f''(\tau) + f(\tau)f'(\tau)$$

Integrating once the above equation, we have

$$f'(\tau) + \frac{1}{2}(f(\tau))^2 + \frac{\tau f(\tau)}{2} = a(\text{say})$$

(v) For the generator $X_5 = -xt\frac{\partial}{\partial x} - t^2\frac{\partial}{\partial t} + (x + tu)\frac{\partial}{\partial u}$, we have the invariants $\frac{x}{t} = \tau$ (say) and $u = -\tau + \frac{f(\tau)}{t}$. Now substituting $u_t = \frac{x}{t} - \frac{f(\tau)}{t^2} - \frac{x}{t^3}f'(\tau)$, $u_x = \frac{-1}{t} + \frac{f'(\tau)}{t^2}$, $u_{xx} = \frac{f''(\tau)}{t^3}$ in the Burger's equation one will obtain the similarity equation as

$$\frac{d^2 f}{d\tau^2} + f(\tau)\frac{df}{d\tau} = 0$$

Which upon integration gives $2\frac{df}{d\tau} + (f(\tau))^2 = C$. This equation gives the following solutions according to the value of C.

(a) When $C = \mu^2$ and $\mu > f(\tau)$, then $f(\tau) = \frac{Ce^{\mu\tau} - 1}{Ce^{\mu\tau} + 1}$ hence $u(x,t) = -\frac{x}{t} + \frac{1}{t}\left(\frac{Ce^{\mu(x/t)} - 1}{Ce^{\mu(x/t)} + 1}\right)$, where C is an arbitrary integration constant

(b) When $C = \mu^2$ and $\mu < f(\tau)$, then $f(\tau) = \frac{Ce^{\mu\tau} + 1}{Ce^{\mu\tau} - 1}$ hence $u(x,t) = -\frac{x}{t} + \frac{1}{t}\left(\frac{Ce^{\mu(x/t)} + 1}{Ce^{\mu(x/t)} - 1}\right)$

(c) When $C = -\mu^2$ then $f(\tau) = \mu \tan\left(c - \frac{\mu\tau}{2}\right)$ hence $u(x,t) = \frac{-x}{t} + \frac{\mu}{t}\tan\left(c - \frac{\mu x}{2t}\right)$.

(d) When $C = 0$, then $f(\tau) = \frac{2}{\tau+c}$ hence $u(x,t) = \frac{-x}{t} + \frac{2}{t\left(\frac{x}{t}+c\right)} = \frac{-x}{t} + \frac{2}{(x+tc)}$, c is an arbitrary integration.

8.10 Symmetry Analysis for Boundary Value Problems

So far the invariant solutions of partial differential equations are constructed without considering the boundary influences on these solutions. Thus, the solutions are not all physically valid. In other words some of the solutions may not exist physically. However, an applied mathematician or a physicist has interest only in those solutions which have physical significance and can be considered in real-life or engineering problems. The boundary value problems are difficult to solve in contrast to the problems without any boundary conditions. In boundary value problems, the differential equation, the surfaces on which the boundary conditions are defined and the conditions themselves must be invariant under the infinitesimal transformations. However, for linear partial differential equations with homogeneous boundary conditions and one nonhomogeneous boundary condition it is not necessary to keep every surface invariant. An infinitesimal generator is admitted by the partial differential equation if it leaves the partial differential equation along with the homogeneous boundary condition invariant. The superposition of the invariant solution for the homogeneous boundary conditions is then used to solve the nonhomogeneous boundary condition. For a detailed theory see Bluman and Kumei [19].

Thus, a boundary value problem admits an infinitesimal generator X if the partial differential equation $\mathcal{F}(x, u, u^{(1)}, u^{(2)}, \ldots, u^{(k)}) = 0, x = (x_1, x_2, \ldots, x_n)$ is invariant under the kth extended form of X, i.e., $X^k\mathcal{F}\big|_{\mathcal{F}=0} = 0$. Again the surfaces $s_\alpha(x) = 0$ are invariant under X, i.e., $Xs_\alpha\big|_{s_\alpha=0} = 0$ and the boundary conditions $\mathcal{B}(x, u, u^{(1)}, \ldots, u^{(k-1)}) = 0$ on $s_\alpha(x) = 0$ are invariant under $(k-1)$th extended form of X, i.e., $X^{k-1}\mathcal{B}\big|_{\mathcal{B}=0} = 0$.

We have already established the Lie point group of transformation for the heat equation. Here, we shall illustrate below the heat equation with different boundary conditions.

Example 8.25. *(Heat equation with Dirichlet's condition)* Solve the equation $u_t = u_{xx}, t > 0, a < x < b$ under the nonhomogeneous boundary conditions

(i) $u(a, t) = u(b, t) = 0, t \geq 0$
(ii) $u(x, t)$ is finite as $t \to \infty$
(iii) $u(x, 0) = f(x), a \leq x \leq b$.

Solution The heat equation is invariant under the infinitesimal generator $X = \frac{\partial}{\partial t} + cu\frac{\partial}{\partial u}$, $c = \frac{c_4}{c_2}$, $c_2 \neq 0$. Again it keeps invariant the surfaces $x = a$, $x = b$ and the boundary conditions (i) and (ii). Since heat equation is linear, it is not necessary that the infinitesimal generator keep every surfaces invariant (see Bluman and Kumei [19]). We have seen that heat equation admits three solutions under the infinitesimal generator X. Now if $c = 0$ then the heat equation admits only trivial solution under given conditions, again if $c = \mu^2$, $\mu \neq 0$ then it violates the condition (ii), hence, we should take $c = -\mu^2$, $\mu \neq 0$ which gives the solution of heat equation as $u(x,t) = e^{-\mu^2 t}(\alpha \cos \mu(x-a) + \beta \sin \mu(x-a))$. After applying the condition (i) we have $\alpha = 0$ and $\mu = \frac{n\pi}{b-a}$, $n = 0, 1, 2, 3, \ldots$ and after superposing all invariant solutions, the solution of the heat equation become $u(x,t) = \sum_{n=1}^{\infty} \beta_n e^{-\mu^2 t} \sin \frac{n\pi}{(b-a)}(x-a)$ where β_n are arbitrary constants. Now the third boundary condition gives $f(x) = \sum_{n=1}^{\infty} \beta_n \sin \frac{n\pi}{(b-a)}(x-a)$, which after multiplying both sides by $\sin \frac{m\pi}{(b-a)}(x-a)$ and then integrating between the limits a and b gives the constants $\beta_n = \frac{2}{b-a} \int_a^b f(x) \sin \frac{n\pi}{b-a}(x-a)dx$ hence, the required solution of the heat equation under Dirichlet's boundary conditions is

$$u(x,t) = \sum_{n=1}^{\infty} \beta_n e^{-\mu^2 t} \sin \frac{n\pi}{(b-a)}(x-a), \quad \text{where } \mu = \frac{n\pi}{b-a}, n = 0, 1, 2, 3, \ldots,$$

and

$$\beta_n = \frac{2}{b-a} \int_a^b f(x) \sin \frac{n\pi}{b-a}(x-a)dx.$$

Again one can easily check that the given boundary conditions are not invariant under all other infinitesimals given in Example 8.22. For instance, the infinitesimal generator $X = \frac{\partial}{\partial x} + c\frac{\partial}{\partial t}$, $c = \frac{c_2}{c_1}$, $c_1 \neq 0$ does not satisfy the surface conditions as $X(x-a)|_{x=a} = 1 \neq 0$ and $X(x-b)|_{x=b} = 1 \neq 0$ hence the surfaces $x = a$ and $x = b$ are not invariant under this generator. So we can conclude that the given boundary value problem is not an invariant under this infinitesimal generator. Again for the generator $\frac{x}{2}\frac{\partial}{\partial x} + t\frac{\partial}{\partial t} + cu\frac{\partial}{\partial u}$, $c = \frac{c_4}{c_3}$, $c_3 \neq 0$ the surface $x = a$ and $x = b$ are invariant only if $x = 0$. Hence, the homogeneous boundary value problem is not an invariant under this generator. Thus, the given Dirichlet's problem is invariant only under translation in time t and dilation in the temperature u.

Example 8.26 (Heat equation with Neumann condition) Solve the equation $u_t = u_{xx}$, $t > 0$, $a < x < b$ under the nonhomogeneous boundary conditions

(i) $u_x(a,t) = u_x(b,t) = 0, t \geq 0$
(ii) $u(x, t)$ is finite as $t \to \infty$
(iii) $u(x, 0) = f(x), a \leq x \leq b.$

Solution The given heat equation is invariant under the infinitesimal generator $X = \frac{\partial}{\partial t} + cu\frac{\partial}{\partial u}$, $c = \frac{c_4}{c_2}, c_2 \neq 0$. Again it keeps invariant the surfaces $x = a, x = b$ and the boundary condition (i) and (ii). Since heat equation is linear, it is not necessary that the infinitesimal generator keep every surfaces invariant (see Bluman and Kumei [19]). We have seen in Example 8.22 that heat equation admits three solutions under the infinitesimal generator X. Now if $c = 0$ then the heat equation admits only trivial solution under given conditions, again if $c = \mu^2, \mu \neq 0$ then it violates the condition (ii), hence we should take $c = -\mu^2, \mu \neq 0$ which gives the solution of heat equation as $u(x, t) = e^{-\mu^2 t}(\alpha \cos \mu(x - a) + \beta \sin \mu(x - a))$. After applying the (i) condition we have $\beta = 0$ and $\mu = \frac{n\pi}{b-a}, n = 0, 1, 2, 3, \ldots$ and after superposing all invariant solutions, the solution of the heat equation become $u(x, t) = \sum_{n=1}^{\infty} \alpha_n e^{-\mu^2 t} \cos \frac{n\pi}{(b-a)}(x - a)$ where α_n is arbitrary constant. Now the third boundary condition gives $f(x) = \sum_{n=0}^{\infty} \alpha_n \cos \frac{n\pi}{(b-a)}(x - a)$ which after multiplying both sides by $\cos \frac{m\pi}{b-a}(x - a)$ and then integrating between the limits a and b gives the constants $\alpha_n = \frac{2}{b-a}\int_a^b f(x) \cos \frac{n\pi}{b-a}(x - a)dx, n \neq 0$ and $\alpha_0 = \frac{1}{b-a}\int_a^b f(x)dx$. Hence the required solution of the heat equation under Neumann boundary conditions is

$$u(x, t) = \sum_{n=0}^{\infty} \alpha_n e^{-\mu^2 t} \cos \frac{n\pi}{b - a}(x - a),$$

where $\mu = \frac{n\pi}{b-a}$, $n = 0, 1, 2, 3, \ldots$ and

$$\alpha_n = \frac{2}{b - a} \int_a^b f(x) \cos \frac{n\pi}{b - a}(x - a)dx$$

for $n \neq 0$ and $\alpha_0 = \frac{1}{b-a}\int_a^b f(x)dx$. As in the previous Example (8.25) for all other infinitesimal generators for which the heat equation admits a solution, the given boundary conditions are not invariant. Thus, the given Neumann problem is invariant only under translation in time t and dilation in temperature u.

Example 8.27 Solve the equation $u_t = u_{xx}, t > 0, -\infty < x < \infty$ under the nonhomogeneous boundary conditions

(i) $u(x, 0) = \delta(x)$
(ii) $u(x, t) \to 0$ as $x \to \pm\infty, t > 0.$

Solution The infinitesimal generator of the heat equation is $X = \xi^1 \frac{\partial}{\partial x} + \xi^2 \frac{\partial}{\partial t} + \eta^1 \frac{\partial}{\partial u}$. The infinitesimal coefficients admitted by the heat equation is given by

$$\xi^1 = c_1 + \frac{c_3}{2}x - 2tc_5 - 2txc_6$$

$$\xi^2 = c_2 + c_3 t + (-2t^2)c_6$$

$$\eta = c_4 u + c_5 ux + c_6\left(tu + \frac{ux^2}{2}\right)$$

The boundary surfaces are $x = \infty, x = -\infty, t = 0$. Now the surface $t = 0$ gives $\xi^2(0) = 0$ which gives $c_2 = 0$. The surfaces $x = \infty, x = -\infty$ do not reduce the parameters. Thus the given boundary value problem admits a five parameter group. Now the boundary condition (i) gives

$$\xi^1(x,0)\frac{\partial \delta(x)}{\partial x} - \eta(x,0,\delta(x)) = 0$$

or

$$\frac{\partial}{\partial x}\left(\xi^1(x,0)\delta(x)\right) - \delta(x)\frac{\partial}{\partial x}\xi^1(x,0) - \left(c_4 + c_5 x + c_6\frac{x^2}{2}\right)\delta(x) = 0.$$

This gives

$$\frac{\partial}{\partial x}\xi^1(0,0)\delta(x) - \delta(x)\frac{\partial}{\partial x}\xi^1(x,0) - c_4\delta(x) = 0 \Rightarrow \xi^1(0,0) = c_1 = 0 \text{ and } c_4 = -\frac{\partial}{\partial x}\xi^1(x,0) = -\frac{c_3}{2}$$

Thus, the given boundary value problem admits a three parameter group given by

$$\xi^1 = \frac{c_3}{2}x - 2tc_5 - 2txc_6, \ \xi^2 = c_3 t + (-2t^2)c_6, \ \eta = -\frac{c_3}{2}u + c_5 ux + c_6\left(tu + \frac{ux^2}{2}\right)$$

This has the infinitesimal generators

$$X_3 = \frac{x}{2}\frac{\partial}{\partial x} + t\frac{\partial}{\partial t} - \frac{u}{2}\frac{\partial}{\partial u}, \ X_5 = -2t\frac{\partial}{\partial x} + ux\frac{\partial}{\partial u}, \ X_6$$

$$= -2tx\frac{\partial}{\partial x} - 2t^2\frac{\partial}{\partial t} + \left(tu + \frac{ux^2}{2}\right)\frac{\partial}{\partial u}$$

Now $X_6 = -2tx\frac{\partial}{\partial x} - 2t^2\frac{\partial}{\partial t} + \left(tu + \frac{ux^2}{2}\right)\frac{\partial}{\partial u} = -2tX_3 + \frac{x}{2}X_5$. Hence any solution invariant under X_3 and X_5 is also an invariant solution under X_6. Now we know that heat equation admits the solution $u(x,t) = \frac{c}{\sqrt{t}} e^{\frac{-x^2}{4t}}$, c is an arbitrary constants under these generators. Now the characteristics equation of X_3 is $\frac{2\mathrm{d}x}{x} = \frac{\mathrm{d}t}{t} = \frac{-2\mathrm{d}u}{u}$ which gives the invariants as $\tau = \frac{x}{\sqrt{t}}$ and $f(\tau)/\sqrt{t} = u$. Since heat equation with the given boundary conditions is a well posed problem, therefore it admits a unique solution. So, $f(\tau) = \frac{c}{\sqrt{t}}$. Now using the boundary condition(i) we have $\lim_{t \to 0} u(x,t) = \delta(x)$ now integrating both sides between the limits $-\infty$ and ∞ we have $\lim_{t \to 0} \int_{-\infty}^{\infty} \frac{c}{\sqrt{t}} e^{\frac{-x^2}{4t}} \mathrm{d}x = \int_{-\infty}^{\infty} \delta(x)\mathrm{d}x = 1$. Now substituting $v = \frac{x}{2\sqrt{t}}$, one would have $\lim_{t \to 0} \int_{-\infty}^{\infty} 2ce^{-v^2} dv = 1$. Since $\int_{-\infty}^{\infty} e^{-v^2} dv = \sqrt{\pi}$, we have $c = \frac{1}{2\sqrt{\pi}}$. Hence, the required solution of the heat equation under the conditions (i) and (ii) is $u(x,t) = \frac{1}{\sqrt{4\pi t}} e^{\frac{-x^2}{4t}}$.

8.11 Noether Theorems and Symmetry Groups

We have discussed in Chap. 7 that symmetries and conservation laws are interrelated and every symmetry group corresponds to a conservation law. We have so far discussed the point symmetries of a system of equations. Point symmetries give the known conservation laws for instance translation of time gives the conservation of energy, space translation gives the conservation of linear momentum, and rotation of space gives the conservation of angular momentum. Conservation laws of a system means that there exists a divergence expression which vanishes for all solutions of the system that is, Div P = 0 where P is a smooth function of the variables of the system. The general theory relating symmetry groups and conservation laws of a variational problem was given by German woman mathematician Amalie Emmy Noether in 1918. Her 1918 paper [21] contained two general theorems that relate the symmetry and conservation laws in variational problems of calculus of variation. While her first theorem gained immense popularity, the second theorem which has equal importance was not much appreciated, and therefore remained unknown to many. The importance of Noether theorem lies in the fact that her theorem is a general correlation connecting conservation laws and symmetry properties and all other means of deducing conservation laws and conservation laws itself are in reality particular instances of her general theory. She gave the general theory by extending Lie's theory of continuous point transformation groups to generalized symmetries also known as Lie-Backlund symmetries under which the Lagrangians depending on higher order derivatives and for systems containing more than one independent variable remain invariant. Generalized symmetries involve transformations in which the infinitesimals are dependent not only upon the independent and dependent variables but also upon the derivatives of the dependent variables up to arbitrary order. The infinitesimal operator of a Lie-Backlund group is the infinite order extension of the infinitesimal generator of the point symmetry group. The theory of Lie-Backlund group is beyond the scope of this book and interested readers may consult the books of Bluman and Kumei [19], Cantwell [18]. Here, we shall describe the invariance of a variational problem for point symmetries only.

A variational integral $\int_V L(x, u, u^{(1)}) dx$ is said to be invariant under the one parameter group of transformation

$$G : \bar{x}^i = f^i(x, u, \epsilon), \ \bar{u}^j = g^j(x, u, \epsilon) \quad i = 1, \ldots, n; j = 1, \ldots, m \tag{8.83}$$

with the infinitesimal generator

$$X = \xi^i(x, u) \frac{\partial}{\partial x^i} + \eta^j(x, u) \frac{\partial}{\partial u^j} \tag{8.84}$$

if $\int_{\bar{V}} L(\bar{x}, \bar{u}, \bar{u}^{(1)}) d\bar{x} = \int_V L(x, u, u^{(1)}) dx$, where $\bar{V} \subset \mathbb{R}^n$ is a volume obtained from V by the transformation (8.83).

Noether's first theorem states that every one parameter symmetry group of a variational problem produces a conservation law of the corresponding Euler–Lagrange equations and conversely, every one parameter symmetry group comes from a conservation law of the Euler–Lagrange equation. Her second theorem states that a variational problem admits an infinite-dimensional symmetry group, that is, the group whose infinitesimal generator consists of infinitesimals that depend on one or more arbitrary functions if and only if the corresponding Euler–Lagrange equation are underdetermined, if the nontrivial combination of their derivatives vanishes identically. Here, we do not go into the details of the second theorem and proceed with the first theorem only.

Noether first theorem is mathematically represented by the following two theorems.

Theorem 8.18 *An integral $I(u) = \int_V L(x, u, u^{(1)}) dx$ is invariant under the group G if and only if $X(L) + LD_i(\xi^i) = 0$, where X is once extended generator of the equation of motion given by the Lagrangian L, i.e.,*

$$X(L) = \xi^i \frac{\partial L}{\partial x^i} + \eta^j \frac{\partial L}{\partial u^j} + \zeta^j_i \frac{\partial L}{\partial u^j_i}, \ \zeta^j_i = D_i(\eta^j) - u^j_\alpha D_i(\xi^\alpha).$$

Theorem 8.19 *If a variational integral $I(u) = \int_V L(x, u, u^{(1)}, u^{(2)}, \ldots, u^{(k)}) dx \quad x = (x_1, x_2, \ldots, x_n)$ is invariant under the group (8.83) with the generator (8.84) then the vector $T = (T^1, T^2, \ldots, T^n)$ defined by $T^i = L\xi^i + W^i[u, \eta - u_j\xi^j], i = 1, \ldots, n,$ where*

$$W^i[u, \eta - u_j\xi^j] = (\eta^j - \xi^\alpha u_\alpha^j) \left[\frac{\partial L}{\partial u_i^j} + \cdots + (-1)^{k-1} D_{i_1} D_{i_2} \cdots D_{i_{k-1}} \frac{\partial L}{\partial u_{ii_1 i_2 \cdots i_{k-1}}^j} \right] + D_{i_1}(\eta^j$$

$$- \xi^\alpha u_\alpha^j) \left[\frac{\partial L}{\partial u_{i_1 i}^j} + \cdots + (-1)^{k-2} D_{i_2} D_{i_3} \cdots D_{i_{k-1}} \frac{\partial L}{\partial u_{i_1 i_2 \cdots i_{k-1}}^j} \right]$$

$$+ \cdots + D_{i_1} D_{i_2} \cdots D_{i_{k-1}} (\eta^j - \xi^\alpha u_\alpha^j) \frac{\partial L}{\partial u_{i_1 i_2 \cdots i_{k-1} i}^j}$$

is a conserved vector of the Euler–Lagrange equations, i.e., $D_i(T^i) = 0$ on the solutions of Euler–Lagrange equations.

Theorem 8.20 If the variational integral $I(u) = \int_V L(x, u, u^{(1)}, u^{(2)}, ..., u^{(k)}) dx$, $x = (x_1, x_2, ..., x_n)$ corresponding to the Euler-Lagrange equation $\frac{\delta L}{\delta u^j} \equiv \frac{\partial L}{\partial u^j} + \sum_{k=1}^{n} (-1)^k$ $D_{i_1 i_2 ... i_l} \left(\frac{\partial L}{\partial u_{i_1 i_2 ... i_l}^j} \right) = 0$, be such that $X^{[k]}(L) + LD_i(\xi^i) = D_i(B_i)$, $i = 1, 2, ..., n$, $B_i(x, u, u^{(1)}, u^{(2)}, ..., u^{(k)})$ is a differential function, where $X^{[k]}$ is the kth extended generator of the equation of motion given by the Lagrangian L then the conservation laws of the Euler-Lagrange equations is given by $D_i(L\xi^i + W^i[u, \eta - u_j\xi^j] - B_i) = 0$.

Example 8.28 Determine the conservation laws of the free motion of a particle.

Solution The kinetic energy of a free particle is $\frac{1}{2}mv^2$, $|v|^2 = \sum_{i=1}^{3} (v_i)^2$ and potential energy is zero. Hence, the Lagrangian of a freely moving particle is $L = \frac{1}{2}m\left((v_1)^2 + (v_2)^2 + (v_3)^2\right)$. The governing equation of motion is $mv_i\dot{v}_i = 0$ or $\ddot{x}_i = 0$, $i = 1, 2, 3$. The integral $I(u) = \int_V L(q, u, u^{(1)}) dq$ is invariant under the four parameter group with the basis of infinitesimal generators given by

$$X_t = \frac{\partial}{\partial t}, \ X_i = \frac{\partial}{\partial x_i}, X_{ij} = x_j \frac{\partial}{\partial x_i} - x_i \frac{\partial}{\partial x_j}, Z_i = t \frac{\partial}{\partial x_i}, \ i, j = 1, 2, 3$$

We will now determine the conservation laws using Noether's theorem to each of the above generators below.

(i) Here B = 0, therefore according to the Noether's theorem the conserved quantities of a variational problem is given by $T^i = L\xi^i + (\eta^j - \xi^\alpha u_\alpha^j) \frac{\partial L}{\partial u_i^j}$, $i = 1, ..., n$

For the generator $X_t = \frac{\partial}{\partial t}$, the infinitesimals are $\xi = 1$ and $\eta^1 = \eta^2 = \eta^3 = 0$. Hence, the conserved quantity is given by

$$T = L + (-v^i)\frac{\partial L}{\partial v^i}$$

$$= \frac{1}{2}m\left((v_1)^2 + (v_2)^2 + (v_3)^2\right) - v_1(mv_1) - v_2(mv_2) - v_3(mv_3)$$

$$= -\frac{1}{2}m\left((v_1)^2 + (v_2)^2 + (v_3)^2\right)$$

Since the conserved quantity represents the kinetic energy of freely moving particle therefore setting $T = -E$, the conserved quantity can be written as $E = \frac{1}{2}m\left((v_1)^2 + (v_2)^2 + (v_3)^2\right) = \frac{1}{2}m|v|^2 = \frac{1}{2}m|v|^2$.

(ii) For the generator $X_i = \frac{\partial}{\partial x_i}, i = 1, 2, 3$, the infinitesimals are $\xi = 0$ and $\eta^1 = 1, \eta^2 = \eta^3 = 0$. Therefore, we have, $T^1 = \frac{\partial L}{\partial v_1} = mv_1 = p_1$. Similarly, one will obtain $T^2 = \frac{\partial L}{\partial v_2} = mv_2 = p_2$ and $T^3 = \frac{\partial L}{\partial v_3} = mv_3 = p_3$. Hence the conserved quantity is the linear momentum.

(iii) We have the rotation generator $X_{ij} = x_j\frac{\partial}{\partial x_i} - x_i\frac{\partial}{\partial x_j}; i, j = 1, 2, 3$. Now consider the rotation around the x_1-axis. Hence the generator becomes $X_{23} = x_3\frac{\partial}{\partial x_2} - x_2\frac{\partial}{\partial x_3}$ and the infinitesimals are $\xi = 0$ and $\eta^1 = 0, \eta^2 = x_3, \eta^3 = -x_2$. Now from Noether's theorem we have

$$T^1 = \eta^2\frac{\partial L}{\partial v_2} + \eta^3\frac{\partial L}{\partial v_3} = x_3(mv_2) + (-x_2)(mv_3)$$

$$= m(x_3v_2 - x_2v_3) = M^1$$

Similarly $M^2 = m(x_3v_1 - x_1v_3)$ and $M^3 = m(x_1v_2 - x_2v_1)$. Hence, the total angular momentum $M = m(x \times v)$ is the required conserved quantity.

Example 8.29 Determine the conservation laws of the Kepler's system.

Solution The equations of Kepler's laws of motion are given by

$$m\frac{d^2x_1}{dt^2} = \frac{\mu x_1}{r^3}$$

$$m\frac{d^2x_2}{dt^2} = \frac{\mu x_2}{r^3}$$

$$m\frac{d^2x_3}{dt^2} = \frac{\mu x_3}{r^3}$$

The Lagrangian of the system is obtained as $L = \frac{m}{2}\left(v_1^2 + v_2^2 + v_3^2\right) - \frac{\mu}{r}, \mu = \text{constant}$. Now we calculate the point symmetries of the Kepler's system. The Kepler's system of equations are second-order equation and therefore the infinitesimal generators are given by $X_1 = \xi\frac{\partial}{\partial t} + \eta^1\frac{\partial}{\partial x_1}$, $X_2 = \xi\frac{\partial}{\partial t} + \eta^2\frac{\partial}{\partial x_2}$, $X_3 = \xi\frac{\partial}{\partial t} + \eta^3\frac{\partial}{\partial x_3}$. The twice extended generators are given by

$X_1 = \xi \frac{\partial}{\partial t} + \eta^1 \frac{\partial}{\partial x_1} + (\eta^1)^2 \frac{\partial}{\partial \dot{x}_1}$, $X_2 = \xi \frac{\partial}{\partial t} + \eta^2 \frac{\partial}{\partial x_2} + (\eta^2)^2 \frac{\partial}{\partial \dot{x}_2}$, $X_3 = \xi \frac{\partial}{\partial t} + \eta^3 \frac{\partial}{\partial x_3} +$ $(\eta^3)^2 \frac{\partial}{\partial \dot{x}_3}$. Now applying the infinitesimal criterion $XF|_{F=0} = 0$ to each of the equation of the Kepler's system with $\frac{d^2 x_1}{dt^2} = \frac{\mu x_1}{mr^3}, \frac{d^2 x_2}{dt^2} = \frac{\mu x_2}{mr^3}, \frac{d^2 x_3}{dt^2} = \frac{\mu x_3}{mr^3}$. The infinitesimals obtained by solving (using Introtosymmetry.m, see Cantwell [18]) the overdetermined system of the Kepler's system is given by $\xi = c_1 + c_2 t, \eta^1 = \frac{2}{3} c_2 x_1 - c_3 x_2 - c_4 x_3, \eta^2 = c_3 x_1 + \frac{2}{3} c_2 x_2 - c_5 x_3, \eta^3 = c_4 x_1 + c_5 x_2 + \frac{2}{3} c_2 x_3$.

Thus, the Kepler's system possesses five point symmetries $X = \frac{\partial}{\partial t}$, $X_{ij} = x_j \frac{\partial}{\partial x_i} - x_i \frac{\partial}{\partial x_j}$, $i, j = 1, 2, 3$ and $Z = 3t \frac{\partial}{\partial t} + 2x_i \frac{\partial}{\partial x_i}$.

We shall now use Noether's theorem to obtain the conservation laws of Kepler's system as follows:

(i) For the generator $X_t = \frac{\partial}{\partial t}$, the infinitesimals are $\xi = 1$ and $\eta^1 = \eta^2 = \eta^3 = 0$. Now according to Noether theorem the conserved quantities of a variational problem is given by

$$T^i = L\xi^i + (\eta^j - \xi^\alpha u_\alpha^j) \frac{\partial L}{\partial u_i^j}, \quad i = 1, \ldots, n$$

In this case, the required conserved quantity is given by

$$T = L + (-v^i) \frac{\partial L}{\partial v^i}$$
$$= \frac{1}{2} m \left((v_1)^2 + (v_2)^2 + (v_3)^2 \right) - \frac{\mu}{r} - v_1(mv_1) - v_2(mv_2) - v_3(mv_3)$$
$$= -\frac{1}{2} m \left((v_1)^2 + (v_2)^2 + (v_3)^2 \right) - \frac{\mu}{r}$$

Since the conserved quantity is the energy of the given system, therefore setting $T = -E$, the conserved quantity can be written as

$$E = \frac{1}{2} m \left((v_1)^2 + (v_2)^2 + (v_3)^2 \right) + \frac{\mu}{r} = \frac{1}{2} m |v|^2 + \frac{\mu}{r}.$$

(ii) We have the rotation generator $X_{ij} = x_j \frac{\partial}{\partial x_i} - x_i \frac{\partial}{\partial x_j}$; $i, j = 1, 2, 3$. Now consider the rotation around the x_1-axis. Hence the generator becomes $X_{23} = x_3 \frac{\partial}{\partial x_2} - x_2 \frac{\partial}{\partial x_3}$ and the infinitesimals are $\xi = 0$ and $\eta^1 = 0$, $\eta^2 = x_3$, $\eta^3 = -x_2$. Now from Noether's theorem the conserved quantity is given by

$$T^1 = \eta^2 \frac{\partial L}{\partial v_2} + \eta^3 \frac{\partial L}{\partial v_3} = x_3(mv_2) + (-x_2)(mv_3)$$
$$= m(x_3 v_2 - x_2 v_3) = M^1$$

Similarly for rotation around x_2-axis and x_3-axis, one will have the conserved quantities $M^2 = m(x_3 v_1 - x_1 v_3)$ and $M^3 = m(x_1 v_2 - x_2 v_1)$, respectively.

Hence, the total angular momentum $M = m(x \times v)$ is the required conserved quantity.

(iii) We have the generator $Z = 3t\frac{\partial}{\partial t} + 2x_1\frac{\partial}{\partial x_1} + 2x_2\frac{\partial}{\partial x_2} + 2x_3\frac{\partial}{\partial x_3}$ and the infinitesimals are $\xi = 3t$, $\eta^1 = 2x_1$, $\eta^2 = 2x_2$, $\eta^3 = 2x_3$. Now the condition for the given system to remain invariant under the given generator is $X(L) + LD_i(\xi^i) = 0$, i.e.,

$$
\begin{aligned}
&\left(3t\frac{\partial}{\partial t} + 2x_1\frac{\partial}{\partial x_1} + 2x_2\frac{\partial}{\partial x_2} + 2x_3\frac{\partial}{\partial x_3} + (\eta_t^1 \right. \\
&\quad + (x_1)_t\eta_{x_1}^1 - (x_1)_t\xi_t - (x_1)_t^2\xi_{x_1})\frac{\partial}{\partial \dot{x}_1} + (\eta_t^2 + (x_2)_t\eta_{x_2}^2 - (x_2)_t\xi_t \\
&\quad - (x_2)_t^2\xi_{x_2})\frac{\partial}{\partial \dot{x}_2} + (\eta_t^3 + (x_3)_t\eta_{x_3}^3 - (x)_t\xi_t \\
&\quad \left. - (x_3)_t^2\xi_{x_3})\frac{\partial}{\partial \dot{x}_3}\right)\left(\frac{m}{2}\left(v_1^2 + v_2^2 + v_3^2\right) - \frac{\mu}{r}\right) + 3\left(\frac{m}{2}\left(v_1^2 + v_2^2 + v_3^2\right) - \frac{\mu}{r}\right) \\
&= 2\mu x_1\frac{x_1}{(x_1^2 + x_2^2 + x_3^2)^{3/2}} + 2\mu x_2\frac{x_2}{(x_1^2 + x_2^2 + x_3^2)^{3/2}} \\
&\quad + 2\mu x_3\frac{x_3}{(x_1^2 + x_2^2 + x_3^2)^{3/2}} - m(v_1)^2 - m(v_2)^2 - m(v_3)^2 \\
&\quad + 3\left(\frac{m}{2}\left(v_1^2 + v_2^2 + v_3^2\right) - \frac{\mu}{r}\right) \\
&= \frac{\mu}{(x_1^2 + x_2^2 + x_3^2)^{1/2}} + \frac{m|v|^2}{2} \neq 0
\end{aligned}
$$

Hence, the dilation group Z is not a variational symmetry of the Kepler's system.

8.12 Symmetry Analysis of Kortweg-de Vries (KdV) Equation

Dutch physicists Kortweg and de Vries [22] studied the phenomena of long waves of an incompressible and inviscid fluid in a rectangular shallow channel in 1895 and gave the following equation known as KdV equation

$$u_{xxx} + uu_x + u_t = 0 \tag{8.85}$$

The above equation is nonlinear in nature, propagating in a dispersive medium. We know that waves on a surface of fluid occur when the fluid is disturbed by some means. These waves may die down immediately after its occurrence or durable for some time. Generally, dispersive waves do not preserve their shape like dispersion less waves during their propagation. But, KdV equation admits solutions which like some other nonlinear dispersive waves maintain their shape and speed, and are permanent in nature for a considerable amount of time. These solutions of KdV equation are well known as solitary wave solution after its discovery by the Scottish naval engineer, John Scott Russel in 1834 while riding on a horseback along the bank of a canal near Edinburgh, Scotland. After that Boussinesque in 1871 and Lord Rayleigh in 1876 gave the mathematical form of these waves. But an exact mathematical relationship which admits the solitary waves solution was given by Kortweg and de Vries.

Dispersive waves when occur in a fluid of less depth, spreads while propagating hence cannot preserve their shape, continuously diminishing in size and ultimately fades. But when the amplitude of these dispersive waves is considerably larger, the nonlinear convection balances the dispersion of waves; as a result the waves do not spread and steeply rises forming a hump of fluid thereby creating a sink. The volume of the fluid forming the hump is equal to the volume of the fluid displaced. This sharp rise of fluid forming a hump above the surface of the fluid is known as solitary wave. These waves are long-lasting which preserve their shape and velocity appreciably, while propagating. These waves are therefore also known as great wave of translation. The solitary wave solution of the above KdV equation is due to the balance of the nonlinear inertial term uu_x and the linear dispersion term u_{xxx}. One of the main requirements for these types of waves to occur is that the ratio $\frac{\lambda}{h} \ll 1$, where λ is the wavelength of the wave and h is the depth of the fluid and $\frac{a}{h} \ll 1$, where a is the amplitude of the wave. These types of waves propagate faster when amplitude becomes larger.

The preservation of shape and size of the solitary waves implies the existence of symmetry underlying the KdV equation. So, we will now try to obtain these invariant solutions using Lie symmetry analysis. The KdV equation has two independent variables x and t and one dependent variable u. So the infinitesimal generator of this equation has the following form

$$X = \xi^1 \frac{\partial}{\partial x} + \xi^2 \frac{\partial}{\partial t} + \eta \frac{\partial}{\partial u} \tag{8.86}$$

Since the governing equation is of third order, the infinitesimal generator (8.86) can be thrice extended or prolonged. The thrice extended generator can be written as

$$X^{[3]} = \xi^1(x,t,u)\frac{\partial}{\partial x} + \xi^2(x,t,u)\frac{\partial}{\partial t} + \eta(x,t,u)\frac{\partial}{\partial u} + \zeta^1 \frac{\partial}{\partial u_x} + \zeta^2 \frac{\partial}{\partial u_t} + \zeta^{111} \frac{\partial}{\partial u_{xxx}} \tag{8.87}$$

The invariant surface condition $X^{[3]}F|_{F=0} = 0$ then gives

$$\eta u_x + \zeta^1 u + \zeta^2 + \zeta^{111} = 0 \tag{8.88}$$

The coefficients ζ^1, ζ^2 and ζ^{111} are given by

$$\zeta^1 = D_x(\eta) - u_x D_x(\xi^1) - u_t D_x(\xi^2), \zeta^2 = D_t(\eta) - u_x D_t(\xi^1) - u_t D_t(\xi^2) \tag{8.89}$$

and

$$\zeta^{111} = D_x(\zeta^{11}) - u_{xxx} D_x(\xi^1) - u_{xxy} D_x(\xi^2)$$
$$\text{where } \zeta^{11} = D_x(\zeta^1) - u_{xx} D_x(\xi^1) - u_{xy} D_x(\xi^2) \tag{8.90}$$

Now putting these coefficients in (8.88), we have obtained following equation as

$$\eta u_x + u(\eta_x + u_x \eta_u - u_x \xi_x^1 - (u_x)^2 \xi_u^1 - u_t \xi_x^2 - u_x u_t \xi_u^2) + (\eta_t + u_t \eta_u - u_x \xi_t^1$$
$$- u_x u_t \xi_u^1 u_t \xi_t^2 - (u_t)^2 \xi_u^2) + (3\eta_{uuu} - \xi_{xxx}^1) u_x + (-\xi_{xxx}^2) u_y + (3\eta_{xu} - 3\xi_{xx}^1) u_{xx}$$
$$+ (3\eta_{xuu} - 3\xi_{xxu}^1)(u_x)^2 + (\eta_{uuu} - 3\xi_{xuu}^1)(u_x)^3 + (3\eta_{uu} - 9\xi_{ux}^1) u_{xx} u_x$$
$$+ (-3\xi_u^1)(u_{xx})^2 + (-6\xi_{uu}^1)(u_x)^2 u_{xx} + (-\xi_{uuu}^1)(u_x)^4 + (-3\xi_{xx}^2) u_{xt} + (-3\xi_{xxu}^2) u_{xx} u_t + (-3\xi_{xu}^2) u_{xx} u_t$$
$$+ (-6\xi_{xu}^2) u_x u_{xt} + (-3\xi_{xuu}^2)(u_x)^2 u_t + (-3\xi_u^2) u_{xx} u_{xt} + (-3\xi_u^2) u_x u_{xxt} + (-3\xi_{uu}^2) u_x u_t u_{xx}$$
$$+ (-3\xi_{uu}^2)(u_x)^2 u_{xt} + (-\xi_{xuu}^2)(u_x)^2 u_t + (-\xi_{uuu}^2)(u_x)^3 u_t + (-3\xi_x^2) u_{xxt}$$
$$+ (\eta_u - 3\xi_x^1 - 4\xi_u^1 u_x + \eta_{xxx} - \xi_u^2 u_t) u_{xxx} = 0$$

Replacing u_{xxx} by $-u_t - u u_x$, and then setting the coefficients of each term equal to zero, the following overdetermined system of equations are accomplished

$$-3\xi_u^1 = 0, -3\xi_u^2 = 0, -6\xi_{uu}^1 = 0, -3\xi_{uu}^2 = 0, \xi_{uuu}^1 = 0, \xi_{uuu}^2$$
$$= 0, -3\xi_x^2 = 0, 3\eta_{uu}^1 - 9\xi_{xu}^1 = 0,$$
$$-6\xi_{xu}^2 = 0, -3\xi_{xu}^2 = 0, \eta_{uuu} - 3\xi_{xuu}^1 = 0, -3\xi_{xuu}^2 = 0, 3\eta_{xu} - 3\xi_{xx}^1 = 0, -3\xi_{xx}^2 = 0,$$
$$3u\xi_u^1 + 3\eta_{xuu} - 3\xi_{xxu}^1 = 0, 3\xi_u^1 - 3\xi_{xxu}^2 = 0, \eta_t + u\eta_x + \eta_{xxx} = 0$$
$$\eta - \xi_t^1 + 2u\xi_x^1 + 3\eta_{xxu} - \xi_{xxx}^1 = 0, -\xi_t^2 + 3\xi_x^1 - u\xi_x^2 - \xi_{xxx}^2 = 0$$

Solving the above equations, the following infinitesimals are found

$$\xi^1 = c_1 + c_2 t - \frac{c_3}{2} x, \; \xi^2 = c_4 - \frac{3}{2} c_3 t, \; \eta = c_2 + c_3 u$$

Since KdV equation admits a four parameter group, hence a four-dimensional Lie algebra is formed. The corresponding generators are given by

$$X_1 = \frac{\partial}{\partial x}, \; X_2 = t\frac{\partial}{\partial x} + \frac{\partial}{\partial u}, \; X_3 = \frac{-x}{2}\frac{\partial}{\partial x} - \frac{3t}{2}\frac{\partial}{\partial t} + u\frac{\partial}{\partial u}, \; X_4 = \frac{\partial}{\partial t}$$

Table 8.12 Commutator table of KdV equation

	X_1	X_2	X_3	X_4
X_1	0	0	0	$\frac{-1}{2}X_1$
X_2	0	0	X_1	$\frac{-3}{2}X_4$
X_3	0	$-X_1$	0	X_2
X_4	$\frac{1}{2}X_1$	$\frac{3}{2}X_4$	$-X_2$	0

Lie algebra is conveniently summarized in the commutator Table 8.12 by evaluating the Lie bracket.

(i)   KdV equation admits two translation groups one in time t and another in space x and their combination are

$$c_1 \frac{\partial}{\partial x} + c_4 \frac{\partial}{\partial t}$$

The corresponding characteristic equation can be formed as

$$\frac{dx}{c_1} = \frac{dt}{c_4} = \frac{du}{0}$$

By solving this equation, the invariants are given as

$$x - \frac{c_1}{c_4} t = x - ct = \eta(\text{say}) \text{ where } c = \frac{c_1}{c_4}$$
$$u = f(\eta) \quad f \text{ is an arbitrary function}$$

Now substituting $u = f(\eta), u_t = -cf'(\eta), u_x = f'(\eta), u_{xxx} = f'''(\eta)$ in KdV equation, it becomes

$$f'''(\eta) + f(\eta)f'(\eta) - cf'(\eta) = 0 \tag{8.91}$$

On integration we have $f''(\eta) + \frac{1}{2}(f(\eta))^2 - cf(\eta) = \text{const} = a$ (say)
Multiplying the above equation by f' and then integrating we have

$$\frac{1}{2}(f'(\eta))^2 + \frac{1}{6}(f(\eta))^3 - \frac{c}{2}(f(\eta))^2 = af + b \tag{8.92}$$

Cnoidal wave solution
When $a, b \neq 0$, then the Eq. (8.92) admits Jacobi elliptic function as solution which corresponds to the cnoidal waves. For $a, b \neq 0$ the Eq. (8.92) can be written as

$$\sqrt{3}\frac{df}{d\eta} = \sqrt{-f^3 + 3cf^2 + 6af + 6b}$$

Separation of variables gives $\dfrac{\sqrt{3}df}{\sqrt{-f^3 + 3cf^2 + 6af + 6b}} = d\eta$.

If r_1, r_2, r_3 are the roots of the algebraic equation $(f(\eta))^3 - 3c(f(\eta))^2 - 6af(\eta) - 6b = 0$ then we can write the above equation as

$$\frac{\sqrt{3}df}{\sqrt{(r_1 - f)(f - r_2)(f - r_3)}} = d\eta$$

Let $f = r_1 + (r_2 - r_1)\sin^2\phi$ then $df = 2(r_2 - r_1)\sin\phi\cos\phi d\phi$.

Therefore $\dfrac{2\sqrt{3}d\phi}{\sqrt{r_1 - r_3}\sqrt{1 - \frac{(r_1 - r_2)}{(r_1 - r_3)}\sin^2\phi}} = d\eta$ this on integration gives

$$\eta = \frac{2\sqrt{3}}{\sqrt{r_1 - r_3}}\int_0^\phi \frac{d\phi}{\sqrt{1 - \frac{r_2 - r_1}{r_1 - r_3}\sin^2\phi}}$$

Hence in terms of Jacobi elliptic function, the above equation is written as $\text{sn}\left(\frac{\sqrt{r_1 - r_3}\eta}{2\sqrt{3}}, k\right) = \sin\phi$, where $k = \sqrt{\frac{r_2 - r_1}{r_1 - r_3}}$ is the modulus of the elliptic integral
or,

$$\sin^2\phi = \text{sn}^2\left(\frac{\sqrt{r_1 - r_3}\eta}{2\sqrt{3}}, k\right)$$

or,

$$\frac{f - r_1}{r_2 - r_1} = \text{sn}^2\left(\frac{\sqrt{r_1 - r_3}\eta}{2\sqrt{3}}, k\right)$$

which gives $u(x, t) = (r_2 - r_1)\text{cn}^2\left(\frac{\sqrt{r_1 - r_3}\eta}{2\sqrt{3}}, k\right) + r_2$ or

$u(x, t) = (r_2 - r_1)\text{cn}^2\left(\frac{\sqrt{r_1 - r_3}(x - ct)}{2\sqrt{3}}, k\right) + r_2$.

Soliton wave solution

Soliton solutions can be obtained as a limiting case of cnoidal wave solution when $k = 1$ i.e. $r_1 - r_3 = r_2 - r_1$. In that case $cn(t, 1) = \text{sech}t$, which gives the soliton solution $u(x, t) = (r_1 - r_3)\text{sech}^2\left(\frac{\sqrt{r_1 - r_3}(x - ct)}{2\sqrt{3}}\right) + r_2$. Again,

Fig. 8.8 One hump soliton

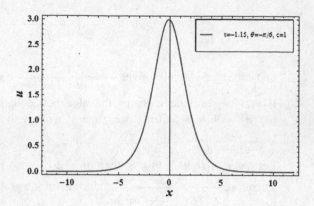

we can obtain the soliton solution from the Eq. (8.91) itself when $a, b = 0$. The Eq. (8.91) is then written as

$$\left(\frac{df}{d\eta}\right)^2 = cf(\eta)^2 - \frac{1}{3}f(\eta)^3 \tag{8.93}$$

Separation of variables gives $\dfrac{df}{f\sqrt{\frac{-1}{3}f(\eta) + c}} = \pm d\eta$

whose solution $\left(\text{by choosing } \sqrt{c - \frac{1}{3}f(\eta)} = v(\text{say})\right)$ is given by

$$f(\eta) = 3c\left(1 - \tanh^2\left(\pm\frac{\sqrt{c}}{2}\eta + \text{const}\right)\right) = 3c\,\text{sech}^2\left(\pm\frac{\sqrt{c}}{2}\eta + \theta\right)$$

or,

$$u(x, t) = 3c\,\text{sech}^2\left(\pm\frac{\sqrt{c}}{2}(x - ct) + \theta\right),$$

where θ is the phase of the system. This is the famous soliton solution (Fig. 8.8).

If in addition $c = 0$, then Eq. (8.92) becomes

$$\left(\frac{df}{d\eta}\right)^2 = -\frac{1}{3}f(\eta)^3$$

The solution of this equation is readily obtained as $f(\eta) = \frac{-12}{(\eta + \theta)^2}$ or $u(x, t) = \frac{-12}{x + \theta}$, singular stationary solution.

(ii) KdV equation also admits the group $t\frac{\partial}{\partial x} + \frac{\partial}{\partial u}$. The corresponding characteristic equation is

$$\frac{dx}{t} = \frac{dt}{0} = \frac{du}{1}$$

The invariants are given by $t = \eta$, and $tu - x = f(\eta)$.
Substituting
$u = (f(\eta) + x) \cdot (\eta)^{-1}, u_x = \eta^{-1}, u_{xxx} = 0, u_t = f'(\eta)\eta^{-1} - f(\eta)\eta^{-2} - x\eta^{-2}$the
KdV equation become

$$\frac{df}{d\eta} = 0$$

which gives the solution as $f(\eta) = \text{const} = \theta$ (say) or $u = \frac{x+\theta}{t}$.

(iii) KdV equation also admits the scaling group $x\frac{\partial}{\partial x} + 3t\frac{\partial}{\partial t} - 2u\frac{\partial}{\partial u}$. The corresponding characteristic equation is

$$\frac{dx}{x} = \frac{dt}{3t} = \frac{du}{-2u}$$

which gives the invariants as $xt^{-1/3} = \eta$(say) and $ut^{2/3} = f(\eta)$.
Substituting

$$u = t^{-2/3}f(\eta), \ u_x = t^{-1}f'(\eta), \ u_{xxx} = t^{-5/3}f'''(\eta), \ u_t$$
$$= \frac{-2}{3}t^{-5/3}f(\eta) - \frac{1}{3}t^{-5/3}\eta f'(\eta),$$

KdV equation becomes

$$\frac{-2}{3}f(\eta) - \frac{1}{3}\eta f'(\eta) + f'''(\eta) + f(\eta)f'(\eta) = 0$$

The above equation is a nonlinear third-order ordinary differential equation whose solution cannot be obtained by the usual procedures.

Conservation Laws of the KdV Equation

The KdV equation has no Lagrangian. Now writing $u = v_x$, the KdV equation becomes

$$v_{xt} + v_x v_{xx} + v_{xxxx} = 0$$

This equation admits the point symmetries and the corresponding groups are

$$X_1 = \frac{\partial}{\partial x}, X_2 = \frac{\partial}{\partial t}, X_3 = \frac{\partial}{\partial u}, X_4 = t\frac{\partial}{\partial x} + x\frac{\partial}{\partial u}, X_5 = -x\frac{\partial}{\partial x} - 3t\frac{\partial}{\partial t} + u\frac{\partial}{\partial u}.$$

The Lagrangian for this equation is given by $L = \frac{1}{2} v_x v_t - \frac{1}{6}(v_x)^3 - \frac{1}{2}(v_{xx})^2$. If we take the generalized coordinate as v then generalized momentum, say p is given by $p = \frac{\partial L}{\partial v_t} = \frac{v_x}{2}$ then, the Hamiltonian of the system is given by

$$H = pq_t - L = \frac{v_x}{2} v_t - \frac{v_x v_t}{2} + \frac{(v_x)^3}{6} + \frac{(v_{xx})^2}{2} = \frac{(v_x)^3}{6} + \frac{(v_{xx})^2}{2}$$

which is a conserved quantity equal to the energy of the system (since the Lagrangian L is explicitly independent of time t).

(i) For $X_1, B = 0$ (see Theorems 8.19 and 8.20) hence the conserved quantities are given by

$$T_1 = L\xi_x + (\eta - \xi_x v_x)\left(\frac{\partial L}{\partial v_x} - \frac{\partial L}{\partial v_{xx}}\right) = L - \frac{1}{2} v_x v_t + \frac{1}{2}(v_x)^3 - v_x v_{xx}$$

$$= \frac{1}{3}(v_x)^3 - \frac{1}{2}(v_{xx})^2 - v_x v_{xx}$$

and

$$T_2 = L\xi_t + (\eta - \xi_x v_x - \xi_t v_t)\frac{\partial L}{\partial v_t} = -v_x \frac{v_x}{2} = -\frac{(v_x)^2}{2}.$$

(ii) For $X_2, B = 0$ hence the conserved quantities are given by

$$T_1 = L\xi_x + (\eta - \xi_x v_x - \xi_t v_t)\left(\frac{\partial L}{\partial v_x} - \frac{\partial L}{\partial v_{xx}}\right) = -v_t\left(\frac{\partial L}{\partial v_x} - \frac{\partial L}{\partial v_{xx}}\right)$$

$$= -\frac{1}{2}(v_t)^2 + \frac{1}{2} v_t (v_x)^2 - v_t v_{xx}$$

and

$$T_2 = L\xi_t + (\eta - \xi_x v_x - \xi_t v_t)\frac{\partial L}{\partial v_t} = L - v_t \frac{v_x}{2} = -\frac{1}{6}(v_x)^3 - \frac{1}{2}(v_t)^2,$$

energy of the system is conserved.

(iii) For $X_3, \xi_x = 0, \xi_t = 0, \eta = 1$ and

$$X_3^{[2]} = \xi_x \frac{\partial}{\partial x} + \xi_t \frac{\partial}{\partial t} + \eta \frac{\partial}{\partial u} + \eta_x^{(1)} \frac{\partial}{\partial u_x} + \eta_t^{(1)} \frac{\partial}{\partial u_t} + \eta_x^{(2)} \frac{\partial}{\partial u_{xx}}.$$

Now

$$X_3^{[2]} L = -\eta_x^{(2)} v_{xx} + \eta_t^{(1)} \frac{v_x}{2} + \eta_x^{(1)} \left(\frac{v_t}{2} - \frac{(v_x)^2}{2} \right)$$

$$= -D_x(D_x\eta) v_{xx} + (D_t\eta) \frac{v_x}{2} + (D_x\eta) \left(\frac{v_t}{2} - \frac{(v_x)^2}{2} \right) = 0.$$

Hence $B = 0$. The conserved quantities are therefore given by

$$T_1 = L\xi_x + (\eta - \xi_x v_x - \xi_t v_t) \left(\frac{\partial L}{\partial v_x} - \frac{\partial L}{\partial v_{xx}} \right) = \left(\frac{\partial L}{\partial v_x} - \frac{\partial L}{\partial v_{xx}} \right)$$

$$= \frac{v_t}{2} - \frac{1}{2} (v_x)^2 + v_{xx}$$

and

$$T_2 = L\xi_t + (\eta - \xi_x v_x - \xi_t v_t) \frac{\partial L}{\partial v_t} = \eta \frac{\partial L}{\partial v_t} = \frac{v_x}{2},$$

the momentum of the system is conserved.

(iv) For $X_4, \xi_x = t, \xi_t = 0, \eta = x$ and

$$X_4^{[2]} = \xi_x \frac{\partial}{\partial x} + \xi_t \frac{\partial}{\partial t} + \eta \frac{\partial}{\partial u} + \eta_x^{(1)} \frac{\partial}{\partial u_x} + \eta_t^{(1)} \frac{\partial}{\partial u_t} + \eta_x^{(2)} \frac{\partial}{\partial u_{xx}}.$$

Now

$$X_4^{[2]} L = -\eta_x^{(2)} v_{xx} + \eta_t^{(1)} \frac{v_x}{2} + \eta_x^{(1)} \left(\frac{v_t}{2} - \frac{(v_x)^2}{2} \right)$$

$$= -D_x(D_x\eta) v_{xx} + (D_t\eta - D_t\xi_x v_x) \frac{v_x}{2} + (D_x\eta) \left(\frac{v_t}{2} - \frac{(v_x)^2}{2} \right)$$

$$= \frac{v_t}{2} - (v_x)^2$$

Now the invariance condition

$$X_4^{[2]} L - L(D_x\xi_x + D_t\xi_t) = \frac{v_t}{2} - \frac{(v_x)^2}{2} - 0 = \frac{v_t}{2} - (v_x)^2$$

(cannot be written in the form $D_i(B)$), hence X_4 is not a variational symmetry.

(v) For

$$X_5 = -x\frac{\partial}{\partial x} - 3t\frac{\partial}{\partial t} + u\frac{\partial}{\partial u}, \xi_x = -x, \xi_t = -3t, \eta = u$$

and

$$X_5^{[2]} = \xi_x\frac{\partial}{\partial x} + \xi_t\frac{\partial}{\partial t} + \eta\frac{\partial}{\partial u} + \eta_x^{(1)}\frac{\partial}{\partial u_x} + \eta_t^{(1)}\frac{\partial}{\partial u_t} + \eta_x^{(2)}\frac{\partial}{\partial u_{xx}}.$$

Now

$$X_5^{[2]}L = -\eta_x^{(2)}v_{xx} + \eta_t^{(1)}\frac{v_x}{2} + \eta_x^{(1)}\left(\frac{v_t}{2} - \frac{(v_x)^2}{2}\right) = (v_{xx})^2 + 2v_tv_x - \frac{(v_x)^3}{2}$$

Now the invariance condition

$$
\begin{aligned}
X_5^{[2]}L - L(D_x\xi_x + D_t\xi_t) &= (v_{xx})^2 + 2v_tv_x \\
&\quad - \frac{(v_x)^3}{2} + 4\left(\frac{1}{2}v_xv_t - \frac{1}{6}(v_x)^3 - \frac{1}{2}(v_{xx})^2\right) \\
&= -(v_{xx})^2 + 4v_tv_x - \frac{7}{6}(v_x)^3
\end{aligned}
$$

(cannot be written in the form $D_i(B)$), Therefore, X_5 is also not a variational symmetry.

Substitution of $u = \int v dx$ gives the conservation laws of the KdV equation.

8.13   Exercises

(1) (a) Explain Lie groups of transformations. What is the advantage of using Lie groups of transformation of a system?

(b) Discuss one parameter group of transformations.

(c) Define continuous groups of transformations. Give geometrical interpretation of continuous group of transformations.

(2) Determine whether the following one parameter group of transformations are Lie group of transformations or not

(i) $\tilde{x} = x + \epsilon$, $\tilde{y} = y$ (ii) $\tilde{x} = e^\epsilon x$, $\tilde{y} = e^{-\epsilon} y$ (iii) $\tilde{x} = x + \epsilon$, $\tilde{y} = \dfrac{xy}{x+\epsilon}$ (iv) $\tilde{x} = x - \epsilon y$, $\tilde{y} = y + \epsilon x$ (v)

$\tilde{x} = -x + \epsilon$, $\tilde{y} = y$ (vi) $\tilde{x} = \dfrac{x}{1-\epsilon x}$, $\tilde{y} = \dfrac{y}{1-\epsilon x}$ (vi) $\tilde{x} = \dfrac{x}{1-\epsilon x}$, $\tilde{y} = \dfrac{y}{1-\epsilon y}$ (vii) $\tilde{x} = x + \epsilon y$, $\tilde{y} = y + \epsilon x$

(3) Define infinitesimal generator of a group of transformation. Find the infinitesimal generator corresponding to the following one parameter group of transformation

(i) $\tilde{x} = x + \epsilon$, $\tilde{y} = y$ (ii) $\tilde{x} = \epsilon x$, $\tilde{y} = \epsilon^2 y$ (iii) $x \cos \epsilon + y \sin \epsilon$, $\tilde{y} = y \cos \epsilon - x \sin \epsilon$

(iv) $\tilde{x} = e^\epsilon (x \cos \epsilon - y \sin \epsilon)$, $\tilde{y} = e^\epsilon (y \cos \epsilon + x \sin \epsilon)$ (v) $\tilde{x} = \dfrac{x}{1-\epsilon x}$, $\tilde{y} = \dfrac{y}{1-\epsilon y}$

(4) Consider the one parameter group of transformation $\tilde{x} = \dfrac{x}{1-ax}$, $\tilde{y} = \dfrac{y}{1-ay}$, where y is dependent, and

x is independent variable. Find the infinitesimals ξ, η and its first, second and third prolongation .

(5) What do you mean by canonical variables of a group of transformations? Find the canonical variables for the following group of transformations:

(i) $\tilde{x} = e^\epsilon x$, $\tilde{y} = e^{2\epsilon} y$ (ii) $\tilde{x} = x$, $\tilde{y} = y + \epsilon$ (iii) $x \cos \epsilon + y \sin \epsilon$, $\tilde{y} = y \cos \epsilon - x \sin \epsilon$ (iv) $\tilde{x} = xe^\epsilon$, $\tilde{y} = ye^{m\epsilon}$ (v)

$\tilde{x} = x \cosh \epsilon + y \sinh \epsilon$, $\tilde{y} = y \cosh \epsilon + x \sinh \epsilon$

(6) Find the one parameter groups of transformation and canonical variables for the following infinitesimal generators

(i) $X = p\dfrac{\partial}{\partial x} - q\dfrac{\partial}{\partial y}$ (ii) $X = x\dfrac{\partial}{\partial x} + y\dfrac{\partial}{\partial y}$ (iii) $X = y\dfrac{\partial}{\partial x} + x\dfrac{\partial}{\partial y}$ (iv) $X = x\dfrac{\partial}{\partial x} + my\dfrac{\partial}{\partial y}$

(v) $X = x^2 \dfrac{\partial}{\partial x} + xy\dfrac{\partial}{\partial y}$ (vi) $X = (y^2 - x^2)\dfrac{\partial}{\partial x} - 2xy\dfrac{\partial}{\partial y}$ (vi) $X = (1+x^2)\dfrac{\partial}{\partial x} + xy\dfrac{\partial}{\partial y}$

(vii) $X = (x - y)\dfrac{\partial}{\partial x} + (x + y)\dfrac{\partial}{\partial y}$

(7) Find the integral curves of $X = \dfrac{\partial}{\partial x} + x^2 \dfrac{\partial}{\partial y} + (3y - x^3)\dfrac{\partial}{\partial z}$.

(8) Determine the infinitesimal transformation of the following first order differential equations and obtain the solution by using both integrating factor and canonical variable methods.

(i) $y' + y = x$ (ii) $\left(y - \dfrac{3}{2}x - 3 \right) y' + y = 0$ (iii) $yy' = \dfrac{8x+1}{y}$ (iv) $xy' = 3y + \dfrac{y^2}{x}$ (v) $y' = \dfrac{2xy}{x^2 - y^2}$

(9)   Find the envelope and also draw the integral curves of the Clairaut's equation

$$x\left(\frac{dy}{dx}\right)^2 - y\left(\frac{dy}{dx}\right) + m = 0$$

(10) Find the invariant solutions, if any, of the following equations

(i) $4x(y')^2 - (3x-1)^2 = 0$ (ii) $x(y')^2 - 2y'y + 4x = 0$ (iii) $y = 2y'x + (y')^2$ (iv) $y' = y^2$

(v) $(x^2-4)(y')^2 - 2xyy' - x^2 = 0$ (vi) $(y')^2(x^2-a^2) - 2y'xy + y^2 - b^2 = 0$ (vii) $y' = -\dfrac{x^3}{y}$

(11) Find the point symmetries and reduce the following differential equation to the standard forms and hence find the solution

(i) $y'' - \dfrac{2}{x^2}y = 0$ (ii) $y'' + y'^2 + xy = 0$ (iii) $y'' - \dfrac{y'}{x} + e^y = 0$ (iv) $y'' + e^{3y}y'^4 + y'^2 = 0$

(v) $y'' + 2\left(y' - \dfrac{y}{x}\right)^3 = 0$ (vi) $y'' = y'^5$ (vii) $y'' = y^{-2}$

(12) Find the solution of the following equation using differential invariant method

(i) $y'' = -4y + \sin 2x$ (ii) $y'' = -a^2y + \sec ax$ (iii) $y'' + 2e^x y' + 2e^x y = x^2$ (iv) $2yy''' + 6y'y'' = -x^{-2}$

(13) Find the point symmetries of the Blasius equation for a flow of an incompressible fluid over a flat plate

$$y''' + yy'' = 0 .$$

Also, form the commutator table. Is it form solvable Lie algebra?

(14) Find the point transformation and the invariant solutions of the following partial differential equation also construct the Commutator table

(i) $u_t = uu_x$ (ii) $u_t = u_{xx} + u_{yy}$ (iii) $u_t = u_{xx} + u_{yy} + u_{zz}$ (iv) $u_{tt} = cu_{xx}$

(v) $u_t = u_{xx} + xu_x + u$ (vi) $u_{tt} = u_{xx} + u$ (vii) $u_{xx} + u_{yy} = e^u$

(15) Show that the Burger's equation written in the form $u_t = u_{xx} + (u_x)^2$ admits the same symmetry group as admitted by Heat equation.

(16) Show that the equation $u_t - \dfrac{1}{2}(u_x)^2 - u_{xx} = 0$ is invariant under the transformations

$\bar{x} = x, \ \bar{t} = t, \ \bar{u} = u - 2\ln(1 - f(x,t)e^{u/2})$ where f is the solution of the heat equation $f_t - f_{xx} = 0$. Also find the solution under this group.

(17) Prove that $[fX,Y] = f[X,Y] - (Yf)X, \forall f \in C^\infty, \forall X, Y \in \mathcal{L}$ (Lie Algebra).

(18) Draw the commutator table of the following infinitesimal generators

(i) $X_1 = \dfrac{\partial}{\partial t}, X_2 = \dfrac{\partial}{\partial x}, X_3 = 2t\dfrac{\partial}{\partial t} + x\dfrac{\partial}{\partial x} - u\dfrac{\partial}{\partial u}$ (ii) $X_1 = \dfrac{\partial}{\partial x}, X_2 = x\dfrac{\partial}{\partial x}, X_3 = x^2\dfrac{\partial}{\partial x}$

(iii) $X_1 = t^2\dfrac{\partial}{\partial t} + xt\dfrac{\partial}{\partial x}, X_2 = t\dfrac{\partial}{\partial t}, X_3 = x\dfrac{\partial}{\partial t}, X_4 = tx\dfrac{\partial}{\partial t} + x^2\dfrac{\partial}{\partial x}$

Also check whether the Lie algebra formed by these operators are solvable or not.

(19) Determine the point symmetry group of the following equation using the mathematica software package IntroToSymmetry.m package(see Cantwell 2001).

(i) $u_{xxx} + (u_x)^2 + uu_{xx} + u_{tt} = 0$ (Boussinesq   equation)   (ii) $u^3 u_{xxx} - u_t = 0$ (Harry   Dim   equation)

(iii) $u_{xx} + u_{yy} + u_{zz} = 0$ (Laplace   equation)   (iv) $u_{xx} + xu_x - u_t = 0$ (Fokker- planck   equation)   (v)

$u_{xxt} + u_{yyt} + u_y u_{xxx} + u_y u_{xyy} - u_x u_{xxy} - u_x u_{yyy} - \nu u_{xxxx} - 2\nu u_{xxyy} - \nu u_{yyyy} = 0$ (ν is the viscosity of the fluid)

References

1. Yaglom, M.: Felix Klein and Sophus Lie: Evolution of the Idea of Symmetry in the Nineteenth Century. Birkhäuser (1988)
2. Oberlack, M.: A unified approach for symmetries in plane parallel turbulent shear flows. J. Fluid Mech. **427**, 299–328 (2001)
3. Birkhoff, G.: Hydrodynamics—A Study in Logic, Fact and Similitude. Princeton University Press, Princeton (1960)
4. Ovsiannikov, L.V.: Groups and group-invariant solutions of differential equations. Dokl. Akad. Nauk. USSR **118**, 439–442 (1958). (in Russian)
5. Ovsiannikov, L.V.: Groups properties of the nonlinear heat conduction equation. Dokl. Akad. Nauk. USSR **125**, 492–495 (1958). (in Russian)
6. Ovsiannikov, L.V.: Groups properties of differential equations. Novosibirsk (1962). (in Russian)
7. Ovsiannikov, L.V.: Groups Analysis of Differential Equations. Academic Press, New York (1982)
8. Bluman, G.W., Cole, J.D.: The general similarity solution of the heat equation. J. Math. Mech. **18**, 1025–1042 (1969)
9. Bluman, G.W., Cole, J.D.: Similarity methods for differential equations. Appl. Math. Sci. **13**. Springer-Verlag, New York (1974)
10. Kumei, S.: Invariance transformations, invariance group transformation, and invariance groups of sine-Gordon equations. J. Math. Phys. **16**, 2461–2468 (1975)
11. Kumei, S.: Group theoretic aspects of conservation laws of nonlinear dispersive waves: KdV-type equations and nonlinear Schrödinger equations, invariance group transformation. J. Math. Phys. **18**, 256–264 (1977)
12. Olver, P.J.: Evolution equations possessing infinitely many symmetries. J. Math. Phys. **18**, 1212–1215 (1977)
13. Olver, P.J.: Symmetry groups and group invariant solutions of partial differential equations. J. Diff. Geom. **14**, 497–542 (1979)
14. Olver, P.J.: Symmetry and explicit solutions of partial differential equations. App. Num. Math. **10**, 307–324 (1992)
15. Ibragimov, N.H.: Transformation Groups Applied to Mathematical Physics. Riedel, Boston (1985)
16. Ibragimov, N.H.: Methods of Group Analysis for Ordinary Differential Equations. Znanie Publ, Moscow (1991). (In Russian)

17. Olver, P.J.: Applications of Lie Groups To Differential Equations. Springer-Verlag, New York (2000)
18. Brian, J.: Cantwell: Introduction to Symmetry Analysis. Cambridge University Press, Cambridge (2002)
19. Bluman, G.W., Kumei, S.: Symmetries and Differential Equations, Applied Mathematical Sciences 81. Springer-Verlag, New York (1996)
20. Ibragimov, N.H.: Elementary Lie Group Analysis and Ordinary Differential Equations. John Wiley and Sons (1999)
21. Noether, E.: Invariante Variationsprobleme, Nachr. König. Gesell. Wissen. Göttingen, Math.-Phys. KL, 235–257 (1918)
22. Korteweg, J.D., de Vries, G.: On the change of form of long waves advancing in a rectangular channel, and on a new type of long stationary waves. Philosophical Magazine, Series 5, **39**, 422–443 (1895)

Chapter 9
Discrete Dynamical Systems

So far we have discussed the dynamics of continuous systems. An evolutionary process may also be expressed mathematically as discrete steps in time. Discrete systems are described by maps (difference equations). The composition of map generates the dynamics or flow of a discrete system. It is a sequence of iterations such as for a given function $f : E \rightarrow E \subseteq \mathbb{R}^n$ with an initial point x_0, the sequence of iterates may be generated as

$$x_0, f(x_0), f(f(x_0)), f(f(f(x_0))), \ldots$$

This sequence may be finite or infinite. It is interesting to know how this sequence behaves after some iterations. The discrete maps cover a much greater range of dynamics than continuous systems. The notion of flow generated by a discrete system, its mathematical representation, compositions of maps, orbits, phase portraits, fixed points, periodic points, periodic cycles, stabilities, hyperbolic, non-hyperbolic fixed points with some important theorems and examples will be discussed sequentially in this chapter.

9.1 Maps and Flows

In general, a map is a function $f : E \rightarrow E$. The state x_{n+1} at the $(n + 1)$th stage is expressed in terms of the previous stage x_n by the relation $x_{n+1} = f(x_n)$. If the initial state or seed state is $x_0 \in E$, the sequence of states is given by $x_0, x_1, \ldots, x_n, \ldots$ in E. A discrete system generates a flow represented by $\phi_t(x)$ on E such that $f(x) = \phi_\tau(x)$, $x \in E$ and τ is a discrete time in $\mathbb{R}$. So the discrete dynamical system is the evolution of family of maps $\{f^n\}$, $n = 0, \pm 1, \pm 2, \ldots$ in E.

© Springer India 2015
G.C. Layek, *An Introduction to Dynamical Systems and Chaos*,
DOI 10.1007/978-81-322-2556-0_9

9.2   Composition of Maps

We need a notation for representing composition of maps. The two times composition of a map $f(x)$ is represented by $f^2(x) = (f \circ f)(x) = f(f(x))$, similarly $f^3(x) = f(f(f(x)))$. In general, we write inductively the n times composition of $f(x)$ with itself as

$$f^n(x) = f(f(f(\cdots f(x) \cdots)) \ (n \ \text{times}).$$

We now define composition of two maps. Let $f : \mathbb{R} \to \mathbb{R}$ and $g : \mathbb{R} \to \mathbb{R}$ be two one-dimensional maps. The composition of f and g is denoted symbolically by $f \circ g : \mathbb{R} \to \mathbb{R}$ and it is defined as

$$(f \circ g)(x) = f(g(x)), \forall x {\in} \mathbb{R}.$$

Setting $f = g$, we have $f^2 = f \circ f$. Similarly, $f^3 = f \circ f \circ f$, ..., $f^n = f \circ f \circ \cdots \circ f(n \ \text{times})$. If the map f is one–one and onto, then its inverse $f^{-1} :$ $\mathbb{R} \to \mathbb{R}$ exists and $(f \circ f^{-1})(x) = f(f^{-1}(x)) = x$ and $(f^{-1} \circ f)(x) = f^{-1}(f(x)) = x$, for all $x \in \mathbb{R}$. Therefore, $(f^{-1})^2 = f^{-1} \circ f^{-1}$; $(f^{-1})^2(x) = f^{-1}(f^{-1}(x))$ and so on. We define $(f^{-1})^n$ by $(f^{-1})^n = (f^n)^{-1} = f^{-1} \circ f^{-1} \circ \cdots \circ f^{-1}(n \ \text{times})$.

For example, consider the map $f(x) = -x^3$, $x \in \mathbb{R}$. Its two-fold composition gives

$$f^2(x) = (f \circ f)(x) = -(-x^3)^3 = x^9.$$

Iterating in this process, the successive compositions are obtained as follows:

$$f^3(x) = (f \circ f \circ f)(x) = -(x^9)^3 = -x^{27}$$
$$f^4(x) = x^{81}$$

$$\vdots$$

$$f^n(x) = (-1)^n x^{3^n}, n \in \mathbb{N}.$$

Here the notation f^n represents the composition of f with itself n times, neither the nth power of f nor its nth order derivative. In this way one can generate the sequence of iterations for any map. Higher dimensional composite maps can be similarly defined.

9.3 Orbits

Given a one-dimensional map $f : \mathbb{R} \to \mathbb{R}$ and a point $x_0 \in \mathbb{R}$ the forward orbit of x_0 is defined by

$$O^+(x_0) = \{f^k(x_0)\}_{k=0}^{\infty} = \{x_0, f(x_0), f^2(x_0), \ldots, f^n(x_0), \ldots\}.$$

Similarly, the backward orbit of x_0 is defined as

$$O^-(x_0) = \{x_0, f^{-1}(x_0), f^{-2}(x_0), \ldots, f^{-n}(x_0), \ldots\}.$$

The backward orbit exists if f is a homeomorphism (continuous and has continuous inverse). In general, the orbit of x_0 under a homeomorphism f is defined as follows:

$$O(x_0) = \{f^k(x_0)\}_{k=-\infty}^{\infty}$$
$$= \{\ldots, f^{-n}(x_0), \ldots, f^{-2}(x_0), f^{-1}(x_0), x_0, f(x_0), f^2(x_0), \ldots, f^n(x_0), \ldots\}.$$

Similarly, orbits of higher dimensional maps can be obtained.

9.4 Phase Portrait

Phase portraits are frequently used in dynamical system to represent the dynamics of a map graphically. A phase portrait consists of a diagram exhibiting possible changing positions of a map function and the arrows indicate the change of positions under iterations of the map. Consider a simple one-dimensional map $f : [0, 2\pi] \to [0, 2\pi]$ defined by $f(\theta) = \theta + 0.3\sin(3\theta)$. The phase portrait of this map is displayed in Fig. 9.1. The figure shows that the six points satisfy the

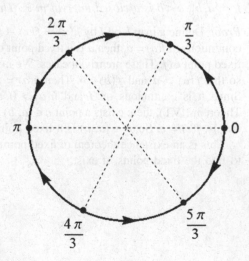

Fig. 9.1 Phase portrait of $f(\theta) = \theta + 0.3\sin(3\theta)$

relationship $f(\theta) = \theta$. The arrows indicate that the flow toward the three points $\frac{\pi}{3}, \pi, \frac{5\pi}{3}$ and the flow moves away from the other three points $0, \frac{2\pi}{3}, \frac{4\pi}{3}$. The points have special interest and we discuss elaborately in the next section.

9.5 Fixed Points

We know that fixed points of a continuous system are basically the constant or equilibrium solutions of the system. The notion of fixed points and their characterizations are very important in discrete systems. A point x^* is said to be a fixed point of a map $f : \mathbb{R} \to \mathbb{R}$ if $f(x^*) = x^*$, that is, if x^* is invariant under f. In other words, x^* is mapped onto itself by the mapping f. The fixed points of a map f can be obtained by finding the roots of the equation $f(x) - x = 0$. For example, consider the map $f : \mathbb{R} \to \mathbb{R}$, $f(x) = 4x(1 - x)$. It has two fixed points, given by

$$f(x^*) = x^* \Rightarrow 4x^*(1 - x^*) = x^* \Rightarrow x^* = 0, 3/4.$$

Note that a map may have infinitely many fixed points. For example, the identity map $I(x) = x$ has infinitely many fixed points in $\mathbb{R}$, that is, every real number is a fixed point of this map. Consider another map $f(x) = x + \sin(\pi x)$, $x \in \mathbb{R}$. In this map, every integer is a fixed point, and there are no others. The map $f(x) = x + 1$, $x \in \mathbb{R}$ has no fixed points. Suppose that x^* is a fixed point of a map f. Then $f(x^*) = x^*$. This yields $f^2(x^*) = f(f(x^*)) = f(x^*) = x^*$. Similarly, $f^3(x^*) = x^*, \dots, f^k(x^*) = x^*, \forall k \in \mathbb{N}$.

Again if $x_n = x^*$, then $x_{n+1} = f(x_n) = f(x^*) = x^*$. Thus, the orbit of f starting from the fixed point x^* remains at x^* for all iterations. Hence the fixed point is a constant solution of the map. This is why the fixed point is also called an equilibrium point. Fixed points of higher dimensional maps can be obtained similarly. Some theorems on the existence of fixed points are given below.

Theorem 9.1 Fixed Point Theorem *Let $f : I \to I$ be a continuous map, where $I = [a, b]$, $a < b$ is a closed interval in $\mathbb{R}$. Then f has at least one fixed point in I.*

Proof Define a map h on I by $h(x) = f(x) - x$, $x \in I$. Since f is continuous, h is also continuous. If $f(a) = a$, then a is a fixed point of f. Similarly, if $f(b) = b$, then b is also fixed point of f. These are trivial cases. We now assume that $f(a) \neq a$ and $f(b) \neq b$ so that $f(a) > a$ and $f(b) < b$. Then $h(a) = f(a) - a > 0$ and $h(b) = f(b) - b < 0$. Since h is continuous on I and $h(a) > 0$ and $h(b) < 0$, by Intermediate Value Theorem (IVT), there exists a point $c \in (a, b)$ such that $h(c) = 0$, that is, $f(c) - c = 0$, that is, $f(c) = c$. So, $c \in (a, b)$ is a fixed point of f. This completes the proof.

This is an existence theorem of fixed point for a map. It says nothing about how to find the fixed points, if exist.

Fig. 9.2 Graphical
representation of fixed point
of a map $f(x)$

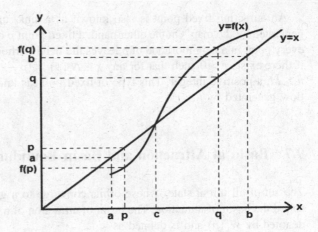

Theorem 9.2 *Let $I = [a, b]$, $a < b$ be a closed interval in $\mathbb{R}$ and $f : I \to \mathbb{R}$ be a continuous map such that $f(I) \supseteq I$. Then f has a fixed point in I.*

Proof Since $f(I) \supseteq I = [a, b]$, $f(I)$ contain points smaller than a and larger than b. That is, there exist points $p, q \in I$ such that $f(p) < a < p$ and $f(q) > b > q$. As before define a map h on I by $h(x) = f(x) - x$. Then h is continuous on I. Also $h(p) < 0$ and $h(q) > 0$. So by IVT there exists $c \in (p, q) \subset (a, b)$ such that $h(c) = 0$, that is, $f(c) = c$. This completes the proof. The graphical representation of the theorem is shown in Fig. 9.2.

Note that every map may not have fixed point. Consider the map $f(x) = x + k$, $x, k \in \mathbb{R}$. Since $x = f(x) = x + k$ has no solution for nonzero values of k, f has no fixed points. If $k = 0$, then $f(x) = x \; \forall x \in \mathbb{R}$, which gives infinite number of fixed points of f. Again fixed points of a map may exist but cannot be found algebraically, for example $f(x) = e^{-x/4}$, $x \in \mathbb{R}$. In this situation we use cobweb diagram (graphical representation of x versus $f(x)$) to find fixed points. The cobweb diagram and its construction will be discussed later.

9.6 Stable and Unstable Fixed Points

A fixed point in some interval I is said to be stable if for any point x_0 in I, the orbit starting from x_0 converges to the fixed point. It is unstable if the orbit moves away from I under iterations. Mathematically, a fixed point p of a map $f : \mathbb{R} \to \mathbb{R}$ is said to be stable or attracting if there exists a neighborhood of p such that iterations of all points in this neighborhood tend to p. This means that if there exists $\varepsilon > 0$ such that for all $x \in N_\varepsilon(p) = (p - \varepsilon, p + \varepsilon)$, the limit $\lim_{n \to \infty} f^n(x) = p$, that is, $\lim_{n \to \infty} |f^n(x) - p| = 0$, holds.

An attracting fixed point is also known as a sink, under the flow imagination generated by the map. On the other hand, a fixed point p of f is said to be unstable if every point in a neighborhood of p leaves the neighborhood under iterations, that is, if there exists $\varepsilon > 0$ such that for any $x \in N_\varepsilon(p) = (p - \varepsilon, p + \varepsilon)$, $f^n(x) \notin N_\varepsilon(p)$ for $n > M$, a positive integer. This type of fixed point is known as a source under the flow generated by f.

9.7   Basin of Attraction and Basin Boundary

The set of all initial states whose orbits converge to a given attractor of a map is called the basin of attraction. The basin of attraction of a fixed point p of a map f is denoted by $W^s(p)$ and is defined as

$$W^s(p) = \left\{ x : \lim_{n \to \infty} f^n(x) = p \right\}.$$

This set is a stable set of the point p. The unstable basin set is defined as

$$W^u(p) = \{ x : f^n(x) \notin N_\varepsilon(p),\ \text{for } n > M,\ \text{a positive integer} \}.$$

The boundary of the set $W^s(p)$ is called the basin boundary of f. Now, consider the function $f(x) = x^3, x \in \mathbb{R}$. It has three fixed points given by,

$$x^3 = x \Rightarrow x(x^2 - 1) = 0 \Rightarrow x = 0, \pm 1.$$

The fixed point 0 is a sink and the other two fixed points are sources. Figure 9.3 depicts them. The dashed line represents the diagonal line $y = x$. Therefore, $W^s(0) = (-1, 1)$, $W^u(1) = (0, \infty)$ is the positive real axis and $W^u(-1) = (-\infty, 0)$ is the negative real axis.

Fig. 9.3 Graphical representation of the map $f(x) = x^3$

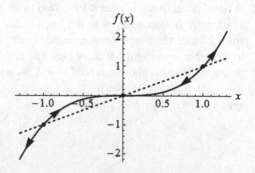

9.8 Linear Stability Analysis

Let x^* be a fixed point of a smooth function $f : \mathbb{R} \to \mathbb{R}$. We would like to determine the stability behavior of the map around the fixed point x^*. In usual procedure stability of a point depends on the first derivative of f. So, we shall assume that the maps are smooth in the vicinity of any fixed point, that is, the map function has continuous derivatives of all orders in the neighborhood of the fixed point. The map is represented by $x_{n+1} = f(x_n)$, where x_n denotes the nth iteration. Let $x_n = x^* + \xi_n$, where ξ_n be a small perturbation in the neighborhood of the fixed point. Then we have

$$
\begin{aligned}
x^* + \xi_{n+1} &= f(x^* + \xi_n) \\
&= f(x^*) + \xi_n f'(x^*) + O(\xi_n^2) \quad \text{[Taylor series expansion]} \\
&= x^* + \xi_n f'(x^*) + O(\xi_n^2) \quad \text{[Since } x^* \text{ is a fixed point of } f] \\
\Rightarrow \quad \xi_{n+1} &= \xi_n f'(x^*) + O(\xi_n^2).
\end{aligned}
$$

Let us choose ξ_n in such a way that $O(\xi_n^2)$ is negligible. Then

$$
\xi_{n+1} = \xi_n f'(x^*) = \lambda \xi_n, n = 0, 1, 2, \ldots,
$$

where $\lambda = f'(x^*)$ is known as the *multiplier*. Putting $n = 0, 1, 2, \ldots$ in the above equation, we get

$$
\begin{aligned}
\xi_1 &= \lambda \xi_0, \\
\xi_2 &= \lambda \xi_1 = \lambda^2 \xi_0, \\
&\vdots \\
\xi_n &= \lambda^n \xi_0.
\end{aligned}
$$

If $|\lambda| = |f'(x^*)| < 1$, then $\xi_n \to 0$ as $n \to \infty$. In this case, the fixed point x^* is *stable*. If $|\lambda| = |f'(x^*)| > 1$, then $\xi_n \to \infty$ as $n \to \infty$, and the fixed point x^* is *unstable*. So the stability of the fixed point x^* is determined by the value of $|f'(x^*)|$. Nothing can be said about the *marginal* case $|\lambda| = |f'(x^*)| = 1$.

Theorem 9.3 (Stability theorem) *Let $f : \mathbb{R} \to \mathbb{R}$ be a smooth function and p be a fixed point of f.*

(i) *If $|f'(p)| < 1$, then p is attracting (sink or stable)*
(ii) *If $|f'(p)| > 1$, then p is repelling (source or unstable)*.

Proof Since $f(x)$ is smooth, $f'(p)$ is well-defined and is given by

$$f'(p) = \lim_{x \to p} \frac{f(x) - f(p)}{x - p}.$$

So,

$$|f'(p)| = \lim_{x \to p} \left| \frac{f(x) - f(p)}{x - p} \right|.$$

(i) Suppose that $|f'(p)| < 1$. Then there exists a point $a \in \mathbb{R}$ with $0 < a < 1$ such that $|f'(p)| < a < 1$. Since $f'(x)$ is continuous at $x = p$, there exists $\varepsilon > 0$ such that

$$\left| \frac{f(x) - f(p)}{x - p} \right| < a \text{ whenever } |x - p| < \varepsilon.$$

So whenever $|x - p| < \varepsilon$,

$$|f(x) - f(p)| < a|x - p|$$

Now,

$$\begin{aligned}
\left| f^2(x) - f^2(p) \right| &= |f(f(x)) - f(f(p))| \\
&< a|f(x) - f(p)| \\
&< a^2|x - p|.
\end{aligned}$$

Similarly, using induction

$$|f^n(x) - f^n(p)| < a^n|x - p|.$$

Since $0 < a < 1$, $a^n \to 0$ as $n \to \infty$. Again p is a fixed point of f implies $f^n(p) = p$ for all $n \in \mathbb{N}$. Therefore, from the above inequality, we get

$$|f^n(x) - p| \to 0 \text{ as } n \to \infty$$

whenever $|x - p| < \varepsilon$, that is, whenever $x \in N_\varepsilon(p) = (p - \varepsilon, p + \varepsilon)$. So from the definition, p is attracting.

(ii) Suppose that $|f'(p)| > 1$. Then there exists a point $a \in \mathbb{R}$ such that $|f'(p)| > a > 1$ and therefore, there exists $\varepsilon > 0$ such that for all $x \in N_\varepsilon(p)$,

$$\left| \frac{f(x) - f(p)}{x - p} \right| > a$$

$$\Rightarrow |f(x) - f(p)| > a|x - p|.$$

Now,

$$\left|f^2(x) - f^2(p)\right| = \left|f(f(x)) - f(f(p))\right|$$
$$> a\left|f(x) - f(p)\right|$$
$$> a^2\left|x - p\right|.$$

Similarly, $\left|f^n(x) - f^n(p)\right| > a^n\left|x - p\right|$. Since $a > 1$, $a^n \to \infty$ as $n \to \infty$. Again, since p is a fixed point of f, $f^n(p) = p$ for all $n \in \mathbb{N}$. Therefore,

$$\left|f^n(x) - p\right| \to \infty \text{ as } n \to \infty, \text{ whenever } x \in N_\varepsilon(p) = (p - \varepsilon, p + \varepsilon).$$

This proves that p is a repelling fixed point of f. This completes the proof.

9.9 Cobweb Diagram

As mentioned earlier fixed points may not find easily for some maps. Also in the linear approximation, we cannot determine the stability of a fixed point p of f when $\left|f'(p)\right| = 1$. In such cases we use *cobweb diagram* to determine the fixed points and their stability characters. Cobweb diagrams are frequently drawn in the xy plane to analyze the stability behaviors of fixed points graphically. We shall first describe cobweb diagram for determining fixed points. Consider a one-dimensional map $f : \mathbb{R} \to \mathbb{R}$. The cobweb diagram of f consists of a diagonal line $y = x$ and the curve $y = f(x)$. The x coordinates of the points where the diagonal line $y = x$ intersect the curve $y = f(x)$ give the fixed points of the map f. If the line $y = x$ does not intersect the curve $y = f(x)$, then f has no fixed point. We shall now discuss the stability of the fixed points of a map using the cobweb diagram.

Let x^* be a fixed point of the map f, which is the x-coordinate of the point $(x^*, f(x^*))$, where the line $y = x$ intersects the curve $y = f(x)$, shown in Fig. 9.4.

Consider an initial point x_0 and draw a vertical line parallel to y-axis from the point x_0 to the curve $y = f(x)$. Suppose it intersects the curve at the point $(x_0, f(x_0))$. Draw a horizontal line parallel to x-axis from this point to the line $y = x$. Suppose it intersects the line at the point $(x_1, f(x_0))$, where $x_1 = f(x_0)$. Repeat this process until the line reaches to the fixed point x^* or it completely moves away from the fixed point. If it finally reaches to the fixed point x^*, then the fixed point is stable. Otherwise, it is unstable (Figs. 9.5 and 9.6).

The importance of the cobweb construction of a map is that it gives the stability of the fixed points globally, that is, using the cobweb diagram we can say that a fixed point (if exists) is either globally stable or globally unstable. But it can be used only in one-dimensional maps.

Fig. 9.4 Cobweb diagram for finding fixed point of a map

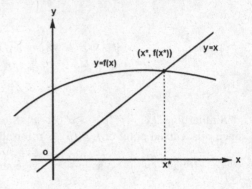

Fig. 9.5 Cobweb diagram for stable fixed point

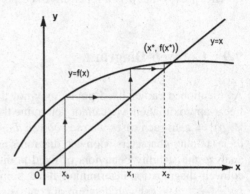

Fig. 9.6 Cobweb diagram for an unstable fixed point

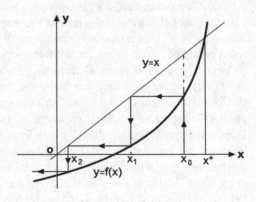

Example 9.1 Consider the map $f(x) = \sin x$, $x \in \mathbb{R}$. Discuss the stability character of the fixed point $x = 0$ of the map.

Solution The derivative of $f(x)$ is $f'(x) = \cos x$. Since $|f'(0)| = |\cos 0| = 1$, we cannot apply linear stability analysis to determine the stability of the fixed point origin. We construct the cobweb diagram as shown in Fig. 9.7.

Fig. 9.7 Cobweb diagram of
$f(x) = \sin(x)$ depicting the
stability character of the fixed
point origin

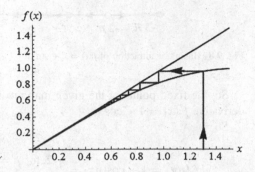

The figure shows that the iterated points move toward the fixed point origin. So the origin is stable.

Example 9.2 Consider the map $f(x) = 1 - \lambda x^2$, where $-1 \le x \le 1$ and $0 \le \lambda \le 2$. Find all fixed points of the map. Also determine their stability characters.

Solution The fixed points of $f(x)$ are the solutions of the equation $f(x) = x$. This gives

$$1 - \lambda x^2 = x \Rightarrow \lambda x^2 + x - 1 = 0 \Rightarrow x = \frac{-1 \pm \sqrt{1+4\lambda}}{2\lambda}.$$

Therefore,

$$x_1^* = \frac{-1 + \sqrt{1+4\lambda}}{2\lambda} \text{ and } x_2^* = \frac{-1 - \sqrt{1+4\lambda}}{2\lambda}$$

are two fixed points of f. Now, $f'(x) = -2\lambda x$. Since $\left| f'(x_2^*) \right| = (1 + \sqrt{1+4\lambda}) > 1 \ \forall \lambda \in [0,2]$, the fixed point x_2^* is unstable for all $\lambda \in [0,2]$. Again, $\left| f'(x_1^*) \right| = (\sqrt{1+4\lambda} - 1)$. Therefore, $\left| f'(x_1^*) \right| < 1$ if $\sqrt{1+4\lambda} - 1 < 1$, that is, if $\lambda < 3/4$. So, the fixed point x_1^* is stable if $0 < \lambda < 3/4$. And $\left| f'(x_1^*) \right| > 1$ if $\sqrt{1+4\lambda} - 1 > 1$, that is, if $\lambda > 3/4$. Hence x_1^* is unstable if $\lambda > 3/4$.

Example 9.3 Find the fixed points of the one-dimensional map $f(x) = x + \sin x, \ x \in \mathbb{R}$. Also find the basins of attraction.

Solution The fixed points of f satisfy

$$f(x) = x \Rightarrow x + \sin x = x \Rightarrow \sin x = 0 \Rightarrow x = n\pi, n = 0, \pm 1, \pm 2, \ldots$$

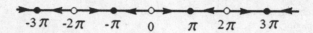

Fig. 9.8 Basins of attraction of $f(x) = x + \sin x$

So, the fixed points of the given map are $x^* = n\pi, n = 0, \pm 1, \pm 2, \ldots$ Now, the derivative $f'(x) = 1 + \cos x$.

So,

$$f'(n\pi) = 1 + \cos(n\pi) = \begin{cases} 2, \text{ for } n = 2m \\ 0, \text{ for } n = 2m + 1 \end{cases}, m = 0, \pm 1, \pm 2, \ldots$$

Therefore, $f'(0) = 2, f'(\pi) = 0, f'(2\pi) = 2$. This shows that $x^* = \pi$ is an attracting fixed point, while $x^* = 0, 2\pi$ are repelling fixed points. So the basin of attraction of π is $W^s(\pi) = (0, 2\pi)$. Similarly, the basin of attraction of 3π is $W^s(3\pi) = (2\pi, 4\pi)$. In general, we get $|f'(2m\pi)| = 2 > 1$ and $|f'((2m+1)\pi)| = 0 < 1, 2m\pi, m \in \mathbb{Z}$ are repelling fixed points, while $(2m+1)\pi$ are attracting fixed points. The basins of attraction of the fixed points $(2m+1)\pi$ are (Fig. 9.8)

$$W^s((2m+1)\pi) = (2m\pi, (2m+2)\pi), m \in \mathbb{Z}.$$

9.10 Periodic Points

A point p is said to be a periodic point of period-k or simply a periodic-k point of a map $f : \mathbb{R} \to \mathbb{R}$ if $f^k(p) = p$. The least positive integer k for which the relation is satisfied is called the prime period of the point p. For example, consider the map $f(x) = x^2 - 1, \ x \in \mathbb{R}$. We see that $f(0) = -1, \ f^2(0) = f(f(0)) = f(-1) = 0, \ f^3(0) = f(f^2(0)) = f(0) = -1$, and $f^4(0) = f(f^3(0)) = f(-1) = 0$. So $x = 0$ is periodic-2 point of the map f.

Note that a periodic-k point of a map f is a fixed point of the map f^k. But the converse is not true. Take $f(x) = x^2, \ x \in \mathbb{R}$. Clearly, $x = 0, 1$ are the fixed points of f. Now, $x = f^2(x) = f(f(x)) = x^4 \Rightarrow x^4 - x = 0 \Rightarrow x(x^3 - 1) = 0 \Rightarrow x(x-1)(x^2 + x + 1) = 0 \Rightarrow x = 0, 1 \ (\because x \in \mathbb{R})$. Therefore, $x = 0, 1$ are the fixed points of f^2. But they are not period-2 points of f, since $f(0) = 0, f(1) = 1, f^2(0) = 0, f^2(1) = 1$.

Maps may have infinite number of periodic points. Consider the map $f(x) = -x$, $x \in \mathbb{R}$. Here $x = 0$ is the only fixed point of f. For all $x \in \mathbb{R}$ we see that $f^2(x) = f(f(x)) = f(-x) = -(-x) = x$. This shows that every real number is a fixed point of $f^2(x)$. But $x = 0$ is the fixed point of f. Hence every nonzero real number is a periodic point of period -2 of the map $f(x) = -x$.

9.11 Periodic Cycles

Maps may have periodic points/periodic orbits. We shall now give definitions of periodic orbits of periods 2, 3, and k for a map. Consider a one-dimensional map $f : \mathbb{R} \to \mathbb{R}$. Let p and q be two points in $\mathbb{R}$ such that $f(p) = q, f(q) = p$. This implies that $f^2(p) = (f \circ f)(p) = f(f(p)) = f(q) = p$ and $f^2(q) = (f \circ f)(q) = f(f(q)) = f(p) = q$. Then $\{p, q\}$ is called a periodic orbit of period-2. A periodic orbit of period-2 is also known as periodic 2-cycle or a periodic 2-orbit or simply a 2-cycle. The points of a periodic 2-cycle give the fixed points of f^2. The 2-cycle represents diagrammatically as

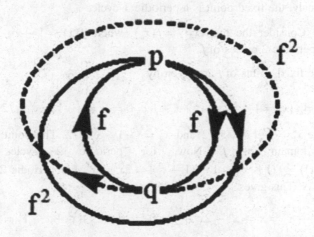

Again, let p, q and r be three points in $\mathbb{R}$ such that $f(p) = q, f(q) = r, f(r) = p$. So, $f^2(p) = r, f^2(q) = p, f^2(r) = q$, but $f^3(p) = f(f^2(p)) = f(r) = p$, $f^3(q) = q$ and $f^3(r) = r$. The set $\{p, q, r\}$ is called a periodic 3-cycle of f. The points p, q, r are the fixed points of f^3. This expresses graphically as In the similar way the periodic k-cycle can be defined as follows:

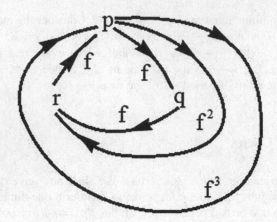

A cycle $\{x_1, x_2, \ldots, x_k\}$ of a map $f : \mathbb{R} \to \mathbb{R}$ is said to be a periodic orbit of period-k (or periodic k-cycle) if $f(x_1) = x_2, f(x_2) = x_3, \ldots, f(x_{k-1}) = x_k, f(x_k) = x_1$, as in above this implies that $f^k(x_1) = x_1, f^k(x_2) = x_2, \ldots, f^k(x_k) = x_k$. The points of the k-cycle are all fixed points of f^k. All the points in a cycle are distinct from each other. Obviously, the fixed point is a periodic 1-cycle.

Example 9.4 Consider the map $x_{n+1} = f(x_n)$, where $f(x) = 1 - x^2$, $x \in [-1, 1]$. Find all the periodic 2-cycles of f.

Solution The fixed points of f are given by

$$x = f(x) = 1 - x^2 \Rightarrow x^2 + x - 1 = 0 \Rightarrow x = \left(-1 \pm \sqrt{5}\right)/2.$$

We denote $x_1^* = (-1 + \sqrt{5})/2$ and $x_2^* = (-1 - \sqrt{5})/2$. The point x_2^* lies outside the domain of f. Now, for periodic 2 cycles we see $f^2(x) = f(f(x)) = f(1 - x^2) = 1 - (1 - x^2)^2 = 2x^2 - x^4$. For periodic 2-cycles, we have $f^2(x) = x$. This gives

$$2x^2 - x^4 = x \Rightarrow x^4 - 2x^2 + x = 0 \Rightarrow x(x-1)(x^2 + x - 1) = 0$$
$$\Rightarrow x = 0, 1, \left(-1 \pm \sqrt{5}\right)/2.$$

So the fixed points of $f^2(x)$ are $x = 0, 1, (-1 \pm \sqrt{5})/2$. But $(-1 \pm \sqrt{5})/2$ are the fixed points of f. Again, $f(0) = 1, f(1) = 0$ and $f^2(0) = 0, f^2(1) = 1$. This shows that the set $\{0, 1\}$ forms the period 2-cycle of the map f.

9.12 Stability of Periodic Point and Periodic Cycle

A periodic-k point of a map $f : \mathbb{R} \to \mathbb{R}$ is said to be

(i) Stable, if it is a stable fixed point of the map f^k;
(ii) Unstable, if it is an unstable fixed point of f^k.

A stable fixed point of f^k is called a *periodic sink* and an unstable fixed point of f^k is called a *periodic source*. From the linear stability analysis the stability/instability criteria may be obtained. This is sometimes called derivative stability test condition. The derivative test for periodic fixed points of f^k is obtained as follows. The chain rule for derivative of function of function $f(g(x))$ gives $\frac{d}{dx}(f(g(x))) = f'(g(x)) \cdot g'(x)$. This implies that the derivative composition function $(f^2)'(x) = f'(f(x)) \cdot f'(x)$. Using linear stability analysis, we have the condition $|(f^2)'(x)| < 1$ for stable, and $|(f^2)'(x)| > 1$ for unstable fixed points of $f^2(x)$. Similarly, for the map $f^k(x)$ we can write the criteria as $\left|(f^k)'(x)\right| < 1$ for stable and $\left|(f^k)'(x)\right| > 1$ for unstable fixed points of f^k. These are all linear stability criteria.

Example 9.5 Find the period of the point $\frac{1}{8}(5 + \sqrt{5})$ for the map $f(x) = 4x(1 - x)$, $x \in [0, 1]$. Also determine its stability.

Solution Given map is $f(x) = 4x(1 - x)$, $x \in [0, 1]$. This is a quadratic map. Now,

$$f\left(\frac{1}{8}\left(5 + \sqrt{5}\right)\right) = 4 \cdot \frac{1}{8}\left(5 + \sqrt{5}\right) \cdot \left(1 - \frac{1}{8}\left(5 + \sqrt{5}\right)\right) = \frac{1}{16}\left(5 + \sqrt{5}\right)\left(3 - \sqrt{5}\right) = \frac{1}{8}\left(5 - \sqrt{5}\right),$$

$$f\left(\frac{1}{8}\left(5 - \sqrt{5}\right)\right) = 4 \cdot \frac{1}{8}\left(5 - \sqrt{5}\right) \cdot \left(1 - \frac{1}{8}\left(5 - \sqrt{5}\right)\right) = \frac{1}{16}\left(5 - \sqrt{5}\right)\left(3 + \sqrt{5}\right) = \frac{1}{8}\left(5 + \sqrt{5}\right)$$

Again, $f^2\left(\frac{1}{8}\left(5 + \sqrt{5}\right)\right) = f\left(f\left(\frac{1}{8}\left(5 + \sqrt{5}\right)\right)\right) = f\left(\frac{1}{8}\left(5 - \sqrt{5}\right)\right) = \frac{1}{8}\left(5 + \sqrt{5}\right)$.

This shows that the point $(5 + \sqrt{5})/8$ is a fixed point of the map f^2 and hence it is a periodic point of period-2 of the given map. We shall now examine the stability of this periodic-2 point. We have

$$f^2(x) = f(f(x)) = f(4x(1 - x)) = 4 \cdot 4x(1 - x)\{1 - 4x(1 - x)\}$$
$$= 16x - 80x^2 + 128x^3 - 64x^4.$$

We shall use the derivative test for finding the stability character of the periodic point of the map. We see that $(f^2)'(x) = 16 - 160x + 384x^2 - 256x^3$. Since

$$\left|(f^2)'\left(\frac{1}{8}\left(5 + \sqrt{5}\right)\right)\right| = 16\left(244 + 105\sqrt{5}\right) > 1,$$

the periodic-2 point $(5 + \sqrt{5})/8$ of f is unstable.

Stability of Periodic Cycles

Stability of periodic cycles is a collective property. Let $\{x_1, x_2, \ldots, x_n\}$ be a periodic n-cycle of a map $f : \mathbb{R} \to \mathbb{R}$. As per definition of periodic cycle each x_i $(i = 1, 2, \ldots, n)$ is a fixed point of the map f^n. The cycle is stable (respectively unstable) if and only if the points x_i $(i = 1, 2, \ldots, n)$ are stable (respectively unstable) fixed points of the map $f^n(x)$. Using the chain rule of differentiation of the composition map $f^n(x_i)$, we get

$$
\begin{aligned}
(f^n)'(x_i) &= (f^{n-1})'(f(x_i))f'(x_i) = (f^{n-1})'(x_{i+1})f'(x_i) \\
&= (f^{n-2})'(f(x_{i+1}))f'(x_{i+1})f'(x_i) \\
&= (f^{n-2})'(x_{i+2})f'(x_{i+1})f'(x_i) \\
&= \cdots = f'(x_{i+n-1})f'(x_{i+n-2})\cdots f'(x_{i+1})f'(x_i) \\
&= f'(x_1)f'(x_2)\cdots f'(x_{n-1})f'(x_n) \quad [\text{Since } \{x_1, x_2, \ldots, x_n\} \text{ is a cycle of } f]
\end{aligned}
$$

Now if x_i is a stable fixed point of $f^n(x)$, then from linear stability analysis $|(f^n)'(x_i)| < 1$. This implies that $|f'(x_1)f'(x_2)\cdots f'(x_n)| < 1$. Similarly, if x_i is an unstable fixed point of f^n, then $|f'(x_1)f'(x_2)\cdots f'(x_n)| > 1$. Hence we have the following definition:

The cycle is said to be *stable* (*sink* or *attracting*) if $|f'(x_1)f'(x_2)\cdots f'(x_n)| < 1$ and it is *unstable* (*source* or *repelling*) if $|f'(x_1)f'(x_2)\cdots f'(x_n)| > 1$. But these criteria are weak in nature.

Example 9.6 Find all fixed points of $f(x) = x^2 - 1$, $x \in \mathbb{R}$. Determine their stabilities. Show that $\{0, -1\}$ is a periodic orbit of period-2. Are the periodic cycle attracting?

Solution The fixed points of f are given by

$$
x = f(x) = x^2 - 1 \Rightarrow x^2 - x - 1 = 0 \Rightarrow x = \frac{1 \pm \sqrt{5}}{2}.
$$

So, the fixed points of the map are $\left(\frac{1+\sqrt{5}}{2}\right)$ and $\left(\frac{1-\sqrt{5}}{2}\right)$. Now, $f'(x) = 2x$. Since $f'\left(\frac{1+\sqrt{5}}{2}\right) = 2\left(\frac{1+\sqrt{5}}{2}\right) = (1 + \sqrt{5}) > 1$ and $f'\left(\frac{1-\sqrt{5}}{2}\right) = 2\left(\frac{1-\sqrt{5}}{2}\right) = (1 - \sqrt{5}) < 1$, both the fixed points are unstable. For periodic orbit, we find $f(0) = -1$, $f(-1) = 0$, $f^2(0) = f(f(0)) = f(-1) = 0$ and $f^2(-1) = f(f(-1)) = f(0) = -1$.

This shows that $\{0, -1\}$ is a periodic orbit of period-2 of the map f. Since $|f'(0)f'(-1)| = |0.(-2)| = 0 < 1$, the cycle is stable.

Example 9.7 Show that $\{-1, 1\}$ is an attracting 2-cycle of the map $f(x) = -x^{1/3}$, $x \in \mathbb{R}$. Find the stability character of the fixed points of f.

Solution Here $f(x) = -x^{1/3}, x \in \mathbb{R}$. We see that $f(-1) = -(-1)^{1/3} = -(-1) = 1$, $f(1) = -(1)^{1/3} = -1$, $f^2(-1) = f(f(-1)) =$

$f(1) = -1, f^2(1) = f(f(1)) = f(-1) = 1$

The points $\{-1, 1\}$ form a cycle. Again we see that.

$$f^3(-1) = f(f^2(-1)) = f(-1) = 1, f^3(1) = f(f^2(1)) = f(1) = -1.$$

This shows that $\{-1, 1\}$ is a 2-cycle of the map f. We shall use the derivative test for stability character of the cycle. The derivative of f gives

$$f'(x) = -\frac{1}{3}x^{-(2/3)} = -\frac{1}{3x^{(2/3)}}$$

$$\therefore f'(-1) = -\frac{1}{3} \text{ and } f'(1) = -\frac{1}{3}$$

The cycle $\{-1, 1\}$ will be linearly stable if $|f'(-1)f'(1)| < 1$. Since $|f'(-1)f'(1)| = |(-\frac{1}{3})(-\frac{1}{3})| = \frac{1}{9} < 1$, so the 2-cycle $\{-1, 1\}$ is stable. We now find the fixed points of the map. The fixed points are obtained by solving the equation

$$f(x) = x \Rightarrow -x^{1/3} = x \Rightarrow -x = x^3 \Rightarrow x^3 + x = 0 \Rightarrow x(x^2 + 1) = 0$$
$$\Rightarrow x = 0 (\because x \in \mathbb{R})$$

So, $x^* = 0$ is the only fixed point of the map f. Since $|f'(x)| > 1$ in the neighborhood of the fixed point 0, the fixed point origin is repelling.

Example 9.8 Consider the map $f(x) = -x^3$, $x \in \mathbb{R}$. Show that the origin is an attracting fixed point and $\{-1, 1\}$ is a repelling 2-cycle of the map.

Solution The fixed points of the map f are given by

$$x = f(x) = -x^3$$
$$\Rightarrow \quad x^3 + x = 0$$
$$\Rightarrow \quad x(x^2 + 1) = 0$$
$$\Rightarrow \quad x = 0.$$

So the origin is the only fixed point of f. Now $f'(x) = -3x^2$. Since $f'(0) = 0 < 1$, the fixed point origin is stable. Again, $f(1) = -1$, $f(-1) = 1$. Now $f^2(x) = f(f(x)) = -(-x^3)^3 = x^9$, $f^3(x) = -x^{27}$ and $f^4(x) = x^{81}$. Therefore, $f^2(1) = 1$, $f^2(-1) = -1$, $f^3(1) = -1$ and $f^3(-1) = 1$, $f^4(1) = 1$ and $f^4(-1) = -1$. This shows that $\{-1, 1\}$ is a periodic 2-cycle. The stability condition of the cycle gives that $|f'(-1)f'(1)| = |(-3)(-3)| = 9 > 1$. Hence $\{-1, 1\}$ is a repelling 2-cycle.

Example 9.9 Show that the map $f(x) = -\frac{3}{2}x^2 + \frac{5}{2}x + 1$, $x \in \mathbb{R}$ has a 3-cycle. Comment about the stability of the cycle.

Solution Take three points $0, 1,$ and $2.$ We see that

$$f(0) = 1, \quad f(1) = -\frac{3}{2} + \frac{5}{2} + 1 = 2,$$
$$f(2) = -\frac{3}{2}(4) + \frac{5}{2}(2) + 1 = -6 + 5 + 1 = 0$$

So $f(0) = 1, f(1) = 2, f(2) = 0.$ And $f^2(0) = f(f(0)) = f(1) = 2,$ $f^2(1) = f(f(1)) = f(2) = 0,$ $f^2(2) = f(f(2)) = f(0) = 1.$ $f^3(0) = f(f^2(0)) = f(2) = 0,$ $f^3(1) = f(f^2(1)) = f(0) = 1, f^3(2) = f(f^2(2)) = f(1) = 2.$ The three points $0,1,$ and 2 are fixed points of $f^3.$ Also, $f^4(0) = f(f^3(0)) = f(0) = 1,$ $f^4(1) = f(f^3(1)) = f(1) = 2, f^4(2) = f(f^3(2)) = f(2) = 0.$

This shows that $\{0, 1, 2\}$ is a periodic 3 cycle of the map $f.$ We shall test the stability of the 3-cycle using derivative test. This gives the condition $|f'(0)f'(1)f'(2)| < 1.$ Now, $f'(x) = -3x + \frac{5}{2}.$ So, $f'(0) = \frac{5}{2},$ $f'(1) = -3 + \frac{5}{2} = -\frac{1}{2},$ $f'(2) = -6 + \frac{5}{2} = -\frac{7}{2}.$ Since $|f'(0)f'(1)f'(2)| = \left|\frac{5}{2}\left(-\frac{1}{2}\right)\left(-\frac{7}{2}\right)\right| = \frac{35}{8} > 1,$ the cycle $\{0, 1, 2\}$ is unstable.

Example 9.10 Find all periodic two orbits of the quadratic map $f(x) = 4x(1 - x),$ $x \in [0, 1].$ Show that they are unstable.

Solution It can be easily shown that the fixed points of $f^2(x),$ where $f(x) = rx(1 - x),$ $x \in [0, 1]$ are

$$x^* = 0, \left(1 - \frac{1}{r}\right), p, q$$

where $p = \frac{r + 1 + \sqrt{(r+1)(r-3)}}{2r}$ and $q = \frac{r + 1 - \sqrt{(r+1)(r-3)}}{2r}.$ But $x^* = 0, \left(1 - \frac{1}{r}\right)$ are the fixed points of $f(x) = rx(1 - x).$ Here $r = 4.$ So

$p = \frac{4 + 1 + \sqrt{(4+1)(4-3)}}{2.4} = \frac{5 + \sqrt{5}}{8}$ and $q = \frac{5 - \sqrt{5}}{8}.$ Now

$$f(p) = 4p(1 - p) = 4\left(\frac{5 + \sqrt{5}}{8}\right)\left(1 - \frac{5 + \sqrt{5}}{8}\right) = \frac{5 + \sqrt{5}}{2} \cdot \frac{3 - \sqrt{5}}{8} = \frac{5 - \sqrt{5}}{8}$$

$$= q.$$

That is, $f(p) = q.$ Similarly, $f(q) = p.$ Thus we get $f(p) = q, f(q) = p$ and $f^2(p) = p$ and $f^2(q) = q.$ Therefore the periodic-2 cycle of f is $\{p, q\},$ i.e., $\left\{\frac{5+\sqrt{5}}{8}, \frac{5-\sqrt{5}}{8}\right\}.$ Here $f'(x) = 4 - 8x.$ So,

$$f'(p) = 4 - 8p = 4 - 8\left(\frac{5 + \sqrt{5}}{8}\right) = -\left(1 + \sqrt{5}\right) \text{ and}$$

$$f'(q) = 4 - 8q = 4 - 8\left(\frac{5 - \sqrt{5}}{8}\right) = -\left(1 - \sqrt{5}\right).$$

The derivative test of the cycle $\{p, q\}$ gives $|f'(p)f'(q)| = |(1+\sqrt{5})(1 - \sqrt{5})| = 4 > 1$. Hence the cycle $\{(5+\sqrt{5})/8, (5 - \sqrt{5})/8\}$ is unstable.

9.13 Eventually Fixed Point, Periodic Point, Periodic Orbit

A point p is said to be an eventually fixed point of a map f if it is not a fixed point of f but reaches to a fixed point of f after a finite number of iterations. In other words, p is a fixed point of f if there exists a positive integer n such that $f^n(p) = x^*$ but $f^{n-1}(p) \neq x^*$, where x^* is a fixed point of f. This implies that $f^{n+1}(p) = f(f^n(p)) = f(x^*) = x^* = f^n(p)$. Thus, p is an eventually fixed point of f if $f^n(p)$ is a fixed point of f for some positive integer n. This can be understood very easily diagrammatically.

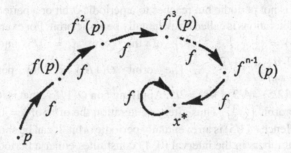

Consider the map $f(x) = 4x(1 - x)$, $x \in [0, 1]$. The fixed points of f are given by $x = f(x) = 4x(1 - x) \Rightarrow x = 0, 3/4$. At $x = 1/4$, $f(x) = 4.\frac{1}{4}\left(1 - \frac{1}{4}\right) = \frac{3}{4}$. At $x = \left(\frac{1}{2} - \frac{\sqrt{3}}{4}\right)$,

$$f(x) = f\left(\frac{1}{2} - \frac{\sqrt{3}}{4}\right) = 4\left(\frac{1}{2} - \frac{\sqrt{3}}{4}\right)\left(\frac{1}{2} + \frac{\sqrt{3}}{4}\right) = \frac{1}{4}$$

and $f^2(x) = f(f(x)) = f(1/4) = 3/4$. Therefore, the points $1/4$ and $\left(\frac{1}{2} - \frac{\sqrt{3}}{4}\right)$ are eventually fixed points of the map f.

Eventually Periodic Point

A point p is said to be an eventually periodic point of a map f if p itself is not a periodic point of f but it reaches to a periodic point of f after finite number of iterations. Mathematically, the point p is said to be an eventually periodic point (of period-k) if there exists a positive integer N such that $f^{n+k}(p) = f^n(p)$ whenever $n \geq N$. That is, if $f^n(p)$ is periodic-k point of f for $n \geq N$. An eventually periodic point of period-2 is shown graphically as below.

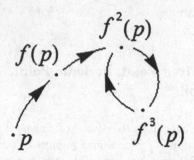

For example, $x = 0$ is an eventually periodic-2 point of $f(x) = 1 - x - x^2$, since $f^3(0) = f(0)$. Similarly, $x = 0.2$ is an eventually periodic point of $f(x) = 3.2x(1 - x)$. Note that eventually periodic points cannot occur for homeomorphic maps.

Eventually Periodic Orbit (or cycle)

An orbit which is not periodic but reaches to a periodic orbit or a periodic cycle after finite number of iterations is called an eventually periodic orbit. For example, consider the orbit $O\left(\frac{1}{5}\right) = \left\{\frac{1}{5}, \frac{2}{5}, \frac{4}{5}, \frac{2}{5}, \frac{4}{5}, \ldots\right\}$ of the point $x = 1/5$ under the map $f(x) = \begin{cases} 2x & \text{if } 0 \leq x \leq 1/2 \\ 2(1-x) & \text{if } 1/2 \leq x \leq 1 \end{cases}$. The orbit $O(1/5)$ is not periodic, since $f(1/5) = 2/5, f(2/5) = 4/5, f(4/5) = 2/5$. Applying f on $O(1/5)$, that is, $O((f1/5))$ we get the periodic orbit $\left\{\frac{2}{5}, \frac{4}{5}\right\}$. Thus after one iteration the orbit of $x = 1/5$ moves to a periodic orbit. Hence $O(1/5)$ is an eventually periodic orbit. It can be shown that every set of rational numbers in the interval $(0, 1)$ constitutes either a periodic orbit or an eventually periodic orbit under the map f. This is not true for the set of irrational numbers in $(0, 1)$.

9.14 Superstable Fixed Point and Superstable Periodic Point

Stability of fixed points/periodic points is determined by the derivative test condition and the smaller magnitude of the derivative gives the faster rate of convergence. But the parameter values at which the derivative is zero are of special importance. A fixed point x^* of a one-dimensional map $f : \mathbb{R} \to \mathbb{R}$ is said to be superattracting (or superstable) if $f'(x^*) = 0$. A superstable fixed point is always stable. Similarly, a period n cycle $\{x_1, x_2, \ldots, x_n\}$ of f is said to be superstable if the derivative $f'(x_1) \cdot f'(x_2) \cdots f'(x_n) = 0$, that is, if at least one $f'(x_i)$ is zero.

Consider the map $f(x) = (3x - x^3)/2$, $x \in \mathbb{R}$. The fixed points of f are given by

$$f(x) = x \Rightarrow (3x - x^3)/2 = x \Rightarrow x = -1, 0, 1.$$

The derivative of $f(x)$ is $f'(x) = 3(1 - x^2)/2$. Since $f'(\pm 1) = 0$, the fixed points $x = \pm 1$ are superstable. Consider another map $f(x) = rx(1 - x)$, $x \in [0, 1]$, $r \geq 0$. Here $f'(x) = r(1 - 2x)$. In this context we define an important map, the unimodal map. It is a very simple nonlinear map with a single point of extremum. The map $f(x)$ is unimodal. For superstable 1-cycle, we must have $f'(x) = 0$. This gives $x = 1/2$. However, 1-cycles are the fixed points of the map. So $f(x) = x$ at $x = 1/2$. This yields the parameter value $r = 2$. Thus a superstable 1-cycle of $f(x) = rx(1 - x)$ exists when $r = 2$ and the cycle contains only the point $x = 1/2$. For superstable 2-cycle $\{p, q\}$ we have the condition $f'(p)f'(q) = 0$. This shows that $x = 1/2$ must be an element of the 2-cycle. Since every element of the 2-cycle of $f(x)$ is a fixed point of $f^2(x)$, $x = 1/2$ is a fixed point of $f^2(x)$. By solving the equation $f^2(x) = x$ for $x = 1/2$ we get three values of r, namely $r = 2, (1 - \sqrt{5}), (1 + \sqrt{5})$. But $r = 2$ corresponds to superstable 1-cycle of f and $r = (1 - \sqrt{5})$ is negative. So for superstable 2-cycle we must have $r = (1 + \sqrt{5})$. With this value of r the superstable 2-cycle is given by $\left\{\frac{1}{2}, \frac{1+\sqrt{5}}{4}\right\}$.

9.15 Hyperbolic Points

A fixed point x^* of a map $f : \mathbb{R} \to \mathbb{R}$ is said to be hyperbolic if $|f'(x^*)| \neq 1$. For example, consider the map $f(x) = x^2$. The fixed points of the map are $x^* = 0, 1$. Since $|f'(0)| = 0 \neq 1$ and $|f'(1)| = 2 \neq 1$, the fixed points $x^* = 0, 1$ are hyperbolic.

Hyperbolic Periodic Point
Let p be a periodic point of period n of a map $f : \mathbb{R} \to \mathbb{R}$. Then p is said to be hyperbolic periodic point if $|(f^n)'(p)| \neq 1$. Let $f(x) = -(x + x^3)/2$, $x \in \mathbb{R}$. Clearly, the points ± 1 are periodic-2 points of f. Now, we see that $|(f^2)'(\pm 1)| = 4 \neq 1$. Hence the periodic points ± 1 are hyperbolic.

Example 9.11 Find the source and sink of the map $f(x) = -(x^2 + x)/2$, $x \in \mathbb{R}$. Show that they are hyperbolic in nature.

Solution The fixed points of the map f are given by $f(x) = x$. Therefore, the fixed points are given by $x^* = 0, -3$. Now, $f'(x) = -(2x + 1)/2$ and the derivatives of $f(x)$ at the points 0 and -3 are $f'(0) = -1/2$ and $f'(-3) = 5/2$, respectively. Since $|f'(0)| = 1/2 < 1$, the fixed point $x = 0$ is stable (sink). Again, $|f'(-3)| = 5/2 > 1$, the fixed point $x = -3$ is unstable (source). We see that $|f'(x)| \neq 1$ for both the fixed points. This implies that the source at $x = -3$ and the sink at $x = 0$ are hyperbolic in nature. These fixed points are known as hyperbolic source and hyperbolic sink under the flow f.

Theorem 9.4 *Let $f : \mathbb{R} \to \mathbb{R}$ be a smooth function and p be a hyperbolic fixed point of the map f with $|f'(p)| < 1$. Then there exists a neighborhood $N_\varepsilon(p) = (p - \varepsilon, p + \varepsilon)$ of p such that if $x \in N_\varepsilon(p)$, then $f^n(x) \to p$ as $n \to \infty$.*

Proof Since f is a smooth function and $|f'(p)| < 1$, there exists a real number $0 < a < 1$ and an ε neighborhood $N_\varepsilon(p) = (p - \varepsilon, p + \varepsilon)$ of p such that $|f'(x)| < a < 1$ for all $x \in N_\varepsilon(p)$. Using Mean Value Theorem of calculus and $f(p) = p$, p being fixed point of f, we have

$$|f(x) - p| = |f(x) - f(p)| = |f'(k)||x - p|, k \in (x, p).$$

Since $|f'(p)| < 1$ and taking $k \in N_\varepsilon(p)$, we have $|f'(k)| < a < 1$. Therefore, $|f(x) - p| < a|x - p| < |x - p|$. This shows that $f(x)$ lies in $N_\varepsilon(p)$. Taking the second composition, we have

$$|f^2(x) - p| = |f(f(x)) - p| < a|f(x) - p| < a^2|x - p|.$$

Continuing this process, we finally get

$$|f^n(x) - p| < a^n|x - p|.$$

Since $a < 1$, $a^n \to 0$ as $n \to \infty$. This shows the nth composition $f^n(x) \to p$ as $n \to \infty$. This completes the proof.

Theorem 9.5 *Let p be a hyperbolic fixed point of a one-dimensional map $f : \mathbb{R} \to \mathbb{R}$ with $|f'(p)| > 1$. Then there exists an open interval U of p such that, if $x \in U$, $x \neq p$, then there exists $n > 0$ such that $f^n(x) \notin U$.*

Proof Left as an exercise.

9.16   Non-Hyperbolic Points

A fixed point x^* of a map $f : \mathbb{R} \to \mathbb{R}$ is said to be non-hyperbolic if $|f'(x^*)| = 1$. Thus, for a non-hyperbolic fixed point x^* either $f'(x^*) = 1$ or $f'(x^*) = -1$.

Consider the map $f(x) = \sin x$, $x \in \mathbb{R}$. Here $x^* = 0$ is a fixed point of f. Since $f'(x^*) = \cos(x^*) = \cos 0 = 1$, the fixed point $x^* = 0$ is non-hyperbolic. Similarly, $x = 0$ is a non-hyperbolic fixed point of the map $g(x) = \tan x$.

It is very difficult to say whether a non-hyperbolic fixed point is attracting or repelling. In this situation we use cobweb diagram to analyze the nature of the fixed point. We can also determine the stability of a non-hyperbolic fixed point by using the following theorem.

Theorem 9.6 *Let p be a non-hyperbolic fixed point of a map* $f : \mathbb{R} \to \mathbb{R}$ *such that* $f'(p) = 1$. *Then the following statements hold:*

(i) *If* $f''(p) \neq 0$, *then p is semi-stable*

(ii) *If* $f''(p) = 0$ *and* $f'''(p) < 0$, *then p is asymptotically stable.*

(iii) *If* $f''(p) = 0$ *and* $f'''(p) > 0$, *then p is unstable.*

Example 9.12 Use Theorem 9.6 to determine the stability of the fixed point origin of the maps (i) $f(x) = \sin x$, $x \in \mathbb{R}$, (ii) $f(x) = x^3 - x^2 + x, x \in \mathbb{R}$.

Solution

(i) Here $f(x) = \sin x$, $x \in \mathbb{R}$. Clearly, the origin is a fixed point of the map. Now, $f'(x) = \cos x$, $f''(x) = -\sin x$, $f'''(x) = -\cos x$. Therefore, $f'(0) = \cos 0 = 1$. Also, the third derivative is continuous and $f'''(0) = -\cos 0 = -1 \neq 0$. Since $f''(0) = -\sin 0 = 0$ and $f'''(0) = -1 < 0$, by Theorem 9.6, the origin is asymptotically stable.

(ii) Here $f(x) = x^3 - x^2 + x$. Clearly, the origin is a fixed point of f. Now, $f'(x) = 3x^2 - 2x$, $f''(x) = 6x$ and $f'''(x) = 6$. Here $f'''(x)$ is continuous and $f'''(0) \neq 0$. Since $f''(0) = 0$ and $f'''(0) = 6 > 0$, by Theorem 9.6, the origin is unstable.

9.17 The Schwarzian Derivative

The Schwarzian derivative is important in determining the stability character of non-hyperbolic fixed points. The concept of Schwarzian derivative was introduced by the German mathematician Hermann Schwarz (1843–1921). But D. Singer (1978) first used it in one-dimensional systems. The Schwarzian derivative $Sf(x)$ of a function f at a point x is defined as follows:

$$Sf(x) = \frac{f'''(x)}{f'(x)} - \frac{3}{2}\left[\frac{f''(x)}{f'(x)}\right]^2,$$

provided the derivatives exist. For example, consider the map $f(x) = 4x(1 - x)$, $x \in [0, 1]$. Its Schwarzian derivative is given by

$$Sf(x) = -\frac{3}{2}\left(\frac{-8}{4 - 8x}\right)^2 = -\frac{6}{(1 - 2x)^2} < 0. \text{ At } x = \frac{1}{2}, Sf(x) = -\infty < 0.$$

Take another function $f(x) = x^n$, where $x \in \mathbb{R}$ and $n \neq 0$. Then $Sf(x) = \frac{1-n^2}{2x^2}$. Clearly, $Sf(x) < 0$ when $n^2 > 1$, $Sf(x) = 0$ when $n = \pm 1$ and $Sf(x) > 0$ when $n^2 < 1$, that is, when $-1 < n < 1$.

Elementary Properties of Schwarzian Derivative

Property 1 Suppose $f(x)$ is a polynomial such that all the roots of $f'(x)$ are real and distinct. Then $Sf(x) < 0$.

Proof Suppose that

$$f'(x) = \prod_{i=1}^{n} (x - x_i) = (x - x_1)(x - x_2) \cdots (x - x_n)$$

where each $x_i (i = 1, 2, \ldots, n)$ is real and distinct. Then the second-order derivative is expressed by

$$f''(x) = (x - x_2)(x - x_3) \cdots (x - x_n) + (x - x_1)(x - x_3) \cdots (x - x_n) + (x - x_1)(x - x_2) \cdots (x - x_{n-1})$$
$$= \frac{f'(x)}{x - x_1} + \frac{f'(x)}{x - x_2} + \cdots + \frac{f'(x)}{x - x_n}$$
$$= \sum_{j=1}^{n} \frac{f'(x)}{(x - x_j)}$$

Similarly,

$$f'''(x) = \sum_{\substack{j, k = 1 \\ j \neq k}}^{n} \frac{f'(x)}{(x - x_j)(x - x_k)}$$

Therefore,

$$Sf(x) = \frac{f'''(x)}{f'(x)} - \frac{3}{2}\left(\frac{f''(x)}{f'(x)}\right)^2 = \sum_{j,k=1j\neq k}^{n} \frac{1}{(x - x_j)(x - x_k)} - \frac{3}{2}\left(\sum_{j=1}^{n} \frac{1}{x - x_j}\right)^2$$
$$= -\frac{1}{2}\sum_{j=1}^{n} \left(\frac{1}{x - x_j}\right)^2 - \left(\sum_{j=1}^{n} \frac{1}{x - x_j}\right)^2 < 0.$$

Property 2 Let f and g be two functions such that $Sf < 0$ and $Sg < 0$. Then $S(f \circ g) < 0$.

Proof Using chain rule of differentiation, we have

$$(f \circ g)'(x) = f'(g(x)) \cdot g'(x)$$
$$(f \circ g)''(x) = f''(g(x)) \cdot (g'(x))^2 + f'(g(x)) \cdot g''(x)$$
$$(f \circ g)'''(x) = f'''(g(x)) \cdot (g'(x))^3 + 3f''(g(x)) \cdot g''(x) \cdot g'(x) + f'(g(x)) \cdot g'''(x)$$

Therefore,

$$S(f \circ g)(x) = \frac{(f \circ g)'''(x)}{(f \circ g)'(x)} - \frac{3}{2}\left(\frac{(f \circ g)''(x)}{(f \circ g)'(x)}\right)^2$$

$$= \frac{f'''(g(x)) \cdot (g'(x))^3 + 3f''(g(x)) \cdot g''(x) \cdot g'(x) + f'(g(x)) \cdot g'''(x)}{f'(g(x)) \cdot g'(x)}$$

$$- \frac{3}{2}\left(\frac{f''(g(x)) \cdot (g'(x))^2 + f'(g(x)) \cdot g''(x)}{f'(g(x)) \cdot g'(x)}\right)^2$$

$$= \left[\frac{f'''(g(x))}{f'(g(x))} - \frac{3}{2}\left(\frac{f''(g(x))}{f'(g(x))}\right)^2\right](g'(x))^2 + \frac{g'''(x)}{g'(x)} - \frac{3}{2}\left(\frac{g''(x)}{g'(x)}\right)^2$$

$$= Sf(g(x)) \cdot (g'(x))^2 + Sg(x)$$

This proves that $S(f \circ g) < 0$.

Property 3 If $Sf < 0$, then for every positive integer n, $Sf^n < 0$.

Proof Hint: Use Property 2 with $f = g$ and then use mathematical induction.

Property 4 If for a function $f(x)$ the Schwarzian derivative $Sf < 0$, then $f'(x)$ cannot have a positive local minimum or a negative local maximum.

Proof Suppose that the function $f'(x)$ has a local extremum at x_0. Then $f''(x_0) = 0$. Since $Sf < 0$, we have $Sf(x_0) < 0$. This gives $\frac{f'''(x_0)}{f'(x_0)} < 0$. This is a contradiction, because for positive local minimum (resp. negative local maximum), we have $f'(x_0) > 0$ and $f'''(x_0) > 0$(resp. $f'(x_0) < 0$ and $f'''(x_0) < 0$). Therefore $f'(x)$ cannot have a positive local minimum or a negative local maximum.

Theorem 9.7 *Let p be a non-hyperbolic fixed point of a map $f : \mathbb{R} \to \mathbb{R}$ with $f'(p) = -1$ and $f''(p)$ is continuous. Then*

(i) *if $Sf(p) < 0$, p is asymptotically stable.*
(ii) *if $Sf(p) > 0$, p is unstable.*

Proof Consider the map $h(x) = f^2(x)$. Then $h(p) = f^2(p) = p$. Now

$$h'(x) = \frac{dh(x)}{dx} = \frac{d(f^2(x))}{dx} = \frac{d}{dx}(f(f(x))) = f'(f(x)) \cdot f'(x).$$

Therefore,

$$h'(p) = f'(f(p)) \cdot f'(p) = f'(p) \cdot f'(p) = (-1) \cdot (-1) = 1.$$

Again, $h''(x) = f''(f(x)) \cdot [f'(x)]^2 + f'(f(x)) \cdot f''(x)$. So,

$$h''(p) = f''(f(p)) \cdot [f'(p)]^2 + f'(f(p)) \cdot f''(p) = f''(p)(-1)^2 + (-1)f''(p) = 0.$$

Also we see that

$$h'''(x) = f'''(f(x)) \cdot [f'(x)]^3 + 2f''(f(x)) \cdot f'(x) \cdot f''(x) + f''(f(x)) \cdot f'(x) \cdot f''(x)$$
$$+ f'(f(x)) \cdot f'''(x)$$
$$= f'''(f(x)) \cdot [f'(x)]^3 + 3f''(f(x)) \cdot f'(x) \cdot f''(x) + f'(f(x)) \cdot f'''(x).$$

Therefore,

$$h'''(p) = f'''(f(p)) \cdot [f'(p)]^3 + 3f''(f(p)) \cdot f'(p) \cdot f''(p) + f'(f(p)) \cdot f'''(p)$$
$$= f'''(f(p))(-1)^3 + 3f''(p) \cdot (-1) \cdot f''(p) + (-1)f'''(p)$$
$$= -2f'''(p) - 3[f''(p)]^2$$
$$= 2Sf(p).$$

Thus p is a non-hyperbolic fixed point of the map h with $h'(p) = 1$ and $h''(p) = 0$. We also see that $h'''(p)$ is continuous.

(i)   Suppose that $Sf(p) < 0$. Then $h'''(p) < 0$. Therefore p is an asymptotically stable fixed point of the map h and hence it is asymptotically stable for f.
(ii) Suppose $Sf(p) > 0$. Then $h'''(p) > 0$ and therefore p is an unstable fixed point of the map h and hence it is an unstable fixed point of f.

This completes the proof.

Example 9.13 Determine the stability behavior of the non-hyperbolic fixed points of the quadratic map $Q(x) = ax^2 + bx + c$, $a \neq 0$.

Solution The fixed points x^* of $Q(x)$ satisfy $Q(x^*) = x^*$. This yields two fixed points $x_{\pm}^* = \frac{-(b-1) \pm \sqrt{(b-1)^2 - 4ac}}{2a}$. If x^* is non-hyperbolic, then either $Q'(x^*) = 1$ or $Q'(x^*) = -1$

(i)   Let $Q'(x^*) = 1$. Then $2ax^* + b = 1$. This gives the fixed point $x^* = (1 - b)/2a$ and it exists when $(b - 1)^2 - 4ac = 0$, that is, $(b - 1)^2 = 4ac$ Since $Q''(x^*) = 2a > 0$, from Theorem 9.6 it follows that the fixed point $x^* = (1 - b)/2a$ is semi-stable.
(ii) Let $Q'(x^*) = -1$. Then $2ax^* + b = -1$. This gives the fixed point $x^* = -(b+1)/2a$ and it exists when $-(b - 1) \pm \sqrt{(b - 1)^2 - 4ac} = -(b+1)$, that is, when $(b - 1)^2 = 4(ac + 1)$ Calculate $Q''(x^*) = 2a$ and $Q'''(x^*) = 0$. Since $SQ(x^*) = -6a^2 < 0$, by Theorem 9.7 the non-hyperbolic fixed point $x^* = -(b+1)/2a$ is asymptotically stable.

Example 9.14 Determine the stability of the fixed point origin of the function $f(x) = -\sin x$.

Solution Clearly, $x = 0$ is a fixed point f. We calculate $f'(x) = -\cos x$, $f''(x) = \sin x$ and $f'''(x) = \cos x$. Obviously, $f'(x)$ is continuous and $f'(0) = -1$. The Schwarzian derivative of f at the origin is given by

$$Sf(0) = \frac{f'''(0)}{f'(0)} - \frac{3}{2}\left(\frac{f''(0)}{f'(0)}\right)^2$$

$$= \frac{1}{(-1)} - \frac{3}{2}\left(\frac{0}{(-1)}\right)^2 = -1 < 0.$$

This shows that the origin is asymptotically stable.

9.18   Exercises

1.   (a) For the following maps find the compositions $f^2(x)$ and $f^3(x)$:
   (i) $f(x) = x^2 - 5$ (ii) $f(x) = \sqrt{x+2}$ (iii) $f(x) = 3^x$ (iv) $f(x) = rx(1-x), x \in [0,1]$
   (v) $f(x) = x + x^3$ (vi) $f(x) = \sin x$ (vii) $f(x) = \tan^{-1}(x)$
   (viii) $f(x) = -\frac{3}{2}x^2 + \frac{5}{2}x + 1$

   (b) Consider the map $x(k+1) = ax(k)$ with $x(0) = b$, a and b are given constants. Find an iterative formula for $x(k)$.

   (c) Find compositions $(f \circ g)$ of the maps given below
   (i) $f(x) = x, g(x) = x^2$; (ii) $f(x) = x^3 - x, g(x) = x^2$

   (d) Draw the phase portraits of the following maps:
   (i) $f(x) = \sin x$, (ii) $f(\theta) = \theta + \varepsilon \sin(2\theta)$, $0 < \varepsilon < 1/2$

2.   (a) What do you mean by a fixed point of a map $f: \mathbb{R}^n \to \mathbb{R}^n$? Give geometrical interpretation of fixed point of a map. How do you find fixed points graphically?

   (b) How do you explain the fixed point of a map in the context of flow in a discrete system?

   (c) Consider the map $f(x) = x^2 + k, x \in \mathbb{R}$. Find the values of k for which the map f has

   (i) two fixed points,
   (ii) only one fixed point,
   (iii) no fixed points.

   For what values of k there will be an attracting fixed point of the map?

   (d) Define periodic point of a map $f: \mathbb{R} \to \mathbb{R}$. Give its significance in the context of dynamics of a map.

   (e) Discuss cycle of a map with example.

   (f) Give mathematical definitions of periodic-2 and periodic-3 cycles of a map $f: \mathbb{R} \to \mathbb{R}$. Show these graphically for a map.

3. Find all periodic points of the following maps:

   (i) $f(x) = -x^3, -\infty < x < \infty$

   (ii) $f(x) = \dfrac{1}{2}(x^3 + x), -1 \le x \le 1$

   (iii) $f(x) = x - x^2, 0 \le x \le 1$

   (iv) $f(x) = \dfrac{\pi}{2} \sin x, 0 \le x \le \pi$

4. Consider the map $f(x) = \tan x, -\pi/2 < x < \pi/2$. Prove that $x = 0$ is a fixed point of f. Also describe the dynamical behavior of the points near $x = 0$.

5. Consider the map $f(x) = 1 + \dfrac{1}{2}\sin x$. Show that it has a unique fixed point and find its stability. [*Hint*: Use Cobweb construction]

6. Show that for each positive integer k, there is an orbit of period-k of the map $f(x) = 4x(1-x), x \in [0,1]$.

7. Show that the map $G_4{}^2(x) = G_4 \circ G_4(x)$ where $G_4(x) = 4x(1-x), x \in [0,1]$ has four fixed points such that $x_1 = G_4(x_1)$, $x_2 = G_4(x_2)$, $x_3 = G_4(x_4)$ and $x_4 = G_4(x_3)$.

8. With the help of three suitable maps show that a non-hyperbolic fixed point of a map may be (i) attracting, (ii) repelling or (iii) neither attracting nor repelling.

   [*Hint:* Consider the maps (i) $\sin x$, (ii) $\tan x$, (iii) $e^x - 1$ and the fixed point origin and then use Cobweb diagram of each map]

9. Define k-cycle of a map. Consider the discrete map $x_{n+1} = f(x_n)$ where $f(x) = 1 - x^2, x \in [-1,1]$. Find all 2-cycles of the map.

10. Let $\{x_1, x_2, ..., x_k\}$ be a k-cycle of a smooth function $f(x)$ on $\mathbb{R}$. Prove that the cycle is attracting if $\left| f'(x_1) f'(x_2) ... f'(x_k) \right| < 1$ and repelling if $\left| f'(x_1) f'(x_2) ... f'(x_k) \right| > 1$

11. Use Cobweb diagram to show that the map $f(x) = e^{-x}$ has a unique fixed point. Also determine the stability character of the fixed point from the diagram.

12. Consider the map $f(x) = \lambda x - x^3$, where λ is a real parameter. Determine the fixed points and their stabilities for different values of the parameter λ.

13. Find the stability of all fixed points and period-2 points for the map $f(x) = 3.05x(1-x)$.

14. Let $f(x) = x - x^2$. Show that $x = 0$ is a fixed point of f. Describe the dynamical behavior of points near the fixed point.

15. Consider the map $f(x) = ax, x \in \mathbb{R}$ where a is a constant. Find the fixed points of the map and their stability for different values of a.

16. Consider the map $f(x) = \begin{cases} 2x+3, & x \le -2 \\ \dfrac{x}{2}, & -2 < x < 2 \\ 2x-3, & x \ge 2 \end{cases}$.

    Show that the map has three fixed points. Determine their stability characters.

17. (a) Give an example of a map f for which $f'(0) = 1$ and $x = 0$ is an attracting fixed point. Justify your answer.

    (b) Give an example of a map f for which $f'(0) = 1$ and $x = 0$ is a repelling fixed point. Justify your answer.

    (c) Give an example of a map f for which $f'(0) = 1$ and $x = 0$ is a fixed point but it is neither attracting nor repelling. Justify your answer.

18. Use the chain rule of differentiation to prove that the derivative of f^n evaluated at any of the fixed points of an n-cycle of map f has the same value.

19. Find all fixed points, periodic points and basins of attraction for the maps $f(x) = 2x+3$ and $g(x) = x^3$.

20. Find all critical points of the map $f(x) = x^2, x \in \mathbb{R}$. Also find the basin of attraction. Show that the point $x = -1$ is an eventually fixed point of the map f.

21. Consider the map $g : S \to S$ defined by $g(\theta) = 2\theta$, where S denotes the unit circle. Show that the periodic points of period-k of the map g are the $(2^k - 1)$th roots of unity. Also show that for each positive integer n, $\theta = 2m\pi/2^n, m \in \mathbb{Z}$ is an eventually fixed point of g.

22. If p is a stable fixed point of a map $f : \mathbb{R} \to \mathbb{R}$, then show that it is also a stable fixed point of f^n. Is the converse true? Justify your answer.

23. Consider the map $f(x) = x^2 + k, x \in \mathbb{R}$, where k is a real parameter. Show that the map f has a priod-2 orbit if $k < -\dfrac{3}{4}$. Also show that the orbit is stable if $-\dfrac{5}{4} < k < -\dfrac{3}{4}$.

24. Using the concept of stability of a periodic point of a map derive the stability criteria of a periodic cycle of the map.

25. Find all fixed points of the map $f(x) = ax(1 - x^3)$, $x \in \mathbb{R}$ and determine their stability characters for different values of the parameter a.

26. Give an example of a map that has infinite number of fixed points. Also give an example of a map which has infinite number of periodic points.

27. (a) Let f be a smooth function on $\mathbb{R}$ and p be a fixed point of f. Find a condition that guarantees that every point x in the neighborhood $N_\varepsilon(p) = (p - \varepsilon, p + \varepsilon)$ satisfies $f^n(x) \to p$ as $n \to \infty$.

    (b) Let $f : \mathbb{R} \to \mathbb{R}$ be a continuous map and let I and J are disjoint closed and bounded intervals such that $f(I) \subset J$ and $f(J) \subset I$. Prove that there is a period-2 point of f in I.

28. For a hyperbolic fixed point p of a one-dimensional map $f: \mathbb{R} \to \mathbb{R}$ with $|f'(p)| > 1$. Show that there exists an open interval U of p such that, if $x \in U, x \neq p$, then there exists $k > 0$ such that $f^k(x) \notin U$.

29. (a) Discuss eventual fixed points and eventual periodic points with examples. Show that $\left(\dfrac{1}{2} - \dfrac{1}{4}\sqrt{2+\sqrt{3}} \right)$ is an eventually fixed point of the map $f(x) = 4x(1-x), x \in [0,1]$.

(b) Show that the map $f(x) = 3x \pmod 1$ on the unit interval is eventually periodic iff x is a rational number.

30. (a) Show that the map $f(x) = rx(1-x^2)/\sqrt{3}$, $0 \leq r \leq 4.5$ is a unimodal map (a single maximum or a single minimum in an interval) of the interval $[0,1]$.

(b) Show that the above map has a symmetric period-2 orbit $\left(\sqrt{1+\sqrt{3}/r}, -\sqrt{1+\sqrt{3}/r} \right)$ and the orbit is unstable for all $r > 0$.

31. Find the symmetric period 2-orbit for the map $f(x) = x - rx + rx^3, 0 \leq r \leq 4$.

32. Find all fixed points of the map $f(x) = a - x^2, x \in \mathbb{R}$ for different values of the parameter $a \in \mathbb{R}$. Also find the value(s) of a for which the map has non-hyperbolic fixed points.

33. Use Theorem 9.6 to determine the stability of the fixed point origin of the following maps:

   (i) $f(x) = \sin x$

   (ii) $f(x) = \tan x$

   (iii) $h(x) = x + x^3$

   (iv) $f(x) = x - x^3$

34. (a) Prove that the sine map $x_{n+1} = r \sin(\pi x_n)$ has a negative Schwarzian derivative over the closed interval $[0,1]$, where r is the parameter of the map.

(b) Show that Schwarzian derivative of a function is identically zero if and only if it is a Mobius transformation.

35. Give an example of a function which has a positive Schwarzian derivative. Find the Schwarzian derivatives for the following functions:

   (i) e^x, (ii) $Sin(x)$, (iii) $Cos(x)$, (iv) $Sin^{-1}(x)$, (v) $\ln x$, (vi) $rx - x^3$, (vii) $x^2 + k$.

36. Using the concept of Schwarzian derivative to find the stability of the non-hyperbolic fixed point of the map $f(x) = 3x(1-x), x \in [0,1]$.

References

1. Devaney, R.L.: A First Course in Chaotic Dynamical Systems: Theory and Experiment. Westview Press, Cambridge (1992)
2. Devaney, R.L.: An Introduction to Chaotic Dynamical Systems. Westview Press, Cambridge (2003)

3. Holmgren, R.A.: A First Course in Discrete Dynamical Systems. Springer, New York (1994)
4. Alligood, K.T., Sauer, T.D., Yorke, J.A.: Chaos: An Introduction to Dynamical Systems. Springer, New York (1997)
5. Sharkovsky, A.N., Kolyada, S.F., Sivak, A.G., Fedorenko, V.V.: Dynamics of One-Dimensional Maps. Springer, B. V. (1997)
6. Medio, A., Lines M. Nonlinear Dynamics: A Primer. Cambridge University Press, Cambridge, (2001)
7. Robinson, R.C.: An introduction to dynamical systems: continuous and discrete. American Mathematical Society, 2012.

Chapter 10
Some Maps

This chapter deals with some important maps and their elementary properties. In particular, we are interested in finding fixed points, their stability behaviors, and formation of periodic cycles, stabilities of the periodic cycles, and bifurcation phenomena of some special maps. Maps and their compositions represent many natural phenomena or engineering processes. For example, dynamical models have been used for the study of population of species over centuries. In general, we would like to know how the size, say at $(n + 1)$th generation of a population model is related to the preceding generations of that model. Often a growth rate or reproductive rate of a population appears in the model. This may be expressed by a relationship $x_{n+1} = f(x_n, r)$, where x_n denotes the population at nth generation and r is the population growth parameter. Simple population model for species can be formulated through mathematical modeling where the reproductive rate is a function $r(x)$ which decreases with increasing population x, from an initial value $r(0) = r_0$ to $r(x) = 0$ at some limiting value of population. A simple population model with a linear decrease of growth rate $r(x)$ with increasing x (known as logistic growth rate) can be expressed mathematically by the function $f(x) = rx(1 - x), x \in [0, 1]$. Starting from some initial population x_0, the sequences of population at successive generations are given by $x_{n+1} = rx_n(1 - x_n)$, $n = 0, 1, 2, \ldots$. Similarly, the tent, Euler's shift, and Hénon maps have importance in many contexts and are discussed in the following sections. The branching of a solution at a critical value of the parameter of a map is called bifurcation. It is a qualitative change of dynamics or orbits for changing values of parameters of a map. Bifurcation theory in discrete systems is vast and we shall introduce few particular bifurcations, viz., saddle-node, period-doubling, and transcritical bifurcations.

10.1 Tent Map

The tent map is a one-dimensional piecewise linear map. Its graph resembles the front view of a tent. It is also called a triangle map or a stretch and fold map. The mapping function is made up of sections of two straight line segments. This is a continuous map but it does not have differentiability at the points where the line

© Springer India 2015
G.C. Layek, *An Introduction to Dynamical Systems and Chaos*,
DOI 10.1007/978-81-322-2556-0_10

Fig. 10.1 Graphical
representation of T
(x) depicting the fixed points
$x^* = 0$ and $x^* = 2/3$

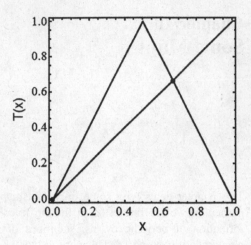

segments of different slopes meet. The tent map has some interesting properties that we explore below. Generally, the tent map $T : [0, 1] \to [0, 1]$ is defined as

$$T(x) = \begin{cases} 2\lambda x, & 0 \le x \le 1/2 \\ 2\lambda(1-x), & 1/2 \le x \le 1 \end{cases} \tag{10.1}$$

where $\lambda (0 \le \lambda \le 1)$ is a control parameter. We can also write the tent map by an iterative sequence as follows:

$$x_{n+1} = f(x_n) = \lambda \left(1 - 2\left| x_n - \frac{1}{2} \right| \right).$$

As we know that a fixed point is an invariant solution of a map, the nature of the fixed points plays an important role in analyzing the dynamical behavior of the map. In this section we shall analyze the fixed points of the tent map for the parameter value $\lambda = 1$. The fixed points satisfy the relation $x = T(x)$. So, we get $x = 2x \Rightarrow x = 0 \in [0, 1/2]$ and $x = 2(1 - x) \Rightarrow x = 2/3 \in [1/2, 1]$. Thus, the only two fixed points of the tent map are $x^* = 0$ and $x^* = 2/3$. The graphical representation of $T(x)$ is shown in Fig. 10.1.

The map $T^2(x)$:

Using the definition of $T(x)$, the twofold composition of the tent map can be obtained as follows:

$$T^2(x) = T(T(x)) = \begin{cases} 2(2x), & 0 \leq 2x \leq 1/2 \\ 2(1-2x), & 1/2 \leq 2x \leq 1 \\ 2 \cdot 2(1-x), & 0 \leq 2(1-x) \leq 1/2 \\ 2(1-2(1-x)), & 1/2 \leq 2(1-x) \leq 1 \end{cases}$$

$$= \begin{cases} 4x, & 0 \leq x \leq 1/4 \\ 2-4x, & 1/4 \leq x \leq 1/2 \\ -2+4x, & 1/2 \leq x \leq 3/4 \\ 4-4x, & 3/4 \leq x \leq 1 \end{cases}$$

For determining the fixed points of $T^2(x)$, we have to solve the equation $x = T^2(x)$. Now,

$$\text{for } 0 \leq x \leq \frac{1}{4}, \ x = T^2(x) \Rightarrow x = 4x \Rightarrow x = 0.$$

$$\text{for } \frac{1}{4} \leq x \leq \frac{1}{2}, \ x = T^2(x) \Rightarrow x = 2 - 4x \Rightarrow x = \frac{2}{5}.$$

$$\text{for } \frac{1}{2} \leq x \leq \frac{3}{4}, \ x = T^2(x) \Rightarrow x = -2 + 4x \Rightarrow x = \frac{2}{3}.$$

$$\text{for } \frac{3}{4} \leq x \leq 1, \ x = T^2(x) \Rightarrow x = 4 - 4x \Rightarrow x = \frac{4}{5}.$$

Therefore, the fixed points of $T^2(x)$ are given by $x^* = 0, \frac{2}{5}, \frac{2}{3}, \frac{4}{5}$. The diagrammatic representation of $T^2(x)$ displaying the fixed points is presented in Fig. 10.2.

Fig. 10.2 The twofold composition $T^2(x)$ depicting four fixed points

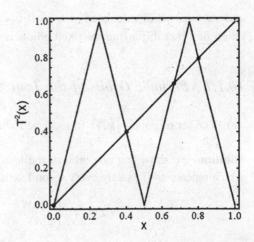

Fig. 10.3 Graphical representation of $T^3(x)$

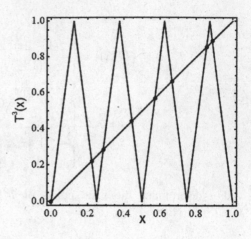

The Map $T^3(x)$

Using $T(x)$ and the twofold composition $T^2(x)$, we can show the representation of $T^3(x)$ as follows:

$$T^3(x) = \begin{cases} 8x, & 0 \le x \le \frac{1}{8} \\ 2-8x, & \frac{1}{8} \le x \le \frac{1}{4} \\ -2+8x, & \frac{1}{4} \le x \le \frac{3}{8} \\ 4-8x, & \frac{3}{8} \le x \le \frac{1}{2} \\ -4+8x, & \frac{1}{2} \le x \le \frac{5}{8} \\ 6-8x, & \frac{5}{8} \le x \le \frac{3}{4} \\ -6+8x, & \frac{3}{4} \le x \le \frac{7}{8} \\ 8-8x, & \frac{7}{8} \le x \le 1 \end{cases}$$

and its fixed point calculated as $x^* = 0, \frac{2}{3}, \frac{2}{7}, \frac{4}{7}, \frac{6}{7}, \frac{2}{9}, \frac{4}{9}, \frac{8}{9}$. The graphical representation of $T^3(x)$ displaying the fixed points is presented in Fig. 10.3.

10.1.1 Periodic Orbits of the Tent Map

(a) The set of points $\{\frac{2}{5}, \frac{4}{5}\}$ forms a periodic-2 cycle of the tent map.

Solution We show that one orbit of the tent map is $\frac{2}{5}, \frac{4}{5}, \frac{2}{5}, \frac{4}{5}, \ldots$, that is, an orbit which repeats itself exactly every second iteration. Now,

$$T(x) = \begin{cases} 2x, & 0 \le x \le \frac{1}{2} \\ 2(1-x) & \frac{1}{2} \le x \le 1 \end{cases}$$

Since $\frac{2}{5} \in \left[0, \frac{1}{2}\right]$, $T\left(\frac{2}{5}\right) = 2\left(\frac{2}{5}\right) = \frac{4}{5}$. Similarly, $T\left(\frac{4}{5}\right) = 2\left(1 - \frac{4}{5}\right) = \frac{2}{5}$.
Also, $T^2\left(\frac{2}{5}\right) = T\left(T\left(\frac{2}{5}\right)\right) = T\left(\frac{4}{5}\right) = \frac{2}{5}$ and $T^2\left(\frac{4}{5}\right) = T\left(T\left(\frac{4}{5}\right)\right) = T\left(\frac{2}{5}\right) = \frac{4}{5}$.
Therefore, $T\left(\frac{2}{5}\right) = \frac{4}{5}$, $T\left(\frac{4}{5}\right) = \frac{2}{5}$, $T^2\left(\frac{2}{5}\right) = \frac{2}{5}$ and $T^2\left(\frac{4}{5}\right) = \frac{4}{5}$.
This shows that $\left\{\frac{2}{5}, \frac{4}{5}\right\}$ forms a periodic-2 cycle of the tent map T. The derivative test, discussed in the previous chapter, gives the stability behavior of the cycle. Now, $\left|T'\left(\frac{2}{5}\right)T'\left(\frac{4}{5}\right)\right| = 4 > 1$. Hence the periodic-2 cycle is unstable.

(b) The cycles $\left\{\frac{2}{7}, \frac{4}{7}, \frac{6}{7}\right\}$ and $\left\{\frac{2}{9}, \frac{4}{9}, \frac{8}{9}\right\}$ form two periodic-3 cycles of the tent map.

Solution We shall show that the points $\left\{\frac{2}{7}, \frac{4}{7}, \frac{6}{7}\right\}$ and $\left\{\frac{2}{9}, \frac{4}{9}, \frac{8}{9}\right\}$ repeat itself every third iterations. Now,

$$T\left(\frac{2}{7}\right) = 2 \cdot \frac{2}{7} = \frac{4}{7}, \; T\left(\frac{4}{7}\right) = 2\left(1 - \frac{4}{7}\right) = \frac{6}{7}, \; \text{and } T\left(\frac{6}{7}\right) = 2\left(1 - \frac{6}{7}\right) = \frac{2}{7}$$

$$T^2\left(\frac{2}{7}\right) = T\left(T\left(\frac{2}{7}\right)\right) = T\left(\frac{4}{7}\right) = \frac{6}{7},$$

$$T^2\left(\frac{4}{7}\right) = T\left(T\left(\frac{4}{7}\right)\right) = T\left(\frac{6}{7}\right) = \frac{2}{7} \text{ and } T^2\left(\frac{6}{7}\right) = T\left(T\left(\frac{6}{7}\right)\right) = T\left(\frac{2}{7}\right) = \frac{4}{7}$$

Again, $T^3\left(\frac{2}{7}\right) = T\left(T^2\left(\frac{2}{7}\right)\right) = T\left(\frac{6}{7}\right) = \frac{2}{7}$. Similarly, $T^3\left(\frac{4}{7}\right) = \frac{4}{7}$ and $T^3\left(\frac{6}{7}\right) = \frac{6}{7}$
This shows that $\left\{\frac{2}{7}, \frac{4}{7}, \frac{6}{7}\right\}$ forms a periodic-3 cycle of T. In the similar manner we can prove that $\left\{\frac{2}{9}, \frac{4}{9}, \frac{8}{9}\right\}$ also forms a periodic-3 cycle of the tent map. The derivative test for 3-cycle gives $\left|T'\left(\frac{2}{7}\right)T'\left(\frac{4}{7}\right)T'\left(\frac{6}{7}\right)\right| = 2^3 > 1$. Similarly, the derivative test for the other cycle gives that value 2^3. So, both the cycles are unstable.

(c) **Periodic cycles of $T^n(x)$**

From the above analysis, it can be shown easily that the n-fold composition, $T^n(x)$, of $T(x)$ has 2^n fixed points. For $n = 1$, there are two period-1 orbits, $x_0^* = 0$ and $x_1^* = 2/3$. In each fold of compositions there is a fixed point $x_0^* = 0$. $T^2(x)$ has four fixed points, of which two are new, $x^* = 2/5, 4/5$. But the new pair of fixed points $\left\{\frac{2}{5}, \frac{4}{5}\right\}$ forms a period-2 orbit of the map. $T^3(x)$ has eight fixed points, among which two fixed points $x_0^* = 0$ and $x_1^* = 2/3$ belong to $n = 1$ and the other fixed points, viz., $\left\{\frac{2}{9}, \frac{4}{9}, \frac{8}{9}\right\}$ and $\left\{\frac{2}{7}, \frac{4}{7}, \frac{6}{7}\right\}$ form two period-3 orbits of the tent map. $T^4(x)$ has 2^4 fixed points, among which two belong to $n = 1$ and another two belong to $n = 2$. So, the other 12 fixed points of $T^4(x)$ must form three distinct period-4 orbits. In general, $T^n(x)$ has 2^n fixed points and we see from above analyses that there is at least one period-n orbit with the points $\left\{\frac{2}{2^n+1}, \frac{2^2}{2^n+1}, \frac{2^3}{2^n+1}, \dots, \frac{2^n}{2^n+1}\right\}$ (Davies [1]).
So, there are periodic orbits of every period. Note that none of these periodic cycles contain the points 0, 1/2 and 1, where the derivatives of the map are not defined. The derivative test can be applied to these cycles for their stability analyses. These

give $T'(x^*) = \pm 2, (T^n)'(x^*) = \pm 2^n \Rightarrow |(T^n)'(x^*)| = 2^n > 1$. This implies that all cycles are unstable in nature. In other words, we can say that infinitely many unstable periodic orbits can exist for the tent map.

10.2 Logistic Map

In the year 1976, the well-known population biologist *Robert M. May (Nature, 261, 459–469 (1976))* analyzed a simple nonlinear one-dimensional map which could have very complicated dynamics. This map is associated with the logistic pattern of population growth (linear growth model, that is, $r(x) = r(1 - x)$, a linear decrease of $r(x)$ with increasing population x) and may be represented by

$$x_{n+1} = f(x_n) = rx_n(1 - x_n) \tag{10.2}$$

which is basically a discrete-time analog of the logistic equation for the population growth model $\dot{x} = rx(1 - x)$, $x \in [0, 1]$. Here $x_n \geq 0$ is a dimensionless measure of the population in the nth generation and $r \geq 0$ is the intrinsic growth rate (population growth parameter). The graph of (10.2) represents a parabola with a maximum value of $(r/4)$ at $x = 1/2$. Figure 10.4 depicts a sketch of $f(x)$ at the parameter value $r = 4$.

We take the control parameter r to be in the range $0 \leq r \leq 4$ so that Eq. (10.2) maps the closed interval [0, 1] onto itself. The logistic map is a one-dimensional quadratic map. It is also a non-invertible map. We may iterate the map forward in time with each x_n leading to a unique subsequent value x_{n+1}, but the reverse is not true. For further reading on mathematical biology, see Robert M. May [2], Leah Edelstein-Keshet [3], and James D. Murray [4].

Besides the logistic map, the following maps also have applications in the study of population dynamics in mathematical biology,

Fig. 10.4 Sketch of $f(x)$ at the parameter value $r = 4$

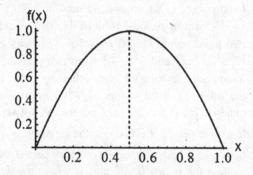

(i) $f(x) = xe^{r(1-x)}, x \in [0, 1]$

(ii) $f(x) = \begin{cases} \lambda x, & \text{if } x \le 1/2 \\ \lambda(1-x), & \text{if } x \ge 1/2 \end{cases}$. This is the general tent map.

(iii) $f(x) = \begin{cases} \lambda x, & \text{if } x < 1 \\ \lambda x^{p-1}, & \text{if } x > 1 \end{cases}$ $(p > 1)$ [Robert May (1976, Nature Vol. 261)].

(iv) The logistic and exponential models with feedback are modeled as $f(x) = rx(1-x) + L$, $L > 0$ for positive feedback, and $f(x) = xe^{r(1-x)} + L$, $L < 0$ for negative feedback, [Somdatta Sinha and Parthasarathy (1995, Phys. Rev. E, Vol. 51)].

We now discuss some properties of the logistic map for different values of population growth parameter r.

10.2.1 Some Properties of the Logistic Map

(a) Find all fixed points of the logistic map $x_{n+1} = rx_n(1 - x_n)$ for $0 \le x_n \le 1$ and $0 \le r \le 4$. Also determine their stability behaviors.

Solution The logistic map function is given by $f(x) = rx(1 - x)$, $0 \le x \le 1$ and $0 \le r \le 4$. For the fixed points of $f(x)$, we have a relation

$$f(x) = x \Rightarrow rx(1 - x) = x \Rightarrow x\{(r-1) - rx\} = 0 \Rightarrow x = 0, \left(1 - \frac{1}{r}\right).$$

So, the map has two fixed points, namely $x_0^* = 0$ and $x_1^* = \left(1 - \frac{1}{r}\right)$. Clearly, $x_0^* = 0$ is a fixed point of the map f for all values of the parameter r. But $x_1^* = \left(1 - \frac{1}{r}\right)$ is a fixed point of f if $r \ge 1$ (since $0 \le x \le 1$). Therefore, the fixed points of the logistic map are $x_0^* = 0$ $\forall r \in [0, 4]$ and $x_1^* = \left(1 - \frac{1}{r}\right)$ for $1 \le r \le 4$. The stability of the fixed points depends on the absolute value of $f'(x) = r - 2rx$. Since $|f'(0)| = |r| = r$, the fixed point $x_0^* = 0$ is stable when $r < 1$ and unstable when $r > 1$. Again, since $\left|f'\left(1 - \frac{1}{r}\right)\right| = |2 - r|$, the fixed point x_1^* is stable if $|2 - r| < 1$, that is, if $1 < r < 3$ and unstable if $|2 - r| > 1$, that is, in the range $3 < r \le 4$.

(b) Prove that the logistic map $f(x) = rx(1 - x)$ has a 2-cycle for all $r > 3$.

Solution The growth rate parameter r is the deciding factor for the evolution of the logistic map. We now find periodic-2 orbit for logistic map which depends upon the parameter r. Now, $f^2(x) = rf(x)\{1 - f(x)\} = r^2 x(1 - \check{x})\{1 - rx(1 - x)\}$. For finding fixed points of twofold composition f^2, we get a relation

$$f^2(x) = x$$

$$\text{or,}\quad r^2 x(1 - x)\{1 - rx(1 - x)\} = x$$

$$\text{or,}\quad x\left[r^2(1 - x)\{1 - rx(1 - x)\} - 1\right] = 0$$

$$\text{or,}\quad x\left[r^2(1 - x) - r^3 x(1 - x)^2 - 1\right] = 0$$

$$\text{or,}\quad x\left[r^2(1 - x) + r^3(1 - x - 1)(1 - x)^2 - 1\right] = 0$$

$$\text{or,}\quad x\left[r^3(1 - x)^3 - r^3(1 - x)^2 + r^2(1 - x) - 1\right] = 0$$

$$\text{or,}\quad x\left\{x - \left(1 - \frac{1}{r}\right)\right\}\left[r^3(1 - x)^2 + r^2(1 - r)(1 - x) + r\right] = 0$$

or,

$$x = 0, 1 - \frac{1}{r} \text{ and } 1 - x = \frac{-r^2(1 - r) \pm \sqrt{r^4(1 - r)^2 - 4r^4}}{2r^3}$$

$$= \frac{r - 1 \pm \sqrt{(r + 1)(r - 3)}}{2r}$$

or,

$$x = 0, 1 - \frac{1}{r} \text{ and } 1 - \frac{r - 1 \pm \sqrt{(r + 1)(r - 3)}}{2r} = \frac{r + 1 \pm \sqrt{(r + 1)(r - 3)}}{2r}$$

$$= p, q \text{ (say)}$$

Therefore, there are four fixed points of f^2, given by $x^* = 0, \frac{1}{r}, p, q$. But the two, $x^* = 0, \frac{1}{r}$, of them are the fixed points of $f(x)$ and the other two are real only for $r \geq 3$. We examine the cycle property of f.

Now,

$$f(p) = rp(1-p) = r \cdot \frac{r+1+\sqrt{(r+1)(r-3)}}{2r}\left\{1 - \frac{r+1+\sqrt{(r+1)(r-3)}}{2r}\right\}$$

$$= \frac{r+1+\sqrt{(r+1)(r-3)}}{2}\left\{\frac{r-1-\sqrt{(r+1)(r-3)}}{2r}\right\}$$

$$= \frac{r^2 - \left\{1+\sqrt{(r+1)(r-3)}\right\}^2}{4r}$$

$$= \frac{r^2 - 1 - 2\sqrt{(r+1)(r-3)} - (r^2 - 2r - 3)}{4r}$$

$$= \frac{2r + 2 - 2\sqrt{(r+1)(r-3)}}{4r}$$

$$= \frac{r+1-\sqrt{(r+1)(r-3)}}{2r} = q$$

$$\Rightarrow f(p) = q.$$

Similarly, we can show that $f(q) = p$. Again, $f^2(p) = f(f(p)) = f(q) = p$ and $f^2(q) = f(f(q)) = f(p) = q$. According to the definition of 2-cycle it is clear that the logistic map $f(x)$ has a periodic-2 orbit or cycle $\{p, q\}$ when $r > 3$.

Note that for $r < 3$ the roots are complex, which means that a cycle does not exist. Hence a 2-cycle of the logistic map appears when $r > 3$. The graph of $f^2(x)$ for $r > 3$ is shown in Fig. 10.5.

(c) Explain bifurcation of a system. Show that the logistic map $f_r(x) = rx(1-x)$, $x \in [0, 1]$ undergoes a transcritical bifurcation at $r = 1$ and a period-doubling bifurcation at $r = 3$.

Fig. 10.5 Graphical representation of the logistic map at the second iteration for $r = 3.7$

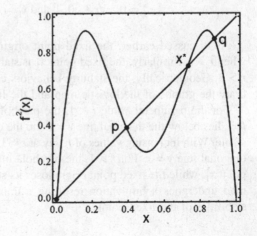

Fig. 10.6 Graphical
representation of the logistic
map for different values of r

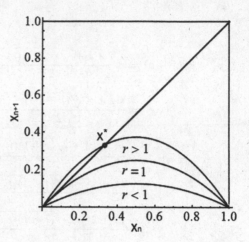

Solution Bifurcation is basically a change in the structure of the orbit as a system parameter (known as control parameter) varies continuously through critical values. Bifurcation theory is concerned with equilibrium solutions of system. The characters of the fixed points and the period orbits are altered (Fig. 10.6).

When the parameter is increased, the two fixed points collide, whereupon they exchange their stabilities. This type of bifurcation is called transcritical bifurcation and this is not a common type of bifurcation occurred in one-dimensional discrete system. Now, the fixed points of the logistic map f are given by

$$x = f(x) = rx(1 - x) \Rightarrow x = 0, \left(1 - \frac{1}{r}\right).$$

The fixed $x = 0$ exists for all values of r, whereas the other fixed point $x = \left(1 - \frac{1}{r}\right)$ exists when $r \geq 1$. So, the fixed points of the logistic map are given by

$$x_0^* = 0 \quad \forall r \in [0, 4] \text{ and } x_1^* = 1 - \frac{1}{r} \text{ for } 1 \leq r \leq 4.$$

As discussed earlier, the fixed point origin is stable when $r < 1$ and unstable when $r > 1$. Similarly, the fixed point x_1^* is stable when $1 < r < 3$ and unstable when $r > 3$. Geometrically, the stability behaviors can be explained in a better way. We draw the graphs of the logistic map and the line $y = x$ in x-$f(x)$ plane.

For the parameter value $r < 1$, the parabola, which represents the logistic map $f(x)$, lies below the diagonal line $y = x$ and the origin is the only fixed point, and it is stable. With increasing values of r, say at $r = 1$, the parabola becomes tangent to the diagonal line $y = x$. For $r > 1$, the parabola intersects the diagonal line at the fixed point x_1^*, while the fixed point origin loses its stability. Thus, we see that at $r = 1$ the map undergoes a bifurcation resulting a transcritical bifurcations by exchanging stabilities of the fixed points.

When r increases beyond 1, the slope of the function f gets increasingly steep and the critical slope is attained when $r = 3$ (at $r = 3$, fixed points are $p = q = 2/3$). This indicates that the logistic map undergoes another bifurcation leading to period-2 cycle. This bifurcation is known as a period-doubling bifurcation or a flip bifurcation. The iterates flip from side to side of the fixed point. So, flip bifurcation is basically a period-doubling bifurcation and occurs at the critical value $r = 3$ for the logistic map.

(d) Prove that the period-2 cycle of the logistic map is linearly stable when $3 < r < (1 + \sqrt{6}) = 3.449\ldots$

Solution We know that the period 2-cycle of the logistic map is $\{p, q\}$, where

$$p, q = \frac{r + 1 \pm \sqrt{(r+1)(r-3)}}{2r}.$$

So, $p + q = \frac{r+1}{r}$ and $pq = \frac{(r+1)^2 - (r+1)(r-3)}{4r^2} = \frac{r+1}{r^2}$.

The derivative of $f(x)$ is given by $f'(x) = r - 2rx$. Therefore, $f'(p) = r - 2rp$ and $f'(q) = r - 2rq$. The linear stability of the 2-cycle $\{p, q\}$ gives the following condition as

$$|f'(p)f'(q)| < 1$$
$$\Rightarrow \quad |(r - 2rp)(r - 2rq)| < 1$$
$$\Rightarrow \quad r^2|(1 - 2p)(1 - 2q)| < 1$$
$$\Rightarrow \quad r^2|1 - 2(p + q) + 4pq| < 1$$
$$\Rightarrow \quad r^2\left|1 - 2\left(\frac{r+1}{r}\right) + 4\left(\frac{r+1}{r^2}\right)\right| < 1$$
$$\Rightarrow \quad |r^2 - 2r(r + 1) + 4(r + 1)| < 1$$
$$\Rightarrow \quad |-r^2 + 2r + 4| < 1$$
$$\Rightarrow \quad |r^2 - 2r - 4| < 1$$
$$\Rightarrow \quad |(r - 1)^2 - 5| < 1$$
$$\Rightarrow \quad -1 < (r - 1)^2 - 5 < 1$$
$$\Rightarrow \quad 4 < (r - 1)^2 < 6$$
$$\Rightarrow \quad 2 < (r - 1) < \sqrt{6}$$
$$\Rightarrow \quad 3 < r < 1 + \sqrt{6}.$$

Hence, the period 2-cycle of the logistic map is linearly stable when $3 < r < (1 + \sqrt{6})$.

10.2.2 Iterative Solutions of the Logistic Equation

We now give successive iterative solutions and corresponding orbits to the logistic map for different values of the parameter r. At each iteration the current value of x is fed back into the map to generate the next value, then this value is again used to generate the next one and so on for a given initial value x_0 (here we take $x_0 = 0.2$).

(a) **For** $r = 0.95$

The successive iteration values of the logistic map are given as follows:

$$x_0 = 0.2, x_1 = 0.152, x_2 = 0.122451, x_3 = 0.102084, x_4 = 0.0870798,$$
$$x_5 = 0.075522, x_6 = 0.0663275, x_7 = 0.0588318, x_8 = 0.0526021,$$
$$x_9 = 0.0473433, x_{10} = 0.0428468, x_{11} = 0.0389604, x_{12} = 0.0355704,$$
$$x_{13} = 0.0325899, x_{14} = 0.0299514, x_{15} = 0.0276016, x_{16} = 0.0254978,$$
$$x_{17} = 0.0236052, x_{18} = 0.0218956, x_{19} = 0.0203454, x_{20} = 0.0189349,$$
$$x_{21} = 0.0176475, x_{22} = 0.0164693, x_{23} = 0.0153882, x_{24} = 0.0143938,$$
$$x_{25} = 0.0134773, x_{26} = 0.0126309, x_{27} = 0.0118478, x_{28} = 0.011122,$$
$$x_{29} = 0.0104484, x_{30} = 0.00982228.$$

This sequence of iterations indicates that the iterations will converge to the value zero. A plot of x_n versus n is presented in Fig. 10.7.

(i) **For** $r = 2.6$

The successive iteration values of the logistic map are given as follows:

0.2, 0.416, 0.631654, 0.604935, 0.621371, 0.6117, 0.61756, 0.614067, 0.616171,
0.614911, 0.615668, 0.615214, 0.615487, 0.615323, 0.615421, 0.615363, 0.615398,
0.615377, 0.615389, 0.615382, 0.615386, 0.615384, 0.615385, 0.615384, 0.615385,
0.615384, 0.615385, 0.615385, 0.615385, 0.615385, 0.615385.

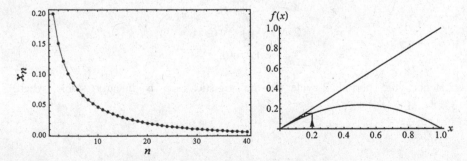

Fig. 10.7 Phase trajectory and cobweb diagram for $r = 0.95$

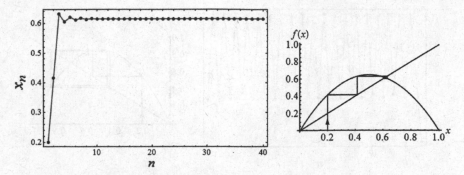

Fig. 10.8 Phase trajectory and cobweb diagram for $r = 2.6$

It indicates that the iterations tend to the value 0.615385, which is the constant solution of the logistic map for this parameter value. This gives a period-1 orbit, because the iterations tend to a fixed value where $x_{n+1} = x_n$ for large n. A plot of x_n versus n is shown in Fig. 10.8.

(ii) **For $r = 3.3$**

The successive iteration values of the logistic map are given as follows:

0.2, 0.528, 0.822413, 0.481965, 0.823927, 0.478736, 0.823508, 0.479631, 0.823631,

0.479368, 0.823595, 0.479444, 0.823606, 0.479422, 0.823603, 0.479428, 0.823603,

$\overline{0.479427}$, $\underline{0.823603}$, 0.479427, 0.823603, 0.479427, 0.823603, 0.479427, 0.823603,

0.479427, 0.823603, 0.479427, 0.823603, 0.479427, 0.823603.

The above data clearly indicate that the sequence of iterations forms a period-2 orbit of the logistic map (marked in the data), that is, $x_{n+2} = x_n$. A plot of x_n versus n is presented in Fig. 10.9 and the cobweb diagram shows the period-2 orbit.

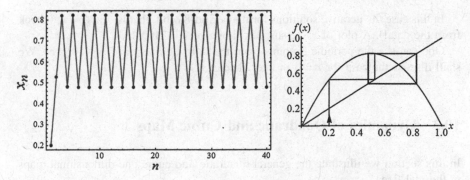

Fig. 10.9 Phase trajectory and cobweb diagram for $r = 3.3$

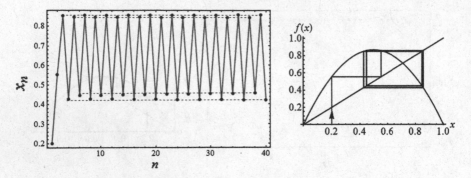

Fig. 10.10 Phase trajectory and cobweb diagram for $r = 3.46$

(iii) **For** $r = 3.46$

The successive iteration values of the logistic map are given as follows:

0.2, 0.5536, 0.85506, 0.428807, 0.847463, 0.447272, 0.85538, 0.428019, 0.847073,
0.44821, 0.85572, 0.427184, 0.846654, 0.449214, 0.856076, 0.426306, 0.84621,
0.450281, 0.856447, 0.425392, 0.84574, 0.451404, 0.856829, 0.424449, 0.84525,
0.452576, 0.857218, 0.423487, 0.844744, 0.453783, 0.857609.

In this case period-4 orbit is formed, that is, $x_{n+4} = x_n$ (check from the data!).
Figure 10.10 depicts a plot of x_n versus n. Also, the cobweb diagram displays
the period-4 orbit of the logistic map for the above parameter value.

(iv) **For** $r = 3.56$

The successive iteration values of the logistic map are given below.

0.2, 0.5696, 0.872755, 0.395352, 0.851014, 0.451371, 0.881581, 0.371649, 0.831353,
0.499132, 0.889997, 0.348531, 0.808324, 0.551573, 0.880531, 0.374498, 0.833928,
0.493033, 0.889827, 0.349004, 0.808832, 0.550456, 0.880937, 0.373398, 0.83294,
0.495377, 0.889924, 0.348735, 0.808544, 0.551091, 0.880707.

In this case the iterative solutions form a period-8 orbit, that is, $x_{n+8} = x_n$ (check
from the data!). A plot of x_n versus n is shown in Fig. 10.11a.

One can see the periodic-8 points from the cobweb diagram (Fig. 10.11b). We
shall discuss the case for $r = 4$ in later chapter.

10.3   Dynamics of Quadratic and Cubic Maps

In this section we illustrate the general quadratic and cubic one-dimensional maps
in the real line.

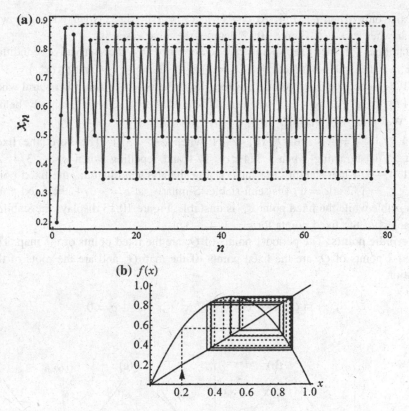

Fig. 10.11 a Phase trajectory for $r = 3.56$. **b** Cobweb diagram for $r = 3.56$

10.3.1 The Quadratic Map

The family of quadratic map is often denoted by Q_c and is defined by $Q_c(x) = x^2 + c$, $x \in \mathbb{R}$, where $c \in \mathbb{R}$ is a parameter. This map is called the Myrbrg family of maps on $\mathbb{R}$ as the domain. Myrbrg was one of the first to study this map extensively, see [5, 6]. The study of dynamics of Q_c with varying c is interesting and we shall discuss it below. First we calculate the fixed points of the quadratic map.

Fixed points: The fixed points of Q_c are simply the roots of the quadratic equation $Q_c(x) - x \equiv x^2 + c - x = 0$. This yields two fixed points $x_+^* = \frac{1}{2}\left(1 + \sqrt{1 - 4c}\right)$ and $x_-^* = \frac{1}{2}\left(1 - \sqrt{1 - 4c}\right)$. Note that the points $x_\pm^*$ depend on the parameter c. Furthermore, they are real if and only if $c \leq 1/4$. So, the fixed points of

Q_c exist only when $c \leq 1/4$. At $c = 1/4$, $x_+^* = x_-^* = 1/2$. No fixed points will appear when $c > 1/4$. Figure 10.12 depicts the three cases.

Stabilities of the fixed points are determined by the derivative condition $Q_c'(x_\pm^*) = 1 \pm \sqrt{1 - 4c}$. Note that $Q_c'(x_+^*) > 1$ when $c < 1/4$ and $Q_c'(x_+^*) = 1$ at $c = 1/4$. Hence the fixed point x_+^* is repelling when $c < 1/4$ and it is neutral when $c = 1/4$. At x_-^*, $Q_c'(x_-^*) = 1$ when $c = 1/4$ and $Q_c'(x_-^*) < 1$ for c slightly below $1/4$. We now see that $|Q_c'(x_-^*)| < 1$ if and only if $-3/4 < c < 1/4$. At $c = -3/4$, $Q_c'(x_-^*) = -1$ and $Q_c'(x_-^*) < -1$ when $c < -3/4$. Therefore, the fixed point x_-^* is attracting when $-3/4 < c < 1/4$ and repelling when $c < -3/4$. At $c = 1/4, -3/4$, the stability test fails. Using the cobweb diagram, the fixed point $x_+^* = x_-^* = 1/2$ at $c = 1/4$ is semi-stable. Similarly, at $c = -3/4$, the fixed point x_-^* is stable while the fixed point x_+^* is unstable. Figure 10.13 displays the stability characters of the fixed points for $c = 1/4, -3/4$.

Periodic points: The period-1 points of Q_c are the fixed points of the map. The period-2 points of Q_c are the fixed points of the map Q_c^2 and are the roots of the equation

$$f(x) \equiv Q_c^2(x) - x = x^4 + 2cx^2 - x + c^2 + c = 0.$$

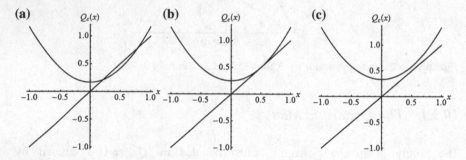

Fig. 10.12 Quadratic map for **a** $c < 1/4$, **b** $c = 1/4$, and **c** $c > 1/4$

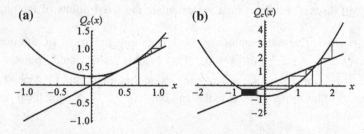

Fig. 10.13 Stability characters of the fixed points at **a** $c = 1/4$ and **b** $c = -3/4$

Since the fixed points of Q_c are also the fixed points of Q_c^2, $(x^2 - x + c)$ is a factor of the polynomial $f(x)$. Dividing $f(x)$ by this factor, we get the other factor as $(x^2 + x + c + 1)$. The solutions $p, q = \frac{-1 \pm \sqrt{-(4c+3)}}{2}$ of the equation $x^2 + x + c + 1 = 0$ yield the other two fixed points of Q_c^2. Note that the points p, q exist only when $c \leq -3/4$, they collide at $(-1/2)$ when $c = -3/4$. Now, calculate

$$Q_c(p) = p^2 + c = \frac{1}{4}\left(1 - 2\sqrt{-4c - 3} - 3 - 4c\right) + c$$

$$= \frac{-1 - \sqrt{-(4c+3)}}{2} = q,$$

and

$$Q_c(q) = q^2 + c = \left(\frac{-1 - \sqrt{-(4c+3)}}{2}\right)^2 + c$$

$$= \frac{-1 + \sqrt{-(4c+3)}}{2} = p.$$

This shows that $\{p, q\}$ forms a period-2 cycle of $Q_c(x)$. Stability of this cycle is determined by the derivative condition of the cycle $Q_c'(p)Q_c'(q) = 4pq = 4(c+1)$. Therefore, $|Q_c'(p)Q_c'(q)| < 1$ when $-1 < 4(c+1) < 1$, that is, when $-5/4 < c < -3/4$. Similarly, $|Q_c'(p)Q_c'(q)| > 1$ when $c < -5/4$ and $|Q_c'(p)Q_c'(q)| = 1$ when $c = -3/4, -5/4$. Thus, the period-2 cycle $\{p, q\}$ of the quadratic map Q_c is stable when $-5/4 < c < -3/4$ and is unstable (repelling) when $c < -5/4$ (Fig. 10.14).

It can be shown that the quadratic map has two period-3 cycles and they occur at $c = -1.75$. Figure 10.5 shows two attracting 3-cycles of the quadratic map at $c = -1.77$.

10.3.2   The Cubic Map

A family of one-dimensional cubic maps is defined by $f(x) = rx - x^3$, where $r \in \mathbb{R}$ is the control parameter. The dynamics of the cubic maps are more complicated than the quadratic maps. The fixed points of f are given by $x = 0, \pm\sqrt{(r-1)}$. Clearly, $x = 0$ is a fixed point of f for every value of r, but the other two fixed points, $x_\pm = \pm\sqrt{(r-1)}$, exist only when $r \geq 1$, and they coincide with $x = 0$ when $r = 1$. Figure 10.16 shows the fixed points of the map at $r = 2, -1$. The fixed points are the intersecting points of the function $f(x)$ with the diagonal line.

Fig. 10.14 Period-2 cycle of
the quadratic map for $c = -0.9$

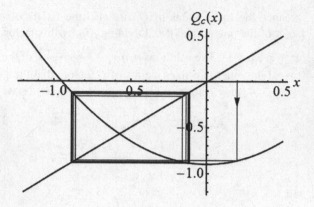

Fig. 10.15 Two period-3
cycles of the quadratic map

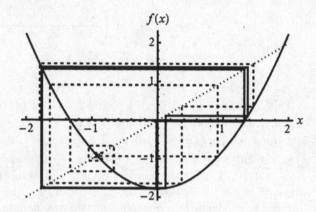

For stability of the fixed points, we calculate the derivative $f'(x) = r - 3x^2$. Since $f'(0) = r$, the fixed point origin is stable when $-1 < r < 1$ and unstable when r lies in the interval $(-\infty, -1) \cup (1, \infty)$. Bifurcations will occur at $r = \pm 1$. Again, since $f'\left(\pm\sqrt{(r-1)}\right) = 3 - 2r$, the fixed points $x_\pm = \pm\sqrt{(r-1)}$ are stable (attracting) when $|3 - 2r| < 1$, that is, when $1 < r < 2$, and are unstable when $r > 2$.

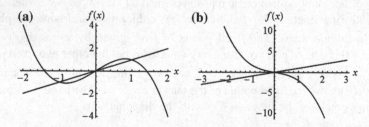

Fig. 10.16 Fixed points of the cubic map at **a** $r = 2$, **b** $r = -1$

A bifurcation may occur at $r = 2$. We shall now determine the period-2 cycle of f, which are simply the solutions of the equation $g(x) \equiv f^2(x) - x = r(rx - x^3) - (rx - x^3)^3 - x = 0$. Since the fixed points of f are also the fixed points of f^2, $x^3 - (r - 1)x$ must be a factor of the polynomial $g(x)$. On division, we obtain the other factor as $(x^2 - r - 1)(x^4 - rx^2 + 1)$ and this gives six 2-cycle points, $a_\pm = \pm\sqrt{r+1}$, $b_\pm = \pm\sqrt{\frac{r+\sqrt{r^2-4}}{2}}$ and $c_\pm = \pm\sqrt{\frac{r-\sqrt{r^2-4}}{2}}$. The points $a_\pm$ exist if $r \geq -1$. At $r = -1$, they coincide with the fixed point $x = 0$. The other four points exist when $r \geq 2$. We see that $f(a_+) = r\sqrt{r+1} - (r+1)\sqrt{r+1} = -\sqrt{r+1} = a_-$, $f(a_-) = -r\sqrt{r+1} + (r+1)\sqrt{r+1} = \sqrt{r+1} = a_+$. Thus, $\{a_+, a_-\}$ forms a period-2 cycle of the cubic map. Similarly, it can be shown that the other two period-2 cycles of the cubic map are $\{b_+, c_-\}$ and $\{b_-, c_-\}$. To determine the stability of the 2-cycles, we evaluate $(f^2)'(x) = (r - 3x^2)\left(r - 3x^2(r - x^2)^2\right)$. Using a little algebraic manipulation, we see that the 2-cycle $\{a_+, a_-\}$ is stable when $1 < r < 2$ and the other two 2-cycles are stable when $2 < r < \sqrt{5}$.

10.4 Symbolic Maps

We first discuss sequence space of 2-symbols and then sequence space of N-symbols. Their few properties are also given.

Sequence space of 2-symbols

We denote $\Sigma_2 = \{s = (s_0 s_1 s_2 \ldots) \mid s_i \in \{0, 1\}\}$ as a sequence space of 2-symbols. So, $(001010\ldots), (111000\ldots) \in \Sigma_2$. Let s and t be two elements of Σ_2, where $s = (s_0 s_1 s_2 \ldots)$ and $t = (t_0 t_1 t_2 \ldots)$. The distance between s and t is defined by

$$d_2(s, t) = \sum_{i=0}^{\infty} \frac{|s_i - t_i|}{2^i}.$$

Since $|s_i - t_i|$ is either 0 or 1, the infinite series is dominated by $\sum_{i=0}^{\infty} \frac{1}{2^i} = 2$. So, $d_2(s, t) \leq 2 \quad \forall s, t \in \Sigma_2$. Hence the space Σ_2 is bounded.

For example, if $s = (10011001 \cdots)$ and $t = (01101110 \cdots)$, then

$$d(s, t) = \sum_{i=0}^{\infty} \frac{|s_i - t_i|}{2^i} = \sum_{i=0}^{\infty} \frac{1}{2^i} = 2.$$

Again, if $s = (00000000 \cdots)$ and $t = (01010101 \cdots)$, then

$$d(s,t) = \sum_{i=0}^{\infty} \frac{|s_i - t_i|}{2^i} = \sum_{i=0}^{\infty} \frac{1}{2^{2i+1}} = \frac{2}{3}.$$

Let s, t, and r be three points in Σ_2, where $s = (s_0 s_1 s_2 \ldots)$, $t = (t_0 t_1 t_2 \ldots)$ and $r = (r_0 r_1 r_2 \ldots)$.

(i)   Since $|s_i - t_i| \geq 0$ for all $i = 0, 1, 2, \ldots$, we have $d_2(s,t) \geq 0$ $\forall s, t \in \Sigma_2$ and $d_2(s,t) = 0$ if and only f $s_i = t_i$ $\forall i$, that is, if only if $s = t$.

(ii) $d_2(s,t) = \sum_{i=0}^{\infty} \frac{|s_i - t_i|}{2^i} = \sum_{i=0}^{\infty} \frac{|t_i - s_i|}{2^i} = d_2(t,s)$ $\forall s, t \in \Sigma_2$.

(iii) For all $i = 0, 1, 2, \ldots$, we have

$$|s_i - t_i| = |(s_i - r_i) + (r_i - t_i)| \leq |s_i - ri| + |ri - ti|$$
$$\Rightarrow \quad d_2(s,t) \leq d_2(s,r) + d_2(r,t)$$

This shows that d_2 is a metric on Σ_2, and therefore (Σ_2, d_2) forms a metric space.

Theorem 10.1 *Suppose* $s = (s_0 s_1 s_2 \ldots)$, *and* $t = (t_0 t_1 t_2 \ldots)$ *are two elements in* Σ_2.

(i)   *If* $s_i = t_i$ *for* $i = 0, 1, 2, \ldots, k$, *then* $d_2(s,t) \leq 1/2^k$.
(ii) *If* $d_2(s,t) < 1/2^k$ *for some* k, *then* $s_i = t_i$ *for* $i = 0, 1, 2, \ldots, k$.

Proof

(i)   Suppose that $s_i = t_i$ for each $i = 0, 1, 2.., k$. Then

$$d_2(s,t) = \sum_{i=0}^{\infty} \frac{|s_i - t_i|}{2^i} = \sum_{i=0}^{k} \frac{|s_i - t_i|}{2^i} + \sum_{i=k+1}^{\infty} \frac{|s_i - t_i|}{2^i}$$

$$= \sum_{i=k+1}^{\infty} \frac{|s_i - t_i|}{2^i}$$

$$\leq \sum_{i=k+1}^{\infty} \frac{1}{2^i} = \frac{1}{2^k}.$$

(ii) Suppose that $d_2(s,t) < 1/2^k$ for some $k \geq 0$. We have to show that $s_i = t_i$ for $i = 0, 1, 2, \ldots, k$. If possible, let there exists an integer j with $j \leq k$, that is, $j \in \{0, 1, 2, \ldots, k\}$ such that $s_j \neq t_j$. Then we get

$$d_2(s,t) = \frac{1}{2^j} + \sum_{i=k+1}^{\infty} \frac{|s_i - t_i|}{2^i} \geq \frac{1}{2^j} \geq \frac{1}{2^k}.$$

This is a contradiction. Hence no such j exists. Therefore, $s_i = t_i$ for each $i = 0, 1, 2, \ldots, k$. This completes the proof.

Sequence space of N-symbols

The sequence space of N-symbols, Σ_N, is defined as

$$\Sigma_N = \{s = (s_0 s_1 s_2 \ldots) | s_i \in \{0, 1, 2, \ldots, N-1\}\}.$$

The distance between two points $s, t \in \Sigma_N$ is defined as

$$d_N(s, t) = \sum_{i=0}^{\infty} \frac{|s_i - t_i|}{N^i}.$$

Since $|s_i - t_i| \leq N - 1$, $d_N(s, t) \leq \sum_{i=0}^{\infty} \frac{N-1}{N^i} = N$, and therefore Σ_N is a bounded space.

10.5 Shift Map

Generally, the shift map $\sigma : \Sigma_2 \to \Sigma_2$ is defined as $\sigma(s_0 s_1 s_2 \ldots) = (s_1 s_2 s_3 \ldots)$ for all $s = (s_0 s_1 s_2 \ldots) \in \Sigma_2$. That is, $(\sigma(s))_i = s_{i+1}, i = 0, 1, 2, \ldots$, where $(\sigma(s))_i$ denotes the ith element of $\sigma(s)$.

Theorem 10.2 *The shift map is continuous.*

Proof To prove that the shift map σ is continuous, we have to prove that for any given $\varepsilon > 0$, there exists a $\delta > 0$ such that $d(\sigma(s), \sigma(t)) < \varepsilon$ whenever $d(s, t) < \delta$. Let $\varepsilon > 0$ be given. We choose a positive integer k such that $1/2^k < \varepsilon$. Set $\delta = 1/2^{k+1}$. Then for any two elements $s = (s_0 s_1 s_2 \ldots), t = (t_0 t_1 t_2 \ldots) \in \Sigma_2$, we have

$$d(s, t) < \delta = \frac{1}{2^{k+1}} \Rightarrow s_i = t_i \text{ for } i = 0, 1, 2, \ldots, (k+1)$$

$$\Rightarrow (\sigma(s))_i = (\sigma(t))_i, \; i = 0, 1, 2, \ldots, k$$

$$\Rightarrow d(\sigma(s), \sigma(t)) \leq \frac{1}{2^k} < \varepsilon$$

Thus, $d(s, t) < \delta \Rightarrow d(\sigma(s), \sigma(t)) < \varepsilon$. Therefore the shift map is continuous on Σ_2.

Periodic points of the shift map

A point p is said to be a periodic point of period n of a map f if $f^n(p) = p$. The least positive integer n for which $f^n(p) = p$ is called the prime period of the periodic point p. The periodic points of a map generate a repeating sequence. This sequence can also be observed for the shift map. For example, consider a point $p_n = (s_0 s_1 s_2 \cdots s_{n-1} s_0 s_1 s_2 \cdots s_{n-1} \cdots)$ in Σ_2. Applying the shift operator σ to p_n repeatedly, we see that

$$\sigma(p_n) = (s_1 s_2 \ldots s_{n-1} s_0 s_1 s_2 \ldots s_{n-1} \ldots)$$
$$\sigma^2(p_n) = (s_2 \ldots s_{n-1} s_0 s_1 s_2 \ldots s_{n-1} \ldots)$$
$$\sigma^3(p_n) = (s_3 \ldots s_{n-1} s_0 s_1 s_2 \ldots s_{n-1} \ldots)$$
$$\vdots$$
$$\sigma^{n-1}(p_n) = (s_{n-1} s_0 s_1 s_2 \ldots s_{n-1} \ldots)$$
$$\sigma^n(p_n) = (s_0 s_1 s_2 \ldots s_{n-1} \ldots) = p_n.$$

This shows that p_n is a periodic point of the shift map σ of prime period-n. Again, it can be shown that $(s_0 s_1 \ldots s_{n-1} 111 \ldots)$ is an eventually fixed point, and $(s_0 s_1 \ldots s_{n-1} 011\,011\,011 \ldots)$ is an eventually periodic point of period-3. In case of Σ_N, s_i may have N different values $(0, 1, \ldots, N-1)$, in each sequence $s_0, s_1, \ldots, s_{n-1}$ in which the s_i's can be chosen in N different ways. Hence there exist N^n different sequences of N-symbols. Thus, the number of periodic points of period-n is equal to N^n but not necessarily of prime period.

10.6 Euler Shift Map

The Euler shift map $S : [0, 1) \rightarrow [0, 1)$ is defined as

$$S(x) = 2x(\text{mod}1) = \begin{cases} 2x, & 0 \leq x < 1/2 \\ 2x - 1, & 1/2 \leq x < 1 \end{cases}$$

Here $2x(\text{mod}1)$ gives the fractional part of $2x$. The map $S(x)$ is a piecewise linear and non-invertible map in which the mapping function has a discontinuity. The function jumps suddenly from one value to another. The graphical representation of the map is shown in Fig. 10.17.

Fig. 10.17 Graphical representation of Euler shift map

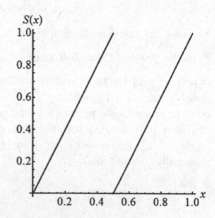

The fixed points of $S(x)$ satisfy the relation $x = S(x)$. This gives $x = 2x \Rightarrow x = 0 \in [0, 1/2)$ and $x = 2x - 1 \Rightarrow x = 1 \notin [1/2, 1)$. Hence the Euler shift map has only one fixed point at $x^* = 0$. Calculate the map $S^2(x)$:

$$S^2(x) = S(S(x)) = \begin{cases} 2(2x), & 0 \le 2x < \frac{1}{2} \\ 2(2x - 1), & 0 \le 2x - 1 < \frac{1}{2} \\ 2(2x - 1), & \frac{1}{2} \le 2x < 1 \\ 2(2x - 1 - 1), & \frac{1}{2} \le 2x - 1 < 1 \end{cases}$$

$$= \begin{cases} 4x, & 0 \le x < \frac{1}{4} \\ 4x - 1, & \frac{1}{4} \le x < \frac{1}{2} \\ 4x - 2, & \frac{1}{2} \le x < \frac{3}{4} \\ 4x - 3, & \frac{3}{4} \le x < 1 \end{cases}$$

It is easy to verify that the fixed points of $S^2(x)$ are $x^* = 0, \frac{1}{3}, \frac{2}{3}$. It is also verified that $\{\frac{1}{3}, \frac{2}{3}\}$ forms a period-2 cycle for $S^2(x)$, and $\{\frac{1}{7}, \frac{2}{7}, \frac{4}{7}\}$ and $\{\frac{3}{7}, \frac{6}{7}, \frac{5}{7}\}$ form two period-3 cycles for $S^3(x)$. The pictorial representations of the maps $S^2(x)$ and $S^3(x)$ are shown in Figs. 10.18 and 10.19, respectively.

Fig. 10.18 Graphical representation of Euler shift map at second iteration

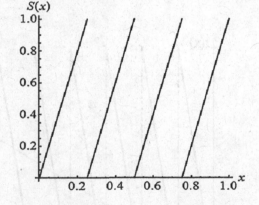

Fig. 10.19 Graphical representation of Euler shift map at third iteration

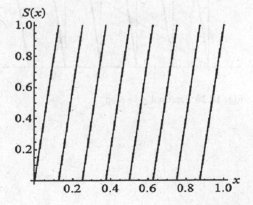

10.7   Decimal Shift Map

The decimal shift map is a piecewise linear map from [0, 1) to itself. The map is defined as

$$f(x) = 10x(\mathrm{mod}1), \quad x \in [0,1).$$

The graph of the map together with the diagonal line $y = x$ is shown in Fig. 10.20.

The fixed points of the decimal shift map satisfy the relation

$$
\begin{aligned}
x^* &= f(x^*) = 10x^*(\mathrm{mod}1) \\
&\Rightarrow \quad x^* = 10x^* - k, \quad k \in \mathbb{Z} \\
&\Rightarrow \quad x^* = k/9
\end{aligned}
$$

But $x^* \in [0,1)$. So, the fixed points of the decimal shift map are given by

$$x^* = 0, \frac{1}{9}, \frac{2}{9}, \frac{1}{3}, \frac{4}{9}, \frac{5}{9}, \frac{2}{3}, \frac{7}{9}, \frac{8}{9}$$

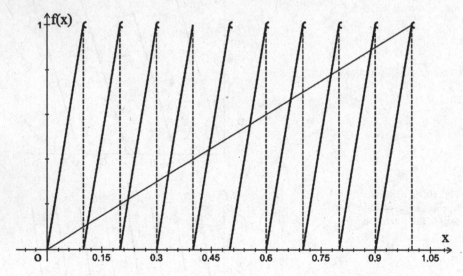

Fig. 10.20 Decimal shift map

10.8   Gaussian Map

The Gaussian map is a one-dimensional nonlinear map in which the mapping function is characterized by two control parameters. The Gaussian map is defined as

$$x_{n+1} = f(x_n) = e^{-bx_n} + c,$$

where b and c are the two control parameters (real).

The fixed points of the map depend on the control parameters. The graphs of the map for $b = 6$, $c = -0.8$ and $b = 6$, $c = -0.1$ are shown in Fig. 10.21.

From these graphs we see that the Gaussian map has three distinct fixed points for $b = 6$, $c = -0.8$ and it has only one fixed point for $b = 6$, $c = -0.1$. Since the map has two control parameters, it is very much difficult to study the behavior of iterates of the map. Some other important maps are (i) the sine map $x_{n+1} = r x_n^2 \sin(\pi x_n)$, (ii) the Baker map $x_{n+1} = 2x_n (\mathrm{mod}\,1)$, (iii) the sine circle map $x_{n+1} = x_n + \omega - \frac{k}{2\pi} \sin(2\pi x_n)(\mathrm{mod}\,1)$, $x_n \in [0, 1)$.

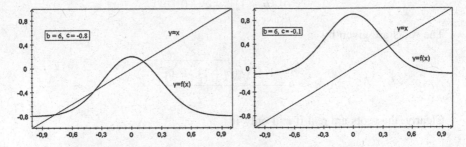

Fig. 10.21 The Gaussian map for different values of the control parameters

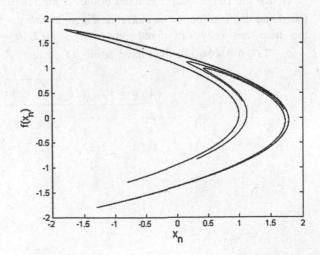

Fig. 10.22 Phase portrait of Hénon map for $a = 1.4$, $b = 0.3$

10.9   Hénon Map

This is a two-dimensional map and was first studied by the French astronomer *Michel Hénon* (1931–2013) in the year 1976. The Hénon map $H_{a,b} : \mathbb{R}^2 \to \mathbb{R}^2$ is defined as

$$(x, y) \to H_{a,b}(x, y) = \left(a - x^2 + by, x\right)$$

$$\text{or,} \quad \begin{bmatrix} x_n \\ y_n \end{bmatrix} \to \begin{bmatrix} x_{n+1} = a - x_n^2 + by_n \\ y_{n+1} = x_n \end{bmatrix}$$

where a, b are two real parameters (Fig. 10.22). The Hénon map has a nonlinear term x^2. The fixed points of this map satisfy the following equations:

$$f(x, y) = (x, y) \Rightarrow a - x^2 + by = x \text{ and } y = x$$

which is equivalent to

$$x^2 - (b - 1)x - a = 0.$$

The roots are given by

$$x = \frac{b - 1 \pm \sqrt{(b - 1)^2 + 4a}}{2}.$$

Clearly, the roots are real if and only if

$$(b - 1)^2 + 4a \geq 0, \text{ that is, } 4a \geq -(b - 1)^2.$$

Hence the Hénon map has fixed points if the relation $4a \geq -(b - 1)^2$ is satisfied and the fixed points lie along the diagonal line $y = x$. When $4a = -(b - 1)^2$, the map has only one fixed point $((b - 1)/2, (b - 1)/2)$ and when $4a > -(b - 1)^2$, it has two distinct fixed points (α, α) and (β, β) where

$$\alpha, \beta = \frac{b - 1 \pm \sqrt{(b - 1)^2 + 4a}}{2}.$$

For example, if $a = 0$ and $b = 0.4$, then the Hénon map has two distinct fixed points $(0, 0)$ and $(-0.6, -0.6)$. The Jacobian matrix of $H_{a,b}$ is given by

$$J(x,y) = \begin{pmatrix} -2x & b \\ 1 & 0 \end{pmatrix}$$

This gives the eigenvalues

$$\lambda = x \pm \sqrt{x^2 + (b/2)^2}.$$

The Jacobian determinant of the Hénon map is constant and it is given by $\det(J) = -b$. Note that $\det(J) \neq 0$ if and only if $b \neq 0$. Hence the Hénon map is invertible if and only if $b \neq 0$ and the inverse is given by

$$H_{a,b}^{-1} = \left(y,\ (x - a + y^2)/b\right).$$

We also see that $|\det(J)| < 1$ if and only if $-1 < b < 1$. Hence the Hénon map is area-contracting (a map $f(x,y)$ is said to be area-contracting if $|\det(J(x,y))| < 1$ everywhere) for $-1 < b < 1$. That is, it contracts the area of any region in each iteration by a constant factor of $|b|$.

Period-2 points

For finding period-2 points we have a relation

$$f^2(x,y) = (x,y) \Rightarrow a - (a - x^2 + by)^2 + bx = x \text{ and } a - x^2 + by = y.$$

Solving the second equation for y, $(1 - b)y = a - x^2$ and then substituting in the first equation, we get

$$\left(x^2 - a\right)^2 + (1 - b)^3 x - (1 - b)^2 a = 0.$$

By factorization we see that $(x^2 + (1 - b)x - a)$ must be a factor of the equation. On division, we get the other factor of the equation as $(x^2 - (1 - b)x - a + (1 - b)^2)$ and the period-2 points are given by the roots of the equation $x^2 - (1 - b)x - a + (1 - b)^2 = 0$, while $(1 - b)y = a - x^2$. The roots are given by

$$x_1, x_2 = \frac{1}{2}\left[(1 - b) \pm \sqrt{4a - 3(1 - b)^2}\right].$$

The period two points lie on $x + y = 1 - b$. It is evident that the Hénon map has a period-2 orbit if and only if $4a > 3(1 - b)^2$.

10.9.1   Skinny Baker Map

This is a two-dimensional map and it is denoted by $B(x, y)$. The map has the property of starching in one direction and contracting it in a cross direction. This is a typical two-dimensional chaotic map. The map function is defined as follows:

$$B(x, y) = \begin{cases} \left(\frac{1}{3}x, 2y\right), & 0 \le y \le \frac{1}{2} \\ \left(\frac{1}{3}x + \frac{2}{3}, 2y - 1\right), & \frac{1}{2} \le y \le 1 \end{cases}$$

This is a discontinuous map, points (x, y) where $y < 1/2$ are mapped to the left of the unit square while points (x, y) where $y > 1/2$ are mapped to the right. In matrix representation, the map is written as

$$\begin{bmatrix} x \\ y \end{bmatrix} \rightarrow B \begin{bmatrix} x \\ y \end{bmatrix} = \begin{bmatrix} \frac{1}{3} & 0 \\ 0 & 2 \end{bmatrix} \begin{bmatrix} x \\ y \end{bmatrix}, \quad 0 \le y \le \frac{1}{2}$$

$$= \begin{bmatrix} \frac{1}{3} & 0 \\ 0 & 2 \end{bmatrix} \begin{bmatrix} x \\ y \end{bmatrix} + \begin{bmatrix} \frac{2}{3} \\ -1 \end{bmatrix}, \quad \frac{1}{2} \le y \le 1$$

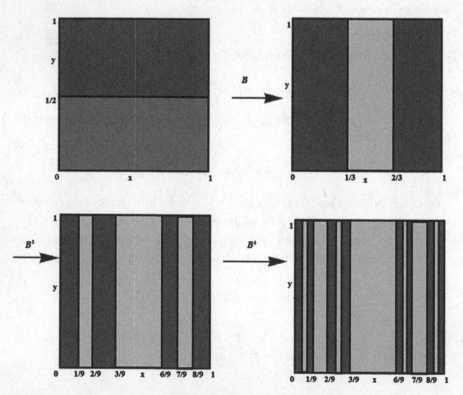

Fig. 10.23 First three iterations of Skinny Baker map

The diagrammatic representations of different compositions are shown in Fig. 10.23.

10.10 Bifurcations in Discrete Systems

Maps represented by $f_\lambda(x) = f(x, \lambda)$ are depending upon the parameter $\lambda \in \mathbb{R}$. It is often found that when λ crosses a critical value, the properties of the orbit, say its stability, periodicity, etc., may change and even new orbit of a discrete system may be created. Such a qualitative change is called bifurcation. Bifurcation theories are vast. In this section we discuss few selective bifurcations in discrete systems.

For example, consider the one-dimensional map $f(x, \lambda) = \lambda \tan^{-1}(x)$; $\lambda, x \in \mathbb{R}$. For $\lambda < 1$, $f_\lambda(x) = \lambda \tan^{-1}(x) = x \Rightarrow x = 0$ is only fixed point since $x/\tan^{-1}(x) > 1$. Since $f_\lambda'(0) = \lambda < 1$, the fixed point origin is stable. For $\lambda > 1, f_\lambda(x) = \lambda \tan^{-1}(x)$ has three fixed points at $x = 0$, $x = x_+ > 0$ and $x = x_- < 0$. The map $f_\lambda(x)$ has three points of intersection. Now, $f_\lambda(x_+) = \lambda \tan^{-1}(x_+) = x_+ \Rightarrow \lambda = \frac{x_+}{\tan^{-1}(x_+)}$. This implies

$$f_\lambda'(x_+) = \frac{\lambda}{1 + (x_+)^2} = \frac{x_+}{1 + (x_+)^2} \frac{1}{\tan^{-1}(x_+)} < 1.$$

The fixed point at x_+ is stable for $\lambda > 1$. Similarly, the fixed point at $x = x_-$ is also stable. But the fixed point $x = 0$ is unstable, since $f_\lambda'(0) = \lambda > 1$. Thus the map undergoes a bifurcation when the parameter λ crosses the value $\lambda = 1$, called bifurcating point. The graphical representations of $f_\lambda(x)$ for $\lambda < 1$ and $\lambda > 1$ are shown in Fig. 10.24.

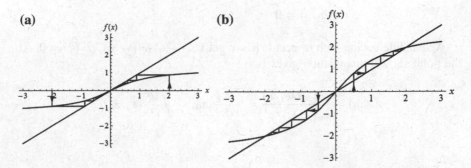

Fig. 10.24 The graphical representations of $f_\lambda(x)$ for **a** $\lambda < 1$ and **b** $\lambda > 1$

Saddle-node bifurcation:

Let $f : \mathbb{R}^2 \to \mathbb{R}$ be a C^r function with $r \geq 2$ and with respect to x and λ, the function $f(x, \lambda) = f_\lambda(x)$ satisfies the following conditions

(i) $f(x_0, \lambda_0) = x_0$, that is, the function f has a fixed point (x_0, λ_0).

(ii) $\frac{\partial}{\partial x} f(x, \lambda)\big|_{(x_0, \lambda_0)} \equiv f'_{\lambda_0}(x_0) = 1$.

(iii) $\frac{\partial^2}{\partial x^2} f(x, \lambda)\big|_{(x_0, \lambda_0)} \equiv f''_{\lambda_0}(x_0) \neq 0$.

(iv) $\frac{\partial}{\partial \lambda} f(x, \lambda)\big|_{(x_0, \lambda_0)} \equiv \frac{\partial f}{\partial \lambda}(x_0, \lambda_0) \neq 0$.

Then there exist intervals I about x_0 and N about λ_0 and a C^r function $\lambda(x): I \to N$ such that

(a) $f_{\lambda(x)}(x) \equiv f(x, \lambda(x)) = x$,

(b) $\lambda(x_0) = \lambda_0$,

(c) The graph of $\lambda(x)$ gives all the fixed points in $I \times N$,

(d) $\lambda'(x_0) = 0$ and $\lambda''(x_0) \neq 0$.

These fixed points are attracting on the one side of x_0 and are repelling on the other side.

Let us now set $C(x, \lambda) = f(x, \lambda) - x$. Then, we get the following:

$C(x_0, \lambda_0) = f(x_0, \lambda_0) - x_0 = 0$, $\frac{\partial C}{\partial x}(x_0, \lambda_0) = \frac{\partial f}{\partial x}(x_0, \lambda_0) - 1 = 0$. So, we cannot solve for x in terms of λ. But $\frac{\partial C}{\partial \lambda}(x_0, \lambda_0) = \frac{\partial f}{\partial \lambda}(x_0, \lambda_0) \neq 0$. Hence by the implicit function theorem, there exists a C^n function $\lambda = \lambda(x) : I \to N$ such that $C(x_0, \lambda_0) = 0$, that is, $f(x_0, \lambda_0) = x_0$ and $C(x, \lambda(x)) = 0 \ \forall x \in I$, that is, $f(x, \lambda(x)) = x$.

Now, differentiating both sides of $C(x, \lambda(x)) = 0$ with respect to x, we get $\frac{\partial C}{\partial x} + \frac{\partial C}{\partial \lambda}\frac{d\lambda}{dx} = 0$. Therefore, at the point (x_0, λ_0), we have

$$\left[\frac{\partial f}{\partial x}(x_0, \lambda_0) - 1\right] + \frac{\partial f}{\partial \lambda}(x_0, \lambda_0)\lambda'(x_0) = 0$$
$$\Rightarrow \lambda'(x_0) = 0$$

Again, differencing with respect to x, we get $C_{xx} + 2C_{x\lambda}\lambda'(x) + C_\lambda \lambda''(x) = 0$. At the point (x_0, λ_0) this relation gives

$$\lambda''(x_0) = -\frac{C_{xx}(x_0, \lambda_0)}{C_\lambda(x_0, \lambda_0)} = -\frac{\partial^2 f}{\partial x^2}(x_0, \lambda_0) \bigg/ \frac{\partial f}{\partial \lambda}(x_0, \lambda_0).$$

For determining the stability of the fixed points, we expand $\frac{\partial f}{\partial x}$ in Taylor series about (x_0, λ_0):

$$\frac{\partial f}{\partial x}(x, \lambda) = 1 + \left.\frac{\partial^2 f}{\partial x^2}\right|_{(x_0,\lambda_0)} (x - x_0) + \left.\frac{\partial^2 f}{\partial x \partial \lambda}\right|_{(x_0,\lambda_0)}$$

$$(\lambda - \lambda_0) + O\left(|x - x_0|^2\right) + O(|x - x_0||\lambda - \lambda_0|) + O\left(|\lambda - \lambda_0|^2\right)$$

Since $\lambda'(x_0) = 0$, $\lambda(x) - \lambda_0 = O\left(|x - x_0|^2\right)$ and therefore,

$$\frac{\partial f}{\partial x}(x, \lambda) = 1 + \left.\frac{\partial^2 f}{\partial x^2}\right|_{(x_0,\lambda_0)} (x - x_0).$$

This implies that $\left\{\frac{\partial f}{\partial x}(x, \lambda) - 1\right\}$ has opposite signs on two sides of $x = x_0$ according for reversal of stability. So, the system is stable for $x > x_0$ if $\frac{\partial^2 f}{\partial x^2}(x_0, \lambda_0) < 0$ and it is stable for $x < x_0$ if $\frac{\partial^2 f}{\partial x^2}(x_0, \lambda_0) > 0$.

Period-doubling or flip bifurcation:

Let $f(x, \lambda) \equiv f_\lambda(x) : \mathbb{R}^2 \to \mathbb{R}$ be a C^r function in variables x, λ with $r \geq 3$. If $f(x, \lambda)$ satisfies the following conditions:

(i) $f_{\lambda_0}(x_0) \equiv f(x_0, \lambda_0) = x_0$.

(ii) $\frac{\partial f}{\partial x}(x_0, \lambda_0) = -1 \Rightarrow \frac{\partial}{\partial x}[f(x, \lambda) - x]_{(x_0, \lambda_0)} \neq 0$,

(iii) The derivative of $\frac{\partial f}{\partial x}$ with respect to λ at the point (x_0, λ_0) is nonzero, that is,

$$\alpha = \left[\frac{\partial^2 f}{\partial \lambda \partial x} + \frac{1}{2}\frac{\partial f}{\partial \lambda}\frac{\partial^2 f}{\partial x^2}\right]_{(x_0,\lambda_0)} \neq 0,$$

(iv) $\beta = \left[\frac{1}{3!}\frac{\partial^3 f}{\partial x^3} + \frac{1}{2!}\left(\frac{\partial f}{\partial x}\right)^2\right]_{(x_0,\lambda_0)} \neq 0.$

Then

(a) there occurs a differentiable curve $x(\lambda)$ of fixed points passing through the point (x_0, λ_0); the stability of the fixed points changes at (x_0, λ_0): the curvature changes from stable to unstable as λ increases past the value λ_0 if $\alpha < 0$ while the converse occurs if $\alpha > 0$.

(b) Moreover, there occurs a period-doubling bifurcation at the point (x_0, λ_0); there is a differentiable curve $\lambda = l(x)$ passing through the point (x_0, λ_0) such that all points on the curve except the point (x_0, λ_0) are hyperbolic period-2 points, $f^2_{\lambda=l(x)}(x) = x$; the curve $\lambda = l(x)$ is tangential to $\lambda = \lambda_0$ at (x_0, λ_0); $l'(x_0) = 0$ and $l''(x_0) = -2\beta/\alpha \neq 0$. Finally, the period 2 orbits are attracting if $\beta > 0$ and are repelling if $\beta < 0$.

We discuss the saddle-node and period-doubling bifurcations through examples based on the above bifurcation theory.

Fig. 10.25 Graph of the given map

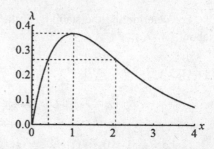

Consider a map $f_\lambda(x) \equiv f(x, \lambda) = \lambda e^x; \lambda > 0$. The fixed points of the map $f_\lambda(x) \equiv f(x, \lambda) = \lambda e^x$ are the solutions of $f_\lambda(x) = x$, which gives $\lambda = xe^{-x}$. The graph of this function is displayed in Fig. 10.25.

From the graph we see that the function $f_\lambda(x) = \lambda e^x; \lambda > 0$ has (i) no fixed points for $\lambda > 1/e$, (ii) one fixed point for $\lambda = 1/e$, and (iii) two fixed points for $\lambda < 1/e$, one on each side of $x = 1$. The following conditions are hold:

(a) $f(x = x_0 = 1, \lambda = \lambda_0 = 1/e) = x_0$

(b) $f'_{\lambda=1/e}(x = x_0 = 1) = 1$

(c) $f''_{\lambda=1/e}(x = x_0 = 1) \neq 0 (= 1)$

(d) $\frac{\partial f_\lambda}{\partial \lambda}\big|_{\lambda=1/e, x=1} = e \neq 0.$

Then there are intervals I about $x_0 = 1$ and N about $\lambda = \lambda_0 = 1/e$ such that there exists a C^τ differentiable function $\lambda = l(x) : I \to N$ with the following properties (Fig. 10.27).

(i) $f_{\lambda=l(x)}(x) = x$ with $\lambda_0 = l(x_0) = 1/e$

(ii) $l'(x = x_0 = 1) = 0$

(iii) $l''(x = x_0 = 1) = -\frac{\partial^2 f}{\partial x^2}\big|_{\substack{x=x_0 \\ \lambda=\lambda_0}} \Big/ \frac{\partial f}{\partial x}\big|_{\substack{x=x_0 \\ \lambda=\lambda_0}} \neq 0$, so that

(iv) $\lambda = l(x) = l(x_0) + l'(x_0)(x - x_0) + l''(x_0)(x - x_0)^2 + O\big((x - x_0)^3\big) = \lambda_0 + l''$
      $(x_0)(x - x_0)^2 + O\big((x - x_0)^3\big)$

Fig. 10.26 Bifurcation diagram of the given map

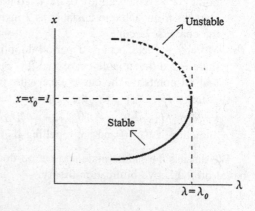

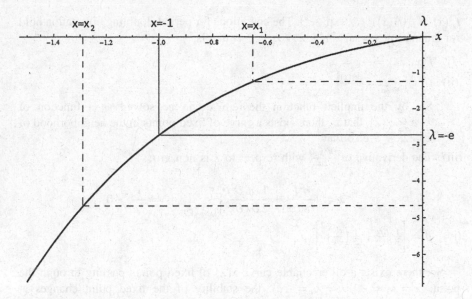

Fig. 10.27 Graph of the given map

This shows that there occurs a saddle-node bifurcation at $(x = x_0 = 1, \lambda = \lambda_0 = 1/e)$. All points on the curve $\lambda = l(x)$ belonging to $I \times N$ are fixed points. At these points, we have

$$\frac{\partial f}{\partial x}\bigg|_{(x, \lambda = l(x))} = \frac{\partial f}{\partial x}\bigg|_{\substack{x=x_0 \\ \lambda=\lambda_0}} + \frac{\partial^2 f}{\partial x^2}\bigg|_{\substack{x=x_0 \\ \lambda=\lambda_0}} (x - x_0) + \frac{\partial}{\partial \lambda}\frac{\partial f}{\partial x}\bigg|_{\substack{x=x_0 \\ \lambda=\lambda_0}} (\lambda - \lambda_0)$$

$$+ O\Big((x - x_0)^2, (x - x_0)(\lambda - \lambda_0), (\lambda - \lambda_0)^2\Big)$$

$$= 1 + (x - x_0) + O\Big((x - x_0)^2, (x - x_0)(\lambda - \lambda_0), (\lambda - \lambda_0)^2\Big) \text{ [Using (b) and(c)]}$$

< 1, for fixed points with $x < x_0 \Rightarrow$ stable.

> 1, forfixed points with $x > x_0 \Rightarrow$ unstable.

Hence, the given map undergoes the saddle-node bifurcation. The bifurcation diagram is presented below in Fig. 10.26.

Again, we consider the map $f_\lambda(x) \equiv f(x, \lambda) = \lambda e^x; \lambda < 0$. The fixed points of the map $f_\lambda(x) \equiv f(x, \lambda) = \lambda e^x$ are the solutions of $f_\lambda(x) = x$, which gives $\lambda = x e^{-x}$.

From the graph it is clear that the function $f_\lambda(x) = \lambda e^x; \lambda < 0$ has only one fixed point for any value of $\lambda(<0)$. At the fixed point at $x = -1, \lambda = x e^{-x} = -e, f_\lambda(x) = \lambda e^x = -1$ and $f'_\lambda(x) = -1$. For the fixed point at $x = x_1 > -1, \lambda = x_1 e^{-x_1} > -e$, we have $|f'_\lambda(x)| = |x_1| < 1$. Similarly, at the fixed point $x = x_2 < -1, \lambda = x_2 e^{-x_2} < -e$, we have $|f'_\lambda(x)| = |x_2| > 1$. So, the fixed point $(x = x_1 > -1, \lambda > -e)$ is attracting, while the other fixed point $(x = x_2 < -1, \lambda < -e)$ is repelling. The stability character changes at $(x = -1, \lambda = -e)$. We now calculate

$f_\lambda^2(x) = f_\lambda(f_\lambda(x)) = \lambda \exp(\lambda e^x)$. The conditions for period-doubling bifurcation hold for this map at $x = x_0 = -1, \lambda = \lambda_0 = -e$, that is,

(i). $f(x_0, \lambda_0) = -1 = x_0,$

(ii) $f_\lambda'(x)\big|_{(x_0,\lambda_0)} = \frac{\partial f_\lambda(x)}{\partial x}\big|_{(x_0,\lambda_0)} = -1$

So, by the implicit function theorem x can be solved as a function of $\lambda : x = x(\lambda)$, that is, there exists a curve of fixed points in the neighborhood of $\lambda = \lambda_0 = -e$ so that

(iii) The derivative of $\frac{\partial f(x,\lambda)}{\partial x}$ with respect to λ is nonzero;

$$\alpha = \left[\frac{\partial^2 f}{\partial \lambda \partial x} + \frac{1}{2} \frac{\partial f}{\partial \lambda} \frac{\partial^2 f}{\partial x^2} \right]_{(x=-1,\lambda=-e)} = \frac{1}{2e} > 0$$

(iv) $\beta = \left[\frac{1}{3!}\frac{\partial^3 f}{\partial x^3} + \frac{1}{2!}\left(\frac{\partial f}{\partial x}\right)^2 \right]_{(x=-1,\lambda=-e)} = \frac{1}{3} > 0.$

So, there exists a differentiable curve $x(\lambda)$ of fixed points passing through the point $(x = x_0 = -1, \lambda = \lambda_0 = -e)$; the stability of the fixed point changes at (x_0, λ_0). Since $\alpha > 0$, the fixed points are attracting for $\lambda > -e$ and repelling for $\lambda < -e$. There also occurs a period-doubling bifurcation at (x_0, λ_0); note that a differentiable curve $\lambda = l(x)$ such that all points on it except (x_0, λ_0) are hyperbolic period-2 points. Since $\beta > 0$, the period-2 points are attracting. The bifurcation diagram is presented in Fig. 10.28.

The period-doubling bifurcation also occurs for the map defined by $f_\lambda(x) \equiv f(x, \lambda) = \lambda \sin x$. The fixed points of the map are given by the solutions of the equation $\lambda \sin x = x$, which gives $\lambda = x / \sin x$ (Fig. 10.29).

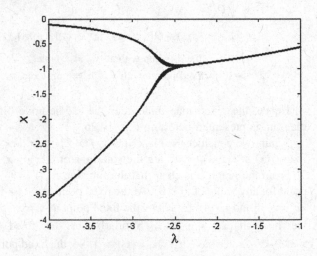

Fig. 10.28 Bifurcation diagram of the given map

Fig. 10.29 Phase trajectory
of the map

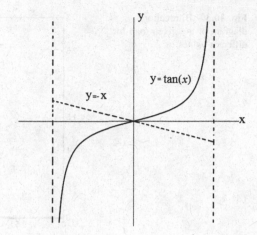

So, the fixed points are (x_0, λ_0), where $\lambda_0 = x_0/\sin x_0$. From the above theorem, we get

$$f'(x_0, \lambda_0) = \lambda_0 \cos x_0 = \frac{x_0}{\tan x_0} = -1.$$

Now, at $x_0 = \frac{\pi}{2} + \xi_0$,

$$\frac{\frac{\pi}{2} + \xi_0}{\tan(\frac{\pi}{2} + \xi_0)} = -1 \Rightarrow \tan(\xi_0) = \frac{1}{(\frac{\pi}{2} + \xi_0)}.$$

Using fixed point iteration $(x_{n+1} = \varphi(x_n)$, where $f(x) = 0 \Rightarrow x = \varphi(x))$, we get $\xi_0 = 0.4580105$, so that $x_0 = \frac{\pi}{2} + 0.4580105 = 2.0288068$,

$$f'_{\lambda_0}(x_0) = \frac{x_0}{\tan x_0} = \frac{2.088068}{\tan(\frac{\pi}{2} + \xi_0)} = -\frac{\frac{\pi}{2} + \xi_0}{\cot \xi_0} = -\left(\frac{\pi}{2} + \xi_0\right) \tan \xi_0 = -1.0001476,$$

and $\lambda_0 = x_0/\sin x_0 = 2.4023656$. Thus

(i) For $x = x_0, \lambda = \lambda_0, f(x_0, \lambda_0) = \lambda_0 \sin x_0 = x_0$,
     where $x_0 = 2.0288068$ and $\lambda_0 = 2.4023656$; x_0 is a fixed point.

(ii) $f'_{\lambda_0}(x_0) = \lambda_0 \cos x_0 = \frac{x_0}{\tan x_0} = -1$, so that there exists a line of fixed points
     $x = x(\lambda)$ through (x_0, λ_0).

(iii) $\alpha = \left[\frac{\partial^2 f}{\partial \lambda \partial x} + \frac{1}{2} \frac{\partial f}{\partial \lambda} \frac{\partial^2 f}{\partial x^2}\right]_{(x=x_0, \lambda=\lambda_0)} = \left[\cos x_0 - (1/2)\lambda_0 \sin^2 x_0\right] < 0.$

(iv) $\beta = \left[\frac{1}{3!} \frac{\partial^3 f}{\partial x^3} + \frac{1}{2!} \left(\frac{\partial f}{\partial x}\right)^2\right]_{(x=x_0, \lambda=\lambda_0)} = \left[-(\lambda_0/6)\cos x_0 + (\lambda_0^2/4)\sin^2 x_0\right] > 0.$

Fig. 10.30 Bifurcation diagram of the given map for different values of λ

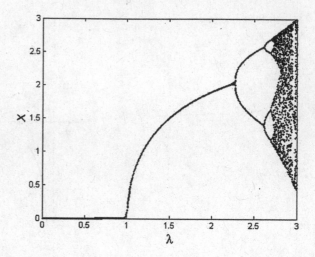

Hence we have

(a) There occurs a period-doubling bifurcation at $(x_0 = \pi/2 + \xi_0, \lambda_0 = x_0/\sin x_0)$

(b) There exists a line of fixed points $x = x(\lambda)$ passing through (x_0, λ_0). As $\alpha < 0$ stability character of the fixed points changes at (x_0, λ_0) in the following manner:

   (i) Fixed points are attracting for $\lambda < \lambda_0$,
   (ii) Fixed points are repelling for $\lambda > \lambda_0$.

(c) $\beta > 0$; so the period-2 orbits are attractors,

(d) In the neighborhood of (x_0, λ_0) the curve of period-2 attractors is $\lambda = l(x)$, where
$l'(x_0) = x_0, l''(x_0) = -2\beta/\alpha$ so that $l(x) = \lambda_0 - \frac{2\beta}{\alpha}(x - x_0)^2 + o\left((x - x_0)^3\right)$.

The bifurcation diagram for this map is presented in Fig. 10.30.
For further reading, see Devaney [7].

10.11   Exercises

1. (a) Find all fixed points of the tent map $T^4(x)$. Show that there must constitute three distinct periodic-4 orbits.
   (b) Prove that the tent map has a periodic 2 cycle. Is the cycle stable or unstable?
   (c) Show that all periodic cycles of the tent map are unstable in nature.
   (d) Show that $x = 1/4$ is an eventually fixed point of the tent map.

2. Show that all periodic points of the logistic map are sources.
3. Show that the logistic map $f_r(x) = rx(1-x)$, $0 \le r \le 4$ has at most one periodic sink.
4. Find all fixed points and classify their stabilities for the cubic map $f(x) = 3x - x^3$.
5. Define superstable fixed point and superstable 2-cycle. Find parameter (r) values so that the logistic map $f_r(x) = rx(1-x)$, $x \in [0,1]$ has a superstable fixed point and a superstable 2-cycle.
6. Show that the logistic map $f_r(x) = rx(1-x)$, $x \in [0,1]$ has negative Schwarzian derivative over the entire interval $[0, 1]$. Are there any restrictions on the parameter r? Justify your answer.
7. (a) Define Euler shift map $S(x)$ and give its importance in symbolic dynamics.
   (b) Find all fixed points of the map $S^2(x)$. Show that $\{\frac{1}{3}, \frac{2}{3}\}$ is a periodic 2-cycle of the map $S(x)$.
   (c) Generate the map $S^3(x)$, where $S(x)$ is the Euler shift map.
   (d) Show that the map $S^3(x)$ has exactly seven fixed points. Show the fixed points graphically.
   (e) Show that $\{\frac{1}{7}, \frac{2}{7}, \frac{4}{7}\}$ and $\{\frac{3}{7}, \frac{6}{7}, \frac{5}{7}\}$ are the two periodic 3-cycles of the map $S(x)$.

8. Show that the logistic map with the parameter $r = 4$ is equivalent to the Euler shift map.
9. Show that the Gaussian map function has two inflection points. Also find them [Inflection points of a map $f(x)$ are those at which $f'(x)$ is either maximum or minimum.].
10. Consider the Hénon map $H_{a,b}(x,y) = (a - x^2 + by, x)$, where a, b are the two real parameters. Prove that the map has a periodic-2 orbit iff $a > \frac{3}{4}(1-b)^2$
11. Show that for $a = 0$ and $b = 0.4$ the Hénon map $H_{a,b}$ has exactly two fixed points, one is stable and another is unstable.
12. Consider the map $f(x,y) = (a - x^2 + 0.4y, x)$, $(x,y) \in \mathbb{R}^2$.

   (a) Prove that for $-0.09 < a < 0.27$, the map f has a stable fixed point and an unstable fixed point.
   (b) Prove that for $-0.27 < a < 0.85$, it has a period-2 sink.

13. Show that the Hénon map is an invertible map and find its inverse. Also show that the map is area-contracting.
14. Let (x_1, y_1) and (x_2, y_2) be two fixed points of the Hénon map $H_{a,b}$ for some fixed parameters a and b. Show that $x_1 - y_1 = x_2 - y_2 = 0$ and $x_1 + y_1 = x_2 + y_2 = b - 1$.

15. Consider the two-dimensional map constructed by Holmes:

$$x_{n+1} = dx_n - x_n^3 - by_n,$$
$$y_{n+1} = x_n$$

Prove that the fixed point $(0, 0)$ is an attractor for $d < (1 - b)$ and is repellor for $d > (1 - b)$.

16. Find all fixed points of the decimal shift map $x_{n+1} = 10x_n(\mathrm{mod}1)$ on $[0, 1)$. Comment on the stability of each of these fixed points. Also, show that the map has periodic points of all periods but they are all unstable.

17. Draw bifurcation diagram for the logistic map for parameter values $r = 2.5$ to $r = 4$. What conclusion can you draw from this diagram?

18. Iterate the logistic map $x_{n+1} = rx_n(1 - x_n)$ for $r = 0.8, 2.8, 3.2, 3.5, 3.8, 4$ taking starting point $x_0 = 0.2$. Tabulate the iteration values to five significant figures. What can you conclude from the results?

19. Find the attractor of the sine map $x_{n+1} = rx_n^2 \sin(\pi x_n)$ for the parameter values $r = 1.5, 1.8$ and 2.2. Iterate the map 20 times using starting points $0.2, 0.5, 0.75,$ and $.9$.

20. Plot the first 20 points of the Hénon attractor for $a = 1.4$, b $= 0.3$ with the initial value $(0, 0)$.

21. Discuss bifurcations and draw bifurcation diagrams for the following maps:

(i) $f(x, r) = rx(1 - x), r = 3, 3.567,$
(ii) $f(x, \lambda) = \lambda x - x^3, \lambda = 1, -1.$

22. Iterate 10 times of the complex maps (i) $z_{n+1} = z_n^2$, (ii) $z_{n+1} = z_n^2 - 1$, (iii) $z_{n+1} = z_n^2 - 1.5$ for initial value $(0 + 0i)$. Find the periodicity of the solutions of the above maps.

References

1. Davies, B.: Exploring Chaos. Westview Press, New York (2004)
2. Robert M. May: Stability and Complexity in Model Ecosystems. Princeton University Press, Princeton (1973)
3. Edelstein-Keshet, L.: Mathematical Biology, SIAM (2005)
4. Murray, J.D.: Mathematical Biology: I. An Introduction, 3rd edn. Springer, Berlin (2002)
5. Myrberg, P.J.: Iteration von Quadratwurzeloperationen. Ann. Acad. Sci. Fenn. A **259**, 1–10 (1958)
6. Abraham, R.H., Gardini, L., Mira, C.: Chaos in Discrete Dynamical Systems. Springer, Berlin (1997)
7. Devaney, R.L.: An Introduction to Chaotic Dynamical Systems. Westview Press, New York (2003)
8. Strogatz, S.H.: Nonlinear Dynamics and Chaos with Application to Physics, Biology, Chemistry and Engineering, Perseus Books. L.L.C, Massachusetts (1994)
9. Lynch, S.: Dynamical Systems with Applications using MATLAB. Birkhauser, Boston (2004)

10. Hubbard, J.H., West, B.H.: Differential Equations: A Dynamical Systems Approach. Springer, New York (1991)
11. Hilborn, R.C.: Chaos and Nonlinear Dynamics: An introduction for Scientists and Engineers. Oxford University Press, Oxford (2000)

Chapter 11
Conjugacy of Maps

The notion of topological conjugacy is very important in connecting the dynamics of different maps. It relates the properties among maps through some conjugacy. In other way, conjugacy is a change of variables that transforms one map into another. The variable changes from one system to another, that is, transition mappings are invertible and continuous, but not necessarily affine (a combination of linear transformation and translation). Two maps are said to be conjugate if they are equivalent to each other and their dynamics are similar. Conjugacy is an equivalence relation among maps. In conjugacy relation, the transformation should be a homeomorphism, so that some topological structures are preserved. Naturally, it is a useful and also a wise trick to find conjugacy between a map and an easier map. Besides conjugacy, the concept of semi-conjugacy is also useful in many contexts with some limitations. It is not always possible to establish conjugacy among maps but may be possible to find semi-conjugacy relation among them. Semi-conjugacy is a less stringent relationship among maps than a conjugacy relationship. In this chapter some definitions and important theorems on conjugacy and semi-conjugacy relations have been given and discussed elaborately with examples.

11.1 Conjugacy

In this section we illustrate the topological semi-conjugacy and conjugacy relations among maps. The conjugacy relations are basically homeomorphic mappings. Here also we discuss homeomorphism with examples.

11.1.1 Topological Semi-conjugacy

Let $f : X \to X$ and $g : Y \to Y$ be two maps. A map $h : Y \to X$ is said to be a topological semi-conjugacy map (or semi-conjugacy in short) if h satisfies the following properties:

© Springer India 2015

G.C. Layek, *An Introduction to Dynamical Systems and Chaos*,
DOI 10.1007/978-81-322-2556-0_11

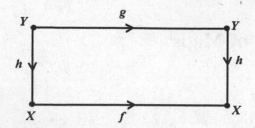

Fig. 11.1 Graphical representation of the composition (iii)

(i) h is continuous,
(ii) h is onto,
(iii) $f \circ h = h \circ g$.

The map g is then called topologically semi-conjugate to the map f through the map h or the map h is called a semi-conjugacy map from g to f. The graphical representation of the composition (iii) is shown in Fig. 11.1.

Consider the maps $f(x) = x : [0, \infty) \to [0, \infty)$ and $g(x) = -x : \mathbb{R} \to \mathbb{R}$. Define $h : \mathbb{R} \to [0, \infty)$ by $h(x) = |x|$. Then the composition relation $f \circ h = h \circ g$ is satisfied. We also see that h is continuous and onto. Hence f and g are semi-conjugate via h. Note that h is not one-one since $h(-1) = h(1)$.

11.1.2   Homeomorphism

Before introducing the concept of conjugacy map we shall familiar with homeomorphism. A function $f : X \to Y$ is said to be homeomorphism if it is continuous, invertible and has continuous inverse, that is, it is a bijective and bi-continuous map. But homeomorphic map may not be differentiable. For example, the map $f(x) = \exp(x)$ is homeomorphic from $\mathbb{R}$ to $\mathbb{R}^+$ but it is not homeomorphic from $\mathbb{R}$ to $\mathbb{R}$ because within this co-domain f is not onto. The map $h(x) = |x| : \mathbb{R} \to [0, \infty)$ is not a homeomorphism, since it is not one-one. The function $f : \mathbb{R} \setminus \{0\} \to \mathbb{R}^+$ defined by $f(x) = 1/x^2$ is continuous in $\mathbb{R} \setminus \{0\}$ but it is not one-one, since $f(-1) = f(1)$. Hence it is not homeomorphic from $\mathbb{R} \setminus \{0\}$ to $\mathbb{R}^+$. Note that the composition of two homeomorphisms is homeomorphism. Also the inverse of homeomorphism is homeomorphism.

11.1.3   Topological Conjugacy

Let $f : X \to X$ and $g : Y \to Y$ be two maps. A map $h : Y \to X$ is said to be a topological conjugacy map (or conjugacy in short) if it satisfies the following properties:

(i) h is a homeomorphism,

(ii) $f \circ h = h \circ g$.

Then the map g is called topologically conjugate to the map f through the map h or the map h is called a conjugacy map from g to f.

In conjugacy, we must find a homeomorphic map h between f and g. But there is no systematic procedure for finding such homeomorphic map. By using conjugacy relation and properties of homeomorphism one can find homeomorphic map. For example, consider the maps $f : \mathbb{R} \to \mathbb{R}$ and $g : \mathbb{R} \to \mathbb{R}$ where $f(x) = 8x$ and $g(x) = 2x$. We should seek conjugacy map for f and g. Let $h : \mathbb{R} \to \mathbb{R}$ be the conjugacy map. Then, the composition relation $f \circ h = h \circ g$ holds. This implies $h(g(x)) = f(h(x))$, that is, $h(2x) = 8h(x)$. The function $h(x) = x^{\log_2 8} = x^3$ satisfies the relation. The map $h(x) = x^3$ is homeomorphic from $\mathbb{R}$ to $\mathbb{R}$. Let us consider another example $f, g : [0, \infty) \to [0, \infty)$ where $f(x) = x/2$ and $g(x) = x/4$. These two maps are conjugate via $h(x) = \sqrt{x} : [0, \infty) \to [0, \infty)$. The homeomorphic mapping $h(x) = \sqrt{x}$ is not differentiable at $x = 0$.

In this connection we now define C^k-conjugacy. The C^k-conjugate map can be defined under the C^r-diffeomorphism. A function $f : X \to Y$ is said to be C^r-diffeomorphism $(r \geq 1)$ if it is a homeomorphism and f and f^{-1} are both C^r-functions. For example, the function $f : \mathbb{R} \to \mathbb{R}, f(x) = x^3$ is a homeomorphism but not a diffeomorphism, because $f^{-1}(x) = x^{1/3}$ exists but not differentiable at $x = 0$. The function $f : \left(-\frac{\pi}{2}, \frac{\pi}{2}\right) \to \mathbb{R}, f(x) = \tan x$ is C^r-diffeomorphism.

A function $g : Y \to Y$ is said to be C^k-conjugate $(k \leq r)$ to a function $f : X \to X$ if there exists a C^r-homeomorphism $h : Y \to X$ such that $f \circ h = h \circ g$. Note that for $k = 0$ the functions f and g are then topologically conjugate.

11.2 Properties of Conjugacy/Semi-conjugacy Relations

Suppose a map $g : Y \to Y$ is topologically conjugate to a map $f : X \to X$. Then there exists a homeomorphism $h : Y \to X$, such that $f \circ h = h \circ g$. Applying h^{-1} on the left we see that $g = h^{-1} \circ f \circ h$. This yields for all $n \in \mathbb{N}$,

$$
\begin{aligned}
g^n &= g \circ g \circ \cdots \circ g \, (n \text{ times}) \\
&= \left(h^{-1} \circ f \circ h\right) \circ \left(h^{-1} \circ f \circ h\right) \circ \cdots \circ \left(h^{-1} \circ f \circ h\right) \\
&= h^{-1} \circ f \circ \left(h \circ h^{-1}\right) \circ f \circ \left(h \circ h^{-1}\right) \circ f \circ \cdots \circ f \circ \left(h \circ h^{-1}\right) \circ f \circ h \\
&= h^{-1} \circ f^n \circ h
\end{aligned}
$$

Thus $g^n = h^{-1} \circ f^n \circ h$ for all $n \in \mathbb{N}$. This implies that the dynamics of two conjugate maps are qualitatively equivalent. In other words, if the dynamics of one map is known, then the dynamics of the other can be obtained through the conjugacy map. Using conjugacy relation the following properties associated with the dynamics can be proved easily.

Using conjugacy relation the following properties associated with the dynamics can be proved easily:

(i)   the orbits under g are similar to the orbits under f via h,
(ii) periodic points of g are mapped to periodic points of f via h with the same prime period,
(iii) dense orbits of g to dense orbits of f,
(iv) asymptotic behavior of the orbits of g to asymptotic behavior of f.

Also, the topological transitive property is preserved under conjugacy. Moreover, the sensitivity property is retained under conjugacy with some conditions. We shall now prove few theorems for conjugacy and semi-conjugacy maps.

Theorem 11.1 *Suppose* $g : Y \to Y$ *is topologically conjugate to* $f : X \to X$. *Then* $g^n : Y \to Y$ *is topologically conjugate to* $f^n : X \to X$ *for all* $n \in \mathbb{N}$.

Proof Since $g : Y \to Y$ is topologically conjugate to $f : X \to X$, there exists a homeomorphism $h : Y \to X$, such that $f \circ h = h \circ g$, that is, $h^{-1} \circ f \circ h = g$. Therefore for all $n \in \mathbb{N}$, we have

$$g^n = g \circ g \circ \cdots \circ g \,(n \text{ times})$$
$$= \left(h^{-1} \circ f \circ h\right) \circ \left(h^{-1} \circ f \circ h\right) \circ \cdots \circ \left(h^{-1} \circ f \circ h\right)$$
$$= h^{-1} \circ f \circ \left(h \circ h^{-1}\right) \circ f \circ \left(h \circ h^{-1}\right) \circ f \circ \cdots \circ f \circ \left(h \circ h^{-1}\right) \circ f \circ h$$
$$= h^{-1} \circ f^n \circ h$$
$$\Rightarrow f^n \circ h = h \circ g^n.$$

Again for all $n \in \mathbb{N}$, $f : X \to X$ and $g : Y \to Y$ imply $f^n : X \to X$ and $g^n : Y \to Y$, respectively. Thus for two maps $f^n : X \to X$ and $g^n : Y \to Y$ there exists a homeomorphism $h : Y \to X$, such that $f^n \circ h = h \circ g^n$. This shows that the map $g^n : Y \to Y$ is topologically conjugate to the map $f^n : X \to X$ through the conjugacy map $h : Y \to X$. This completes the proof.

Theorem 11.2 *Let* $f : X \to X$ *and* $g : Y \to Y$ *be two maps. Then*

(i)   *f and g are topologically conjugate to themselves.*
(ii) *If $g : Y \to Y$ is conjugate to $f : X \to X$ through $h : Y \to X$, then $f : X \to X$ is conjugate to $g : Y \to Y$ through $h^{-1} : X \to Y$.*
(iii) *If $g : Y \to Y$ is conjugate to $f : X \to X$ through $h : Y \to X$ and if $p : Z \to Z$ is conjugate to $g : Y \to Y$ through $k : Z \to Y$, then $p : Z \to Z$ is conjugate to $f : X \to X$ through the map $h \circ k : Z \to X$.*

Proof

(i) We now prove that the map $f : X \to X$ is topologically conjugate to itself. We can establish a relation easily.

$$f \circ I_X = I_X \circ f = f$$

where $I_X : X \to X$ is the identity map. Since the map I_X is a homeomorphism, the map f is topologically conjugate to itself. Similarly, we can prove that the map $g : Y \to Y$ is conjugate to itself through the identity map $I_Y : Y \to Y$.

(ii) Since $g : Y \to Y$ is conjugate to $f : X \to X$ through $h : Y \to X$, h is a homeomorphism and $f \circ h = h \circ g$. Since $h : Y \to X$ is a homeomorphism, $h^{-1} : X \to Y$ is also a homeomorphism. Again, $f \circ h = h \circ g \Rightarrow h^{-1} \circ f = g \circ h^{-1}$, that is, $g \circ h^{-1} = h^{-1} \circ f$. This shows that the map $f : X \to X$ is conjugate to $g : Y \to Y$ through the map $h^{-1} : X \to Y$.

(iii) Since $g : Y \to Y$ is conjugate to $f : X \to X$ through the map $h : Y \to X$ and $p : Z \to Z$ is conjugate to $g : Y \to Y$ through the map $k : Z \to Y$. The maps $h : Y \to X$ and $k : Z \to Y$ are both homeomorphisms. Also, $f \circ h = h \circ g$ and $g \circ k = k \circ p$. Since $h : Y \to X$ and $k : Z \to Y$ are homeomorphisms, so $h \circ k : Z \to X$ is a homeomorphism.

Now,

$$
\begin{aligned}
f \circ (h \circ k) &= (f \circ h) \circ k && [\text{Since } \circ \text{ is associative}] \\
&= (h \circ g) \circ k && [\because f \circ h = h \circ g] \\
&= h \circ (g \circ k) && [\text{Since } \circ \text{ is associative}] \\
&= h \circ (k \circ p) && [\because g \circ k = k \circ p] \\
&= (h \circ k) \circ p && [\text{Since } \circ \text{ is associative}]
\end{aligned}
$$

This shows that the map $p : Z \to Z$ is conjugate to the map $f : X \to X$ through the map $h \circ k : Z \to X$. Therefore, we have

(i) The functions f and g are conjugate to themselves (reflexive property).
(ii) g is conjugate to $f \Rightarrow f$ is conjugate to g (symmetric property).
(iii) g is conjugate to f and p is conjugate to $g \Rightarrow p$ is conjugate to f (transitive property).

So the conjugacy is an equivalence relation.

Theorem 11.3 *Suppose $g : Y \to Y$ is conjugate to $f : X \to X$ via $h : Y \to X$. Then h maps the orbits under g to the orbits under f.*

Proof Let $\{y_0, y_1, y_2, \ldots\}$ be the orbit of a point $y_0 \in Y$ under the map g where $y_{n+1} = g(y_n)$, $n = 0, 1, 2, 3, \ldots$. Since h is a homeomorphism, there exist points $x_n \in X$ such that $x_n = h(y_n)$, $n = 0, 1, 2, 3, \ldots$. To show that $\{x_0, x_1, x_2, \ldots\}$ forms an orbit under f it is sufficient to show that $x_{n+1} = f(x_n)$. Using the conjugacy relation, it follows that

$$x_{n+1} = h(y_{n+1}) = h(g(y_n)) = f(h(y_n)) = f(x_n)$$

This completes the proof.

Theorem 11.4 *Suppose $g : Y \to Y$ is conjugate to $f : X \to X$ via $h : Y \to X$. Then the followings hold:*

(i) *If x^* is a fixed point of g, then $h(x^*)$ is a fixed point of f.*
(ii) *If y is an eventually fixed point of g, then $h(y)$ is an eventually fixed point of f.*

Proof (i) Since x^* is a fixed point of g, we have $g(x^*) = x^*$. This implies $f(h(x^*)) = h(g(x^*)) = h(x^*)$. Hence $h(x^*)$ is a fixed point of f.

(ii) By the definition of an eventually fixed point y of g, there exists a positive integer k such that $g^{k+1}(y) = g^k(y)$. But $f^n \circ h = h \circ g^n$ for all $n \in \mathbb{N}$. Using this relation, we see that $f^{k+1}(h(y)) = h(g^{k+1}(y)) = h(g^k(y)) = f^k(h(y))$. Hence $h(y)$ is an eventually fixed point of f.

Theorem 11.5 *Suppose $g : Y \to Y$ is semi-conjugate to $f : X \to X$ via $h : Y \to X$. If $y \in Y$ is a periodic point of g with prime period n, then $h(y)$ is a periodic point of f with prime period m where $m|n$, that is, m divides n.*

Proof Suppose y is a periodic point of g with prime period n. Then $g^n(y) = y$. We now see that

$$h \circ g^n = (h \circ g) \circ g^{n-1} = (f \circ h) \circ g^{n-1}$$
$$= f \circ (h \circ g^{n-1}) = f \circ ((h \circ g) \circ g^{n-2}) = f \circ ((f \circ h) \circ g^{n-2})$$
$$= f^2 \circ (h \circ g^{n-2}) = \cdots = f^n \circ h.$$

Therefore, $f^n(h(y)) = h(g^n(y)) = h(y)$. This implies that $h(y)$ is a periodic point of f. Suppose that $h(y)$ is a periodic point of f with prime period m. Then by division algorithm there exist integers $q(>0)$ and r such that $n = mq + r$, where $0 \leq r < m$. Therefore

$$h(y) = f^n(h(y)) = f^{mq+r}(h(y)) = f^r \{f^{mq}(h(y))\} = f^r(h(y)).$$

This leads to a contradiction unless $r = 0$. Hence $n = mq$, that is, m divides n.

Theorem 11.6 *Suppose $g : Y \to Y$ is conjugate to $f : X \to X$ via $h : Y \to X$. Then $y \in Y$ is a periodic point of g with prime period n if and only if $h(y)$ is a periodic point of f with prime period n.*

Proof If y is a periodic point of g with prime period n, then $h(y)$ is a periodic point of f with prime period m, where $m|n$. Since h is a homeomorphism, operating with $h^{-1} : X \to Y$, we see that $h^{-1}(x)$ is a periodic point of g with prime period n, where $n|m$ such that there is a point $x \in X \mapsto y = h^{-1}(x) \in Y$. Combining the two results it follows that $m = n$. The converse part can be proved in the similar way.

Theorem 11.7 *Suppose $g : Y \to Y$ is conjugate to $f : X \to X$ through $h : Y \to X$. If y is an eventually periodic point of g, then $h(y)$ is an eventually periodic point of f.*

Proof For proof use Theorem 11.4(ii) and Theorem 11.6.

Theorem 11.8 *Let $h : Y \to X$ be a semi-conjugacy from a map $g : Y \to Y$ to another map $f : X \to X$. If $g : Y \to Y$ is topologically transitive on Y, then $f : X \to X$ is also topologically transitive on X.*

Proof Let us first define the topological transitivity property of a map. A map $f : X \to X$ is said to be topologically transitive on the set X if for any two open sets $U, V \subset X$, there exists an integer $k > 0$ such that $f^k(U) \cap V \neq \varphi$.

Let U_1 and U_2 be two open subsets of X. Since $h : Y \to X$ is onto, there exist open subsets V_1 and V_2 of Y such that h projects V_1 and V_2 to U_1 and U_2, respectively, that is, $h(V_1) = U_1$ and $h(V_2) = U_2$. Now, the semi-conjugacy property implies that

$$(h \circ g^k)(V_i) = (f^k \circ h)(V_i) = f^k(h(V_i)) = f^k(U_i), i = 1, 2.$$

Since $g : Y \to Y$ possesses transitivity property on Y, there exists an integer $k > 0$, such that $g^k(V_1) \cap V_2 \neq \phi$.

From above relations, for arbitrary $U_1, U_2 \subset X$, $\exists k > 0$ such that $f^k(U_1) \cap U_2 \neq \phi$.

This implies that the map f is topologically transitive on X. This completes the proof.

Theorem 11.9 *Suppose that $g : Y \to Y$ is topologically conjugate to $f : X \to X$, where both X and Y are compact spaces. Assume that g has sensitive dependence on Y. Then f has sensitive dependence on X.*

Proof A map $f : X \to X$ is said to be sensitive dependence on X if there exists a $\delta > 0$ such that for any $x \in X$ and any neighborhood $N_\varepsilon(x) = (x - \varepsilon, x + \varepsilon)$ of x, there exists $y \in N_\varepsilon(x)$ and $k > 0$ such that $\left| f^k(x) - f^k(y) \right| > \delta$. We now prove the theorem for two conjugate maps f and g where X and Y are compact spaces. Since $g : Y \to Y$ is topologically conjugate to the map $f : X \to X$, there exists a homeomorphism $h : Y \to X$ such that $f \circ h = h \circ g$. Now, $h : Y \to X$ is a homeomorphism implies $h^{-1} : X \to Y$ exists and both h and h^{-1} are continuous. Since X and Y are compact spaces, h and h^{-1} are uniformly continuous. Let $x_1, x_2 \in X$ and $y_1, y_2 \in Y$ such that $y_1 = h^{-1}(x_1)$ and $y_2 = h^{-1}(x_2)$. Then $x_1 = h(y_1)$ and $x_2 = h(y_2)$. Since h^{-1} is uniformly continuous, given any $r > 0$ there exists $\delta > 0$ (depending on r) such that for all $x_1, x_2 \in X$,

$$|x_1 - x_2| < \delta \Rightarrow \left| h^{-1}(x_1) - h^{-1}(x_2) \right| = |y_1 - y_2| < r.$$

Consequently, for every pair $y_1, y_2 \in Y$ and $x_1 = h(y_1)$, $x_2 = h(y_2) \in X$,

$$|y_1 - y_2| \geq r \Rightarrow |h(y_1) - h(y_2)| = |x_1 - x_2| \geq \delta \tag{11.1}$$

Let $x = h(y) \in X$, where $y \in Y$. Since $h : Y \to X$ is uniformly continuous, given any $\varepsilon > 0$, there exists $\varepsilon' > 0$ (depending on ε) such that $p = h(q) \in N_\varepsilon(x)$, the ε-nbd of x, whenever $q \in N_{\varepsilon'}(y)$, the ε'-nbd of y. That is,

$$|x - p| = |h(y) - h(q)| < \varepsilon \text{ whenever } |y - q| < \varepsilon' \qquad (11.2)$$

We now derive the sensitive dependence of f on X. Since g has sensitive dependence on initial conditions in Y, for $q \in \varepsilon'$-nbd of y and given $r > 0$, there exists $k > 0$ such that

$$\left| g^k(y) - g^k(q) \right| \geq r \text{ for } |y - q| < \varepsilon' \qquad (11.3)$$

Setting $g^k(y) = y_1$ and $g^k(q) = y_2$ and then using (11.1), we have

$$\begin{aligned}
\left| g^k(y) - g^k(q) \right| = |y_1 - y_2| \geq r &\Rightarrow |h(y_1) - h(y_2)| \geq \delta \\
&\Rightarrow \left| h(g^k(y)) - h(g^k(q)) \right| \geq \delta
\end{aligned} \qquad (11.4)$$

Now,

$$f \circ h = h \circ g \ \Rightarrow f^k \circ h = h \circ g^k \ \forall k \in \mathbb{N}.$$

So,

$$\begin{aligned}
h\big(g^k(y)\big) - h\big(g^k(q)\big) &= f^k(h(y)) - f^k(h(q)) \\
&= f^k(x) - f^k(p)
\end{aligned} \qquad (11.5)$$

Substituting (11.5) into (11.4), we get

$$\left| g^k(y) - g^k(q) \right| \geq r \Rightarrow \left| f^k(x) - f^k(p) \right| \geq \delta \qquad (11.6)$$

Combining (11.3) and (11.6), we see that

$$\left| f^k(x) - f^k(p) \right| \geq \delta \text{ for } |y - q| < \varepsilon' \qquad (11.7)$$

Finally, using (11.2) and (11.7), we can say that there exists $k > 0$ such that

$$\left| f^k(x) - f^k(p) \right| \geq \delta \text{ when } |x - p| < \varepsilon$$

This shows that the map f has sensitive dependence on X. This completes the proof.

Note that if X and Y are not compact, then the sensitivity property is not preserved under conjugacy. For example, consider the maps $f : (1, \infty) \to (1, \infty)$ and $g : (0, \infty) \to (0, \infty)$, where $f(x) = 2x$ and $g(x) = 1 + x$. The map f has sensitivity property on initial conditions but g does not exhibit the sensitivity property. The conjugacy relation yields $h(f(x)) = g(h(x)) \Rightarrow h(2x) = 1 + h(x)$. By inspection, we choose $h(x) = \log_2 x, x \in (1, \infty)$. Clearly, the map h is a homeomorphism from $(1, \infty)$ into $(0, \infty)$. Here X and Y are two open intervals and are not compact.

11.3   Conjugacy Between Tent and Logistic Maps

We now show the conjugacy relation between the two important maps, the tent map $T : [0, 1] \rightarrow [0, 1]$ and the logistic map $f(x) = 4x(1 - x), x \in [0, 1]$. The tent map $T : [0, 1] \rightarrow [0, 1]$ is expressed by

$$T(x) = \begin{cases} 2x, & x \leq 1/2 \\ 2(1 - x), & x \geq 1/2 \end{cases}$$

The absolute value of the slope of $T(x)$ is 2 at every point except the point of non-differentiability whereas the slope of logistic map varies from iteration to iteration. Two maps have an extreme point at $x = 1/2$, which maps to the point 1 and then to 0. Each has a fixed point at 0. The other fixed point of $T(x)$ is at $x = 2/3$ and the logistic map is at $x = 3/4$. The tent map has a single period-2 orbit $\{0.4, 0.8\}$ whereas the logistic map has a period-2orbit $\left\{(5 - \sqrt{5})/8, (5 + \sqrt{5})/8\right\}$. In each map, the period-2 orbit lies in the same relation to the other points. The left-hand point lies between the origin and the extreme point and the right-hand point lies between the other fixed point and the point 1. We now examine the stability of the period-2 orbits for both the maps. The derivative of $T(x)$ at the fixed point $x = 2/3$ and the derivative of $f(x)$ at its fixed point $x = 3/4$ are same, both are -2 and so the fixed points are unstable. The derivative of $T^2(x)$ at the two-cycle is $T'(0.4)T'(0.8) = -4$ and the derivative of $f^2(x)$ at its two-cycle is $f'\left(\frac{5-\sqrt{5}}{8}\right)f'\left(\frac{5+\sqrt{5}}{8}\right) = -4$. So, both are equal and unstable in nature. How far can we expect the similarity of two maps? Can it be possible to establish the similarity through a mapping? Yes, this is the conjugacy map that relates the qualitative properties between two maps. We now find the conjugacy map $h(x)$ for the tent and logistic maps. Consider the map $h : [0, 1] \rightarrow [0, 1]$ defined by

$$h(x) = (1 - \cos \pi x)/2 = \sin^2(\pi x/2).$$

Clearly, this is a one-to-one transformation of $[0, 1]$ onto itself. First we establish the relation $f \circ h = h \circ T$. For the case of logistic map $f(x), x \in [0, 1]$, we have

$$f \circ h = f(h(x)) = 4h(x)(1 - h(x)) = 4\left(\frac{1 - \cos \pi x}{2}\right)\left(1 - \frac{1 - \cos \pi x}{2}\right) = \sin^2 \pi x.$$

Again, for the tent map $(x \leq 1/2)$, we get

$$h(T(x)) = h(2x) = \frac{1 - \cos 2\pi x}{2} = \sin^2 \pi x$$

and for $x \geq 1/2$, we get

$$h(T(x)) = h(2(1-x)) = \frac{1 - \cos(2\pi - 2\pi x)}{2} = \frac{1 - \cos 2\pi x}{2} = \sin^2 \pi x.$$

Therefore, $f \circ h = h \circ T \Rightarrow f(h(x)) = h(T(x))$. In the context of dynamics the evolution of x values under the tent map $T(x)$ is equivalent to the evolution of h (x) under the logistic map $f(x)$. We now show that h is a homeomorphic mapping from [0, 1] to [0, 1]. We have $h(x) = \frac{1 - \cos \pi x}{2} = (p \circ q)(x)$, where $p(x) = \frac{1 - \cos x}{2}$ and q $(x) = \pi x$. Since h is the composition of two continuous maps, it is also continuous. Again, since $h(0) = 0$ and $h(1) = 1$, by continuity and Mean Value Theorem it assumes all values in [0, 1]. So, h is onto. Again, since $h'(x) = \pi \sin\left(\frac{\pi x}{2}\right) \cos\left(\frac{\pi x}{2}\right) > 0$ for $0 < x < 1$, h is monotone and so injective. Hence h is invertible and its inverse is also continuous. Thus h is a homeomorphism from [0, 1] to [0, 1].

This shows that the tent map $T(x), 0 \leq x \leq 1$ is conjugate to the logistic map for $r = 4$ through the conjugacy map $h(x) = \frac{1 - \cos \pi x}{2}, x \in [0, 1]$. We now relate the second composition between the tent map and the logistic map as follows:

$$h \circ T^2 = h(T^2(x)) = h(T(T(x))) = f(h(T(x))) = f(f(h(x))) = f^2(h(x)).$$

This indicates that the evolution of x under second iterations of the tent map $T(x)$ is equivalent to the evolution of $h(x)$ under second iterations of the logistic map. Inductively, we get $h(T^n(x)) = f^n(h(x))$. It is interesting to note that the dynamics of the tent map which is linear but not differentiable at $x = 1/2$ is similar to the logistic map via the conjugacy map $h(x)$. See Devaney [1], Davies [2], Wiggins [3].

Example 11.1 Show that the map $T : [-1, 1] \rightarrow [-1, 1]$ defined by

$$T(x) = \begin{cases} 1 + 2x, & -1 \leq x \leq 0 \\ -1 \leq x \leq 0 & 0 \leq x \leq 1 \end{cases}$$

is topologically conjugate to the map $Q_2(x) = 1 - 2x^2, \ -1 \leq x \leq 1$.

Solution Consider the map $h : [-1, 1] \rightarrow [-1, 1]$ defined by

$$h(x) = \sin\left(\frac{\pi x}{2}\right), x \in [-1, 1]$$

We first show that $Q_2 \circ h = h \circ T$.
For all $x \in [-1, 1]$, we have

$$Q_2(h(x)) = 1 - 2(h(x))^2 = 1 - 2\sin^2\left(\frac{\pi x}{2}\right) = \cos(\pi x).$$

For $-1 \leq x \leq 0$,

$$h(T(x)) = h(1 + 2x) = \sin\left(\frac{\pi}{2}(1 + 2x)\right) = \cos(\pi x).$$

For $0 \leq x \leq 1$,

$$h(T(x)) = h(1 - 2x) = \sin\left(\frac{\pi}{2}(1 - 2x)\right) = \cos(\pi x).$$

Therefore, $Q_2 \circ h = h \circ T$. The map $h : [-1, 1] \to [-1, 1]$ is bijective and bi-continuous. Hence it is a homeomorphism. Therefore, $T(x)$ is topologically conjugate to $Q_2(x)$ via $h(x) = \sin\left(\frac{\pi x}{2}\right)$, $x \in [-1, 1]$.

Example 11.2 Show that the tent map $T : [0, 1] \to [0, 1]$ is topologically conjugate to the map $Q_2(x) = 1 - 2x^2, x \in [-1, 1]$

Solution The tent map $T : [0, 1] \to [0, 1]$ is defined already.

Consider the map $h : [0, 1] \to [-1, 1]$ defined by $h(x) = -\cos(\pi x)$. We now show that $Q_2 \circ h = h \circ T$. For all $x \in [0, 1]$, we have

$$Q_2(h(x)) = 1 - 2(h(x))^2 = 1 - 2\cos^2(\pi x) = -\cos(2\pi x).$$

For $0 \leq x \leq \frac{1}{2}$,

$$h(T(x)) = h(2x) = -\cos(2\pi x).$$

For $\frac{1}{2} \leq x \leq 1$,

$$h(T(x)) = h(2 - 2x) = -\cos(\pi(2 - 2x)) = -\cos(2\pi x).$$

Therefore, $Q_2 \circ h = h \circ T$. Here the map h is continuous and has a continuous inverse. So, it is a homeomorphism. Therefore, the tent map $T(x)$ is topologically conjugate to the map $Q_2(x)$ through the map $h(x) = -\cos(\pi x), x \in [0, 1]$.

Example 11.3 If x is a periodic k point of the tent map $T : [0, 1] \to [0, 1]$, prove that $C(x)$ is also a periodic k point of the map $G(x) = 4x(1 - x)$, $x \in [0, 1]$, where C is the conjugacy map from T to G.

Solution Since x is a periodic k point of the tent map T, k is the least positive integer such that $T^k(x) = x$. Again, since T is topologically conjugate to G through the conjugacy map C, so $G \circ C = C \circ T$. Therefore,

$$G^2 \circ C = (G \circ G) \circ C = G \circ (G \circ C) = G \circ (C \circ T) = (G \circ C) \circ T = (C \circ T) \circ T = C \circ T^2$$
$$\Rightarrow G^2 \circ C = C \circ T^2.$$

Similarly,

$$G^3 \circ C = C \circ T^3, G^4 \circ C = C \circ T^4, \ldots, G^k \circ C = C \circ T^k.$$

Therefore, $G^k(C(x)) = C(T^k(x)) = C(x)$. Thus, k is the least positive integer such that $G^k(C(x)) = C(x)$. This shows that $C(x)$ is a periodic k point of the map G.

Example 11.4 Calculate a piece-wise linear map on the interval $[0, 1]$ which is topologically conjugate to the map $g(x) = 4x^3 - 3x$, $-1 \leq x \leq 1$.

Solution Let the piece-wise linear map $f : [0, 1] \to [0, 1]$ be defined as

$$f(x) = \begin{cases} 3x, & 0 \leq x \leq \frac{1}{3} \\ 2 - 3x, & \frac{1}{3} \leq x \leq \frac{2}{3} \\ 3x - 2, & \frac{2}{3} \leq x \leq 1 \end{cases}$$

Now, consider the map $h : [0, 1] \to [-1, 1]$ defined by

$$h(x) = \cos(\pi x), \, x \in [0, 1]$$

Obviously, the map h is one-one, onto, and h^{-1} is continuous. So, h is a homeomorphism.

Now, for all $x \in [0, 1]$, we have

$$g(h(x)) = g(\cos \pi x) = 4\cos^3(\pi x) - 3\cos(\pi x) = \cos(3\pi x).$$

For $0 \leq x \leq \frac{1}{3}$,

$$h(f(x)) = h(3x) = \cos(3\pi x).$$

For $\frac{1}{3} \leq x \leq \frac{2}{3}$,

$$h(f(x)) = h(2 - 3x) = \cos(2\pi - 3\pi x) = \cos(3\pi x).$$

For $\frac{2}{3} \leq x \leq 1$,

$$h(f(x)) = h(3x - 2) = \cos(3\pi x - 2\pi) = \cos(3\pi x).$$

Therefore, $g \circ h = h \circ f$. This shows that the piece-wise linear map f on $[0, 1]$ is topologically conjugate to the map $g(x) = 4x^3 - 3x$, $-1 \leq x \leq 1$ through the map $h(x) = \cos(\pi x), x \in [0, 1]$.

Example 11.5 Find the relation between c and $r(> 0)$ such that the logistic map $f(x) = rx(1 - x)$ is topologically conjugate to the quadratic map $g(x) = x^2 + c$. Also find the conjugacy map.

Solution We try to find a linear conjugacy map of the form $h(x) = \alpha x + \beta$, where α and β are certain constants to be determined. Since f is conjugate to g through the map h, we have

$$h \circ f = g \circ h$$

Now,

$$h \circ f = g \circ h \Leftrightarrow h(f(x)) = g(h(x)) \forall x$$
$$\Leftrightarrow \alpha(rx(1-x)) + \beta = (\alpha x + \beta)^2 + c$$
$$\Leftrightarrow -r\alpha x^2 + r\alpha x + \beta = \alpha^2 x^2 + 2\alpha\beta x + \beta^2 + c$$
$$\Leftrightarrow -r = \alpha, \ r = 2\beta, \ c + \beta^2 = \beta$$

This shows that the map $f(x) = rx(1-x)$ is topologically conjugate to the map $g(x) = x^2 + c$, where $c = \frac{r}{2} - \left(\frac{r}{2}\right)^2 = \frac{1}{4}r(2-r)$. Here the conjugacy map is given by

$$h(x) = -rx + r/2.$$

Example 11.6 Show that the shift map $f(x) = 2x \pmod 1$, $x \in [0,1]$ and the logistic map $g(x) = 4x(1-x)$, $x \in [0,1]$ are topologically semi-conjugate.

Solution Consider the map $h : [0,1] \to [0,1]$ defined by

$$h(x) = \frac{1 - \cos(2\pi x)}{2} = \sin^2(\pi x)$$

We now show that $g \circ h = h \circ f$
For all $x \in [0,1]$, we have

$$g(h(x)) = 4h(x)(1 - h(x))$$
$$= 4 \cdot \frac{1 - \cos(2\pi x)}{2}\left(1 - \frac{1 - \cos(2\pi x)}{2}\right)$$
$$= 1 - \cos^2(2\pi x) = \sin^2(2\pi x)$$

We now calculate $h \circ f$. It should be noted that $\cos(2\pi(n+x)) = \cos(2\pi x) \forall n \in \mathbb{N}$.

So, $\cos(2\pi f(x)) = \cos(2\pi(2x))$.

Therefore, $h(f(x)) = \frac{1 - \cos(2\pi f(x))}{2} = \frac{1 - \cos(2\pi(2x))}{2} = \sin^2(2\pi x)$. This shows that $g \circ h = h \circ f$.

Now, $h(x) = \frac{1 - \cos(2\pi x)}{2} = (p \circ q)(x)$, where $p(x) = \cos(2\pi x)$ and $q(x) = \frac{1-x}{2}$. Since p maps $[0, 1]$ onto $[-1, 1]$ and q is one-one from $[-1, 1]$ to $[0, 1]$, the map h is onto. Moreover, since $p(x)$ and $q(x)$ are continuous, $h(x)$ is also continuous. Hence h is a topological semi-conjugacy from f to g.

11.4   Exercises

1. Explain the concept of topological conjugacy and semi-conjugacy between two maps. Write down the importance of conjugacy relation in dynamical evolution of a map.
2. Prove that the conjugacy is an equivalence relation. Is it true for semi-conjugacy? Give example in support of your answer.
3. Define homeomorphic map. Show that the composition of two homeomorphic maps is also homeomorphic.
4. Which of the following functions are homeomorphisms in their domains of definition? Justify your answers.
    (a) $f(x) = \sin x$
    (b) $f(x) = 1/x$
    (c) $f(x) = e^x$
    (d) $f(x) = ax + b, a(\neq 0), b \in \mathbb{R}$
    (e) $f(x) = x^3 + 3x^2 + 4x + 1$
    (f) $f(x) = x^{1/5}$
    (g) $f(x) = x^{4/5}$
5. Show that homeomorphic map on the real line cannot have eventually periodic points.
6. Let $g :$ unit circle $S \to S$ be given by $g(\theta) = \theta + \alpha + \beta \sin(\theta)$, α, β are constants. Show that g is a homeomorphism on S if $|\beta| < 1$.
7. Show that a homeomorphism $f : \mathbb{R} \to \mathbb{R}$ has no periodic points of period greater than 2.
8. Show that for each positive integer k there exists a homeomorphism $f : \mathbb{R}^2 \to \mathbb{R}^2$ with a period-k point.
9. Find the conjugacy map $h : \mathbb{R} \to \mathbb{R}^+$ for the maps $f : \mathbb{R}^+ \to \mathbb{R}^+$ and $g : \mathbb{R} \to \mathbb{R}$ where $f(x) = 1/x$ and $g(x) = -x$.
10. (a) Show that the maps $f, g : \mathbb{R} \to \mathbb{R}$ with $f(x) = 4x$ and $g(x) = 2x$ are not conjugate but they are conjugate if their domain and co-domain are restricted as $\mathbb{R}^+$. Also find the conjugacy map.
    (b) Examine the conjugacy between the maps $f : (1, \infty) \to (1, \infty)$ $f(x) = 2x$ and $g : (0, \infty) \to (0, \infty)$, $g(x) = 2 + x$.
11. Prove that the maps $f, g : \mathbb{R} \to \mathbb{R}$ where $f(x) = 27x$ and $g(x) = 3x$ are topologically conjugate.
12. For two topologically conjugate maps $f : X \to X$ and $g : Y \to Y$ prove that
    (a) f is one-to-one if and only if g is one-to-one.
    (b) f is onto if and only if g is onto.
13. Under what condition the maps $f : \mathbb{R} \to \mathbb{R}$ and $g : \mathbb{R} \to \mathbb{R}$ defined by $f(x) = ax$ and $g(x) = bx$, where a and b are positive real numbers, are topologically conjugate? Also find the homeomorphism.

14. Define $f:(1,\infty)\to(1,\infty)$ and $g:(0,\infty)\to(0,\infty)$ by $f(x)=3x$ and $g(x)=2+x$. Find the conjugacy map between f and g, if exists.

15. Let $h:Y\to X$ be a conjugacy from $g:Y\to Y$ to $f:X\to X$. If $\overset{*}{x}$ is a fixed point of g, then show that $h\left(x^{*}\right)$ is a fixed point of f.

16. If f and g are conjugate via h and x_0 is a topologically stable fixed point of f, then show that $h(x_0)$ is a topologically stable fixed point of g. [A point x_0 is said to be a topologically stable fixed point of f when there exists an interval I containing x_0 such that every orbit under f which starts in I converges to x_0.]

17. Show that a conjugacy relation between two maps sends orbits of one map into the orbits of the other map. Also show that it sends periodic orbits to periodic orbits of the same period.

18. Let a map $g:Y\to Y$ be topologically conjugate to a map $f:X\to X$ through $h:Y\to X$. If x is a period-k point for g, show that $h(x)$ is a period-k point of f. Also, prove that $\left(f^{k}\right)'(h(x))=\left(g^{k}\right)'(x)$ provided h' is non-zero on the periodic orbits of g.

19. Show by example that conjugacy map does not preserve differentiability.

20. Suppose $g:Y\to Y$ is semi-conjugate to the map $f:X\to X$ through $h:Y\to X$. If $y\in Y$ is a periodic point of g with prime period n, then find the period of the periodic point $h(y)$ of the map f.

21. Show that the stability characters of a periodic orbit of two topologically conjugate maps are identical. Is it true for semi-conjugate maps? Justify.

22. Show that for a general quadratic map $f(x)=ax^2+bx+c$, $a\neq 0$ is conjugate to the quadratic map $g(x)=x^2+d$ through a linear conjugacy map.

23. Show that the tent map $T:[0,1]\to[0,1]$ is topologically conjugate to the map $g(x)=2x^2-1$, $x\in[-1,1]$.

24. Let f and g be two C^k-conjugate maps. Show that the orbits of f are same to the orbits of g under a C^k-diffeomorphism.

25. For C^k-conjugate ($k\geq 1$) of two maps f and g show that the eigenvalues of $Dg(x_0)$ are equal to the eigenvalues of $Df(h(x_0))$ under the diffeomorphism h, x_0 being the fixed point of the map g.

References

1. Devancy, R.L.: An Introduction to Chaotic Dynamical Systems. Westview Press (2003)
2. Davies, B.: Exploring Chaos. Westview Press (2004)
3. Wiggins, S.: Introduction to Applied Nonlinear Dynamical Systems and Chaos, 2nd edn. Springer, New York (2003)

Chapter 12
Chaos

In the development of science in the twentieth century, philosophers and scientists convinced that there could be a motion even for a simple system which is erratic in nature, not simply periodic or quasiperiodic. Moreover, the behaviors of the motion may be unpredictable and therefore long-range prediction is impossible. The science of unpredictability has immense interest. The debate on the cause of unpredictability is continuing over centuries. The great physicist *Albert Einstein* wrote a letter to *Max Born* regarding unpredictability in the cosmos. He wrote: "You believe in the God who plays dice, and I in complete law and order." In fact, nothing in the universe behaves in a way that is predictable totally forever. The perceptions of the infinite, infimum, and their connections with finite are a matter of great concern in science and philosophy. Even there is a order in unpredictable motions. But how order and chaos coexist. What are the laws underlying in chaotic motion? With the advancement of science and computing power it is believed that a simple deterministic system can have very complicated dynamics which is inherently present in the system itself. On the other hand, for an infinitesimal change in system's initial setup the dynamics as a whole may completely change. While studying the unpredictable behaviors of a system, the American mathematician *James Alan Yorke* had introduced the term 'Chaos' for random looking dynamics of simple deterministic systems. In Greek mythology, 'Chaos' is defined as an infinite formless structure. However, the precise definition of chaos either literally or mathematically is lacking behind till date because of its multi-length scales motions with formless structures at infinitum. The appearance of chaotic motion has no definite routes. In this book we shall present a basic understanding of what chaos is and its mathematical theory under some assumptions. In mathematical framework chaos is a phenomenon exhibiting "sensitive dependence on infinitesimally different initial set-ups" and topologically "mixing". The chaotic orbits are generally aperiodic and named as "strange" by Rulle and Takens in 1971 that have fractional dimensions, a new discovery in the twentieth century's nonlinear science. There are connections with chaos and fractal objects. Nonlinearity and dimensionality (≥ 3) are the key requirements for chaos in continuous systems while in discrete systems, even a one-dimensional linear system may exhibit chaotic motion provided the system has lack of differentiability. Chaotic phenomena abound in Nature and in manmade devices.

© Springer India 2015
G.C. Layek, *An Introduction to Dynamical Systems and Chaos*,
DOI 10.1007/978-81-322-2556-0_12

The perception of unpredictable behavior in deterministic systems had been conceived and reported in the works of the French mathematician *Henri Poincaré*. George David Birkhoff, A.M. Lyapunov, M.L. Cartwright, J.E. Littlewood, Andrey Kolmogorov, Stephen Smale, and coresearchers were noteworthy workers at the early stage of developments in chaos theory, particularly for mathematical foundation of chaos. The chaotic motion in atmospheric flows was first described by the American meteorologist, *Edward Lorenz* in the year 1963 from numerical experiments on convective patterns for a very simplified model. He found that the solutions never settled down to fixed points or periodic orbits of this simple system. Trajectories oscillate in an irregular, nonperiodic pattern with completely different behaviors for infinitesimal small change of initial conditions. The solution structure when plotted in three-dimensional Eucledian space resembles as a surface of two wings of a butterfly. Lorenz pointed out that the solution set contained an infinite number of sheets, known as strange attractor with fractional dimension. The sensitive dependence of dynamical evolution for infinitesimal change of initial conditions is called the butterfly effect. In 1972, Lorenz talked on the butterfly effect in the American Association for the Scientific Progress and questioned: "Does the flap of a butterfly's wings in Brazil set off a tornado in Texas?" He claimed that it was difficult to predict the long-range climate conditions correctly. The unpredictability is inherently present in the atmospheric flow itself. Therefore, the long-range prediction would be uncertain.

Quantifying chaos is a central issue for understanding chaotic phenomena. Experimental evidence and theoretical studies predict some qualitative and quantitative measures for quantifying chaos. In this chapter we discuss some measures such as universal sequence (U-sequence), Lyapunov exponent, renormalization group theory, invariant measure, etc., for quantifying chaotic motions. On the other hand, there are some universal numbers applicable for particular class of systems, for example, the Feigenbaum number, Golden mean, etc. The universality is an important feature in chaotic dynamics.

Chaotic nonlinear dynamics is a rapidly expanding field and has now been proved to have potential applications in many manmade devices, social sciences, chemical and biological processes, and computer science. The unexpected fluctuations in sudden occurrence of diseases may be explained with the help of chaos theory. Chaos theory is much helpful in designing true economic and monetary modeling in resource distribution, financial and policy-making decisions. Chaos synchronization theory has been used nowadays for sending secret messages and also in other areas.

12.1   Mathematical Theory of Chaos

Chaos is ubiquitous. Chaotic motions are unpredictable. Philosophers and scientists are trying to understand logically how unpredictability occurs. How it can be expressed in mathematical setup. The unpredictability in chaos and its mathematical foundation are still not well established. The simple looking phenomenon such as the smoke column rising in still air from a cigarette, the oscillations and their patterns in the smoke column are so complicated to defy understanding. Similarly, the weather forecasting and the world stock market prices are the systems that fluctuate with time in a random, irregular ways that the long-term predictions do not often match with reality. Chaos is a deterministically unpredictable phenomenon. In the evolution of chaotic orbit there are trajectories which do not settle down to fixed points or periodic orbits or quasiperiodic orbits as time tends to infinity. Even a deterministic system has no random or noisy inputs; an irregular behavior may appear due to presence of nonlinearity, dimensionality, or nondifferentiability of the system. Although the time evolution obeys strict deterministic laws, the system seems to behave according to its own free will. The mathematical definition of chaos introduces two notions, viz., the topological transitive property implying the mixing and the metrical property measuring the distance. Chaotic orbit may be expressed by fractals. Before defining chaos under the mathematical framework we discuss some preliminary concepts and definitions of topological and metric spaces which are essential for chaos theory.

Let X be a nonempty set and $\tau \subseteq P(X)$, the power set of X. Then τ is said to form a topology on X if

(i)   the null set φ and the whole set X both belong to τ,
(ii) union of any collection of subsets of τ belongs to τ, and
(iii) intersection of finite collection of subsets of τ belongs to τ.

If τ is a topology on X, then the couple (X, τ) is called a topological space. The subsets of τ are called open sets. Some examples of topological spaces are given below:

(a) Let X be a nonempty set and $\tau = P(X)$. Then (X, τ) forms a topological space. In this space, the topology τ is called a discrete topology on X.
(b) Let X be a nonempty set and $\tau = \{\varphi, X\}$. Then (X, τ) forms a topological space. In this space, the topology τ is called a trivial topology or an indiscrete topology on X.
(c) Let $X = \{a, b, c, d\}$. Take $\tau_1 = \{\varphi, X, \{a\}, \{b, c\}, \{a, b, c\}\}$ and $\tau_2 = \{\varphi, X, \{a\}, \{b, c\}\}$. Then (X, τ_1) forms a topological space, since all the axioms of a topological space are satisfied. But (X, τ_2) does not form a topological space, because $\{a\}, \{b, c\} \in \tau_2$ but their union $\{a\} \cup \{b, c\} = \{a, b, c\} \notin \tau_2$.

In a topological space (X, τ), a subset A of X is said to be a neighborhood of a point $p \in X$ if there exists an open set G such that $p \in G \subset A$. A subset A in (X, τ) is a neighborhood of each point $x \in X$ if and only if A is an open set. Complement of an open set is a closed set. For every subset A in a topological space there always exists a smallest closed set, containing A, which is the intersection of all closed sets that contain A. This smallest closed set is known as the closure of A and it is denoted by $\bar{A}$. In other words, closure of a set A is the intersection of all closed supersets of A. For example, in the topological space (X, τ_1), as given above, the closed subsets of X are X, φ, $\{b, c, d\}$, $\{a, d\}$, and $\{d\}$. Therefore,

$$\overline{\{a\}} = \{a,d\} \cap X = \{a,d\}, \ \overline{\{d\}} = \{d\} \cap \{a,d\} \cap \{b,c,d\} \cap X = \{d\}, \text{ and}$$
$$\overline{\{a,b\}} = X.$$

A topology τ on X is said to be finer than another topology τ' on the same set X if $\tau' \subset \tau$. The topology τ' is then called weaker or coarser than τ. A point $p \in X$ is said to be an accumulation point or a limiting point of a subset A of X if every neighborhood of p intersects A at least at one point other than p. The set formed by the accumulation points of a set A is called the derived set of A and it is denoted by A' (or $D(A)$). For every subset A in a topological space, $\bar{A} = A \cup A'$. A topological space (X, τ) is said to be connected if and only if A cannot be expressed as the union of its two nonempty disjoint open subsets. If not, then it is called disconnected. In a topological space (X, τ), let S be a collection of subsets of X. Then S is said to be a cover of a subset $A \subseteq X$ if the union of the sets in S contains A. Moreover, if each set in S is open, then S is called an open cover of A. If S covers A and S' is a subcollection of S that also covers A, then S' is called a subcover of A. A topological space (X, τ) is said to be a separated space if any two distinct points in X always possess two disjoint neighborhoods. A separated space is said to be compact if its every open cover has a finite subcover. Let (X, τ) and (Y, τ') be two topological spaces. A function $f : X \rightarrow Y$ is said to be continuous if for every open set A in Y, $f^{-1}(A)$ is open in X. For details on topological spaces, see the books Simmons [1] and Munkres [2].

A metric space (X, d) contains a nonempty set X and a distance function $d :$ $X \times X \rightarrow \mathbb{R}^+ \cup \{0\}$ ($\mathbb{R}^+$ is the set of all positive real numbers) such that for all $x, y, z \in X$, the following properties hold

(i) $d(x,y) = d(y,x)$, (symmetry)
(ii) $d(x,y) = 0 \Leftrightarrow x = y$, (identity)
(iii) $d(x,y) \le d(x,z) + d(z,y)$, (triangle inequality).

Some examples of metric spaces are listed below:

(a) Let X be a nonempty set and $d : X \times X \rightarrow \mathbb{R}$ be defined as $d(x,y) =$
$$\begin{cases} 0 \text{ if } x = y \\ 1 \text{ if } x \neq y \end{cases} \text{ for } x, y \in X. \text{ Then } (X,d) \text{ is a metric space. This metric space is}$$
known as 'discrete metric space'.

(b) Let $X = \mathbb{R}^n$, the n-dimensional Euclidean space and for $\underset{\sim}{x}, \underset{\sim}{y} \in X$ define
$$d(\underset{\sim}{x}, \underset{\sim}{y}) = \sqrt{\sum_{i=1}^{n} (x_i - y_i)^2}, \qquad\qquad \text{where}$$
$\underset{\sim}{x} = (x_1, x_2, \ldots, x_n)$, $\underset{\sim}{y} = (y_1, y_2, \ldots, y_n)$. Then (X, d) is a metric space.

(c) Let $X = C[a, b]$, the set of all continuous functions over the closed and bounded interval $[a, b]$. For $f, g \in C[a, b]$ define $d(f, g) = \sup_{x \in [a,b]} |f(x) - g(x)|$.

Then $(C[a, b], d)$ forms a metric space.

In a metric space (X, d), an open ball (or an open sphere) with center at $a \in X$ and radius $r > 0$ is the set $B(a, r) = \{x \in X : d(x, a) < r\}$, and a closed ball (or a closed sphere) with center at $a \in X$ and radius $r > 0$ is the set $B[a, r] = \{x \in X : d(x, a) \leq r\}$. For example, in the discrete metric space (X, d), for any $a \in X$, we have

(i)   if $0 < r < 1$, then $B(a, r) = \{x \in X : d(x, a) < r\} = \{x \in X : d(x, a) = 0\} = \{a\}$ and $B[a, r] = \{x \in X : d(x, a) \leq r\} = \{x \in X : d(x, a) = 0\} = \{a\}$

(ii) if $r = 1$, then $B(a, r) = \{x \in X : d(x, a) < r\} = \{x \in X : d(x, a) < 1\} = \{a\}$ and $B[a, r] = \{x \in X : d(x, a) \leq r\} = \{x \in X : d(x, a) \leq 1\} = X$, and

(iii) if $r > 1$, then $B(a, r) = B[a, r] = X$.

Let (X, d) be a metric space. Then the open subsets of X form a topology on X. This topology is called a metric topology on X. A sequence of points $\{x_n\}$ in a metric space (X, d) is said to be convergent to a point $p \in X$ if for any $\varepsilon > 0$ there exists a positive integer N (depending on ε) such that $d(x_n, x) < \varepsilon$ whenever $n \geq N$. In a metric space (X, d), a sequence of points $\{x_n\}$ is said to be a Cauchy sequence if for any $\varepsilon > 0$, there exists a positive integer N (depending on ε) such that $d(x_m, x_n) < \varepsilon$, whenever $m > n \geq N$. Every convergent sequence in a metric space is a Cauchy sequence. But the converse is not true, in general. For example, consider the set $X = (0, 1]$, an interval in $\mathbb{R}$. Then, with the usual metric defined by $d(x, y) = |x - y|$, $x, y \in X$, it forms a metric space. In this space, the sequence $\{x_n\} = \{\frac{1}{n}\}$ of points of X is a Cauchy sequence but it does not converge to a point of X, since $\lim_{n \to \infty} x_n = 0 \notin X$. A metric space (X, d) in which every Cauchy sequence in X converges to a point in X is called a complete metric space. For example, the metric space $(\mathbb{R}, d)$, where d is the usual metric, is complete. Let (X, d) and (Y, d') be two metric spaces. A function $f : X \to Y$ is said to be continuous at a point $a \in X$ if for any $\varepsilon > 0$, there exists a $\delta > 0$ such that $d(x, a) < \delta$ implies $d(f(x), f(a)) < \varepsilon$, for any point $x \in X$. If f is continuous at every point of X, then f is said to be continuous on X. For metric spaces, see the books Copson [3] and Reisel [4].

A dynamical system can be viewed as a couple (X, f), where $f : X \to X$ is a function from the topological space (or metric space) X to itself. The system (X, f) is said to be reversible if f is a homeomorphism from X to X.

Definition 12.1 (*Invariant set*) Let $f : X \to X$ be a map. A set $A \subseteq X$ is said to be invariant under the map f if for any $x \in A$, $f^n(x) \in A$, $\forall n$. Specifically, the set A is invariant if $f(A) = A$. Let (X, f) be a discrete dynamical system. A subset A of X is said to be a *positively invariant* set if $f(A) \subset A$. If $f(A) = A$, then A is *strictly positively invariant*. The set of periodic points of a map is always an invariant set. The unit interval $[0, 1]$ of the logistic map $f_4(x) = 4x(1 - x)$, $x \in [0, 1]$ is an invariant set. For $r > 4$, the Cantor set is an invariant set of $f_r(x) = rx(1 - x), x \in [0, 1]$.

Definition 12.2 (*Dense set*) In a topological space (X, τ), a subset A of X is said to be a dense set (or an everywhere dense set) if $\bar{A} = X$. In other words, A is said to be dense subset of X if for any $x \in X$, any neighborhood of x contains at least one point of A. For example, the set of all rational numbers is dense subset of the set of all real numbers. Again, in the topological space (X, τ_1), given above, we have seen that $\overline{\{a, b\}} = X$. Therefore, $\{a, b\}$ is a dense subset of X. But $\{a\}$ is not a dense subset of X, since $\overline{\{a\}} = \{a, d\} \neq X$. The existence of dense set (or dense orbit) is a feature of chaotic map.

Definition 12.3 (*Perfect systems*) A dynamical system (X, f) is said to be a perfect system if every point of X is its limit point.

Definition 12.4 (*Regular systems*) A dynamical system (X, f) is said to be regular if the set of the periodic points of f is dense in X. Note that in a metric space (X, d), the system (X, f) is regular if and only if for every $x \in X$ and for any $\varepsilon > 0$, there exists a point y in the set of the periodic points of f such that $d(x, y) < \varepsilon$. For example, the system (X, f), where $X = [-2, 2]$ and $f(x) = x^2 - 2$, is regular.

Definition 12.5 (*Sensitive dependence on initial conditions*) An attribute for a chaotic system is to exhibit exponentially fast separation of nearby trajectories for infinitesimally changed initial conditions. Mathematically, this can be expressed in (X, f) as follows:

A map $f : X \rightarrow X$ is said to have sensitive dependence on initial conditions (SDIC—) property if there exists a $\delta > 0$ such that for any $x \in X$ and any neighborhood $N_\varepsilon(x) = (x - \varepsilon, x + \varepsilon)$ of x, there exists $y \in N_\varepsilon(x)$ and an integer $k > 0$ such that the property $|f^k(x) - f^k(y)| > \delta$ holds good.

This relation indicates that for $x \in X$ there are points in X arbitrarily close to x which separate from x by at least δ under iterations of the map f. Let us explain the concept by considering an example. The doubling map $g : S \rightarrow S$ on the unit circle S is defined by $g(\theta) = 2\theta$. Let $\theta_1 \in S$ and $N_\varepsilon(\theta_1) = (\theta_1 - \varepsilon, \theta_1 + \varepsilon)$ be an ε-neighborhood of θ_1. Let $\delta > 0$, then there exists $\theta_2 \in N_\varepsilon(\theta_1)$ and $k > 0$ such that

$$|g^k(\theta_1) - g^k(\theta_2)| = |2^k\theta_1 - 2^k\theta_2|$$
$$= 2^k|\theta_1 - \theta_2| > \delta, \ \forall \theta_1, \theta_2 \in N_\varepsilon(\theta_1).$$

This implies that the map g has sensitive dependence property. It can also be verified by considering two neighboring points, say $x = 0.25$ and $y = 0.2501$ and then calculating the difference $|g^k(x) - g^k(y)|$ for increasing k, as shown in Table 12.1.

Table 12.1 Iterations of the map g

k (Number of iterations)	$g^k(x)$	$g^k(y)$	$\left|g^k(x) - g^k(y)\right|$
0	0.25	0.2501	0.0001
1	0.5	0.5002	0.0002
2	1.0	1.0004	0.0004
3	2.0	2.0008	0.0008
4	4.0	4.0016	0.0016
5	8.0	8.0032	0.0032
6	16.0	16.0064	0.0064
7	32.0	32.0128	0.0128
8	64.0	64.0256	0.0256
9	128	128.0512	0.0512
10	256	256.1024	0.1024
11	512	512.2048	0.2048
12	1024	1024.4096	0.4096
13	2048	2048.8192	0.8192
14	4096	4097.6384	1.6384
15	8192	8195.2768	3.2768
16	16384	16390.5536	6.5536
17	32768	32781.1072	13.1072
18	65536	65562.2144	26.2144
19	131072	131124.4288	52.4288
20	262144	262248.8576	104.8576

The data in the table show that for increasing k the difference $\left|g^k(x) - g^k(y)\right|$ becomes more and more larger than the initial separation $|x - y| = 0.0001$. Hence, the map g satisfies the sensitive dependence property. Also, the difference $\left|g^k(x) - g^k(y)\right|$ increases linearly with the composition number k. The linearity property does not hold for all maps showing SDIC property. One such map is the doubling map on the real line,

$$f(x) = 2x(\mathrm{mod}1) = \begin{cases} 2x, & 0 \leq x < 1/2 \\ 2x - 1, & 1/2 \leq x < 1 \end{cases}$$

It represents the fractional part of $2x$ for $x \in [0, 1)$. Table 12.2 represents first 20 iterations of f taking $x = 0.45$ and $y = 0.4501$ as two seeds.

Clearly, the map f follows the sensitivity property of two neighboring points. But after 12 iterations the difference $\left|f^k(x) - f^k(y)\right|$ becomes completely uncorrelated with increasing values of k. Graphical representations of the iterations are shown in Fig. 12.1.

Table 12.2 Iterations of the map $f(x) = 2x(\mathrm{mod}1)$

k	$f^k(x)$	$f^k(y)$	$\left| f^k(x) - f^k(y) \right|$
0	0.45	0.4501	0.0001
1	0.9	0.9002	0.0002
2	0.8	0.8004	0.0004
3	0.6	0.6008	0.0008
4	0.2	0.2016	0.0016
5	0.4	0.4032	0.0032
6	0.8	0.8064	0.0064
7	0.6	0.6128	0.0128
8	0.2	0.2256	0.0256
9	0.4	0.4512	0.0512
10	0.8	0.9024	0.1024
11	0.6	0.8084	0.2048
12	0.2	0.6096	0.4096
13	0.4	0.2192	0.1808
14	0.8	0.4384	0.3616
15	0.6	0.8768	0.2768
16	0.2	0.7536	0.5536
17	0.4	0.5072	0.1072
18	0.8	0.0144	0.7856
19	0.6	0.0288	0.5712
20	0.2	0.0576	0.1424

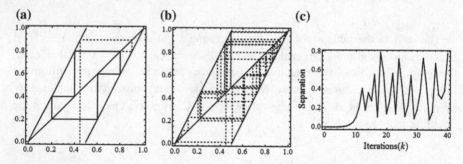

(a) **(b)** **(c)**

Fig. 12.1 First 40 iterations of the angle-doubling map $f(x) = 2x(\mathrm{mod}1)$ with the initial points **a** $x_0 = 0.45$, **b** $x_0 = 0.4501$, and **c** their separations with iterations

Similarly, the tent map

$$T(x) = \begin{cases} 2x, & 0 \le x \le 1/2 \\ 2(1-x), & 1/2 \le x \le 1 \end{cases}$$

satisfies the sensitivity property, as represented graphically in Fig. 12.2. In this figure we take two slightly different initial points $x_0 = 0.2$ (Fig. 12.2a) and $x_0 =$

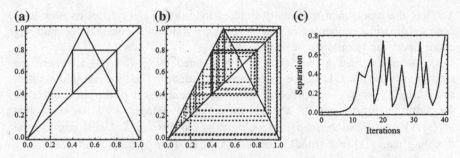

Fig. 12.2 First 40 iterations of the tent map taking the initial points **a** $x_0 = 0.2$, **b** $x_0 = 0.2001$, and **c** their separations with iterations

0.2001 (Fig. 12.2b) and plot their iterations under the map T. The graphs show that for the first case the iterated solutions form a periodic orbit, whereas for the second the solutions form an irregular orbit which is not periodic. Figure 12.2c displays separations of the two iterations.

For any map, the source (unstable fixed point) is always sensitive dependence on initial conditions.

Definition 12.6 A map $f : X \rightarrow X$ is said to be expansive provided there is a $r > 0$ (independent of the points) such that for each pair of points $x, y \in X$, there exists a positive integer k such that $|f^k(x) - f^k(y)| > r$. The quantity r is called the constant of expansiveness. This is an effect in which any initial error is always magnified when iterations are continued. The expansive property of a map cannot alone define a system to be chaotic. There are systems which are expansive but not chaotic. Expansiveness gives only amplification at some constant rate. For example, the doubling map on the real line is expansive with $r = 2$. Again, consider a finite space X with the trivial metric function d defined by $d(x,y) = \begin{cases} 0 & \text{if } x = y \\ 1 & \text{if } x \neq y \end{cases}$. Let $f : X \rightarrow X$ be a bijective map. Then (X, f) is expansive with $r = 1$ but it is not sensitive to initial conditions.

Definition 12.7 (*Topological transitivity*) Transitivity is one of the fundamental properties in the mathematical theory of chaos. A map $f : X \rightarrow X$ is said to be topologically transitive on X if for any two open sets $U, V \subset X$ there exists $k \in \mathbb{N}$ such that $f^k(U) \cap V \neq \varphi$, the null set. The function f is called total transitivity when the composition f^n is topologically transitive for all integer $n \geq 1$. A topologically transitive map has fixed points which eventually move under iterations from one arbitrarily small neighborhood to the other. Hence, the orbit cannot be decomposed into two disjoint open sets which are invariant under the map. A discrete dynamical system is decomposable if there exists a finite open cover (with at least two elements) of X such that each open set of the cover is positively invariant under the map f. On the other hand, the system is indecomposable if and only if it cannot be expressed as the union of two nonempty, closed, and positively invariant subsets of

X. Thus the topological transitivity implies indecomposability. Let us now concentrate on some properties of transitive maps which are essential for studying chaos under the framework of mathematics.

Consider the map g : unit circle $S \to S$ defined by $g(\theta) = \theta + \alpha$, where the rotation α is irrational. Let U be an open arc of length l. Then we can find a $n \in \mathbb{N}$ such that $g^n(U) = U + n\alpha$ will cover the entire circumference of the circle S. Hence for arbitrary open sets U and V, there exists integer $k > 0$ such that $g^k(U) \cap V \neq \varphi$. Thus the map g is topologically transitive on S. Again, consider the doubling map $f(x) = 2x(\mathrm{mod}1)$ defined on [0,1). Take two open intervals, $U = \left(0, \frac{1}{4}\right)$, $V = \left(\frac{2}{3}, \frac{3}{4}\right)$ in [0,1). We calculate $f(U) = \left(0, \frac{1}{2}\right), f^2(U) = (0, 1)$. This implies $f^2(U) \cap V \neq \varphi$. Therefore, the map f is topologically transitive on [0,1).

Proposition 12.1 *A dynamical system. (X, f) is topologically transitive if and only if for any non-empty open subset A of X, the set $\bigcup_{n \in \mathbb{N}} f^n(A)$ is dense in X.*

Proposition 12.2 *The system (X, f) is topologically transitive if and only if for each element $(x, y) \in X \times X$ and for each open ball B_x and B_y with center at x and y, respectively, there exist an element $z \in B_x$ and a positive integer k such that $f^k(z) \in B_y$.*

Proposition 12.3 *In a topologically transitive system (X, f), the set X can be expressed as $X = \overline{f(X)}$.*

Proposition 12.4 *In a topologically transitive system (X, f) if the metric space (X, d) is compact (that is, if every open cover of the separated space (X, d) has a finite subcover), then $X = f(X)$.*

Proposition 12.5 *In a compact space, topological transitivity is equivalent to the existence of dense orbit.*

Proposition 12.6 *Let (X, f) be a dynamical system where the set X is compact. Then the following properties are equivalent:*

(a) *(X, f) is topologically transitive.*
(b) *for every nonempty open sets $U, V \subset X$ and for all $n \in \mathbb{N}$, $f^{-n}(U) \cap V \neq \varphi$.*
(c) *for every nonempty open set $A \subset X$, $\overline{\bigcup_{n \in \mathbb{N}} f^{-n}(A)} = X$.*

Proposition 12.7 *If a perfect system possesses a dense orbit, then the system is topologically transitive.*

Interested reader can try to prove the propositions, (see Bahi and Guyex [5]).

Definition 12.8 (*Topological mixing*) Topological mixing is a stronger notion of topological transitivity property. A map $f : X \to X$ is said to be *strongly transitive* if for every $x, y \in X$ and for all real $r > 0$ there exist $n \in \mathbb{N}$ and $z \in (x - r, x + r)$ such that $f^n(z) = y$, that is, if for every $x, y \in X$ we can find at least one point z very near to x that moves under iterations to small neighborhood of y. If X is compact, then one can prove easily that topological transitivity and strong transitivity are equivalent. In topological transitivity, the two open sets U and V are either disjoint or not. A map $f : X \rightarrow X$ is said to be topologically mixing on X if for any two open sets $U, V \subset X$ with $U \cap V = \varphi$, there exists a positive integer N such that $f^n(U) \cap V \neq \varphi, \forall n \geq N$.

Definition 12.9 (*Chaotic map*) A map $f : X \rightarrow X$ (X is either a topological space or a metric space) is said to be a chaotic map on an invariant subset $A \subseteq X$ if the following conditions are satisfied:

(i) the map function f has sensitive dependence on initial conditions on A;
(ii) f is topologically transitive on A;
(iii) the periodic points of f are dense in A.

The condition (iii) introduced by *Devaney* implies a trait of regularity on the dynamic behavior of the map. However, the first two properties are closed to the chaotic phenomena and the property (iii) implies the dense set on A. So, we may drop the third condition and take the first two as the defining properties of chaos.

Example 12.1 Show that the doubling map $g : S \rightarrow S$ defined by $g(\theta) = 2\theta, \theta \in S$ is chaotic on the unit circle S.

Solution As shown above, the map g is sensitive to initial conditions. To prove the topologically transitivity property consider an open arc A of the unit circle S of length l. Since the map g doubled a point in S in every iteration, we can find $k \in \mathbb{N}$ such that $g^k(A)$ will cover the entire circumference of the circle S. Thus for any two open sets U and V of S, one must find a $k \in \mathbb{N}$ such that $g^k(U) \cap V \neq \varphi$. Hence g is topologically transitive on S. We now find the periodic points of period-n. According to the definition, we get

$$g^n(\theta) = \theta = \theta + 2k\pi$$
$$\Rightarrow \quad 2^n\theta = \theta + 2k\pi$$
$$\Rightarrow \quad \theta = \frac{2k\pi}{2^n - 1}, k = 1, 2, 3, \ldots, 2^n.$$

This relation indicates that the period points of period-n of the map g are the $(2^n - 1)$th roots of unity. For large n, the set of periodic-n points forms a dense subset of S. So, according to definition of chaos, the doubling map g is chaotic on the invariant set S.

Proposition 12.8 *Let $f : X \rightarrow X$ be a continuous map on an infinite dimensional metric space X. If f is chaotic, then it has sensitive dependence on initial conditions.*

Proposition 12.9 *Let $f : X \rightarrow X$ be a continuous map. Then f is chaotic if and only if for any two open subsets U and V of X there exists a periodic point $x \in U$ and a positive integer n such that $f^n(x) \in V$.*

Proposition 12.10 *Suppose $g : Y \rightarrow Y$ is conjugate to $f : X \rightarrow X$. Then g is chaotic on Y if and only if f is chaotic on X.*

12.2   Dynamics of the Logistic Map

(a)   Period-doubling cascade

We have studied the logistic map for different values of the growth parameter $r(<4)$. The logistic map is a simple one-dimensional map which has chaotic property depending upon r. The logistic map can be expressed iteratively as $x_{n+1} = rx_n(1 - x_n), n = 0, 1, 2, \ldots$, where $x_n \in [0, 1]$ and $r \geq 0$. Taking x_0 as the initial point the successive iterations x_n of x can be calculated as follows:

$$x_1 = rx_0(1 - x_0)$$
$$x_2 = rx_1(1 - x_1)$$
$$x_3 = rx_2(1 - x_2)$$
$$\vdots$$
$$x_n = rx_{n-1}(1 - x_{n-1})$$

and so on. Thus we see that each iterated value x_n (a dimensionless measure of the population of the species in the nth generation) depends on the control parameter r as well as on the previous iterates. As discussed earlier, the equilibrium points (constant solutions) are obtained from the relation $x_{n+1} = x_n = x^*$. This gives two equilibrium points, namely $x_1* = 0$ and $x_2^* = (r - 1)/r$. The origin is an equilibrium point for all values of r whereas the other equilibrium point exists for $r \geq 1$. From linear stability analysis the equilibrium point origin is stable for $0 \leq r < 1$ and unstable for $r > 1$, and the second equilibrium point $x_2^* = (r - 1)/r$ is stable for $1 < r < 3$ and unstable for $r \geq 3$. Therefore, there is an exchange of stability as the parameter r passes through the value 1 resulting a bifurcation, known as the transcritical bifurcation. However, one can obtain solutions which repeat after every two iterations, that is, $x_{n+2} = x_n$, or three iterations $(x_{n+3} = x_n)$, so on, or N iterations $(x_{n+N} = x_N)$ for different values of r. The solutions are called period-2 orbit, or period-3 orbit, ..., or period-N orbit (or cycles). With increasing values of r the period-doubling sequence or period-doubling cascade for the logistic map is formed as follows.

The period-1 solution, that is, the equilibrium solution at $x_2^* = (r - 1)/r$ is unstable for $r > 3$ and the slope f' becomes (-1) at $r = 3$. The period-2 solution is born at $r = 3$ and exists in the range $3 \leq r < 3.449\ldots$ and this period-2 cycle becomes unstable at the critical value $r = (1 + \sqrt{6}) = 3.449\ldots$ and again the slope f' is (-1). This results the birth of period-4 solution at $r = 3.449 \cdots \simeq 3.5$. The period-4 bifurcation exists in the range $(1 + \sqrt{6}) < r < 3.544091$ resulting in the birth of period-8 bifurcation at the critical value $r = 3.544112$ with the range of existence $3.544091\ldots < r < 3.564408\ldots$. The bifurcation sequence occurred faster and faster with infinitesimal small increasing values of r. It is interesting to note that the range of existence of the cycle decreases very fast with increasing values of r. We are approaching at the infinitum state. The bifurcation sequence reaches a dense state (resulting a dense orbit). The sequence of parameters r_n, at which

Fig. 12.3 A sketch of period-doubling cascade for logistic map at different values of r

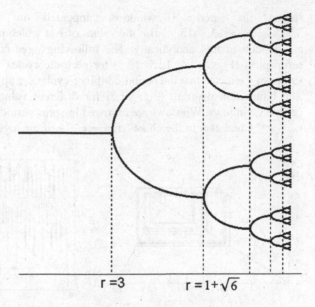

$r = 3$ $r = 1 + \sqrt{6}$

period-2^n bifurcation occurs, converges to a limiting value and we denote it by r_∞ and it is given by $r_\infty = 3.57$(approximately). This limiting state of bifurcation sequence is called the boundary of chaos and the whole mechanism is called the period-doubling cascade. A sketch of the period-doubling cascade is shown graphically in Fig. 12.3.

The figure clearly shows that the map undergoes first bifurcation at $r = 3$ (period-1). The second or period-2 bifurcation takes place at the value $r = 1 + \sqrt{6}$ and this sequence of bifurcation continues upon the critical value $r_\infty(<4)$. The numerical values of r at different stages of period-doubling bifurcation sequence are presented in tabular form below.

$0 < r < 1$	orbits coverge to the equilibrium point origin
$1 < r < 3 (r_c = 3)$	period-1(equilibrium point)
$3 < r < 3.449490 \cdots (r_{c=3.449489\ldots})$	period-2
$3.449490 \cdots < r < 3.544091 \cdots$	period-4
$3.544091 \cdots < r < 3.564408 \cdots$	period-8
$\vdots$	$\vdots$

(b) Periodic windows

Another pattern, called windows appeared as in period-doubling bifurcation sequence. Windows are followed in some periodic sequence. Graphically, it can be shown that period-3 windows appeared in the range of r as $3.8284\ldots \le r \le 3.8415\ldots$. But the values of r at which period-3 windows are born are difficult to find analytically. The following three cobweb diagrams and time series plots (Figs. 12.4, 12.5, 12.6) for periodic cycles for three slightly different values of r indicate that the period-doubling cycles are approaching to the dense set. The bifurcation diagram (Fig. 12.7) for different values of r shows clearly the period-3 windows. Windows are followed in some periodic pattern within the range $r_\infty \le r \le 4$ and also in the chaotic range $r \ge 4$ of the logistic map. Period-doubling

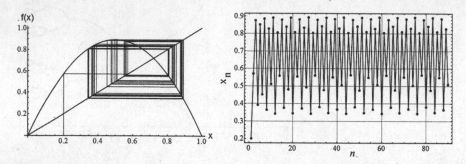

Fig. 12.4 Cobweb diagram and time series plot of the logistic map at $r = 3.57$

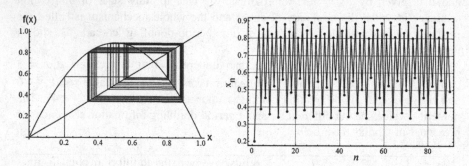

Fig. 12.5 Cobweb diagram and time series plot of the logistic map at $r = 3.578$

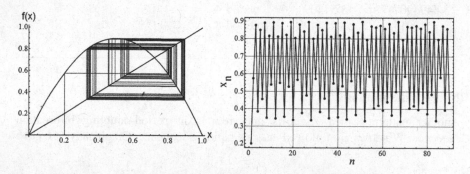

Fig. 12.6 Cobweb diagram and time series plot of the logistic map at $r = 3.59$

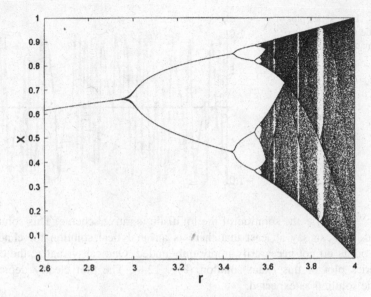

Fig. 12.7 Bifurcation diagram of the logistic map for different values of r

cascade of periods 3×2^n successively occurred and becomes chaotic. The sequence of bifurcation of windows followed like 3×2^n bands $\rightarrow 3 \times 2^{n-1}$ bands and so on, until a range of r where chaos apparently appears in three bands of windows (Ott [6]). It was shown by Yorke et al. [7] that the high periodic window attained a universal value which is independent for particular class of maps, more specifically the class one-dimensional dissipative maps undergoing period-doubling bifurcation sequence. The ratio of the width in the parameter of a periodic window to the parameter difference between the starting point of the periodic window and the occurrence of the first period doubling in the window approaches the value 9/4 universally for one-dimensional maps with periodic window cascade (Grebogi et al. [8]), the value is approximately matched for the periodic-3 windows. Periodic windows are seen in all dynamical processes undergoing period doubling.

(c)   Dynamics of logistic map for $r \geq 4$

The dynamics of logistic map for $r \geq 4$ has interesting features. For $r = 4$, the map has exact solution. We take $x_n = \sin(2^n), n = 0, 1, 2, \ldots$, and substitute it in the iterative logistic map. We then get

$$
\begin{aligned}
rx_n(1 - x_n) &= 4\sin^2(2^n)\left(1 - \sin^2(2^n)\right) \\
&= 4\sin^2(2^n)\cos^2(2^n) \\
&= \sin^2(2^{n+1}) = x_{n+1}.
\end{aligned}
$$

Fig. 12.8 Time series plot of the chaotic solution for the logistic map at $r = 4$

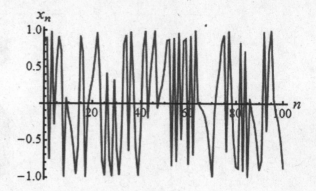

Hence $\sin(2^n)$ is the solution of the logitistic iterative scheme. This solution is bounded. We can say at least that there is an analytical solution for chaos and chaotic orbits do not necessarily imply unbounded. One can visualize the chaotic time series plot of this exact solution (Fig. 12.8). The plot clearly depicts the aperiodic solution as expected.

For $r > 4$, we can find a point $p \in I = [0, 1]$, in particular $p = 1/2$ such that $f_r(p) > 1$, the second iterates $f_r^2(p) < 0$ and $f_r^n(p) \to -\infty$ as $n \to \infty$. For $r > 0$, there exists an open interval $S_0 = (p_-, p_+)$ centered about the point $p = 1/2$ such that $f_r(x) > 1$, $f_r^2(x) < 0$ and $f_r^n(x) \to -\infty$ as $n \to \infty$, for all $x \in S_0$. In the previous section, we have seen that the map undergoes period-doubling sequence of bifurcations and periodic windows on the unit interval [0,1] for $0 < r < 4$. So, for $r > 4$, there exists $x_0 \in [0, 1]$ such that $f_r^k(x_0) > 1$ for certain integer k. Thereafter, the iterates $f_r^{n+k}(x_0) \to -\infty$ as $n \to \infty$. There exists a set $\Lambda_r (r > 4) \subset [0, 1]$ which has a Cantor set structure. $f_r(x)$ is chaotic on Λ_r. We now establish this by mathematical theory of the set Λ_r.

Thus $S_0 = \{x \in I : f_r(x) > 1\}$ is a single open interval such that $x \in I \backslash S_0$, which is a closed interval that does not fly off the interval I after the first iteration (Fig. 12.9).

Let $S_1 = \{x \in I : f_r(x) \leq 1, f_r^2(x) > 1\}$. Then $S_1 = f_r^{-1}(S_0)$ can be expressed as the union of two distinct open intervals (see Fig. 12.10) and for any $x \in I \backslash (S_0 \cup S_1)$ that does not fly off the interval I after two iterations. Hence the set $S_0 \cup S_1$ is the union of $1 + 2 = 3 (= 2^2 - 1)$ open intervals, indicating that $I \backslash (S_0 \cup S_1)$ is a closed set (Fig. 12.11).

Proceeding in this way, we can find an interval

$$S_n = f_r^{-1}(S_{n-1}) = \left\{ x \in I : f_r^k(x) \leq 1 \; \forall k \leq n \text{ and } f_r^{n+1}(x) > 1 \right\}$$

that can be expressed as the union of 2^n open intervals and for any $x \in I \backslash \bigcup_{k=0}^{n} S_k$ does not fly off I after n iterations. Note that $\bigcup_{k=0}^{n} S_k$ is the union of $1 + 2 + 2^2 + \cdots + 2^n = (2^{n+1} - 1)$ open intervals. So, the set $\Lambda_r = I - \bigcup_{n=0}^{\infty} S_n$ is closed. We now prove some theorems for the set Λ_r which are important in connection with the dynamics of logistic map when $r > 4$.

Fig. 12.9 The logistic map
and the open interval
$S_0 = (p_-, p_+)$

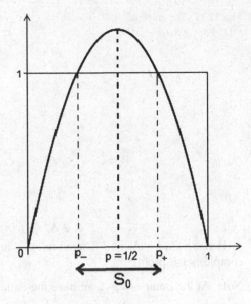

Fig. 12.10 The logistic map
and the intervals S_0 and S_1

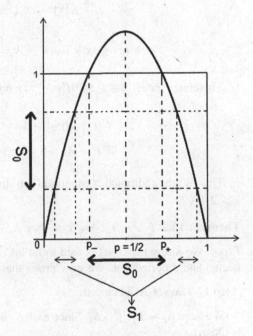

Theorem 12.1 *The set Λ_r is an invariant set for the logistic map $f_r(x)$ on I.*

Proof If $x \in I$ be any point such that $f_r(x) \notin \Lambda_r$. Then $x \in \bigcup_{n=0}^{\infty} S_n \Rightarrow x \notin \Lambda_r$.
Again, if $x \notin \Lambda_r$, then $x \in \bigcup_{n=0}^{\infty} S_n$. This implies that some iterates of x will fly off
the interval I and so, $f_r(x) \notin \Lambda_r$. Therefore, $x \notin \Lambda_r \Leftrightarrow f_r(x) \notin \Lambda_r$. This implies

Fig. 12.11 The graph of $f_r^2(x)$ for $r > 4(r = 4.3)$

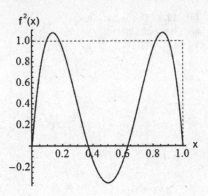

$$x \notin \Lambda_r \Leftrightarrow f_r(x) \notin \Lambda_r.$$

Hence $f_r(\Lambda_r) = \Lambda_r$ on I. Therefore, Λ_r is an invariant set for the map f_r on I. This completes the proof.

Note At the point $x = p_\pm$, we have the equation given by

$$f_r(x) = rx(1 - x) = 1$$

$$\Rightarrow rx^2 - rx + 1 = 0 \Rightarrow x = p_\pm = \frac{r \pm \sqrt{r^2 - 4r}}{2r}$$

Therefore, for all $x \in I \backslash S_0$, $\left|f_r'(x)\right| > 1$ provided $\left|f_r'(p_\pm)\right| > 1$

$$\Rightarrow |r - 2rp_\pm| > 1 \Rightarrow \left|\sqrt{r^2 - 4r}\right| > 1$$

$$\Rightarrow r^2 - 4r > 1 \Rightarrow r > 2 + \sqrt{5}$$

This is an additional assumption on the control parameter r that satisfies $r > 2 + \sqrt{5}$.

Theorem 12.2 *For $r > 2 + \sqrt{5}$, the set Λ_r is a Cantor set.*

Proof We know that a set is said to be a Cantor set if it is closed, totally disconnected and a perfect set. We now prove these three properties for the set Λ_r.

Step I   **The set Λ_r is closed.**

We have $\Lambda_r = I \backslash \bigcup_{n=0}^{\infty} S_n$. Since each S_n is open, $\bigcup S_n$ is also open and hence Λ_r is closed.

Step II   **The set Λ_r is totally disconnected.**

A set is said to be totally disconnected if it contains no intervals. Assume that the parameter satisfies $r > 2 + \sqrt{5}$. Then $\left|f_r'(x)\right| > 1$, for all $x \in \Lambda_r$. Hence there exists

$\lambda > 1$ such that $|f'_r(x)| > \lambda \ \forall x \in \Lambda_r$. Using chain rule of differentiation, we see that $|(f^n_r)'(x)| > \lambda^n$, for all $x \in \Lambda_r$. We shall now show that Λ_r does not contain any interval. If possible, let $x, y \in \Lambda_r$ be two distinct points such that the closed set (interval) $[x, y] \subset \Lambda_r$. Then, by the Mean Value Theorem (MVT) of differential calculus, there exists $\xi \in (x, y)$ such that $|f^n_r(y) - f^n_r(x)| = |(f^n_r)'(\xi)||y - x|$. Since $\xi \in (x, y)$, it is in Λ_r and so $|(f^n_r)'(\xi)| > \lambda^n$. Therefore, $|f^n_r(y) - f^n_r(x)| > \lambda^n|y - x|$. Let us now choose n sufficiently large so that $\lambda^n|y - x| > 1$. Then $|f^n_r(y) - f^n_r(x)| > 1$. This shows that one of $f^n_r(x)$ and $f^n_r(y)$ lies outside the interval I. That is, at least one of x and y is not in the set Λ_r. This is a contradiction. Hence Λ_r cannot contain an interval and therefore it is totally disconnected.

Step III The set Λ_r is a perfect set (that is, every point of Λ_r is its limit point).

From Step I, Λ_r is closed. Now, we can express Λ_r as

$$\Lambda_r = I \setminus \bigcup_{n=0}^{\infty} S_n = \bigcap_{k=1}^{\infty} C_k$$

where each closed set C_k is formed by the union of k closed intervals. The set C_1 is formed by removing the set S_0 from the closed interval I. That is,

$$C_1 = I \setminus S_0 = I_0 \cup I_1 = \bigcup_{i=0,1} I_i,$$

where I_0 and I_1 are two closed subintervals of I. Thus, C_1 is the union of two closed subintervals I_0 and I_1. Also, it is easy to verify that $f_r(I_0) = f_r(I_1) = I$. If a and b be two end points of I_0 (or I_1), then by MVT, there exists $\xi \in (a, b)$ such that $|f_r(b) - f_r(a)| = |f'_r(\xi)||b - a| > \lambda|b - a|$. Hence

$$1 = |f_r(b) - f_r(a)| = L(I) > \lambda L(I_i)$$
$$\Rightarrow \quad L(I_i) < \frac{1}{\lambda}L(I) = \frac{1}{\lambda}$$

where $L(I_i)$ denotes the length of the interval $I_i (i = 0, 1)$. The set C_2 is obtained by removing S_1 from C_1. That is,

$$C_2 = C_1 \setminus S_1 = I \setminus (S_0 \cup S_1) = \bigcup_{i_0, i_1 = 0}^{1} I_{i_0 i_1}$$
$$= (I_{00} \cup I_{01}) \cup (I_{10} \cup I_{11})$$

which is the union of four closed subintervals and hence a closed set. Let a, b be two end points of I_{00} (or I_{01} or I_{10} or I_{11}). Then by MVT $\exists \xi \in (a, b)$ such that

$$|f_r(b) - f_r(a)| = |f_r'(\xi)||b - a| > \lambda |b - a|$$

$$\Rightarrow \quad L(I_{i_0 i_1}) < \frac{1}{\lambda} L(I_i) < \frac{1}{\lambda^2}.$$

Proceeding in this way, we can generate the closed intervals C_k by removing S_{k-1} from $C_{k-1}(k = 1, 2, \ldots)$ and we can express C_k as

$$C_k = C_{k-1} \backslash S_{k-1}$$

$$= I - \bigcup_{m=0}^{k-1} S_m$$

$$= \cup I_{i_0 i_1 \cdots i_{k-1}} \, (i_0, i_1, \ldots, i_{k-1} \in \{0,1\})$$

where $L(i_0 i_1 \cdots i_{k-1}) < \lambda^{-n}$. Thus we see that $\Lambda_r = \bigcap_{k=1}^{\infty} C_k$, where each C_k is given by

$$C_k = \cup \, I_{i_0 i_1 \cdots i_{k-1}} (i_0, i_1, \ldots, i_{k-1} \in \{0, 1\})$$

Observe that each $C_k, k \in \mathbb{N}$ is closed and $C_k \subset C_{k-1}$. Next, we prove that every $x \in \Lambda_r$ is a limit point of it. Let $x \in \Lambda_r$. Then $x \in C_k, k \in \mathbb{N}$. Let for each $k \in \mathbb{N}$, $a_k = a_k(x)$ be an end point of the closed set (interval) C_k that contains the point x. Then $a_k \in \Lambda_r$ and x is a limit point of the sequence $\{a_k\}$. Hence, every $x \in \Lambda_r$ is a limit point of Λ_r and therefore the set Λ_r is perfect. Thus, we see that the set Λ_r is closed, totally disconnected, and perfect. Hence the invariant hyperbolic set Λ_r is a Cantor set for the logistic map f_r on $I = [0, 1]$. This completes the proof.

In this connection we remark that the Cantor set is a self-similar fractal with fractional dimension 0.6309. The proof will be given in the next chapter. So the chaotic dynamics or the orbit of the logistic map for $r > 4$ resembles Cantor set which is a self-similar fractal. Again, we reach at an infinitum state where the dynamical evolution attained a formless structure, an important feature of chaotic orbit.

12.3   Symbolic Dynamics

The periodic points are dense in chaotic motion. The regularity in chaotic dynamics is an intrinsic feature. But it is very difficult to prove mathematically this property for maps. Restricted to dynamics of a map in an invariant set and its relation with other easier maps, the properties of a chaotic map may establish quite easily. The idea of characterizing the orbit structure of a dynamical system with the help of sequences of symbols, say 0 and 1, letters of the alphabet, Chinese characters, etc., known as symbolic dynamics is interesting in which many chaotic maps can be related with symbolic maps. Originally, this technique came from the work of Hadamard (1898) in the study of geodesics on surfaces of negative curvature and then *Birkhoff* (1927, 1935) used this idea in the study of dynamical systems. This section contains the properties of sequence spaces of two symbols and the dynamics of shift map.

Theorem 12.3 *The periodic points of the shift map $\sigma : \Sigma_2 \to \Sigma_2$ form a dense set.*

Proof To prove the theorem we have to show that for an arbitrary point $s = (s_0 s_1 s_2 \cdots) \in \Sigma_2$ there exist periodic points $p_n \in \Sigma_2$ which converge to s. Choose $p_n = (s_0 s_1 s_2 \cdots s_{n-1} s_0 s_1 s_2 \cdots s_{n-1} \cdots)$. Then $p_n \in \Sigma_2$. From above discussion p_n is a periodic-n point of σ. Also, we see that

$$d(p_n, s) < \frac{1}{2^{n-1}} \to 0 \text{ as } n \to \infty$$

$$\Rightarrow p_n \to s \text{ as } n \to \infty.$$

This completes the proof.

Itinerary map: Let $x \in \Lambda_r \subset I_0 \cup I_1$. Then the orbits of x lie in Λ_r and $f_r^n(x) \in I_0$ or I_1, for all $n \in \mathbb{N}$. Consider a map $S : \Lambda_r \to \Sigma_2$ defined by

$$S(x) = (s_0 s_1 s_2 \cdots s_n \cdots)$$

where $s_i = 0$ if $f_r^i(x) \in I_0$ and $s_i = 1$ if $f_r^i(x) \in I_1$. This map is called the *itinerary* map or simply *itinerary* of $x \in \Lambda_r$.

Theorem 12.4 *For $r > 2 + \sqrt{5}$, the map $S : \Lambda_r \to \Sigma_2$ is a homeomorphism.*

For proof, see Devaney [9].

Theorem 12.5 *The map $\sigma : \Sigma_2 \to \Sigma_2$ is topologically conjugate to the logistic map $f_r : \Lambda_r \to \Lambda_r$ through the map $S : \Lambda_r \to \Sigma_2$.*

Proof The map $S : \Lambda_r \to \Sigma_2$ is a homeomorphism. We shall verify only the conjugacy relation $S \circ f_r = \sigma \circ S$. Let $x \in \Lambda_r$, then there exists a sequence $\{I_{s_0 s_1 \cdots s_n}\}$ of nested intervals such that x can be expressed uniquely as

$$x = \bigcap_{n=0}^{\infty} I_{s_0 s_1 \cdots s_n}$$

and is determined by the itinerary $S(x)$. From definition, we can show

$$I_{s_0 s_1 \cdots s_n} = I_{s_0} \cap f_r^{-1}(I_{s_1}) \cap f_r^{-2}(I_{s_2}) \cap \cdots \cap f_r^{-n}(I_{s_n})$$

$$\Rightarrow f_r(I_{s_0 s_1 \cdots s_n}) = f(I_{s_0}) \cap I_{s_1} \cap f_r^{-1}(I_{s_2}) \cap \cdots \cap f_r^{-(n-1)}(I_{s_n})$$

$$= I_{s_1} \cap f_r^{-1}(I_{s_2}) \cap \cdots \cap f_r^{-(n-1)}(I_{s_n}) \quad [\because f_r(I_{s_0}) = I]$$

$$= I_{s_1 \cdots s_n}.$$

Therefore,

$$(S \circ f_r)(x) = S(f_r(x)) = S\left(f_r\left(\bigcap_{n=0}^{\infty} I_{s_0 s_1 \cdots s_n}\right)\right)$$

$$= S\left(\bigcap_{n=1}^{\infty} I_{s_1 \cdots s_n}\right)$$

$$= (s_1 s_2 \cdots s_n)$$

$$= \sigma(s_0 s_1 s_2 \cdots) = \sigma(S(x)) \quad [\because S(x) = (s_0 s_1 s_2 \cdots)]$$

$$= (\sigma \circ S)(x)$$

$$\Rightarrow \quad S \circ f_r = \sigma \circ S.$$

Hence the map σ is topologically conjugate to f_r through the map S. Through conjucacy relation we can able to connect the logistic and the symbolic maps.

Theorem 12.6 *For $r > 2 + \sqrt{5}$, the logistic map $f_r : \Lambda_r \to \Lambda_r$ is chaotic.*

Proof From Theorem 12.5, the shift map $\sigma : \Sigma_2 \to \Sigma_2$ is topologically conjugate to the logistic map $f_r : \Lambda_r \to \Lambda_r$ through the map $S : \Lambda_r \to \Sigma_2$. Note that the shift map σ has a dense orbit on Σ_2. Thus the map σ is topologically transitive on Σ_2. This implies that $f_r : \Lambda_r \to \Lambda_r$ is topologically transitive on Λ_r through topological conjugacy. We shall now show that f_r has sensitive dependence on initial conditions on Λ_r. Let us choose a positive real quantity δ such that $\delta < \text{diam}(S_0)$, where $\text{diam}(S_0)$ is the diameter of the open set S_0. Let x and y be two distinct elements in Λ_r. Then $S(x) \neq S(y)$ and therefore the itineraries $S(x)$ and $S(y)$ of x and y, respectively, must differ by a positive real number, however small at least one iteration, say the kth iteration. This implies that the two iterations $f_r^k(x)$ and $f_r^k(y)$ lie on the opposite sides of S_0. Therefore, $\left|f_r^k(x) - f_r^k(y)\right| > \delta$. This shows that the logistic map f_r has SDIC on Λ_r. Thus we see that $f_r(x)$ is topologically conjugate and possesses SDIC property on the set Λ_r for $r > 2 + \sqrt{5}$. Hence $f_r(x)$ is a chaotic map for $r > 2 + \sqrt{5}$. This completes the proof.

12.3.1 The Sawtooth Map

In this subsection we have focused our attention on the sawtooth map (a shift map with sawtooth structure) which is represented in terms of binary decimal form. Generally, the sawtooth map S is defined on the symbolic space as

$$S(s_0 s_1 s_2 \cdots) = s_1 s_2 s_3 \cdots.$$

The map S chops the leftmost symbol in the sequence of symbols. For any number, say $x \in [0, 1)$, the binary decimal representation of x is

$$x = 0.a_1 a_2 a_3 \cdots a_n \cdots$$
$$= \frac{a_1}{2} + \frac{a_2}{2^2} + \frac{a_3}{2^3} + \cdots + \frac{a_n}{2^n} + \cdots$$

where $a_i = 0$ or 1, $\forall i \in \mathbb{N}$. When x is a rational number, then it is either a terminating decimal number or a recurring decimal number. The sawtooth map $S(x)$: $[0, 1) \to [0, 1)$ can also be defined as

$$S(x) = \mathrm{frac}(2x)$$
$$= .a_2 a_3 \cdots a_n \cdots, \quad \text{when } x = 0.a_1 a_2 \cdots a_n \cdots, \text{ where } a_i = 0 \text{ or } 1, \ \forall i \in \mathbb{N}.$$

This justified the definition of the sawtooth map as Euler shift map (here shift from a_1 to a_2). We see the character of the fixed points of the map as below.

$$S(.a_1 a_2 a_3 \cdots a_n \cdots) = .a_2 a_3 \cdots a_n$$
$$S^2(.a_1 a_2 a_3 \cdots a_n) = .a_3 a_4 \cdots a_n$$
$$\vdots$$
$$S^n(.a_1 a_2 a_3 \cdots a_n) = 0$$
$$S^{n+1}(.a_1 a_2 a_3 \cdots a_n) = 0$$

Again, $S^n \left(.a_1 a_2 a_3 \cdots a_n \overline{b_1 b_2 \cdots b_k} \right) = \left(.\overline{b_1 b_2 \cdots b_k} \right)$

and $S^{n+k} \left(.a_1 a_2 a_3 \cdots a_n \overline{b_1 b_2 \cdots b_k} \right) = \left(.\overline{b_1 b_2 \cdots b_k} \right)$.

This shows that the sawtooth map has eventually fixed and periodic points. In eventual periodic points we have the relation

$$S^{n+k}(x) = S^n(x).$$

If x is rational, then either $x = 0.a_1 a_2 \cdots a_n$ (terminating decimal) or $x = 0.a_1 a_2 \cdots a_n \overline{b_1 b_2 \cdots b_k}$ (recurring decimal). In terminating decimal $S^n(x) = 0$. So when $x \neq 0$, the terminating decimal point x is an eventually fixed point whereas the point $x = 0$ is a fixed point of the map. On the other hand, when x is a recurring decimal, we have the relation

$$S^{n+k}(x) = S^n(x).$$

This implies that when $n \neq 0$, the recurring decimal point x is eventually periodic-k point of the map S. When $n = 0$, it is a periodic-k point of S. Thus all rational numbers in the interval $[0,1)$ are either fixed points or periodic points or eventually fixed points or eventually periodic points of the map S. When the number $x \in [0, 1)$ is irrational, its binary decimal representation is neither recurring nor terminating, that is, it is a nonterminating, nonrecurring binary decimal number. We shall now show that the dynamics of the sawtooth map is chaotic. In usual notation the sawtooth map can be expressed by

$$S(x) = 2x \pmod 1 = \begin{cases} 2x, & 0 \le x < \frac{1}{2} \\ 2x - 1, & \frac{1}{2} \le x < 1 \end{cases}$$

For binary decimal number $x = .a_1 a_2 a_3 a_4 \cdots$, $S(x)$ gives $S(x) = .a_2 a_3 a_4 \cdots$, where $a_i = 0$ or 1, for all $i \in \mathbb{N}$. We shall now prove that the map $S(x)$ is chaotic.

(i) Sensitive dependence property

Suppose x and y be two points in $I = [0, 1)$, where $x = .a_1 a_2 a_3 a_4 \cdots a_n a_{n+1} a_{n+2} \cdots$ and $y = .a_1 a_2 a_3 a_4 \cdots a_n b_{n+1} b_{n+2} \cdots$ such that if $a_{n+i} = 0$, then $b_{n+i} = 1$ and if $a_{n+i} = 1$, then $b_{n+i} = 0$, for all $i \in \mathbb{N}$. The difference of x and y gives $|x - y| = \frac{1}{2^{n+1}} < \frac{1}{2^n}$. This implies $|S(x) - S(y)| = \frac{1}{2}$. Hence for any preassigned positive number ε, we can choose n sufficiently large such that $|x - y| < \varepsilon \Rightarrow |S(x) - S(y)| = \frac{1}{2} = \delta$, say. This implies $|S^n(x) - S^n(y)| > \delta$. Therefore, the map $S(x)$ follows the SDIC property.

(ii) Topologically transitive property

Let A and B be two subintervals of I such that $|A| =$ length of $A = |B| = \frac{1}{2^{n+1}}$. Since the periodic points of $S(x)$ are dense in I, there exists a periodic point $b = .\overline{b_1 b_2 \cdots b_k} \in B$. Let $c = .a_1 a_2 a_3 \cdots$ be the midpoint of A. Let $z = .a_1 a_2 a_3 \cdots a_{n+2} \overline{b_1 b_2 \cdots b_k}$. Then $|c - z| < \frac{1}{2^{n+2}}$. This implies that $z \in A$ and $S^{n+2}(z) = .\overline{b_1 b_2 \cdots b_k} = b \in B$. Hence the map $S(x)$ is topologically transitive on I. Since the sets A and B are disjoint, the map $S(x)$ has topologically mixing.

(iii) Desity of periodic points

Let $x = .a_1 a_2 a_3 a_4 \cdots$ be any number in I and ε be any preassigned small positive number. The numbers of the form $t_n = .a_1 a_2 a_3 \cdots a_n$ are all periodic points of the map $S(x)$ and the difference

$$|x - t_n| = |.a_1 a_2 a_3 a_4 \cdots - .a_1 a_2 a_3 \cdots a_n|$$

$$= \sum_{i=0}^{\infty} \frac{a_{n+i}}{2^{n+i}}, \ a_i = 0 \text{ or } 1$$

$$\le \sum_{i=0}^{\infty} \frac{1}{2^{n+i}} = \frac{1}{2^n} < \varepsilon \quad ,$$

where n is sufficiently large. This shows that the periodic points of $S(x)$ are dense in I. Hence, $S(x)$ is a chaotic map.

12.4 Quantifying Chaos

The laws for chaotic motions are not well understood. Naturally, one way for understanding chaos is to seek the qualitative and quantitative measures at least applicable for some systems. We discuss some qualitative and quantitative measures for chaotic systems, viz., the universal sequence, Feigenbaum universal number, Lyapunov exponent, invariant set theoretic measure, etc. Note that most of the quantifying chaotic measures are applicable for one-dimensional chaotic maps of particular types. But it is interesting to say that some of higher dimensional systems can be expressed in one-dimensional maps. Thus the study of quantifications of one-dimensional chaotic systems is useful in analyzing the complicated higher dimensional systems.

12.4.1 Universal Sequence

We have seen an infinite sequence of period-doubling bifurcations and periodic windows for logistic map for varying control parameter r. This bifurcation sequence is same for all unimodal maps. This is a qualitative property and was discovered by Metropolis et al. [10] in the year 1973. It was noticed that the relative ordering of periodic windows for one-dimensional unimodal maps follows the sequence as 1, 2, 4, 6, 5, 3, 6, 5, 6, 4, 6, 5, 6 (up to period 6). This sequentially pattern is interesting and termed as the Universal sequence or simply the U-sequence. The pattern of appearance of windows is same for all unimodal maps undergoing bifurcations when the parameter is varying. Sometimes it also called the MSS sequence. The U-sequence of windows is seen in nonlinear electrical circuits, the Belousov–Zhabotinsky chemical reaction, and all experiments undergoing period doublings. The bifurcation diagrams (Fig. 12.12a–c) for the logistic and sine maps $(x_{n+1} = r \sin(\pi x_n),\ x, r \in [0, 1])$ depict the bifurcation sequence indicating the U-sequence. The depicted numbers in the diagrams indicate the U-sequence for both the maps.

12.4.2 Feigenbaum Number

The logistic map is a prototype map representing the population of species from generation to generation. We have seen the period-doubling bifurcation sequence of the map. This type of bifurcation scenario for varying parameters appeared in the large class of systems like mechanical, electrical, chemical, and biological processes. Period doubling is a mechanism or a characteristic route for a class of systems to reach the complex aperiodic motion. In 1975 the American theoretical Physicist, *Mitchel Feigenbaum* while working at the *Los Alamos National*

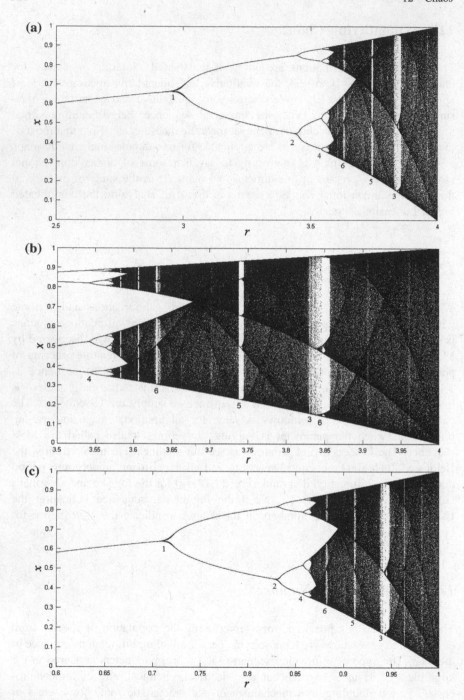

Fig. 12.12 Bifurcation diagrams of **a** the logistic map in [0, 4], **b** a blowup of the bifurcation diagram (**a**) in [3.5, 4], and **c** the sine map in [0, 1]

Laboratory, New Mexico discovered a remarkable feature in period-doubling sequence of bifurcations for the logistic map. In the limit of this bifurcation cascade, he obtained a unique, definite number which is universal to all one-dimensional dissipative maps undergoing period-doubling bifurcations. From the bifurcation diagram of the logistic map it is well understood that as the control parameter $r \in [0, 4]$ increases, the period-doubling sequence occurs in the range $3 \le r \le 3.57$(approx.), appears in a faster rate than that of the previous cascade, and the periodic regions become smaller and smaller. Within this critical range of the period-doubling cascade, a small change in the control parameter may produce a significant change in the system's behavior. The period-doubling bifurcation sequence is as follows:

$$\text{Period-1} \rightarrow \text{Period-2} \rightarrow \text{Period-}2^2 \rightarrow \cdots \rightarrow \text{Period-infinitum}.$$

Suppose r_n denotes the value of r at which the period-2^n bifurcation occurs and r_∞ is the value of r for which the period infinitum (n tending to ∞) attains. The sequence $\{r_n\}$ converges to r_∞ geometrically. This feature can be written as $(r_\infty - r_n) \propto \delta^{-n}$ for a fixed value of δ when n is very large. That is, $r_n = r_\infty - c\delta^{-n}$ for large n, where c is a constant varies from map to map. The value of c can be determined from the relation $c = \lim\limits_{n \to \infty} \frac{r_\infty - r_n}{\delta^{-n}}$. According to Figenbaum the value δ is estimated as follows. From the scaling relation, we have for large n

$$(r_n - r_{n-1}) = -c\delta^{-n}(1 - \delta) \text{ and } (r_{n+1} - r_n) = -c\delta^{-n}(\delta^{-1} - 1).$$

On division, we can estimate δ as $\delta = \frac{(r_n - r_{n-1})}{(r_{n+1} - r_n)}$ as $n \to \infty$. Thus we see that the number δ_n defined by the ratio of $(r_n - r_{n-1})$ and $(r_{n+1} - r_n)$, that is, $\delta_n = \frac{r_n - r_{n-1}}{r_{n+1} - r_n}$ approaches to the constant value δ as the number of period-doubling bifurcation becomes larger and larger. Bearing his name, this limiting constant value is known as the Feigenbaum number or the Feigenbaum constant. Remarkably, this universal number appears in many natural systems like oscillators, fluid turbulence, population dynamics, etc. Note that the Feigenbaum number δ is only an approximate value. No matter how large the period-2^n is, we cannot calculate its exact value. With the aid of δ we can formulate the estimation of the value of r_∞ in terms of three successive values of r_n for large n. From the scaling law, we have an equation of r_∞ for large n as follows:

$$(r_{n-1} - r_\infty) - \delta(r_n - r_\infty) = 0 \Rightarrow (r_{n-1} - r_\infty) - \frac{(r_n - r_{n-1})}{(r_{n+1} - r_n)}(r_n - r_\infty) = 0$$

$$\Rightarrow r_\infty = \frac{r_{n-1}r_{n+1} - r_n^2}{r_{n+1} - 2r_n + r_{n-1}}.$$

These estimate the values of r_∞ and δ in terms of three successive r_n for enough large n. The numerical values of r_∞ and δ are calculated as follows:

Period of the n-cycle	Critical value of r
2	$r_0 = 3$
2^2	$r_1 = 3.449489...$
2^3	$r_2 = 3.544112...$
2^4	$r_3 = 3.564445...$
2^5	$r_4 = 3.568809...$
2^6	$r_5 = 3.569745...$
...	....
2^∞	$r_\infty = 3.57$

Therefore, $\delta_1 = \frac{(r_1-r_0)}{(r_2-r_1)} \simeq 4.75031$, $\delta_2 = \frac{(r_2-r_1)}{(r_3-r_2)} \simeq 4.65367$, $\delta_3 = \frac{(r_3-r_2)}{(r_4-r_3)} \simeq 4.65926$, $\delta_4 = \frac{(r_4-r_3)}{(r_5-r_4)} \simeq 4.66239$, and so on. The limiting value of δ or Feigenbaum constant is $\delta \simeq 4.6692016148$.

12.4.3 Renormalization Group Theory and Superstable Cycle

Consider the quadratic function $f_r(x) = rx(1 - x), x \in [0, 1]$ with $r > 2$. The map f_r has two fixed points, $x^* = 0$ and $x_r^* = (1 - 1/r)$. Consider the point $x_r = 1/r$. The points x_r^* and x_r situate symmetrically about the point $x = 1/2$, the point at which $f_r(x)$ is maximum. This is because $\left|x_r^* - \frac{1}{2}\right| = \left|x_r - \frac{1}{2}\right|$. Also, $f_r(x_r) = x_r^*$. Draw the graphs of $f_r(x)$ and $f_r^2(x)$ in [0,1] (Fig. 12.13).

The graph of $f_r^2(x)$ in $\left[x_r, x_r^*\right]$ looks similar to that of $f_r(x)$ in [0,1] with the following common key traits:

(a) f_r is unimodal in [0,1] and f_r^2 is unimodal in $\left[x_r, x_r^*\right]$,
(b) Both the maps f_r and f_r^2 have the single critical point at $x = 1/2$ (critical in the sense of extremum) in [0, 1] and $\left[x_r, x_r^*\right]$, respectively. f_r is maximum at $x = 1/2 \in [0, 1]$ and f_r^2 is minimum at $x = 1/2 \in \left[x_r, x_r^*\right]$.,
(c) The map f_r has two fixed points in [0,1], one at $x = 0$, an end point of [0,1] and the other in (0,1). Similarly, the map f_r^2 has two fixed points in $\left[x_r, x_r^*\right]$, one at the end point $x = x_r^*$ and the other in $\left(x_r, x_r^*\right)$.

Feigenbaum renormalization group method

Feigenbaum nicely explained mathematically the universal qualitative property of one-dimensional unimodal maps by introducing the renormalization group theory

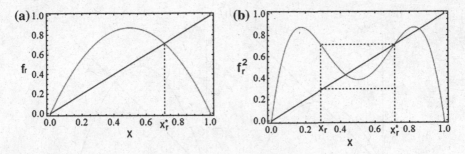

Fig. 12.13 Graphs of **a** $f_r(x)$ and **b** $f_r^2(x)$ for $r = 3.5$

which leads to a universal function. Consider the linear transformation $L_r(x) = \frac{x - x_r^*}{x_r - x_r^*}$. It is a combination of translation and scaling. Clearly, L_r expands the smaller interval $[x_r, x_r^*]$ onto [0,1] with a reversal of orientation. It is easy to verify that the transformation L_r is continuous and invertible with the continuous inverse $L_r^{-1}(x) = (x_r - x_r^*)x + x_r^*$. Thus, $L_r : [x_r, x_r^*] \to [0, 1]$ is a homeomorphism. Define a map Rf_r on [0,1] such that $Rf_r \circ L_r(x) = L_r \circ f_r^2(x) \Rightarrow Rf_r(x) = L_r \circ f_r^2 \circ L_r^{-1}(x)$. Since L_r is a homeomorphism, Rf_r is topologically conjugate to f_r^2 and therefore the dynamical properties of these two maps are topologically equivalent. The map Rf_r is called the **renormalization** of the map f_r.

Properties of renormalization Rf_r:

(i) $Rf_r(0) = L_r \circ f_r^2 \circ L_r^{-1}(0) = L_r \circ f_r^2(x_r^*) = L_r(x_r^*) = 0.$

(ii) $Rf_r(1) = L_r \circ f_r^2 \circ L_r^{-1}(1) = L_r \circ f_r^2(x_r) = L_r(x_r^*) = 0.$

(iii) Put $x_1 = L_r^{-1}(x) = (x_r - x_r^*)x + x_r^*$, $x_2 = f(x_1) = rx_1(1 - x_1)$, $x_3 = f(x_2) = rx_2(1 - x_2)$, and $x_4 = L_r(x_3) = \frac{x_3 - x_r^*}{x_r - x_r^*}$. Then $Rf_r(x) = L_r \circ f_r^2 \circ L_r^{-1}(x) = x_4$.
This implies that

$$(Rf_r)'(x) = \frac{d}{dx}(Rf_r(x)) = \frac{dx_4}{dx} = \frac{dx_4}{dx_3}\frac{dx_3}{dx_2}\frac{dx_2}{dx_1}\frac{dx_1}{dx}$$

$$= \frac{1}{x_r - x_r^*} \cdot r(1 - 2x_2) \cdot r(1 - 2x_1) \cdot (x_r - x_r^*)$$

$$= r^2(1 - 2x_1)(1 - 2x_2).$$

Therefore, $(Rf_r)'(x) = 0$ at $x_1 = 1/2$ and $x_2 = 1/2$. But

$$x_1 = 1/2 \Rightarrow (x_r - x_r^*)x + x_r^* = 1/2 \Rightarrow x = 1/2, \text{ and}$$

$$x_2 = 1/2 \Rightarrow rx_1(1 - x_1) = 1/2 \Rightarrow x_1 = \frac{1}{2}\left[1 \pm \sqrt{\left(1 - \frac{2}{r}\right)}\right] = x_1^\pm \text{ (say)}.$$

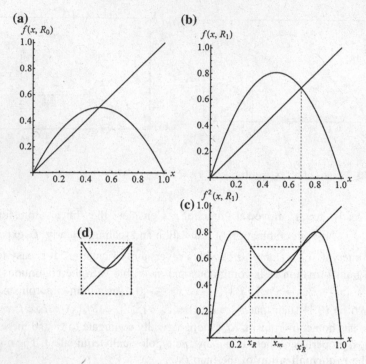

Fig. 12.14 Graphical representations of $f(x, R_0)$, $f(x, R_1)$, and $f^2(x, R_1)$

The points $x_1^{\pm}$ lie outside of the interval $\left[x_r, x_r^*\right]$ or equivalently outside of the interval $[0, 1]$. This is because the distance of x_r^* from the midpoint $\frac{1}{2}\left(x_r + x_r^*\right) = \frac{1}{2}$ of $\left[x_r, x_r^*\right]$ is $\left|\frac{1}{2} - x_r^*\right| = \left|\frac{1}{2} - 1 + \frac{1}{r}\right| = \frac{1}{2}\left(1 - \frac{2}{r}\right) < \frac{1}{2}\sqrt{\left(1 - \frac{2}{r}\right)} = \left|\frac{1}{2} - x_1^{\pm}\right|$, the distance of $x_1^{\pm}$ from $x = 1/2$. Similarly, $\left|\frac{1}{2} - x_r\right| < \left|\frac{1}{2} - x_1^{\pm}\right|$. Therefore, $(Rf_r)'(x) = 0$ only at $x = 1/2$.

(iv) If $f_r^2(x) = x$, that is, if x is a period-2 point of the map f_r, then $Rf_r(L_r(x)) = L_r(x)$, that is, Rf_r has a fixed point at $\xi = L_r(x)$.

We shall now apply the renormalization technique to the superstable cycles of the quadratic map $x_{n+1} = f(x_n, r), n = 0, 1, 2, \ldots$, where $f(x, r) = rx(1 - x), x \in [0, 1], r \in [0, 4]$. In Chap. 9 we have calculated the values of r at which superstable 1 and 2-cycles are born at $R_0 = 2$ and $R_1 = \left(1 + \sqrt{5}\right)$, respectively. Similarly, we can calculate the other values of R_n at which the superstable 2^n-cycle occurs. Now, consider the functions $f(x, R_0)$ and $f^2(x, R_1)$. Figure 12.14a, c depict them. The point $x_m (= 1/2)$ is a point of extremum of the functions and it is also a member of the superstable 2^n-cycle. The function $f^2(x, R_1)$ has four fixed points. Two of them (e.g., 0 and x_R^1) are the fixed points of $f(x, R_1)$ and the other two form a 2-cycle. Taking the point x_R^1 as a corner we draw a square as shown in Fig. 12.14c.

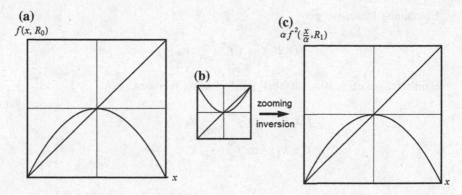

Fig. 12.15 Scaling operations of $f(x, R_0)$

The graph of $f^2(x, R_1)$ on the interval $[x_R, x_R^1]$ resembles similar to the graph of f (x, R_0) on [0,1]. These similarity patterns suggest that both the maps have similar dynamical behaviors and are connected under some scaling. We now shift the origin of x to x_m using the transformation $x \mapsto x - x_m$ (see Fig. 12.15a, b). Now, if we enlarge Fig. 12.15b by a factor $\beta > 1$ [e.g., $\beta \approx 2.5029$] and then invert it, that is, if we apply the transformations $x \mapsto \beta x$ and then $x \mapsto -x$, it looks like Fig. 12.15c.

These transformations convert $f^2(x, R_1)$ into $\alpha f^2\left(\frac{x}{\alpha}, R_1\right)$, where $\alpha = -\beta$. From Fig. 12.15a, c we see that under these transformations

$$f(x, R_0) \approx \alpha f^2\left(\frac{x}{\alpha}, R_1\right).$$

Similarly, taking $f^2(x, R_1)$ and $f^4(x, R_2)$, $f^4(x, R_2)$ and $f^8(x, R_3)$, and so on, we have

$$f^2(x, R_1) \approx \alpha f^4\left(\frac{x}{\alpha}, R_2\right),$$

$$f^4(x, R_2) \approx \alpha f^8\left(\frac{x}{\alpha}, R_3\right),$$

$$f^8(x, R_3) \approx \alpha f^{16}\left(\frac{x}{\alpha}, R_4\right),$$

and so on. In general,

$$f^{2^{n-1}}(x, R_{n-1}) \approx \alpha f^{2^n}\left(\frac{x}{\alpha}, R_n\right), n = 1, 2, 3, \dots$$

Combining these, we get

$$f(x, R_0) \approx \alpha^n f^{2^n}\left(\frac{x}{\alpha^n}, R_n\right).$$

Similarly, we can easily establish the following relations:

$$f(x, R_1) \approx \alpha^n f^{2^n}\left(\frac{x}{\alpha^n}, R_{n+1}\right),$$
$$f(x, R_2) \approx \alpha^n f^{2^n}\left(\frac{x}{\alpha^n}, R_{n+2}\right),$$

$$\vdots$$

Proceeding in this way, we have

$$f(x, R_k) \approx \alpha^n f^{2^n}\left(\frac{x}{\alpha^n}, R_{n+k}\right), k = 0, 1, 2, \ldots.$$

For large n, the function $\alpha^n f^{2^n}\left(\frac{x}{\alpha^n}, R_{n+k}\right)$ yields a universal function $g_k(x) = \lim_{n\to\infty} \alpha^n f^{2^n}\left(\frac{x}{\alpha^n}, R_{n+k}\right)$ with a superstable 2^k-cycle. When $k \to \infty$, R_k tends to the accumulation point r_∞. At this point, we have the universal function $g(x) = \lim_{n\to\infty} \alpha^n f^{2^n}\left(\frac{x}{\alpha^n}, r_\infty\right)$. From the compositions of functions, we have

$$f^{2^n}\left(\frac{x}{\alpha^n}, R_{n+k}\right) = f^{2^{n-1}}\left(f^{2^{n-1}}\left(\frac{x}{\alpha^n}, R_{n+k}\right), R_{n+k}\right).$$

This implies

$$\alpha^n f^{2^n}\left(\frac{x}{\alpha^n}, R_{n+k}\right) = \alpha^n f^{2^{n-1}}\left(f^{2^{n-1}}\left(\frac{x}{\alpha^n}, R_{n+k}\right), R_{n+k}\right)$$
$$= \alpha\alpha^{n-1} f^{2^{n-1}}\left(\frac{1}{\alpha^{n-1}}\alpha^{n-1} f^{2^{n-1}}\left(\frac{(x/\alpha)}{\alpha^{n-1}}, R_{n+k}\right), R_{n+k}\right)$$
$$= \alpha\alpha^m f^{2^m}\left(\frac{1}{\alpha^m}\alpha^m f^{2^m}\left(\frac{(x/\alpha)}{\alpha^m}, R_{m+k+1}\right), R_{m+k+1}\right),$$

where $m = n - 1$. Taking the limit as $n \to \infty$ in the above relation, we get

$$g_k(x) = \alpha g_{k+1}\left(g_{k+1}\left(\frac{x}{\alpha}\right)\right),$$

which, in the limiting sense, yields the functional equation

$$g(x) = \alpha g\left(g\left(\frac{x}{\alpha}\right)\right) \tag{12.1}$$

Note that at the new origin (Fig. 12.15a), the function $f(x, r)$ has maximum at the point $x = 0$. This suggest that $g'(0) = 0$. Without loss of generality we take $g(0) = 1$. We now solve the functional equation (12.1) subject to the boundary conditions $g(0) = 1$ and $g'(0) = 0$. At $x = 0$, the equation yields $\alpha = 1/g(1)$. Since the function $g(x)$ is quadratic in x, we can expand it in the polynomial form

$$g(x) = 1 + a_1 x^2 + a_2 x^4 + a_3 x^6 + \cdots$$

so that the boundary conditions are satisfied. Substituting this into Eq. (12.1) and then equating the coefficients of like powers of x from both sides we can obtain the values of $a_i (i = 1, 2, 3, \ldots)$. But this task is very difficult analytically. In the year 1979, Feigenbaum [11] taking the polynomial of degree 14, calculated the coefficients $a_i (i = 1, 2, \ldots, 7)$ numerically correct up to ten significant figures as follows:

$$a_1 = -1.527632997,\ a_2 = 1.048151943 \times 10^{-1},$$
$$a_3 = 2.670567349 \times 10^{-2},\ a_4 = -3.52413864 \times 10^{-3},$$
$$a_5 = 8.158191343 \times 10^{-5},\ a_6 = 2.536842339 \times 10^{-5},$$
$$a_7 = -2.687772769 \times 10^{-6}.$$

This gives $\alpha = 1/g(1) = -2.5029$ (approx.). We can also calculate the value of α as follows:

Consider a 2^n-period cycle of the quadratic map $f(x, r)$. The successive approximate values of R_n are calculated as follows:

$$R_0 = 2$$
$$R_1 = 3.236079775$$
$$R_2 = 3.4985616993$$
$$R_3 = 3.5546408628$$
$$R_4 = 3.5666673799$$
$$R_5 = 3.5692435316$$

and so on. Suppose that d_n denotes the difference between the point $x = 1/2$ and the other point on the 2^n-cycle nearest to $x = 1/2$. Then

$$d_n = f^{2^{n-1}}\left(\frac{1}{2}, R_n\right) - \frac{1}{2}, n = 1, 2, 3, \ldots$$

The successive approximate values of d_n are obtained as follows:

$$d_1 = 0.30902$$
$$d_2 = -0.116402$$
$$d_3 = 0.0459752$$
$$d_4 = -0.0183262$$
$$d_5 = 0.00731843$$

and so on. We see that

$$\frac{d_1}{d_2} = -2.65477, \frac{d_2}{d_3} = -2.53184, \frac{d_3}{d_4} = -2.50871, \frac{d_4}{d_5} = -2.50412, \ldots.$$

Continuing this process we see that for large n, the ratio $\frac{d_n}{d_{n+1}}$ converges to the number (-2.5029) (approximately). The quantity $\alpha = \lim_{n \to \infty} \left(\frac{d_n}{d_{n+1}} \right) = -2.5029$ is a universal number, independent of the class of one-dimensional maps. The tremendous practical importance of the quadratic map lies in the existence of the universal numbers δ and α which are general characteristics of the period-doubling route to chaos (Fig. 12.16).

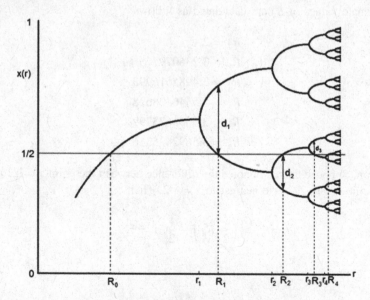

Fig. 12.16 Superstable cycles for the quadartic map

12.4.4 Lyapunov Exponent

The Lyapunov exponent of a map is an important quantitative measure to be chaotic. The basic idea of Lyapunov is that the rate of attraction or repulsion of nearby trajectories of a fixed point or any other point is exponentially fast. This gives an average measure at the exponential rate of which nearby orbits of a map $f : X \rightarrow X$ move apart. If the map possesses the property of sensitive dependence on initial conditions, the neighboring orbits should move apart exponentially. This would require a positive value of the average exponential rate which is diverging from neighboring orbits and hence a positive value of the Lyapunov exponent is a signature of chaos.

Consider a one-dimensional map $x_{n+1} = f(x_n)$ with two neighboring initial points x_0 and $x_0 + \varepsilon$. The Nth step of iteration takes the form $x_0 \rightarrow f^N(x_0)$ and $(x_0 + \varepsilon) \rightarrow f^N(x_0 + \varepsilon)$, ε being a small quantity. A number λ depending on the initial point x_0 is defined in the limit $N \rightarrow \infty$ as

$$\lim_{N \to \infty} e^{N\lambda} = \lim_{N \to \infty} \frac{|f^N(x_0 + \varepsilon) - f^N(x_0)|}{\varepsilon} \qquad (12.2)$$

Obviously, $\varepsilon \rightarrow 0$ as $N \rightarrow \infty$. The number λ is called the Lyapunov exponent. Obviously, the dynamics of a system is chaotic if $\lambda > 0$ and it is nonchaotic, that is, regular if $\lambda \leq 0$. Taking logarithm in the limiting Eq. (12.2), we get the following expression for λ:

$$\lambda = \lim_{\substack{N \to \infty \\ \varepsilon \to 0}} \frac{1}{N} \ln \frac{|f^N(x_0 + \varepsilon) - f^N(x_0)|}{\varepsilon}$$

$$= \lim_{N \to \infty} \frac{1}{N} \ln \left| \frac{df^N}{dx}(x_0) \right| = \lim_{N \to \infty} \frac{1}{N} \ln \left| \left((f^N)'(x_0) \right) \right|$$

Using chain rule of differentiation, we obtain the expression

$$\left(f^N\right)'(x_0) = \left(f(f^{N-1})\right)'(x_0)$$

$$= \left(f'(f^{N-1})\right)(x_0) \cdot \left(f^{N-1}\right)'(x_0)$$

$$= f'\left(f^{N-1}(x_0)\right) \cdot \left(f'(f^{N-2})\right)(x_0) \cdot \left(f^{N-2}\right)'(x_0)$$

$$= f'\left(f^{N-1}(x_0)\right) f'\left(f^{N-2}(x_0)\right) f'\left(f^{N-3}(x_0)\right) \ldots f'(f(x_0)) f'(x_0)$$

Since $f^k(x_0) = x_k \forall k \in \mathbb{Z}$, that is, $x_1 = f(x_0), x_2 = f^2(x_0) = f(x_1), x_3 = f^3(x_0) = f(x_2), \ldots$, we write the above expression as

$$\left(f^N\right)'(x_0) = f'(x_{N-1}).f'(x_{N-2}).f'(x_{N-3})\ldots f'(x_1).f'(x_0)$$

$$= \prod_{i=0}^{N-1} f'(x_i).$$

Therefore,

$$\lambda = \lim_{N \to \infty} \frac{1}{N} \ln \left| \prod_{i=0}^{N-1} f'(x_i) \right| = \lim_{N \to \infty} \frac{1}{N} \sum_{i=0}^{N-1} \ln|f'(x_i)|. \qquad (12.3)$$

Note that the Lyapunov exponent λ depends on the initial condition x_0. Thus for a given initial condition x_0, the Lyapunov exponent λ or $\lambda(x_0)$ of f is given by $\lambda = \lim\limits_{N \to \infty} \frac{1}{N}\sum_{i=0}^{N-1} \ln|f'(x_i)|$, provided the limit exists. We now find Lyapunov exponent for some important cases.

(I) Lyapunov exponent for stable periodic and superstable cycles:

Let f be a one-dimensional map with a stable k-cycle containing the initial point x_0. Since x_0 is an element of the k-cycle, it must be a fixed point of f^k. Again, since the cycle is stable, the multiplier $\left|\left(f^k\right)'(x_0)\right| < 1$. That implies $\ln\left|\left(f^k\right)'(x_0)\right| < \ln(1) = 0$. Therefore, the Lyapunov exponent is given by

$$\lambda = \lim_{N \to \infty} \frac{1}{N} \sum_{i=0}^{N-1} \ln|f'(x_i)| = \frac{1}{k} \sum_{i=0}^{k-1} \ln|f'(x_i)|$$

(Since the same k terms will come in the infinite sum)
Using the chain rule of differentiation in reverse, we get

$$\lambda = \frac{1}{k} \sum_{i=0}^{k-1} \ln|f'(x_i)| = \frac{1}{k}\ln\left|\left(f^k\right)'(x_0)\right| < 0.$$

Hence for stable periodic cycle the Lyapunov exponent is less than zero. So, stable periodic cycle is a regular property of a map. For superstable k-cycle, along with the k-cycle conditions the condition $\left|\left(f^k\right)'(x_0)\right| = 0$ must hold. Thus for superstble cycles the Lyapunov exponent λ is obtained as $\lambda = \frac{1}{k}\ln(0) = -\infty$. Thus the superstable cycle is also a regular property and we say that these properties are nonchaotic.

(II) Lyapunov exponent for the tent map

For the general tent map $T(x) = \begin{cases} 2rx, & 0 \leq x \leq 1/2 \\ 2r(1-x), & 1/2 \leq x \leq 1 \end{cases}$, we calculate $|T'(x)| = 2r, \forall x \in [0, 1]$, except at $x = 1/2$, the point of nondifferentiability, Here the parameter r lies in the interval $0 \leq r \leq 1$. Thus the Lyapunov exponent of the tent map is given by

$$\lambda = \lim_{N \to \infty} \frac{1}{N} \sum_{i=0}^{N-1} \ln|T'(x_i)| = \lim_{N \to \infty} \frac{1}{N} \sum_{i=0}^{N-1} \ln 2r = \lim_{N \to \infty} \frac{1}{N} \cdot N \ln 2r$$

$$= \ln 2r$$

Since $\lambda > 0$ for $2r > 1$, that is, for $r > 1/2$, the tent map is chaotic for $r > 1/2$. It is nonchaotic for $r \leq 1/2$. The transition from nonchaotic to chaotic behavior occurs at $r = r_c = 1/2$.

(III) Lyapunov exponent for the Bernoulli's shift map

The Bernoulli's shift is defined as

$$B(x) = \begin{cases} 2x & 0 \leq x \leq \frac{1}{2} \\ 2x - 1 & \frac{1}{2} \leq x \leq 1 \end{cases}$$

We have $|B'(x)| = 2$, $\forall x \in [0, 1]$. So, the Lyapunov exponent is obtained as

$$\lambda = \lim_{N \to \infty} \frac{1}{N} \sum_{i=0}^{N-1} \ln|B'(x_i)| = \lim_{N \to \infty} \frac{1}{N} \sum_{i=0}^{N-1} \ln 2$$

$$= \lim_{N \to \infty} \frac{1}{N} \cdot N \ln 2 = \ln 2 \simeq 0.693147.$$

Since the Lyapunov exponent is positive, Bernoulli's shift is chaotic.

(IV) Lyapunov exponent of the logistic map

As we know the logistic map is expressed as $f_r(x) = rx(1-x)$, $0 \leq r \leq 4$, $x \in [0, 1]$. We have $f_r'(x) = r(1 - 2x)$. Therefore, the Lyapunov exponent is obtained as

$$\lambda = \lim_{N \to \infty} \frac{1}{N} \sum_{i=0}^{N-1} \ln|f_r'(x_i)| = \lim_{N \to \infty} \frac{1}{N} \sum_{i=0}^{N-1} \ln|r(1 - 2x_i)|$$

Taking the initial point x_0 and the iterations $x_{n+1} = f_r(x_n), n = 0, 1, 2, \ldots$ one can calculate the Lyapunov exponent of the logistic map for different parameter values r. Figure 12.17 presents the Lyapunov exponent of the logistic map for $2.5 \leq r \leq 4$ taking 1000 iterations and $x_0 = 0.2$ as the initial point. The points at

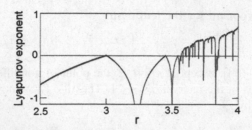

Fig. 12.17 Lyapunov exponent of the logistic map in the range $2.5 \leq r \leq 4$

which the graph touches the x-axis represent the bifurcation points. The points at which the graph moves to $-\infty$ are the points of superstability and they occur between two bifurcation points. The Lyapunov exponent for $r = 4$ is $\ln 2$. One can easily obtain this value because the tent map is conjugate to the logistic map $f_4(x)$ through the conjugacy map $h(x) = \sin^2(\pi x/2)$.

So far we have discussed the technique for calculating Lypunov exponents of one-dimensional maps. Our next target is to determine the Lyapunov exponents for higher dimensional systems. Consider a n-dimensional system represented by

$$\underset{\sim}{x}_{n+1} = \underset{\sim}{f}(\underset{\sim}{x}_n); \underset{\sim}{x}_n \in \mathbb{R}^n. \tag{12.4}$$

Take a point $\underset{\sim}{x}_0$ in the phase space of the system. Let $N(\underset{\sim}{x}_0)$ represent a small neighborhood of $\underset{\sim}{x}_0$. Choose another point $\underset{\sim}{x}_0 + \Delta \underset{\sim}{x}_0$ in $N(\underset{\sim}{x}_0)$, where $\Delta \underset{\sim}{x}_0$ represents the separation of the neighboring two points $\underset{\sim}{x}_0$ and $\underset{\sim}{x}_0 + \Delta \underset{\sim}{x}_0$. These two points correspond to two orbits of the system under iterations. Let the successive points on the orbits be $\underset{\sim}{x}_0, \underset{\sim}{x}_1, \underset{\sim}{x}_2, \ldots, \underset{\sim}{x}_N, \ldots$ and $\underset{\sim}{x}_0 + \Delta \underset{\sim}{x}_0, \underset{\sim}{x}_1 + \Delta \underset{\sim}{x}_1, \underset{\sim}{x}_2 + \Delta \underset{\sim}{x}_2, \ldots, \underset{\sim}{x}_N + \Delta \underset{\sim}{x}_N, \ldots$, respectively, where $\underset{\sim}{x}_i = \underset{\sim}{f}(\underset{\sim}{x}_{i-1})$ and $\underset{\sim}{x}_i + \Delta \underset{\sim}{x}_i = \underset{\sim}{f}(\underset{\sim}{x}_{i-1} + \Delta \underset{\sim}{x}_{i-1})$, $i = 1, 2, \ldots, N, \ldots$. Therefore, the separations between the successive neighboring points on the orbits are

$$\Delta \underset{\sim}{x}_1 = \underset{\sim}{f}(\underset{\sim}{x}_0 + \Delta \underset{\sim}{x}_0) - \underset{\sim}{f}(\underset{\sim}{x}_0) \simeq Df(\underset{\sim}{x}_0)\Delta \underset{\sim}{x}_0,$$

$$\Delta \underset{\sim}{x}_2 = \underset{\sim}{f}(\underset{\sim}{x}_1 + \Delta \underset{\sim}{x}_1) - \underset{\sim}{f}(\underset{\sim}{x}_1) \simeq Df(\underset{\sim}{x}_1)\Delta \underset{\sim}{x}_1,$$

$$\Delta \underset{\sim}{x}_3 = \underset{\sim}{f}(\underset{\sim}{x}_2 + \Delta \underset{\sim}{x}_2) - \underset{\sim}{f}(\underset{\sim}{x}_2) \simeq Df(\underset{\sim}{x}_2)\Delta \underset{\sim}{x}_2,$$

$$\vdots$$

$$\Delta \underset{\sim}{x}_N = \underset{\sim}{f}(\underset{\sim}{x}_{N-1} + \Delta \underset{\sim}{x}_{N-1}) - \underset{\sim}{f}(\underset{\sim}{x}_{N-1}) \simeq Df(\underset{\sim}{x}_{N-1})\Delta \underset{\sim}{x}_{N-1},$$

$$\vdots$$

where $Df(x_i)$ represents the $n \times n$ Jacobian matrix of the map f at the point x_i. Therefore, we can easily determine the separation Δx_N at the Nth iteration as follows:

$$\Delta x_N = Df(x_{N-1})Df(x_{N-2})\cdots Df(x_1)Df(x_0)\Delta x_0 = D_N \Delta x_0 \qquad (12.5)$$

where $D_N = Df(x_{N-1})Df(x_{N-2})\cdots Df(x_1)Df(x_0)$, a $n \times n$ matrix depends on both N and x_0. From this discussion, it follows that the separation of two neighboring orbits at the Nth iteration is completely determined by the matrix D_N and the initial separation Δx_0. Since the phase space is multidimensional, we have to introduce the eigenvectors of the matrix D_N as a natural basis for the decomposition of the vectors in the phase space. The eigenvectors, $e(N)$, of the matrix D_N are obtained from the relation

$$D_N e_i(N) = v_i(N) e_i(N), \; i = 1, 2, \ldots, n \qquad (12.6)$$

where both $v_i(N)$ and $e_i(N)$ depend on N. The eigenvectors are generally chosen as orthonormal. If they are not orthonornal, we make them orthonormal using Gram–Schmidt orthonormalization technique. The quantities $v_i(N)$ are known as the eigenvalues of the matrix D_N and are determined from the characteristic equation

$$\det(D_N - v(N)I) = 0 \qquad (12.7)$$

where I is the identity matrix of order n. The characteristic Eq. (12.7) gives n eigenvalues. Suppose in terms of the basis vectors (eigenvectors) $e_i(N)$, the initial separation Δx_0 is expressed as

$$\Delta x_0 = \sum_{i=1}^{n} c_i e_i(N) \qquad (12.8)$$

where c_i's are the coordinates of Δx_0 in the basis $e(N)$. Substituting (12.8) into (12.5), we get

$$\Delta x_N = D_N \sum_{i=1}^{n} c_i e_i(N) = \sum_{i=1}^{n} c_i D_N e_i(N) = \sum_{i=1}^{n} c_i v_i(N) e_i(N) \qquad (12.9)$$

From (12.8) and (12.9) it is clear that the separation of the orbits along the ith direction is characterized by the eigenvalue $v_i(N)$ of the coefficient matrix D_N. The exponential rate of separation of orbits leads to the Lyapunov exponents. Let λ_i denote the Lyapunov exponent along the ith direction. Then

$$|v_i(N)| \approx e^{\lambda_i N} \text{ for large } N.$$

This implies

$$\lambda_i = \lim_{N\to\infty} \frac{1}{N} \ln(|v_i(N)|), \; i = 1, 2, \ldots, n \tag{12.10}$$

Equation (12.10) gives an estimation of the Lyapunov exponents for multidimensional systems. Note that the number of Lyapunov exponents of a multidimensional system is equal to the dimensionality of the phase space of that system. The Lyapunov exponents are positive, zero as well as negative. A positive Lyapunov exponent signifies chaotic behavior.

Example 12.2 Determine the Lyapunov exponents for the Skinny-Baker map B (x, y) on the unit square $[0, 1] \times [0, 1]$.

Solution The Skinny-Baker map $B(x, y)$ is defined by

$$B(x,y) = \begin{cases} \begin{pmatrix} 1/3 & 0 \\ 0 & 2 \end{pmatrix} \begin{pmatrix} x \\ y \end{pmatrix}, & 0 \le y \le 1/2 \\ \begin{pmatrix} 1/3 & 0 \\ 0 & 2 \end{pmatrix} \begin{pmatrix} x \\ y \end{pmatrix} + \begin{pmatrix} 2/3 \\ -1 \end{pmatrix}, & 1/2 < y \le 1 \end{cases}$$

The Jacobian matrix of $B(x, y)$ at a point $(x, y) \in [0, 1] \times [0, 1]$ is given by $J = \begin{pmatrix} 1/3 & 0 \\ 0 & 2 \end{pmatrix}$. After Nth iterations the Jacobian matrix is $D_N = \begin{pmatrix} \frac{1}{3^N} & 0 \\ 0 & 2^N \end{pmatrix}$. The matrix D_N has two distinct eigenvalues, namely $v_1 = 1/3^N$ and $v_2 = 2^N$. Therefore, the map has two distinct Lyapunov exponents and are given by

$$\lambda_1 = \lim_{N\to\infty} \frac{1}{N} \ln(1/3^N) = -\ln 3 < 0 \text{ and } \lambda_2 = \lim_{N\to\infty} \frac{1}{N} \ln(2^N) = \ln 2 > 0.$$

Hence the Skinny-Baker map is chaotic. This can be shown easily that the dynamics of Skinny-Baker map has Cantor set-like structure.

Now if we start with the unit square $[0, 1] \times [0, 1]$ and then applying the given map $B(x, y)$ to each of the rectangles $[0, 1] \times [0, 1/2]$ and $[0, 1] \times [1/2, 1]$, it gives a Cantor-like structure, that is, yields two rectangles $[0, 1/3] \times [0, 1]$ and $[2/3, 1] \times [0, 1]$, each rectangle is of base length 3^{-1} and height of unit length, creating a gap of rectangular column of cross section $[1/3, 2/3] \times [0, 1]$, that is, the area of the square get reduced by 1/3 square unit. Applying once again the transformation $B(x, y)$ to each of the rectangles $[0, 1/3] \times [0, 1/2] \cup [0, 1/3] \times [1/2, 1]$ and $[2/3, 1] \times [0, 1/2] \cup [2/3, 1] \times [1/2, 1]$ would yield four rectangles $[0, 1/9] \times [0, 1]$, $[2/3, 7/9] \times [0, 1]$, $[2/9, 1/3] \times [0, 1]$, and $[8/9, 1] \times [0, 1]$, each rectangle is of base length 3^{-2} and height 1. This process of composition continues in this way and eventually yields 2^n rectangles each with a base length 3^{-n} and height 1. For each composition the area of the cross section becomes 2/3 of the previous. Since the successive composition of Skinny-Baker map gives Cantor set-like structure, the given Skinny-Baker map is self-similar.

12.4.5 *Invariant Measure*

We have studied the dynamics of logistic map for $r \geq 4$. It has been observed that the frequency of visits of an orbit falls in the set $\Lambda_r (r > 4) \subset [0, 1]$ which is a Cantor set in [0,1]. Specifically, for $r > (2 + \sqrt{5})$ the orbit has Cantor set structure. Frequency of visits of an orbit is also described mathematically by a function known as invariant density denoted by $\rho(x)$. The invariant density function is defined for any interval $[a,b]$ in which the orbits visit in the interval $[a,b]$ of the fraction of time given by $\int_a^b \rho(x) \mathrm{d}x$. For the tent map $\rho(x) = 1$ in [0,1], since the map divides the interval symmetrically into two halves such that the oribit visits each half exactly once. The probabilistic view of frequency of visits of orbits in an interval is called the natural invariant density function $\rho(x)$. We find the natural invariant density function $\rho(x)$ for the one-dimensional logistic map for $r = 4$ for randomly chosen initial conditions x_0 in the interval [0,1]. Let $\rho(y)$ be the natural invariant density function for the one-dimensional tent map $T(y)$. As in the definition for density function, $\bar{\rho}(y) = 1, 0 \leq y \leq 1$, for tent map. We know that the tent map is topologically conjugate to the logistic map at $r = 4$ through the conjugacy map $h(y) = \sin^2(\pi y/2)$. So we consider that two maps must visit at same frequency to the respective intervals $[y, y + dy]$ and $[x, x + dx]$. So, we have

$$\bar{\rho}(y)|\mathrm{d}y| = \rho(x)|\mathrm{d}x| \Rightarrow \rho(x) = \left| \frac{\mathrm{d}y}{\mathrm{d}x} \right| \bar{\rho}(y)$$

Now, $x = h(y) = \sin^2(\pi y/2) \Rightarrow y = h^{-1}(x) = 2\pi^{-1} \sin^{-1}(\sqrt{x})$. So, $\frac{\mathrm{d}y}{\mathrm{d}x} = \pi^{-1} \frac{1}{\sqrt{x(1-x}}$ and $\bar{\rho}(y) = 1$. The natural density function $\rho(x)$ is obtained as $\rho(x) = \frac{1}{\pi\sqrt{x(1-x)}}$. Note that $\rho(x)$ has singularities at $x = 0$ and $x = 1$. The density functions for logistic map at $r = 3.8$ and at $r = 4$ are shown in Fig. 12.18.

The natural density functions for general and higher dimensional maps have some limitations. It canot be defined properly. Instead of density function one can equivalently deal with the set theoretic measure denoted by μ. Measure is basically just a way of measuring the size of a set. We now briefly remind the notion of measurable sets, outer and inner measure of a set in $\mathbb{R}$, where length of a closed and bounded interval $[a, b] = b - a$. The outer measure of any bounded subset, say $A \subset \mathbb{R}$ is the infimum of the length of all open sets which contain A and the inner measure is the supremum taken over the lengths of all the closed sets B contained in A. The set A is measurable if its outer measure is equal to the inner measure. In this

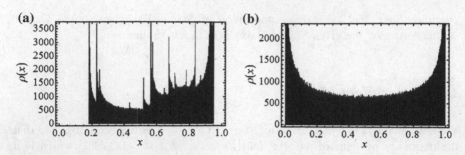

Fig. 12.18 The natural density function $\rho(x)$ at **a** $r = 3.8$ and **b** $r = 4$

case the measure of A is the same as its outer and inner measures. Mathematically, the measure μ is defined as follows:

A function $\mu : \mathcal{A} \to \mathbb{R}$ is called a *measure* if it assigns a nonnegative number to each subset of X such that

(i) $\mu(\phi) = 0$, ϕ is an empty set;
(ii) For $B \subset C, \mu(B) \leq \mu(C)$
(iii) If E_n is a countable collection of pairwise disjoint sets in $\mathcal{A}$ (i.e., $E_n \cap E_m = \phi$ for $n \neq m$) then $\mu\left(\bigcup_{n=1}^{\infty} E_n\right) = \sum_{n=1}^{\infty} \mu(E_n)$.

The triplet $(X, \mathcal{A}, \mu)$ is then called a measure space. Again, if $\mu(X) < \infty$, then μ is called as finite measure. We now define invariant measure in context of dynamical system.

Let X be a metric space and $f : X \times \mathbb{R} \to X$ be a flow generated by the system. Now for $A \subset X$, we consider the set $I(A) = \{x \in A : x(\mathbb{R}) \subset A\}$ in which the set of points of A remains in A for all positive and negative times under the flow f. The set A is called invariant if $I(A) = A$.

We now briefly define the σ-algebra.

A set $\mathcal{B}$ of subsets of a set X is called σ-algebra of subsets of X if the following properties hold:

(i) The empty set $\phi \in \mathcal{B}$, and $X \in \mathcal{B}$
(ii) If $E \in \mathcal{B}$ then its complement $X \backslash E \in \mathcal{B}$,
(iii) If $E_n \in \mathcal{B}, n = 1, 2, \ldots$, is a countable sequence of sets S in $\mathcal{B}$ then their union $\bigcup_{n=1}^{\infty} E_n \in \mathcal{B}$.
    This implies the following property:
(iv) For finite collection $E_1, E_2, \ldots, E_n \in \mathcal{B}$, $\bigcap_{i=1}^{n} E_n \in \mathcal{B}$

The space $(X, \mathcal{B})$ is said to be a measurable space if X is a set and $\mathcal{B}$ a σ-algebra of subsets of X.

In general, it is very difficult to find the measure of any arbitrary subset of X. So, for studying chaos one should consider the set theoretic measure in *Borel* σ-algebra of subsets of X.

For a compact metric space X, the *Borel* σ-*algebra* $\mathcal{B}(X)$ is defined as the smallest σ-algebra of subsets of X which contains all the open subsets of X. The members of the *Borel* σ-algebra are called the Borel subsets of X (see Royden [12]).

Probability measure

The probability measure on a measurable space $(X, \mathcal{B})$ is a function $\mu : \mathcal{B} \to [0, 1]$ if it satisfies the following properties:

(i) The measure of union of all pairewise disjoint sets B_n equals the sum of the measure of these sets, that is, $\mu\left(\bigcup_{n \in \mathbb{N}} B_n\right) = \sum_{n \in \mathbb{N}} \mu(B_n)$, if $B_i \cap B_j = \phi, \forall i \neq j$,

(ii) The measure of any empty set $\phi \in \mathcal{B}$ is zero, that is, $\mu(\phi) = 0$,

(iii) $\mu(X) = 1$.

So, we can say that a measure μ of a set X is probability measure if measure of X, that is, $\mu(X) = 1$. Moreover, the measure of any subset of X is in between 0 and 1, that is, $0 < \mu(A) < 1$, $A \neq \phi \subset X$. The triplet $(X, \mathcal{B}, \mu)$ is referred as a probability space, where $\mathcal{B}$ is the event space and X is the sample space with probability measure μ.

A transformation $f : X \to X$ of a probability space $(X, \mathcal{B}, \mu)$ is said to be measurable if $f^{-1}B \in \mathcal{B}$, for all $B \in \mathcal{B}$. Now we give the definition of measure preserving map.

Definition 12.10 A measure μ for a map $f : X \to X$ of the probability space $(X, \mathcal{B}, \mu)$ is said to be an invariant probability measure if $\mu(f^{-1}B) = \mu(B)$ for all measurable sets $B \in \mathcal{B}$ and the map or transformation f is called measure preserving.

In context of dynamic evolution for a map f, the invariant probability measure means that the probability distribution of $f(x)$ is unaltered in X.

For instance, Leabsgue measure L^1 of a set is an invariant measure. Leabsgue measure is an extension of the notion of length, area, volume to the Borel subsets of $\mathbb{R}, \mathbb{R}^2, \mathbb{R}^3$. Leabsgue measure for open and closed subsets of $\mathbb{R}$, that is, $L(a, b)$ and $L[a, b]$ is $b - a$, again $L(A) = \sum (b_i - a_i)$ where $A = \bigcup_{n \in \mathbb{N}} [a_n, b_n]$ such that $[a_n, b_n] \cap [a_m, b_m] = \phi, n \neq m$. So for any arbitrary set A in $\mathbb{R}^n$, the Leabsgue measure $L^n(A)$ is defined as

$$L^n(A) = \inf\left\{ \sum_{n=1}^{\infty} I_n : A \subset \bigcup_{n=1}^{\infty} I_n \right\},$$

where I_n is the cover of A.

Definition 12.11 (*Ergodic map*) A map $f : X \rightarrow X$ is called an ergodic map on a probability measure space $(X, \mathcal{B}, \mu)$ if for every $A \in \mathcal{B}$ either $\mu(A) = 0$ or $\mu(A) = 1$.

Ergodic hypothesis states that for an ergodic map f with respect to certain invariant measure μ, many smooth functions $g : X \rightarrow \mathbb{R}$, and many states $x = (q, p)$, the time average of g, that is, $\lim\limits_{T \rightarrow \infty} \int_0^T g(f^n(x_0))dt$ exists and equals to the space average of g which is $\int_X g(x)d\mu(x)$.

An invariant probability measure μ is called *ergodic* if it cannot be decomposed in the form $\mu = P\mu_1 + (1 - P)\mu_2$, where μ_1 and μ_2 are invariant probability measures such that $\mu_1 \neq \mu_2$ and $0 < P < 1$.

Theorem 12.7 (Ergodic theorem) *For an integrable function g(x), the ergodic probability measure μ has value 1 for all those values of $x_0 \in A(\in \mathcal{B}$, a Borel σ-algebra) which satisfies the ergodic hypothesis and has value 0 for all those values of x_0 which do not satisfy the hypothesis.*

The ergodic theorem was proved by G.D. Birkhoff in 1931. More formally, this theorem can be stated as follows:

If $T : X \rightarrow X$ is an ergodic measure preserving map on a probability measure space $(X, \mathcal{B}, \mu)$ with invariant probability measure $\mu(x)$, and $f \in L^1(\mu)$ then

$$\lim_{n \rightarrow \infty} \frac{1}{n} \sum_{k=0}^{n-1} f(T^k x) = \int_X f(x)d\mu(x) \qquad \text{for almost all } x.$$

For an ergodic continuous flow $(T_t)_{t \geq 0}$

$$\lim_{n \rightarrow \infty} \frac{1}{T} \int_0^T f(T_t x)dt = \int_X f(x)d\mu(x) \quad \text{for almost all } x.$$

The left- and right-hand sides give the time average and ensemble average, respectively. The ergodic theorem gives the result that the temporal mean coincides with the spatial mean. Now we give some examples in connection with the ergodicity and invariant measure.

(a) Invariant measure of logistic and tent maps

Consider the logistic map $f_4(x) = 4x(1 - x)$ which is chaotic on the set $X = [0, 1]$. If x_0 is not a periodic point of f_4, orbit $O(x_0)$ of x_0 will visit every open set $U \subset [0, 1]$, that is, visit the neighborhood, however small of every $y \in [0, 1]$ (due to transitivity property of the chaotic map). The periodic points are enumerable. Hence for every measurable set A either $\mu(A) = 0$, where A is a set of measurable points or $\mu(A) = \mu(X)$, where A contains a chaotic orbit. So, the logistic map f_4 is ergodic on $(X, \mathcal{B}, \mu)$ and hence the tent map T is also ergodic because both maps are conjugate. The chaotic orbit $O(x)$ of any point x has an invariant measure, that is,

$\mu(A) = \mu(f^{-1}(A))$. Here the map f is the logistic map f_4 or the tent map T. If μ is the probability measure of X then we have

$$\mu[x, x + \mathrm{d}x] = \mu(f^{-1}[x, x + \mathrm{d}x])$$
$$= \mu([x_1, x_1 + \mathrm{d}x_1]) + \mu([x_2 - \mathrm{d}x_2, x_2])$$
$$\Rightarrow \mu(x)\mathrm{d}x = \mu(x_1)\mathrm{d}x_1 + \mu(x_2)\mathrm{d}x_2,$$

$\mu(x)\mathrm{d}x$, $\mu(x_1)\mathrm{d}x_1$, and $\mu(x_2)\mathrm{d}x_2$ are the probabilities that a trajectory visit the interval between x and $x + dx$, x_1 and $x_1 + dx_1$, and x_2 and $x_2 + dx_2$, respectively. Here $x = f$ (x_1) when $x = x_1$ and $x = f(x_2)$ when $x = x_2$ for the unimodal map. This follows that

$$\mu(x) = \frac{\mu(x_1)}{\left|\frac{\mathrm{d}f}{\mathrm{d}x}\right|_{x=x_1}} + \frac{\mu(x_2)}{\left|\frac{\mathrm{d}f}{\mathrm{d}x}\right|_{x=x_2}}.$$

The tent map $T(x) : [0, 1] \rightarrow [0, 1]$ is defined as

$$T(x) = \begin{cases} 2x & \text{for } 0 \leq x \leq \frac{1}{2} \\ 2(1 - x) & \text{for } \frac{1}{2} \leq x \leq 1 \end{cases}$$

This gives $\frac{\mathrm{d}f}{\mathrm{d}x} = \frac{\mathrm{d}T}{\mathrm{d}x} = \pm 2$. So for the tent map

$$\mu(x) = \frac{1}{2}[\mu(x_1) + \mu(x_2)] \quad (x_2 = 1 - x_1 \text{ by symmetry}).$$

Since for an ergodic map in a probability measure space, probability measure μ is either 0 or 1. Now since the tent map is chaotic and divides the phase space into two symmetrical halves where the orbit occurs exactly once in each halves, we have $\mu(x_1) = 1$ and $\mu(x_2) = 1$. Hence the solution of the functional equation gives $\mu(x) = 1$ (normalization on [0,1]). The logistic map $f_4(x)$ is topologically conjugate to the tent map $T(x)$ through the conjugacy map $h(y) = \sin^2\left(\frac{\pi}{2}y\right)$. This implies that $x = h(y) = \sin^2\left(\frac{\pi}{2}y\right) \Rightarrow y = h^{-1}(x) = \frac{2}{\pi}\sin^{-1}(\sqrt{x})$. The invariant measure for the logistic map f_4 is therefore given by $\mu(x) = \mu(y)\frac{\mathrm{d}y}{\mathrm{d}x} = \frac{1}{\pi\sqrt{x(1-x)}}$[since $\mu(y) = 1$ is the probability density for the tent map $T(y)$].

We have calculated the Lyapunov exponent for the logistic map for different values of r. The Lyapunov exponent of $f_4(x)$ can be obtained easily through probability measure function $\rho(x)$. The Lyapunov exponent is obtained as

$$\lambda = \lim_{n\to\infty} \sup \frac{1}{n} \sum_{i=0}^{n-1} \ln|f_4'(x_i)| = \int_X \ln|f_4'(x)d\mu(x)|$$

$$= \int_0^1 \frac{\ln|4(1-2x)|}{\pi\sqrt{x(1-x)}}dx \ (\text{put } x = \sin^2\theta)$$

$$= \frac{2}{\pi} \int_0^{\pi/2} [2\ln 2 + \ln\cos 2\theta]d\theta$$

$$= \ln 2.$$

Note that the values of Lyapunov exponents for the logistic map with $r = 4$, the tent map with $r = 1$, and the Bernoulli shift map are same, positive, and equal to $\ln 2$. These three maps are conjugate to each other through conjugacy maps. This suggests that these three cojugacy maps have same Lyapunov exponent and they are chaotic.

(b) **Invariant measure of doubling map**

Consider the doubling map $T(x) : X \to X,$ $\begin{cases} 2x & 0 \le x < \dfrac{1}{2} \\ 2x-1 & \dfrac{1}{2} \le x < 1 \end{cases}$ $, X = \mathbb{R}\backslash\mathbb{Z}$ is a circle.

The doubling map is also known as Bernoulli's shift map and can also be written as $x_{n+1} = 2x_n$ modulo 1. The doubling map is a map on a circle, as the variable x increasing from 0 to 1 acts like an angle because of modulo 1, and is analogus to one round around the circle. The given map therefore can be regarded as a stretch–twist and fold operations. Under these operations, the given map at first expands the circle twice, which means that the circumference will be twice to that of the original circle. This expanded circle is then twisted into two lobes upper and lower, each of them are circles of original length. And eventually the two circles are pressed together by folding down the upper circle on to the lower circle. By this operation a point in the original circle is mapped to a point in the final compressed double circle. The foregoing discussion thus clears the reason of calling the given map, doubling map. We now show that this map is measure preserving.

Let μ be the measure, then to prove that the doubling map is μ-invariant we have to prove that $\mu(T^{-1}[a,b]) = \mu[a,b]$. Now $T^{-1}[a,b] = \{x \in \mathbb{R}\backslash\mathbb{Z} : T(x) \in [a,b] \subset \mathbb{R}\backslash\mathbb{Z}\} = \left[\frac{a}{2}, \frac{b}{2}\right] \cup \left[\frac{a+1}{2}, \frac{b+1}{2}\right]$. Therefore,

$$\mu(T^{-1}[a,b]) = \mu\left(\left[\frac{a}{2}, \frac{b}{2}\right] \cup \left[\frac{a+1}{2}, \frac{b+1}{2}\right]\right) = \mu\left[\frac{a}{2}, \frac{b}{2}\right] + \mu\left[\frac{a+1}{2}, \frac{b+1}{2}\right]$$

$$= \frac{b}{2} - \frac{a}{2} + \frac{b+1}{2} - \frac{a+1}{2} = b - a = \mu[a,b]$$

Hence, the doubling map is a measure preserving map.

The doubling map is similar to the tent map, and it also divides the phase space into intervals of equal length where the orbit occurs exactly once. And the histogram of the fraction of times an orbit of finite length originating from a particular initial condition is equal for each interval along the x-axis. So, if the phase space is divided into N equal intervals then the probability distribution in each interval is equal to $1/N$ when the length of the orbit tends to infinity. Hence we have the probability measure for the doubling map, that is, $\mu(x) = 1/2 + 1/2 = 1$ in $[0,1]$.

12.4.6   Sharkovskii Order

The common well-known order in natural numbers is $1, 2, 3, 4, 5, 6, 7, 8, 9, \ldots$. Let us consider another ordering of natural numbers presented below.

$$3 \rhd 5 \rhd 7 \rhd \cdots \rhd 2 \cdot 3 \rhd 2 \cdot 5 \rhd 2 \cdot 7 \rhd \cdots \rhd 2^2 \cdot 3 \rhd 2^2 \cdot 5 \rhd 2^2 \cdot 7 \rhd \cdots$$
$$\cdots \rhd 2^n \cdot 3 \rhd 2^n \cdot 5 \rhd 2^n \cdot 7 \rhd \cdots \rhd 2^n \rhd \cdots \rhd 2^4 \rhd 2^3 \rhd 2^2 \rhd 2 \rhd 1$$

This ordering is known as the Sharkovskii order of natural numbers and is constructed as follows:

First, list all positive odd integers greater than 1 in increasing order. Then list the integers that are 2 times the odd integers greater than 1 in increasing order and then the numbers which are 2^2 times of the odd integers, and so on. Finally, list the integers in decreasing order that are the integral powers of 2. In the Sharkovskii order, the integer 3 is the smallest number and 1 is the largest.

Sharkovskii's theorem

In 1964, the Russian (Ukrainian) mathematician, *A.N. Sharkovskii* in his paper *"Coexistence of Cycles of a Continuous Map of a line into itself,"* published in *Ukrainian Mathematical Journal* (1964), proved a remarkable theorem in discrete dynamical system. According to his name the theorem is called Sharkovskii's theorem. The theorem plays an important role in verifying the existence of periodic cycles of certain periods of a one-dimensional real-valued map from the existence of periodic cycles of different periods of the map. In this section we only state the theorem and then deduce some important results from it. See Devaney [9] for proof.

Theorem 12.8 *Let $f : [a, b] \to \mathbb{R}$ be a continuous map. Suppose f has periodic cycle of period n. If $n \rhd k$ in the Sharkovskii order, then the map f also has a periodic cycle of period k.*

For explanation, if a map f has a 8-cycle, then it also has at least one 4-cycle, one 2-cycle and one 1-cycle (equilibrium point). Similarly, if f has a periodic-3 cycle, then it has all cycles according to Sharkovskii order. The converse of Sharkovskii's theorem is also true and is given below.

Theorem 12.9 *For each natural number n there exists a continuous map* $f : \mathbb{R} \to \mathbb{R}$ *that has a periodic cycle of period n but no cycles of period k for any k appearing before n in the Sharkovskii order.*

Some remarks:

(i)   Sharkovskii's theorem is applicable only for real numbers.

(ii) If f has a periodic cycle whose period is not a power of 2, then f has infinitely many periodic cycles. On the other hand, if f has only a finite number of periodic cycles, then they all necessarily have periods equal to a power of 2.

(iii) If f has a periodic-3 cycle, then it has all cycles. This is simply the theorem of Li and Yorke, "Period-3 implies all periods."

(iv) It does not give any result about the stability of the cycles. It just shows the existence of cycles of certain periods.

Example 12.3 Use the converse of Sharkovskii's theorem to show that there is a continuous map $f : \mathbb{R} \to \mathbb{R}$ with periodic points of order 5, but no period-3 orbits.

Solution Consider the piecewise linear map $f: [1, 5] \to [1, 5]$ satisfying

$$f(1) = 3, f(2) = 5, f(3) = 4, f(4) = 2, f(5) = 1.$$

The graphical representation of the map f is shown in Fig. 12.19.

From the figure we see that $f^5(k) = k$ for $k = 1, 2, 3, 4, 5$. Therefore, the points 1, 2, 3, 4, and 5 are the periodic-5 points of the map f. Now,

$$f([1,2]) = [3,5], f([3,5]) = [1,4], f([1,4]) = [2,5]$$
$$\Rightarrow \quad f^3([1,2]) = [2,5].$$

Thus the only fixed point of the map f^3 in [1,2] may be the point 2. But it is a periodic-5 point of f. So f^3 has no fixed point in [1,2]. Similarly, f^3 has no fixed points in both the intervals [2,3] and [4,5]. Now, $f([3, 4]) = [2, 4], f([2, 4]) = [2, 5], f([2, 5]) = [1, 5]$ imply $f^3([3,4]) = [1,5]$. Again, since $f : [3,4] \to [2,4], f : [2,4] \to [2,5]$, and $f: [2, 5] \to [1, 5]$ are monotonically decreasing, the function $f^3 : [3,4] \to [1,5]$ is also monotonically decreasing on [3,4]. Therefore, f^3 has a

Fig. 12.19 Graphical
representation of the map f

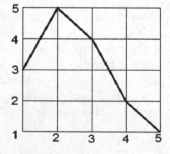

unique fixed point on [3,4]. That is, there exists a point $\alpha \in [3, 4]$ such that $f^3(\alpha) = \alpha$.

Now, on [3,4], $f(x)$ is given by $f(x) = -2x + 10$. Clearly, $\beta = 10/3$ is the unique fixed point of f on [3,4]. Again, on [3,4], $f^2(x)$ and $f^3(x)$ are, respectively, given as $f^2(x) = -10 + 4x$ and $f^3(x) = 30 - 8x$, and they all have unique fixed points at $\beta = 10/3$. Since, $f^3(\beta) = f^2(\beta) = f(\beta) = \beta$, α is not a period-3 point of f. Hence f has no 3-periodic points and therefore it has no 3 cycles.

12.4.7   Period 3 Implies Chaos

In the year 1975, *James Alan Yorke* and his doctoral student *Tien-Yien Li* in a paper entitled "Period 3 Implies Chaos" [13], proved a remarkable theorem which states that if a continuous function on some closed interval in the real line has a point of period 3, then it has points of all periods. Before proceeding to the proof of the theorem we state the following two lemmas.

Lemma 12.1 *Let $f : \mathbb{R} \to \mathbb{R}$ be a continuous map. If $f(I) \supseteq I$, then f has a fixed point in I.*

Lemma 12.2 *Let $f : \mathbb{R} \to \mathbb{R}$ be a continuous map, and I and J be two closed intervals such that $f(I) \supseteq J$. Then there exists a closed interval $K \subseteq I$ such that $f(K) = J$.*

Theorem 12.10 *Let $f : X \to X$ be a continuous map defined on some interval $X \subseteq \mathbb{R}$. If f has a periodic point of period 3, then it has a periodic point of period n for each $n \in \mathbb{N}$.*

Proof Since the map f has a periodic point of period 3, there exists a cycle $\{a,b,c\}$ such that either $f(a) = b$ or $f(a) = c$. Without loss of generality, we assume that $f(a) = b$. Then $f(b) = c$ and $f(c) = a$. We also assume that $a < b < c$. Let $I = [a, b]$ and $J = [b, c]$. Then by the continuity of f (Fig. 12.20), we have

(i) $f(a) = b, f(b) = c \Rightarrow f(I) \supseteq J$ and
(ii) $f(b) = c, f(c) = a \Rightarrow f(J) \supseteq [a,c] = I \cup J$

That is, $f(I) \supseteq J$ and $f(J) \supseteq I \cup J$.

Step I f has a periodic-1 point, that is, a fixed point

Since $f(J) \supseteq I \cup J \supseteq J$, by Lemma 12.1 f has a fixed point in J. Thus the theorem follows for $n = 1$.

Step II f has a periodic-2 point

Since $f(J) \supseteq I \cup J \supseteq I$, by Lemma 12.2 there exists a closed interval $K \subseteq J$ such that $f(K) = I$. Since $f^2(K) = f(I) \supseteq J \supseteq K$, by Lemma 12.1 the map f^2 has a fixed point, say, α in K. We see that $f(\alpha) \in f(K) = I$ and $\alpha \in K \subseteq J$. So, $f(\alpha) \neq \alpha$. Therefore, α is a periodic-2 point of f. Thus the theorem follows for $n = 2$.

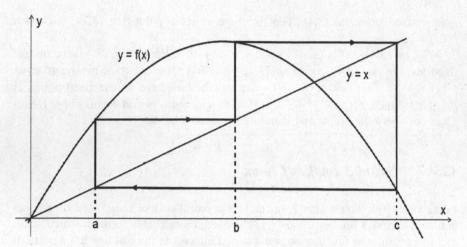

Fig. 12.20 Cobweb diagram for a periodic-3 cycle

Step III f has a periodic point of period $n(>3)$

Since $f(J) \supseteq J$, Lemma 12.2 implies that there exists a closed interval $K_1 \subseteq J$ such that $f(K_1) = J$. Again, since $f(K_1) = J \supseteq K_1$, by Lemma 12.2 there exists a closed interval $K_2 \subseteq K_1 \subseteq J$ such that $f(K_2) = K_1$. Then $f^2(K_2) = f(K_1) = J$. Continuing in this process $(n-2)$ times, we get a closed interval $K_{n-2} \subseteq K_{n-3}$ such that $f(K_{n-2}) = K_{n-3}$ and $f^{n-2}(K_{n-2}) = J$. Since $f(J) \supseteq I$, we have $I \supseteq f^{n-1}(K_{n-1})$. So, by Lemma 12.2 there exists a closed interval a closed interval $K_{n-1} \subseteq K_{n-2}$ such that $f^{n-1}(K_{n-1}) = I$. Again, since $f(I) \supseteq J$, we have $J \subseteq f^n(K_{n-1})$. Thus we obtain a nested collection of closed intervals $K_{n-1} \subseteq K_{n-2} \subseteq K_{n-3} \subseteq \cdots \subseteq K_2 \subseteq K_1 \subseteq K_0 = J$ such that

(i) $f(K_i) = K_{i-1}$, $i = 1, 2, ..., n-2$
(ii) $f^i(K_i) = J$, $i = 1, 2, ..., n-2$
(iii) $f^{n-1}(K_{n-1}) = I$
(iv) $J \subseteq f^n(K_{n-1})$

From (iv) it follows that $K_{n-1} \subseteq J \subseteq f^n(K_{n-1})$ and therefore by Lemma 12.1 f^n has a fixed point, say α in K_{n-1} (that is, in J). This implies that α is periodic point of f. We claim that n is the prime period of the priodic point α. Then from (ii) it follows that the first $(n-2)$ iterates of α lie in J and from (iii) the $(n-1)$th iterate lies in I. Suppose $\alpha = c$. Then $f(\alpha) = a$ lies on the boundary of I. This implies $n = 2$, since $f^{n-1}(\alpha) \in I$. Similarly, $\alpha = b$, implies $n = 3$. But we assume that $n > 3$. So α must lie in the open interval (b, c). Since $f^{n-1}(\alpha)$ lies in I, $f^{n-1}(\alpha) \neq \alpha$. That is, α cannot have prime period $(n-1)$. If α has a prime period less than $(n-1)$, then

Fig. 12.21 Graphical
description of period-3 cycle

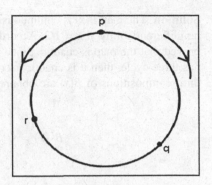

from (ii) it follows that the orbit of α will never cross the interval (b, c). This contradicts (iii). Hence α cannot have prime period less than $(n - 1)$. Therefore α is a periodic pint of f with prime period n. This completes the proof.

Theorem 12.11 *Suppose $f : [a, b] \rightarrow \mathbb{R}$ is a continuous map. If f has an orbit of period 3, then f is chaotic.*

Proof Let $\{p, q, r\}$ be a period 3 orbit of f such that $p < q < r$. Then either $f(p) = q, f(q) = r, f(r) = p$ or $f(p) = r, f(r) = q, f(q) = p$, (see Glendinning [14]).

From Fig. 12.21 we see that these two cases are equivalent after reversing the direction of movement. Without loss of generality, we choose the first case.

Since $f(q) = r > q$ and $f(r) = p < r$, by the fixed point theorem there exists $y \in (q, r)$ such that $f(y) = y$. Again, since $f(p) = q(<y)$ and $f(q) = r(>y)$, by intermediate value theorem (IVT) there exists $x \in (p, q)$ such that $f(x) = y$. Thus using the continuity condition of the map f and the condition of existence of periodic orbit of period 3 of f we find five points $p < x < q < y < r$ such that $f(p) = q$, $f(x) = y, f(q) = r, f(y) = y$, and $f(r) = p$. Now, consider the map $f^2([y, z])$. Since f is continuous, f^2 is also continuous. Again, from the above relations we see that

$$f^2(x) = f(f(x)) = f(y) = y, f^2(q) = f(r) = p < x \text{ and } f^2(y) = y.$$

Therefore by IVT, there exist points $c \in (x, q)$ and $d \in (q, y)$ such that $f^2(c) = x$ and $f^2(d) = x$. Let $M = (x, a)$ and $N = (a, y)$. Then M and N are two disjoint open subsets of $X = (x, y)$. We also see that $f^2(M) = X = f^2(N)$. Hence by definition the map f^2 has a horseshoe, and so f is chaotic. This completes the proof.

Example 12.4 Show that the map $B:[0, 1] \rightarrow [0, 1]$ defined by

$$B(x) = \begin{cases} 2x, & 0 \leq x \leq \frac{1}{2} \\ 2x - 1, & \frac{1}{2} \leq x \leq 1 \end{cases}$$

is chaotic on $[0,1]$.

Solution The map $B(x)$ is continuous on [0,1]. This is a one-dimensional map with nondifferentiability at $x = 1/2$. According to the definition of chaotic map we have proved that the map is chaotic. Since the map is continuous on [0,1], so if the map has three cycle, then it is chaotic according to the above theorem. The second and third compositions of $B(x)$ are obtained easily by composition formula as

$$B^2(x) = \begin{cases} 4x, & 0 \le x \le \frac{1}{4} \\ 4x - 1, & \frac{1}{4} \le x \le \frac{1}{2} \\ 4x - 2, & \frac{1}{2} \le x \le \frac{3}{4} \\ 4x - 3, & \frac{3}{4} \le x \le 1 \end{cases}$$

and

$$B^3(x) = B^2(B(x)) = \begin{cases} 8x, & 0 \le x \le \frac{1}{8} \\ 8x - 1, & \frac{1}{8} \le x \le \frac{1}{4} \\ 8x - 2, & \frac{1}{4} \le x \le \frac{3}{8} \\ 8x - 3, & \frac{3}{8} \le x \le \frac{1}{2} \\ 8x - 4, & \frac{1}{2} \le x \le \frac{5}{8} \\ 8x - 5, & \frac{5}{8} \le x \le \frac{3}{4} \\ 8x - 6, & \frac{3}{4} \le x \le \frac{7}{8} \\ 8x - 7, & \frac{7}{8} \le x \le 1 \end{cases}$$

The fixed points of $B(x)$ are $x^* = 0, 1$. $B^2(x)$ has four fixed points, viz., $x^* = 0, \frac{1}{3}, \frac{2}{3}, 1$. On the other hand $B^3(x)$ has eight fixed points, viz., $x^* = 0, \frac{1}{7}, \frac{2}{7}, \frac{3}{7}, \frac{4}{7}, \frac{5}{7}, \frac{6}{7}, 1$. We shall examine now its 3 cycles. We see that $B\left(\frac{1}{7}\right) = \frac{2}{7}$, $B\left(\frac{2}{7}\right) = \frac{4}{7}$, $B\left(\frac{4}{7}\right) = \frac{1}{7}$. Again, $B^3\left(\frac{1}{7}\right) = \frac{1}{7}$, $B^3\left(\frac{2}{7}\right) = \frac{2}{7}$, and $B^3\left(\frac{4}{7}\right) = \frac{4}{7}$. According to the definition of 3 cycles for a map, the set $\left\{\frac{1}{7}, \frac{2}{7}, \frac{4}{7}\right\}$ forms a 3-cycle of the map $B(x)$. So, by the theorem of Li and Yorke, the given map is chaotic on the interval [0,1].

12.5 Chaotic Maps

In this section we discuss three important maps. These three maps are useful for studying chaos.

12.5.1 Poincaré Map

The analytical solutions of most of the nonlinear continuous systems are very cumbersome and sometimes they become impossible. Also, their dynamics are very complicated for analysis. Instead of studying continous system, we may look at its discrete times in which the flow trajectory will be represented by a sequence of dots in the phase plane or space. Henri Poincaré devised this technique which set up a

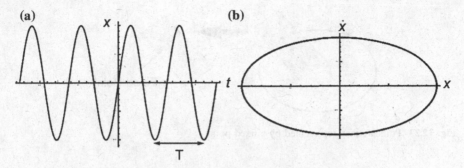

Fig. 12.22 a Time history and **b** phase portrait of a typical periodic solution

bridge between continuous system and its discrete analog. The discrete version is sometimes easy for analysis in particular chaotic motion. The technique is known as Poincaré section or map. This representation does not provide any information on the time history of evolution. In this section we shall study elaborately the Poincaré map for periodic solutions and then for an arbitrary attractor. A periodic solution is that repeats continuously after a certain time interval. The most common and useful tools for analyzing a periodic solution graphically are the representations of (i) displacement x versus time t and (ii) velocity $\dot{x}$ versus displacement x, in which the former is known as the time history of motion and the later as the phase portrait of the solution. The plane $x - \dot{x}$ is called the phase plane. Figure 12.22 depicts the time history and the phase portrait of a typical periodic solution.

Note that here the periodic solution is a trigonometric function, so the phase trajectory is an ellipse. The great advantage of periodic solutions is that instead of observing the whole motion in the phase plane or space, we may contemplate only at discrete time intervals. This is because the motion is periodic. Within this small time interval the trajectory in the phase plane will seem to appear as a sequence of dots (points). For periodic motions it is convenient to assume the quantity $T(=2\pi/\omega)$ as the sampling time for the period of oscillation, so that their periods are proportional to T, that is, the periods are of the form nT, n is a positive integer. With this sampling time T, we now estimate the displacement and velocity as follows:

$$x(0), x(T), x(2T), \ldots, x(nT), \ldots \text{ and } \dot{x}(0), \dot{x}(T), \dot{x}(2T), \ldots, \dot{x}(nT), \ldots.$$

Each point $(x(nT), \dot{x}(nT))$ represents a dot in the phase plane $x - \dot{x}$. For solutions of period-T, all the points $(x(nT), \dot{x}(nT))$ coincide with the point $(x(0), \dot{x}(0))$, so that the solution trajectory will appear as a dot (a single point) in the phase plane and this single point corresponds to the Poincaré map of the period-T solution and the plane $(x - \dot{x}$ plane) is referred to as the Poincaré section. Again, for period-$2T$ solutions, the points $(x(2kT), \dot{x}(2kT))$ coincide with the point

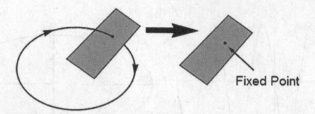

Fixed Point

Fig. 12.23 Poincaré map represented by a fixed point

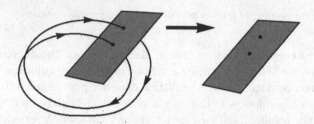

Fig. 12.24 Poincaré map represented by a period-2 cycle

$(x(0), \dot{x}(0))$ and $(x((k+1)T), \dot{x}((k+1)T))$ with $(x(T), \dot{x}(T))$, imply that the solution will appear as two dots in the phase plane which correspond to the Poincaré map of the period-$2T$ solution. Similarly, for period-nT solutions the phase portrait contains n distinct points that constitute the Poincaré map of the solution. This is shown graphically below in Figs. 12.23, 12.24.

Thus, we have seen that the Poincaré map reduces the periodic solutions with continuous time to solutions, constant in time. We shall now focus our attention to Poincaré maps of attractors. It is quite similar to that of periodic solutions. Let us consider an attractor of a N-dimensional continuous system. The concept of Poincaré map reduces the N-dimensional flow into a $(N–1)$-dimensional map. To construct the Poincaré map associated with the flow we consider a $(N–1)$-dimensional surface in the N-dimensional phase space in such a way that the trajectory of the attractor intersects the surface. This surface is known as the surface of section or simply the Poincaré section. To understand the construction we consider the flow in the three-dimensional phase space, the minimum dimensional phase space required for chaotic motions in a continuous system. Now, place a two-dimensional Poincaré section in the phase space transverse to the trajectory of the attractor. That is, the trajectories intersect the surface. As time goes on, a trajectory of the attractor may intersect the surface at more than one point. Suppose that initially the trajectory intersects the surface at some point, say x_0 (see Fig. 12.25). Starting with this point on the surface we follow the trajectory until it reintersects the surface. Denote this point on the surface by x_1. In a similar manner, starting with x_1, we can find the next intersecting point, x_2, and so on. On the surface, we can now introduce a function that maps x_0 into x_1, x_1 into x_2, and so on. We denote this function by f. Thus, on the

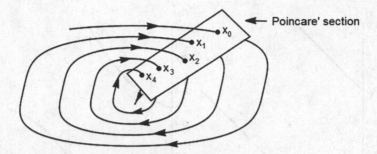

Fig. 12.25 Graphical representation of Poincaré section

surface, $x_1 = f(x_0)$, $x_2 = f(x_1) = f(f(x_0))$, and so on. The function f is known as a *first return map*. A return map is a function that allows us to know where a trajectory of an attractor, starting from some previous intersecting point on the Poincaré section, will reintersect the surface. A first return map takes every point on the surface into the next one and is sometimes called the Poincaré return map or simply the Poincaré map.

For construction of Poincaré map from continuous system see Gluckenheimer and Holmes [15].

12.5.2   Circle Map

The circle map is a one-dimensional iterated map which leaves a circle invariant. The most general form of a circle map is given as

$$\theta_{n+1} = f(\theta_n) = \theta_n + \omega - \frac{k}{2\pi} g(\theta_n) \tag{12.11}$$

where θ_{n+1} is calculated over mod 1. Here ω and k are two parameters and ω represents the rational frequency and is restricted to the interval [0,1]. The parameter k denotes the strength of the nonlinearity of the map. The circle map (12.11) is linear if $k = 0$ and nonlinear when $k \neq 0$. The function g is a nonlinear periodic function of period-1: $g(\theta + 1) = g(\theta)$. Different choices of g give rise to different circle maps. In this book, we shall consider only $g(\theta) = \sin(2\pi\theta)$, the nonlinear, 1-periodic sine function. Then the circle map (12.11) looks like

$$\theta_{n+1} = f(\theta_n) = \theta_n + \omega - \frac{k}{2\pi} \sin(2\pi\theta_n) \tag{12.12}$$

and this map is called the sine circle map. In the absence of the nonlinear term (obtained by setting $k = 0$), the map (12.12) reduces to the bare circle map (linear circle map)

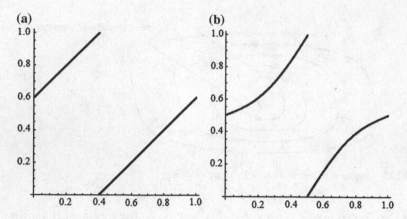

Fig. 12.26 Graphical representation of **a** the linear circle map at $w = 0.6$, $k = 0$ and **b** the sine circle map at $w = 0.5$, $k = 0.6$

$$\theta_{n+1} = f(\theta_n) = \theta_n + \omega. \qquad (12.13)$$

Grapical representations of the linear and sine circle maps are shown in Fig. 12.26.

Starting with some initial point $\theta_0 \in [0, 1)$, the successive iterations of the linear circle map (12.13) are calculated as follows:

$$\theta_1 = \theta_0 + \omega$$
$$\theta_2 = \theta_1 + \omega = \theta_0 + 2\omega$$
$$\theta_3 = \theta_2 + \omega = \theta_0 + 3\omega$$

and so on. In general, the nth iteration of the linear circle map is calculated as $\theta_n = \theta_0 + n\omega$. Suppose $\omega = m/n$ is a rational number. Then $\theta_n = \theta_0 + m = \theta_0$, since $m = 0 \pmod 1$. This shows that the linear circle map (12.13) has a period-n cycle, and the graph of the map contains exactly n number of points in total, of which m number of points are located on the lower branch (below the diagonal line) of the map. Figure 12.27a shows the period-5 orbit of the linear circle map with $\omega = 0.6$ and $\theta_0 = 0.3$. Note that the cycle does not depend on the initial point θ_0. If we plot the orbits, for the same value of ω, with a different initial point θ_0, then we will get the same result (see Fig. 12.27b).

Note that if ω is irrational, then this phenomenon is no longer valid. In such case, the orbit of the map will never reach to its initial value and hence we will never obtain an exact periodic cycle. Such an orbit is called a quasiperiodic orbit. Figure 12.28 depicts a quasiperiodic orbit of the linear circle map with $\omega = 1/\sqrt{2}$ and $\theta_0 = 0.3$.

For linear circle map, the quantity ω is also called the 'winding number'. In general, the winding number, W, for a circle map (12.11) is defined as

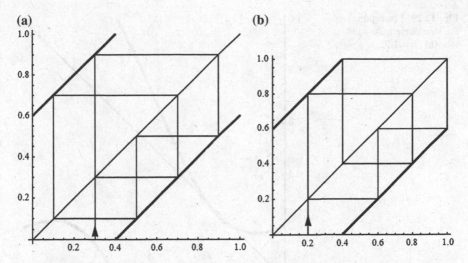

Fig. 12.27 Iterations of the linear circle map at $\omega = 0.6$ with the starting values **a** $\theta_0 = 0.3$, and **b** $\theta_0 = 0.2$

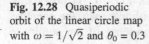

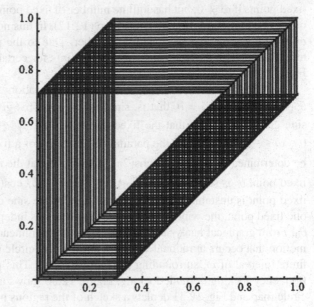

Fig. 12.28 Quasiperiodic orbit of the linear circle map with $\omega = 1/\sqrt{2}$ and $\theta_0 = 0.3$

$$W = \lim_{n \to \infty} \frac{f^n(\theta_0) - \theta_0}{n} = \lim_{n \to \infty} \frac{\theta_n - \theta_0}{n}$$

The 'winding number' W measures the average increase in θ per iteration, and is calculated without taking 'mod 1' on θ. For the linear circle map (12.13), $\theta_n = \theta_0 + n\omega$ and so $W = \lim_{n \to \infty} \frac{\theta_0 + n\omega - \theta_0}{n} = \omega$. Note that the linear circle map has no

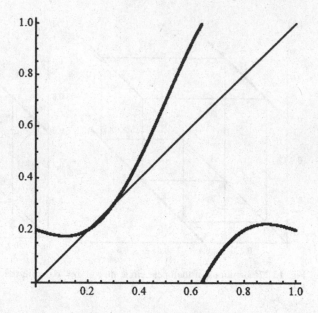

Fig. 12.29 Fixed points of the sine circle map with $w = 0.1$, $k = 1.32$

fixed points if $\omega \neq 0$, but has infinite number of fixed points if $\omega = 0$. We now focus our concentration to the sine circle map (12.12). In this nonlinear map, the quantity ω itself does not give the winding number. Due to the presence of the nonlinear term, the sine circle map can have fixed points for certain values of ω and k (see Fig. 12.29).

The fixed points are the solutions of equation $f(\theta) = \theta$, which gives $\theta + \omega - \frac{k}{2\pi}\sin(2\pi\theta) = \theta$, that is, $\sin(2\pi\theta) = \frac{2\pi\omega}{k}$. This gives all fixed points of the sine circle map. Note that the fixed points exist if $\frac{2\pi\omega}{k} \leq 1$, that is, if the relation $0 \leq \omega \leq \frac{k}{2\pi}$ is satisfied by the parameters. Stability of a fixed point θ^*, if exists, can be determined by $\frac{\partial f}{\partial \theta} = 1 - k\cos(2\pi\theta^*)$, evaluated at the fixed point. Therefore, the fixed point θ^* is stable if $-1 < \frac{\partial f}{\partial \theta} < 1$, that is, if $0 < k\cos(2\pi\theta^*) < 2$. Otherwise, the fixed point is unstable. It can be proved that for the sine circle map, having at least one fixed point, the winding number is zero and it is independent of the initial point θ_0. From graphical analysis, it is shown that for nonlinear circle maps, the periodic motion that occurs at rational values of ω for linear circle map, can be locked over a finite interval of ω, surrounding the rational point. This phenomenon is termed as frequency-locking or mode-locking. Figure 12.30 shows a period motion of the sine circle map and Fig. 12.31 depicts a sketch of the regions of frequency-locking in the $(k–w)$ plane.

The figure also shows that the frequency-locking regions are widened as k increases. These regions are called 'Arnold tongues', named after the Russian mathematician *Vladimir Igorevich Arnold* (1937–2010). Inside the 'Arnold tongues' the motions are periodic but they are quasiperiodic outside the tongues. For $k = 0$, that is, for linear circle map, the Arnold tongues are an isolated set of rational numbers. As k increases from zero, the width of the tongues increases and they form

Fig. 12.30 Periodic motion
of the sine circle map for
$\omega = 0.65$, $k = 0.98$, and
$\theta_0 = 0.3$

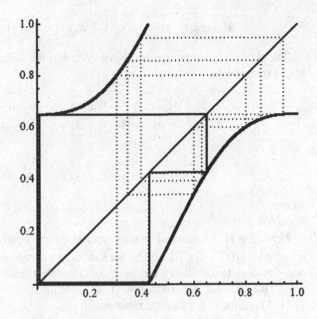

Fig. 12.31 Demonstration of
frequency-locking
phenomena

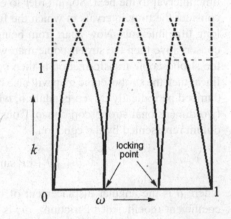

a Cantor set at $k = 1$. For further increase in k, the tongues start to overlap each other, implying that several periodic motions will occur for a given ω and k. Also, at $k > 1$, the map (12.13) has a maximum and minimum value at $\theta = 1 - \frac{1}{2\pi}\cos^{-1}\left(\frac{1}{k}\right)$ and $\theta = \frac{1}{2\pi}\cos^{-1}\left(\frac{1}{k}\right)$, respectively, and so the map is not invertible for $k > 1$ (see Fig. 12.29). Therefore, chaotic phenomenon may occur for $k > 1$.

A continuous system may express its discrete analog uniquely with the help of stroboscopic map for a periodically driven continuous system.

Let us consider a perodically driven continuous system represented by a system of differential equations given by

$$\dot{x} = \upsilon(x,t),\ x = (x_1, x_2, x_n)\ \text{and}\ \upsilon(x,t) = (\upsilon_1, \upsilon_2, \ldots, \upsilon_n),$$

The stroboscopic map for the above system is the n-dimensional iterated discrete map expressed by

$$x_{1,N+1} = \mathcal{G}_1(x_{1,N}, x_{2,N}, \ldots, x_{n,N}; \tau)$$
$$x_{2,N+1} = \mathcal{G}_2(x_{1,N}, x_{2,N}, \ldots, x_{n,N}; \tau)$$
$$\vdots$$
$$x_{n,N+1} = \mathcal{G}_n(x_{1,N}, x_{2,N}, \ldots, x_{n,N}; \tau),$$

where $\qquad\qquad x_{1,N} = x_1(t), \ldots, x_{n,N} = x_n(\tau) \qquad\qquad$ and
$x_{1,N+1} = x_1(t+\tau), \ldots, x_{n,N+1} = x_n(t+\tau)$.

Now if ϕ is a time evolutionary operator then $\phi(x_i(t))$ is a function of all the n variables $x_1(t), \ldots, x_n(t)$, and by the successive composition of ϕ, that is, $\phi, \phi^2, \ldots$ one would go to the times $t+\tau, t+2\tau, \ldots$ implying that the functions $\mathcal{G}_1, \mathcal{G}_2, \ldots, \mathcal{G}_n$ will be the same for all iterations $N = 1, 2, \ldots$. If the system is strobed at time intervals $T, 2T, 3T, \ldots$ for $T \neq \tau$ then the functions $\mathcal{G}_1, \mathcal{G}_2, \ldots, \mathcal{G}_n$ will not be the same from one time interval to the next. So, in order to construct the stroboscopic map one should consider the time intervals in which the functions $\mathcal{G}_1, \mathcal{G}_2, \ldots, \mathcal{G}_n$ remains same over long time intervals. Now apart from being periodically driven if the system is also conservative then the stroboscopic map will be area preserving and the Jacobian of the stroboscopic transformation is then equal to one. Furthermore, if the system is linear then the stroboscopic map will also be linear. We shall now give the example of damped periodically driven pendulum, where the circle map comes from a discrete two-dimensional stroboscopic map. Consider the damped, periodically driven pendulum represented by the equation

$$\ddot{\theta} + \beta\dot{\theta} + \omega_0^2 \sin\theta = A\cos(\omega t),$$

where θ is the angular displacement of the pendulum at time t, β is the damping coefficient (coefficient of friction), ω_0 is the frequency of natural oscillation, and A is the strength of the forcing (external). Setting $x = \theta, y = \dot{\theta}, z = \omega t$, the above equation can be written as

$$\left.\begin{array}{l} \dot{x} = y \\ \dot{y} = -\beta y - \omega_0^2 \sin x + A\cos z \\ \dot{z} = \omega \end{array}\right\}$$

Now $\nabla \cdot (\dot{x}, \dot{y}, \dot{z}) = \frac{\partial \dot{x}}{\partial x} + \frac{\partial \dot{y}}{\partial y} + \frac{\partial \dot{z}}{\partial z} = -\beta < 0$. Therefore, the Jacobian J of the system is $e^{-\beta t}$. Since the given system is periodic, therefore it can be replaced by the two-dimensional stroboscopic map for a time shift $t + \tau$, and has the following form:

$$\theta_{n+1} = \mathcal{G}_1(\theta_n, \dot{\theta}_n)$$
$$\dot{\theta}_{n+1} = \mathcal{G}_2(\theta_n, \dot{\theta}_n)$$

where the iterations $n = 0, 1, 2, \ldots$ are equivalent to the times $0, 2T, 3T, \ldots$, and $T = 2\pi/\omega$, the period of the external force which make sure that the functions $\mathcal{G}_1(\theta_n, \dot{\theta}_n)$ and $\mathcal{G}_2(\theta_n, \dot{\theta}_n)$ remain unchanged with time. The functions $\mathcal{G}_1(\theta_n, \dot{\theta}_n)$ and $\mathcal{G}_2(\theta_n, \dot{\theta}_n)$ are, respectively, given by

$$\mathcal{G}_1(\theta, \dot{\theta}) = \theta + \Omega + g(\theta, \dot{\theta}) + (1 - J)\dot{\theta}/b \text{ and } \mathcal{G}_2(\theta, \dot{\theta}) = J\dot{\theta} + \Omega' + h(\theta, \dot{\theta}),$$

where $b = \exp(-2\pi\beta/\omega)$ and g, h are periodic functions with period 2π in θ, and Ω, Ω' are constants. J is the Jacobian of the map which is equal to the Jacobian of the equation of the damped-driven pendulum evaluated at $t = \tau$. If $\dot{\theta}_n \to h(\theta_n)$ as $n \to \infty$, then the two-dimensional stroboscopic map will lead to the one-dimensional circle map given by

$$\theta_{n+1} = \mathcal{G}_1(\theta_n, h(\theta_n)) = f(\theta_n),$$

where $f(\theta_n + 2\pi) = f(\theta_n) + 2\pi$. Thus the nonlinear damped-driven pendulum can be analyzed by studying the discrete one-dimensional circle map. For the chaotic motion of the nonlinear forced damped pendulum, see Jordan and Smith [16], Gluckenheimer and Holmes [15], Grebogi et al. [8], McCauley [17].

12.5.3   Smale Horseshoe Map

In the year 1967 the American mathematician, Stephen Smale built the horseshoe map in the two-dimensional plane $\mathbb{R}^2$ as an illustrative example of the symbolic map or dynamics, his another discovery in the context of discrete dynamical systems. The construction of the horseshoe map and its connection with the shift map are illustrated as follows.

Take a unit square S (see Fig. 12.32a). Now, expand S uniformly in the vertical direction (y-direction) by a factor λ, greater than 2 and then uniformly shorten it horizontally (x direction) by a factor μ, less than 1/2 (in our figures we take $\lambda = 3$ and $\mu = 1/3$), so that it appears as a long thin strip as shown in Fig. 12.32b. Now, bend this strip in clockwise direction into a horseshoe shape as presented in Fig. 12.32c. Place this horseshoe on the unit square S (see Fig. 12.32d).

Through the geometrical construction of the horseshoe structure from the square S we can formulate a map, known as the horseshoe map, which maps the unit square S into the horseshoe. We denote the map by h. We can then express h as the

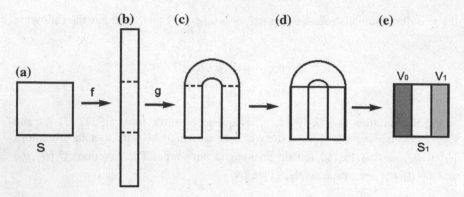

Fig. 12.32 Construction of the horseshoe map

composition of two maps, $h = g \circ f$, where $f(x, y) = (\mu x, \lambda y)$ is a linear transformation and g is a nonlinear transformation that bends the strip into a horseshoe in clockwise direction. From Fig. 12.32d we see that certain fraction of the area of the unit square S is mapped to the region outside S. Let this fraction be $(1-a)$, where a denotes the area of S that mapped inside S under h. Removing the portion of the horseshoe outside S we obtain a unit square, S_1, with two vertical strips labeled by V_0 and V_1 (see Fig. 12.32e). In the similar manner, applying h on S_1 we get another horseshoe as shown in Fig. 12.33c.

As previous, if we place the horseshoe on the original square S, we then again see that the fraction $(1-a)$ of the area of S_1 (or equivalently the area of S) is mapped to the region outside S. Removing them again we obtain another unit square S_2 with four vertical strips, labeled by V_{00}, V_{01}, V_{10}, V_{11}, in which the strips V_{00} and V_{01} are contained in V_0, and V_{10}, V_{11} are in V_1. Also, the area of S that mapped into S under h^2 is a^2. Proceeding in this way, we can construct the unit square S_n, with 2^n vertical strips. Note that after n successive applications of h on S the area of S that will map into S is $a^n \to 0$ as $n \to \infty$. Thus, after enough large iterations we can find points

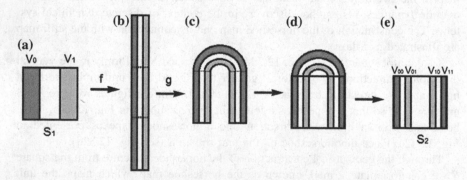

Fig. 12.33 Construction of the square S_2 from the square S_1

Fig. 12.34 Step-by-step construction of the horizontal strips H_0, H_1 from the vertical strips V_0, V_1

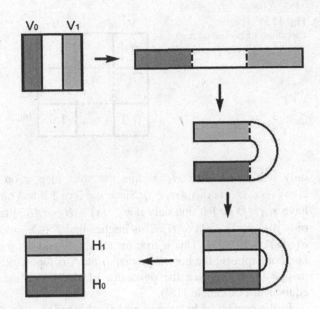

(very small squares) that will never leave the square S. We denote the set of all those points by Λ. Then Λ constitutes an invariant set of the map f.

If we apply h^{-1}, the inverse of h, on the square S_1, we then get the same square S_1 with two horinzontal strips H_0, H_1, instead of the two vertical strips V_0, V_1. A step-by-step construction of the horinzontal strips is presented in Fig. 12.34. Thus, the horizontal strips H_0, H_1 and the vertical strips V_0, V_1 are connected by the relation $H_0 = h^{-1}(V_0)$ and $H_1 = h^{-1}(V_1)$. That is, the vertical strips are the h-images of the horinzontal strips. Simlarly, $H_{00} = h^{-2}(V_{00}), H_{01} = h^{-2}(V_{01}), H_{10} = h^{-2}(V_{10})$ and $H_{11} = h^{-2}(V_{11})$, and so on. Since all the tiny vertical (respectively, horinzontal) strips in S_n are contained in $V_0 \cup V_1$ (respectively, $H_0 \cup H_1$), the invariant set Λ must contain in both $V_0 \cup V_1$ and $H_0 \cup H_1$, and so, it must contain in their intersection, $(V_0 \cup V_1) \cap (H_0 \cup H_1)$. Similarly, Λ must contain in $(V_{00} \cup V_{01} \cup V_{10} \cup V_{11}) \cap (H_{00} \cup H_{01} \cup H_{10} \cup H_{11})$, and so on. Note that the intersection $(V_0 \cup V_1) \cap (H_0 \cup H_1)$ contains 4 squares and the intersection $(V_{00} \cup V_{01} \cup V_{10} \cup V_{11}) \cap (H_{00} \cup H_{01} \cup H_{10} \cup H_{11})$ contains $16(= 4^2)$ squares smaller than the previous 4 squares. It can be easily shown that the intersection of union of the vertical and horizontal strips of the square S_n will contain 4^n small squares inside S (Fig. 12.35).

Thus, we see that under every iteration, each square splits into four small squares. If we set $n \to \infty$, then we get the invariant set Λ. Note that for parameters $\lambda = 3$ and $\mu = 1/3$ the set Λ gives the structure like a Cantor dust which is formed by the intersection of the Cantor sets of the vertical and the horinzontal strips. This is a fractal structure. The Cantor dust will be discussed elaborately in the next chapter.

We now correspond the elements of Λ with the elements of Σ_2, the set of all bi-infinte sequences of two symbols 0's and 1's, by the correspondence $\varphi : \Lambda \to \Sigma_2$ defined by $s = \varphi(x)$, where $s = (\cdots s_{-3}s_{-2}s_{-1}.s_0s_1s_2s_3 \cdots)$ with $s_i = 0$ or 1 if and

Fig. 12.35 Folding
operations of horseshoe map

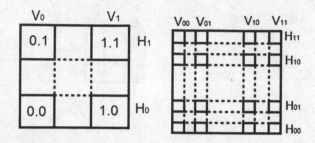

only if $h^i(x) \in H_0$ or H_1. Define the shift map σ on Σ_2 by $\sigma(s) = s'$, where $s' = (\cdots s_{-3} s_{-2} s_{-1} s_0 s_1 s_2 s_3 \cdots)$. Since $s_i = 0$ or 1 if and only if $h^i(x) \in H_0$ or H_1, we have $s_{i+1} = 0$ or 1 if and only if $h^{i+1}(x) \in H_0$ or H_1, that is, $s_{i+1} = 0$ or 1 if and only if $h^i(h(x)) \in H_0$ or H_1. This implies that $s' \in \Sigma_2$ and $s' = \varphi(h(x))$. Therefore, $\sigma(\varphi(x)) = \varphi(h(x))$. This is true for all $x \in \Lambda$ and so, $\sigma \circ \varphi = \varphi \circ h$. Since φ is a homeomorphism, the horseshoe map h on Λ is topologically conjugate to the shift map σ on Σ_2. Hence the dynamics of the horseshoe and the shift maps are equivalent (see Smale [18]).

In the context of horseshoe map the following definitions and lemma are very important for studying chaotic dynamics.

Definition 12.12 Let $f : I \to \mathbb{R}$ be a continuous map on the interval I. Then the map f has a horseshoe if and only if there exists a interval J contained in I and two disjoint open subintervals K_1 and K_2 such that $f(K_1) = f(K_2) = J$.

Definition 12.13 A continuous map $f : I \to I$ is chaotic iff the map f^k has a horseshoe for some $k \geq 1$.

Lemma 12.3 *Suppose a continuous map $f : I \to I$ has a horseshoe. Then*

  (i) *f^k has at least 2^k fixed points.*
 (ii) *f has periodic points of every period.*
(iii) *f has an uncountable number of aperiodic orbits.*

12.6   Routes of Chaos

There is no unique route in which the evolution of a system reaches chaotic regime. Chaos is a multiscale and multiplicative phenomena. Chaotic system has extremely sensitive dependence on infinitesimal change of initial conditions with topologically mixing property. It has infinitely many periodic cycles and attained a dense orbit at the infinitum. We have discussed some of qualitative and quantitative measures for chaos. However, there is an order in chaotic motion. The routes from order or regular to chaos are called routes of chaos. There are infinite routes. Some of the well-known routes are period-doubling cascade, periodic windows, intermittency, crises, etc. These routes are observed in systems like fluid turbulence convection, electric circuits, lasers, population dynamics, chemical and biological processes, etc. We shall discuss few routes briefly.

(i) **Period-doubling route**: This is the most well-known route of chaos and occurs in many one-dimensional systems. Many natural systems exhibit this route. The systems are dissipative in nature. The route begins with limit cycle. The limit cycle may be formed from a bifurcation involving a node or other type of fixed points of the system. When the parameter of the system varies, this limit cycle becomes unstable and gives the birth of a period-2 limit cycle. For further changes of parameter, the period-2 limit cycle becomes unstable and gives birth to period-4 cycles. The period-doubling cascades may continue until the period becomes infinite. This implies that the trajectory never repeats itself. The orbit is then chaotic orbit (see Cvitanović [24] for details).

(ii) **Intermittency**: Period doubling is a common route to chaos. Intermittency is another type of chaotic behavior that appears in deterministic systems. This is characterized by long periods of regular motion interrupted by short duration chaotic bursts. A number of experimental investigations on the convective motion of a fluid layer heated underneath had shown a transition to turbulence via intermittency. In these experiments the reduced Rayleigh number, say r plays the role of the control parameter. For values of r below a critical value, say r_T the motion is periodic. But as soon as r reaches the value r_T the periodic motion losses its stability and a random apparently periodic fluctuating motion starts in some region of this periodic motion. For r slightly larger than r_T, the time interval of the fluctuations expands. But the periodic motion is still observed in some long time intervals (known as 'laminar phases'). Thus for $r > r_T$, the regular motion is intermittently interrupted by random aperiodic bursts. Each burst has finite duration. After the duration period of the burst a periodic motion is observed again and then another finite duration burst starts and so on. The time durations of the bursts are randomly different but one can determine the average time duration between the bursts. In 1980, Berge et al. observed this intermittency behavior in the convectional motion of fluid. Pomeau and Manneville [19] nicely represents the intermittency phenomena by solving the Lorenz model numerically. Figure 12.36 displays the time history of the z coordinate of the Lorenz model for the Prandtl number $\sigma = 10$ and $b = 8/3$. It can be proved, at least graphically, that a period-3 cycle of the quadratic map borns at the critical parameter value $r = r_{c3} = 1 + 2\sqrt{2} \approx 3.82843$. Note that at $r = r_{c3}$ the graph of f_r^3 touches the diagonal line at three distinct points, the points where the diagonal line is tangent to the curve $y = f_r^3(x)$. A slight increase in the parameter value r will produce 3-cycles of the map, but no 3-cycles will occur for slightly decreased r from the critical value r_{c3}. For example, we take the parameter values $r = 1 + 2\sqrt{2} + 0.0001$

(a)   (b)

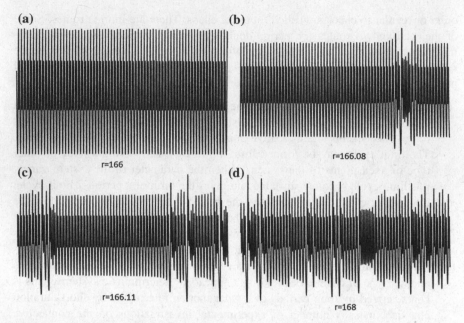

r=166

(c)   (d)

r=166.11

r=168

Fig. 12.36 **a** represents the stable periodic motion corresponds to the value $r = 166$. **b–d** are obtained for r slightly greater than the critical value $r_T \simeq 166.06$. The graphs clearly show that the bursts become more frequent as r increases. In intermittency we observed that as r increases r_T the stable periodic orbit becomes unstable but never settle down to a stable periodic orbit. This is because the orbit moves far away from the original periodic orbit. This can be nicely observed in maps also. For example, we consider the quadratic map $f_4(x) = rx(1 - x)$

and $r = 1 + 2\sqrt{2} - 0.0001$. At $r = 1 + 2\sqrt{2} + 0.0001$, the graph of $f_r^3(x)$ intersects the diagonal line at eight distinct points 0, 0.738803, 0.158988, 0.511916, 0.956588, 0.160861, 0.516792, and 0.956052, in which the first two are the fixed points of $f_r(x)$ and the other six points form two 3-cycles of $f_r(x)$. At $r = r_{c3}$, $f_r^3(x) = x$ produces two fixed points of $f_r(x)$ and three other points, 0.159929, 0.514355, and 0.956318, where $f_r^3(x)$ touches the diagonal line. But, at $r = 1 + 2\sqrt{2} - 0.0001$, $f_r^3(x) = x$ gives 5 solutions, two real and three complex conjugates. The real roots represent the constant solutions of $f_r(x)$. The appearance of the complex roots signifies that period-3 cycles will never occur at $r < (1 + 2\sqrt{2})$. Taking $x = 0.158988$ as initial seed or starting point, which is an element of the 3-cycles of $f_r(x)$ at $r = 1 + 2\sqrt{2} + 0.0001$, we plot the successive iterations of $f_r(x)$ and the difference $(x_{i+3} - x_i)$, $i = 1, 2, 3, \ldots$ for the three parameter values (see Figs. 12.37, 12.38, 12.39). Figure 12.37b shows that the difference $(x_{i+3} - x_i)$ converges rapidly to zero. Figure 12.38b shows the similar behavior but quite different picture: the difference $(x_{i+3} - x_i)$ converges to zero but very slowly with iterations. On the other hand, Fig. 12.39b shows a completely different scenario. From this figure, we see that the deviation $(x_{i+3} - x_i)$ is small in certain intervals of i,

(a)   **(b)**

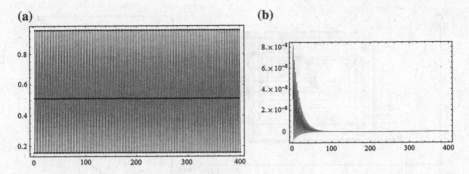

Fig. 12.37 Plots of **a** successive iterations and **b** the difference $(x_{i+3} - x_i)$ for $r = 1 + 2\sqrt{2} + 0.0001$

(a)   **(b)**

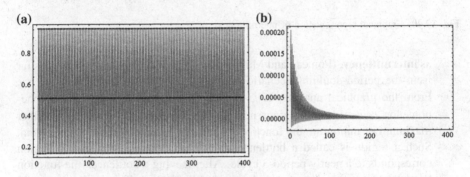

Fig. 12.38 Plots of **a** successive iterations and **b** the difference $(x_{i+3} - x_i)$ for $r = 1 + 2\sqrt{2}$

(a)   **(b)**

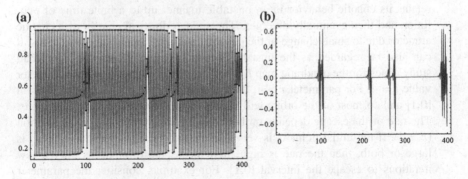

Fig. 12.39 Plots of **a** successive iterations and **b** the difference $(x_{i+3} - x_i)$ for $r = 1 + 2\sqrt{2} - 0.0001$

Fig. 12.39a reveals that the periodic patterns of Figs. 12.37a and 12.38a are observed in these intervals and between two successive almost periodic behaviors the system exhibits a chaotic behavior. This phenomenon is known

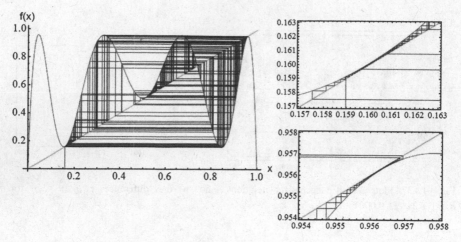

Fig. 12.40 Graphical representation of intermittency

as **intermittency,** (Pomeau and Manneville [19]). It is also a route to chaos far from the period-doubling scenario and it occurs only in the vicinity of r_{c3}. From the graphical analysis of f_r^3 we can understand easily why this phenomenon occurs. In the regions near to the diagonal line, where f_r^3 is either maximum or minimum, the function takes many iterative steps to leave them. Such a region is called a bottleneck and each iterative step in this region corresponds to a nearly period-3 cycle. After leaving a bottleneck the function takes large iterative steps to reach the next bottleneck. This is shown graphically in Fig. 12.40.

(iii)   **Crises**: This is another important route in which the orbit of a chaotic region retains its chaotic behavior in an unstable manner up to a finite time of evolution and then eventually diverges. The sudden drastical change of a chaotic attractor due to small change of the parameter of a system is known as crisis. It can also be regarded as the unstable situation of chaotic motion. We shall study crises for the quadratic map $f_r(x) = rx(1 - x), x \in [0, 1]$ at the parameter value $r = 4$. For parameter $r > 4$, the function $f_r(x)$ leaves the closed interval [0,1] and so most of the orbits will escape the interval [0,1] under iterations. The rate of the escape depends on both the initial point x_0 and the separation $(r - 4)$. If the initial point x_0 is very close to 0.5 or the separation $(r - 4)$ is large or both, then the rate is high, that means the orbit needs only a few iterations to escape the interval [0,1]. For example, consider the parameter value $r = 4.01$ and two distinct initial points $x_0 = 0.3$ and $x_0 = 0.6$. Figures 12.41 and 12.42 display the graphical iteration and the time series plot of the orbits of the points $x_0 = 0.3$ and $x_0 = 0.6$, respectively. From these figures we see that both the orbits eventually diverges to $(-\infty)$. We also see that the orbit of $x_0 = 0.3$ takes more iterations than the orbit of $x_0 = 0.6$ to escape the interval.

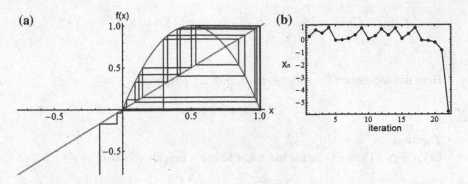

Fig. 12.41   a Graphical iteration and **b** time series plot of the orbit of $x_0 = 0.3$ under the map $f_r(x)$

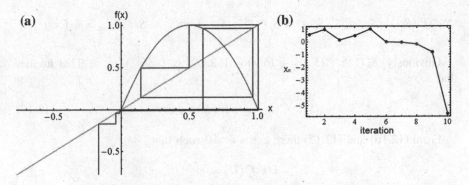

Fig. 12.42   a Graphical iteration and **b** time series plot of the orbit of $x_0 = 0.6$ under the map $f_r(x)$

Example 12.5 Let $\Omega(f)$ denote the nonwandering set of a map $f : X \to X$. Show that

(i) the set $\Omega(f)$ is closed.

(ii) $\Omega(f_r) = \Lambda_r$, where f_r is the quadratic map with $r > 2 + \sqrt{5}$.

Solution

(i) Let $\bar{p}$ be the limiting point of the set $\Omega(f)$. Then every neighborhood N_ε of $\bar{p}$ contains at least one point $p \in \Omega(f)$. Since p is a nonwandering point, an arbitrary open neighborhood U of p contains a point x such that $f^n(x) \in U$. Since the choice of U is arbitrary, we can select U such that $U \subset N_\varepsilon$. Then $f^n(x) \in U \Rightarrow f^n(x) \in N_\varepsilon$, an arbitrary N_ε of $\bar{p}$. Hence $\bar{p} \in \Omega(f)$. This shows that the set $\Omega(f)$ is closed.

(ii) Let $\{S_n\}$ be a sequence of open intervals constructed in the following manner:

$$f_r(S_0) \cap I = \varphi \text{ and } S_n = fr^{-1}(S_{n-1}) \text{ so that} f_r(S_n) = S_{n-1}, n = 1, 2, \ldots$$
$$(12.14)$$

Here the sequence $\{S_n\}$ satisfies the following property:

$$S_i \cap S_j = \varphi \, \forall \, i \neq j \tag{12.15}$$

Therefore, $\Lambda_r = I - \bigcup_{n=0}^{\infty} S_n$.

Let $p \notin \Lambda_r$. Then $p \in S_k$ for some k. Choose a neighborhood U of p such that

$$U \subset S_k \tag{12.16}$$

From (12.14) and (12.15), we have

$$S_k \cap f_r^n(S_k) = S_k \cap f_r^{n-1}(S_{k-1}) = S_k \cap f_r^{n-2}(S_{k-2}) = \cdots = S_k \cap S_{k-n} = \varphi \text{ for } n \leq k.$$

Obviously, $S_k \cap S_r^{k+m}(S_k) = \varphi$ for $m = 1, 2, \ldots$ as $f_r(S_0) \cap I = \varphi$. That implies that

$$S_k \cap f_r^n(S_k) = \varphi \text{ for all } n \in \mathbb{N}. \tag{12.17}$$

From (12.16) and (12.17) there exists $n > 0$ such that

$$U \cap f_r^n(U) = \varphi$$

for every neighborhood U of p. Therefore, $p \notin \Omega(f_r)$.

Thus $p \notin \Lambda_r \Rightarrow p \notin \Omega(f_r)$. That is, $p \notin \overline{\Lambda_r} \Rightarrow p \notin \overline{\Omega(f_r)}$, where $\overline{\Lambda_r}$ represents the complement of Λ_r. This implies that

$$\overline{\Lambda_r} \subseteq \overline{\Omega(f_r)} \tag{12.18}$$

Again, let $p \notin \Omega(f_r)$. Then there exists a neighborhood U of p such that

$$f_r^n(U) \cap U = \varphi \quad \forall n > 0 \tag{12.19}$$

Let $x \in \Lambda_r$. Then some iterate, say $f_r^n(x)$ will come sufficiently close to x (this follows from the topological conjugacy between $f_r : \Lambda_r \to \Lambda_r$ and $\sigma : \sum_2 \to \sum_2$). Thus if $x \in \Lambda_r$, then (12.19) cannot be true. Hence U cannot contain any $x \in \Lambda_r$. That is, $U \cap \Lambda_r = \varphi$, that is, $U \subset S_k$ for some k. Since U is a neighborhood of p, we have $p \in S_k$. So $p \notin \Lambda_r$. Thus $p \notin \Omega(f_r) \Rightarrow p \notin \Lambda_r$. That is, $p \in \overline{\Omega(f_r)} \Rightarrow p \in \overline{\Lambda_r}$ and so

$$\overline{\Omega(f_r)} \subseteq \overline{\Lambda_r}. \tag{12.20}$$

Combining (12.18) and (12.20), we get

$$\overline{\Omega(f_r)} = \overline{\Lambda_r} \Rightarrow \Omega(f_r) = \Lambda_r.$$

Example 12.6 Show that the quadratic map $Q_{-2}(x) = x^2 - 2$ is chaotic on [–2, 2].

Solution Consider the one-dimensional map $f(x) = 2|x| - 2$, defined on [–2, 2]. Clearly, the interval [–2, 2] is invariant under these two maps. The map $Q_{-2}(x) = x^2 - 2$ is topologically semiconjugate to $f(x) = 2|x| - 2$ under the map $h(x) = -2\cos\left(\frac{\pi}{2}x\right)$, $x \in [-2, 2]$. Therefore, if we show that the map $f(x) = 2|x| - 2$ is chaotic in [–2, 2], we can reach our goal. The map $f(x) = 2|x| - 2$ can be written as

$$f(x) = \begin{cases} 2x - 2, & \text{if } x > 0 \\ -2, & \text{if } x = 0 \\ -2x - 2, & \text{if } x < 0 \end{cases}$$

The fixed points of $f(x)$ are $x^* = 0, -\frac{2}{3}$. We shall calculate the third composition $f^3(x)$ of $f(x)$ which is semiconjugate to the one-dimensional quadratic map $Q_{-2}(x)$. It is semiconjugate to a similar linear map whose dynamics can be studied easily. The third composition of $f(x)$ is obtained as

$$f^3(x) = \begin{cases} 8x - 14, & x > \frac{3}{2} \\ -8x + 10, & x < \frac{3}{2} \\ 8x - 6, & x > \frac{1}{2} \\ -8x + 2, & x < \frac{1}{2} \\ 8x + 2, & x > -\frac{1}{2} \\ -8x - 6, & x < -\frac{1}{2} \\ 8x + 10, & x > -\frac{3}{2} \\ -8x - 10, & x < -\frac{3}{2} \end{cases}$$

$f^3(x)$ has eight fixed points, viz., $x^* = 2, \frac{10}{9}, \frac{6}{7}, \frac{2}{9}, -\frac{2}{7}, -\frac{2}{3}, \frac{10}{7}, -\frac{14}{9}$. But $x^* = 0, -\frac{2}{3}$ are the fixed points of $f(x)$. We shall search for periodic cycles of $f^3(x)$. We see that

$$f\left(\frac{10}{9}\right) = \frac{2}{9}, f\left(\frac{2}{9}\right) = -\frac{14}{9}, f\left(-\frac{14}{9}\right) = \frac{10}{9}, f^3\left(\frac{2}{9}\right) = \frac{2}{9}, f^3\left(-\frac{14}{9}\right) = -\frac{14}{9}.$$

This shows that $\left\{\frac{10}{9}, \frac{2}{9}, -\frac{14}{9}\right\}$ forms a period-3 cycle of f. Similarly, it can be shown that $\left\{\frac{6}{7}, \frac{2}{7}, -\frac{10}{7}\right\}$ also forms a period-3 cycle of f. Hence the map f is chaotic on [–2,2] and therefore the quadratic map $Q_{-2}(x)$ is also chaotic on [–2,2].

Example 12.7 Show that the map $f(x) = 4x(1-x), x \in [0,1]$ is a chaotic map.

Solution Consider the maps $g :$ unit circle $S \rightarrow S$ and $h_1 : S \rightarrow [-1,1]$ defined by $g(\theta) = 2\theta$ and $h_1(\theta) = \cos\theta$, respectively. Clearly, the map h_1 is onto but not one–one, because $\cos(2\pi + \theta) = \cos\theta$. Let $q(x) = 2x^2 - 1, x \in [-1,1]$. Then for all $\theta \in S$, we have $(h_1 \circ g)(\theta) = h_1(g(\theta)) = \cos 2\theta$ and $(q \circ h_1)(\theta) = q(\cos\theta) = 2\cos^2\theta - 1 = \cos 2\theta$. So, $q \circ h_1 = h_1 \circ g$. This shows that g is topologically semiconjugate to the map q through the map h_1. Next, define the map $h_2 : [-1,1] \rightarrow [0,1]$ by $h_2(x) = \frac{1-x}{2}$. Clearly, the map h_2 is a homeomorphism. Now, for all $x \in [-1,1]$,

$$h_2(q(x)) = \frac{1 - (2x^2 - 1)}{2} = 1 - x^2$$

and

$$f(h_2(x)) = f\left(\frac{1-x}{2}\right) = 4\left(\frac{1-x}{2}\right)\left(1 - \frac{1-x}{2}\right) = 1 - x^2.$$

So, $f \circ h_2 = (h_2 \circ q)$. This shows that q is conjugate to f through h_2. We also see that

$$f \circ (h_2 \circ h_1) = (f \circ h_2) \circ h_1$$
$$= (h_2 \circ q) \circ h_1$$
$$= (h_2 \circ q) \circ h_1 = h_2 \circ (h_1 \circ g) = (h_2 \circ h_1) \circ g.$$

Again, h_2 is homeomorphism and h_1 is onto together imply $(h_2 \circ h_1)$ is onto. So, $(h_2 \circ h_1)$ semiconjugates g to f. Hence f is transitive on [0,1], since g is transitive on the circle S. We now establish the sensitive dependence on initial conditions property of f. Let U be an open arc of the circle S and $(h_2 \circ h_1)(U) = V \subset [0,1]$. Let $x \in V$. Since $(h_2 \circ h_1)$ is onto, for sufficiently large n, $g^n(U)$ covers the circle S. So, $f(V) = f(h_2 \circ h_1(U)) = (h_2 \circ h_1) \circ g^n(U)$ covers [0,1]. So, there exists point $y \in N_\varepsilon(x) \subset V$ and $n \in \mathbb{N}$ such that

$$|f^n(x) - f^n(y)| > \delta$$

This shows that the map f has SDIC property on [0,1]. Hence f is chaotic on [0,1].

12.7 Universality in Chaos

The remarkable fact for chaotic motion is that the Feigenbaum constant, superstable ratio, subharmonic spectra peak in Fourier spectra and other quantitative and qualitative measures retain their respective values constant from transitions to chaos. These constant values do not change from map to map but remain unaltered for a particular class of maps. This is why these numbers are called universal numbers. The renormalization group method explains nicely the universality for the class of unimodal one-dimensional maps undergoing period doublings. This scaling behavior has been noticed in many physical systems like fluid flows, nonlinear circuits, laser, etc. This indicates that there is an order or regular pattern in the chaotic regime. Metropilos et al. [10] noticed that when the control parameter is varied for different hump maps, the sequence of symbols is same. This is another indication of the existence of the order or regularity in chaotic regime for which deterministic systems exhibit chaotic behavior. For conservative systems one may obtain a universal number which is different from Feigenbaum number. Consider a two-dimensional map (Benettin et al. [20])

$$\left. \begin{array}{c} x_{n+1} = y_n \\ y_{n+1} = -x_n + 2y_n(c+y_n) \end{array} \right\}$$

Helleman constructed this map in 1980 from the motion of a proton particle in a storage ring with periodic impulses. This map is conservative in nature and so the area is preserved under evolution (since det(Jacobian matrix) = 1). Iterations of this map exhibit a multiperiodic orbit for $c \geq -1$. The transitions from one to two periodic orbits and from two to four periodic, and so on, are obtained for the values

$$c_1 = -1, c_2 = (1 - \sqrt{5}), c_3 = 1 - \sqrt{4 + \sqrt{5/4}}.$$

For large iterations, say n, the sequence of bifurcations of this two-dimensional conservative map occurs for the values of c as

$$c_n - c_\infty + A/\delta_c^n, \text{ where } \delta_c = 8.72\ldots.$$

It has been observed that δ_c is universal for all conservative maps undergoing bifurcations sequence. This number is differed from the Feigenbaum number 4.6692(approx.). This simple analysis reveals that even for conservative system there is an order maintained in the chaotic regime. However, there does not exist a single universal number covering all dissipative or conservative systems.

Large class of nonlinear systems display transitions to chaos that is universal and quantitatively measurable. The universal number δ (Feigenbaum number) has been measured in many experiments like in hydrodynamics, electrical circuits, laser, chemical reaction, Brusselator systems, etc. The results confirmed that for all dissipative systems undergoing period-doubling bifurcations including infinite-dimensional systems (e.g., fluid flows), are agreed well with this universal number. The universality feature is commonly exhibited in many biological systems also.

12.8   Exercises

1.(a) Discuss predictability and unpredictability in natural phenomena with examples.
  (b) Write some properties of chaos. Express chaotic properties in terms of mathematics.
  (c) Discuss chaos and its importance.
  (d) Define invariant set with example. Write its significance in chaotic dynamics.
  (e) What are the roles of positive invariant set in chaotic dynamics?
  (f) Give the definition of expansive map with example. State its roles in chaotic systems.
  (g) Define dense set and dense periodic orbit. How do you explain in the context of chaotic orbit?
  (h) Prove that topological transitive set is a dense set. Is the reverse true? Give example in support of your answer.
  (i) Let X be a finite space with a trivial distance function $d(x, y) = 0$ when $x = y$ and $d(x, y) = 1$ else. Show that (X, f) is expansive but not sensitive to initial conditions, where $f : X \rightarrow X$ is a bijective map.
  (j) Show that for any map the fixed point source is sensitive dependence on initial points.

2. Discuss period-doubling cascade in logistic map for different values of growth parameter.
3. Discuss the period-doubling bifurcation sequences for the maps
   (i) $x_{n+1} = rx_n(1 - x_n)$, $3 < r < 4$,
   (ii) $x_{n+1} = r\sin(\pi x_n)$ for different values of r.
       Also, draw bifurcation diagrams for the above two maps.

4.(a) Explain periodic windows in chaotic regime. What does it signify?
  (b) Prove that the following maps are chaotic: (i) $nx(\mathrm{mod}1)$, n is an integer, (ii) $\exp(-\alpha x^2) + \beta$, $\alpha, \beta \in \mathbb{R}$.

5. Explain the idea of universality in chaos.
6. Define super stable fixed point and super stable cycle of a map. Discuss the significance of these fixed points in chaotic dynamics. Find the equation for r in which the quadratic map has a super stable 3-cycle.
7. By drawing cobweb diagram, starting from $x_0 = 1/2$ obtain the period of super stable for the logistic map at $r = 3.739$.

8. Discuss Feigenbaum number. Calculate this number for the case of logistic map.

9. Calculate the Feigenbaum number for the sine map.

10.(a) What do you mean by boundary of chaos? Explain it in the context of dynamics of different one-dimensional maps.

   (b) Explain the universality feature in nonlinear systems. Formulate a physical system which shows the universality feature.

11. Show that the periodic points of the map g : unit circle $S \to S$ defined by $g(\theta) = 2\theta$ are dense in S. Also show that the eventually fixed points of g are dense in S.

12. Find all fixed points and periodic points of the tent map $T : [0, 1] \to [0, 1]$. Prove that the tent map T has infinitely many periodic orbits and hence show that T is chaotic.

13. Find the fixed points of the cubic map $x_{n+1} = rx_n - x_n^3$ and determine the stabilities. Also, find period-doubling bifurcations sequence of the map.

14. Find the equation for r in which the map $x_{n+1} = 1 - rx_n^2$ has a super stable three cycle.

15. Consider the points in Σ_2
    $$s = (101011101011\cdots), \quad t = (010100010100\cdots) \qquad \text{and}$$
    $r = (111000111000\cdots)$. Find $d(s,t)$, $d(t,r)$, and $d(r,s)$. Also, verify that $d(r, s) + d(t, r) = d(s, t)$.

16. Show that d is a metric on Σ_N, where d represents the distance between two elements $s, t \in \Sigma_N$.

17. Let $s, t \in \Sigma_N$, the sequence space of N symbols $0, 1, 2, \ldots, N-1$. Show that

   (a) if $s_i = t_i$ for $i = 0, 1, 2\ldots, k$, then $d(s, t) < 1/N^k$.
   (b) if $d(s, t) < 1/N^k$ for some $k \geq 0$, then $s_i = t_i$ for $i \leq k$.

18. The shift map $\sigma : \Sigma_N \to \Sigma_N$ is defined by

    $$\sigma(s_0 s_1 s_2 \cdots) = (s_1 s_2 s_3 \cdots) \quad \forall s = (s_0 s_1 s_2 \cdots) \in \Sigma_N.$$

   (a) Show that σ is continuous on Σ_N.
   (b) Show that the periodic points of σ forms a dense set. Also show that σ has N^n periodic points of period n (not necessarily prime period).
   (c) Show that the tent map is conjugate to the two symbols L and R.

19. For Cantor set C prove the following:

   19. $$C_n = C_{n-1} - \bigcup_{k=0}^{\infty} \left(\frac{1+3k}{3^n}, \frac{2+3k}{3^n} \right),$$

   (ii) C is closed and compact,
   (iii) C is uncountable,

   (iv) C is perfect,
   (v) C is totally disconnected,
  (vi) C has Lebesgue measure zero.

20.(a) Show that for $r > 2 + \sqrt{5}$, the map $f_r(x) = rx(1-x)$ is chaotic on the Cantor set Λ_r

  (b) Prove that the period 3 implies all periods.

21. Show that period 3 implies chaos.

22. Prove that the map $S(x) = 2x(\mathrm{mod}1)$ on the real line exhibits chaotic orbit.

23.(a) What do you mean by Lyapunov exponent? Derive its mathematical expression.

  (b) Write down geometrical significance of Lyapunov exponent of a map.

24. Find Lyapunov exponent for the logistic map and draw the graph for different values of r.

25. Calculate Lyapunov exponent for the following maps (i) $x_{x+1} = 1 - rx_n^2$, (ii) $x_{x+1} = 10x_n(\mathrm{mod}1)$, (iii) $x_{x+1} = r\sin(\pi x_n), r, x_n \in [o, 1]$, (iv) $x_{x+1} = re^{x_n}$, $r > 0$, (v) $x_{x+1} = rx_n(1 - x_n^2)$, (vi) $x_{x+1} = (x_n + q)(\mathrm{mod}1)$, q being irrational number.

26. Define invariant measure and ergodic map. State ergodic hypothesis. Discuss invariant measure of the logistic map for $r = 4$.

27. Find the probability density functions for the tent and logistic maps.

28. Plot the probability density functions or orbital densities for the logistic map with $r = 3.8$ and $r = 4.0$. Explain the significant difference between the two plots.

29. Define horseshoe map. Discuss periodic cascade of horseshoe map.

30. For the Smale horseshoe map f, prove the following: (i) a countable infinity of periodic orbits of arbitrarily high period, (ii) an uncountable infinity of non-periodic orbits, (iii) a dense orbit.

31. Let $f : [a, b] \to \mathbb{R}$ be a continuous map. Prove that f has periodic points of period $2n$ for all $n \in \mathbb{N}$.

32. Show that a continuous map $f : \mathbb{R} \to \mathbb{R}$ is chaotic if and only if f^n has a horseshoe for some $n \geq 1$.

33. Explain Li and Yorke theorem on 'period-3 implies chaos' and hence show that the tent map is chaotic.

34. Show that the map $f_4(x) = 4x(1-x), x \in [0, 1]$ is chaotic.

35. Show that there exists a continuous map f with periodic point of period 7, but no periodic-5 orbits.

36. Prove that for each natural number $n > 1$, there exists a continuous map f with a cycle of period $(2n + 1)$ but not a cycle of period $(2n-1)$.

37. Show that the tent map has exactly 2^n periodic points of period n.

38. Consider a forced damped nonlinear pendulum represented by the equation $\ddot{\theta} + \alpha\dot{\theta} + \sin\theta = A\cos t$ with $\alpha = .22$ and $A = 2.7$. Graphically show that the system has a chaotic behavior. Show that the orbit is a Cantor-like structure by zooming a part of the attractor. Also, plot the Poincaré section. (Grebogi et al. [8]).

References

1. Simmons, G.F.: Introduction to Topology and Modern Analysis. Tata McGraw-Hill (2004)
2. Munkres, J.R.: Topology. Prentice Hall (2000)
3. Copson, E.T.: Metric Spaces. Cambridge University Press (1968)
4. Reisel, R.B.: Elementary Theory of Metric Spaces. Springer-Verlag (1982)
5. Bahi, J.M., Guyeux, C.: Discrete Dynamical Systems and Chaotic Mechanics: Theory and Applications. CRC Press (2013)
6. Ott, E.: Chaos in Dynamical Systems, 2nd edition. Cambridge University press (2002)
7. Yorke, J.A., Grebogi, C., Ott, E., Tedeschini-Lalli, L.: Scaling behavior of windows in dissipative dynamical systems. Phys. Rev. Lett. **54**, 1095–1098 (1985)
8. Grebogi, C., Ott, E., Yorke, J.A.: Chaos, Strange attractors, and fractal basin boundaries in nonlinear dynamics. Science **238**, 632–638 (1987)
9. Devaney, R.L.: An Introduction to Chaotic Dynamical systems. Westview Press (2003)
10. Metropolis, N., Stein, M.L., Stein, P.R.: On finite limit sets for transformations on the unit interval. J. Comb. Theory (A) **15**, 25–44 (1973)
11. Feigenbaum, M.J.: The universal metric properties of nonlinear transformations. J. Stat. Phys. **21**, 669–706 (1979)
12. Royden, H.L.: Real Analysis. Pearson Education (1988)
13. Li, T., Yorke, J.A.: Period three implies chaos. Am. Math. Monthly **82**, 985–992 (1975)
14. Glendinning, P.: Stability, Instability and Chaos: An Introduction to the Theory of Nonlinear Differential Equations. Cambridge University Press (1994)
15. Gluckenheimer, J., Holmes, P.: Nonlinear Oscillations, Dynamical Systems, and Bifurcations of Vector Fields. Springer (1983)
16. Jordan, D.W., Smith, P.: Non-linear Ordinary Differential Equations. Oxford University Press (2007)
17. McCauley, J.L.: Chaos, Dynamics and Fractals. Cambridge University Press (1993)
18. Smale, S.: Differentiable dynamical systems. Bull. Am. Math. Soc. **73**, 747–817 (1967)
19. Pomeau, Y., Manneville, P.: Intermittent transition to turbulence in dissipative dynamical systems. Commun. Math. Phys. **74**, 189 (1980)
20. Benettin, G., Cercignani, C., Galgani, L., Giorgilli, A.: Universal properties in conservative dynamical systems. Lett. Nuovo Cimento **28**, 1–4 (1980)
21. Gleick, J.: Chaos: Making a new science. Viking, New York (1987)
22. Devaney, R.L.: A First Course in Chaotic Dynamical Systems: Theory and Experiment. Westview Press (1992)
23. Davies, B.: Exploring Chaos. Westview Press (2004)
24. Cvitanović, P.: Universality in Chaos, 2nd edn. IOP, Bristol and Philadelphia (1989)
25. Arrowsmith, D.K., Place, L.M.: Dynamical Systems: Differential Equations, Maps and Chaotic Behavior. Chapman and Hall/CRC (1992)
26. Addison, P.S.: Fractals and Chaos: An Illustrated Course. Overseas Press (2005)
27. Manneville, P.: Instabilities, An introduction to Nonlinear Dynamics and Complex Systems. Imperial College Press, Chaos and Turbulence (2004)
28. Lorenz, E.: The Essence of Chaos. University of Washington Press (1993)
29. Tél, T., Gruiz, M.: Chaotic Dynamics: An Introduction Based on Classical Mechanics. Cambridge University Press (2006)
30. Sternberg, S.: Dynamical Systems. Dover Publications (2010)
31. Parker, T.S., Chua, L.O.: Practical Numerical Algorithms for Chaotic Systems. Springer-Verlag New York Inc. (1989)
32. Stewart, I.: Does God play dice?: The Mathematics of Chaos. Oxford (1990)
33. Gulick, D.: Encounters with Chaos and Fractals. CRC Press (2012)
34. Bhi-lin, H.: Elementary Symbolic Dynamics and Chaos in Dissipative Systems. World Scientific (1989)

35. Baker, G.L., Gollub, J.P.: Chaotic Dynamics: An Introduction. Cambridge University Press (1990)
36. McMullen, C.T.: Complex Dynamics and Renormalization. Princeton University Press (1994)
37. Medio, A., Lines, M.: Nonlinear Dynamics: A Primer. Cambridge University Press (2001)
38. Feigenbaum, M.J.: Universal behavior in nonlinear systems. Los Alamos Science **1**, 4–27 (1980)
39. Hilborn, R.C.: Chaos and Nonlinear Dynamics: An Introduction for Scientists and Engineers. Oxford University Press (2000)
40. Lichtenberg, A.J., Lieberman, M.A.: Regular and Stochastic Motion. Springer, Berlin (1983)
41. Schuster, H.G.: Deterministic Chaos, Chaps. 1, 2. Physik-Verlag, Weinheim (1984)

Chapter 13
Fractals

With the advent of civilization human mind always tend to unravel the wealth of knowledge in nature, whether it is his curiosity to know the universe or to measure the length of the coastlines of the earth. However, despite discovering modern technological tools, most of the knowledge remains unknown. Nature possesses objects that are irregular and erratic in shape. With the aid of Euclidean geometry it became possible to give detail descriptions of length, area, volume of objects like line, square, cube, etc., but for many natural objects like cauliflower, leaves patterns of trees, shape of mountains, clouds, etc., the idea of length, area remains vague until Benoit Mandelbrot (1924–2010). In the year 1975, he introduced a new branch of geometry known as fractal geometry which finds order in chaotic shapes and processes, and also wrote a book on "The fractal geometry of nature" in 1977. More specifically, fractal geometry describes the fractals, which are complex geometric structures prevalent in natural and physical sciences. In the earliest civilizations these complex structures now known as fractals were regarded as formless, consequently the study of fractals was rejected until Mandelbrot who claimed that these formless objects can be described by fractal geometry. Most of fractal objects are self-similar in nature. By self-similarity we mean that if a tiny portion of a geometric structure is magnified, an analogous structure of the whole is obtained. The self-similarity does not always mean the usual geometric self-similarity in which self-similarity of the shape is considered, but it may also be statistical in which the degree of irregularity or fragmentation of the shape is same at all scales. However the converse is not true, that is not every self-similar object is a fractal. The geometry of fractal is different from Euclidean geometry. Among the differences first comes the dimension of the objects, the objects of Euclidean geometry always have an integer dimension like line has dimension 1, square has dimension 2 and a cube has dimension 3. Wherein fractal objects usually have fractional dimension for instance 1.256, even though there are exceptions such as the Brownian motion, Peano curve, which have fractal dimension 2 and devil's staircase has fractal dimension 1. The perimeter and surface area of a fractal object is not unique like the regular objects and changes its values with finer resolution. Thus the perimeter and area of fractals are undefined; this means that the object cannot be well approximated with regular geometry like square and cubes of Euclidean geometry. Natural objects like ferns, trees, snowflakes, seashells, lightning bolts, cauliflower, or broccoli are fractals. The natural processes which grow with the

© Springer India 2015
G.C. Layek, *An Introduction to Dynamical Systems and Chaos*,
DOI 10.1007/978-81-322-2556-0_13

evolution of time such as sea coast, surface of moon, clouds, mountains, veins, and lungs of humans and animals are all approximately described by fractal geometry. As we know chaotic orbits are highly irregular. Fractals are useful to study chaotic orbits and may be represented by fractals.

Fractals are not just a matter of geometry but have a number of applications for the well-being of life. Fractal properties are useful in medical science. Its applicability in medical science paves the way to identify fatal diseases, for instance, the fractal properties of the blood vessels in the retina may be useful in diagnosing the diseases of the eye or in determining the severity of the disease. Herein we begin with a detailed study of fractals including its difference with the Euclidean geometry. Next, some construction of mathematical fractals will be given and finally the various methods of calculating the dimension of fractal structures are discussed in the subsequent sections.

13.1   Fractals

Euclidean geometry is a magnificent concept to describe natural objects like roads, houses, books, etc. But the methods of Euclidean geometry and calculus are not suitable to describe all natural objects. There are objects, such as trees, coastlines, cloud boundaries, monument ranges, river meanders, coral structures, etc., for which the knowledge of Euclidean geometry is insufficient to describe them as they lacks order and are erratic in shape. Such objects are called fractal objects or simply fractals and are described by fractal geometry with fractional dimensions. The name 'fractal' was invented by Polish-French-American mathematician *Benoit Mandelbrot* in the year 1975, from the Latin adjective *fractus*, which means 'broken' or 'fractured.' Mandelbrot created a beautiful fractal represented by a set of complex numbers named after him as Mandelbrot set. Since then he has been regarded as the originator of fractal geometry and fractals. However, at that time there already existed some mathematically developed fractals such as Cantor set by *George Cantor* in 1872, Peano's curve by *Giuseppe Peano* in 1890, Koch's curve by *Helge von Koch* in 1904, Sierpinsky's fractals such as carpet, triangle, etc., by *Waclaw Sierpinski* in 1916, Julia set by *Gaston Julia* in 1918 but Mandelbrot was the man who gave all of these structures a common name and a tool to describe the properties and complexities underlying these irregular and erratic structures. However, there is a huge difference between the natural fractals and mathematically developed fractals because natural fractals are always growing with time, they are generally dynamical processes while those developed mathematically are regarded as static (do not change with time). Fractal is an object that appears self-similar with varying degrees of magnification. It does not have any characteristic length scale to measure and details of its structure would reveal if looked at finer resolution. Furthermore, it possesses symmetry across the scale with each small part of the object reproducing the structure of the whole. As a matter of fact, fractals are complex geometric shapes with fine structure at arbitrarily small scales. Usually the

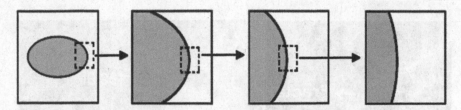

Fig. 13.1 A non-fractal object (*ellipse*)

orbits of a system are regular objects belonging to Euclidean geometry, but this is not true for chaotic orbits. The fundamental difference between fractal and non-fractal objects is that when a non-fractal object is magnified, it cannot reveal the original feature. For instance, if a section of an ellipse is magnified, it lost its feature of being an ellipse (shown in Fig. 13.1); but a fractal object always reveals its original feature under magnifications. Fractals may also be analyzed statistically. The statistical properties of fractal objects are quite different than those of non-fractal objects for instance, for non-fractal objects the average of the data reaches a limiting value even if more data is added but for a fractal the average does not reach a limiting value and keeps on changing with resolution. This implies that average does not exist for a fractal object. The self-similarity property suggests that the moment such as mean, variance do not exist for a fractal object, that is, the moment either increases or decreases with finer resolution. Thus, the statistical properties are measured by analyzing how these properties depend on the resolution used to measure them. Figures 13.1 and 13.2 depict non-fractal and fractal objects.

From above discussion it is clear that fractals have no proper definition, but there exists some properties which help in identifying and analyzing fractal objects. We shall now describe main properties of fractal objects below.

The main properties of fractal objects are

(i) Fractals appear with some degrees of self-similarity that is, if a tiny portion of a fractal object is enlarged, features reminiscent of the whole will be discerned. This means that the fractal objects could be broken into even finer pieces having same features as the original;

(ii) The topological dimension of fractal objects is always less than its fractal dimension (both of these notions are discussed later in this chapter). The dimension of a fractal object is usually not an integer but fractional;

(iii) Usually, the perimeter and area of a fractal object is undefined with changing values depending upon the resolution taken to measure it. In some cases, the perimeter is immeasurable (being infinite), but area is finite;

(iv) Fractal geometry can be represented by iterative algorithm for instance the Mandelbrot set is the set of points (say z_0) in the complex plane for which the iteration is given by $z_{n+1} = z_n^2 + z_0$ which remains bounded. Another example is the Julia set represented by the function $f(z) = z^2 + c$.

(a)   (b)

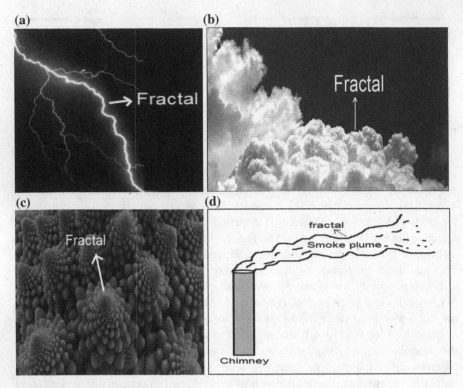

Fig. 13.2 Some fractal objects: **a** forked lighting in the sky, **b** cloud boundaries, **c** broccali, **d** smoke plume from chimney

The self-similarity is an intrinsic property of natural objects. And many properties of fractal object can be established by knowing the self-similar or similarity pattern of the object. In particular, the self-similar dimension or fractal dimension is very convenient to express the fractal object mathematically. In the following section we shall discuss the self-similarity and scaling.

13.2   Self-similarity and Scaling

Self-similarity and scaling are two bases for describing fractal geometry. Basically, the concept of self-similarity is an extension of the concept of similarity in mathematics or physical processes. Two objects are said to be similar if they are of the same shape, regardless of their size. The solid circles as shown in the Fig. 13.3 are similar, but the rectangles shown in the Fig. 13.4 are not similar.

Fig. 13.3 Two similar *circles*

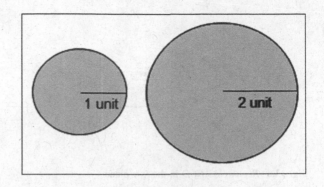

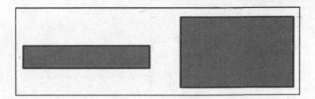

Fig. 13.4 Two non-similar *rectangles*

Normally, the similarity between two objects or among objects is studied under the light of transformations. Two objects are said to be similar if one is obtained from the other by a transformation, known as similarity transformation. Similarity transformations are the combinations of translation, rotation, and scaling. We shall now discuss these transformations in detail in $\mathbb{R}^2$.

Let $P(x, y)$ be any point in $\mathbb{R}^2$. A translation operation, denoted by T, translates the point P into a new point $P'(x', y')$ by the rule $(x', y') = T(x, y) = (x + c, y + d)$, that is, $x' = x + c$, $y' = y + d$, where c and d are real numbers. Here c represents the displacement of the point P in the horizontal direction (x-axis) and d represents the displacement of P in the vertical direction (y-axis). For example, consider the rectangle $ABCD$ shown in Fig. 13.5.

Taking $c = -1$ and $d = 1$, we can find the translated rectangle $A'B'C'D'$ shown in Fig. 13.6.

Thus under translation T the shape and the size of an object remain invariant. When we apply a rotational operation R on the point $P(x, y)$, it yields a new point, say $P'(x', y')$ satisfying the transformation relation as

$$(x', y') = T(x, y) = (x \cos \theta - y \sin \theta, x \sin \theta + y \cos \theta),$$

i.e., $x' = x \cos \theta - y \sin \theta$, and $y' = x \sin \theta + y \cos \theta$, where θ measures the angle of rotation (in the anti-clock wise direction). For example, consider the rectangle in the Fig. 13.5. Under a rotation of 90° the rectangle $ABCD$ yields a new rectangle $A'B'C'D'$ shown in the Fig. 13.7.

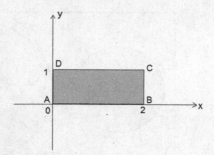

Fig. 13.5 *Rectangle ABCD* in the *x*–*y* plane

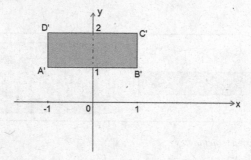

Fig. 13.6 Translation of the *rectangle ABCD* −1 unit along the *x*-axis and 1 unit along the *y*-axis

Fig. 13.7 Rotation of the
rectangle ABCD by 90° angle

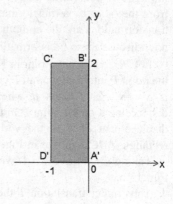

A scaling operation S takes the point $P(x, y)$ to a new point $P'(x', y')$ by the formula

$$(x', y') = T(x, y) = (\alpha x, \alpha y), \text{that is,} \, x' = \alpha x, y' = \alpha y,$$

where $\alpha > 0$ is a real number, known as the scaling factor. Under the scaling operation S an object will either contract or expand. For $\alpha < 1$, it contracts and for

Fig. 13.8 Scaling of the *rectangle ABCD*

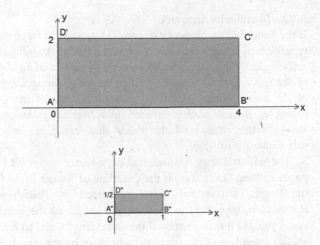

$\alpha > 1$, it expands. Consider the rectangle given in the Fig. 13.8, for the scaling factors $\alpha = 2$ and $\alpha = 1/2$, the rectangle *ABCD* reduces to the rectangles *A'B'C'D'* and *A″B″C″D″*, respectively, as shown in Fig. 13.8.

This illustrates that the transformed figures are associated with some transformations. We will now try to understand the concept of self-similarity by the following example of the self-similarity of a square.

Consider a self-similar square. This square can be divided into four equal small squares as shown in Fig. 13.9.

Taking any one of these four squares and magnifying (scale) it by the scaling factor $\alpha = 2$, we will obtain the original square. This is actually the property of self-similar object. Self-similarity usually does not mean that a magnified view is identical to the whole object, but instead the character of patterns is same on all scales. Normally, self-similarity of fractal objects, natural, physical, and biological processes may occur in different types, viz., self-similarity in space, self-similarity in time, and also statistical self-similarity. We shall illustrate these different types of self-similarities as follows:

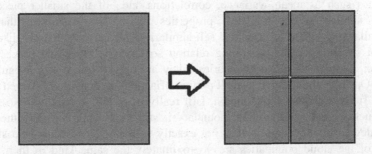

Fig. 13.9 Self-similarity of a *square*

(a) Self-similarity in space

If the pattern of an object or dynamical process in large structures obtains under the repetition of smaller structure then the similarity exhibited by the object is called as self-similarity in space. The measurement of area of an object depends on the length of the space used to measure it. The length of the space gives the spatial resolution of the measurement. For finer resolution, area of the object gets increased. For examples, the self-similar patterns in biological systems such as the arteries and the veins in the retina and the tubes that bring air into the lungs maintain the self-similarity in space.

The self-similarity in space had been better described by Mandelbrot in his 1967 paper entitled "How long is the coastline of Britain" (see [20]), based on measuring the length of coastlines first considered by British meteorologist Lewis Fry Richardson, who tried to measure the length of the coastline of Britain by laying small straight line segments of the same length, end to end, along the coastline. The spatial resolution of the measurement is set by the length of these line segments and the combined length of these line segments give the total length of the coastline. He observed that the length of coastline increased each time when the length of the line segment measuring the coastline decreased, i.e., whenever the measurement was taken at ever finer resolutions since the smaller line segments included the smaller bays and peninsulas that were not included in the measurement of larger line segments. The total length of coastlines thus varies with the length of line segment measuring it. Coastlines are approximately described by fractal geometry. Thus the length of fractals is immeasurable.

(b) Self-similarity in time

If the pattern of the smaller fluctuations of a process over short times is repeated in the larger fluctuations over longer times then the self-similarity in the process is found in terms of time and is known as self-similarity in time. The electrical signals generated by the contraction of the heart, the volumes of breaths over time drawn into the lung, etc. are some examples of self-similarity in time.

(c) Statistical self-similarity

As mentioned earlier, self-similarity does not mean that smaller pieces of an object are exact smaller copies of the whole object, but instead of the same kind of their larger pieces, that is, approximately similar to the larger pieces. When the statistical properties such as mean, variance, correlations, etc., of the smaller pieces are exactly similar to the statistical properties of the larger pieces, then the self-similarity is called as statistical self-similarity. We can say that just like geometrical structure satisfies a scaling relation so does the stochastic process. For instance, consider the cloud boundaries. They are rugged (uneven or irregular) in nature. One may think that if a small portion of a cloud boundary is enlarged it will lose its ruggedness and look smooth. Is it really true? No, it does not. No matter how much we enlarge the cloud boundary it still looks approximately the same (same degree of ruggedness). It is not exactly self-similar. Actually, the smaller copies of the cloud boundaries are approximately the same kind of their larger portions, but some of the statistical properties remain invariant. This type of

self-similarity is known as **statistical self-similarity.** The blood vessels in the
retina, the tubes that bring air into the lungs, the electrical voltage across the cell
membrane, wall crakes, sea coast lines, and ECG report of heartbeat of a normal
man possess the statistical self-similar property.

The probability density function (pdf) of an object or process gives the number of
pieces of different sizes that constitute the object. We can therefore say an object as
statistical self-similar if the pdf measured at resolution say $\mathcal{R}$ is geometrically
similar, that is, it has the same shape as the pdf measured at resolution $a\mathcal{R}$. Under
statistical self-similarity a statistical property of an object at a resolution is in proportion to that at another resolution. This means that if $S(\alpha)$ represents a statistical
property of an object at resolution α then $S(k\alpha)$ has the same statistical property of the
object at resolution $k\alpha$. In other words, $S(k\alpha) = aS(\alpha)$, where a is the proportionality
constant. It is important to say that many natural and biological systems exhibit the
statistical self-similarity. For instance, the fluid turbulence possesses statistical
self-similarity. In fact, early turbulent flow research after Osborne Reynolds, pioneered by G.I. Taylor, led into the path in which statistical self-similarity of various
flow quantities in homogeneous and isotropic turbulent flow such as the velocity
correlation function, pressure–velocity correlation function, etc., were considered. It
would not be altogether wrong if we say that this concept opens a new perspective to
look at the problem of turbulence. In recent times under Sophus Lie's symmetry
group of transformations, it is speculated that certain statistical properties of turbulence can be determined with the help of symbolic computational tools for
symmetry. For example, the mean velocity and its multiple layer structures of steady
turbulent plane parallel boundary layer flows are established mathematically, Martin
Oberlack [21]. Moreover, the scaling relation of the correlation function is also
determined whose structure remains invariant with the scale of the flow.

13.3   Self-similar Fractals

Fractal objects are omnipresent in nature and can be well-approximated mathematically. Mandelbrot set, Koch snowflake, and Sierpinski triangle are all examples
of mathematically generated fractals. All fractal objects in nature are self-similar at
least approximately or statistically if not exactly. Natural objects are usually
approximately fractal, and we are well aware of the fact that one cannot find any
natural object or phenomena which are exactly self-similar. In contrast the mathematical fractals are perfect fractals where each smaller copy is an exact copy of the
original. For example, consider a triangle S presented in Fig. 13.10, the details of its
construction are given in the later subsection.

The triangle S known as Sierpinski triangle may be considered as the composition of three small equal triangles, each of which is exactly half the size of the
original triangle S. Thus if we magnify any of these three triangles by a factor of 2,
we will get the triangle S. So, the triangle S consists of three small triangles, which
are self-similar copies of S. Again, each of these three small self-similar triangles

Fig. 13.10 *Sierpinski triangle*

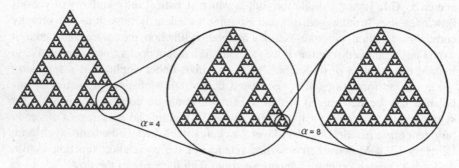

Fig. 13.11 Magnification of a portion of the *Sierpinski triangle* first by scaling factor 4 and then by factor 8

can be considered as the combination of another three small self-similar triangles, each of which is the scaled-down copy with scaling factor $\alpha = 1/2$ and so on.

Thus we see that if we enlarge a small portion of the Sierpinski triangle by a suitable scaling factor (Fig. 13.11), it will give the original feature. This type of self-similarity is known as **exact self-similarity**. It is the strongest self-similarity occurred in fractal images. Likewise Sierpinski gasket, Cantor set, von Koch curve, fern tip, Manger sponge, etc., possess exact self-similarity property.

It is to be noted that most of the fractal objects are self-similar, but a self-similar object is not necessarily a fractal. For example, a solid square can be divided into four small solid squares that resemble the original large square, and each small square can be divided into four smaller solid squares resembling the large square, and so on. Hence a square is self-similar. But it is not a fractal, and is a regular geometric shape whose properties can be well described by the Euclidean geometry. Besides, its fractal dimension coincides with its topological dimension, and it also does not possess any other fractal property.

13.4   Constructions of Self-similar Fractals

We shall now give some examples of constructing mathematical fractals. The mathematical fractals can be designed using mathematical functions and their compositions.

(i) Cantor set

The Cantor set was invented by the German mathematician *Georg Cantor* (1845–1918) in the year 1883 (see Cantor [1]) as an example of perfect, nowhere dense set. The Cantor set is an infinite set of points in the unit interval [0, 1]. At first glance, Cantor set does not seem a fractal. However, it is the most important of all as it modeled many of the other fractals such as Julia set. Moreover, many fractals such as Cantor dust; Baker's attractor etc., are the product fractals of Cantor set with itself or some other set. It is constructed as follows:

Consider the unit interval $S_0 = [0, 1]$. Remove its open middle third, that is, delete the open interval $\left(\frac{1}{3}, \frac{2}{3}\right)$ and leave the end points behind. This produces the pair of closed intervals shown below (Fig. 13.12).

Then we remove the open middle third of the above two closed intervals to produce S_2. Repeat this process several times. The limiting set $C = S_\infty$ is called the **middle-third Cantor set** or simply the **Cantor set**. It is difficult to visualize the Cantor set but it consists of an infinite number of infinitesimal pieces, separated by gaps of various sizes. The formation of Cantor set is shown in Fig. 13.13.

The sequences of closed subintervals $S_0, S_1, S_2, \ldots$ are given as follows.

$$S_0 = [0, 1]$$

$$S_1 = \left[0, \frac{1}{3}\right] \cup \left[\frac{2}{3}, 1\right]$$

$$S_2 = \left[0, \frac{1}{9}\right] \cup \left[\frac{2}{9}, \frac{1}{3}\right] \cup \left[\frac{2}{3}, \frac{7}{9}\right] \cup \left[\frac{8}{9}, 1\right] \text{ and so on.}$$

The length L of the Cantor set at step 0 is 1, at first step it is $\frac{2}{3}$, at second step it becomes $\left(\frac{2}{3}\right)^2$ and at the nth stage the length becomes $\left(\frac{2}{3}\right)^n$ which tends to 0 as $n \to \infty$ and hence the Cantor set has zero measure.

The points in [0, 1] which belongs to the Cantor set C can be represented by triadic expansions, by triadic expansions we mean that a number say x is expanded with respect to base 3,

$$x = 0 \cdot a_1 a_2 a_3 \cdots = a_1 \cdot 3^{-1} + a_2 \cdot 3^{-2} + a_3 \cdot 3^{-3} + \cdots$$

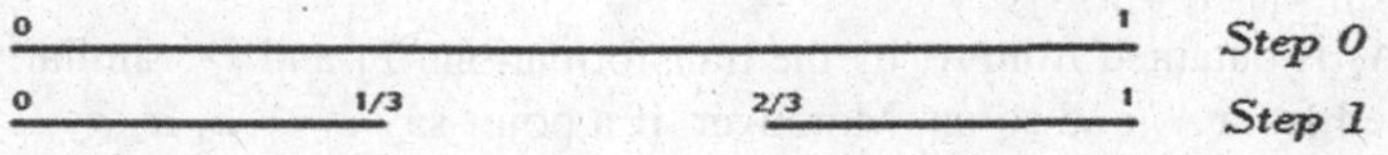

Fig. 13.12 First two steps of formation of cantor set

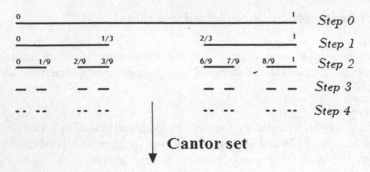

Cantor set

Fig. 13.13 Cantor set

where the numbers $a_1, a_2, a_3, \ldots$ takes the values from the set$\{0, 1, 2\}$, but Cantor set contains only those element in the unit interval which consists entirely of 0 and 2 for instance, 1 belongs to the Cantor set has the expansion 0.222222 as 1 can be written as $1 = \frac{2}{3} \cdot (1/(1 - 1/3)) = \frac{2}{3} + \frac{2}{3^2} + \frac{2}{3^3} + \cdots$, likewise $\frac{1}{3}$ has the triadic expansion 0.0222222, as $\frac{1}{3}$ can be written as $\frac{1}{3} = \frac{2}{9}(1/(1 - 1/3)) = \frac{0}{3} + \frac{2}{3^2} + \frac{2}{3^3} + \cdots$, likewise $\frac{2}{3}$ has the triadic expansion 0.2000... as $\frac{2}{3} = \frac{2}{3} + \frac{0}{3^2} + \frac{0}{3^3} + \cdots$, $\frac{1}{9}$ has the triadic expansion 0.00222 as $\frac{1}{9} = \frac{0}{3} + \frac{0}{3^2} + \frac{2}{3^3} + \frac{2}{3^4} + \cdots = \frac{2}{3^3}(1/(1 - 1/3))$ and so on. Moreover, each element belonging to the Cantor set has a unique ternary representation with 0 and 2, except the end points of the interval. For instance, 1/3 has the ternary expansion $0.1 = \frac{1}{3} + \frac{0}{3^2} + \cdots$ as well as the ternary expansion 0.0222... shown above, but this too has the ternary expansion with 0 and 2 in the latter expansion. So all those points which has ternary expansion with 0 and 2 in [0, 1] belongs to the Cantor set C. Again there exist an infinite number of possibilities of the representation of ternary expansion with 0 and 2 in [0, 1], therefore all the elements of the Cantor set cannot fit into a single list. Hence Cantor set is uncountable. Generally, construction of self-similar fractals can be related with the affine transformation $T :$ $\mathbb{R}^n \to \mathbb{R}^n$ of the form $T : x \to a_i x + b_i$, where a_i is a linear transformation on $\mathbb{R}^n$ and b_i belongs to some subset of a vector space. In fact, affine transformation is a combination of a translation, rotation, dilation and a reflection contracting with differing ratios in different directions. Affine transformations particularly maps circles to ellipses, squares to parallelograms, etc. For the construction of the Cantor set, the affine transformation $T : \mathbb{R} \to \mathbb{R}$, a_i is a linear transformation in $\mathbb{R}$, $b_i \in [0, 1]$, and the transformations are defined as $T_1 : x \to \frac{x}{3}, T_2 : x \to \frac{x}{3} + \frac{2}{3}$. If we start with the closed and bounded interval $A_0 = [0, 1]$ then under these transformations we will have $A_1 = T_1[0, 1] \cup T_2[0, 1] = \left[0, \frac{1}{3}\right] \cup \left[\frac{2}{3}, 1\right]$. Applying these transformations once again to each of the above intervals will give $A_2 = T_1\left[0, \frac{1}{3}\right] \cup T_2\left[0, \frac{1}{3}\right] \cup T_1\left[\frac{2}{3}, 1\right] \cup T_2\left[\frac{2}{3}, 1\right] = \left[0, \frac{1}{9}\right] \cup \left[\frac{2}{3}, \frac{7}{9}\right] \cup \left[\frac{2}{9}, \frac{1}{3}\right] \cup \left[\frac{8}{9}, 1\right]$.

Thus A_2 is obtained from A_1 by the transformations T_1 and T_2, similarly A_3 will be obtained from A_2 and so on. Moreover, if a point say $a = \cdot a_1 a_2 a_3 \ldots \in C$ has the triadic expansion with only 0 or 2 then the points obtained after applying the transformations T_1 and T_2 have also the triadic expansion with 0 or 2 thus $TC \subset C$. This implies that the Cantor set is invariant under T. Also $T_1(\cdot a_2a_3...) = \cdot 0a_2a_3..., T_2(\cdot a_2a_3...) = \cdot 2a_2a_3...$, this shows that $a = \cdot a_1a_2a_3...$ is the image of T_1 if $a_1 = 0$ and is the image of T_2 if $a_1 = 2$ thus $TC = C$. Hence the Cantor set is self-similar under the contraction transformations T_1 and T_2.

The main properties of Cantor set are summarized below:

 (i)   The Cantor set C has structure at arbitrarily small scales;
 (ii) It is self-similar;
 (iii) The dimension of C is a fraction;
 (iv) It is totally disconnected therefore nowhere dense;
 (v)   C is perfect;
 (vi) C has measure zero and consists of uncountably many points.

The topological dimension (defined later in this chapter) of Cantor set is 0 as it is made up of disconnected points and does not contain any line segment.

(ii) Even-fifth Cantor set

We have described the construction of middle-third Cantor set and its various properties. But note that it is not the only possibility of constructing a Cantor set and there exists infinite way of constructing Cantor like sets. For instance instead of removing the middle third of an interval one can try with removal of middle half of each subinterval thus creating a middle-halves Cantor set. Likewise, Cantor set of the type in which each interval is scaled down by an odd factor followed by the removal of even subinterval can be created. Here we will now see how an even-fifth Cantor set is created. For this consider an interval $S_0 = [0,1]$. Now scale down this interval into five equal pieces. Then remove the second and fourth subinterval. This produces

$$S_1 = [0, 1/5) \cup (2/5, 3/5) \cup (4/5, 1]$$

Now repeating the same procedure with each subinterval in S_1, we will obtain

$$S_2 = [0, 1/25) \cup (2/25, 3/25) \cup (4/25, 1/5) \cup (2/5, 11/25) \cup (12/25, 13/25)$$
$$\cup (14/25, 3/5) \cup (4/5, 21/25) \cup (22/25, 23/25) \cup (24/25, 1]$$

and so on. This construction of Cantor set is called even fifth Cantor set (see Strogatz [19]). The construction is shown in Fig. 13.14a.

(iii) von Koch curve

The Swedish mathematician *Helge von Koch* (1870–1924) introduced the Koch curve in the year 1904 (von Koch [2, 3]), as an example of a curve that is continuous but nowhere differentiable. We know that tangent at a corner of a curve is not uniquely defined. The Koch curve is made out of corners everywhere, therefore it is not possible to draw tangent at any of its point hence nowhere differentiable. This curve can be constructed geometrically by successive iterations as follows.

We start with a line segment, say, S_0 of length L_0. To generate S_1, divide it into three equal line segments. Then replace the middle segment by an equilateral

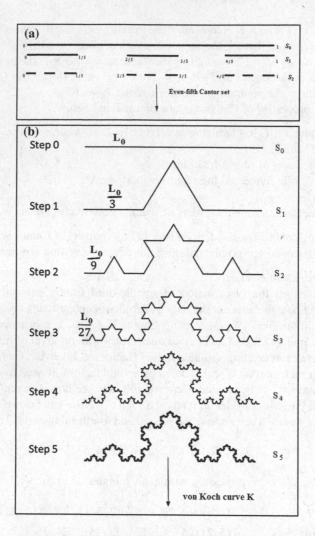

Fig. 13.14 **a** Construction of even-fifth Cantor set. **b** Construction of von *Koch curve*

triangle without a base. This completes the first step (S_1) of the construction, giving a curve of four line segments, each of length $l = \frac{L_0}{3}$ and the total length is $\frac{4L_0}{3}$ in this stage of construction. Subsequent stages are generated recursively by a scaled-down of the generator l. Thus, at the nth stage number of copies is equal to $N = 4^n$ line segments, each of length $\frac{L_0}{3^n}$. The limiting set $K = S_\infty$ is known as the von Koch curve. The graphical representation is presented in Fig. 13.14b. At the nth stage of iterations, we will get 4^n line segments each of length $\frac{L_0}{3^n}$. The length increases by a factor of $\frac{4}{3}$ at each stage of the construction. So, at the nth stage the length of the segments are given by

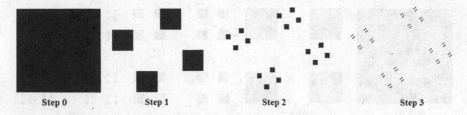

Step 0   Step 1   Step 2   Step 3

Fig. 13.15 Construction of Cantor dust

$$S_n = 4^n \cdot \frac{L_0}{3^n} = L_0 \left(\frac{4}{3}\right)^n \to \infty \text{ as } n \to \infty.$$

Hence, the length of the Koch curve is infinite.

(iv) Cantor dust

Cantor dust is analogs to the Cantor set in a plane. Cantor dust is created by dividing a square into m^2 smaller squares of side lengths $1/m$ of which one square in each column is retained and the rests are discarded at each stage. At first stage a square is divided into 16 squares, out of these 16 squares only four are kept and the other discarded. Thus at first stage four squares are obtained each of side length $l_1 = \frac{1}{4}$. At second stage each of these four squares are divided into 16 squares of which only four are kept and the rest are discarded. Thus at second stage 4^2 squares are obtained each of side length $l_2 = \frac{1}{4^2}$. Proceeding in this way at the kth stage 4^k squares will be obtained each of side length $\frac{1}{4^k}$. The construction of Cantor dust is shown in Fig. 13.15. This arrangement of the Cantor dust has the sequences of closed sub-squares as $S_0, S_1, S_2, \ldots$ are given as follows.

$$S_0 = [0,1] \times [0,1]$$

$$S_1 = \left[0, \frac{1}{4}\right] \times \left[0, \frac{1}{4}\right] \cup \left[0, \frac{1}{4}\right] \times \left[\frac{3}{4}, 1\right] \cup \left[\frac{3}{4}, 1\right] \times \left[0, \frac{1}{4}\right] \cup \left[\frac{3}{4}, 1\right] \times \left[\frac{3}{4}, 1\right] \text{ and so on.}$$

Note that the Cantor set thus obtained is not the only one and one can use other arrangements or number of squares to get different sets. For instance, if the unit square is divided into nine equal squares and then retaining the four outer squares and discarding the other squares at each stage, one will get a different Cantor dust. For this arrangement of Cantor dust, one will get 4^k squares at each stage each of side length 3^{-k}. Figure 13.16 display this Cantor dust. This later arrangement has the sequences of closed sub-squares as $S_0, S_1, S_2, \ldots$ are given as follows.

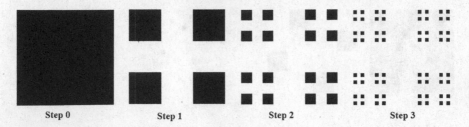

Step 0   Step 1   Step 2   Step 3

Fig. 13.16 Another construction of Cantor dust

$$S_0 = [0,1] \times [0,1]$$

$$S_1 = \left[0,\frac{1}{3}\right] \times \left[0,\frac{1}{3}\right] \cup \left[\frac{2}{3},1\right] \times \left[0,\frac{1}{3}\right] \cup \left[0,\frac{1}{3}\right] \times \left[\frac{2}{3},1\right] \cup \left[\frac{2}{3},1\right] \times \left[\frac{2}{3},1\right]$$

$$S_2 = \left[0,\frac{1}{9}\right] \times \left[0,\frac{1}{9}\right] \cup \left[0,\frac{1}{9}\right] \times \left[\frac{2}{9},\frac{1}{3}\right] \cup \left[\frac{2}{9},\frac{1}{3}\right] \times \left[0,\frac{1}{9}\right] \cup \left[\frac{2}{9},\frac{1}{3}\right] \times \left[\frac{2}{9},\frac{1}{3}\right] \cup \left[\frac{2}{3},\frac{7}{9}\right]$$

$$\times \left[0,\frac{1}{9}\right] \cup \left[\frac{8}{9},1\right] \times \left[0,\frac{1}{9}\right] \cup \left[\frac{2}{3},\frac{7}{9}\right] \times \left[\frac{2}{9},\frac{1}{3}\right] \cup \left[\frac{8}{9},1\right] \times \left[\frac{2}{9},\frac{1}{3}\right] \cup \left[0,\frac{1}{9}\right] \times \left[\frac{2}{3},\frac{7}{9}\right] \cup \left[0,\frac{1}{9}\right]$$

$$\times \left[\frac{8}{9},1\right] \cup \left[\frac{2}{9},\frac{1}{3}\right] \times \left[\frac{2}{3},\frac{7}{9}\right] \cup \left[\frac{2}{9},\frac{1}{3}\right] \times \left[\frac{8}{9},1\right] \cup \left[\frac{2}{3},\frac{7}{9}\right] \times \left[\frac{2}{3},\frac{7}{9}\right] \cup \left[\frac{8}{9},1\right]$$

$$\times \left[\frac{2}{3},\frac{7}{9}\right] \cup \left[\frac{2}{3},\frac{7}{9}\right] \times \left[\frac{8}{9},1\right] \cup \left[\frac{8}{9},1\right] \times \left[\frac{8}{9},1\right].$$

and so on.

(v) Sierpinski fractals

Sierpinski gasket and carpet were discovered by the Polish mathematician Waclaw Sierpinski (1882–1969) in 1916 (Sierpinski [4, 5]). We now give the construction of these fractals one by one:

(a) Sierpinski triangle (gasket)

Consider a solid (filled) equilateral triangle, say S_0, with each side of unit length. Now divide this triangle into four equal small equilateral triangles using the midpoints of the three vertices of the original triangle S_0 as new vertices. Then remove the interior of the middle triangle. This generates the stage S_1. Repeat this process in each of the remaining three equal solid equilateral triangles to produce the stage S_2. Repeat this process continuously for further evolution and finally the Sierpinski triangle is formed. The Fig. 13.17 displays five steps of Sierpinski triangle.

From the figure we see that S_1 is covered by $N = 3$ small equal equilateral triangles each of side $\varepsilon = \frac{1}{2}$. Similarly, S_2 is covered by $N = 3^2$ triangles of side $\varepsilon = \frac{1}{2^2}$. In general, S_n at the nth stage is covered by $N = 3^n$ equal equilateral triangles each of side $\varepsilon = \frac{1}{2^n}$. The area at the nth stage is obtained as $3^n \frac{\sqrt{3}}{4} \left(\frac{1}{2}\right)^{2n} = \frac{\sqrt{3}}{4} \left(\frac{3}{4}\right)^n \to 0$ as $n \to \infty$.

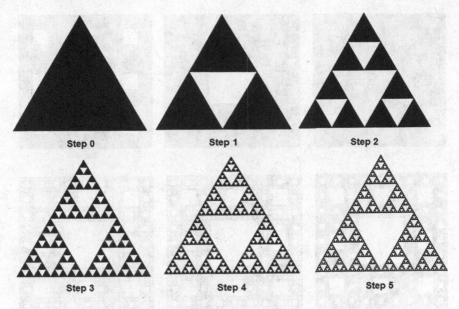

Fig. 13.17 Construction of *Sierpinski triangle*

(b) Sierpinski carpet

Consider a solid (filled) square of unit side length. Now, divide it into nine small equal squares and then remove the interior of the middle square. Repeat this process in each of the remaining eight squares and so on. The limiting set is known as the Sierpinski carpet. The Fig. 13.18 displays the Sierpinski carpet up to five iterations.

At the nth stage (nth iteration) of the Sierpinski carpet, we will get 8^n small squares each of length (side) $\frac{1}{3^n}$. Hence $N = 8^n$ and $\varepsilon = \frac{1}{3^n}$.

Many students have a wrong idea about the self-similarity of fractal objects. The fractal objects, viz., von Koch curve, Sierpinski triangle, Sierpinski carpet, Manger sponge, etc., are not self-similar at any finite stage of their constructions. For example, consider the Sierpinski triangle as constructed in step 3. If we scale a part of the triangle by a factor of 2, we will get the triangle as in step 2, not equal to the triangle as in step 3 (Fig. 13.19).

We shall illustrate a remarkable correspondence of self-similarity with Sierpinski triangle. Consider the geometric series of real numbers as

$$\sum_{k=0}^{\infty} a^k = 1 + a + a^2 + a^3 + \cdots, \text{ where } -1 < a < 1.$$

Let $S_n = \sum_{k=0}^{n} a^k = 1 + a + a^2 + \cdots + a^n$ be the sum of first n terms of the series. Then we have

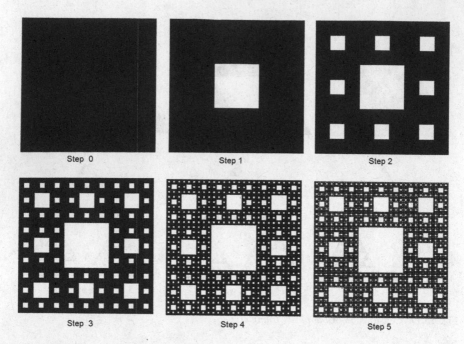

Fig. 13.18 Construction of Sierpinski carpet

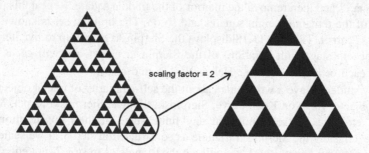

Fig. 13.19 Magnification of a part of *Sierpinski triangle* by scaling factor 2

$$S_n - aS_n = 1 - a^{n+1} \Rightarrow S_n = \frac{1 - a^{n+1}}{1 - a}.$$

Since $|a| < 1$, the sum converges to $1/(1 - a)$ as n becomes very large. Now, multiplying the geometric series by a, we get

$$a \sum_{k=0}^{\infty} a^k = a + a^2 + a^3 + a^4 + \cdots$$

$$\Rightarrow \sum_{k=0}^{\infty} a^k = 1 + a + a^2 + a^3 + \cdots = 1 + a \sum_{k=0}^{\infty} a^k$$

$$\Rightarrow \sum_{k=0}^{\infty} a^k = 1 + a \sum_{k=0}^{\infty} a^k.$$

This represents the self-similarity of the geometric series, in which the value of the sum of the series is one plus the scaled-down version of the series. This self-similarity relation does not hold for finite terms. For example, we take the sum $S_n = \sum_{k=0}^{n} a^k$. In this case we see that

$$1 + aS_n = 1 + a \sum_{k=0}^{n} a^k = 1 + \sum_{k=0}^{n} a^{k+1} = S_{n+1} \neq S_n.$$

Thus the self-similarity of the geometric series does hold for any finite term, but holds for the limiting case, containing an infinite number of terms. Like this one can prove that the self-similarity of the Sierpinski triangle does not hold in any finite stage of constructions, but holds in the limiting case. We hope that the reader should try to find an analogy between the construction of the sum of the geometric series and the construction of the Sierpinski triangle, Sierpinski carpet, von Koch curve and other known self-similar fractals.

(vi) Borderline fractals

Borderline fractals are those objects which may be either called as fractals or non-fractals and have integer fractal dimensions. Devil's staircase is an example of borderline fractal whose fractal dimension is 1. Devil's staircase has finite length and lacks self-similar properties of fractal. The reason of its being regarding as fractal is that it represents a very strange function, which is constant everywhere except in those points belonging to the Cantor set. Devil's staircase is also called as Cantor function as it is closely related to the Cantor set and its construction. The devil's staircase is constructed as follows.

Consider a square with side of length 1 and then begin the construction just like that of Cantor set. The only difference is that Cantor set is constructed on a line segment of unit length; here the construction is on a square of unit area. In this process middle third of length $\frac{1}{3^k}$ of the square is removed and pasted in a rectangular column of width $\frac{1}{3^k}$ and a certain height. This means that the square is filled with rectangles whose base is of length $1/3^k$ and height is of some different length. In this process, at first a rectangular column of height 1/2 is created, standing over the middle third of the base side of the interval [1/3, 2/3] of the square, in the next step two columns one of height 1/4 and the other of height 3/4 are created, standing over the interval [1/9, 2/9] and [7/9, 8/9], respectively. Similarly in the third step four columns of height 1/8, 3/8, 5/8, 7/8 are erected and in the kth step 2^{k-1} number of columns of height $\frac{1}{2^k}, \frac{3}{2^k}, \cdots, \frac{(2^k-1)}{2^k}$ are erected. In the limit an area is obtained, whose upper boundary is called the devil's staircase or Cantor function. The name devil's staircase is given because it has uncountably many steps whose step heights become infinitely small, moving upwards from left to right as the process proceeds. The devil's staircase thus splits the square into two halves, fractionally. The length of the curve is exactly 2!

It is called as Cantor function because it is constructed in the same way as that of Cantor set. By construction this function can be expressed mathematically as $P(x) = 0 \cdot \frac{\varepsilon_1}{a-1} \frac{\varepsilon_2}{a-1} \cdots \frac{\varepsilon_N}{a-1} \cdots$ a binary number $\left(\text{i.e.} \frac{\varepsilon_i}{a-1} = 0 \text{ or } 1\right)$, where $x = 0 \cdot \varepsilon_1 \varepsilon_2 \cdots \varepsilon_N \cdots$ belongs to the Cantor set ($\varepsilon_i = 0$ or $a - 1$). If x_1 and x_2 are the end points of a removed open interval then $P(x_1) = P(x_2)$. For instance, if $x_1 = a^{-1} = 0.0a - 1a - 1\ldots$ then $P(x_1) = 0.01111 = 0.1 = \frac{1}{2}$ and $x_2 = 1 - a^{-1}$ for which $P(x_2) = 0.1 = \frac{1}{2}$. This means that half of the iterates of the map lies in the interval $[0, a^{-1}]$ and half in $[1 - a^{-1}, 1]$. Similarly, if $x_1 = a^{-2} = 0.00a - 1a - 1\ldots$ then $P(x_1) = 0.00111\ldots = 0.01\ldots = \frac{1}{4}$ (decimal representation) and $x_2 = a^{-1} - a^{-2}$ gives $P(x_2) = 0.01 = \frac{1}{4}$ which means that $\frac{1}{4}$ of the iterations lies in the interval $[0, 1/a^{-2}]$. Again if $x_1 = 1 - a^{-1} - a^{-2}$ then $P(x_1) = 0.1011\ldots = 0.11 = \frac{3}{4}$ and $P(x_2) = 0.11\ldots = \frac{3}{4}$ which means that $\frac{3}{4}$ of the iterates fall into three intervals $[0, a^{-2}]$, $[a^{-1} - a^{-2}, a^{-1}]$ and $[1 - a^{-1}, 1 - a^{-1} + a^{-2}]$ and the remaining $\frac{1}{4}$ lies in the interval $[1 - a^{-2}, 1]$. Continuing this process of extending the function $P(x) = $ constant on the closure of every removed open interval, we will see that the function $P(x)$ is continuous but non-decreasing and its slope vanishes almost everywhere. Cantor function is therefore a probability distribution for the iteration of the tent map on the Cantor set (Figs. 13.20 and 13.21).

Another example of borderline fractal is the Peano curve whose fractal dimension is 2. Peano curve is constructed as follows.

A straight line piece of the curve is taken, and in the first stage nine line segments scaled down by a factor of 3 replace this curve. In the first stage, the curve intersects itself at exactly two points. In the second stage, each line segment obtained in the first stage is taken and replaced by nine line segments scaled down

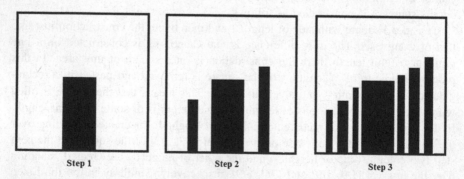

Step 1   Step 2   Step 3

Fig. 13.20 Construction of devil's staircase

Fig. 13.21 Devil's staircase

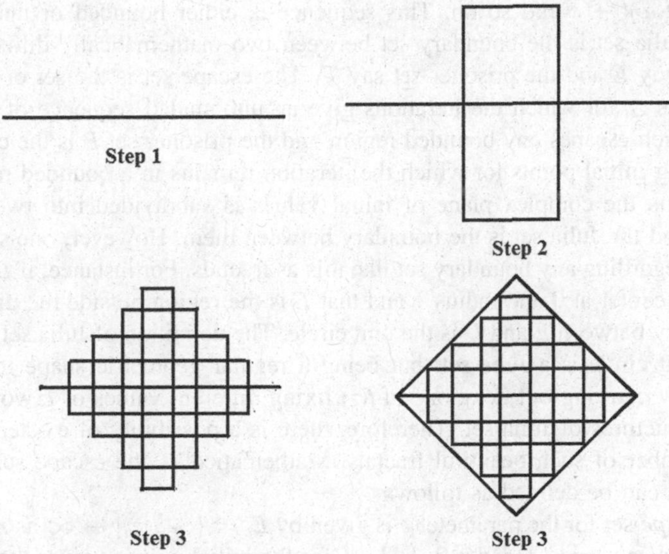

Step 1

Step 2

Step 3

Step 3

Fig. 13.22 Construction of *Peano curve*

by a factor of 3. In this stage, the curve intersects itself at about 32 points in the curve. Repeating this procedure of scaling down the line segments by a factor of 3 in each stage of construction, one will obtained in the kth step, a line segment of length $1/3^k$ which finally embeds into a square (Fig. 13.22). Since in each step of construction of the Peano curve, each line segment is replaced by nine line segments of one-third the length of the line segments obtained in the previous stage so, the length of the Peano curve can be easily calculated. Therefore, if we presume that the length of the original line segment was 1 unit, then after first stage its length will be $9 \times 1/3 = 3$ unit, and after second stage the length will be $9 \times 9 \times \frac{1}{3^2} = 9$ unit. Hence, the resulting curve increases in length by a factor of 3 in each step of the construction. Therefore, the length of the curve after kth stage is 3^k, see Peitgen, Jürgens, Saupe [16].

(vii) Julia sets

Julia set was discovered by Gaston Julia (1893–1975) in the year 1918 (Julia [6]). Julia set represents the transformation function which is either a complex polynomial or complex rational function from one state of the system to the next, and is a source of the majority of attractive fractals known at present. Julia set is obtained by iterating the quadratic function $f(z) = z^2 + c$ for a complex initial value say z_0, where c is an arbitrary fixed complex constant. Thus, fixing the value of c and taking an initial value z_0 of z, one will obtain $f(z) = z_0^2 + c$ after first iteration. The next iteration will give $f(z) = (z_0^2 + c)^2 + c$. Thus for a fixed value of c, the successive iterations give a sequence of complex numbers, i.e., $z_0^2 + c$, $(z_0^2 + c)^2 + c$, $((z_0^2 + c)^2 + c)^2 + c$ and so on. This sequence is either bounded or unbounded. Actually, Julia set is the boundary set between two mathematically different sets, escape set say E and the prisoner set say $\mathcal{P}$. The escape set is the set of all those initial points z_0 for which the iterations give an unbounded sequence of complex number which escapes any bounded region and the prisoner set P is the collection of remaining initial points for which the iteration remains in a bounded region for always. Thus the complex plane of initial values is subdivided into two subsets E and $\mathcal{P}$, and the Julia set is the boundary between them. However, one should be careful in regarding any boundary set like this as fractals. For instance, if D denotes a disk with center at 0 and radius 1 and that E is the region outside the disk. Then the boundary between E and D is the unit circle. The definition of Julia set suggests that the unit circle is a Julia set, but being a regular geometric shape it is not a fractal. Now carrying out iterations of $f(z)$ fixing different values of c would yield different structures of Julia set. Therefore, there is a possibility of existence of an infinite number of such beautiful fractals. Mathematically, the escape set and the prisoner set can be defined as follows

The *escape* set for the parameter c is given by $E_c = \{z_0 : |z_n| \to \infty \text{ as } n \to \infty\}$. The iterations $z_{n+1} = z_n^2 + c$, $n = 0, 1, 2, \ldots$ for the initial point z_0 gives the orbit z_0, z_1, z_2, $\ldots$

The *prisoner* set for parameter c is given by $\mathcal{P}_c = \{z_0 : z_0 \notin E_c\}$.

Both escape set E and the prisoner set $\mathcal{P}$ fill some part of the complex plane and complement each other. Thus, the boundary of the prisoner set is simultaneously the boundary of the escape set, which is the Julia set for c. When points from the escape set in the neighborhood of the Julia set are iterated using the iterated map $z \to z^2 + c$, they move towards infinity, away from the Julia set. Thus, the Julia set may be called a repeller with respect to the transformation $z \to z^2 + c$. Now if the transformation is inverted, then the Julia set will cease to act as a repeller and becomes an attractor. Some Julia set structures are given in Fig. 13.23 (figures are drawn using Maple software).

Properties of the Julia set

(1) The Julia set J being the closure of the repelling periodic points of the polynomial f, is a dynamical repeller.

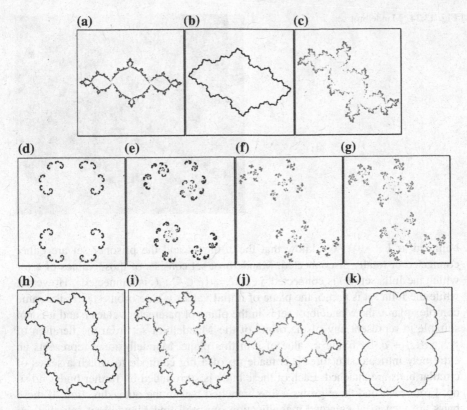

Fig. 13.23 Julia set of the quadratic function $f(z) = z^2 + c$ for **a** $c = -1 + 0.5i$, **b** $c = -0.5 + 0.3i$, **c** $c = -0.2 + 0.75i$, **d** $c = 0.5$, **e** $0.5 + 0.3i$, **f** $c = 0.25 + 0.8i$, **g** $c = 0.25 + 0.7i$, **h** $c = 0.25 + 0.4i$, **i** $c = 0.25 + 0.5i$, **j** $c = -0.8 + 0.156i$, **k** $c = 0.25$

(2) The set J is an uncountable compact set containing no isolated points and is invariant under the polynomial f and its inverse f^{-1}.

(3) The Julia set J has either periodic or chaotic orbit.

(4) All unstable periodic points are on J.

(5) The set J is either totally connected or totally disconnected.

(6) The Julia set has fractal structure almost every time.

We have described the Julia set only for a quadratic polynomial in the complex plane and one may try the Julia sets for higher degree polynomials of some other form. For a detailed description of Julia set, see the book of Falconer [17].

(viii) Mandelbrot set

Mandelbrot set is attributed to Benoit Mandelbrot, who discovered this set in 1979 (Mandelbrot [7]), while looking for the dichotomy in the Julia set produced by the variation of the parameter value c in the entire complex plane instead of the initial value z_0. The Mandelbrot set is the region in the complex plane comprising the values of c for which the trajectories defined by $z_{k+1} = z_k^2 + c$; $k = 0, 1, 2, \ldots$, remain

Fig. 13.24 Mandelbrot set

bounded as $k \to \infty$. We know that the Julia set and the prisoner set are either connected or totally disconnected. Mandelbrot set consists of those values of c for which the Julia set (J_c) is connected, i.e., $M = \{c \in \mathbb{C} : J_c$ is connected$\}$. However, while the Julia set is part of the plane of initial values whose orbits stay in the same complex plane, the Mandelbrot set is in the plane of parameter values c and it is not suitable to represent any of the orbits of the Mandelbrot set from the iteration of $z_{k+1} \to z_k^2 + c$, by fixing a value of c in this plane. Mandelbrot set represents an extremely intricate structure. It is made up of a big cardiode to which a series of circular buds are attached. Each of these buds is surrounded by further buds and so on. From each bud there grows a fine branched hair in the outer direction. If these hairs are viewed at enlarged magnification one will find Mandelbrot sets that are self-similar with the actual Mandelbrot set (Fig. 13.24). Mandelbrot set contains an enormous amount of information about the structure of Julia set as when the boundaries of Mandelbrot set are magnified an infinite structures of Julia sets are revealed for some values of c.

13.5   Dimensions of Fractals

The main goal in fractal geometry is to find dimension because dimension contains much information about the geometrical properties of an object. The dimension gives a quantitative measure of the fractal properties of self-similarity and scaling. It is thus important as it gives the measure of the appearance of additional pieces of an object when the object is looked at larger magnification or finer resolutions.

Dimension of an object can be classified into three different types each measuring different properties of the object. They are: (1) Fractal dimension (2) Topological dimension, and (3) Embedding dimension.

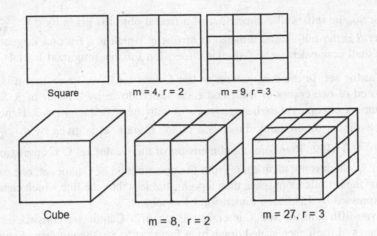

Square   m = 4, r = 2   m = 9, r = 3

Cube   m = 8, r = 2   m = 27, r = 3

Fig. 13.25 Number of copies for classical similar objects *square* and *cube*

(1) Fractal dimension

Fractal dimension describes the space-filling property of an object. The value of fractal dimension gives the quantitative measure of self-similarity of an object. The fractal dimension of an object gives the number of smaller pieces obtained when an object is looked at a smaller length scale or finer resolution. Moreover, it measures correlations between small and large pieces and also among the smaller pieces themselves. The larger value of fractal dimension means that upon viewing at finer resolution results into the revelation of larger number of smaller pieces. There exist different techniques for calculating the fractal dimension each having different characteristic way of filling the space of an object for instance self-similar dimension, box dimension, Hausdorff dimension, pointwise and correlation dimensions. Among these different fractal dimensions, the self-similar dimension is the simplest one. A through description of these dimensions with examples is given as follows.

(i) Similar dimension

We know that self-similar fractals are made up of scaled-down copies of themselves of any arbitrarily small scales. The dimension of this type of fractal can be obtained simply by observing the classical self-similar objects like square or cube. If we scale down a square by a factor of 2 in each side of it, then we get $4(= 2^2)$ small squares similar to each other and the whole. Again if we scale down the original square by a factor of 3, then we get $9(= 3^2)$ small squares. In general, the number of small squares is r^2, where r is the scaling factor. Similarly, for a cube the number of copies (small cubes) is r^3. The power indices 2 and 3 of the scale factor r are coming due to the two dimensionality of the square and the three dimensionality of the cube, respectively (shown in the Fig. 13.25). In general, for all self-similar fractal objects the number of copies (m) depends on scale factor (r) and dimension of the object (d). Naturally, the relation $m = r^d$ holds good for self-similar fractal objects. So, the self-similar dimension of a fractal object is given by $d = \frac{\log m}{\log r}$. The quantity d is the self-similar dimension of fractal which is a fraction in general.

We shall now calculate self-similar dimension for few important fractals.

(a) **Cantor set**: In the construction of the Cantor set C we know that it is composed of two copies of itself and each scaled down by a factor of 3. So, the number of copies at each stage is $m = 2$ with the scale factor $r = 3$. Hence, the self-similar or similarity dimension of the Cantor set is given by $d = \frac{\log m}{\log r} = \frac{\log 2}{\log 3} \approx 0.6309$. Therefore, the dimension of the Cantor set C is approximately 0.63, which is not an integer. From the construction of Cantor set, one can say that the set fills more space than a point, but less than the line which cannot be expressed by Euclidean dimension of an object.

(b) **Even-fifth Cantor set**: Cantor set: Even fifth Cantor set consists of three copies of itself each scaled down by a factor of 5. So, the number of copies at each stage is $m = 3$ with the scale factor $r = 5$. Therefore, the self-similarity dimension of even-fifth Cantor set is

$$d = \frac{\log m}{\log r} = \frac{\log 3}{\log 5} \approx 0.682606$$

(c) **von Koch curve**: The von Koch curve is a self-similar fractal. The curve is made up of four equal pieces, each of which is similar to the original curve and is scaled down by a factor of 3 for each segment of its construction. Hence, the number of copies is $m = 4$, and $r = 3$. The similar dimension of the self-similar von Koch curve is obtained as $d = \frac{\log m}{\log r} = \frac{\log 4}{\log 3} \approx 1.26$. The value 1.26 of similar dimension of von Koch curve tells us that the curve has dimension larger than 1 as it has infinite length and it is smaller than 2 because the curve has zero area.

(d) **Sierpinski gasket**: The Sierpinski gasket is made up of three equilateral triangles, each of which is similar to the initial equilateral triangle and is scaled down by a factor of 2 at each stage of its construction. Hence the number of copies is $m = 3$ and the scaling factor is $r = 2$. Thus, the self-similar dimension of Sierpinski gasket is given as $d = \frac{\log 3}{\log 2} = 1.58496$.

(e) **Peano curve**: The Peano curve is made up of nine equal pieces, each of which is scaled down by a factor of 3 at each stage of its construction. Hence the number of copies is $m = 3^2$ and the scaling factor is 3. Hence the dimension is given as $d = \frac{2\log 3}{\log 3} = 2$.

(f) **Cantor dust**: The dimension of the Cantor dust for the arrangement where unit square is divided into 16 smaller squares retaining 4^k squares of side length 4^{-k} at kth stage is given as $d = \frac{\log 4}{\log 4} = 1$ and for the arrangement where the unit square is divided into nine smaller squares retaining 4^k squares of side length 3^{-k} at kth stage is given as $d = 2\frac{\log 2}{\log 3} = 2 \times 0.6309 = 1.2618$ approximately.

Note that the calculation of dimension of a self-similar object is worthwhile only for a small class of strictly self-similar objects. So we need any other methods for calculating dimension in which we can calculate the dimension of self-similar objects.

(ii) Box dimension

Box dimension is one of the most widely used dimensions. This is also often called as capacity dimension. The reason of its widespread use is mainly because it is comparatively easy for mathematical calculation and empirical estimation. The concept of box dimension came in 1930s. All fractals are not self-similar. For finding dimension of non-similar objects approximately, the idea of measurement at a scale (ε) has been introduced. The irregularities of size less than ε are ignored. The boxes of length ε are used to cover the object. The number of boxes $N(\varepsilon)$ depends on ε and the dimension of the object. Naturally, the relation $N(\varepsilon) \propto 1/\varepsilon^d$ holds good, where d is the dimension of the object. In the limiting sense we find the dimension d of the object.

Let us consider a geometric object and ε be the length of cells which covers the space occupied by the object. The number $N(\varepsilon)$ is the minimum number of cells required to cover the space. Now, for a line segment of length L, $N(\varepsilon) \propto L/\varepsilon$ and for a plane area A, $N(\varepsilon) \propto A/\varepsilon^2$.

In general, we can take $N(\varepsilon)$ as $N(\varepsilon) = \frac{1}{\varepsilon^d}$.

Taking logarithm of both sides, we get

$$\log N(\varepsilon) = -d \log \varepsilon$$
$$\Rightarrow d = -\frac{\log N(\varepsilon)}{\log \varepsilon} = \frac{\log N(\varepsilon)}{\log(1/\varepsilon)}$$

If the limit of the above expression exists for $\varepsilon \to 0$, then d is called the capacity dimension or the box dimension of the non-similar fractal. So, the box dimension (or capacity dimension) of a fractal is given by

$$d = \lim_{\varepsilon \to 0} \frac{\log N(\varepsilon)}{\log(1/\varepsilon)}, \text{ provided the limit exists.}$$

The capacity or box dimension of a point in a two-dimensional space is 0, since in this case $N(\varepsilon) = 1$ for all ε.

Again, consider a line segment of length L. Here $N(\varepsilon) = \frac{L}{\varepsilon}$. So the box dimension is given by

$$d = \lim_{\varepsilon \to 0} \frac{\log N(\varepsilon)}{\log(1/\varepsilon)} = -\lim_{\varepsilon \to 0} \frac{\log L - \log \varepsilon}{\log \varepsilon} = 1.$$

The box dimension has been calculated for few objects below.

(a) Box dimension of Cantor set

The Cantor set is covered by each of the sets S_n which are used in its construction. Each S_n consists of 2^n intervals of length $\left(\frac{1}{3}\right)^n$. We take $\varepsilon = \left(\frac{1}{3}\right)^n$. We need all 2^n of these intervals to cover the Cantor set. Hence $N(\varepsilon) = 2^n$. Therefore, the box dimension of the Cantor set is obtained as

$$
\begin{aligned}
d &= \lim_{\varepsilon \to 0} \frac{\log N(\varepsilon)}{\log(1/\varepsilon)} \\
&= \lim_{n \to \infty} \frac{\log(2^n)}{\log(3^n)} \; [\because \varepsilon \to 0 \Rightarrow n \to \infty] \\
&= \frac{\log 2}{\log 3} \approx 0.6309.
\end{aligned}
$$

The box dimension of the Cantor set is equal to the similarity dimension.

(b) Box (capacity) dimension of the Sierpinski triangle

We have seen that the Sierpinski triangle S_1 is covered by $N = 3$ small equal equilateral triangles each of side $\varepsilon = \frac{1}{2}$. Similarly, the stage S_2 is covered by $N = 3^2$ triangles of side $\varepsilon = \frac{1}{2^2}$. In general, S_n at the nth stage is covered by $N = 3^n$ equal equilateral triangles each of side $\varepsilon = \frac{1}{2^n}$. Therefore the box (capacity) dimension of the Sierpinski triangle is given by

$$
d = \lim_{\varepsilon \to 0} \frac{\log N}{\log(1/\varepsilon)} = \lim_{n \to \infty} \frac{\log 3^n}{\log 2^n} = \frac{\log 3}{\log 2} \approx 1.58
$$

(c) Box (capacity) dimension of the Sierpinski carpet

In the construction of the Sierpinski carpet we have seen that at the nth stage (nth iteration) of the Sierpinski carpet, we will get 8^n small squares each of length (side) $\frac{1}{3^n}$. Hence $N = 8^n$ and $\varepsilon = \frac{1}{3^n}$. Therefore, the fractal dimension of the Sierpinski carpet is given by

$$
d = \lim_{\varepsilon \to 0} \frac{\log N}{\log(1/\varepsilon)} = \lim_{n \to \infty} \frac{\log 8^n}{\log 3^n} = \frac{\log 8}{\log 3} \approx 1.89
$$

(d) Box dimension of the von Koch curve

According to the construction of the von Koch curve, the minimum number of cells to cover the space at the nth iteration is given by $N(\varepsilon) = 4^n$, where $\varepsilon = $ length of the cells $= \frac{L_0}{3^n}$.

Therefore, the box dimension of the von Koch curve with $L_0 = 1$ is obtained as

$$
d = \lim_{\varepsilon \to 0} \frac{\log N(\varepsilon)}{\log(1/\varepsilon)} = \lim_{n \to \infty} \frac{\log(4^n)}{\log(3^n)} = \frac{\log 4}{\log 3} \approx 1.26
$$

(e) Box dimension of von Koch snowflake curve

Consider an equilateral triangle, say S_0 of side L_0. To generate S_1, divide each of its three sides into three equal line segments and then replace the middle segment by an equilateral triangle without a base. This completes the first step (S_1) of the construction, giving a curve of twelve line segments, each of length $l = \frac{L_0}{3}$ and the total length is $3 \cdot 4 \cdot \left(\frac{L_0}{3}\right)$. Subsequent stages are generated recursively by a scaled-down of the generator. Thus, at the nth stage number of copies is equal to $N = 3 \cdot 4^n$ line segments, each of length $\frac{L_0}{3^n}$. The limiting set $S = S_\infty = \lim_{n \to \infty} S_n$ is known as von Koch snowflake curve. At the nth stage, we will get 3.4^n line segments each of length $\frac{L_0}{3^n}$. So, $S_n = 3.4^n \left(\frac{L_0}{3^n}\right) = 3L_0 \left(\frac{4}{3}\right)^n \to \infty$ as $n \to \infty$. Hence the length of the von Koch snowflake curve is infinite.

At the nth stage (nth iteration) of the von Koch snowflake curve, we get 3.4^n small line segments each of length $\frac{L_0}{3^n}$. Hence $N = 3.4^n$ and $\varepsilon = \frac{L_0}{3^n}$.

Therefore, the fractal dimension of the von Koch snowflake curve for $L_0 = 1$ is given by

$$d = \lim_{\varepsilon \to 0} \frac{\log N}{\log(1/\varepsilon)} = \lim_{n \to \infty} \frac{\log 3.4^n}{\log(3^n)} = \lim_{n \to \infty} \frac{1}{n} \frac{\log 3}{\log 3} + \frac{\log 4}{\log 3} = \frac{\log 4}{\log 3} \approx 1.26.$$

(iii) Hausdorff dimension

Hausdorff definition of fractal object is much more general then the previous two dimensions as it can be defined for any set and is mathematically suitable, as it is based on measures, which are comparatively easy to calculate. However, in many cases it is hard to calculate or estimate using computational techniques. According to the concept, the shapes of the members of the cover $\{N_i\}$ of a set F that is, $\cup N_i \supseteq F$ are chosen arbitrarily. Moreover, the shape of any two covers N_i may be different and they are not necessarily box shaped. The only restriction is that $\dim(N_i) \le \varepsilon \to 0$ for each i, where $\varepsilon(>0)$ is a very small quantity. To establish the definition of Hausdorff dimension certain topological concepts such as the diameter of a set, cover of a set, etc., are required. We will also give these concepts in the process of defining Hausdorff dimension.

Let $F \subset \mathbb{R}^n$ and s is a non-negative number then for any $\delta > 0$ we can define $H_\delta^s = \inf\left\{\sum_{i=1}^\infty |N_i|^s : \{N_i\} \text{ is a } \delta \text{ cover of } F\right\}$ where $|N_i|$ is the diameter of the non-empty set $N \subset \mathbb{R}^n$. By δ cover we mean that the set F is covered by a countable collection of the sets $\{N_i\}$ with diameter at most δ that is, $F \subset \cup_{i=1}^\infty N_i$ with $0 \le |N_i| \le \delta$. The diameter for any non-empty subset N of the Euclidean space $\mathbb{R}^n$ is defined as $|N| = \sup\{|x - y| : x, y \in N\}$. The class of possible covers of F reduces with the decrease in the diameter δ while the infimum H_δ^s increases and approaches a limiting value as $\delta \to 0$ i.e., $H^s(F) = \lim_{\delta \to 0} H_\delta^s(F)$. The limiting value $H^s(F)$ can be 0 or ∞, called as s-dimensional Hausdorff measure of F. For an empty set the Hausdorff measure is zero, again $H^1(F)$ is the measure of length of a smooth curve and $H^2(F)$ is the measure of area of a surface.

Now, for $F \subset \mathbb{R}^n$, if there be a mapping $f : F \to \mathbb{R}^n$ such that $|f(x) - f(y)| \le c|x - y|^{\alpha}$, x, $y \in F$, c, $\alpha > 0$ then for each s, $H^{s/\alpha}(f(F)) \le c^{s/\alpha}H^s(F)$. The former inequality is known as Holder's inequality. As a special case when $\alpha = 1$ then the latter inequality becomes $H^s(f(F)) \le c^s H^s(F)$.

Housdorff dimension say $d_H(F)$ is defined as that critical value of s below which the Housdorff measure $H^s(F)$ assumes 0 value and above which it takes the value infinity that is, Housdorff dimension $d_H(F) = \alpha$ such that

$$H^s(F) = \left. \begin{array}{l} 0 \text{ for } s < \alpha \\ \infty \text{ for } s > \alpha \end{array} \right\}$$

Some properties of Housdorff dimension

(1) If the fractal set $F \subset \mathbb{R}^n$ is open then the Housdorff dimension $d_H(F) = n$, since F contains ball of positive n-dimensional volume.
(2) If F is smooth that is, continuously differentiable m-dimensional surface of $\mathbb{R}^n$ then $d_H(F) = m$.
(3) If $F \subset G$, then $d_H(F) < d_H(G)$.
(4) For a countable set F, Housdorff dimension $d_H(F) = 0$.
(5) For $F \subset \mathbb{R}^n$, F is totally disconnected if $d_H(F) < 1$.

Calculation of Hausdorff dimensions for some objects

(a) If F is a flat disk of radius 1unit in $\mathbb{R}^3$ then from the properties of length, area, volume $\quad H^1(F) = \text{length}(F) = \infty$, $\quad 0 < H^2(F) = \frac{\pi}{4} \times \text{area}(F) < \infty \quad$ and $H^3(F) = \frac{4\pi}{3} \times \text{volume}(F) = 0$. Thus Housdorff dimension of the disk is $\alpha_H = 2$ with $H^s(F) = \infty$ if $s < 2$ and $H^s(F) = 0$ if $s > 2$.
(b) For the Cantor dust the Housdorff dimension is 1. The Cantor set is constructed from the unit square, at each stage of which the squares are divided into 16 squares with a length sides of sizes ¼, of which the same pattern of four squares is repeated. Thus $1 \le H^1(F) \le \sqrt{2}$, so dimension is 1.
(c) For the Cantor set Housdorff dimension $d_H(C_\infty) = 0.630929$. However, the calculation is somewhat Heuristic as here we assume that $0 < H^s(C_\infty) < \infty$. We know that the Cantor set C_∞ divides into two parts $C_\infty \cap [0, 1/3]$ and $C_\infty \cap [2/3, 1]$, which are similar to the Cantor set C_∞ scaled by a factor 1/3. Thus the Housdorff measure is given by

$$H^s(C_\infty) = H^s(C_\infty \cap [0, 1/3]) + H^s(C_\infty \cap [2/3, 1]) = c^s H^s(C_\infty) + c^s H^s(C_\infty)$$

Now, dividing both sides by H^s(it is permissible as $H^s \ne 0$, ∞), we will obtain $2c^s = 1$, applying logarithm on both sides we will have $s = \frac{\log 2^{-1}}{\log c}$. For Cantor set, c = 1/3, therefore $s = \frac{\log 2}{\log 3} = 0.630929$.

(d) For the Sierpinski gasket Housdorff dimension $d_H(\mathcal{S}) = 1.58496$. The calculation is done by assuming that $0 < H^s(\mathcal{S}) < \infty$. We know that the Sierpinski gasket divides into three equal parts which are similar to the gasket scaled by a factor 1/2. Thus the Housdorff measure is given by

$$H^s(\mathcal{S}) = H^s(\mathcal{S}_1) + H^s(\mathcal{S}_2) + H^s(\mathcal{S}_3) = 3\left(\frac{1}{2}\right)^s H^s(\mathcal{S})$$

Now, dividing both sides by H^s (it is permissible as $H^s \neq 0, \infty$), we will obtain $3\left(\frac{1}{2}\right)^s = 1 \Rightarrow 2^{-s} = 3^{-1}$, applying logarithm on both sides we will have $s = \frac{\log 3}{\log 2} = 1.58496$ approximately.

(iv) Pointwise and correlation dimensions

We have already learnt that the chaotic orbits or strange attractors are fractals. The dimension of strange attractor can be found out using point wise and correlation dimensions. Correlation dimension is not defined for a general set, but for an orbit of a dynamical system and therefore differs from other fractal dimensions. In the process of calculating the correlation dimension, a set of points say $\{x_i, i = 1, \ldots\}$ or a trajectory formed by these points is generated on the attractor by letting the system evolve for a long time. By repeating this procedure for several distinct trajectories would give better statistics. Since all trajectories give the same statistic, it is sufficient to run one trajectory for a long time. Let $N_x(\epsilon)$ is the number of points inside an open set with center at x (a point fixed on the attractor) and radius ϵ. When the radius ϵ is varied, the number of points in the open set $N_x(\epsilon)$ increases as a power law with the increase of ϵ, i.e., $N_x(\epsilon) \propto \epsilon^d$, where d is the pointwise dimension of the attractor. The pointwise dimension of the attractor depends on the point x and its value will be small in thin regions of the attractor. The overall dimension of the attractor can be obtained by averaging $N_x(\epsilon)$ over a number of points x that is,

$$C(\epsilon) = \lim_{N \to \infty} \frac{\{\text{pairs}\{x_1, x_2\} : x_1, x_2 \in N_x(\epsilon), |x_1 - x_2| < \epsilon\}}{\{\text{pairs}\{x_1, x_2\} : x_1, x_2 \in N_x(\epsilon)\}}$$

The average quantity $C(\epsilon)$ thus obtained is found empirically to scale as $C(\epsilon) \propto \epsilon^d$, where d is the correlation dimension of the attractor. More precisely,

Correlation dimension, $d = \lim_{\epsilon \to 0} \frac{\log C(\epsilon)}{\log(\epsilon)}$, provided the limit exists.

Since correlation dimensions depends upon the density of points on the attractor, it is different from box dimension which does not consider the density of points. It was obtained by Grassberger and Procaccia [8] in 1983 that correlation dimension $\leq$ box dimension.

(2) Topological dimension

We know that topology deals qualitatively with the form and shape of an object. Topology analyzes the way in which shapes can be deformed into other shapes. For instance a circle can be continuously deformed into a triangle, a triangle can be deformed into a Koch curve. Topologically these shapes are identical. Topological dimension describes the way the points of an object connected to each other. The topological dimension always assumes an integer value as opposed to fractal dimension. Topological dimension of edges, surfaces and volumes are, respectively, 1, 2, and 3. Two topologically same structures may be fractally different. For instance the fractal dimension of the von Koch curve describing the space-filling properties of the parameter is approximately 1.2619. Now whatever be the resolution taken to measure the parameter of this curve, it remains a line whose dimension is 1. Therefore, topologically a line and the Koch curve cannot be distinguishable. Thus, from topological viewpoint dimension of the parameter of the von Koch curve is 1. Similarly the Koch Island is identical to a circle. The topological dimension of an object therefore aids us in identifying whether the object is a line or a surface or a solid. However, it is not necessary that fractal objects have always a fractional dimension and it may have integer fractal dimension for instance the irregular and erratic motion of small particles of solid matter suspended in a liquid seen under a microscope called as Brownian motion is a fractal whose fractal dimension is 2. Actually, Brownian motion's trail is topologically a curve of dimension 1. However, in reality it fills a plane; hence its fractal dimension is 2. Any curve whose fractal dimension is greater than 1 (topological dimension of a curve) or which has space-filling properties is called as fractal curve. Another mathematically developed fractal whose fractal dimension is 2 is the Peano curve in which line segments are replaced by a generator curve consisting of nine segments, each one being one-third as long. Topological dimension are determined by two ways viz. (i) Covering dimension (ii) Iterative dimension.

(i) Covering dimension

The measurement of covering dimension of an object is obtained by determining the least number of sets required to cover all the parts of an object. These sets may overlap each other. If each point of the object is covered by less than equal to say n number of sets than the covering dimension d will be $d = n - 1$.

For instance the covering dimension of a plane is 2 as if circles are used to cover the plane, then its each point will be covered by less than equal to three circles (Fig. 13.26).

(ii) Iterative dimension

A D dimensional space has borders of dimension D-1. For example borders of a two-dimensional space are lines whose dimension is 1. To get the measure of iterative dimension of an object there is a need to take the borders of the borders of the borders of the object until one reaches a point whose dimension is zero. If the border is repeated m times than the iterative dimension $d = m$. For the plane the border are lines, whose endpoints are points which has zero dimension. Thus one

Fig. 13.26 Covering a *plane* by *circles*

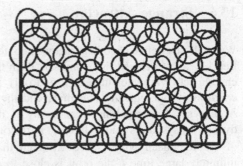

Fig. 13.27 Iterative dimension of a plane

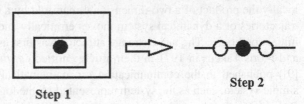

needs to take the border of the borders twice to reach the points. Therefore, the iterative dimension of plane is 2 (Fig. 13.27).

Fractal dimension of a fractal object is always greater than the topological dimension. When the fractal dimension is greater than the topological dimension it means that the object has finer (smaller) pieces than one would have presumed with its topological dimension. The additional pieces at finer resolutions mean that the object covers extra space (i.e., topological dimension plus an additional fraction) than the topological space. This means that the von Koch curve having fractal dimension 1.2619 (1 + 2.619) covers a space more than a one-dimensional line but less than a two-dimensional surface. Thus a fractal object can be defined more formally as an object in space or a process in time that has a fractal dimension greater than its topological dimension.

(3) Embedding dimension

The dimension of the space that contains the fractal object is described by the embedding dimension. The fractal objects may be contained in a non-integer dimensional space as well as in an integer dimensional space. Usually the fractal objects are present in one-, two-, or three-dimensional spaces. For example the tubes that bring air into the lung are spread out into a three-dimensional space, the nerves and the blood vessels in the retina are embedded in a two-dimensional space, a chemical reaction can occur in one-, two-, or three-dimensional space.

Thus far some constructions of fractals are given which are obtained by iterating a single length scale or some polynomial functions. We shall now give some examples which are generated by smooth discrete maps of dynamical systems such as tent map, baker map, etc.

13.6   Strange Attractor

We shall first define chaotic attractor in context of dynamical evolution of a system. Let $A \subset X$ be a closed, invariant set. Then the action of $f: X \to X$ on A is called a chaotic set or attractor under f if and only if it is sensitively dependent on initial conditions, topologically transitive, and the set of all periodic points is dense in A. The set A is repelling if and only if it is not attracting. In other words, a dynamical system is said to have an attractor if there exists a proper subset say, $\mathcal{M}$ of the Euclidean phase space $\mathbb{R}^E$ such that for almost all initial state and for infinitely large time t, the orbit is close to some point of $\mathcal{M}$. strange attractor is locally the product of a two-dimensional manifold by a Cantor set inside which the trajectories of a dynamical system moves erratically and are highly sensitive to the initial conditions. The name strange attractor was first used jointly by David Ruelle and Floris Takens in 1971 in their article entitled *On the nature of turbulence* (see [9]) published in the communication on mathematical physics. The attractors of simple system, such as the system represented by the logistic map $y \to ry(1 - y)$ for real r (growth rate of a population) and $y \in [0, 1]$, are strange and are the characteristic of complex and realistic system. Strange attractors are patterns which describe the final state of dissipative systems that are extremely complex and chaotic. The geometry of strange attractor set is very strange as it resembles the infinite complex surface and cannot be represented in Euclidean geometry with integral dimensions. In fact chaos and fractals inevitably occur simultaneously at the points of strange attractors. Geometrically strange attractors are fractals, and dynamically they are chaotic. Strange attractors are generated by iteration of a map or the solution of a system of initial-value differential equations that exhibit chaos. Scientists now believe that many unsolved problems like climate of earth, activity of human brain, the phenomena of turbulence etc., can be resolved by strange attractors. Strange attractors are fractals as almost all of the strange attractors known at present have fractal dimensions greater than their topological dimensions.

The simplest planar dynamical system that is, the dissipative baker's transformation acts as a model of strange attractor. The attractor generated by baker map is called strange as it has the Cantor set structure and therefore has fractal properties. The dissipative baker's transformation is given by

$$y_{n+1} = \begin{cases} \varepsilon y_n, & 0 \le x_n < \frac{1}{2}, \\ \varepsilon y_n + \frac{1}{2}, & \frac{1}{2} \le x_n \le 1. \end{cases}$$

$\varepsilon > 0$ with the Bernoulli shift $x_{n+1} = 2x_n (\mathrm{mod}\ 1)$. The name baker is given to the transformation because it is similar to the process of constantly stretching and doubling the dough. In Fig. 13.28, a cross section of the square of dough is shown having the head of a cat (see Arnold and Avez [10]). The given transformation first rolls this dough, as a result the head of the dough becomes half its height and twice its length. This stretched and flattened figure of dough is then cut in the center. The initial square is then reconstructed by placing the right segment of the dough above

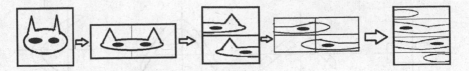

Fig. 13.28 Baker's transformation on Arnold's cat

the left segment. This transformation is repeated once which then gives a structure in which the cat's head is embedded in four layers of dough. The baker's transformation is a piecewise linear and discontinuous transformation. The attractor for the baker's transformation is a unit square $[0, 1] \times [0, 1]$. For each iteration, the total area of the attractor is reduced by a factor 2ε. In the first iteration two rectangles of side length $l_1 = \varepsilon$ are generated with a gap of $\frac{1}{2} - \varepsilon$ in the middle. In the next iteration four rectangles each of side length $l_2 = \varepsilon^2$ will be formed separated from each other by a gap of $\left(\frac{1}{2} - \varepsilon\right)\varepsilon^{-1}$ and in the nth iteration the unit square is contracted onto 2^n rectangles each of side length $l_n = \varepsilon^n$ with a gap of at least $\left(\frac{1}{2} - \varepsilon\right)\varepsilon^{1-n}$. The fractal dimension of the attractor is the sum of D_x and D_y where $D_x = 1$ follows from the Bernoulli shift when x_0 is irrational. While D_y is obtained from the baker's map which transforms the region $0 \le x_0 \le \frac{1}{2}, 0 \le y_0 \le 1$ onto the region $0 \le x_0 \le 1, 0 \le y_0 \le \varepsilon < \frac{1}{2}$ and the region $\frac{1}{2} \le x_0 \le 0, 0 \le y_0 \le 1$ onto $0 \le x_0 \le 1, \frac{1}{2} \le y_0 \le \frac{1}{2} + \varepsilon$. From the construction of the attractor we have $D_y = \frac{\ln 2}{|\ln \varepsilon|}$, $D_y < 1$ if $\varepsilon < \frac{1}{2}$. The fractal dimension is therefore between 1 and 2. Furthermore, baker's transformation has sensitive dependence on initial conditions and the periodic points of the transformation are dense in the fractal thus obtained, hence this attractor is a chaotic attractor for the baker's transformation. Thus the fractal obtained from baker's transformation is the product of $[0, 1]$ and the uniform Cantor set which is formed by continuously replacing intervals say I by a pair of subintervals of lengths $\varepsilon|I|$ (Fig. 13.29).

The Skinny-baker map which is described in Chap. 12 also gives a Cantor-like structure on a unit square and hence is a fractal which also exhibits chaos on the invariant set $[0, 1] \times [0, 1]$.

In natural science, the Lorenz attractor discovered in 1962 was the first, which was recognized as strange attractor. The fractal dimension of Lorenz attractor also

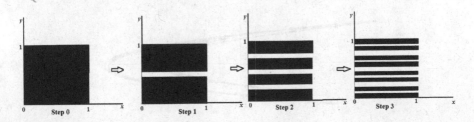

Fig. 13.29 Attractor due to dissipative baker transformation

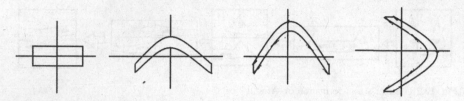

Fig. 13.30 Construction of the Hénon attractor by first bending a rectangle followed by contracting and then reflecting in the line $y = x$

known as Saltzman-Lorenz attractor is 2.06. This dimension was obtained by Ya. G. Sinai [11].

In 1976 French astronomer Michel Hénon suggested a model [12] which is a simplified form of the complex Lorenz attractor named after him as Hénon attractor. It provides another example of strange attractor. Hénon attractor is not only simple in comparison to the Lorenz system but exhibits characteristic inherent in the Lorenz attractor and other complex attractor and are obtained from a stretching and folding map known as Hénon map given by $f(x, y) = (y + 1 - ax^2, bx)$, a, b are constants. This mapping has Jacobian $-b$ for all x, y. This implies a contraction of area at a constant rate in the entire $\mathbb{R}^2$ plane. This mapping may be decomposed into an area preserving bend $(x_1, y_1) = (x, 1 - ax^2 + y)$, a contraction $(x_2, y_2) = (bx_1, y_1)$ in the x-direction and a reflection $(x_3, y_3) = (y_2, x_2)$ in the line $y = x$ shown in Fig. 13.30, which results into a horseshoe like map which is locally the product of a line segment and a Cantor-like set. Hence the attractor thus obtained by the map is a strange attractor whose fractal dimension is 1.28.

Hénon system occurs in a two-dimensional discrete dynamical system, with coordinates as x and y and an orbit of the system consists of an initial point say (x_0, y_0) and its iterated images i.e. $(x_{r+1}, y_{r+1}) = (y_{r+1} + 1 - ax_r^2, bx_r)$, $r = 0, 1, 2, \ldots$ (Fig. 13.31). The dynamics of the Hénon system depend on the choice of the constants a and b and the initial point (x_0, y_0). The dimensions of Hénon attractor being a strange attractor cannot be easily found. The most widely used box-counting dimension is not suitable for Hénon attractor since it is constructed by iterating many times the Hénon transformation. The number of iterations for measuring the total number of boxes $N(s)$ of box size s increases with the decrease in the value of s. So for the Hénon attractors the number of boxes cannot be computed directly and

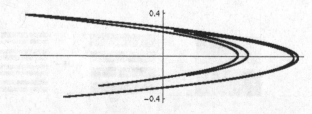

Fig. 13.31 Hénon attractor

depends upon the iteration n. However, the number of iterations enough for measuring the dimension is not clear. In 1983 Peter Grassberger on the basis of his tabulated data in his article *On the fractal dimension of the Hénon attractor* published in *Physics letters*, suggested that $N(s, n) \approx N(s) \approx$ constant $\cdot$ $s^{-\alpha} n^{-\beta}$, for large number of iterations. Accordingly the rate is given by $\frac{\Delta N(s,n)}{\Delta n} \approx$ constant $\cdot$ $s^{-\alpha} n^{-\beta-1}$. Grassberger roughly calculated the values of α and β as $2 \cdot 42 \pm 0 \cdot 15$ and $0 \cdot 89 \pm 0 \cdot 03$, respectively, after performing the analysis with five runs of 7.5 million iterations and using five different box sizes s. This led him to report the box dimension of the Hénon attractor as $1 \cdot 28 \pm 0 \cdot 01$. On the other hand the correlation dimensions of the Hénon attractor is found to be $\approx 1 \cdot 23 < 1 \cdot 28 \pm 0 \cdot 01$, box-counting dimension (see the book of Alligood et al. [13]) by taking an orbit of the attractor formed by some 1000 points. The proportion that lies within the radius ϵ is then counted for $\epsilon = 2^{-2}, ..., 2^{-8}$ of $(1000)(999)/2$ possible pairs of the points on the orbit. According to the definition of correlation dimension estimating the slope of the graph of $\log C(\epsilon)$ versus $\log \epsilon$ as $\epsilon \to 0$, one would obtain the correlation dimensions $\approx 1 \cdot 23$.

The Hénon attractor has the following characteristic features:

(i)   The attractor has a fractal structure and is therefore called as strange attractor.
(ii) There is a trapping region within which all orbits starting at some initial point lead to the attractor.
(iii) Hénon attractor is chaotic as the orbit generated from the Hénon map exhibits sensitive dependence to initial conditions.

Lorenz attractor is a continuous dynamical system described by E. N. Lorenz in a model of turbulence (see [18]) consisting of three coupled ordinary differential equation which contains two nonlinear terms. But the qualitative behavior of the system was not sufficiently found. So, Otto E. Rössler proposed a system of equation in 1976 (see [14]) which is simpler than Lorenz system of equation and generates a similar flow forming a single spiral. This system gives a strange attractor known as Rössler attractor. Rössler called his equations as a prototype of the Lorenz model of turbulence that has no physical interpretation. The system of differential equation which generates the Rössler attractor is given by

$$\left.\begin{array}{l} \dot{x} = -y - z \\ \dot{y} = x + ay \\ \dot{z} = b + xz - cz \end{array}\right\} \tag{13.2}$$

with $a = 0.2$, $b = 0.2$ and $c = 5.7$. The coefficients a, b, and c are constants and can be modified. The Rössller system has only one nonlinearity in the form of xz. A unique trajectory is defined by the system for any given initial coordinates parameterized by time t that satisfies the equations at all times. Rössller attractor represents a strange attractor simpler than the Lorenz attractor. Orbits on the attractor spiral out along the xy-plane from the origin, most of the time stays in the vicinity of the xy-plane. And after reaching some critical distance from the origin,

(a)   **(b)**

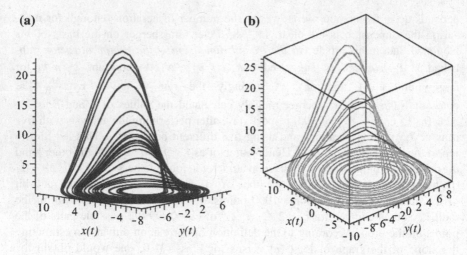

Fig. 13.32 Rosselor attractor for **a** $a = 0.2$, $b = 0.2$, $c = 5.7$; **b** $a = 0.2$, $b = 0.2$, $c = 6.3$

Fig. 13.33 Rösseler attractor
for $a = 0.2$, $b = 0.2$, $c = 8.0$
which are trajectories of the
system for attracting periodic
point

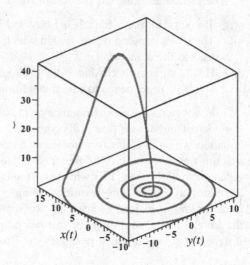

an orbit is first lifted away from the xy-plane. Then, after reaching some maximal value of z, it re-enters into the spiraling portion of the attractor near the plane. The orbit will lie closer to the origin when the z amplitude becomes larger and the spiraling process followed by excursion in the z-direction and then re-insertion into the xy-plane continues. The nonlinear stretch-and-fold operation is hidden in the system and the attractor has the structure of a folded band. A function modeling the plot of Rössller attractor is called a *Lorenz map* whose dynamics is also a stretch-and-fold operation. It provides a connection between the dynamics of a continuous system and the discrete dynamics of transformations of an interval. Figures 13.32 and 13.33 show the behavior of Rössller attractor for different values of c with fixed values of a and b.

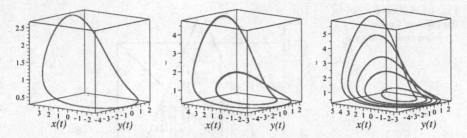

Fig. 13.34 Rosseller attractor for three different values of a (0.3, 0.375, 4) for fixed values of $b = 2$ and $c = 4$

The fractal dimension of a chaotic attractor from the Rössller family must be $2 + \varepsilon$ where ε is a very small number. It is this fractal character that makes the Rössller's attractor a *strange* attractor. It is numerically estimated that the dimension of the Rössller attractor is slightly above 2, i.e., 2.01 or 2.02.

For fixed values of $b = 2$, $c = 4$ and varying the value of a, we will see that the nature of attractor changes. For small value of a, the attractor is a simple closed curve. After increasing the value of a, it will be seen that the attractor splits into a double loop, then a quadruple loop and so on. Thus, a period doubling of the attractor takes place shown in Fig. 13.34.

13.7 Strange Repeller

A dynamical system is said to have a repeller if there exists a proper subset say, $\mathcal{M}$ of the Euclidean phase space $\mathbb{R}^E$ such that for almost all initial state and for infinitely large time t, the orbit seems to be pushed away from $\mathcal{M}$ before converging to stable equilibrium elsewhere. The set of all unstable equilibrium states along with its limit points is called repeller. When this set has fractal structures then it is named as strange repeller. Here we provide an example of strange repeller given by the discrete dynamical map, tent map.

The well known tent map can be defined as

$$f(x, \varepsilon) = \begin{cases} \varepsilon x & x < \frac{1}{2} \\ \varepsilon(1 - x), & x > \frac{1}{2} \end{cases}$$

Tent map splits the unit interval dynamically into two sets; one is 'escape set' which occupies all the space and the other is a fractal strange repeller set which is a uniform invariant Cantor set of Housdorff and box dimension log 2/log 3. Here the Cantor sets are called strange repellors because the points which lie in the region $x_n > 1$ escape to infinity as the iteration $n \to \infty$ i.e. if $x < 0$ then $f(x) < \varepsilon x$, so $f^k(x) \to -\infty$ as $k \to \infty$. if $x > 1$ then $f(x, \varepsilon) < 0$ and $f^n(x, \varepsilon) \to -\infty$. This means

Fig. 13.35 Construction of
the uniform Cantor set
invariant under tent map

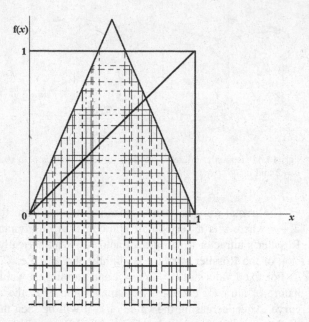

that the invariant Cantor set repels these points. After first iteration, i.e., for the
initial point say $x_0 \in (\varepsilon^{-1}, 1 - \varepsilon^{-1})$, $x_1 > 1$ tent map breaks the phase space into two
equal closed intervals $[0, \varepsilon^{-1}]$, $[1 - \varepsilon^{-1}, 1]$, each of length ε^{-1}. In the next iteration
the set of initial conditions give $x_2 > 1$, tent map breaks the phase space into the four
closed intervals $[0, \varepsilon^{-2}]$, $[\varepsilon^{-1}(1 - \varepsilon^{-1}), \varepsilon^{-1}]$, $[1 - \varepsilon^{-1}, 1 - \varepsilon^{-1} + \varepsilon^{-2}]$ and $[1 - \varepsilon^{-2}, 1]$
each with length ε^{-2}. Repeating this process, after nth iterations, the map breaks the
phase space into $N = 2^n$ closed intervals each of length $l_n = \varepsilon^{-n}$. For $\varepsilon = 3$, the tent
map $f(x)$ maps the middle third Cantor set onto itself shown in Fig. 13.35.
Furthermore, the iterates of the map have sensitive dependence on initial condition
i.e. for a slight change in the initial conditions give large fluctuations in the later
stages, For instance, $f^k(x_1) \in \left[0, \frac{1}{3}\right]$ but $f^k(x_2) \in \left[\frac{2}{3}, 1\right]$, where x_2 is an approximate
value of x_1. For this reason the tent map f exhibits chaotic behavior on the uniform
Cantor set.

Let us consider a model for the well-known Navier-Stokes equations in fluid
mechanics. The model is known as poor man's Navier-Stokes equations referred in
the book "Turbulence" by Uriel Frisch, Cambridge University Press, 1995. Fluid
turbulence is a chaotic phenomenon. It is one of the oldest unsolved problems in
classical mechanics. Turbulent flows are an irregular, unpredictable, nonlinear, and
vortical. The flow phenomenon is dissipative and multi-length scales in nature.
Turbulence also exhibits mixing and thereby enhances diffusivity. The flow can be
described mathematically by the Navier-Stokes equations of motion under the
framework of continuum mechanics. The Navier-Stokes equations of motion for an
viscous incompressible fluid (fluid density ρ = constant) may be written in usual
notations as

$$v_t + v \cdot \nabla v = -\nabla p + v \nabla^2 v + f,$$

where v_t denotes the partial derivative with respect to t, $(v \cdot \nabla v)$ is the nonlinear convection term, ∇_p is the pressure gradient, $v\nabla^2 v$ is the viscous dissipation term, f denotes the body force, and $v = \mu/\rho$ the kinematic fluid viscosity, μ being the fluid viscosity. However, the Navier-Stokes equation of motion is deterministic, that is, a unique solution is expected for all times for given initial conditions (There is no mathematical proof for this!). The behaviors of turbulent signals are unpredictable in nature but the flow statistics are reproducible. The signals can be quantified through statistics. This leads the mathematicians/physicists, viz. Taylor, Kolmogorov, von Karman, and others to look for a probabilistic description of turbulence. How can chaos appear in a deterministic system? This can be analyzed by constructing a simple one-dimensional quadratic map as a model for the Navier-Stokes equations, known as poor man's model considering only temporal evolution of the motion. Obviously, this is a crude model but has significance for understanding turbulence.

Consider the map $L(v_t) = rv_t(1 - v_t)$, $r \in [0,4]$ and $v_t \in [0,1]$, known as logistic map described extensively in previous chapters, v_t being the variable representing the temporal growth of fluid velocity. The Navier-Stokes equations of motion above can be modeled very crudely by a logistic map as

$$v_{t+1} - v_t = -av_t^2 + vv_t + c$$
$$\Rightarrow v_{t+1} = f(v_t) = -av_t^2 + (1+v)v_t + c$$

in which the approximations are $O(v_{t+1} - v_t) \simeq O(\partial_t v)$, $O(av_t^2) \simeq O$ $(v \cdot \nabla v + \nabla p)$, $O(vv_t) \simeq O(v\nabla^2 v)$ and $f \simeq O(c)$. As we know conjugacy is an important relation among maps and the dynamics of conjugacy maps are similar. It is easy to find that the map above (which is a logistic map) is conjugate to the map $v_{t+1} = g(v_t) = 1 - 2v_t^2$, $t = 0, 1, 2, ...,$ $v_t \in [-1, 1]$ with given v_0, via the linear conjugacy map $h_1 : [0,1] \rightarrow [-1,1]$, $h_1(v_t) = \frac{a}{2}v_t - \frac{(1+v)}{4}$ subject to the condition $v^2 + 4ac = 9$, the kinematic viscosity $v > 0$. Again, the tent map $T(x)$ in $[0, 1]$ is conjugate to the map g via the conjugacy map $h_2 : [0,1] \rightarrow [-1,1]$, $h_2(x) = -\cos(\pi x)$. In other words, g is conjugate to T via h_2^{-1} (see Chap. 11). Thus the three maps are conjugate to each other through their respective conjugacy maps and their dynamics are similar. The evolution of tent map, that is, $T^n(x)$ have been discussed previously. The function $T^n(x)$ intersects twice with the line $y = x$ in each of the subintervals $[(i-1)/2^{n-1}, i/2^{n-1}]$, $i = 1, 2, ..., 2^{n-1}$. The points of intersections are the periodic points of T which get increased in number when number of iterations n gets larger and larger creating smaller and even smaller intervals. It follows that the periodic points of the tent map $T(x)$ are dense in $[0,1]$. The tent map is also topologically transitive and has sensitivity property in $[0,1]$ which is an invariant set. It gives a Cantor set structure as $n \rightarrow \infty$. The dynamics of tent map is similar to the poor man's Navier-Stokes equation via the conjugacy map $h_1^{-1} \circ h_2$.

The Cantor set has a self-similar fractal structure with dimension 0.6309. From this analysis, following observations are inferred:

(i) Transition to turbulence occurred through period-doubling bifurcation,
(ii) Periodic points are dense,
(iii) Unpredictability occurs after exceeding a critical value.
(iv) Order (regularity) and chaos co-exist in turbulence.
(v) It has self-similar fractal structure.
(vi) Self-similar fractal has Cantor set like structure

13.8 Exercises

1. What is fractal? Give examples fractal objects both in natural and manmade. How do you distinguish between fractal and non-fractal objects?
2. Give the definition of 'self-similar' fractal with examples.
3. Show that similar object does not mean fractal.
4. Give few non-similar fractal objects.
5. Discuss self-similarity in time and the statistically self-similarity with examples.
6. Explain different types of similar transformations.
7. Define von Koch curve. Describe the procedure of constructing von Koch curve. Sketch the curve for five steps.
8. Show that the length of the Koch curve is infinite.
9. Construct the Sierpinski triangle. Show that the area is fixed but infinite in length.
10. Construct the Sierpinski carpet. Show mathematically that it is a self-similar fractal.
11. Define Cantor set. Construct Cantor set for 8 steps. State and prove the properties of Cantor set.
12. How do you find the dimension of a self-similar fractal? Give it general formula.
13. Find the self-similar dimensions of (i) von Koch curve, (ii) Sierpinski triangle, (iii) Sierpinski carpet, and (iv) Cantor set.
14. Explain the concept of box dimension and correlation dimensions for fractal objects. Find the box dimensions and correlation dimension of the Hénon attractor.
15. Find the box dimension of (i) von Koch curve, (ii) Sierpinski triangle, (iii) Sierpinski carpet, and (iv) Cantor set.
16. Show that the Cantor set is invariant under the contracting mapping $T_1 : x \rightarrow \frac{x}{3}, T_2 : x \rightarrow \frac{x}{3} + \frac{2}{3}$.
17. Show that the box dimension of a finite number of isolated points is one.

18. Prove that the measure of a Cantor set is zero and then find the box dimension of Cantor set.

19. Construct a Cantor set by deleting the middle subinterval of length 4^{-k} from each remaining interval at step k, starting at the interval $[0, 1]$ at step. Also determine the length and box-counting dimension of the set. Is the set is of measure zero.

20. Construct a middle third Cantor-like set by deleting the middle half i.e. 2^{-k} from each remaining interval at step k, starting at the interval $[0, 1]$ at step. Determine the length and box-counting dimension of the set.

21. Construct a set F by replacing the unit interval $[0, 1]$ by two intervals; one of one-fourth of the length at the left end and the other of one half the length at the right-hand end that is, $F_0 = [0, 1]$, $F_1 = [0, 1/4] \cup [1/2, 1]$ and so on. Find the Housdorff and box dimension of this set.

22. Draw the graphics of the Koch snowflake. Also calculate its area.

23. Prove that the similarity dimension of Koch curve is 1.26.

24. Construct a curve by repeatedly replacing the middle portion $\epsilon, 0 < \epsilon \le \frac{1}{3}$ of each interval by the other two sides of an equilateral triangle. Show that the Housdorff dimension of this curve is the solution of $2\epsilon^s + 2\left(\frac{1}{2}(1 - \epsilon)^s\right) = 1$.

25. Construct a curve by repeatedly dividing a line segment into three equal parts and then replacing the middle portion of each interval by the other three sides of a square. Find the Housdorff dimension of this curve.

26. Find the box dimension of a bounded set in $\mathbb{R}^n$.

27. Show that the box-counting dimension of a disk (a circle together with its interior) is 2.

28. Find the box-counting dimension of the set of integers $\{0, 1, \ldots, 100\}$.

29. Find the box-counting dimension of the set of all rational numbers in $[0, 1]$.

30. Show that the capacity dimension of the von Koch curve is equal to the similarity dimension of the curve.

31. Consider a solid (filled) unit square. Divide it into nine equal small squares and then delete the interior of any one of these nine squares, selected arbitrarily. Repeat this process on each of the remaining eight small squares and so on. Find the box dimension of the limiting set.

32. Consider a solid (filled) unit equilateral triangle. Divide it into three small equal equilateral triangles and then delete the interior of any one of these three triangles, selected arbitrarily. Repeat this process for each of the remaining two equilateral triangles and so on. Find the box dimension of the limiting set.

33. Show that the area of the Sierpinski carpet is zero.

34. Let $S = F \times [0, 1] \subset \mathbb{R}^2$ be the product of the middle third Cantor set F and the unit interval. Find the box and Housdorff dimension of S.

35. Let $S = \{(x, y) \in \mathbb{R}^2 : x \in F \text{ and } 0 \le y \le x^2\}$, F is the middle third Cantor set. Find the Hausdorff dimension of S.

36. Find a fractal invariant set F for the tent-like map $f : \mathbb{R} \to \mathbb{R}$ given by $f(x) = 2(1 - |2x - 1|)$. Show that F is a repeller for f and that f is chaotic on F.

37. Describe the Julia set of $f(z) = z^2 + 4z + 2$.

38. Show that if $|c| < 1$ then the Julia set of $f(z) = z^3 + cz$ is a simple closed curve.
39. Obtain an estimate of the Hausdorff dimension for the Julia set of f $(z) = z^3 + c$ when $|c|$ is large.
40. Generate computer graphics for the Julia set $f(z) = z^3 + c$, z is complex.
41. Prove that the Hénon map (i) is invertible (ii) is dissipative and (iii) has a trapping zone.

References

1. Cantor, G.: Über unendliche, lineare Punktmannigfaltigkeiten V. Math. Ann. **21**, 545–591 (1883)
2. von Koch, H.: Sur une courbe continue sans tangente, obtenue par une construction géometrique élémentaire. Arkiv för Matematik **1**, 681–704 (1904)
3. von Koch, H.: Une méthode géométrique élémentaire pour l'étude de certaines questions de la théorie des courbes planes. Acta Mathematica **30**, 145–174 (1906)
4. Sierpinski, W.: Sur une courbe cantorienne dont tout point est un point de ramification. C. R. Acad. Paris **160**, 302 (1915)
5. Sierpinski, W.: Sur une courbe cantorienne qui contient une image biunivoquet et continue detoute courbe donnée. C. R. Acad. Paris **162**, 629–632 (1916)
6. Julia, G.: Mémoire sur l'iteration des fonctions rationnelles. J. de Math. Pure et Appl. **8**, 47–245 (1918)
7. Benoit, B.: Mandelbrot: Fractal aspects of the iteration $z \longmapsto \lambda\, z(1 - z)$ of for complex λ and z and Annals NY Acad. Sciences **357**, 249–259 (1980)
8. Grassberger, P., Procaccia, I.: Characterization of strange attractors. Phys. Rev. Lett. **50**, 346 (1983)
9. Ruelle, D., Takens, F.: On the nature of turbulence. Commun. Math. Phys. **20**, 167–192; **23**, 343–344 (1971)
10. Arnold, V.I., Avez, A.: Ergodic problems in classical mechanics. Benjamin, New York (1968)
11. Ya, G.: Sinai: Self-similar probability distributions. Theor. Probab. Appl. **21**, 64–80 (1976)
12. Hénon, M.: A Two-dimensional Mapping with a Strange Attractor. Commun. Math. Phys. **50**, 69–77 (1976)
13. Alligood, K.T., Sauer, T.D, Yorke, J.A.: Chaos: An Introduction to Dynamical Systems. Springer (1997)
14. Otto, E.: Rössler: An equation for continuous chaos. Phys. Lett. **57A**, 5 (1976)
15. Benoit, B.: Mandelbrot: The fractal geometry of nature. W. H. Freeman and Company, New York (1977)
16. Peitgen, Heinz-Otto: Hartmut Jürgens, Dietmar Saupe: Chaos and fractals. Springer-Verlag, New York (2004)
17. Falconer, K.: Fractal geometry, Mathematical foundations and applications. Wiley, New York (1990)
18. Lorenz, E.N.: Deterministic non-periodic flow. J. Atmos. Sci. **20**, 130–141 (1963)
19. Strogatz, S.H.: Nonlinear Dynamics and Chaos with application to physics, biology, chemistry and engineering. Perseus books, L.L.C, Massachusetts (1994)
20. Benoit, B.: Mandelbrot: How long is the coast of Britain? Science **156**, 636–638 (1967)
21. Oberlack, M.: A unified approach for symmetries in plane parallel turbulent shear flows. J. Fluid Mech. **427**, 299–328 (2001)

Index

A

Absorbing set, 31, 34
Affine transformation, 586
Algorithm to calculate Lie symmetry, 332
Asymptotic stability, 91, 95, 96, 122, 131, 132, 157
Attracting set, 31, 32, 154, 235, 237

B

Basin boundary, 154, 156, 158, 414
Basin of attraction, 32, 154, 158, 414, 420
Bendixson's negative criterion, 175
Bifurcation
  codimension-1, 203
  codimension-2, 203
  global, 169
  hard, 223
  local, 21, 204
  soft, 223
Bifurcation diagram, 203, 205, 207, 209, 210, 217, 219
Bifurcation point, 203, 205, 206, 214, 216, 219, 221–223, 229, 534
Birkhoff's ergodic theorem, 540
Borderline fractal, 593, 594
Box dimension, 599, 601–603, 611, 616
Brusselator chemical reaction, 126, 168

C

Canonical form, 85, 89, 93, 96, 334, 357
Canonical parameter, 334–336
Cantor dust, 559, 585, 589, 600, 604
Cantor set, 502, 512, 514, 516
Center manifold theorem, 153
Circle map, 465, 551–554, 557
Cobweb diagram, 413, 417, 430, 453, 454, 456, 510, 546
Configuration space, 257, 287
C^k-conjugacy, 483

Conservative system, 5, 26, 28, 34, 133, 292, 301, 569
Continuous group, 323, 324, 343, 350
Correlation dimension, 599, 605, 611, 616
Covering dimension, 606
Crises, 561, 564
Critical point, 14, 20, 33, 34

D

Damping
  critical, 108, 110, 111
  strong, 23, 109
  weak, 108–110
Decimal shift map, 464, 478
Decomposability, 505
Degrees of freedom, 259, 264, 271, 274, 275, 278, 290, 291, 303
Dense orbit, 502, 506, 508, 560
Dense set, 502, 507, 510, 571, 585
Devil's staircase, 575, 593–595
Dilation invariance, 319
Dissipative system, 26, 28, 29, 34, 106, 236, 569, 608
Dulac's criterion, 175, 176, 198

E

Embedding dimension, 607
Equilibrium point
  asymptotic stable, 91, 131, 140
  hyperbolic, 90, 101, 140–142, 149, 152, 227
  neutrally stable, 95, 96, 98, 114, 120
  non-hyperbolic, 207
  stable, 22, 93–95, 100, 130, 131, 133, 192, 222, 226, 508
  unstable, 22, 91, 93, 95, 96, 102, 138, 207, 223
Ergodic map, 540, 541, 572
Euler shift map, 463, 477, 519
Eventually fixed point, 427, 462, 476, 519, 571

© Springer India 2015
G.C. Layek, *An Introduction to Dynamical Systems and Chaos*,
DOI 10.1007/978-81-322-2556-0

Printed in the United States
By Bookmasters